Lecture Notes in Computer Science 9290

Commenced Publication in 1973
Founding and Former Series Editors:
Gerhard Goos, Juris Hartmanis, and Jan van Leeuwen

More information about this series at http://www.springer.com/series/7410

Javier Lopez · Chris J. Mitchell (Eds.)

Information Security

18th International Conference, ISC 2015
Trondheim, Norway, September 9–11, 2015
Proceedings

Springer

Editors
Javier Lopez
University of Málaga
Málaga
Spain

Chris J. Mitchell
Royal Holloway, University of London
Egham
UK

ISSN 0302-9743 ISSN 1611-3349 (electronic)
Lecture Notes in Computer Science
ISBN 978-3-319-23317-8 ISBN 978-3-319-23318-5 (eBook)
DOI 10.1007/978-3-319-23318-5

Library of Congress Control Number: 2015947114

LNCS Sublibrary: SL4 – Security and Cryptology

Springer Cham Heidelberg New York Dordrecht London

Printed on acid-free paper

Springer International Publishing AG Switzerland is part of Springer Science+Business Media
(www.springer.com)

Preface

The Information Security Conference (ISC) is an annual international conference dedicated to research on the theory and applications of information security. It started as a workshop in 1997, changed to a conference in 2001, and has been held on five different continents. ISC 2015, the 18th in the series, was held in the delightful and historic city of Trondheim in Norway, September 9–11, 2015. The conference was organized by Colin Boyd and Danilo Grigoroski of the Norwegian University of Science and Technology (NTNU) in Trondheim.

This year we received 103 submissions with authors from 35 different countries. Each submission was reviewed by at least two and in almost all cases by three or even four Program Committee members, and the review process was conducted in a 'double-blind' fashion. After detailed and careful discussions, the committee decided to accept 30 papers, yielding an acceptance rate of 29%. The conference program also included two fascinating invited keynote talks, given by Professors Tor Helleseth ("Sequences, Codes and Cryptography") and Kenny Paterson ("Authenticated Encryption and Secure Channels: There and Back Again").

The success of ISC 2015 depended critically on the help and hard work of many people, whose assistance we gratefully acknowledge.

First, we would like to thank all the authors for submitting their papers to us. We further thank the authors of accepted papers for revising papers according to the various reviewer suggestions and for returning the source files in good time. The revised versions were not checked by the Program Committee, and so authors bear final responsibility for their contents.

We heartily thank the 54 members of the Program Committee (from 19 different countries) and the 84 external reviewers, listed on the following pages, for their careful and thorough reviews. Thanks must also go to the hard-working shepherds for their guidance in improving a number of papers.

Huge thanks are due to Colin Boyd and Danilo Grigoroski for acting as general chairs, and taking care of every detail, large and small. We are grateful to the ISC Steering Committee for their advice and support. The local administrator was Mona Nordaune; this conference would not have been successful without her vital assistance with a multitude of details. We are also very grateful to Prof. Yuming Jiang for his generous advice regarding many administrative and organizational matters. We must also warmly thank Carmen Fernandez-Gago for getting the word out, and enabling us to have such a healthy number of submissions. In our expressions of gratitude we must not forget Slartibartfast, the planetary coastline designer who was responsible for the fjords of Norway including, of course, Trondheimsfjord.

Last, but not least, we would like to thank EasyChair for providing the user-friendly management system we used for managing the submission and review phases, and Springer for, as always, providing a meticulous service for the timely production of the proceedings.

September 2015

Javier Lopez
Chris J. Mitchell

Organization

Information Security Conference 2015, Trondheim, Norway
September 9–11, 2015

General Chairs

Colin Boyd	NTNU, Trondheim, Norway
Danilo Grigoroski	NTNU, Trondheim, Norway

Program Chairs

Javier Lopez	University of Málaga, Spain
Chris J. Mitchell	Royal Holloway, University of London, UK

Steering Committee

Sherman S.M. Chow	Chinese University of Hong Kong, Hong Kong, SAR China
Ed Dawson	Queensland University of Technology, Australia
Javier Lopez	University of Málaga, Spain
Masahiro Mambo	Kanazawa University, Japan
Eiji Okamoto	University of Tsukuba, Japan
Susanne Wetzel	Stevens Institute of Technology, USA
Yuliang Zheng	University of North Carolina at Charlotte, USA

Local Administrator

Mona Nordaune	NTNU, Trondheim, Norway

Publicity Chair

Carmen Fernandez-Gago	University of Málaga, Spain

Program Committee

Habtamu Abie	Norsk Regnesentral – Norwegian Computing Center, Norway
Gail-Joon Ahn	Arizona State University, USA
Claudio A. Ardagna	Università degli Studi di Milano, Italy
Tuomas Aura	Aalto University, Finland
Feng Bao	Huawei, Singapore
Alex Biryukov	University of Luxembourg, Luxembourg
Marina Blanton	University of Notre Dame, USA

Liqun Chen	Hewlett-Packard Laboratories, UK
Sherman S.M. Chow	Chinese University of Hong Kong, Hong Kong, SAR China
Bruno Crispo	University of Trento, Italy
Paolo D'Arco	University of Salerno, Italy
Robert Deng	Singapore Management University, Singapore
Yvo Desmedt	University of Texas at Dallas, USA
Roberto Di Pietro	Bell Labs, France
Josep Domingo-Ferrer	Universitat Rovira i Virgili, Spain
Christian Gehrmann	Swedish Institute of Computer Science, Sweden
Dieter Gollmann	Hamburg University of Technology, Germany
Dimitris Gritzalis	Athens University of Economics and Business, Greece
Stefanos Gritzalis	University of the Aegean, Greece
Tor Helleseth	University of Bergen, Norway
Xinyi Huang	Fujian Normal University, China
Sushil Jajodia	George Mason University, USA
Audun Jøsang	University of Oslo, Norway
Sokratis Katsikas	University of Piraeus, Greece
Stefan Katzenbeisser	Technische Universität Darmstadt, Germany
Di Ma	University of Michigan-Dearborn, USA
Fabio Martinelli	IIT-CNR, Italy
Leonardo Martucci	Karlstad University, Sweden
Catherine Meadows	NRL, USA
Jean-François Misarsky	Orange, France
Stig Mjølsnes	NTNU, Norway
Refik Molva	Eurecom, France
Yi Mu	University of Wollongong, Australia
David Naccache	École Normale Supérieure, France
Eiji Okamoto	University of Tsukuba, Japan
Josef Pieprzyk	Queensland University of Technology, Australia
Michalis Polychronakis	Stony Brook University, USA
Joachim Posegga	University of IT-Security and Security Law, Passau, Germany
Bart Preneel	KU Leuven, iMinds, Belgium
Indrajit Ray	Colorado State University, USA
Kui Ren	State University of New York, Buffalo, USA
Rei Safavi-Naini	University of Calgary, Canada
Pierangela Samarati	Università degli Studi di Milano, Italy
Miguel Soriano	Universitat Politècnica de Catalunya, Spain
Claudio Soriente	ETH Zürich, Switzerland
Ketil Stølen	SINTEF, Norway
Nils K. Svendsen	Gjøvik University College, Norway
Carmela Troncoso	Gradiant, Spain
Vijay Varadharajan	Macquarie University, Australia
Guilin Wang	Huawei International, China
Susanne Wetzel	Stevens Institute of Technology, USA

Stephen Wolthusen	RHUL, UK and Gjøvik University College, Norway
Moti Yung	Google, USA
Jianying Zhou	Institute for Infocomm Research, Singapore

External Reviewers

Ahmad Ahmadi Mamunur
Asghar Monir Azraoui
Ero Balsa
Matthias Beckerle
Paul Black
Alberto Blanco-Justicia
Fábio Borges
Arcangelo Castiglione
Luigi Catuogno
Rongmao Chen
Davide D'Arenzo
Dumitru Daniel Dinu
Patrick Derbez
Kaoutar Elkhiyaoui
Gencer Erdogan
Fabian Foerg
Maria Isabel Gonzalez Vasco
Fuchun Guo
Matt Henricksen
John Iliadis
Mohammad Jafari
Cees Jansen
Mahavir Jhawar
Maria Karyda
Jongkil Kim
Elisavet Konstantinou
Russell W.F. Lai
Meixing Le
Hao Lei
Yan Li
Kaitai Liang
Jianghua Liu
Jiqiang Lu
Oussama Mahjoub
Tobias Marktscheffel
Sergio Martinez
Weizhi Meng
Bart Mennink
Bo Mi
Tarik Moataz
Subhojeet Mukherjee
Alexios Mylonas
Panagiotis Nastou
Ana Nieto
David Nuñez
Aida Omerovic
Melek Önen
Simon Oya
Juan David Parra
Tobias Pulls
Baodong Qin
David Rebollo
Jordi Ribes-González
Akand Muhammad Rizwan
Sondre Rønjom
Olivier Sanders
Andrea Saracino
Fredrik Seehusen
Setareh Sharifian
Matteo Signorini
Jordi Soria-Comas
Yannis Stamatiou
Ralf Staudemeyer
Vasilis Stavrou
George Stergiopoulos
Mikhail Strizhov
Nikolaos Tsalis
Aleksei Udovenko
Vesselin Velichkov
Sridhar Venkatesan
Nick Virvilis
Pengwei Wang
Qingju Wang
Zhan Wang
Zhuo Wei
Shuang Wu
Yongdong Wu
Jia Xu
Yanjiang Yang
Artsiom Yautsiukhin
Tsz Hon Yuen
Lei Zhang
Ye Zhang
Yongjun Zhao

Contents

Cryptography I: Signatures

Black-Box Separations on Fiat-Shamir-Type Signatures in the Non-Programmable Random Oracle Model

Masayuki Fukumitsu[1(✉)] and Shingo Hasegawa[2]

[1] Faculty of Information Media, Hokkaido Information University, Nishi-Nopporo 59-2, Ebetsu, Hokkaido 069-8585, Japan
fukumitsu@do-johodai.ac.jp

[2] Graduate School of Information Sciences, Tohoku University, 41 Kawauchi, Aoba-ku, Sendai, Miyagi 980–8576, Japan
hasegawa@cite.tohoku.ac.jp

Abstract. In recent years, Fischlin and Fleischhacker showed the impossibility of proving the security of specific types of FS-type signatures, the signatures constructed by the Fiat-Shamir transformation, via a single-instance reduction in the non-programmable random oracle model (NPROM, for short).

In this paper, we pose a question whether or not the impossibility of proving the security of *any* FS-type signature can be shown in the NPROM. For this question, we show that each FS-type signature cannot be proven to be secure via a key-preserving reduction in the NPROM from the security against the impersonation of the underlying identification scheme under the passive attack, as long as the identification scheme is secure against the impersonation under the active attack.

We also show the security incompatibility between the discrete logarithm assumption and the security of the Schnorr signature via a single-instance key-preserving reduction, whereas Fischlin and Fleischhacker showed that such an incompatibility cannot be proven via a non-key-preserving reduction.

Keywords: Fiat-Shamir transformation · The Schnorr signature · Non-programmable random oracle model · Meta-reduction

1 Introduction

The Fiat-Shamir (FS, for short) transformation [18] is a general method to construct secure and efficient signature schemes from identification (ID, for short) schemes. It is known that there are many FS-type signatures which are signatures derived by using this method. For example, the Schnorr signature [37] and the Guillou-Quisquater (GQ, for short) signature [27] are constructed by using the FS transformation.

J. Lopez and C.J. Mitchell (Eds.): ISC 2015, LNCS 9290, pp. 3–20, 2015.
DOI: 10.1007/978-3-319-23318-5_1

The security of FS-type signatures is discussed in several literature. Pointcheval and Stern [36] first showed that an FS-type signature is existential unforgeable against the chosen-message attack (EUF-CMA, for short) in the random oracle model (ROM, for short) if the underlying ID scheme is an honest-verifier zero-knowledge proof of knowledge. By employing their result, in the ROM, one can show that the Schnorr signature is proven to be EUF-CMA from the discrete logarithm (DL, for short) assumption, and the GQ signature is proven to be EUF-CMA from the RSA assumption, respectively. Subsequently, Abdalla, An, Bellare and Namprempre [1] relaxed the condition of the honest-verifier zero-knowledge proof of knowledge. More precisely, they proved the equivalence between the EUF-CMA security of an FS-type signature and the security of the underlying ID scheme against the impersonation under the passive attack (imp-pa security, for short) in the ROM. This result indicates that the imp-pa security of the underlying ID schemes is essential for proving the security of FS-type signatures in the ROM.

On the other hand, Paillier and Vergnaud [34] gave a negative circumstantial evidence on proving the security of FS-type signatures in the *standard model*. More specifically, they showed that the Schnorr signature cannot be proven to be EUF-CMA via an algebraic reduction from the DL assumption, as long as the One-More DL (OM-DL, for short) assumption [4] holds. In a similar manner to the Schnorr signature, they also showed the impossibility of proving the security of the GQ signature in the standard model.

The security of FS-type signatures can be proven in the ROM, whereas it may not be proven in the standard model. The main reason is the programmable property of the random oracle. Informally, this property allows a reduction, which aims to prove the security of the designated cryptographic scheme, to program outputs of the random oracle. Although the programmable property is valuable, it is known that this property is strong. This is because concrete hash functions in the standard model seem not to satisfy such a property completely. As an intermediate model between the ROM and the standard model, the non-programmable random oracle model (NPROM, for short) was proposed. The concept of the NPROM was formalized by Nielsen [31]. Subsequently, the NPROM was first applied to a security proof in [20]. In the NPROM, the random oracle outputs a random value as in the ROM, but it is dealt with an independent party in the security proof. Namely, the reduction is prohibited to program outputs of the random oracle in the NPROM.

Recently, Fischlin and Fleischhacker [19] showed that the Schnorr signature cannot be proven to be EUF-CMA via a single-instance reduction in the NPROM from the DL assumption as long as the OM-DL assumption holds. Such a single-instance reduction would invoke a forger against the Schnorr signature only once, but it is allowed to rewind the forger many times. They mentioned that this impossibility result can be extended to cover any FS-type signature satisfying the following two conditions: Its secret key consists of one component, and the one-more assumption related to the cryptographic assumption from which the security of the signature is proven in the ROM holds. Therefore, their impossibility result seems not to be applied to the other FS-type signatures such as the signatures derived from the Okamoto ID [32] and the standard protocol for proving equality of DLs [11,12,25] by using the FS transformation.

1.1 Our Results

In this paper, we pose a question whether or not the impossibility of proving the security of *any* FS-type signature can be shown in the NPROM. In order to apply their result to other FS-type signatures, one strategy is to find concrete conditions corresponding to the target signatures. Indeed, Fischlin and Fleischhacker [19] showed their impossibility of proving the security of the Schnorr signature by employing such a strategy. Another one is to consider abstract conditions which can apply to any FS-type signatures. In this paper, we employ the latter strategy. As a candidate of such abstract conditions, we consider the security property of the underlying ID scheme and the type of the security reductions. More precisely, we show that any FS-type signature cannot be proven to be existential unforgeable against even the key-only attack (EUF-KOA, for short) via a key-preserving reduction in the NPROM from the imp-pa security of the underlying ID scheme, as long as the ID scheme is secure against the impersonation under the *active* attack (imp-aa secure, for short).

Our result is proven by employing the meta-reduction technique. This technique was often used to give impossibility results on the security proofs including FS-type signatures and on relationships among cryptographic assumptions [2,3,8–10,13–17,19,21–24,28–30,33–35,38–40].

As the first condition employed in our result, we restrict the reduction to being key-preserving. The *key-preserving reduction* means that a reduction is limited to invoke a forger with the same public key as the public key given to the reduction [35]. This setting was introduced by Paillier and Villar [35] to give an impossibility of proving the security of factoring-based encryptions in the standard model. Subsequently, the key-preserving property was considered to discuss the provable security of the full domain hash [14,29] and the other cryptographic schemes [10,16,17] and to investigate the strength of security models [30].

On the other hands, Fischlin and Fleischhacker [19] restricted the reduction to being *group-preserving* implicitly. This means that the reduction $\mathcal{R}$ invokes the forger with a public key which contains the same group description as that input to $\mathcal{R}$. Comparing the key-preserving setting and the group-preserving one, the key-preserving setting is stronger in a sense that the entire components of a public key are preserved in the key-preserving reduction, whereas the partial ones are only preserved in the group-preserving reduction. Nevertheless, we should note that the key-preserving property seems not to be unreasonable. This is because the security of many cryptographic schemes including FS-type signatures in the ROM [1,36] is proven via a key-preserving reduction. In particular, one can employ the forking technique [36], which was utilized to prove the security of FS-type signatures in the ROM, under even such a setting.

As the second condition, we also require the imp-aa security of the ID scheme in our impossibility result. This requirement is likely to be reasonable, because ID schemes were generally proved to satisfy the security stronger than the imp-aa security, namely the security against the impersonation under the concurrent attack (imp-ca security, for short) [6]. In fact, many ID schemes including the Schnorr ID, the GQ ID and the Okamoto ID were proven to be imp-ca secure, and hence imp-aa secure, respectively [6,32].

By the above observations, our result indicates that the security of FS-type signatures may not be proven from the imp-pa security of the underlying ID schemes by employing ordinary proof techniques in the NPROM. Note that we do not rule out the possibility that the security of FS-type signatures other than ones to which the Fischlin-Fleischhacker's result can be applied is proven from the imp-pa security of the underlying ID schemes via a non-key-preserving reduction in the NPROM.

We also consider the question whether or not the security incompatibility between the DL assumption and the security of the Schnorr signature can be proven in the NPROM. The *security incompatibility* means that the security of the Schnorr signature in the NPROM is not compatible with the DL assumption. This question was first discussed by Fischlin and Fleischhacker [19]. They showed that this incompatibility cannot be proven via a non-key-preserving reduction. On the other hand, in this paper, we give such an incompatibility via a reduction that is different from the non-key-preserving one. More precisely, we show that the Schnorr signature cannot be proven to be EUF-CMA via a single-instance key-preserving reduction in the NPROM from the DL assumption as long as the DL assumption holds. Our incompatibility result means that the EUF-CMA security of the Schnorr signature is proven from the DL assumption via a single-instance key-preserving reduction in the NPROM if and only if the DL assumption does not hold.

Recall that the Schnorr signature cannot be proven to be EUF-CMA via a single-instance reduction in the NPROM from the DL assumption, as long as the OM-DL assumption holds [19]. Therefore, it is not known whether or not this impossibility holds in the case where the OM-DL assumption does not holds. On the other hand, our incompatibility result implies that such an impossibility via a single-instance key-preserving reduction holds even when the OM-DL assumption does not hold, but the DL assumption remains to hold.

Our incompatibility result is proven by employing the invoking twin reductions technique proposed in [19]. It should be noted that our result does not contradict the one in [19], because the reduction concerned in our result differs from theirs. In [19], the non-key-preserving reduction is concerned, whereas we consider the single-instance key-preserving one. Note also that the single-instance property is used in the ordinary security proofs as well as the key-preserving one.

2 Preliminaries

In this section, we introduce some notions and notations used in this paper. Let λ denote the empty string. We denote by $x \in_{\mathrm{U}} D$ that the element x is chosen uniformly at random from the finite set D. By $x := y$, we mean that x is defined or substituted by y. For any algorithm $\mathcal{A}$, $y \leftarrow \mathcal{A}(x)$ indicates that the algorithm $\mathcal{A}$ outputs y on input x. Note that when $\mathcal{A}$ is a probabilistic algorithm, $y \leftarrow \mathcal{A}(x)$ is a shorten notation of $y \leftarrow \mathcal{A}(x; r)$ with a randomly chosen random coins r,

---

The EF-ATK game between a challenger $\mathcal{C}$ and a forger $\mathcal{F}$

Init: $\mathcal{C}$ generates $(pk, sk) \leftarrow \mathsf{KGen}(1^k)$, and then $\mathcal{C}$ submits pk to the forger $\mathcal{F}$.
Signing Oracle: This phase is only provided when ATK = CMA. When $\mathcal{F}$ hands a message m_i, $\mathcal{C}$ returns a valid signature σ_i on the pair (pk, m_i).
Challenge: When $\mathcal{F}$ outputs a pair (m^*, σ^*), $\mathcal{C}$ outputs 1 if and only if $m^* \notin \{m_i\}_i \wedge \mathsf{Ver}(pk, m^*, \sigma^*) = 1$.

Fig. 1. The description of the EF-ATK game

and y is distributed according to such random coins. A function $\nu(k)$ is *negligible* if for any polynomial μ, there exists a constant k_0 such that $\nu(k) < 1/\mu(k)$ for any $k \geq k_0$. Let $\mathtt{negl}$ denote a negligible function. We use k to denote a security parameter.

2.1 Digital Signature Scheme

A *signature scheme* SIG consists of the following three polynomial-time algorithms $(\mathsf{KGen}, \mathsf{Sign}, \mathsf{Ver})$. KGen is a probabilistic polynomial-time (PPT, for short) key generation algorithm that on input 1^k, generates a public key pk and the corresponding secret key sk. Sign is a PPT signing algorithm that on input (pk, sk, m), issues a signature σ on the message m. Ver is a deterministic verification algorithm that on input (pk, m, σ), outputs 1 if σ is a signature on the message M under the public key pk, or 0 otherwise.

We consider the EUF-KOA security and the EUF-CMA security, respectively [26]. Let ATK $\in$ {KOA, CMA}. We depict in Fig. 1 the descriptions of both the existentially forgeable game against the key only attack (EF-KOA game, for short) and the existentially forgeable game against the chosen message attack (EF-CMA game, for short). Note that when ATK = KOA, $\mathcal{C}$ outputs 1 in the **Challenge** phase if and only if $\mathsf{Ver}(pk, m^*, \sigma^*) = 1$. Then the forger $\mathcal{F}$ is said to *win the EF-ATK game* if the challenger $\mathcal{C}$ finally outputs 1 in the corresponding game. A signature scheme $\mathsf{SIG} = (\mathsf{KGen}, \mathsf{Sign}, \mathsf{Ver})$ is *EUF-ATK* if for any forger $\mathcal{F}$, $\mathcal{F}$ wins the EF-ATK game of SIG with at most negligible probability in k. The probability is taken over the coin flips of $\mathcal{C}$ and $\mathcal{F}$.

2.2 Canonical Identification Scheme

A *canonical identification scheme* ID (ID scheme, for short) [1,36] consists of $(K, \mathcal{CH}, P_1, P_2, V)$. K is a PPT key generator that on input 1^k, issues a pair (pk, sk) of a public key and the corresponding secret key. $\mathcal{CH} := \{\mathcal{CH}_{pk}\}_{pk}$ is a polynomial-time samplable family indexed by public keys of sets, namely given a public key pk generated by $K(1^k)$, one can sample an element uniformly at random from the set $\mathcal{CH}_{pk}$ in PPT. (P_1, P_2) are prover algorithms. Specifically, P_1 outputs a pair (st, cmt) of a state and a commitment in PPT on input (pk, sk), and P_2 outputs a response res on input a key pair (pk, sk), a pair (st, cmt)

$$\begin{array}{rcl} \text{Prover } \mathcal{P} & & \text{Verifier } \mathcal{V} \\ (\mathrm{st}, \mathrm{cmt}) \leftarrow P_1(pk, sk) & \xrightarrow{\mathrm{cmt}} & \\ & \xleftarrow{\mathrm{cha}} & \mathrm{cha} \in_{\mathrm{U}} \mathcal{CH}_{pk} \\ \mathrm{res} \leftarrow P_2(pk, sk, \mathrm{st}, \mathrm{cmt}, \mathrm{cha}) & \xrightarrow{\mathrm{res}} & \text{output } V(pk, \mathrm{cmt}, \mathrm{cha}, \mathrm{res}) \end{array}$$

Fig. 2. ID scheme

and a challenge $\mathrm{cha} \in \mathcal{CH}_{pk}$. V is a deterministic polynomial-time verification algorithm that outputs either 0 or 1 on input $(pk, \mathrm{cmt}, \mathrm{cha}, \mathrm{res})$. The protocol between a prover and a verifier is described as in Fig. 2.

We now define the imp-pa security and the imp-aa security, respectively [1,5,6]. Let $\mathrm{atk} \in \{\mathrm{pa}, \mathrm{aa}\}$. These are formalized by the imp-atk game depicted in Fig. 3. These games represent the situation where an impersonator $\mathcal{I}$ aims to *impersonate* a honest prover, namely $\mathcal{I}$ tries to find a commitment $\widehat{\mathrm{cmt}}$ and a response $\widehat{\mathrm{res}}$ in the **Challenge** phase such that $V\left(pk, \widehat{\mathrm{cmt}}, \widehat{\mathrm{cha}}, \widehat{\mathrm{res}}\right) = 1$ holds for a challenge $\widehat{\mathrm{cha}}$ given by $\mathcal{C}$. For this purpose, $\mathcal{I}$ is given an oracle access in the **Oracle Query** phase as a hint. In the case where $\mathrm{atk} = \mathrm{pa}$, $\mathcal{I}$ can adaptively obtain transcriptions $(\mathrm{cmt}, \mathrm{cha}, \mathrm{res})$ of conversations between a honest prover and a honest verifier. On the other hand, $\mathcal{I}$ is allowed to obtain such transcripts by

imp-atk game between a challenger $\mathcal{C}$ and an impersonator $\mathcal{I}$

Init: $\mathcal{C}$ generates a key pair $(pk, sk) \leftarrow K(1^k)$, feeds the public key pk to $\mathcal{I}$, and then sets $\mathrm{PS} := \lambda$ and $\mathrm{ST} := \lambda$.

Oracle Query: On a t-th query from $\mathcal{I}$, $\mathcal{C}$ returns $M_{\mathrm{OUT}} \leftarrow \mathsf{Tr}^{\mathsf{ID}}_{pk,sk}()$ if $\mathrm{atk} = \mathrm{pa}$, or $M_{\mathrm{OUT}} \leftarrow \mathsf{Prov}^{\mathsf{ID}}_{pk,sk}(i, M_{\mathrm{IN}})$ if $\mathrm{atk} = \mathrm{aa}$, where (i, M_{IN}) is queried from $\mathcal{I}$.

Challenge: When $\mathcal{I}$ outputs a commitment $\widehat{\mathrm{cmt}}$, $\mathcal{C}$ sends a randomly challenge $\widehat{\mathrm{cha}} \in_{\mathrm{U}} \mathcal{CH}_{pk}$ to $\mathcal{I}$. After receiving a response $\widehat{\mathrm{res}}$ from $\mathcal{I}$, $\mathcal{C}$ finally outputs $V(pk, \widehat{\mathrm{cmt}}, \widehat{\mathrm{cha}}, \widehat{\mathrm{res}})$, and then halts.

$\mathsf{Tr}^{\mathsf{ID}}_{pk,sk}()$

(Tr-1) $(\mathrm{st}_t, \mathrm{cmt}_t) \leftarrow P_1(pk, sk)$;

(Tr-2) $\mathrm{cha}_t \in_{\mathrm{U}} \mathcal{CH}_{pk}$;

(Tr-3) $\mathrm{res}_t \leftarrow P_2(pk, sk, \mathrm{st}_t, \mathrm{cmt}_t, \mathrm{cha}_t)$;

(Tr-4) output $M_{\mathrm{OUT}} := (\mathrm{cmt}_t, \mathrm{cha}_t, \mathrm{res}_t)$;

$\mathsf{Prov}^{\mathsf{ID}}_{pk,sk}(i, M_{\mathrm{IN}})$

When $\mathrm{PS} \neq i$**,** proceeds as follows:
1. choose random coins r_i;
2. $(\mathrm{st}_i, \mathrm{cmt}_i) \leftarrow P_1(pk, sk; r_i)$;
3. set $\mathrm{PS} := i$ and $\mathrm{ST} := (r_i, (\mathrm{st}_i, \mathrm{cmt}_i))$; and
4. output $M_{\mathrm{OUT}} := \mathrm{cmt}_i$.

When $\mathrm{PS} = i$**,** proceeds as follows:
1. $\mathrm{res}_i \leftarrow P_2(pk, sk, \mathrm{st}_i, \mathrm{cmt}_i, M_{\mathrm{IN}}; r_i)$;
2. set $\mathrm{PS} := \lambda$, $\mathrm{ST} := \lambda$; and
3. output $M_{\mathrm{OUT}} := \mathrm{res}_i$.

Fig. 3. Description of the imp-atk game

directly interacting with honest provers when atk is aa. In this case, $\mathcal{I}$ plays the role of a verifier. Note that $\mathcal{I}$ is prohibited to move to the **Oracle Query** phase once $\mathcal{I}$ is in the **Challenge** phase. An impersonator $\mathcal{I}$ *wins the imp-atk game* if $\mathcal{C}$ outputs 1 in the imp-atk game. Then, an ID scheme $\mathsf{ID} = (K, \mathcal{CH}, P_1, P_2, V)$ is *imp-atk secure* if for any PPT impersonator $\mathcal{I}$, $\mathcal{I}$ wins the imp-atk game against ID with at most negligible probability in k. The probability is taken over the coin flips of $\mathcal{C}$ and $\mathcal{I}$.

Note that we define the *game-based* imp-pa security (imp-aa security, resp.), whereas it was given in [6] is experiment-based. Observe that the game-based definition is equivalent to the experiment-based one.

2.3 Fiat-Shamir Transformation

Let $\mathsf{ID} = (K, \mathcal{CH}, P_1, P_2, V)$ be an ID scheme, and let PK_k denote the set of all public keys which could be generated by $\mathsf{KGen}\left(1^k\right)$ for each k. We denote by $\left\{H_{pk} : \{0,1\}^* \to \mathcal{CH}_{pk}\right\}_{k, pk \in \mathsf{PK}_k}$ a family of hash functions indexed by security parameters k and public keys $pk \in \mathsf{PK}_k$. Then, the signature $\mathsf{FS\text{-}Sig}$ is given by the Fiat-Shamir transformation [18] as in Fig. 4. We call the signatures derived from the Fiat-Shamir transformation *FS-type signatures*. For the security of each FS-type signature, it is known that the signature $\mathsf{FS\text{-}Sig}$ is EUF-CMA in the ROM if and only if ID is imp-pa secure [1].

3 Impossibility of Proving the Security of FS-Type Signatures in the NPROM

In this section, we show that an FS-type signature cannot be proven to be EUF-KOA via a key-preserving reduction in the NPROM from the imp-pa security of the underlying ID scheme. We fix an ID scheme $\mathsf{ID} = (K, \mathcal{CH}, P_1, P_2, V)$, and the FS-type signature $\mathsf{FS\text{-}Sig} = (\mathsf{KGen}, \mathsf{Sign}, \mathsf{Ver})$ derived by ID. We first describe the situation where $\mathsf{FS\text{-}Sig}$ is proven to be EUF-KOA from the imp-pa security of ID. This is formalized by the contrapositive setting as in [19, 34]. Namely, this statement holds if there exists a black-box reduction $\mathcal{R}$ that wins the imp-pa game against ID with at least non-negligible probability by black-box access to any forger $\mathcal{F}$ which wins the EF-KOA game against $\mathsf{FS\text{-}Sig}$. Through the black-box access, $\mathcal{R}$ would play the EF-KOA game with a forger $\mathcal{F}$ in which $\mathcal{R}$ is placed at the challenger's position.

Let pk be a public key given by the imp-pa challenger $\mathcal{C}$ to $\mathcal{R}$. Then, the imp-pa impersonator $\mathcal{R}$ aims to impersonate a honest prover without the secret key sk corresponding to the public key pk. Namely $\mathcal{R}$ attempts in the **Challenge** phase of the imp-pa game to find a commitment $\hat{\mathrm{cmt}}$ and a response $\hat{\mathrm{res}}$ such that $V\left(pk, \hat{\mathrm{cmt}}, \hat{\mathrm{cha}}, \hat{\mathrm{res}}\right) = 1$ holds for a challenge $\hat{\mathrm{cha}}$ given by $\mathcal{C}$. Here, the imp-pa impersonator $\mathcal{R}$ is allowed to adaptively query to the transcript oracle $\mathsf{Tr}^{\mathsf{ID}}_{pk,sk}$ to obtain a valid transcript $(\mathrm{cmt}_t, \mathrm{cha}_t, \mathrm{res}_t)$. Moreover, $\mathcal{R}$ is also able to invoke the winning EF-KOA forger $\mathcal{F}$ polynomially many times. More precisely,

The signature scheme FS-Sig given by applying the Fiat-Shamir transformation to an ID scheme $(K, \mathcal{CH}, P_1, P_2, V)$

KGen coincides with K.
Sign on input a key pair (pk, sk) and a message m, proceeds as follows:
(1) obtain $(\mathrm{st}, \mathrm{cmt}) \leftarrow P_1(pk, sk)$;
(2) set $\mathrm{cha} := H_{pk}(\mathrm{cmt}, m)$;
(3) issue a response $\mathrm{res} \leftarrow P_2(pk, sk, \mathrm{st}, \mathrm{cmt}, \mathrm{cha})$; and then
(4) output $\sigma := (\mathrm{cmt}, \mathrm{res})$ as a signature.
Ver on input (pk, m, σ), sets $c := H_{pk}(\mathrm{cmt}, m)$, and then outputs $V(pk, \mathrm{cmt}, c, \mathrm{res})$.

Fig. 4. Fiat-Shamir transformation

$\mathcal{R}$ can obtain a message/signature pair (m_i, σ_i) with non-negligible probability by handing an i-th public key pk_i to $\mathcal{F}$. $\mathcal{R}$ eventually sends a commitment $\hat{\mathrm{cmt}}$ to $\mathcal{C}$. After receiving a challenge $\hat{\mathrm{cha}}$ from $\mathcal{C}$, $\mathcal{R}$ finally outputs a response $\hat{\mathrm{res}}$. For the transcript $\left(\hat{\mathrm{cmt}}, \hat{\mathrm{cha}}, \hat{\mathrm{res}}\right)$, the probability that $V\left(pk, \hat{\mathrm{cmt}}, \hat{\mathrm{cha}}, \hat{\mathrm{res}}\right) = 1$ would be non-negligible in k.

We force the reduction $\mathcal{R}$ to be *key-preserving*. Namely, each public key pk_i fed by $\mathcal{R}$ is always pk which is given by the imp-pa challenger $\mathcal{C}$. In the NPROM, $\mathcal{R}$ obtains a hash value of H_{pk} from an external random oracle, whereas $\mathcal{R}$ simulates the random oracle in the ROM. $\mathcal{F}$ invoked by $\mathcal{R}$ is also allowed to make random oracle queries. On a random oracle query from $\mathcal{F}$, $\mathcal{R}$ replies a hash value to $\mathcal{F}$ by forwarding its query to own random oracle. Here $\mathcal{R}$ is prohibited to simulate a random oracle for $\mathcal{F}$, although it allows to observe any query given from $\mathcal{F}$. This rule captures that one cannot adopt the programming techniques used in [1,7,36] in the NPROM.

Theorem 1. *Assume that* FS-Sig *is proven to be EUF-KOA via a key-preserving reduction in the NPROM from the imp-pa security of* ID. *Then,* ID *is not imp-aa secure.*

Proof (*Sketch*). Assume that FS-Sig is proven to be EUF-KOA via a key-preserving reduction in the NPROM from the imp-pa security of ID. Then there exists a PPT reduction algorithm $\mathcal{R}$ that is key-preserving and wins the imp-pa game with at least non-negligible probability ϵ by black-box access to any forger $\mathcal{F}$ which wins the EF-KOA game with non-negligible probability in the NPROM.

We shall construct a meta-reduction $\mathcal{M}$ that wins the imp-aa game against ID with the reduction $\mathcal{R}$. Recall that $\mathcal{R}$ can impersonate a honest prover in the imp-pa game if a winning EF-KOA forger $\mathcal{F}$ and a valid transcript oracle $\mathsf{Tr}^{\mathsf{ID}}_{pk,sk}$ are provided for $\mathcal{R}$. Below, we first describe a hypothetical and specific unbounded EF-KOA forger $\widetilde{\mathcal{F}}$. Note that the reduction $\mathcal{R}$ should win the imp-pa game with probability at least ϵ even when such a forger $\widetilde{\mathcal{F}}$ is provided. Next, we give the description of $\mathcal{M}$. $\mathcal{M}$ executes $\mathcal{R}$ with the simulations of $\widetilde{\mathcal{F}}$ and $\mathsf{Tr}^{\mathsf{ID}}_{pk,sk}$ for $\mathcal{R}$. We also show that $\mathcal{M}$ succeeds in such simulations in polynomial time.

$\widetilde{\mathcal{F}}$ ← pk_i

(F-1) find cmt_i such that $(\mathrm{st}_i, \mathrm{cmt}_i) \leftarrow P_1(pk_i, sk_i)$;

(F-2) choose a polynomially-length message m_i at random;

(F-3) $\mathrm{cha}_i := H_{pk_i}(\mathrm{cmt}_i, m_i)$; and

(F-4) find res_i such that $\mathrm{res}_i \leftarrow P_2(pk_i, sk_i, \mathrm{st}_i, \mathrm{cmt}_i, \mathrm{cha}_i)$. → $(m_i, (\mathrm{cmt}_i, \mathrm{res}_i))$

Fig. 5. The description of an (unbounded) EF-KOA forger $\widetilde{\mathcal{F}}$

Description of Unbounded Forger $\widetilde{\mathcal{F}}$ We depict in Fig. 5 the unbounded EF-KOA forger $\widetilde{\mathcal{F}}$, where sk_i denotes the secret key corresponding to the public key pk_i given to $\widetilde{\mathcal{F}}$. We should note that the processes (F-1) and (F-4) are not necessarily done in polynomial time here. However, we will construct an imp-aa impersonator $\mathcal{M}$ that can simulate $\widetilde{\mathcal{F}}$ in polynomial time under the key-preserving property of $\mathcal{R}$. Moreover, $\widetilde{\mathcal{F}}$ computes $H_{pk_i}(\mathrm{cmt}_i, m_i)$ in the NPROM by querying (cmt_i, m_i) to the random oracle. Since the tuple $(\mathrm{cmt}_i, \mathrm{cha}_i, \mathrm{res}_i)$ is issued through the same processes of $\mathsf{Sign}(pk_i, sk_i, m_i)$ as in Fig. 4, the tuple $(m_i, (\mathrm{cmt}_i, \mathrm{res}_i))$ output by $\widetilde{\mathcal{F}}$ always satisfies that $\mathsf{Ver}(pk_i, m_i, (\mathrm{cmt}_i, \mathrm{res}_i)) = 1$.

Description of Meta-Reduction $\mathcal{M}$ In Fig. 6, we depict the meta-reduction $\mathcal{M}$ which wins the imp-aa game, where I denotes the upper bound of the total number of invoking an EF-KOA forger by $\mathcal{R}$ and rewinding it, and q denotes the upper bound of the number of queries to $\mathsf{Tr}^{\mathsf{ID}}_{pk,sk}$, respectively.

We show that $\mathcal{M}$ wins the imp-aa game with probability at least ϵ in PPT. In (M-5), $\mathcal{M}$ just intermediates between $\mathcal{C}$ and $\mathcal{R}$. This implies that $\mathcal{M}$ can impersonate a honest prover in the imp-aa game with probability at least ϵ without the secret key sk corresponding to the public key pk given to $\mathcal{M}$ if $\mathcal{R}$ impersonates the honest prover in the imp-pa game. On the other hand, as mentioned above, $\mathcal{R}$ can impersonate the honest prover in the imp-pa game with probability at least ϵ when $\mathcal{M}$ succeeds in the simulations of the forger $\widetilde{\mathcal{F}}$ and the transcript oracle $\mathsf{Tr}^{\mathsf{ID}}_{pk,sk}$ for $\mathcal{R}$. For the simulations in (M-4), the following claims hold. Here, we show that $\mathcal{M}$ perfectly simulates these in (M-4).

Claim 2. *$\mathcal{M}$ perfectly simulates $\widetilde{\mathcal{F}}$ in the $\mathcal{R}$'s viewpoint.*

Proof. We fix an i-th invocation of the EF-KOA forger by $\mathcal{R}$. $\mathcal{R}$ would invoke such a forger on an i-th public key pk_i in (M-4). It should be noted that pk_i always coincides with the public key pk. This is because $\mathcal{R}$ is supposed to be key-preserving. On the i-th public key pk by $\mathcal{R}$, $\mathcal{M}$ queries the pair (cmt_i, m_i) to the random oracle and then returns the message/signature pair $(m_i, (\mathrm{cmt}_i, \mathrm{res}_i))$ in (M-4). The pair $(m_i, (\mathrm{cmt}_i, \mathrm{res}_i))$ is issued in (M-1). Therefore, it suffices that $\mathcal{M}$ issues such a pair $(m_i, (\mathrm{cmt}_i, \mathrm{res}_i))$ in (M-1) in the same way as $\widetilde{\mathcal{F}}$.

In (a), the imp-aa impersonator $\mathcal{M}$ obtains the i-th commitment $\mathrm{cmt}_i \leftarrow P_1(pk, sk)$ by querying to the prover oracle provided by the imp-aa challenger $\mathcal{C}$. It follows from $pk = pk_i$ that cmt_i is issued as in (F-1). The processes (b)

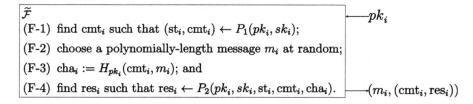

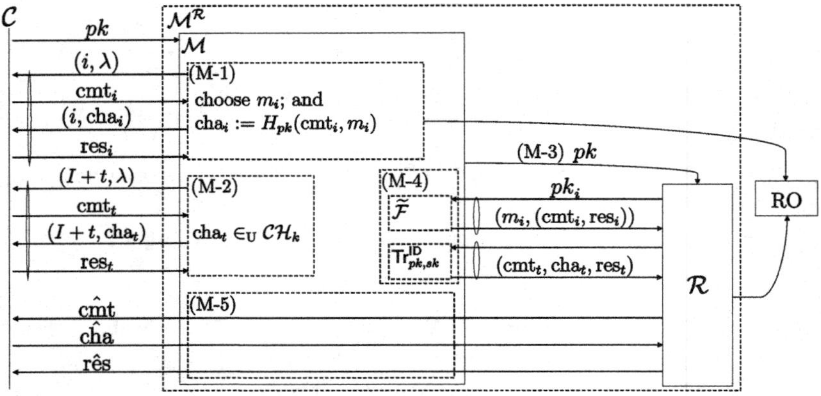

Meta-Reduction $\mathcal{M}(pk)$

(M-1) For each $1 \le i \le I$, $\mathcal{M}$ proceeds as follows:
 (a) obtain cmt_i by querying (i, λ) to $\mathcal{C}$;
 (b) choose a polynomially-length message m_i at random;
 (c) obtain the hash value cha_i of (cmt_i, m_i) by querying to the random oracle; and
 (d) obtain res_i by querying (i, cha_i) to $\mathcal{C}$.

(M-2) For each $1 \le t \le q$, $\mathcal{M}$ proceeds as follows:
 (a) obtain cmt_t by querying $(I+t, \lambda)$ to $\mathcal{C}$;
 (b) choose $\mathrm{cha}_t \in_{\mathrm{U}} \mathcal{CH}_{pk}$; and
 (c) obtain res_t by querying $(I+t, \mathrm{cha}_t)$ to $\mathcal{C}$.

(M-3) $\mathcal{M}$ runs $\mathcal{R}$ on input pk;

(M-4) For each query from $\mathcal{R}$, $\mathcal{M}$ responds in the following way;
 For an i-th invocation of $\mathcal{F}$, $\mathcal{M}$ queries (cmt_i, m_i) to the random oracle provided by $\mathcal{R}$, and then returns $(m_i, (\mathrm{cmt}_i, \mathrm{res}_i))$ issued in (M-1).
 For a t-th query to $\mathsf{Tr}^{\mathsf{ID}}_{pk,sk}$, $\mathcal{M}$ returns $(\mathrm{cmt}_t, \mathrm{cha}_t, \mathrm{res}_t)$ issued in (M-2).

(M-5) When $\mathcal{R}$ eventually outputs $\widehat{\mathrm{cmt}}$, $\mathcal{M}$ moves to the **Challenge** phase by sending $\widehat{\mathrm{cmt}}$ to $\mathcal{C}$. After receiving $\widehat{\mathrm{cha}} \in_{\mathrm{U}} \mathcal{CH}_{pk}$ from $\mathcal{C}$, $\mathcal{M}$ forwards it to $\mathcal{R}$. Once $\mathcal{R}$ outputs $\widehat{\mathrm{res}}$, $\mathcal{M}$ also outputs it and then halts.

Fig. 6. Configuration of $\mathcal{M}$

and (c) are identical to (F-2) and (F-3), respectively. This is because both $\mathcal{M}$ and $\widetilde{\mathcal{F}}$ obtain the i-th challenge cha_i by choosing m_i at random, and then querying the pair (cmt_i, m_i) to the random oracle. In (d), $\mathcal{M}$ obtains the i-th response res_i by querying (i, cha_i) to the prover oracle. Since $\mathcal{C}$ answers $\mathrm{res}_i \leftarrow P_2(pk, sk, \mathrm{st}_i, \mathrm{cmt}_i, \mathrm{cha}_i)$ and $pk_i = pk$, res_i is also issued in the same way as (F-4). Thus $\mathcal{M}$ perfectly simulates $\widetilde{\mathcal{F}}$. □

In a similar manner to **Claim** 2, the following claim is proven.

Claim 3. *$\mathcal{M}$ perfectly simulates $\mathsf{Tr}^{\mathsf{ID}}_{pk,sk}$ in the $\mathcal{R}$'s viewpoint.*

Proof. We fix a t-th $\mathcal{R}$'s query to the transcript oracle. On the t-th query by $\mathcal{R}$, $\mathcal{M}$ returns the transcript $(\mathrm{cmt}_t, \mathrm{cha}_t, \mathrm{res}_t)$ in (M-4). This transcript is issued in (M-2). Therefore, it suffices that $\mathcal{M}$ issues such a transcript $(\mathrm{cmt}_t, \mathrm{cha}_t, \mathrm{res}_t)$ in (M-2) in the same way as the transcript oracle $\mathsf{Tr}^{\mathsf{ID}}_{pk,sk}$ under the pair (pk, sk) of the public key pk given to $\mathcal{R}$ and the corresponding secret key sk.

In (a), the imp-aa impersonator $\mathcal{M}$ obtains the t-th commitment $\mathrm{cmt}_t \leftarrow P_1(pk, sk)$ by querying to the prover oracle provided by the imp-aa challenger $\mathcal{C}$. Since $\mathcal{C}$ generates this commitment cmt_t under the public key pk given to $\mathcal{R}$, cmt_t is issued in the same way as in (Tr-1) in the $\mathcal{R}$'s viewpoint. The process (b) is identical to (Tr-2), because both $\mathcal{M}$ and $\mathsf{Tr}^{\mathsf{ID}}_{pk,sk}$ choose the t-th challenge $\mathrm{cha}_t \in_{\mathrm{U}} \mathcal{CH}_{pk}$ in (b) and in (Tr-2), respectively. In (c), $\mathcal{M}$ obtains the t-th response res_t by querying $(I + t, \mathrm{cha}_t)$ to the prover oracle. Since $\mathcal{C}$ answers $\mathrm{res}_t \leftarrow P_2(pk, sk, \mathrm{st}_t, \mathrm{cmt}_t, \mathrm{cha}_t)$, res_t is also issued in the same way as (Tr-3). Thus $\mathcal{M}$ perfectly simulates $\mathsf{Tr}^{\mathsf{ID}}_{pk,sk}$. □

We need to consider the case where $\mathcal{R}$ rewinds the EF-KOA forger during an i-th invocation. We now show that such a rewind can be replaced with the newly invocation of the forger. Since $\mathcal{M}$ simulates the specific forger $\widetilde{\mathcal{F}}$, it returns the final output $(m_i, (\mathrm{cmt}_i, \mathrm{res}_i))$ soon after $\mathcal{R}$ gives the hash value of the random oracle query (cmt_i, m_i). Therefore, $\mathcal{R}$ would rewind the forger soon after $\mathcal{M}$ makes a random oracle query. Recall that the key-preserving reduction $\mathcal{R}$ always feeds the same public key pk_i as the public key pk given to $\mathcal{R}$. In this case, $\mathcal{M}$ aborts the i-th simulation of $\widetilde{\mathcal{F}}$, and then it starts the simulation of the $(i + 1)$-th invocation. Then $\mathcal{M}$ hands the random oracle query $(\mathrm{cmt}_{i+1}, m_{i+1})$ to $\mathcal{R}$, and then it proceeds to the simulation as in Fig. 6. In the $\mathcal{R}$'s viewpoint, an EF-KOA forger makes a new random oracle query soon after $\mathcal{R}$ rewinds the forger.

We evaluate the running time of $\mathcal{M}$. $\mathcal{M}$ only chooses polynomially many messages and challenges in (M-1) and (M-2), respectively, and $\mathcal{M}$ makes queries to the prover oracle and the random oracle, and invokes $\mathcal{R}$ once. Therefore, $\mathcal{M}$ runs in polynomial time. By the correctness of the simulations, it follows that with probability at least ϵ, $\mathcal{R}$ can find $(\widehat{\mathrm{cmt}}, \widehat{\mathrm{res}})$ in (M-5) such that

---

The Schnorr Signature [37]

KGen on input 1^k, outputs $(pk, sk) := (Y, x) \leftarrow \mathsf{IGen}(1^k)$.

Sign on a tuple (pk, sk, m), where $pk = (\mathbb{G}, p, g, y)$, issues a signature $\sigma = (\mathrm{cmt}, \mathrm{res})$ in the following way:

(1) $\mathrm{st} \in_{\mathrm{U}} \mathbb{Z}_p$;

(2) $\mathrm{cmt} := g^{\mathrm{st}}$;

(3) $\mathrm{cha} := H_{pk}(\mathrm{cmt}, m)$; and

(4) $\mathrm{res} := \mathrm{st} + sk \cdot \mathrm{cha} \bmod p$.

Ver on a tuple $(pk, m, (\mathrm{cmt}, \mathrm{res}))$, sets $c := H_{pk}(\mathrm{cmt}, m)$, and then outputs 1 if and only if $g^{\mathrm{res}} = \mathrm{cmt} \cdot y^c$.

Fig. 7. The Schnorr signature

$V\left(\hat{\mathsf{cmt}}, \hat{\mathsf{cha}}, \hat{\mathsf{res}}\right) = 1$. Thus $\mathcal{M}$ can win the imp-aa game with probability at least ϵ, and hence ID is not imp-aa secure. □

4 Security Incompatibility Between the DL Assumption and the EUF-CMA Security of the Schnorr Signature in the NPROM

In this section, we show the security incompatibility between the DL assumption and the EUF-CMA security of the Schnorr signature via the single-instance key-preserving reduction, whereas Fischlin and Fleischhacker [19] showed that one cannot prove such an incompatibility via the non-key-preserving reduction. More specifically, we show that the DL assumption does not hold if the Schnorr signature is proven to be EUF-CMA via a single-instance key-preserving reduction in the NPROM from the DL assumption. Our incompatibility result can be applied to any FS-type signature whose security proof in the ROM is given from a non-interactive cryptographic assumption, such as the GQ signature [27].

Let $\mathbb{G}$ be a group of prime order p with a generator g. For any natural number N, we use $\mathbb{Z}_N$ to stand for the residue ring $\mathbb{Z}/N\mathbb{Z}$. We denote by IGen the DL instance generator. On input 1^k, IGen generates a pair (Y, x) of an instance $Y := (\mathbb{G}, p, g, y)$ and the solution x corresponding to Y, where $\mathbb{G}$ denotes the description of a group, p is a polynomial-length prime in k which represents the order of $\mathbb{G}$, g is a generator of $\mathbb{G}$, $y = g^x$ and $x \in_U \mathbb{Z}_p$. An algorithm $\mathcal{R}$ is said to solve the *DL problem* if $\mathcal{R}$ outputs the solution x on input Y, where $(Y, x) \leftarrow \mathsf{IGen}\left(1^k\right)$ for some k. The *DL assumption* holds if for any PPT algorithm $\mathcal{R}$, $\mathcal{R}$ solves the DL problem with at most negligible probability. The probability is taken over the coin flips of IGen and $\mathcal{R}$. The Schnorr signature [37] consists as in Fig. 7.

As in the previous section, we formalize the situation where the Schnorr signature is proven to be EUF-CMA via a single-instance key-preserving reduction in the NPROM from the DL assumption. Namely this situation holds if there

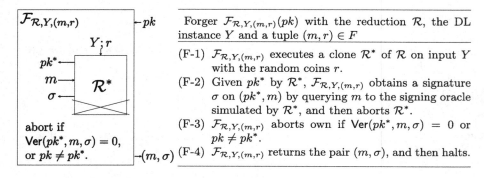

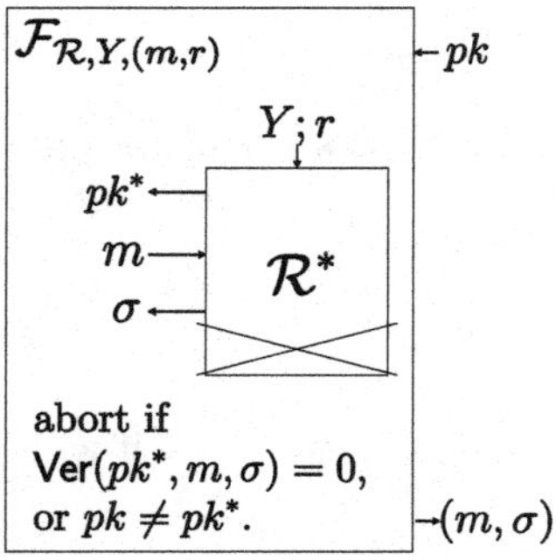

Forger $\mathcal{F}_{\mathcal{R},Y,(m,r)}(pk)$ with the reduction $\mathcal{R}$, the DL instance Y and a tuple $(m, r) \in F$

(F-1) $\mathcal{F}_{\mathcal{R},Y,(m,r)}$ executes a clone $\mathcal{R}^*$ of $\mathcal{R}$ on input Y with the random coins r.
(F-2) Given pk^* by $\mathcal{R}^*$, $\mathcal{F}_{\mathcal{R},Y,(m,r)}$ obtains a signature σ on (pk^*, m) by querying m to the signing oracle simulated by $\mathcal{R}^*$, and then aborts $\mathcal{R}^*$.
(F-3) $\mathcal{F}_{\mathcal{R},Y,(m,r)}$ aborts own if $\mathsf{Ver}(pk^*, m, \sigma) = 0$ or $pk \neq pk^*$.
(F-4) $\mathcal{F}_{\mathcal{R},Y,(m,r)}$ returns the pair (m, σ), and then halts.

Fig. 8. The description of a hypothetical forger $\mathcal{F}_{\mathcal{R},Y,(m,r)}$

exists a single-instance key-preserving reduction $\mathcal{R}$ that solves the DL problem with at least non-negligible probability by black-box access to a forger $\mathcal{F}$ that wins the EF-CMA game against the Schnorr signature with non-negligible probability in the NPROM. The reduction $\mathcal{R}$ is forced to be *single-instance* in addition to the key-preserving property. The single-instance reduction means that $\mathcal{R}$ is limited to invoke $\mathcal{F}$ only once, but it is allowed to rewind $\mathcal{F}$ many times [19]. Another different point from the previous section is that $\mathcal{R}$ needs to simulate the signing oracle for the EF-CMA forger $\mathcal{F}$. As the condition implicitly considered in [19], the reduction $\mathcal{R}$ is supposed to succeed in such a simulation with at most negligible error probability. Most reduction given in security proofs such as [1, 19, 36] satisfies this condition.

Theorem 4. *Assume that the Schnorr signature can be proven to be EUF-CMA via a single-instance key-preserving reduction in the NPROM from the DL assumption. Then the DL assumption does not hold.*

Proof (*Sketch*). We assume that there exists a single-instance key-preserving reduction $\mathcal{R}$ that solves the DL problem with black-box access to any EF-CMA forger against the Schnorr signature in the NPROM. Then we shall construct a meta-reduction $\mathcal{M}$ that solves the DL problem. On any DL instance $Y = (\mathbb{G}, p, g, y)$, $\mathcal{R}$ would find the solution x with probability at least ϵ if an EF-CMA forger against the Schnorr signature is provided for $\mathcal{R}$. We first describe a hypothetical and specific forger $\mathcal{F}_{\mathcal{R},Y,(m,r)}$. It depends on the reduction $\mathcal{R}$, the DL instance Y given to $\mathcal{R}$, a message m and random coins r. $\mathcal{F}_{\mathcal{R},Y,(m,r)}$ exploits a clone $\mathcal{R}^*$ of the reduction $\mathcal{R}$ on input Y with the random coins r. In a similar manner to Theorem 1, the reduction $\mathcal{R}$ should solve the DL problem with probability at least ϵ even when such a forger $\mathcal{F}_{\mathcal{R},Y,(m,r)}$ is provided. Next, we construct $\mathcal{M}$ that solves the DL problem by utilizing $\mathcal{R}$ with a simulation of such a forger $\mathcal{F}_{\mathcal{R},Y,(m,r)}$. We also show that $\mathcal{M}$ perfectly simulates the hypothetical forger $\mathcal{F}_{\mathcal{R},Y,(m,r)}$ in the view of the single-instance key-preserving reduction $\mathcal{R}$.

Description of $\mathcal{F}_{\mathcal{R},Y,(m,r)}$ Let F be a set of pairs (m, r) of a message m and random coins r. For the reduction $\mathcal{R}$, the DL instance Y and each $(m, r) \in F$, we depict a hypothetical forger $\mathcal{F}_{\mathcal{R},Y,(m,r)}$ in Fig. 8.

We show a fact that for each $(m, r) \in F$, if $pk = pk^*$, then $\mathcal{F}_{\mathcal{R},Y,(m,r)}$ wins the EF-CMA game of the Schnorr signature between the challenger $\mathcal{C}$ and $\mathcal{F}_{\mathcal{R},Y,(m,r)}$ with at least the probability that the clone reduction $\mathcal{R}^*$ correctly answers a signature $\sigma = (\mathrm{cmt}, \mathrm{res})$ on (pk^*, m). Assume that $pk = pk^*$. In order to show this fact, it suffices that $\mathcal{F}_{\mathcal{R},Y,(m,r)}$ wins the EF-CMA game of the Schnorr signature when $\mathcal{R}^*$ correctly answers a signature $\sigma = (\mathrm{cmt}, \mathrm{res})$ on (pk^*, m). Therefore, we assume that $\mathcal{R}^*$ correctly answers a signature $\sigma = (\mathrm{cmt}, \mathrm{res})$ on (pk^*, m). Since $\mathcal{F}_{\mathcal{R},Y,(m,r)}$ makes no query in the **Signing oracle** phase of the EF-CMA game between $\mathcal{C}$ and it, it wins this game if it merely returns a pair (m, σ) such that $\mathsf{Ver}(pk, m, \sigma) = 1$ on the public key pk given to $\mathcal{F}_{\mathcal{R},Y,(m,r)}$. The assumption on $\mathcal{R}^*$ implies that σ satisfies that $\mathsf{Ver}(pk^*, m, \sigma) = 1$. Note that $\mathcal{F}_{\mathcal{R},Y,(m,r)}$ queries the pair (cmt, m) to the external random oracle in order to obtain the hash value of (cmt, m) in the verification in (F-3). In addition, $\mathcal{F}_{\mathcal{R},Y,(m,r)}$ also asks all random oracle queries made by $\mathcal{R}^*$ to the external random oracle. It follows from $pk = pk^*$ that $\mathsf{Ver}(pk, m, \sigma) = \mathsf{Ver}(pk^*, m, \sigma) = 1$. Thus, $\mathcal{F}_{\mathcal{R},Y,(m,r)}$ wins the EF-CMA game of the Schnorr signature with at least the probability that the clone reduction $\mathcal{R}^*$ correctly answers a signature $\sigma = (\mathrm{cmt}, \mathrm{res})$ on (pk^*, m). Note that we will show that the condition $pk = pk^*$ is guaranteed under a key-preserving reduction in **Claim** 5.

Description of $\mathcal{M}$ We depict in Fig. 9 the description of $\mathcal{M}$ that solves the DL problem. Note that $\mathcal{M}$ just outputs x once $\mathcal{R}^*$ outputs the solution x without the invocation of an EF-CMA forger. Hereafter, we only consider the other case. The following claim can be shown.

Claim 5. *$\mathcal{M}$ perfectly simulates $\mathcal{F}_{\mathcal{R},Y,(m,r)}$ in the $\mathcal{R}$'s viewpoint.*

Proof. We show that $\mathcal{M}$ behaves in (M-2) in the same way as $\mathcal{F}_{\mathcal{R},Y,(m,r)}$. On a public key pk fed by $\mathcal{R}$, $\mathcal{M}$ executes a clone $\mathcal{R}^*$ of $\mathcal{R}$ on the DL instance Y given to $\mathcal{M}$ with the random coins r in (a) as in (F-1). In the same manner to (F-2), $\mathcal{M}$ obtains a signature $\sigma = (\mathrm{cmt}, \mathrm{res})$ on the message m under the public key pk^* given by $\mathcal{R}^*$ in (b). As the behavior of $\mathcal{F}_{\mathcal{R},Y,(m,r)}$, $\mathcal{M}$ asks (cmt, m) and a sequence Q of all random oracle queries made by $\mathcal{R}^*$ to the random oracle

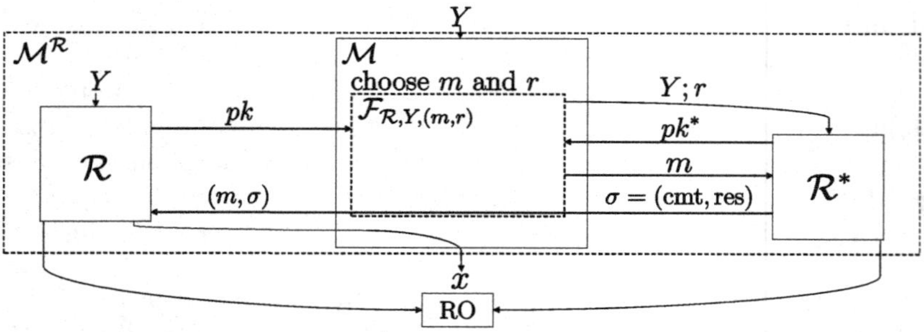

Meta-Reduction $\mathcal{M}(Y)$

(M-1) $\mathcal{M}$ executes $\mathcal{R}$ on the DL instance Y.
(M-2) When $\mathcal{R}$ invokes an EF-CMA forger on pk, $\mathcal{M}$ chooses a message m and random coins r at random, and then proceeds as follows:
  (a) execute a clone $\mathcal{R}^*$ of $\mathcal{R}$ on Y with r;
  (b) given pk^* by $\mathcal{R}^*$, obtain a signature $\sigma = (\mathrm{cmt}, \mathrm{res})$ on m by querying m to $\mathcal{R}^*$, and then abort $\mathcal{R}^*$;
  (c) query the pair (cmt, m) and a sequence Q of all random oracle queries made by $\mathcal{R}^*$ to the random oracle provided by $\mathcal{R}$;
  (d) abort $\mathcal{M}$ if $\mathsf{Ver}(pk^*, m, \sigma) = 0$, or proceed to (e) otherwise; and
  (e) return (m, σ) to $\mathcal{R}$.
(M-3) When $\mathcal{R}$ outputs x as the final output, $\mathcal{M}$ also outputs x and then halts.

Fig. 9. Configuration of $\mathcal{M}$

provided by $\mathcal{R}$ in (c). In (d), $\mathcal{M}$ aborts own if $\mathsf{Ver}(pk^*, m, \sigma) = 0$. It should be noted that $pk = pk^* = Y$. This is because the key-preserving property of $\mathcal{R}$ leads that any public key fed by $\mathcal{R}$ always coincides with the DL instance Y input to $\mathcal{R}$. This implies that the process (d) is identical to (F-3). $\mathcal{M}$ eventually returns the pair (m, σ) to $\mathcal{R}$ in (e). Therefore, $\mathcal{M}$ indeed behaves in the same way as $\mathcal{F}_{\mathcal{R},Y,(m,r)}$.

We also show that the pair (m, σ) returned by $\mathcal{M}$ satisfies that $\mathsf{Ver}(pk, m, \sigma) = 1$ when $\mathcal{R}^*$ correctly answers a signature σ on (pk^*, m) in (b). Assume that $\mathcal{R}^*$ correctly answers a valid signature $\sigma = (\mathrm{cmt}, \mathrm{res})$ on (pk^*, m). $\mathcal{R}$ can know the same input/output pairs of hash values which are issued from $\mathcal{R}^*$, because $\mathcal{M}$ asks the pair (cmt, m) and the sequence Q to the random oracle provided by $\mathcal{R}$ in (d). It follows from $pk = pk^*$ that $\mathcal{R}$ can ensure that $g^{\mathrm{res}} = \mathrm{cmt} \cdot y^c$ for $c = H_{pk}(\mathrm{cmt}, m)$ on $pk = (\mathbb{G}, p, g, y)$, and hence $\mathsf{Ver}(pk, m, (\mathrm{cmt}, \mathrm{res})) = 1$.

Thus $\mathcal{M}$ perfectly simulates $\mathcal{F}_{\mathcal{R},Y,(m,r)}$ in the $\mathcal{R}$'s viewpoint. □

It should be noted that $\mathcal{R}$ may rewind the forger $\mathcal{F}_{\mathcal{R},Y,(m,r)}$. Since the single-instance reduction $\mathcal{R}$ is limited to invoke the forger only once, the rewind would be occurred during its invocation. Observe that the behavior of $\mathcal{F}_{\mathcal{R},Y,(m,r)}$ is deterministic for the reduction $\mathcal{R}$, the DL instance Y given to $\mathcal{M}$ and the pair (m, r) which is fixed by $\mathcal{M}$. Therefore, rewinding $\mathcal{F}_{\mathcal{R},Y,(m,r)}$ does not affect the simulation of $\mathcal{F}_{\mathcal{R},Y,(m,r)}$.

We evaluate that $\mathcal{M}$ runs in polynomial time, because $\mathcal{M}$ just chooses a polynomial-length message and polynomial-length random coins in (M-2), makes at most polynomially many queries to the random oracle, and $\mathcal{M}$ invokes $\mathcal{R}$ and $\mathcal{R}^*$ once. Moreover, we show that $\mathcal{M}$ can output the correct solution x of the DL instance Y with probability at least $\epsilon - \mathtt{negl}(k)$. Recall that $\mathcal{R}$ invoked by $\mathcal{M}$ would output the solution x with probability at least ϵ if a winning EF-CMA forger $\mathcal{F}$ is provided for $\mathcal{R}$. On the other hand, as shown in **Claim** 5, $\mathcal{M}$ indeed simulates the EF-CMA forger $\mathcal{F}_{\mathcal{R},Y,(m,r)}$ for $\mathcal{R}$. Note that $\mathcal{M}$ aborts with negligible probability in (d) of (M-2). This is because the reduction $\mathcal{R}^*$ is supposed to simulate the signing oracle with negligible error probability. These imply that $\mathcal{M}$ can solve the DL problem with at least $\epsilon - \mathtt{negl}(k)$, and hence the DL assumption does not hold. □

Acknowledgements. We would like to thank anonymous reviewers for their valuable comments and suggestions. A part of this work is supported by JSPS KAKENHI Grant Number 15K16001.

References

1. Abdalla, M., An, J.H., Bellare, M., Namprempre, C.: From identification to signatures via the Fiat-Shamir transform: necessary and sufficient conditions for security and forward-security. IEEE Trans. Inf. Theory **54**(8), 3631–3646 (2008). Conference Ver.: Proc. EUROCRYPT 2002, LNCS, vol. 2332, pp. 418–433, 2002
2. Abe, M., Groth, J., Ohkubo, M.: Separating short structure-preserving signatures from non-interactive assumptions. In: Lee, D.H., Wang, X. (eds.) ASIACRYPT 2011. LNCS, vol. 7073, pp. 628–646. Springer, Heidelberg (2011)

3. Baldimtsi, F., Lysyanskaya, A.: On the security of one-witness blind signature schemes. In: Sako, K., Sarkar, P. (eds.) ASIACRYPT 2013, Part II. LNCS, vol. 8270, pp. 82–99. Springer, Heidelberg (2013)
4. Bellare, M., Namprempre, C., Pointcheval, D., Semanko, M.: The one-more-RSA-inversion problems and the security of Chaum's blind signature scheme. J. Cryptology **16**(3), 185–215 (2003). Conference Ver.: Proc. Financial Cryptography 2001, LNCS, vol. 2339, 2002
5. Bellare, M., Namprempre, C., Neven, G.: Security proofs for identity-based identification and signature schemes. J. Cryptology **22**(1), 1–61 (2009)
6. Bellare, M., Palacio, A.: GQ and Schnorr identification schemes: proofs of security against impersonation under active and concurrent attacks. In: Yung, M. (ed.) CRYPTO 2002. LNCS, vol. 2442, pp. 162–177. Springer, Heidelberg (2002)
7. Bellare, M., Rogaway, P.: Random oracles are practical: a paradigm for designing efficient protocols. In: Proceedings of ACM CCS 1993, Fairfax, Virginia, USA, pp. 62–73. ACM Press, New York (1993)
8. Boneh, D., Venkatesan, R.: Breaking RSA may not be equivalent to factoring. In: Nyberg, K. (ed.) EUROCRYPT 1998. LNCS, vol. 1403, pp. 59–71. Springer, Heidelberg (1998)
9. Bresson, E., Monnerat, J., Vergnaud, D.: Separation results on the "one-more" computational problems. In: Malkin, T. (ed.) CT-RSA 2008. LNCS, vol. 4964, pp. 71–87. Springer, Heidelberg (2008)
10. Brown, D.R.L.: What hashes make RSA-OAEP secure? Cryptology ePrint Archive, Report 2006/223 (2006). http://eprint.iacr.org/
11. Camenisch, J., Stadler, M.: Proof systems for general statements about discrete logarithms. Technical report (1997)
12. Chaum, D., Evertse, J.-H., van de Graaf, J.: An improved protocol for demonstrating possession of discrete logarithms and some generalizations. In: Price, W.L., Chaum, D. (eds.) EUROCRYPT 1987. LNCS, vol. 304, pp. 127–141. Springer, Heidelberg (1988)
13. Chen, Y., Huang, Q., Zhang, Z.: Sakai-Ohgishi-Kasahara identity-based non-interactive key exchange scheme, revisited. In: Susilo, W., Mu, Y. (eds.) ACISP 2014. LNCS, vol. 8544, pp. 274–289. Springer, Heidelberg (2014)
14. Coron, J.-S.: Optimal security proofs for PSS and other signature schemes. In: Knudsen, L.R. (ed.) EUROCRYPT 2002. LNCS, vol. 2332, pp. 272–287. Springer, Heidelberg (2002)
15. Dagdelen, Ö., Fischlin, M., Gagliardoni, T.: The Fiat–Shamir transformation in a quantum world. In: Sako, K., Sarkar, P. (eds.) ASIACRYPT 2013, Part II. LNCS, vol. 8270, pp. 62–81. Springer, Heidelberg (2013)
16. El Aimani, L.: On generic constructions of designated confirmer signatures. In: Roy, B., Sendrier, N. (eds.) INDOCRYPT 2009. LNCS, vol. 5922, pp. 343–362. Springer, Heidelberg (2009)
17. El Aimani, L.: Efficient confirmer signatures from the "signature of a commitment" paradigm. In: Heng, S.-H., Kurosawa, K. (eds.) ProvSec 2010. LNCS, vol. 6402, pp. 87–101. Springer, Heidelberg (2010)
18. Fiat, A., Shamir, A.: How to prove yourself: practical solutions to identification and signature problems. In: Odlyzko, A.M. (ed.) CRYPTO 1986. LNCS, vol. 263, pp. 186–194. Springer, Heidelberg (1987)
19. Fischlin, M., Fleischhacker, N.: Limitations of the meta-reduction technique: the case of schnorr signatures. In: Johansson, T., Nguyen, P.Q. (eds.) EUROCRYPT 2013. LNCS, vol. 7881, pp. 444–460. Springer, Heidelberg (2013). Full Ver.: Cryptology ePrint Archive, Report 2013/140

20. Fischlin, M., Lehmann, A., Ristenpart, T., Shrimpton, T., Stam, M., Tessaro, S.: Random oracles with(out) programmability. In: Abe, M. (ed.) ASIACRYPT 2010. LNCS, vol. 6477, pp. 303–320. Springer, Heidelberg (2010)
21. Fleischhacker, N., Jager, T., Schröder, D.: On tight security proofs for schnorr signatures. In: Sarkar, P., Iwata, T. (eds.) ASIACRYPT 2014. LNCS, vol. 8873, pp. 512–531. Springer, Heidelberg (2014)
22. Fukumitsu, M., Hasegawa, S., Isobe, S., Koizumi, E., Shizuya, H.: Toward separating the strong adaptive pseudo-freeness from the strong RSA assumption. In: Boyd, C., Simpson, L. (eds.) ACISP 2013. LNCS, vol. 7959, pp. 72–87. Springer, Heidelberg (2013)
23. Fukumitsu, M., Hasegawa, S., Isobe, S., Shizuya, H.: On the impossibility of proving security of strong-RSA signatures via the RSA assumption. In: Susilo, W., Mu, Y. (eds.) ACISP 2014. LNCS, vol. 8544, pp. 290–305. Springer, Heidelberg (2014)
24. Garg, S., Bhaskar, R., Lokam, S.V.: Improved bounds on security reductions for discrete log based signatures. In: Wagner, D. (ed.) CRYPTO 2008. LNCS, vol. 5157, pp. 93–107. Springer, Heidelberg (2008)
25. Goh, E.J., Jarecki, S., Katz, J., Wang, N.: Efficient signature schemes with tight reductions to the Diffie-Hellman problems. J. Cryptology **20**(4), 493–514 (2007)
26. Goldwasser, S., Micali, S., Rivest, R.L.: A digital signature scheme secure against adaptive chosen-message attacks. SIAM J. Comput. **17**(2), 281–308 (1988)
27. Guillou, L.C., Quisquater, J.-J.: A practical zero-knowledge protocol fitted to security microprocessor minimizing both transmission and memory. In: Barstow, D., et al. (eds.) EUROCRYPT 1988. LNCS, vol. 330, pp. 123–128. Springer, Heidelberg (1988)
28. Hanaoka, G., Matsuda, T., Schuldt, J.C.N.: On the impossibility of constructing efficient key encapsulation and programmable hash functions in prime order groups. In: Safavi-Naini, R., Canetti, R. (eds.) CRYPTO 2012. LNCS, vol. 7417, pp. 812–831. Springer, Heidelberg (2012)
29. Kakvi, S.A., Kiltz, E.: Optimal security proofs for full domain hash, revisited. In: Pointcheval, D., Johansson, T. (eds.) EUROCRYPT 2012. LNCS, vol. 7237, pp. 537–553. Springer, Heidelberg (2012)
30. Kawai, Y., Sakai, Y., Kunihiro, N.: On the (im)possibility results for strong attack models for public key cryptsystems. JISIS **1**(2/3), 125–139 (2011)
31. Nielsen, J.B.: Separating random oracle proofs from complexity theoretic proofs: the non-committing encryption case. In: Yung, M. (ed.) CRYPTO 2002. LNCS, vol. 2442, pp. 111–126. Springer, Heidelberg (2002)
32. Okamoto, T.: Provably secure and practical identification schemes and corresponding signature schemes. In: Brickell, E.F. (ed.) CRYPTO 1992. LNCS, vol. 740, pp. 31–53. Springer, Heidelberg (1993)
33. Paillier, P.: Impossibility proofs for RSA signatures in the standard model. In: Abe, M. (ed.) CT-RSA 2007. LNCS, vol. 4377, pp. 31–48. Springer, Heidelberg (2006)
34. Paillier, P., Vergnaud, D.: Discrete-log-based signatures may not be equivalent to discrete log. In: Roy, B. (ed.) ASIACRYPT 2005. LNCS, vol. 3788, pp. 1–20. Springer, Heidelberg (2005)
35. Paillier, P., Villar, J.L.: Trading one-wayness against chosen-ciphertext security in factoring-based encryption. In: Lai, X., Chen, K. (eds.) ASIACRYPT 2006. LNCS, vol. 4284, pp. 252–266. Springer, Heidelberg (2006)
36. Pointcheval, D., Stern, J.: Security arguments for digital signatures and blind signatures. J. Cryptology **13**(3), 361–396 (2000)
37. Schnorr, C.: Efficient signature generation by smart cards. J. Cryptology **4**(3), 161–174 (1991)

38. Seurin, Y.: On the exact security of schnorr-type signatures in the random oracle model. In: Pointcheval, D., Johansson, T. (eds.) EUROCRYPT 2012. LNCS, vol. 7237, pp. 554–571. Springer, Heidelberg (2012)
39. Villar, J.L.: Optimal reductions of some decisional problems to the rank problem. In: Wang, X., Sako, K. (eds.) ASIACRYPT 2012. LNCS, vol. 7658, pp. 80–97. Springer, Heidelberg (2012)
40. Zhang, J., Zhang, Z., Chen, Y., Guo, Y., Zhang, Z.: Black-box separations for one-more (static) CDH and its generalization. In: Sarkar, P., Iwata, T. (eds.) ASIACRYPT 2014, Part II. LNCS, vol. 8874, pp. 366–385. Springer, Heidelberg (2014)

The Generic Transformation from Standard Signatures to Identity-Based Aggregate Signatures

Bei Liang[1,2,3](✉), Hongda Li[1,2,3], and Jinyong Chang[1,3]

[1] State Key Laboratory of Information Security,
Institute of Information Engineering of Chinese Academy of Sciences,
Beijing, China
{liangbei,lihongda,changjinyong}@iie.ac.cn
[2] Data Assurance and Communication Security Research Center
of Chinese Academy of Sciences, Beijing, China
[3] University of Chinese Academy of Sciences, Beijing, China

Abstract. Aggregate signature system allows a collection of signatures can be compressed into one short signature. Identity-based signature schemes (IBS) allow a signer to sign a message, in which the signature can be verified by his identity. The notion of identity-based aggregate signatures (IBAS) were formally introduced by Gentry and Ramzan (PKC'06). Over the past decade, several constructions of IBAS have been proposed, which are restricted to share a common token or require sequential additions. The problem about how to achieve IBAS from standard signatures still is not resolved.

In this work, we present a generic transformation that yields IBAS schemes starting with standard signature schemes. Specifically, we provide a generic construction of an n-bounded IBAS scheme that can be proven selectively secure in the standard model from any secure signature scheme by using indistinguishability obfuscation and selective one-time universal parameters scheme. The complexity leveraging requires sub-exponential hardness assumption of indistinguishability obfuscation, puncturable PRF and one-way functions.

Keywords: Aggregate signature · Identity-based signature · Identity-based aggregate signature · Indistinguishability obfuscation · Universal parameters.

1 Introduction

Aggregate signatures, as introduced by Boneh et al. [2], are digital signatures that allow n users (whose verification and secret signing key pair is

This research is supported by the Strategy Pilot Project of Chinese Academy of Sciences (Grant No. Y2W0012203).

J. Lopez and C.J. Mitchell (Eds.): ISC 2015, LNCS 9290, pp. 21–41, 2015.
DOI: 10.1007/978-3-319-23318-5_2

$\{(vk_i, sk_i)\}_{i\in[n]}$) of a given group of potential signers to sign n different messages $\{m_i\}_{i\in[n]}$ respectively, and all the signatures of those users on those messages can be aggregated into a single short signature σ. This single signature σ and the n original verification key/message pairs $\{(vk_i, m_i)\}_{i\in[n]}$ are enough to convince the verifier that the n signers did indeed sign the n original messages m_i respectively. Aggregate signatures are useful in many real-world applications where one needs to simultaneously verify several signatures from different signers on different messages in environments with communication or storage resource constraints, such as secure route attestation.

Identity-Based Aggregate Signatures. In 1984, Shamir proposed a new model for public key cryptography, the identity-based cryptography and constructed an identity-based signature scheme (IBS) [11]. The idea of identity based cryptography is to simplify the public key of the user by using user's identity, which uniquely defines the user. In an identity based signature scheme, each user is provided with a secret signing key corresponding to his identity and he/she signs their messages using the secret signing key. The signature can be verified by using the identity of the signer and public parameters of the system.

The features of an identity-based signature scheme make it particularly appealing for use in conjunction with aggregate signature schemes. Gentry and Ramzan first formally introduced the notion of identity-based aggregate signatures (IBAS) and corresponding security model [7]. In an identity-based aggregate signature scheme, a trusted private key generator generates a private signing key sk_{id} corresponding to user's identity id. Using private signing key sk_{id} user can obtain a signature σ_{id} for message corresponding to identity id. Furthermore a signature σ_1 on identity/message pair (id_1, m_1) can be combined with a signature σ_2 on (id_2, m_2) to produce a new signature $\tilde{\sigma}$ on the set $\{id_1, m_1), (id_2, m_2)\}$. Crucially, the size of aggregated signature $\tilde{\sigma}$ should be independent of the number of signatures aggregated. The aggregated signature $\tilde{\sigma}$ can be verified by using the identity/message pair of the signer and public parameters of the system. The system will be secure in the sense that it is hard to produce an aggregate signature on a identity/message list L that contains some (id_i, m_i) never queried before — i.e., for all the adversary's queries L', $(id_i, m_i) \notin L'$.

Current State of the Art. Standard signatures imply identity-based signatures following the "certification paradigm", e.g. [6] , i.e. by simply attaching signer's public key and certificate to each signature. However, it is not clear how to convert standard signatures into identity-based aggregate signatures. Although over the past decade many identity-based aggregate signature schemes have been proposed [3,4,7,10], all of these constructions are restricted to share a common token [7] (e.g., where a set of signatures can only be aggregated if they were created with the same common token) or require sequential additions [3] (e.g., where a group of signers sequentially form an aggregate by each adding their own signature to the aggregate-so-far).

In 2013, Hohenberger, Sahai and Waters [10] implemented the Full Domain Hash with a Naor-Reingold-type structure that is publicly computable by using leveled multilinear maps. And departing from this result they constructed the first identity-based aggregate signature scheme that admits unrestricted aggregation. However, since their solution to identity-based aggregate signature scheme is firstly to build a BLS type-signature that admit unrestricted aggregation, the problem about how to achieve identity-based aggregate signatures from standard signatures still is not resolved.

Our Results in a Nutshell. In this work, we present a generic transformation that yields identity-based aggregate signature schemes based on standard signature schemes. Specifically, we provide a generic construction of an n-bounded identity-based aggregate signature scheme that can be proven selectively secure in the standard model from any secure signature scheme by using indistinguishability obfuscation and selective one-time universal parameters scheme. Although our IBAS scheme requires an a-priori bound n on the number of signatures that can be aggregated, the size of the public parameters and aggregated signatures are independent of it.

Before we describe our construction we briefly overview the underlying primitive: universal parameters scheme. Intuitively, a universal parameters (UP) scheme allows multiple paries to sample a consistent elements from arbitrary distributions while insuring that an adversary cannot learn the randomness that yields this element. In UP there is a universal parameter generation algorithm, UniversalGen, which takes as input a security parameter and output "universal parameters" U. In addition, there is a second algorithm InduceGen which takes as input universal parameters U and a distribution specified by a circuit d, and outputs the induced parameters $d(z)$ for hidden random coins z that are pseudorandomly derived from U and d. The security definition states that it is computationally difficult to distinguish an honest execution of U from that generated by a simulator SimUGen that has access to the parameters oracle.

To transform standard signature scheme into identity-based aggregate signature, we proceed in two steps. In the first step, we show how to obtain IBS from standard signature scheme (SIG). The basic idea is to use one signature instance for each identity id of IBS by universal parameter, which is inspired by the application of universal parameter for transforming public key encryption (PKE) into identity-based encryption (IBE) [8]. Precisely, choose a universal parameter U and a key pair $(pk_{\mathsf{PKE}}, sk_{\mathsf{PKE}})$ of PKE. Let $\mathsf{Prog}\{pk_{\mathsf{PKE}}\}$ be a circuit that taking a random string $r = r_1 \| r_2$ as input, first samples $(vk_{\mathsf{SIG}}, sk_{\mathsf{SIG}}) \leftarrow \mathsf{SIG.Setup}(1^\lambda; r_1)$, then encrypts sk_{SIG} under pk_{PKE} via $c' \leftarrow \mathsf{PKE.Enc}(pk_{\mathsf{PKE}}, sk_{\mathsf{SIG}}; r_2)$, and finally outputs (vk_{SIG}, c'). Here we view pk_{PKE} as a constant hardwired into the circuit $\mathsf{Prog}\{pk_{\mathsf{PKE}}\}$ and $r = r_1 \| r_2$ as input, where we make the random coins of the SIG.Setup and PKE.Enc explicit. For identity id we compute $(vk_{id}, c'_{id}) \leftarrow \mathsf{InduceGen}(U, \mathsf{Prog}\{pk_{\mathsf{PKE}}\} \| id)$. This way, we can use sk_{PKE} as a master trapdoor to extract the signing key sk_{id} from c'_{id} and thus obtain individual user secret signing key for identity id. Using that secret signing key sk_{id} corresponding to id, user can sign their messages.

The second step is to make this IBS support aggregation. Our main solution idea departs fundamentally from Hohenberger, Koppula and Waters's method of aggregating signatures using indistinguishability obfuscation [9]. Basing on their idea we provide an approach to make this IBS support aggregation. Our construction relies on the puncturable PRF and additively homomorphic encryption HE.

Our IBAS scheme is setup as follows. Randomly choose a key K of puncturable PRF and a key pair $(pk_{\mathsf{HE}}, sk_{\mathsf{HE}})$ of HE and obtain n encryption of 0 under pk_{HE}, e.g. $\mathsf{ct}_i \leftarrow \mathsf{HE.Enc}(pk_{\mathsf{HE}}, 0)$. Create program $\mathsf{Prog}\{K, \mathsf{ct}_1, \ldots, \mathsf{ct}_n\}$ which taking as inputs $\{vk_{id_i}, (id_i, m_i), \sigma_i\}_{i\in[n]}$, firstly verifies (m_i, σ_i) is valid under verification key vk_{id_i}, then computes $t = \sigma_1 \cdot \mathsf{ct}_1 + \ldots + \sigma_n \cdot \mathsf{ct}_n$ and $s_i = F(K, vk_{id_i}\|id_i\|m_i\|i\|t)$, and finally outputs $\sigma_{\mathsf{agg}} = (t, \oplus_i s_i)$. In addition, create program $\mathsf{Prog}\{K\}$ that taking as inputs $\{vk_{id_i}, (id_i, m_i)\}_i$ and $\sigma_{\mathsf{agg}} = (t, s)$, computes $s' = \oplus_i F(K, vk_{id_i}\|id_i\|m_i\|i\|t)$ and outputs 1 if $s' = s$, else outputs 0. Set obfuscated programs $P_1 = i\mathcal{O}(\mathsf{Prog}\{K, \mathsf{ct}_1, \ldots, \mathsf{ct}_n\})$ and $P_2 = i\mathcal{O}(\mathsf{Prog}\{K\})$ as public parameters. To aggregate the signatures σ_i of identity/message pair (id_i, m_i) for all $i \in [n]$, firstly obtain verification key vk_{id_i} corresponding identity id_i by universal parameter U and run program $P_1(\{vk_{id_i}, (id_i, m_i), \sigma_i\}_i)$ to get $\sigma_{\mathsf{agg}} = (t, s)$. To verify an aggregate signature, $\sigma_{\mathsf{agg}} = (t, s)$, on $\{(id_i, m_i)\}_{i\in[n]}$, firstly obtain verification key vk_{id_i} corresponding identity id_i by universal parameter U and return the output of $P_2(\{vk_{id_i}, (id_i, m_i)\}_i, \sigma_{\mathsf{agg}})$.

We prove the selective security where the attacker declares before seeing the public parameters a idenntity/message pair (id^*, m^*) by performing a sequence of games. In game 1 challenger first guesses an index i^* (incurring a $1/n$ loss) where the forgery occurs. In game 2, we change ct_{i^*} to be an encryption of 1. This causes an honestly computed value t to be an encryption of the i^*-th signature that we will eventually use for extraction. In game 3 we use the programmed generated algorithm $\mathsf{SimUGen}$ to produce U such that $(vk_{id_{i^*}}, c_{id_{i^*}}) \leftarrow \mathsf{InduceGen}(U, \mathsf{Prog}\{pk_{\mathsf{PKE}}\}\|id_{i^*})$ where $c_{id_{i^*}} = \mathsf{PKE.Enc}(pk_{\mathsf{PKE}}, sk_{id_{i^*}})$. In game 4 we replace $c_{id_{i^*}}$ with an encryption of 1^λ. At this time the simulator cannot answer the $\mathsf{KeyGen}(\mathsf{msk}, \cdot)$ and $\mathsf{Sign}(\cdot, \cdot)$ queries, since it cannot decrypt the ciphertext c_{id_i}. We overcome this obstacle by employing a wCCA-secure PKE that requires that the attacker has access to decryption oracle only after seeing the challenge ciphertext. When using wCCA-secure PKE, simulator can use the wCCA decryption oracle to answer $\mathsf{KeyGen}(\mathsf{msk}, \cdot)$ and $\mathsf{Sign}(\cdot, \cdot)$ queries. For forgery $\sigma^*_{\mathsf{agg}} = (t^*, s^*)$, since t^* is an encryption of σ_{i^*} under sk_{HE}, if $\mathsf{SIG.Vefy}(vk_{id_{i^*}}, m_{i^*}, \mathsf{HE.Dec}(sk_{\mathsf{HE}}, t^*)) = 1$, then $(m_{i^*}, \mathsf{HE.Dec}(sk_{\mathsf{HE}}, t^*))$ is a forgery for basic signature scheme SIG, which contradicts with the existential unforgeability of SIG. Therefore when $\mathsf{SIG.Vefy}(vk_{id_{i^*}}, m_{i^*}, \mathsf{HE.Dec}(sk_{\mathsf{HE}}, t^*)) = 0$, we can use the punctured key $K\{y\}$ at punctured point $y = vk_{id_{i^*}}\|id_{i^*}\|m_{i^*}\|i^*\|t^*$ to replace program $\mathsf{Prog}\{K, \mathsf{ct}_1, \ldots, \mathsf{ct}_n\}$ with $\mathsf{Prog}\{K\{y\}, \mathsf{ct}_1, \ldots, \mathsf{ct}_n\}$. In addition, we replace $i\mathcal{O}(\mathsf{Prog}\{K\})$ with $i\mathcal{O}(\mathsf{Prog}\{y, z = F(K, y), K\{y\}\})$, where program $\mathsf{Prog}\{y, z = F(K, y), K\{y\}\}$ employs an one-way function to check the correctness for $F(K, vk_{id_{i^*}}\|id_{i^*}\|m_{i^*}\|i^*\|t^*)$ and $F(K\{y\}, vk_{id_{i^*}}\|id_{i^*}\|m_{i^*}\|i^*\|t)$ that in turn can be computed by $\oplus_{i\neq i^*} F(K\{y\}, vk_{id_i}\|id_i\|m_i\|i\|t) \oplus s$. By the pseudorandomness property of puncturable PRF we replace $F(K, y)$ with a random strings z. This perfectly simulates the game.

Since there will be an exponential number of intermediate hybrid games, we will be using stronger security for the indistinguishability obfuscation, the puncturable PRF and the one way function, which requires sub-exponential hardness assumption.

Organization. The rest of this paper is organized as follows. In Sect. 2 we describe the basic tools which will be used in our construction. In Sect. 3 we introduce the notions of identity-based aggregate signatures that are considered in this work. In Sect. 4 we present our generic transformation to build IBAS from standard signature scheme and prove the security of our IBAS scheme.

2 Preliminaries

In this section, we give the definitions of cryptographic primitives that will be used in our constructions. Below, we recall the notions of indistinguishability obfuscation, puncturable pseudorandom functions and universal parameters.

2.1 Indistinguishability Obfuscation

Here we recall the notion of indistinguishability obfuscation which was originally proposed by Barak et al. [1]. The formal definition we present below is from [5].

Definition 1 (Indistinguishability Obfuscation [5]**).** *A PPT algorithm $i\mathcal{O}$ is said to be an indistinguishability obfuscator for a circuits class $\{\mathcal{C}_\lambda\}$, if the following conditions are satisfied:*

- *For all security parameters $\lambda \in \mathbb{N}$, for all $C \in C_\lambda$, for all inputs x, we have that*
$$\Pr[C'(x) = C(x) : C' \leftarrow i\mathcal{O}(\lambda, C)] = 1.$$
- *For any (not necessarily uniform) PPT adversaries (Samp, D), there exists a negligible function $negl(\cdot)$ such that the following holds: if $\Pr[\forall x, C_0(x) = C_1(x) : (C_0, C_1, \sigma) \leftarrow \mathsf{Samp}(1^\lambda)] > 1 - negl(\lambda)$, then we have:*
$$\big|\Pr[D(\sigma, i\mathcal{O}(\lambda, C_0)) = 1 : (C_0, C_1, \sigma) \leftarrow \mathsf{Samp}(1^\lambda)]$$
$$-\Pr[D(\sigma, i\mathcal{O}(\lambda, C_1)) = 1 : (C_0, C_1, \sigma) \leftarrow \mathsf{Samp}(1^\lambda)]\big| \leq negl(\lambda).$$

In a recent work, Garg et al. [5] gave the first candidate construction of indistinguishability obfuscator $i\mathcal{O}$ for all polynomial size circuits under novel algebraic hardness assumptions. In this paper, we will take advantage of such indistinguishability obfuscators for all polynomial size circuits.

2.2 Puncturable PRFs

Puncturable PRFs, as introduced by Sahai and Waters [12], are PRFs that a punctured key can be derived to allow evaluation of the PRF on all inputs, except for any polynomial-size set of inputs. The definition is formulated as in [12].

Definition 2. *A puncturable family of PRFs F mapping is given by a triple of Turing Machines (Key_F, $\mathsf{Puncture}_F$, and Eval_F), and a pair of computable functions $\tau_1(\cdot)$ and $\tau_2(\cdot)$, satisfying the following conditions:*

- ***[Functionality preserved under puncturing].*** *For every PPT adversary $\mathcal{A}$ such that $\mathcal{A}(1^\lambda)$ outputs a set $S \subseteq \{0,1\}^{\tau_1(\lambda)}$, then for all $x \in \{0,1\}^{\tau_1(\lambda)}$ where $x \notin S$, we have that:*

$$\Pr[\mathsf{Eval}_F(K,x)=\mathsf{Eval}_F(K_S,x) : K \leftarrow \mathsf{Key}_F(1^\lambda), K_S=\mathsf{Puncture}_F(K,S)]=1$$

- ***[Pseudorandom at punctured points].*** *For every PPT adversary $(\mathcal{A}_1, \mathcal{A}_2)$ such that $\mathcal{A}_1(1^\lambda)$ outputs a set $S \subseteq \{0,1\}^{\tau_1(\lambda)}$ and state σ, consider an experiment where $K \leftarrow \mathsf{Key}_F(1^\lambda)$ and $K_S = \mathsf{Puncture}_F(K,S)$. Then we have*

$$\left|\Pr[\mathcal{A}_2(\sigma, K_S, S, \mathsf{Eval}_F(K,S)) = 1] - \Pr[\mathcal{A}_2(\sigma, K_S, S, U_{\tau_2(\lambda)\cdot|S|}) = 1]\right| = negl(\lambda),$$

where $\mathsf{Eval}_F(K,S)$ denotes the concatenation of $\mathsf{Eval}_F(K,x_1),\dots,\mathsf{Eval}_F(K,x_k)$ where $S=\{x_1,\dots,x_k\}$ is the enumeration of the elements of S in lexicographic order, $negl(\cdot)$ is a negligible function, and $U_{\tau_2(\lambda)\cdot|S|}$ denotes the uniform distribution over $\tau_2(\lambda)\cdot|S|$ bits.

Theorem 1 [12]. *If one-way functions exist, then for all efficiently computable functions $\tau_1(\lambda)$ and $\tau_2(\lambda)$, there exists a puncturable PRFs family that maps $\tau_1(\lambda)$ bits to $\tau_2(\lambda)$ bits.*

2.3 Universal Parameters

In a recent work, Hofheinz et al. [8] introduced the notion of universal parameters. A universal parameters scheme UP, parameterized by polynomials ℓ_{ckt}, ℓ_{inp} and ℓ_{out}, consists of algorithms $\mathsf{UniversalGen}$ and $\mathsf{InduceGen}$ defined below.

- $\mathsf{UniversalGen}(1^\lambda)$ takes as input the security parameter λ and outputs the universal parameters U.
- $\mathsf{InduceGen}(U,d)$ takes as input the universal parameters U and a circuit d which takes as input ℓ_{inp} bits and outputs ℓ_{out} bits. The size of circuit d is at most ℓ_{ckt} bits.

Definition 3 (Selectively-Secure One-Time Universal Parameters Scheme). *Let ℓ_{ckt}, ℓ_{inp}, ℓ_{out} be efficiently computable polynomials. A pair of efficient algorithms ($\mathsf{UniversalGen}$, $\mathsf{InduceGen}$) is a selectively-secure one-time universal parameters scheme if there exists an efficient algorithm $\mathsf{SimUGen}$ such that:*

- *There exists a negligible function $negl(\cdot)$ such that for all circuits d of length ℓ_{ckt}, taking ℓ_{inp} bits of input, and outputting ℓ_{out} bits, and for all strings $p_d \in \{0,1\}^k$, we have that:*

$$\Pr[\mathit{InduceGen}(\mathit{SimUGen}(1^\lambda, d, p_d), d) = p_d] = 1 - negl(\lambda).$$

- *For every efficient adversary $\mathcal{A} = (\mathcal{A}_1, \mathcal{A}_2)$, where $\mathcal{A}_2$ outputs one bit, there exists a negligible function $negl(\cdot)$ such that the following holds. Consider the following two experiments:*

The experiment $\mathit{Real}(1^\lambda)$ is as follows:	*The experiment $\mathit{Ideal}(1^\lambda)$ is as follows:*
1. $(d^*, \sigma) \leftarrow \mathcal{A}_1(1^\lambda)$	*1. $(d^*, \sigma) \leftarrow \mathcal{A}_1(1^\lambda)$*
2. Output $\mathcal{A}_2(\mathit{UniversalGen}(1^\lambda), \sigma)$.	*2. Choose r uniformly from $\{0,1\}^{\ell_{\mathsf{inp}}}$.*
	3. Let $p_d = d^*(r)$.
	4. Output $\mathcal{A}_2(\mathit{SimUGen}(1^\lambda, d^*, p_d), \sigma)$.

Then we have:

$$\left|\Pr[\mathit{Real}(1^\lambda) = 1] - \Pr[\mathit{Ideal}(1^\lambda) = 1]\right| = negl(\lambda).$$

Hofheinz et al. [9] construct a selectively secure one-time universal parameters scheme, assuming a secure indistinguishability obfuscator and a selectively secure puncturable PRF.

3 Identity-Based Aggregate Signatures

Syntax. An identity-based aggregate signatures (IBAS) scheme can be described as a tuple of polynomial time algorithms $\mathsf{IBAS} = (\mathsf{Setup}, \mathsf{KeyGen}, \mathsf{Sign}, \mathsf{Aggregate}, \mathsf{Verify})$ as follows:

- $\mathsf{Setup}(1^\lambda)$ The setup algorithm takes as input the security parameter and outputs the public parameters PP of the scheme and master secret key msk.
- $\mathsf{KeyGen}(\mathsf{msk}, id \in \{0,1\}^{\ell_{\mathsf{id}}})$ The key generation algorithm run by the master entity, takes as input the master secret key msk and an identity id, and outputs a secret signing key sk_{id} corresponding to id.
- $\mathsf{Sign}(\mathsf{sk}_{id}, m \in \{0,1\}^{\ell_{\mathsf{msg}}})$ The signing algorithm takes as input a secret signing key sk_{id} as well as a message $m \in \{0,1\}^{\ell_{\mathsf{msg}}}$), and outputs a signature $\sigma \in \ell_{\mathsf{sig}}$ for identity id on m.
- $\mathsf{Aggregate}(\mathsf{PP}, \{(id_i, m_i), \sigma_i\}_{i=1}^t)$ The aggregation algorithm takes as input t tuples $\{(id_i, m_i), \sigma_i\}$ (for some arbitrary t) where each tuple is $(\ell_{\mathsf{id}}, \ell_{\mathsf{msg}}, \ell_{\mathsf{sig}})$-length. It outputs an aggregate signature σ_{agg} whose length is polynomial in λ, but independent of t.
- $\mathsf{Verify}(\mathsf{PP}, \{(id_i, m_i)\}_{i=1}^t, \sigma_{\mathsf{agg}})$ The verification algorithm takes as input the public parameters PP, t tuples $\{(id_i, m_i)\}$ that are $(\ell_{\mathsf{id}}, \ell_{\mathsf{msg}})$-length, and an aggregate signature σ_{agg}. It outputs 0 or 1 to indicate whether verification succeeded.

Correctness. For all $\lambda, n \in \mathbb{N}$, $(\mathsf{PP}, \mathsf{msk}) \leftarrow \mathsf{Setup}(1^\lambda, n)$, t tuples $\{(id_i, m_i)\}$ that are $(\ell_{\mathsf{id}}, \ell_{\mathsf{msg}})$-length, for all $i \in [t]$, $sk_{id_i} \leftarrow \mathsf{KeyGen}(\mathsf{msk}, id_i)$, $\sigma_i \leftarrow \mathsf{Sign}(\mathsf{sk}_{id_i}, m_i)$, and $\sigma_{\mathsf{agg}} \leftarrow \mathsf{Aggregate}(\mathsf{PP}, \{(id_i, m_i), \sigma_i\}_{i=1}^t)$, we require that $\mathsf{Verify}(\mathsf{PP}, \{(id_i, m_i)\}_{i=1}^t, \sigma_{\mathsf{agg}}) = 1$.

Selective Security. We consider a weaker attack (selective in both the identity and the message) where a forger is challenged on a given identity/message pair (id^*, m^*) chosen by the adversary before receiving the public parameters. More formally, the selective experiment $\mathbf{Exp}_{\mathsf{IBAS},\mathcal{A}}^{\mathsf{sel\text{-}uf}}(\lambda)$ between a challenger and an adversary $\mathcal{A}$ with respect to scheme $\mathsf{IBAS} = (\mathsf{Setup}, \mathsf{KeyGen}, \mathsf{Sign}, \mathsf{Aggregate}, \mathsf{Verify})$ is defined as follows:

Experiment $\mathbf{Exp}_{IBAS,\mathcal{A}}^{sel\text{-}uf}(\lambda)$

1. $(id^*, m^*) \leftarrow \mathcal{A}(\mathsf{PP})$;
2. $(\mathsf{PP}, \mathsf{msk}) \leftarrow \mathsf{Setup}(1^\lambda)$;
3. $(L^* = \{(id_i, m_i)\}_{i=1}^t, \sigma_{\mathsf{agg}}^*) \leftarrow \mathcal{A}^{\mathsf{KeyGen}(\mathsf{msk},\cdot),\mathsf{Sign}(\cdot,\cdot)}(\mathsf{PP})$;
   - $\mathsf{KeyGen}(\mathsf{msk}, \cdot)$ oracle: on input an identity id, returns secret keys for arbitrary identities.
   - $\mathsf{Sign}(\cdot, \cdot)$ oracle: on input an identity id and a message m, sets $sk_{id} \leftarrow \mathsf{KeyGen}(\mathsf{msk}, id)$ and returns $\mathsf{Sign}(sk_{id}, m)$.
4. The adversary $\mathcal{A}$ wins or the output of this experiment is 1 if the following hold true:
   (a) $\mathsf{Verify}(\mathsf{PP}, \{(id_i, m_i)\}_{i=1}^t, \sigma_{\mathsf{agg}}^*)=1$,
   (b) $\exists i^* \in [t]$ such that
      i. $(id^*, m^*) = (id_{i^*}, m_{i^*}) \in L^*$,
      ii. id^* has not been asked to the $\mathsf{KeyGen}(\mathsf{msk}, \cdot)$ oracle,
      iii. (id^*, m^*) has not been submitted to the $\mathsf{Sign}(\cdot, \cdot)$ oracle.

The advantage of an adversary $\mathcal{A}$ in the above game is defined to be

$$\mathsf{Adv}_{\mathsf{IBAS},\mathcal{A}}^{\mathsf{sel\text{-}uf}} = \Pr[\mathbf{Exp}_{\mathsf{IBAS},\mathcal{A}}^{\mathsf{sel\text{-}uf}}(\lambda) = 1],$$

where the probability is taken over all coin tosses of the Setup, KeyGen, and Sign algorithm and of $\mathcal{A}$.

Definition 4 *(Selective Unforgeability). An identity-based aggregate signature scheme IBAS is existentially unforgeable with respect to selectively chosen identiry/message pair attacks if for all probabilistic polynomial time adversaries $\mathcal{A}$, the advantage $Adv_{IBAS,\mathcal{A}}^{sel\text{-}uf}$ in the experiment $\mathbf{Exp}_{IBAS,\mathcal{A}}^{sel\text{-}uf}(\lambda)$ is negligible in λ.*

In our setting, we define an n-bounded identity-based aggregate signatures scheme, which means that at most n signatures can be aggregated.

Definition 5. *An n-bounded identity-based aggregate signatures scheme IBAS=(Setup, KeyGen, Sign, Aggregate, Verify) is an IBAS in which* Setup *algorithm takes an additional input 1^n and* Aggregate *algorithm takes in t tuples $\{(id_i, m_i), \sigma_i\}_{i \in [t]}$ satisfying $t \leq n$. The public parameters output by* Setup *have size bounded by some polynomial in λ and n. However, the aggregated signature has size bounded by a polynomial in λ, but is independent of n.*

Comparison to Previous Definitions. Our definition of IBAS make the requirement that there is an a-priori bound n on the number of signatures that can be aggregated. It is different from the definition described in [10], which allows any two aggregate signatures can be combined into a new aggregate signature.

4 Generic Construction of Identity-Based Aggregate Signatures

In this section, we present our generic transformation to build n-bounded IBAS from length-bounded signature scheme, which can be proven selectively secure in the standard model. Besides indistinguishability obfuscator $i\mathcal{O}$, we will use the following primitives.

- A selectively one-time secure $(\ell_{\mathsf{ckt}}, \ell_{\mathsf{inp}}, \ell_{\mathsf{out}})$ universal parameter scheme UP=(UniversalGen,InduceGen).
- A wCCA-secure public-key encryption scheme $\mathsf{PKE}_{\mathsf{wCCA}} = (\mathsf{PKE.Setup}_{\mathsf{wCCA}}, \mathsf{PKE.Enc}_{\mathsf{wCCA}}, \mathsf{PKE.Dec}_{\mathsf{wCCA}})$. Let the randomness space of $\mathsf{PKE.Enc}_{\mathsf{wCCA}}$ be $\{0,1\}^{\ell_{\mathsf{inp}}/2}$. We give a formal definition of wCCA-secure PKE scheme in Appendix 1.
- A $(\ell_{\mathsf{vk}}, \ell_{\mathsf{msg}}, \ell_{\mathsf{sig}})$-length signature scheme $\mathsf{SIG} = (\mathsf{SIG.Setup}, \mathsf{SIG.Sign}, \mathsf{SIG.Vefy})$ that the verification keys output by SIG.Setup have length at most $\ell_{\mathsf{vk}}(\lambda)$, SIG.Sign takes as input messages of length at most $\ell_{\mathsf{msg}}(\lambda)$ and outputs signatures of length bounded by $\ell_{\mathsf{sig}}(\lambda)$. Let the randomness space of SIG.Setup be $\{0,1\}^{\ell_{\mathsf{inp}}/2}$. We give a formal definition of signature scheme in Appendix 2.
- An additively homomorphic encryption scheme HE=(HE.Setup, HE.Enc,HE.Dec, HE.Add) with message space $\mathbb{F}_p$ for some prime $p > 2^{\ell_{\mathsf{sig}}}$ and ciphertext space $\mathcal{C}_{\mathsf{HE}}$, where each ciphertext in $\mathcal{C}_{\mathsf{HE}}$ can be represented using ℓ_{HEct} bits. We give a formal definition of additively homomorphic encryption scheme in Appendix 3.
- A puncturable PRF F with key space $\mathcal{K}$, input space $\{0,1\}^{\ell_{\mathsf{vk}}+\ell_{\mathsf{id}}+\ell_{\mathsf{msg}}+\log n+\ell_{\mathsf{HEct}}}$ and range $\{0,1\}^{\ell}$.
- An injective one-way function $f : \{0,1\}^{\ell} \to \{0,1\}^{2\ell}$.

Our n-bounded identity-based aggregate signature scheme consists of algorithms IBAS.Setup, IBAS.KeyGen, IBAS.Sign, IBAS.Aggregate and IBAS.Verify described below.

IBAS.Setup$(1^\lambda, n)$: On input 1^λ, the IBAS.Setup algorithm works as follows.

1. It runs $(pk_{\mathsf{HE}}, sk_{\mathsf{HE}}) \leftarrow \mathsf{HE.Setup}(1^\lambda)$ and computes ciphertext $\mathsf{ct}_i \leftarrow \mathsf{HE.Enc}(pk_{\mathsf{HE}}, 0)$ for all $i \in [n]$.
2. It runs $(pk_{\mathsf{wCCA}}, sk_{\mathsf{wCCA}}) \leftarrow \mathsf{PKE.Setup}_{\mathsf{wCCA}}(1^\lambda)$ to generate a key pair for $\mathsf{PKE}_{\mathsf{wCCA}}$.
3. Then it creates a program $\mathsf{Prog}\{pk_{\mathsf{wCCA}}\}$ which is defined below as Fig. 1.

4. Choose a puncturable PRF key K, create programs $\mathsf{Prog}\{K, \mathsf{ct}_1, \ldots, \mathsf{ct}_n\}$ and $\mathsf{Prog}\{K\}$ described below as Figs. 2 and 3 respectively, and set $P_1 = i\mathcal{O}(\mathsf{Prog}\{K, \mathsf{ct}_1, \ldots, \mathsf{ct}_n\})$ and $P_2 = i\mathcal{O}(\mathsf{Prog}\{K\})$.
5. Finally it computes $U \leftarrow \mathsf{UniversalGen}(1^\lambda)$.

The public parameters PP is $(pk_{\mathsf{HE}}, U, \mathsf{Prog}\{pk_{\mathsf{wCCA}}\}, P_1, P_2)$ and the master secret key msk is sk_{wCCA}.

IBAS.KeyGen(**msk**, id) : On input the master secret key msk and $id \in \{0,1\}^{\ell_{\mathsf{id}}}$, it computes $(vk_{id}, c_{id}) \leftarrow \mathsf{InduceGen}(U, \mathsf{Prog}\{pk_{\mathsf{wCCA}}\}\|id)$ and returns $\mathsf{sk}_{id} \leftarrow \mathsf{PKE.Dec}_{\mathsf{wCCA}}(\mathsf{msk}, c_{id})$.

Remark. For any program $\mathsf{Prog}\{pk_{\mathsf{wCCA}}\}$, we let $\mathsf{Prog}\{pk_{\mathsf{wCCA}}\}\|id$ denote the program $\mathsf{Prog}\{pk_{\mathsf{wCCA}}\}$ extended with an additional string $id \in \{0,1\}^{\ell_{\mathsf{id}}}$. Although their description is different, program $\mathsf{Prog}\{pk_{\mathsf{wCCA}}\}\|id$ has the same functionality as program $\mathsf{Prog}\{pk_{\mathsf{wCCA}}\}$. We require that this extension is performed in some standard and deterministic way, for instance by always adding the id string at the end of the code.

IBAS.Sign(**sk**$_{id}$, m): On input a secret signing key sk_{id} and a message $m \in \{0,1\}^{\ell_{\mathsf{msg}}}$, it runs $\sigma \leftarrow \mathsf{SIG.Sign}(\mathsf{sk}_{id}, m)$ and returns σ.

IBAS.Aggregate(**PP**, $\{(id_i, m_i), \sigma_i\}_i$): On input public parameters PP and tuples $\{(id_i, m_i), \sigma_i\}_i$, if tuples $\{(id_i, m_i), \sigma_i\}$ are not distinct, the algorithm outputs $\perp$. Else, it computes $(vk_{id_i}, c_{id}) \leftarrow \mathsf{InduceGen}(U, \mathsf{Prog}\{pk_{\mathsf{wCCA}}\}\|id_i)$ and outputs $P_1(\{vk_{id_i}, (id_i, m_i), \sigma_i\}_i)$.

IBAS.Verify(**PP**, $\{(id_i, m_i)\}_i, \sigma_{\mathsf{agg}} = (t, s)$): The verification algorithm checks if the tuples $\{(id_i, m_i)\}_i$ are distinct. If not, it outputs 0. Else, it computes $(vk_{id_i}, c_{id}) \leftarrow \mathsf{InduceGen}(U, \mathsf{Prog}\{pk_{\mathsf{wCCA}}\}\|id_i)$, and outputs $P_2(\{(vk_{id_i}, id_i, m_i)\}_i, \sigma_{\mathsf{agg}} = (t, s))$.

$\mathsf{Prog}\{pk_{\mathsf{wCCA}}\}$

Hardwired into the circuit: pk_{wCCA}.
Input to the circuit: $r = r_0\|r_1 \in \{0,1\}^{\ell_{\mathsf{inp}}}$.
Algorithm:
1. Compute $(vk_{\mathsf{SIG}}, sk_{\mathsf{SIG}}) \leftarrow \mathsf{SIG.Setup}(1^\lambda; r_0)$ and $c \leftarrow \mathsf{PKE.Enc}_{\mathsf{wCCA}}(pk_{\mathsf{wCCA}}, sk_{\mathsf{SIG}}; r_1)$.
2. Output (vk_{SIG}, c).

Fig. 1. Program Prog$\{pk_{w\mathsf{CCA}}\}$.

The correctness of this scheme follows immediately from the correctness of SIG, $\mathsf{PKE}_{\mathsf{wCCA}}$, HE and (UniversalGen, InduceGen).

Remark. The setup algorithm is parameterized by a polynominal n that gives an a-priori bound on the number of signatures that can be aggregated. The size of the parameters and aggregate signatures are independent of it.

Prog$\{K, \mathsf{ct}_1, \ldots, \mathsf{ct}_n\}$

Hardwired into the circuit: $K, \mathsf{ct}_1, \ldots, \mathsf{ct}_n$.
Input to the circuit: $\{vk_{id_i}, (id_i, m_i), \sigma_i\}_i$.
Algorithm:

1. If $\exists i$ such that $\mathsf{SIG.Vefy}(vk_{id_i}, m_i, \sigma_i) = 0$, then output $\perp$.
   End if.
2. Compute $t = \sum_i \sigma_i \cdot \mathsf{ct}_i$.
3. Compute $s_i = F(K, vk_{id_i} \| id_i \| m_i \| i \| t)$.
4. Output $\sigma_{\mathsf{agg}} = (t, \oplus_i s_i)$.

Fig. 2. Program Prog$\{K, \mathsf{ct}_1, \ldots, \mathsf{ct}_n\}$.

Prog$\{K\}$

Hardwired into the circuit: K.
Input to the circuit: $\{(vk_{id_i}, id_i, m_i)\}_i, \sigma_{\mathsf{agg}} = (t, s)$.
Algorithm:

1. Compute $s' = \oplus_i F(K, vk_{id_i} \| id_i \| m_i \| i \| t)$.
2. Output 1 if $s' = s$, else output 0.

Fig. 3. Program Prog$\{K\}$.

Theorem 2. *Let $\mathcal{A}$ be any PPT adversary, and SIG a $(\ell_{vk}, \ell_{msg}, \ell_{sig})$-length secure signature scheme. Let $\mathsf{Adv}^{\mathsf{sel\text{-}uf}}_{\mathsf{IBAS},\mathcal{A}}$ denote the advantage of $\mathcal{A}$ in the identity-based aggregate signatures. Let $\mathsf{Adv}_{\mathsf{UP}}$, $\mathsf{Adv}_{\mathsf{SIG}}$, $\mathsf{Adv}_{\mathsf{HE}}$, $\mathsf{Adv}_{\mathsf{PKE}_{\mathsf{wCCA}}}$, $\mathsf{Adv}_{i\mathcal{O}}$, $\mathsf{Adv}_{\mathsf{PRF}}$ and Adv_f denote the maximum advantage of a PPT adversary against universal parameters scheme UP, signature scheme SIG, additively homomorphic encryption scheme HE, wCCA secure public key encryption $\mathsf{PKE}_{\mathsf{wCCA}}$, indistinguishability obfuscator $i\mathcal{O}$, selectively secure puncturable PRF F and one way function f respectively. Then,*

$$\mathsf{Adv}^{\mathsf{sel\text{-}uf}}_{\mathsf{IBAS},\mathcal{A}} \leq n\big(\mathsf{Adv}_{\mathsf{HE}} + \mathsf{Adv}_{\mathsf{UP}} + \mathsf{Adv}_{\mathsf{PKE}_{\mathsf{wCCA}}} + 2^{\ell_{\mathsf{HEct}}}(6\mathsf{Adv}_{i\mathcal{O}} + 2\mathsf{Adv}_{\mathsf{PRF}} + \mathsf{Adv}_f) + \mathsf{Adv}_{\mathsf{SIG}}\big)$$

where ℓ_{HEct} is the length of ciphertexts in $\mathcal{C}_{\mathsf{HE}}$.

We now prove via a sequence of exponential number hybrid games Game 0, Game 1, Game 2, Game 3, Game 4, Game 5,0, Game 5,0-1, Game 5,0-2,..., Game 5,0-6, Game 5,1, Game 5,1-1, Game 5,1-2,..., Game 5,1-6, Game 5,2,..., Game $5,2^{\ell_{\mathsf{HEct}}}$, each of which we prove to be indistinguishable from the previous one.

Sequence of Games.

Game 0. This game is the original selective security game $\mathbf{Exp}^{\mathsf{sel\text{-}uf}}_{\mathsf{IBAS},\mathcal{A}}(\lambda)$ in Sect. 3 instantiated by our construction.

1. $\mathcal{A}$ first choose a challenge (id^*, m^*).
2. The challenger chooses $(pk_{\mathsf{HE}}, sk_{\mathsf{HE}})$, $(pk_{\mathsf{wCCA}}, sk_{\mathsf{wCCA}})$, $U \leftarrow \mathsf{UniversalGen}(1^\lambda)$ and $K \leftarrow \mathsf{PRF.Setup}(1^\lambda)$. Compute $\mathsf{ct}_i \leftarrow \mathsf{HE.Enc}(pk_{\mathsf{HE}}, 0)$ for all $i \in [n]$, $P_1 \leftarrow i\mathcal{O}(\mathsf{Prog}\{K, \mathsf{ct}_1, \ldots, \mathsf{ct}_n\})$ and $P_2 \leftarrow i\mathcal{O}(\mathsf{Prog}\{K\})$. Let $\mathsf{Prog}\{pk_{\mathsf{wCCA}}\}$ be circuit as defined in the Fig. 1. Set $\mathsf{PP} = (pk_{\mathsf{HE}}, U, \mathsf{Prog}\{pk_{\mathsf{wCCA}}\}, P_1, P_2)$ and $\mathsf{msk} = sk_{\mathsf{wCCA}}$, and send PP to $\mathcal{A}$.
3. On attacker's $\mathsf{KeyGen}(\mathsf{msk}, \cdot)$ queries and $\mathsf{Sign}(\cdot, \cdot)$ queries, the challenger responds as follows:
   - On $\mathsf{KeyGen}(\mathsf{msk}, \cdot)$ query for id, the challenger computes $(vk_{id}, c_{id}) \leftarrow \mathsf{InduceGen}(U, \mathsf{Prog}\{pk_{\mathsf{wCCA}}\} \| id)$ and returns $\mathsf{sk}_{id} \leftarrow \mathsf{PKE.Dec}_{\mathsf{wCCA}}(\mathsf{msk}, c_{id})$.
   - On $\mathsf{Sign}(\cdot, \cdot)$ query for identity id and message m, the challenger first runs $\mathsf{sk}_{id} \leftarrow \mathsf{KeyGen}(\mathsf{msk}, id)$ and then returns $\sigma \leftarrow \mathsf{SIG.Sign}(\mathsf{sk}_{id}, m)$.
4. Finally the adversary outputs $(L^* = \{(id_i, m_i)\}_i, \sigma^*_{\mathsf{agg}})$. The adversary $\mathcal{A}$ wins or the output of this experiment is 1 if the following hold true:
   (a) $\mathsf{IBAS.Verify}(\mathsf{PP}, L^* = \{(id_i, m_i)\}_i, \sigma^*_{\mathsf{agg}}) = 1$,
   (b) $\exists i^*$ such that
      i. $(id^*, m^*) = (id_{i^*}, m_{i^*}) \in L^*$,
      ii. id^* has not been asked to the $\mathsf{KeyGen}(\mathsf{msk}, \cdot)$ oracle,
      iii. (id^*, m^*) has not been submitted to the $\mathsf{Sign}(\cdot, \cdot)$ oracle.

Game 1. This game is exactly similar to the previous one, except that the challenger guesses a position $i^* \leftarrow [n]$, and the attacker wins if $id^* = id_{i^*}$ and $m^* = m_{i^*}$.

Game 2. This game is similar to the previous one, except that ct_{i^*} is an encryption of 1 under pk_{HE}, instead of 0. That is, compute $\mathsf{ct}_i \leftarrow \mathsf{HE.Enc}(pk_{\mathsf{HE}}, 0)$ for all $i \in [n]$ and $i \neq i^*$, and $\mathsf{ct}_{i^*} \leftarrow \mathsf{HE.Enc}(pk_{\mathsf{HE}}, 1)$.

Game 3. This game is identical to Game 2, except for the following. The experiment generates parameters as $U \leftarrow \mathsf{SimUGen}(1^\lambda, \mathsf{Prog}\{pk_{\mathsf{wCCA}}\} \| id_{i^*}, (vk_{id_{i^*}}, c_{id_{i^*}}))$, where $(vk_{id_{i^*}}, c_{id_{i^*}}) \leftarrow \mathsf{Prog}\{pk_{\mathsf{wCCA}}\} \| id_{i^*}(r)$ for uniformly random $r \in \{0,1\}^{\ell_{\mathsf{inp}}}$. And on attacker's $\mathsf{KeyGen}(\mathsf{msk}, \cdot)$ query for id, the challenger computes $(vk_{id}, c_{id}) \leftarrow \mathsf{InduceGen}(U, \mathsf{Prog}\{pk_{\mathsf{wCCA}}\} \| id)$, where $U \leftarrow \mathsf{SimUGen}(1^\lambda, \mathsf{Prog}\{pk_{\mathsf{wCCA}}\} \| id_{i^*}, (vk_{id_{i^*}}, c_{id_{i^*}}))$, and returns $\mathsf{sk}_{id} \leftarrow \mathsf{PKE.Dec}_{\mathsf{wCCA}}(\mathsf{msk}, c_{id})$.

Game 4. The only difference between this game and the previous one is in the behavior of evaluation on the $d_{id_{i^*}} = \mathsf{Prog}\{pk_{\mathsf{wCCA}}\} \| id_{i^*}$. In Game 3, the entry corresponding to $d_{id_{i^*}}$ is of the form $(d_{id_{i^*}}, (vk_{id_{i^*}}, c_{id_{i^*}}))$ where $r = r_0 \| r_1 \in \{0,1\}^{\ell_{\mathsf{inp}}}$, $(vk_{id_{i^*}}, sk_{id_{i^*}}) \leftarrow \mathsf{SIG.Setup}(1^\lambda; r_0)$ and $c_{id_{i^*}} \leftarrow \mathsf{PKE.Enc}_{\mathsf{wCCA}}(pk_{\mathsf{wCCA}}, sk_{id_{i^*}}; r_1)$. In this game, the entry corresponding to $d_{id_{i^*}}$ is $(d_{id_{i^*}}, (vk_{id_{i^*}}, c_{id_{i^*}}))$, where $c_{id_{i^*}} \leftarrow \mathsf{PKE.Enc}_{\mathsf{wCCA}}(pk_{\mathsf{wCCA}}, 1^\lambda; r_1)$.
We will now describe an exponential number of hybrid experiments Game 5,j for $j \leq 2^{\ell_{\mathsf{HEct}}}$. Let us define some notations. Recall $\mathsf{Prog}\{K\}$ takes as input tuples of the form $(\{(vk_{id_i}, id_i, m_i)\}_i, (t, s))$. We say tuple $(\{(vk_{id_i}, id_i, m_i)\}_i, (t, s))$ is (i^*, sk_{HE})-rejecting if $\mathsf{SIG.\ Vefy}(vk_{id_{i^*}}, m_{i^*}, \mathsf{HE.Dec}(sk_{\mathsf{HE}}, t)) = 0$.

Game 5, j. In this game, the adversary does not win if the forgery input $(\{(vk_{id_i}, id_i, m_i)\}_i, (t^*, s^*))$ is (i^*, sk_{HE})-rejecting and $t^* \leq j$. That is, finally the adversary $\mathcal{A}$ outputs $(L^* = \{(id_i, m_i)\}_i, \sigma^*_{\mathsf{agg}} = (t^*, s^*))$, and $\mathcal{A}$ wins if the following hold true:

1. $\mathsf{IBAS.Verify}(\mathsf{PP}, L^* = \{(id_i, m_i)\}_i, \sigma^*_{\mathsf{agg}}) = 1$,
2. $(\{(vk_{id_i}, id_i, m_i)\}_i, (t^*, s^*))$ is not (i^*, sk_{HE})-rejecting or $t^* > j$,
3. id_{i^*} has not been asked to the $\mathsf{KeyGen}(\mathsf{msk}, \cdot)$ oracle,
4. (id_{i^*}, m_{i^*}) has not been submitted to the $\mathsf{Sign}(\cdot, \cdot)$ oracle.

Game 5, j-1. In this game, the challenger replace P_2 with obfuscation of program $\mathsf{Prog}\text{-}1\{K\}$ instead of $\mathsf{Prog}\{K\}$. That is $P_2 = i\mathcal{O}(\mathsf{Prog}\text{-}1\{K\})$. In program $\mathsf{Prog}\text{-}1\{K\}$ as described in Fig. 4, instead of checking whether $s = \oplus_i s_i$, it uses an injective one way function f to check if $f(s \oplus (\oplus_{i \neq i^*} s_i)) = f(s_{i^*})$.

Prog-1$\{K\}$

Hardwired into the circuit: K.
Input to the circuit: $\{(vk_{id_i}, id_i, m_i)\}_i, \sigma_{\mathsf{agg}} = (t, s)$.
Algorithm:
1. Compute $s' = \oplus_{i \neq i^*} F(K, vk_{id_i} \| id_i \| m_i \| i \| t) \oplus s$.
2. Output 1 if $f(F(K, vk_{id_{i^*}} \| id_{i^*} \| m_{i^*} \| i^* \| t)) = f(s')$, else output 0.

Fig. 4. Program Prog-1$\{K\}$.

Game 5, j-2. In this game, $\mathsf{Prog}\{K, \mathsf{ct}_1, \ldots, \mathsf{ct}_n\}$ and $\mathsf{Prog}\text{-}1\{K\}$ are replaced by $\mathsf{Prog}\text{-}1\{K\{y\}, \mathsf{ct}_1, \ldots, \mathsf{ct}_n\}$ (described in Fig. 5) and $\mathsf{Prog}\text{-}2\{y, z, K\{y\}\}$ (described in Fig. 6) respectively. Both the replaced programs use the punctured key at punctured point $y = vk_{id_{i^*}} \| id_{i^*} \| m_{i^*} \| i^* \| (j+1)$. More precisely, the challenger computes $y = vk_{id_{i^*}} \| id_{i^*} \| m_{i^*} \| i^* \| (j+1)$, $K\{y\} \leftarrow \mathsf{PRF.Puncture}(K, y)$ and $z = f(F(K, y))$. Let $P_1 = i\mathcal{O}(\mathsf{Prog}\text{-}1\{K, \mathsf{ct}_1, \ldots, \mathsf{ct}_n\})$ and $P_2 = i\mathcal{O}(\mathsf{Prog}\text{-}2\{K\})$.

Game 5, j-3. This game is similar to the previous one, except that z is a uniformly random string. That is, the challenger randomly chooses z', and computes $z = f(z')$.

Game 5, j-4. In this game, the challenger modifies the winning condition. That is, $\big(\{(vk_{id_i}, id_i, m_i)\}_i, (t^*, s^*)\big)$ is not (i^*, sk_{HE})-rejecting or $t^* > j+1$.

Game 5, j-5. In this game, the challenger sets $z = f(F(K, y))$ as in Game 4-j-2.

Game 5, j-6. In this game, the challenger changes program $\mathsf{Prog}\text{-}1\{K, \mathsf{ct}_1, \ldots, \mathsf{ct}_n\}$ and program $\mathsf{Prog}\text{-}2\{K\}$ back to $\mathsf{Prog}\{K, \mathsf{ct}_1, \ldots, \mathsf{ct}_n\}$ and $\mathsf{Prog}\text{-}1\{K\}$ respectively.

Game 6. This game is identical to Game $5, 2^{\ell_{\mathsf{HEct}}}$.

Prog-1$\{K\{y\}, \mathbf{ct}_1, \ldots, \mathbf{ct}_n\}$

Hardwired into the circuit: $K\{y\}, \mathsf{ct}_1, \ldots, \mathsf{ct}_n$.
Input to the circuit: $\{vk_{id_i}, (id_i, m_i), \sigma_i\}_i$.
Algorithm:

1. If $\exists i$ such that $\mathsf{SIG.Vefy}(vk_{id_i}, m_i, \sigma_i) = 0$, then output $\perp$.
   End if.
2. Compute $t = \sigma_1 \cdot \mathsf{ct}_1 + \ldots + \sigma_n \cdot \mathsf{ct}_n$.
3. Compute $s_i = F(K\{y\}, vk_{id_i} \| id_i \| m_i \| i \| t)$.
4. Output $\sigma_{\mathsf{agg}} = (t, \oplus_i s_i)$.

Fig. 5. Program Prog-1$\{K\{y\}, \mathsf{ct}_1, \ldots, \mathsf{ct}_n\}$.

Prog-2$\{y, z, K\{y\}\}$

Hardwired into the circuit: $y, z, K\{y\}$.
Input to the circuit: $\{(vk_{id_i}, id_i, m_i)\}_i, \sigma_{\mathsf{agg}} = (t, s)$.
Algorithm:

1. Compute $s' = \oplus_{i \neq i^*} F(K\{y\}, vk_{id_i} \| id_i \| m_i \| i \| t) \oplus s$.
2. If $vk_{id_{i^*}} \| id_{i^*} \| m_{i^*} \| i^* \| t = y$, then output 1 if $z = f(s')$, else output 0.
3. Else output 1 if $F(K\{y\}, vk_{id_{i^*}} \| id_{i^*} \| m_{i^*} \| i^* \| t) = f(s')$, else output 0.

Fig. 6. Program Prog-2$\{y, z, K\{y\}\}$.

Analysis. Let $\mathsf{Adv}^i_{\mathcal{A}}$ denote the advantage of adversary $\mathcal{A}$ in Game i. We now establish the difference of the attacker's advantage between each adjacent game via a sequence of lemmas.

Lemma 1. *For any adversary* $\mathcal{A}$, $\mathsf{Adv}^1_{\mathcal{A}} = \mathsf{Adv}^0_{\mathcal{A}}/n$.

Proof. This follows from the definitions of Game 0 and Game 1. The only difference between the two experiments is the change in winning condition, which now includes the guess i^*. This guess is correct with probability $1/n$.

Lemma 2. *For any PPT adversary* $\mathcal{A}$, $\mathsf{Adv}^1_{\mathcal{A}} - \mathsf{Adv}^2_{\mathcal{A}} \leq \mathsf{Adv}_{\mathsf{HE}}(\lambda)$.

Proof. Suppose there exists an adversary $\mathcal{A}$ such that $\mathsf{Adv}^1_{\mathcal{A}} - \mathsf{Adv}^2_{\mathcal{A}} = \epsilon$. We will construct a PPT algorithm $\mathcal{B}$ that breaks the semantic security of HE scheme using $\mathcal{A}$.

$\mathcal{B}$ receives the public key pk_{HE}. It sends $0, 1$ as challenge messages to the HE challenger, and receives ct in response. On receiving (id^*, m^*) from $\mathcal{A}$, $\mathcal{B}$ chooses $i^* \leftarrow [n]$, $(pk_{\mathsf{wCCA}}, sk_{\mathsf{wCCA}})$, $U \leftarrow \mathsf{UniversalGen}(1^\lambda)$ and computes ciphertext $\mathsf{ct}_i \leftarrow \mathsf{HE.Enc}(pk_{\mathsf{HE}}, 0)$ for all $i \neq i^*$. It sets $\mathsf{ct}_{i^*} = \mathsf{ct}$. Let $\mathsf{Prog}\{pk_{\mathsf{wCCA}}\}$ be circuit as defined in the Fig. 1. It chooses $K \leftarrow \mathsf{PRF.Setup}(1^\lambda)$ and computes $P_1 = i\mathcal{O}(\mathsf{Prog}\{K, \mathsf{ct}_1, \ldots, \mathsf{ct}_n\})$ and $P_2 = i\mathcal{O}(\mathsf{Prog}\{K\})$. $\mathcal{B}$ sends $\mathsf{PP} = (pk_{\mathsf{HE}}, U, P_1, P_2, \mathsf{Prog}\{pk_{\mathsf{wCCA}}\})$ to $\mathcal{A}$.

$\mathcal{A}$ then asks for $\mathsf{KeyGen}(\mathsf{msk},\cdot)$, $\mathsf{Sign}(\cdot,\cdot)$ queries, which $\mathcal{B}$ can simulate perfectly. Finally, $\mathcal{A}$ outputs a forgery $\sigma^*_{\mathsf{agg}} = (t^*, s^*)$ and tuples $\{(id_i, m_i)\}_i$. If $\mathcal{A}$ wins as per the winning conditions (which are the same in both Games 1 and 2), output 0, else output 1.

Clearly, if ct is an encryption of 0, then this corresponds to Game 1, else it corresponds to Game 2. This completes our proof.

Lemma 3. *For any PPT adversary $\mathcal{A}$, $\mathsf{Adv}^2_{\mathcal{A}} - \mathsf{Adv}^3_{\mathcal{A}} \leq \mathsf{Adv}_{UP}(\lambda)$.*

Proof. Suppose there exists a PPT adversary $\mathcal{A}$ such that $\mathsf{Adv}^2_{\mathcal{A}} - \mathsf{Adv}^3_{\mathcal{A}} = \epsilon$. We will construct a PPT algorithm $\mathcal{B}$ such that $\left|\Pr[\mathsf{Real}^{\mathcal{B}}(1^\lambda) = 1] - \Pr[\mathsf{Ideal}^{\mathcal{B}}_{\mathsf{SimUGen}}(1^\lambda) = 1]\right| = \epsilon$.

$\mathcal{B}$ interacts with $\mathcal{A}$ and participates in either the Real or Ideal game. On receiving (id^*, m^*) from $\mathcal{A}$, $\mathcal{B}$ chooses $i^* \leftarrow [n]$, and sets $d_{id_{i^*}} = \mathsf{Prog}\{pk_{\mathsf{wCCA}}\}\|id_{i^*}$. $\mathcal{B}$ sends $d_{id_{i^*}}$ to the challenger of universal parameters UP. The challenger of universal parameters UP computes $(vk_{id_{i^*}}, c_{id_{i^*}}) \leftarrow d_{id_{i^*}}(r)$ for uniformly random $r = r_0\|r_1 \in \{0,1\}^{\ell_{\mathsf{inp}}}$, where $(vk_{id_{i^*}}, sk_{id_{i^*}}) \leftarrow \mathsf{SIG.Setup}(1^\lambda; r_0)$ and $c_{id_{i^*}} \leftarrow \mathsf{PKE.Enc}_{\mathsf{wCCA}}(pk_{\mathsf{wCCA}}, sk_{id_{i^*}}; r_1)$. $\mathcal{B}$ receives U from the challenger of universal parameters. $\mathcal{B}$ then chooses $(pk_{\mathsf{wCCA}}, sk_{\mathsf{wCCA}})$, $(pk_{\mathsf{HE}}, sk_{\mathsf{HE}})$ and compute ciphertext $\mathsf{ct}_i \leftarrow \mathsf{HE.Enc}(pk_{\mathsf{HE}}, 0)$ for all $i \in [n]$ and $i \neq i^*$, $\mathsf{ct}_{i^*} \leftarrow \mathsf{HE.Enc}(pk_{\mathsf{HE}}, 1)$. Let $\mathsf{Prog}\{pk_{\mathsf{wCCA}}\}$ be circuit as defined in the Fig. 1. It chooses $K \leftarrow \mathsf{PRF.Setup}(1^\lambda)$ and computes $P_1 = i\mathcal{O}(\mathsf{Prog}\{K, \mathsf{ct}_1, \ldots, \mathsf{ct}_n\})$ and $P_2 = i\mathcal{O}(\mathsf{Prog}\{K\})$. $\mathcal{B}$ sends $\mathsf{PP} = (pk_{\mathsf{HE}}, U, \mathsf{Prog}\{pk_{\mathsf{wCCA}}\}, P_1, P_2)$ to $\mathcal{A}$.

For the $\mathsf{KeyGen}(\mathsf{msk},\cdot)$ and $\mathsf{Sign}(\cdot,\cdot)$ queries, $\mathcal{B}$ computes $(vk_{id}, c_{id}) \leftarrow \mathsf{InduceGen}(U, d_{id})$, where $d_{id} = \mathsf{Prog}\{pk_{\mathsf{wCCA}}\}\|id$, and returns $\mathsf{sk}_{id} \leftarrow \mathsf{PKE.Dec}_{\mathsf{wCCA}}(sk_{\mathsf{wCCA}}, c_{id})$ by using sk_{wCCA}. Finally, it receives a forgery $\sigma^*_{\mathsf{agg}} = (t^*, s^*)$ and tuples $\{(id_i, m_i)\}_i$. Note that since there is no $\mathsf{Honest\ Parameter\ Violation}$, $\mathsf{InduceGen}(U, \mathsf{Prog}\{pk_{\mathsf{wCCA}}\}\|id_{i^*}) = (vk_{id_{i^*}}, c_{id_{i^*}})$. Therefore, Game 2 corresponds to $\mathsf{Real}^{\mathcal{B}}(1^\lambda)$ experiment, while Game 3 corresponds to $\mathsf{Ideal}^{\mathcal{B}}_{\mathsf{SimUGen}}$. Hence, $\left|\Pr[\mathsf{Real}^{\mathcal{B}}(1^\lambda) = 1] - \Pr[\mathsf{Ideal}^{\mathcal{B}}_{\mathsf{SimUGen}}(1^\lambda) = 1]\right| = \epsilon$.

Lemma 4. *For any PPT adversary $\mathcal{A}$, $\mathsf{Adv}^3_{\mathcal{A}} - \mathsf{Adv}^4_{\mathcal{A}} \leq \mathsf{Adv}_{\mathsf{PKE}_{\mathsf{wCCA}}}(\lambda)$.*

Proof. Note that the only difference between Games 3 and 4 is in the behavior of evaluation on the $d_{id_{i^*}} = \mathsf{Prog}\{pk_{\mathsf{wCCA}}\}\|id_{i^*}$. Suppose there exists an adversary $\mathcal{A}$ such that $\mathsf{Adv}^3_{\mathcal{A}} - \mathsf{Adv}^4_{\mathcal{A}} = \epsilon$. We will construct a PPT algorithm $\mathcal{B}$ that breaks the wCCA security of $\mathsf{PKE}_{\mathsf{wCCA}}$ scheme using $\mathcal{A}$.

$\mathcal{B}$ receives the public key pk_{wCCA}. On receiving (id^*, m^*) from $\mathcal{A}$, $\mathcal{B}$ chooses $i^* \leftarrow [n]$, $(pk_{\mathsf{HE}}, sk_{\mathsf{HE}})$ and compute ciphertext $\mathsf{ct}_i \leftarrow \mathsf{HE.Enc}(pk_{\mathsf{HE}}, 0)$ for all $i \in [n]$ and $i \neq i^*$, $\mathsf{ct}_{i^*} \leftarrow \mathsf{HE.Enc}(pk_{\mathsf{HE}}, 1)$. Let $\mathsf{Prog}\{pk_{\mathsf{wCCA}}\}$ be circuit as defined in the Fig. 1. It chooses $K \leftarrow \mathsf{PRF.Setup}(1^\lambda)$ and computes $P_1 = i\mathcal{O}(\mathsf{Prog}\{K, \mathsf{ct}_1, \ldots, \mathsf{ct}_n\})$ and $P_2 = i\mathcal{O}(\mathsf{Prog}\{K\})$.

$\mathcal{B}$ chooses $r = r_0\|r_1 \in \{0,1\}^{\ell_{\mathsf{inp}}}$ and computes $(vk_{id_{i^*}}, sk_{id_{i^*}}) \leftarrow \mathsf{SIG.Setup}(1^\lambda; r_0)$. $\mathcal{B}$ sends $sk_{id_{i^*}}, 1^\lambda$ as the challenge messages to the wCCA challenger. It receives in response a ciphertext $\mathsf{ct}^*_{\mathsf{wCCA}}$. Let $d_{id_{i^*}}(r) = (vk_{id_{i^*}}, \mathsf{ct}^*_{\mathsf{wCCA}})$, where $d_{id_{i^*}} = \mathsf{Prog}\{pk_{\mathsf{wCCA}}\}\|id_{i^*}$. Then $\mathcal{B}$ compute $U \leftarrow \mathsf{SimUGen}(1^\lambda)$ and sends $\mathsf{PP} = (pk_{\mathsf{HE}}, U, \mathsf{Prog}\{pk_{\mathsf{wCCA}}\}, P_1, P_2)$ to $\mathcal{A}$.

On receiving $\mathsf{KeyGen}(\mathsf{msk},\cdot)$ query for id, $\mathcal{B}$ computes $(vk_{id}, c_{id}) \leftarrow \mathsf{InduceGen}(U, d_{id})$, where $d_{id} = \mathsf{Prog}\{pk_{\mathsf{wCCA}}\}\|id$, then submits a decryption query to oracle $\mathsf{PKE.Dec}_{\mathsf{wCCA}}(sk_{\mathsf{wCCA}},\cdot)$ for c_{id}, and returns whatever $\mathsf{PKE.Dec}_{\mathsf{wCCA}}(sk_{\mathsf{wCCA}},\cdot)$ returns. Finally, $\mathcal{A}$ outputs a forgery $\sigma^*_{\mathsf{agg}} = (t^*, s^*)$ and tuples $\{(id_i, m_i)\}_i$. If $\mathcal{A}$ wins as per the winning conditions (which are the same in both Games 3 and 4), output 0, else output 1.

If $\mathsf{ct}^*_{\mathsf{wCCA}}$ is an encryption of $sk_{id_{i^*}}$, then this is a perfect simulation of Game 3, while if $\mathsf{ct}^*_{\mathsf{wCCA}}$ is an encryption of 1^λ, then this is a perfect simulation of Game 4. This completes our proof.

Observation 1. *For any PPT adversary $\mathcal{A}$, $\mathsf{Adv}^4_{\mathcal{A}} = \mathsf{Adv}^{5,0}_{\mathcal{A}}$.*

Lemma 5. *For any j, any PPT adversary $\mathcal{A}$, $\mathsf{Adv}^{5,j}_{\mathcal{A}} - \mathsf{Adv}^{5,j\text{-}1}_{\mathcal{A}} \leq \mathsf{Adv}_{i\mathcal{O}}(\lambda)$.*

Proof. To prove this lemma, we need to show that the programs $\mathsf{Prog}\{K\}$ and $\mathsf{Prog\text{-}1}\{K\}$ are functionally identical. This follows from the observation that f is an injective function, and hence, for any t, s,

$$\begin{aligned} s = \oplus_i F(K, vk_{id_i}\|id_i\|m_i\|i\|t) &= \oplus_{i\neq i^*} F(K, vk_{id_i}\|id_i\|m_i\|i\|t) \oplus F(K, vk_{id_{i^*}}\|id_{i^*}\|m_{i^*}\|i^*\|t) \\ \iff \oplus_{i\neq i^*} F(K, vk_{id_i}\|id_i\|m_i\|i\|t) \oplus s &= F(K, vk_{id_{i^*}}\|id_{i^*}\|m_{i^*}\|i^*\|t) \\ \iff f(\oplus_{i\neq i^*} F(K, vk_{id_i}\|id_i\|m_i\|i\|t) \oplus s) &= f(F(K, vk_{id_{i^*}}\|id_{i^*}\|m_{i^*}\|i^*\|t)). \end{aligned}$$

Lemma 6. *For any j, any PPT adversary $\mathcal{A}$, $\mathsf{Adv}^{5,j\text{-}1}_{\mathcal{A}} - \mathsf{Adv}^{5,j\text{-}2}_{\mathcal{A}} \leq 2\mathsf{Adv}_{i\mathcal{O}}(\lambda)$.*

Proof. Let $K \leftarrow \mathsf{PRF.Setup}(1^\lambda)$, $y = vk_{id_{i^*}}\|id_{i^*}\|m_{i^*}\|i^*\|(j+1)$, $K\{y\} \leftarrow \mathsf{PRF.Puncture}(K, y)$ and $z = f(F(K,y))$. As in the previous proof, it suffices to show that $\mathsf{Prog}\{K, \mathsf{ct}_1, \ldots, \mathsf{ct}_n\}$ and $\mathsf{Prog\text{-}1}\{K\{y\}, \mathsf{ct}_1, \ldots, \mathsf{ct}_n\}$ have identical functionality, and $\mathsf{Prog\text{-}1}\{K\}$ and $\mathsf{Prog\text{-}2}\{y, z, K\{y\}\}$ have identical functionality.

Let us first consider $\mathsf{Prog}\{K, \mathsf{ct}_1, \ldots, \mathsf{ct}_n\}$ and $\mathsf{Prog\text{-}1}\{K\{y\}, \mathsf{ct}_1, \ldots, \mathsf{ct}_n\}$. Consider input $\{vk_{id_i}, (id_i, m_i), \sigma_i\}_i$. Let $t = \sigma_1 \cdot \mathsf{ct}_1 + \ldots + \sigma_n \cdot \mathsf{ct}_n$. From the correctness property of puncturable PRFs, it follows that the only case in which $\mathsf{Prog}\{K, \mathsf{ct}_1, \ldots, \mathsf{ct}_n\}$ and $\mathsf{Prog\text{-}1}\{K\{y\}, \mathsf{ct}_1, \ldots, \mathsf{ct}_n\}$ can possibly differ is when $\mathsf{SIG.Vefy}(vk_{id_i}, m_i, \sigma_i)=1$ for all $i \in [n]$, and $id^*=id_{i^*}$, $m^*=m_{i^*}$ and $t=j+1$. But this case is not possible, since $\mathsf{SIG.Vefy}(vk_{id_{i^*}}, m_{i^*}, \mathsf{HE.Dec}(sk_{\mathsf{HE}}, t))=\mathsf{SIG.Vefy}(vk_{id_{i^*}}, m_{i^*}, \sigma_{i^*})=1$, while $\mathsf{SIG.Vefy}(vk_{id_{i^*}}, m_{i^*}, \mathsf{HE.Dec}(sk_{\mathsf{HE}}, j+1))=0$.

Next, let us consider the programs $\mathsf{Prog\text{-}1}\{K\}$ and $\mathsf{Prog\text{-}2}\{y, z, K\{y\}\}$. Both programs have identical functionality, because $z = f(F(K,y))$ and for all $y' \neq y$, $F(K, y') = F.\mathsf{eval}(K\{y\}, y')$. This concludes our proof.

Lemma 7. *For any j, any PPT adversary $\mathcal{A}$, $\mathsf{Adv}^{5,j\text{-}2}_{\mathcal{A}} - \mathsf{Adv}^{5,j\text{-}3}_{\mathcal{A}} \leq \mathsf{Adv}_{\mathsf{PRF}}(\lambda)$.*

Proof. We will construct a PPT algorithm $\mathcal{B}$ such that $\mathsf{Adv}^{\mathcal{B}}_{\mathsf{PRF}} = \mathsf{Adv}^{5,j\text{-}2}_{\mathcal{A}} - \mathsf{Adv}^{5,j\text{-}3}_{\mathcal{A}}$. On receiving (id^*, m^*) from $\mathcal{A}$, $\mathcal{B}$ chooses $i^* \leftarrow [n]$, $(pk_{\mathsf{wCCA}}, sk_{\mathsf{wCCA}})$, $(pk_{\mathsf{HE}}, sk_{\mathsf{HE}})$ and compute ciphertext $\mathsf{ct}_i \leftarrow \mathsf{HE.Enc}(pk_{\mathsf{HE}}, 0)$ for all $i \in [n]$ and $i \neq i^*$, $\mathsf{ct}_{i^*} \leftarrow \mathsf{HE.Enc}(pk_{\mathsf{HE}}, 1)$. Let $\mathsf{Prog}\{pk_{\mathsf{wCCA}}\}$ be circuit as defined in the

Fig. 1. It chooses $K \leftarrow \mathsf{PRF.Setup}(1^\lambda)$, and let $y = vk_{id_{i^*}}\|id_{i^*}\|m_{i^*}\|i^*\|(j+1)$. $\mathcal{B}$ sends y to the PRF challenger, and receives $K\{y\}, z'$, where either $z' = F(K,y)$ or $z' \leftarrow \{0,1\}^\ell$. It computes $z = f(z')$, $P_1 = i\mathcal{O}(\mathsf{Prog}\{K\{y\}, \mathsf{ct}_1, \ldots, \mathsf{ct}_n\})$ and $P_2 = i\mathcal{O}(\mathsf{Prog}\{y, z, K\{y\}\})$. $\mathcal{B}$ computes $(vk_{id_{i^*}}, c_{id_{i^*}}) \leftarrow \mathsf{Prog}\{pk_{\mathsf{wCCA}}\|id_{i^*}(r)$ for uniformly random $r = r_0\|r_1 \in \{0,1\}^{\ell_{\mathsf{inp}}}$, where $(vk_{id_{i^*}}, sk_{id_{i^*}}) \leftarrow \mathsf{SIG.Setup}(1^\lambda; r_0)$ and $c_{id_{i^*}} \leftarrow \mathsf{PKE.Enc_{wCCA}}(pk_{\mathsf{wCCA}}, 1^\lambda; r_1)$. Then $\mathcal{B}$ compute $U \leftarrow \mathsf{SimUGen}(1^\lambda)$ and sends $\mathsf{PP} = \big(pk_{\mathsf{HE}}, U, \mathsf{Prog}\{pk_{\mathsf{wCCA}}\}, P_1, P_2\big)$ to $\mathcal{A}$.

$\mathcal{A}$ then asks for $\mathsf{KeyGen}(\mathsf{msk}, \cdot)$, $\mathsf{Sign}(\cdot,\cdot)$ queries, which $\mathcal{B}$ can simulate perfectly by using sk_{wCCA}. Finally, $\mathcal{A}$ outputs a forgery $\sigma^*_{\mathsf{agg}} = (t^*, s^*)$ and tuples $\{(id_i, m_i)\}_i$. If $\mathcal{A}$ wins as per the winning conditions (which are the same in both Game 5, j-2 and Game 5, j-3), output 0, else output 1.

Clearly, if $z' = F(K,y)$, then this corresponds to Game 5, j-2; if $z' \leftarrow \{0,1\}^\ell$, it corresponds to Game 5, j-3. This completes our proof.

Lemma 8. *For any j, any PPT adversary $\mathcal{A}$, $\mathsf{Adv}_{\mathcal{A}}^{5,j\text{-}3} - \mathsf{Adv}_{\mathcal{A}}^{5,j\text{-}4} \leq \mathsf{Adv}_f(\lambda)$.*

Proof. Suppose there exists a PPT adversary $\mathcal{A}$ such that $\mathsf{Adv}_{\mathcal{A}}^{5,j\text{-}3} - \mathsf{Adv}_{\mathcal{A}}^{5,j\text{-}4} = \epsilon$. We will construct a PPT algorithm $\mathcal{B}$ that inverts the one way function f using $\mathcal{A}$.

Note that the only way an adversary can distinguish between Game 5, j-3 and Game 5, j-4 is by submitting a forgery $\sigma^*_{\mathsf{agg}} = (t^* = j+1, s^*)$ and tuples $\{(id_i, m_i)\}_i$ such that $\big(\{(vk_{id_i}, id_i, m_i)\}_i, (t^* = j+1, s^*)\big)$ is (i^*, sk_{HE})-rejecting and $\mathsf{Prog\text{-}2}\{y, z, K\{y\}\}(\{(vk_{id_i}, id_i, m_i)\}_i, (t^* = j+1, s^*)) = 1$. From the definition of $\mathsf{Prog\text{-}2}\{y, z, K\{y\}\}$, it follows that $f(\oplus_{i \neq i^*} F(K, vk_{id_i}\|id_i\|m_i\|i\|t) \oplus s^*) = z$.

$\mathcal{B}$ receives z from the OWF challenger. On receiving (id^*, m^*) from $\mathcal{A}$, $\mathcal{B}$ chooses $i^* \leftarrow [n]$, $(pk_{\mathsf{wCCA}}, sk_{\mathsf{wCCA}})$, $(pk_{\mathsf{HE}}, sk_{\mathsf{HE}})$ and compute ciphertext $\mathsf{ct}_i \leftarrow \mathsf{HE.Enc}(pk_{\mathsf{HE}}, 0)$ for all $i \in [n]$ and $i \neq i^*$, $\mathsf{ct}_{i^*} \leftarrow \mathsf{HE.Enc}(pk_{\mathsf{HE}}, 1)$. Let $\mathsf{Prog}\{pk_{\mathsf{wCCA}}\}$ be circuit as defined in the Fig. 1. It chooses $K \leftarrow \mathsf{PRF.Setup}(1^\lambda)$, and let $y = vk_{id_{i^*}}\|id_{i^*}\|m_{i^*}\|i^*\|(j+1)$, $K\{y\} \leftarrow F.\mathsf{Puncture}(K, y)$. It computes $P_1 = i\mathcal{O}(\mathsf{Prog}\{K\{y\}, \mathsf{ct}_1, \ldots, \mathsf{ct}_n\})$ and $P_2 = i\mathcal{O}(\mathsf{Prog}\{y, z, K\{y\}\})$. $\mathcal{B}$ computes $(vk_{id_{i^*}}, c_{id_{i^*}}) \leftarrow \mathsf{Prog}\{pk_{\mathsf{wCCA}}\|id_{i^*}(r)$ for uniformly random $r = r_0\|r_1 \in \{0,1\}^{\ell_{\mathsf{inp}}}$, where $(vk_{id_{i^*}}, sk_{id_{i^*}}) \leftarrow \mathsf{SIG.Setup}(1^\lambda; r_0)$ and $c_{id_{i^*}} \leftarrow \mathsf{PKE.Enc_{wCCA}}(pk_{\mathsf{wCCA}}, 1^\lambda; r_1)$. Then $\mathcal{B}$ compute $U \leftarrow \mathsf{SimUGen}(1^\lambda)$ and sends $\mathsf{PP} = \big(pk_{\mathsf{HE}}, U, \mathsf{Prog}\{pk_{\mathsf{wCCA}}\}, P_1, P_2\big)$ to $\mathcal{A}$.

$\mathcal{A}$ then asks for $\mathsf{KeyGen}(\mathsf{msk}, \cdot)$, $\mathsf{Sign}(\cdot,\cdot)$ queries, which $\mathcal{B}$ can simulate perfectly by using sk_{wCCA}. Finally, $\mathcal{A}$ outputs a forgery $\sigma^*_{\mathsf{agg}} = (t^* = j+1, s^*)$ and tuples $\{(id_i, m_i)\}_i$. $\mathcal{B}$ sends $\oplus_{i \neq i^*} F(K, vk_{id_i}\|id_i\|m_i\|i\|t) \oplus s^*$ as inverse of z to the OWF challenger, and clearly, $\mathcal{B}$ wins if $\mathcal{A}$ wins. This completes our proof.

Lemma 9. *For any j, any PPT adversary $\mathcal{A}$, $\mathsf{Adv}_{\mathcal{A}}^{5,j\text{-}4} - \mathsf{Adv}_{\mathcal{A}}^{5,j\text{-}5} \leq \mathsf{Adv}_{\mathsf{PRF}}(\lambda)$.*

Proof. Similar to the proof of Lemma 7.

Lemma 10. *For any j, any PPT adversary $\mathcal{A}$, $\mathsf{Adv}_{\mathcal{A}}^{5,j\text{-}5} - \mathsf{Adv}_{\mathcal{A}}^{5,j\text{-}6} \leq 2\mathsf{Adv}_{i\mathcal{O}}(\lambda)$.*

Proof. Similar to the proof of Lemma 6.

Lemma 11. *For any j, any PPT adversary $\mathcal{A}$, $\mathsf{Adv}_{\mathcal{A}}^{5,j\text{-}6} - \mathsf{Adv}_{\mathcal{A}}^{5,(j+1)} \leq \mathsf{Adv}_{i\mathcal{O}}(\lambda)$.*

Proof. Similar to the proof of Lemma 5.

Lemma 12. *For any PPT adversary $\mathcal{A}$, $\mathsf{Adv}_{\mathcal{A}}^{6} \leq \mathsf{Adv}_{\mathsf{SIG}}(\lambda)$.*

Proof. Suppose $\mathsf{Adv}_{\mathcal{A}}^{6} = \epsilon$. We will construct a PPT algorithm $\mathcal{B}$ that breaks the security of SIG with advantage ϵ.

$\mathcal{B}$ receives vk from the challenger of signature scheme SIG. On receiving (id^*, m^*) from $\mathcal{A}$, $\mathcal{B}$ chooses $i^* \leftarrow [n]$, $(pk_{\mathsf{wCCA}}, sk_{\mathsf{wCCA}})$, $(pk_{\mathsf{HE}}, sk_{\mathsf{HE}})$ and compute ciphertext $\mathsf{ct}_i \leftarrow \mathsf{HE.Enc}(pk_{\mathsf{HE}}, 0)$ for all $i \in [n]$ and $i \neq i^*$, $\mathsf{ct}_{i^*} \leftarrow \mathsf{HE.Enc}(pk_{\mathsf{HE}}, 1)$. Let $\mathsf{Prog}\{pk_{\mathsf{wCCA}}\}$ be circuit as defined in the Fig. 1. It chooses $K \leftarrow \mathsf{PRF.Setup}(1^\lambda)$ and computes $P_1 = i\mathcal{O}(\mathsf{Prog}\{K, \mathsf{ct}_1, \ldots, \mathsf{ct}_n\})$ and $P_2 = i\mathcal{O}(\mathsf{Prog}\{K\})$. $\mathcal{B}$ computes $c_{id_{i^*}} \leftarrow \mathsf{PKE.Enc}_{\mathsf{wCCA}}(pk_{\mathsf{wCCA}}, 1^\lambda; r_1)$ for uniformly random $r = r_0 \| r_1 \in \{0,1\}^{\ell_{\mathsf{inp}}}$ and sets $\mathsf{Prog}\{pk_{\mathsf{wCCA}} \| id_{i^*}(r) = (vk, c_{id_{i^*}})$. Then $\mathcal{B}$ compute $U \leftarrow \mathsf{SimUGen}(1^\lambda)$ and sends $\mathsf{PP} = (pk_{\mathsf{HE}}, U, \mathsf{Prog}\{pk_{\mathsf{wCCA}}\}, P_1, P_2)$ to $\mathcal{A}$.

$\mathcal{A}$ then asks for $\mathsf{KeyGen}(\mathsf{msk}, \cdot)$ queries for $id \neq id_{i^*}$, which $\mathcal{B}$ can simulate perfectly by using sk_{wCCA}. On $\mathcal{A}$'s $\mathsf{Sign}(\cdot,\cdot)$ queries, if $id = id_{i^*}$ AND $m \neq m_{i^*}$, $\mathcal{B}$ forwards m to the signing oracle of signature scheme SIG, and receives σ, which is sent to $\mathcal{A}$ as response; if $id \neq id_{i^*}$, $\mathcal{B}$ generates the signature of m by using $\mathsf{sk}_{id} \leftarrow \mathsf{PKE.Dec}_{\mathsf{wCCA}}(sk_{\mathsf{wCCA}}, c_{id})$.

Finally, $\mathcal{A}$ outputs a forgery $\sigma_{\mathsf{agg}}^* = (t^*, s^*)$ and tuples $\{(id_i, m_i)\}_i$. $\mathcal{A}$ wins if id^* has not been asked to the $\mathsf{KeyGen}(\mathsf{msk}, \cdot)$ oracle, (id^*, m^*) has not been submitted to the $\mathsf{Sign}(\cdot,\cdot)$ oracle, and $\mathsf{SIG.Vefy}(vk, m^*, \mathsf{HE.Dec}(sk_{\mathsf{HE}}, t^*)) = 1$. It sends $(m^*, \mathsf{HE.Dec}(sk_{\mathsf{HE}}, t^*))$ as forgery. Note that $\mathcal{B}$ wins the signature game if $\mathcal{A}$ wins Game 6. This concludes our proof.

5 Conclusions

In this work, we consider n-bounded identity-based aggregate signatures (IBAS), which requires at most n signatures can be aggregated. We also provide a generic transformation to build n-bounded IBAS scheme from any secure signature scheme by using indistinguishability obfuscation and selective one-time universal parameters scheme. Based on the sub-exponential hardness of indistinguishability obfuscation, puncturable PRF and one-way functions, we prove that our n-bounded IBAS scheme is selectively secure in the standard model.

A Appendix

1 Public Key Encryption

Definition 6. *A public-key encryption scheme (PKE) consists of PPT algorithms $\mathsf{PKE} = (\mathsf{PKE.Setup}, \mathsf{PKE.Enc}, \mathsf{PKE.Dec})$.*

- **Key Generation**. *PKE.Setup takes as input security parameter* 1^λ *and returns a key pair* (pk, sk).
- **Encryption**. *PKE.Enc takes as input public key* pk *and message* m, *and returns a ciphertext* $c \leftarrow \mathsf{PKE.Enc}(pk, m)$.
- **Decryption**. *PKE.Dec takes as input secret key* sk *and ciphertext* c, *and returns a message* $m \leftarrow \mathsf{PKE.Dec}(sk, c)$.

Correctness. *For all* $\lambda \in \mathbb{N}$, $(pk, sk) \leftarrow \mathsf{PKE.Setup}(1^\lambda)$, *messages* $m \in \mathcal{M}(\lambda)$, *we require that* $\mathsf{PKE.Dec}(sk, \mathsf{PKE.Enc}(pk, m)) = m$.
We say that public-key encryption scheme PKE is wCCA secure, if

$$\left|\Pr[\mathsf{Exp}^{\mathsf{wCCA\text{-}0}}_{\mathsf{PKE},\mathcal{A}}(\lambda) = 1] - \Pr[\mathsf{Exp}^{\mathsf{wCCA\text{-}1}}_{\mathsf{PKE},\mathcal{A}}(\lambda) = 1]\right| \leq negl(\lambda)$$

for some negligible function negl and for all PPT attackers $\mathcal{A}$, *where* $\mathsf{Exp}^{\mathsf{wCCA\text{-}b}}_{\mathsf{PKE},\mathcal{A}}(\lambda)$ *is the following experiment with scheme PKE and attacker* $\mathcal{A}$:

1. $(pk, sk) \leftarrow \mathsf{PKE.Setup}(1^\lambda)$.
2. $(m_0, m_1) \leftarrow \mathcal{A}(1^\lambda, pk)$.
3. $b \leftarrow \{0, 1\}$ *and compute* $c^* \leftarrow \mathsf{PKE.Enc}(pk, m_b)$.
4. $b' \leftarrow \mathcal{A}^{\mathcal{O}_{\mathsf{wCCA}}}(1^\lambda, c^*)$.
   Here $\mathcal{O}_{\mathsf{wCCA}}$ *is an oracle that on input c returns* $\mathsf{PKE.Dec}(sk, c)$ *for all* $c \neq c^*$.

Note that this is a weakened version of standard IND-CCA security, because the attacker has access to $\mathcal{O}_{\mathsf{wCCA}}$ only after seeing the challenge ciphertext.

2 Signature Schemes

Definition 7. *A signature scheme with message space* $M(\lambda)$, *signature key space* $\mathcal{SK}(\lambda)$ *and verification key space* $\mathcal{VK}(\lambda)$ *consists of PPT algorithms* $\mathsf{SIG} = (\mathsf{SIG.Setup}, \mathsf{SIG.Sign}, \mathsf{SIG.Vefy})$:

- **Key Generation**. *SIG.Setup is a randomized algorithm that takes as input security parameter* 1^λ *and outputs signing key* $\mathsf{sk} \in \mathcal{SK}$ *and verification key* $\mathsf{vk} \in \mathcal{VK}$.
- **Signature Generation**. *SIG.Sign takes as input the signing key* $\mathsf{sk} \in \mathcal{SK}$ *and a message* $m \in M$ *and outputs a signature* σ.
- **Verification**. *SIG.Vefy takes as input a verification key* $\mathsf{vk} \in \mathcal{VK}$, *message* $m \in \mathcal{M}$ *and signature* σ *and outputs either 0 or 1.*

Correctness. *For all* $\lambda \in \mathbb{N}$, $(\mathsf{vk}, \mathsf{sk}) \leftarrow \mathsf{SIG.Setup}(1^\lambda)$, *messages* $m \in \mathcal{M}(\lambda)$, *we require that* $\mathsf{SIG.Vefy}(\mathsf{vk}, \mathsf{SIG.Sign}(\mathsf{sk}, m)) = 1$.
We say that signature scheme $\mathsf{SIG} = (\mathsf{SIG.Setup}, \mathsf{SIG.Sign}, \mathsf{SIG.Vefy})$ *is existentially unforgeable under a chosen message attack if*

$$\Pr[\mathsf{Exp}^{\mathsf{uf\text{-}cma0}}_{\mathsf{SIG},\mathcal{A}}(\lambda) = 1] \leq negl(\lambda)$$

for some negligible function negl and for all PPT attackers $\mathcal{A}$, *where* $\mathsf{Exp}^{\mathsf{uf\text{-}cma}}_{\mathsf{SIG},\mathcal{A}}(\lambda)$ *is the following experiment with scheme SIG and attacker* $\mathcal{A}$:

1. $(\mathsf{vk}, \mathsf{sk}) \leftarrow \mathsf{SIG.Setup}(1^\lambda)$.
2. $(m, \sigma) \leftarrow \mathcal{A}^{\mathsf{Sign}(\mathsf{sk}, \cdot)}(1^\lambda, pk)$.
   If $\mathsf{SIG.Vefy}(\mathsf{vk}, m, \sigma) = 1$ *and* m *was not queried to* $\mathsf{Sign}(\mathsf{sk}, \cdot)$ *oracle*
   Then return 1 else return 0.

3 Additively Homomorphic Encryption

Definition 8. *An additively homomorphic encryption scheme with message space* $\mathbb{F}_p$ *and ciphertext space* $\mathcal{C}_{HE}$ *consists of PPT algorithms* HE=($\mathsf{HE.Setup}$, $\mathsf{HE.Enc}$, $\mathsf{HE.Dec}$, $\mathsf{HE.Add}$).

- $\mathsf{HE.Setup}(1^\lambda)$ *takes the security parameter* 1^λ *as input and outputs public key* pk, *secret key* sk.
- $\mathsf{HE.Enc}(\mathsf{pk}, m)$ *takes as input a public key* pk *and message* $m \in \mathbb{F}_p$ *and outputs a ciphertext* $\mathsf{ct} \in \mathcal{C}_{HE}$.
- $\mathsf{HE.Dec}(\mathsf{sk}, \mathsf{ct})$ *takes as input a secret key* sk, *a ciphertext* $\mathsf{ct} \in \mathcal{C}_{HE}$ *and either outputs an element in* $\mathbb{F}_p$ *or* $\perp$.
- $\mathsf{HE.Add}(\mathsf{pk}, \mathsf{ct}_1, \mathsf{ct}_2)$ *takes as input a public key* pk *and two ciphertexts* $\mathsf{ct}_1, \mathsf{ct}_2 \in \mathcal{C}_{HE}$ *and outputs a ciphertext* ct.

Correctness. *Let* p *be any prime and* q *any polynomial in* λ. *For all* $\lambda \in \mathbb{N}$, $(\mathsf{pk}, \mathsf{sk}) \leftarrow \mathsf{HE.Setup}(1^\lambda)$, q *messages* $m_1, \ldots, m_q \in \mathbb{F}_p$, *the following holds*

$$\mathsf{HE.Dec}(\mathsf{sk}, \mathsf{HE.Enc}(\mathsf{pk}, m_1) + \ldots + \mathsf{HE.Enc}(\mathsf{pk}, m_q)) = m_1 + \ldots + m_q.$$

We say that additively homomorphic encryption scheme HE *is IND-CPA secure, if*

$$\left|\Pr[\mathsf{Exp}_{\mathsf{HE},\mathcal{A}}^{\mathsf{CPA\text{-}0}}(\lambda) = 1] - \Pr[\mathsf{Exp}_{\mathsf{HE},\mathcal{A}}^{\mathsf{CPA\text{-}1}}(\lambda) = 1]\right| \leq negl(\lambda)$$

for some negligible function negl and for all PPT attackers $\mathcal{A}$, *where* $\mathsf{Exp}_{\mathsf{HE},\mathcal{A}}^{\mathsf{CPA\text{-}b}}(\lambda)$ *is the following experiment with scheme* HE *and attacker* $\mathcal{A}$:

1. $(\mathsf{pk}, \mathsf{sk}) \leftarrow \mathsf{HE.Setup}(1^\lambda)$.
2. $(m_0, m_1) \leftarrow \mathcal{A}(1^\lambda, \mathsf{pk})$.
3. $b \leftarrow \{0, 1\}$ *and compute* $c^* \leftarrow \mathsf{HE.Enc}(\mathsf{pk}, m_b)$.
4. $b' \leftarrow \mathcal{A}(1^\lambda, c^*)$.

References

1. Barak, B., Goldreich, O., Impagliazzo, R., Rudich, S., Sahai, A., Vadhan, S.P., Yang, K.: On the (im)possibility of obfuscating programs. In: Kilian, J. (ed.) CRYPTO 2001. LNCS, vol. 2139, p. 1. Springer, Heidelberg (2001)
2. Boneh, D., Gentry, C., Lynn, B., Shacham, H.: Aggregate and verifiably encrypted signatures from bilinear maps. In: Biham, E. (ed.) EUROCRYPT 2003. LNCS, vol. 2656, pp. 416–432. Springer, Heidelberg (2003)
3. Boldyreva, A., Gentry, C., ONeill, A., Yum, D.H.: Ordered multisignatures and identity-based sequential aggregate signatures, with applications to secure routing. Cryptology ePrint Archive, report 2007/438 (2010). Revised 21 February 2010
4. Bagherzandi, A., Jarecki, S.: Identity-based aggregate and multi-signature schemes based on RSA. In: Nguyen, P.Q., Pointcheval, D. (eds.) PKC 2010. LNCS, vol. 6056, pp. 480–498. Springer, Heidelberg (2010)

5. Garg, S., Gentry, C., Halevi, S., Raykova, M., Sahai, A., Waters, B.: Candidate indistinguishability obfuscation and functional encryption for all circuits. In: FOCS (2013)
6. Galindo, D., Herranz, J., Kiltz, E.: On the generic construction of identity-based signatures with additional properties. In: Lai, X., Chen, K. (eds.) ASIACRYPT 2006. LNCS, vol. 4284, pp. 178–193. Springer, Heidelberg (2006)
7. Gentry, C., Ramzan, Z.: Identity-based aggregate signatures. In: Yung, M., Dodis, Y., Kiayias, A., Malkin, T. (eds.) PKC 2006. LNCS, vol. 3958, pp. 257–273. Springer, Heidelberg (2006)
8. Hofheinz, D., Jager, T., Khurana, D., Sahai, A., Waters, B., Zhandry, M.: How to generate and use universal parameters. Cryptology ePrint Archive, report 2014/507 (2014). http://eprint.iacr.org/
9. Hohenberger, S., Koppula, V., Waters, B.: Universal signature aggregators. Cryptology ePrint Archive, report 2014/745 (2014). http://eprint.iacr.org/
10. Hohenberger, S., Sahai, A., Waters, B.: Full domain hash from (leveled) multilinear maps and identity-based aggregate signatures. In: Canetti, R., Garay, J.A. (eds.) CRYPTO 2013, Part I. LNCS, vol. 8042, pp. 494–512. Springer, Heidelberg (2013)
11. Shamir, A.: Identity-based cryptosystems and signature schemes. In: Blakely, G.R., Chaum, D. (eds.) CRYPTO 1984. LNCS, vol. 196, pp. 47–53. Springer, Heidelberg (1985)
12. Sahai, A., Waters, B.: How to use indistinguishability obfuscation: deniable encryption, and more. In: STOC, pp. 475–484 (2014)

Leveled Strongly-Unforgeable Identity-Based Fully Homomorphic Signatures

Fuqun Wang[1,2,3](✉), Kunpeng Wang[1,2], Bao Li[1,2], and Yuanyuan Gao[1,2,4]

[1] State Key Laboratory of Information Security, Institute of Information Engineering, Chinese Academy of Sciences, Beijing, China
{fqwang,kpwang,lb,yygao13}@is.ac.cn
[2] Data Assurance and Communication Security Research Center, Chinese Academy of Sciences, Beijing, China
[3] University of Chinese Academy of Sciences, Beijing, China
[4] School of Science, Sichuan University of Science and Engineering, Zigong, China

Abstract. Recently, Gorbunov, Vaikuntanathan and Wichs proposed a new powerful primitive: (fully) homomorphic trapdoor function (HTDF) based on *small integer solution* (SIS) problem in standard lattices, from which they constructed the first leveled existentially-unforgeable fully homomorphic signature (FHS) schemes.

In this paper, we first extend the notion of HTDF to identity-based setting with stronger security and better parameters. The stronger security requires that the identity-based HTDF (IBHTDF) is not only *claw-free*, but also *collision-resistant*. And the maximum noise comparing to Gorbunov-Vaikuntanathan-Wichs' HTDF roughly reduces from $O(m^d\beta)$ to $O(4^d m\beta)$, which will result in polynomial modulus $q = \text{poly}(\lambda)$ when $d = O(\log\lambda)$, where λ is the security parameter and d is the depth bound of circuit. We then define and construct the first leveled strongly-unforgeable identity-based fully homomorphic signature (IBFHS) schemes.

Keywords: Identity-based homomorphic trapdoor function · Identity-based fully homomorphic signature · Small integer solution · Strong unforgeability

1 Introduction

Following the fast development of cloud computing, cryptographic schemes with homomorphic property attract a large number of researchers' sights. They allow a client to securely upload his/her encrypted/signed data to a remote server. Meanwhile they also allow the server to run computation over the data. The seminal study of fully homomorphic encryption (FHE) [17] demonstrates how

This work is supported in part by the National Nature Science Foundation of China under award No. 61272040 and No. 61379137, and in part by the National 973 Program of China under award No. 2013CB338001.

J. Lopez and C.J. Mitchell (Eds.): ISC 2015, LNCS 9290, pp. 42–60, 2015.
DOI: 10.1007/978-3-319-23318-5_3

to perform homomorphic computation over encrypted data without the knowledge of secret key. The recent works [6,18,23] of (leveled) fully homomorphic signatures demonstrate how to perform homomorphic computation on signed data.

In this work, we focus on the latter question: public authenticity of the result of homomorphic computation over signed data. In a homomorphic signature scheme, a client signs some data $\mathbf{x} = (x_1, \ldots, x_N)$ using his/her signing key and outsources the signed data $\boldsymbol{\sigma} = (\sigma_1, \ldots, \sigma_N)$ to a remote server. At any later point, the server can perform homomorphically some operation $y = g(\mathbf{x})$ over the signed data $\boldsymbol{\sigma}$ and produce a short signature σ_g certifying that y is the correct output of the operation g over the data $\mathbf{x}$. Anyone can verify the tuple (g, y, σ_g) using the client's public verification key and be sure of this fact without the knowledge of the underlying data $\mathbf{x}$.

Linear Homomorphic Signatures. A number of works discussed signatures with linear functions [2,4,10,16]. Such linear homomorphic signature schemes have meaningful applications in network coding and proofs of retrievability.

Somewhat Homomorphic Signatures. Boneh and Freeman [5] were the first to define and construct homomorphic signature schemes beyond linear functions, but limited to constant-degree polynomials based on ring SIS assumption in the random oracle model. Not long ago, Catalano, Fiore and Warinschi [11] gave an alternative scheme from multi-linear maps in the standard model.

Leveled Fully Homomorphic Signatures. Gorbunov, Vaikuntanathan and Wichs [18] proposed the first leveled FHS schemes based on SIS assumption. To this end, they drew on the ideas of constructing attribute-based encryption from standard lattices [7] and proposed a new primitive: HTDF. They required that HTDF functions have claw-freeness property, which is sufficient to show their FHS schemes (constructed directly from the HTDF functions) are existentially unforgeable in the static chosen-message-attack (EU-sCMA) model. Additionally, they showed that one can transform an EU-sCMA secure FHS to an EU-aCMA (existential-unforgeability under adaptive chosen-message-attack) secure FHS via homomorphic chameleon hash function. Recently, Boyen, Fan and Shi [6] also proposed EU-aCMA secure FHS schemes using vanishing trapdoor technique [1]. In the meantime, Xie and Xue [23] showed that leveled FHS schemes can be constructed if indistinguishability obfuscation and injective one way function exist.

1.1 Motivation

We observe that all schemes with homomorphism above are existentially unforgeable. In this model, a verifiable forgery (g, y', σ') such that g is admissible on messages $\mathbf{x}$ and $y' \neq y$ $(y = g(\mathbf{x}))$ captures two facts. One is that σ' is a usual existential-forgery corresponding to the usual notion of signature forgery if $g(\mathbf{x}) = \pi_i(\mathbf{x}) = x_i$ is a special projection function. The other is that σ' is a homomorphic existential-forgery if g is a generally admissible function (defined in Sect. 3.1); in other words, the forgery σ' authenticates y' as $g(\mathbf{x})$ but in fact this is not the case.

However, as is well-known, security of signature schemes without homomorphism also can reach up to strong-unforgeability. In the stronger model, a forger can not give a forgery of message x_i, even he has a message-signature pair (x_i, σ_i). As a matter of course, we have a question: *can we define and construct strongly-unforgeable* (IB)FHS?

In this paper, we will give a positive response. Our main observation is that homomorphic computations on signed data are deterministic in all above schemes. In this scenario, we can define meaningful strong-unforgeability. In this model, given message-signature pairs $(\mathbf{x}, \boldsymbol{\sigma})$, a forger produce a verifiable strong-forgery (g, y', σ') such that $y' = y = g(\mathbf{x})$ and $\sigma' \neq \sigma_g$ that captures two facts. One is that $\sigma' \neq \sigma_i$ is a usual strong-forgery corresponding to the usual notion of strong-forgery if $g(\mathbf{x}) = \pi_i(\mathbf{x}) = x_i$. The other is that $\sigma' \neq \sigma_g$ is a homomorphic strong-forgery if g is a generally admissible function; in other words, the forgery σ' authenticates y' as $g(\mathbf{x})$ but in fact any forger can not produce $\sigma' \neq \sigma_g$.

Furthermore, as we all know, identity-based signature (IBS) is a nontrivial extension of signature [22]. In an IBS system, in order to verify a signature σ_i of a message x_i, the verifier requires only the global public parameters and the target identity *id*. Therefore, there is no need to issue a verification key for each user in an IBS system, which greatly simplifies the key management. Naturally, constructing an IBS with homomorphism is interesting. As far as we know, there is no construction of identity-based FHS. In fact, we will propose the first strongly-unforgeable IBFHS as a response to above question.

1.2 Contribution

We define and construct the first leveled strongly-unforgeable IBFHS schemes. To this end, we extend HTDF, the underlying primitive of FHS, to IBHTDF with stronger security and better parameters, the underlying primitive of IBFHS using the trapdoor technique in [1,12,19]. The stronger security requires that IBHTDFs are not only *claw-free*, but also *collision-resistant* to show the strong-unforgeability of IBFHS. We use Barrington's theorem to reduce the parameters as done in FHE world [9]. The maximum noise-level comparing to Gorbunov-Vaikuntanathan-Wichs' FHS roughly reduces from $O(m^d\beta)$ to $O(4^d m\beta)$, which will result in polynomial modulus $q = \text{poly}(\lambda)$ when $d = O(\log\lambda)$, where λ is the security parameter and d is the maximum depth of admissible circuit.

1.3 Paper Organization

In Sect. 2, we give some background on lattices and related tools as used in this paper. We propose formally the IBHTDF functions in Sect. 3 and demonstrate how to homomorphically evaluate a permutation branching program in Sect. 4. In Sect. 5, we define and construct the leveled strongly-unforgeable IBFHS. Finally, we conclude in Sect. 6.

2 Preliminaries

We use the bold upper-case letters (e.g., $\mathbf{A}, \mathbf{B}$) to represent matrices and bold lower-case letters (e.g. $\mathbf{a}, \mathbf{b}$) to represent column vectors. Let $\|\mathbf{A}\|_\infty = \max_{i,j}\{|a_{i,j}|\}$ denote the infinite norm and a_i or $\mathbf{a}[i]$ represent the i-entry of $\mathbf{a}$. Let $[\mathbf{A}||\mathbf{B}]$ denote the concatenation of two matrices and $(\mathbf{A}, \mathbf{B}) = [\mathbf{A}^T||\mathbf{B}^T]^T$. We use λ to denote the *security parameter* and $\mathrm{negl}(\lambda)$ to denote a negligible function that grows slower than λ^{-c} for any constant $c > 0$ and any large enough value of λ.

2.1 Entropy and Statistical Distance

For discrete random variables $X \leftarrow \mathcal{X}, Y \leftarrow \mathcal{Y}$, we define the *statistical distance* $\triangle(X, Y) \triangleq \frac{1}{2}\sum_{\omega\in\mathcal{X}\cup\mathcal{Y}} |\mathbf{Pr}[X = \omega] - \mathbf{Pr}[Y = \omega]|$. We say that two random variables X, Y are statistically indistinguishable, denoted as $X \approx_s Y$, if $\triangle(X, Y) = \mathrm{negl}(\lambda)$. The *min-entropy* of a random variable X, denoted by $\mathbf{H}_\infty(X)$, is defined as $\mathbf{H}_\infty(X) \triangleq -\log(\max_x \mathbf{Pr}[X = x])$. The *average min-entropy* of X conditioned on Y, denoted with $\widetilde{\mathbf{H}}_\infty(X|Y)$, is defined as

$$\widetilde{\mathbf{H}}_\infty(X|Y) \triangleq -\log(\mathbf{E}_{y\leftarrow\mathcal{Y}}[\max_x \mathbf{Pr}[X = x|Y = y]]) = -\log(\mathbf{E}_{y\leftarrow\mathcal{Y}}[2^{-\mathbf{H}_\infty(X|Y=y)}]).$$

The optimal probability of an unbounded attacker surmising X given the correlated value Y is $2^{-\widetilde{\mathbf{H}}_\infty(X|Y)}$.

Lemma 2.1 ([15]). *Let $X \leftarrow \mathcal{X}, Y \leftarrow \mathcal{Y}$ be two (correlated) random variables. It then holds that $\widetilde{\mathbf{H}}_\infty(X|Y) \geq \mathbf{H}_\infty(X) - \log(|\mathcal{Y}|)$.*

2.2 Background on Lattices and Hard Problems

Lattices. Lattices-based cryptography usually use so-called q-ary integer lattices, which contain $q\mathbb{Z}^m$ as a sublattice for some modulus q. Let n, m, q be positive integers. For a matrix $\mathbf{A} \in \mathbb{Z}_q^{n\times m}$ we define the following q-ary integer lattice:

$$\Lambda^\perp(\mathbf{A}) = \{\mathbf{u} \in \mathbb{Z}^m : \mathbf{A}\mathbf{u} = \mathbf{0} \mod q\}.$$

For a vector $\mathbf{v} \in \mathbb{Z}_q^n$, we define the coset (or "shifted" lattice):

$$\Lambda_\mathbf{v}^\perp(\mathbf{A}) = \{\mathbf{u} \in \mathbb{Z}^m : \mathbf{A}\mathbf{u} = \mathbf{v} \mod q\}.$$

SIS. Let n, m, q, β be integers. The short integer solution ($\mathrm{SIS}_{n,m,q,\beta}$) problem is, given a uniformly random matrix $\mathbf{A} \xleftarrow{\$} \mathbb{Z}_q^{n\times m}$, to find a nonzero vector $\mathbf{u} \in \mathbb{Z}_q^n$ with $||\mathbf{u}||_\infty \leq \beta$ such that $\mathbf{A}\mathbf{u} = \mathbf{0}$ (i.e., $\mathbf{u} \in \Lambda^\perp(\mathbf{A})$). For $q \geq \beta \cdot \omega(\sqrt{n\log n})$, solving $\mathrm{SIS}_{n,m,q,\beta}$ in the average case is as hard as solving $\mathrm{GapSVP}_{\tilde{O}(\beta\cdot\sqrt{n})}$ in the worst case in standard lattices [20,21].

Discrete Gaussian Distribution. Let $\mathcal{D}_{\mathbb{Z}^m,r}$ be the truncated discrete Gaussian distribution over $\mathbb{Z}^m$ with parameter r. Namely, for $\mathbf{u} \leftarrow \mathcal{D}_{\mathbb{Z}^m,r}$, if $\|\mathbf{u}\|_\infty$ is larger than $r \cdot \sqrt{m}$, then the output is replaced by $\mathbf{0}$. In other words, $\|\mathbf{u}\|_\infty \le r \cdot \sqrt{m}$ with probability 1 if $\mathbf{u} \leftarrow \mathcal{D}_{\mathbb{Z}^m,r}$.

Lattices Trapdoor. Here we recall the MP12-trapdoor generation algorithm and Gaussian sampling algorithm [19]. We ignore all details of implementation which are not strictly necessary in this work.

For integers n, q and $\ell = \lceil \log q \rceil$, let $\mathbf{G} = \mathbf{I}_n \otimes \mathbf{g}^T \in \mathbb{Z}_q^{n \times n\ell}$, where $\mathbf{g}^T = (1, 2, 2^2, \ldots, 2^{\ell-1})$ and $\mathbf{I}_n$ denotes the n-dimensional identity matrix.

Lemma 2.2 ([19]). *Let n, q, ℓ, m_0, m_1 be integers such that $n = \text{poly}(\lambda)$, $q = q(n), \ell = \lceil \log q \rceil, m_0 = n(\ell + O(1))$, $m_1 = n\ell$. For $\mathbf{A}_0 \xleftarrow{\$} \mathbb{Z}_q^{n \times m_0}$ and $\mathbf{H} \in \mathbb{Z}_q^{n \times n}$, there exists an randomized algorithm* TrapGen*$(\mathbf{A}_0, \mathbf{H})$ to generate a matrix $\mathbf{A}$ $(= [\mathbf{A}_0 \| \mathbf{H}\mathbf{G} - \mathbf{A}_0\mathbf{R}]) \in \mathbb{Z}_q^{n \times (m_0+m_1)}$ with trapdoor $\mathbf{R}$ such that $\mathbf{R} \leftarrow \mathcal{D}_{\mathbb{Z}^{m_0 \times m_1},r}$ for large enough r $(\ge \omega(\sqrt{\log n}))$ and $\mathbf{A}$ is $\text{negl}(\lambda)$-far from $(\mathbf{V}_0, \mathbf{V}_1) \xleftarrow{\$} \mathbb{Z}_q^{n \times m_0} \times \mathbb{Z}_q^{n \times m_1}$. Here, $\mathbf{R}$ is called an* MP12*-trapdoor (or $\mathbf{G}$-trapdoor) of $\mathbf{A}$ with tag $\mathbf{H}$.*

Furthermore, for any non-zero $\mathbf{u} = (\mathbf{u}_0, \mathbf{u}_1) \in \mathbb{Z}_q^{m_0+m_1}$, the average min-entropy of $\mathbf{R}\mathbf{u}_1$ given $\mathbf{A}_0$ and $\mathbf{A}_0\mathbf{R}$ is at least $\Omega(n)$.

Lemma 2.3 ([19]). *Given parameters in above lemma and a uniformly random vector $\mathbf{v} \in \mathbb{Z}_q^n$, for some s $(\ge O(\sqrt{n \log q})) \in \mathbb{R}$ and a fixed function $\omega(\sqrt{\log n})$ growing asymptotically faster than $\sqrt{\log n}$, if the tag matrix $\mathbf{H}$ is invertible, there then exists an efficient algorithm* SamplePre$(\mathbf{A}_0, \mathbf{R}, \mathbf{H}, \mathbf{v}, s)$ *that samples a vector $\mathbf{u}$ from $\mathcal{D}_{\Lambda_{\mathbf{v}}^{\perp}(\mathbf{A}), s \cdot \omega(\sqrt{\log n})}$ such that $\mathbf{A} \cdot \mathbf{u} = \mathbf{v}$. Note that $\|\mathbf{u}\|_\infty \le s\sqrt{m_0 + m_1} \cdot \omega(\sqrt{\log n})$ with probability 1.*

Furthermore, for $\mathbf{u}' \leftarrow \mathcal{D}_{\mathbb{Z}^m, s \cdot \omega(\sqrt{\log n})}$ and $\mathbf{v}' = \mathbf{A}\mathbf{u}'$, we have $(\mathbf{A}, \mathbf{R}, \mathbf{u}, \mathbf{v}) \approx_s (\mathbf{A}, \mathbf{R}, \mathbf{u}', \mathbf{v}')$.

Lemma 2.4 ([7,18,19]). *Let $m = m_0 + 2m_1$ and $\widetilde{\mathbf{G}} = [\mathbf{G} \| \mathbf{0}] \in \mathbb{Z}_q^{n \times m}$. For any matrix $\mathbf{V} \in \mathbb{Z}_q^{n \times m}$ there exists deterministic algorithm to output a $\{0,1\}$-matrix $\widehat{\mathbf{V}} \in \mathbb{Z}_q^{m \times m}$ such that $\widetilde{\mathbf{G}}\widehat{\mathbf{V}} = \mathbf{V}$ (or denoted by $\widetilde{\mathbf{G}}^{-1}(\mathbf{V}) = \widehat{\mathbf{V}}$[1]).*

2.3 Permutation Branching Program.

In this section, we define permutation branching program closely following [9]. A width-w permutation branching program Π of length L with input space $\{0,1\}^t$ is a sequence of L tuples of the form $(h(k), \sigma_{k,0}, \sigma_{k,1})$ where

- $h : [L] \to [t]$ is a function associates the k-th tuple with an input bit $x_{h(k)}$.
- $\sigma_{k,0}, \sigma_{k,1}$ are permutations over $[w] = \{1, 2, \ldots, w\}$.

[1] Here $\widetilde{\mathbf{G}}^{-1}$ is not the inverse matrix of $\widetilde{\mathbf{G}}$ but a deterministic algorithm.

A permutation branching program Π performs evaluation on input $\mathbf{x} = (x_1, x_2, \ldots, x_t)$ as follows. Let the initial state be $\eta_0 = 1$ and the k-th state be $\eta_k \in [w]$. We compute the state η_k recursively as

$$\eta_k = \sigma_{k,x_{h(k)}}(\eta_{k-1}).$$

Finally, after L steps, the end state is η_L. The output of Π is 1 if $\eta_L = 1$, and 0 otherwise.

To slow the growth of noise in homomorphic operations, we represent the states to bits, as demonstated in [9]. More specially, we replace the state $\eta_k \in [w]$ with some w-dimensional unit vector $\mathbf{v}_k$, e.g., $\mathbf{v}_0 = (1, 0, 0, \ldots, 0)$ institutes for $\eta_0 = 1$. The idea is that $\mathbf{v}_k[i] = 1$ if and only if $\sigma_{k,x_{h(k)}}(\eta_{k-1}) = i$. A more important equivalent relation is that $\mathbf{v}_k[i] = 1$ if and only if either:

- $x_{h(k)} = 1$ and $\mathbf{v}_{k-1}[\sigma_{k,1}^{-1}(i)] = 1$; or
- $x_{h(k)} = 0$ and $\mathbf{v}_{k-1}[\sigma_{k,0}^{-1}(i)] = 1$.

Hence, for $k \in [L], i \in [w]$, we have

$$\begin{aligned} \mathbf{v}_k[i] &= \mathbf{v}_{k-1}[\sigma_{k,1}^{-1}(i)] \cdot x_{h(k)} + \mathbf{v}_{k-1}[\sigma_{k,0}^{-1}(i)] \cdot (1 - x_{h(k)}) \\ &= \mathbf{v}_{k-1}[\gamma_{k,i,1}] \cdot x_{h(k)} + \mathbf{v}_{k-1}[\gamma_{k,i,0}] \cdot (1 - x_{h(k)}) \end{aligned} \tag{1}$$

where $\gamma_{k,i,1} \triangleq \sigma_{k,1}^{-1}(i)$ and $\gamma_{k,i,0} \triangleq \sigma_{k,0}^{-1}(i)$ are fully determined by the description of Π and can be computed easily and publicly. Thus, $\{(h(k), \gamma_{k,i,0}, \gamma_{k,i,1})\}_{k \in [L], i \in [w]}$ is an alternative description of a permutation branching program and is the form that we will work with under homomorphic computations.

3 Identity-Based Homomorphic Trapdoor Functions

We give the definition, construction and security proof of IBHTDFs in this section. In next section we will show how to homomorphically compute a circuit. Looking ahead, we will homomorphically compute a permutation branching program instead of a (boolean) circuit to reduce the parameters and increase the efficiency and security.

3.1 Definition

An *identity-based homomorphic trapdoor function* (IBHTDF) consists of six poly-time algorithms ($\mathsf{IBHTDF.Setup}, \mathsf{IBHTDF.Extract}, f, \mathsf{Invert}, \mathsf{IBHTDF.Eval}^{in}$, $\mathsf{IBHTDF.Eval}^{out}$) with syntax as follows:

- $(mpk, msk) \leftarrow \mathsf{IBHTDF.Setup}(1^\lambda)$: A master key setup procedure. The security parameter λ defines the identity space $\mathcal{I}$, the index space $\mathcal{X}$, the input space $\mathcal{U}$, the output space $\mathcal{V}$ and some efficiently samplable input distribution $\mathcal{D}_\mathcal{U}$ over $\mathcal{U}$. We require that elements in $\mathcal{I}, \mathcal{U}, \mathcal{V}$ or $\mathcal{X}$ can be efficiently certified and that one can efficiently sample elements from $\mathcal{V}$ uniformly at random.

- $(pk_{id}, sk_{id}) \leftarrow \mathsf{IBHTDF.Extract}(mpk, msk, id)$: An identity-key extraction procedure. As a matter of course, we require that pk_{id} can be extracted deterministically from mpk and $id \in \mathcal{I}$ without using the knowledge of msk.
- $f_{pk_{id},x} : \mathcal{U} \rightarrow \mathcal{V}$: A deterministic function indexed by pk_{id} and $x \in \mathcal{X}$.
- $\mathsf{Invert}_{sk_{id},x} : \mathcal{V} \rightarrow \mathcal{U}$: A probabilistic inverter indexed by sk_{id} and $x \in \mathcal{X}$.
- $u_g = \mathsf{IBHTDF.Eval}^{in}(g, (x_1, u_1, v_1), \ldots, (x_t, u_t, v_t))$: A deterministic *input* homomorphic evaluation algorithm. It takes as input some function $g : \mathcal{X}^t \rightarrow \mathcal{X}$ and values $\{x_i \in \mathcal{X}, u_i \in \mathcal{U}, v_i \in \mathcal{V}\}_{i \in [t]}$ and outputs $u_g \in \mathcal{U}$.
- $v_g = \mathsf{IBHTDF.Eval}^{out}(g, v_1, \ldots, v_t)$: A deterministic *output* homomorphic evaluation algorithm. It takes as input some function $g : \mathcal{X}^t \rightarrow \mathcal{X}$ and values $\{v_i \in \mathcal{V}\}_{i \in [t]}$ and outputs $v_g \in \mathcal{V}$.

Correctness of Homomorphic Computation. Let algorithm $(pk_{id}, sk_{id}) \leftarrow \mathsf{IBHTDF.Extract}$ extracts the identity-key for id. Let $g : \mathcal{X}^t \rightarrow \mathcal{X}$ be a function on $x_1, \ldots, x_t \in \mathcal{X}$ and set $y = g(x_1, \ldots, x_t)$. Let $u_1, \ldots, u_t \in \mathcal{U}$ and set $v_i = f_{pk_{id},x}(u_i)$ for $i = 1, \ldots, t$. Set $u_g = \mathsf{IBHTDF.Eval}^{in}(g, (x_1, u_1, v_1), \ldots, (x_t, u_t, v_t))$, $v_g = \mathsf{IBHTDF.Eval}^{out}(g, v_1, \ldots, v_t)$. We require that $u_g \in \mathcal{U}$ and $f_{pk_{id},x}(u_g) = v_g$.

Relaxation Correctness of Leveled IBHTDFs. In a leveled IBHTDF, every input $u_i \in \mathcal{U}$ will carry with noise $\beta_i \in \mathbb{Z}$. The initial samples chosen from the input-distribution $\mathcal{D}_{\mathcal{U}}$ carry with small noise β_0 and the noise β_g of the homomorphically evaulation u_g depends on the noise β_i of u_i, the indices x_i and the function g. In fact, if the noise $\beta_g > \beta_{max}$, where β_{max} is a threshold of noise, there is no guarantee of the correctness. Therefore, we should restrict the class of functions that can be computed. We say a function g is *admissible* on indices $x_1, \ldots, x_t$ if $\beta_g \leq \beta_{max}$ whenever u_i carries with noise $\beta_i \leq \beta_0$.

Distributional Equivalence of Inversion. To show the security of our main construction IBFHS in next section, we require the following statistical indistinguishability:

$$(pk_{id}, sk_{id}, x, u, v) \approx_s (pk_{id}, sk_{id}, x, u', v')$$

where $(pk_{id}, sk_{id}) \leftarrow \mathsf{IBHTDF.Extract}$, $x \in \mathcal{X}$, $u \leftarrow \mathcal{D}_{\mathcal{U}}, v = f_{pk_{id},x}(u), v' \xleftarrow{\$} \mathcal{V}, u' \leftarrow \mathsf{Invert}_{sk_{id},x}(v')$.

IBHTDF Security. Gorbunov *et al.* [18] required *claw-freeness* for HTDF security to provide *existential-unforgeability* for FHS. Here, we require not only *claw-freeness* but also *collision-resistance* for IBHTDF security to guarantee *strong-unforgeability* for IBFHS.

The experiment $\mathbf{Exp}^{\mathrm{sID}}_{\mathcal{A},\mathsf{IBHTDF}}(1^\lambda)$ defined in Fig. 1 describes the *selective-identity* security, where the adversary has to appoint a target identity id^* to attack before seeing the master public-key. Moreover, the adversary can query identity-keys for all identities except id^*. He is then forced to find $u \neq u' \in \mathcal{U}$, $x, x' \in \mathcal{X}$ such that $f_{pk_{id^*},x}(u) = f_{pk_{id^*},x'}(u')$. Remark that if $x = x'$, then (u, u') is a collision, a claw otherwise.

$$\mathbf{Exp}^{\mathrm{sID}}_{\mathcal{A},\mathsf{IBHTDF}}(1^\lambda)$$

- $(id^*, state) \leftarrow \mathcal{A}(1^\lambda)$
- $(mpk, msk) \leftarrow \mathsf{IBHTDF.Setup}(1^\lambda)$
- $(u, u', x, x') \leftarrow \mathcal{A}^{\mathsf{IBHTDF.Extract}(mpk, msk, \cdot)\backslash\{id^*\}}(mpk, state)$
- $\mathcal{A}$ wins if $u \neq u' \in \mathcal{U},\ x, x' \in \mathcal{X}$ are such that $f_{pk_{id^*}, x}(u) = f_{pk_{id^*}, x'}(u')$.

Fig. 1. Definition of *selective-identity* security for IBHTDF

We say that an identity-based homomorphic trapdoor function is *selective-identity* secure if $\Pr[\mathbf{Exp}^{\mathrm{sID}}_{\mathcal{A},\mathsf{IBHTDF}}(1^\lambda)] \leq \mathrm{negl}(\lambda)$.

In the stronger model of *adaptive-identity security*, the adversary can not find $u \neq u' \in \mathcal{U},\ x, x' \in \mathcal{X}$ such that $f_{pk_{id}, x}(u) = f_{pk_{id}, x'}(u')$ for any identity id, for which he has never queried identity-key sk_{id}. We note that one may construct *adaptive-identity* secure IBHTDF using the *vanishing trapdoor* techniques [1,8, 12] in the cost of both efficiency and security.

3.2 Construction: Basic Algorithms and Security

Recall that λ is the security parameter. To describe the IBHTDF functions succinctly, we give some public parameters as follows.

- Let flexible d be the circuit depth such that $d \leq \mathrm{poly}(\lambda)$ and set $L = 4^d$.
- Choose an integer $n = \mathrm{poly}(\lambda)$ and a sufficiently large prime $q = q(n)$. Let $\ell = \lceil \log q \rceil$, $m_0 = n(\ell + O(1))$, $m_1 = n\ell$ and $m = m_0 + 2m_1$. Set $\beta_0 = O((n \log q)^{3/2})$, $\beta_{max} = O(4^d m \beta_0)$, $\beta_{SIS} = O(m_1 \beta_0)\beta_{max} < q$.
- $\mathbf{G} = \mathbf{I}_n \otimes \mathbf{g}^T \in \mathbb{Z}_q^{n \times n\ell}$ is the primitive matrix, where $\mathbf{g}^T = (1, 2, 2^2, \ldots, 2^{\ell-1})$. Set $\widetilde{\mathbf{G}} = [\mathbf{G} \| \mathbf{0}] \in \mathbb{Z}_q^{n \times m}$ be the garget matrix used below.
- We assume that identities are elements in $\mathrm{GF}(q^n)$, and say $\mathbf{H} : \mathrm{GF}(q^n) \rightarrow \mathbb{Z}_q^{n \times n}$ is an invertible difference, if $\mathbf{H}(id_1) - \mathbf{H}(id_2)$ is invertible for any two different identities id_1, id_2 and $\mathbf{H}$ is computable in polynomial time in $n\ell$ (see an example in [1]).
- Set $\mathcal{X} = \mathbb{Z}_2, \mathcal{I} = \mathbb{Z}_q^n, \mathcal{V} = \mathbb{Z}_q^{n \times m}$ and $\mathcal{U} = \{\mathbf{U} \in \mathbb{Z}_q^{m \times m} : \|\mathbf{U}\|_\infty \leq \beta_{max}\}$. Define the distribution $\mathcal{D}_\mathcal{U}$ is a truncated discrete Gaussian distribution over $\mathcal{U}$, so that $\|\mathbf{U}\|_\infty \leq \beta_0$ if $\mathbf{U} \leftarrow \mathcal{D}_\mathcal{U}$.

Now we describe the basic algorithms of IBHTDF function $\mathcal{F}$.

- $\mathsf{IBHTDF.Setup}(1^\lambda)$: On input a security parameter λ, set $d, L, n, m_0, m_1, m, q, \beta_0, \beta_{max}, \beta_{SIS}$ as specified above. Then do:
  1. Choose $\mathbf{A}_0 \stackrel{\$}{\leftarrow} \mathbb{Z}_q^{n \times m_0}$. Run $\mathsf{TrapGen}(\mathbf{A}_0, \mathbf{0})$ to generate a matrix $\mathbf{A} = [\mathbf{A}_0 \| \mathbf{A}_1] = [\mathbf{A}_0 \| -\mathbf{A}_0\mathbf{R}] \in \mathbb{Z}_q^{n \times (m_0 + m_1)}$ and a trapdoor $\mathbf{R}$ such that $\mathbf{R} \leftarrow \mathcal{D} \triangleq \mathcal{D}_{\mathbb{Z}^{m_0 \times m_1}, \omega(\sqrt{\log n})}$ and $\mathbf{A}$ is $\mathrm{negl}(\lambda)$-far from uniform. Set the master secret key as $msk = \mathbf{R}$. Note that $\mathbf{A} \cdot (\mathbf{R}, \mathbf{I}_{m_1}) = \mathbf{0}$, namely $\mathbf{R}$ is a $\mathbf{G}$-trapdoor of $\mathbf{A}$ with tag $\mathbf{0}$.

2. Choose $\mathbf{A}_2 \xleftarrow{\$} \mathbb{Z}_q^{n\times m_1}$ and set the master public key as $mpk = \{\mathbf{A}, \mathbf{A}_2\}$.

– $\mathsf{IBHTDF.Extract}(mpk, \mathbf{R}, id)$: On input a master public key mpk, a master secret key $\mathbf{R}$ and an identity $id \in \mathcal{I}$, do:
  1. Compute $\mathbf{H}(id)$ for $id \in \mathcal{I}$ and let $\mathbf{A}'_{id} = [\mathbf{A}_0 || \mathbf{H}(id)\cdot\mathbf{G} + \mathbf{A}_1]$ (Note that $\mathbf{R}$ is a $\mathbf{G}$-trapdoor of $\mathbf{A}'_{id}$ with tag $\mathbf{H}(id)$). Set user-specific public-key $pk_{id} = \mathbf{A}_{id} = [\mathbf{A}'_{id}||\mathbf{A}_2]$.
  2. Run algorithm $\mathsf{SamplePre}(\mathbf{A}_0, \mathbf{R}, \mathbf{H}(id), \mathbf{G} - \mathbf{A}_2, O(\sqrt{n\log q}))$ to output $\mathbf{R}_{id} \in \mathbb{Z}^{(m_0+m_1)\times m_1}$ such that $\mathbf{A}'_{id}\cdot\mathbf{R}_{id} = \mathbf{G} - \mathbf{A}_2$ (Note that $\mathbf{R}_{id}$ is a $\mathbf{G}$-trapdoor of $\mathbf{A}_{id}$ with tag $\mathbf{I}_n$). Set secret key $sk_{id} = \mathbf{R}_{id}$.

– $f_{pk_{id},x}(\mathbf{U})$: On input $mpk, id \in \mathcal{I}, x \in \mathcal{X}$ and $\mathbf{U} \in \mathcal{U}$, do:
  1. Compute $pk_{id} = \mathbf{A}_{id} = [\mathbf{A}_0||\mathbf{H}(id)\cdot\mathbf{G} + \mathbf{A}_1||\mathbf{A}_2]$ as above.
  2. For $id \in \mathcal{I}, x \in \mathcal{X}$ and $\mathbf{U} \in \mathcal{U}$, define $f_{pk_{id},x}(\mathbf{U}) \triangleq \mathbf{A}_{id}\cdot\mathbf{U} + x\cdot\widetilde{\mathbf{G}}$.

– $\mathsf{Invert}_{sk_{id},x}(\mathbf{V})$: On input an identity $id \in \mathcal{I}$, an identity-key $\mathbf{R}_{id}$, an index $x \in \mathcal{X}$ and $\mathbf{V} \in \mathcal{V}$, run $\mathsf{SamplePre}(\mathbf{A}'_{id}, \mathbf{R}_{id}, \mathbf{I}_n, \mathbf{V} - x\cdot\widetilde{\mathbf{G}}, O(n\log q))$ to output $\mathbf{U}$ (such that $\mathbf{A}_{id}\cdot\mathbf{U} = \mathbf{V} - x\cdot\widetilde{\mathbf{G}}$).

Distributional Equivalence of Inversion. Let $x \in \mathcal{X}$ and $(pk_{id} = \mathbf{A}_{id}, sk_{id} = \mathbf{R}_{id}) \leftarrow \mathsf{IBHTDF.Extract}(mpk, \mathbf{R}, id)$. Let $\mathbf{U} \in \mathcal{U}$, $\mathbf{V} = f_{pk_{id},x}(\mathbf{U}) = \mathbf{A}_{id}\cdot\mathbf{U} + x\widetilde{\mathbf{G}}$, $\mathbf{V}' \xleftarrow{\$} \mathcal{V}$, $\mathbf{U}' \leftarrow \mathsf{SamplePre}(\mathbf{A}'_{id}, \mathbf{R}_{id}, \mathbf{I}_n, \mathbf{V}' - x\widetilde{\mathbf{G}}, O(n\log q))$. By Lemma 2.3 and the fact that $(\mathbf{V}' - x\widetilde{\mathbf{G}})$ is uniformly random, using a simple hybrid argument, we have

$$(\mathbf{A}_{id}, \mathbf{R}_{id}, \mathbf{U}, \mathbf{A}_{id}\cdot\mathbf{U}) \approx_s (\mathbf{A}_{id}, \mathbf{R}_{id}, \mathbf{U}', \mathbf{V}' - x\widetilde{\mathbf{G}}).$$

Then, we have

$$(\mathbf{A}_{id}, \mathbf{R}_{id}, x, \mathbf{U}, \mathbf{V} = \mathbf{A}_{id}\cdot\mathbf{U} + x\widetilde{\mathbf{G}}) \approx_s (\mathbf{A}_{id}, \mathbf{R}_{id}, x, \mathbf{U}', \mathbf{V}') \quad (2)$$

by applying the same function to both sides: put in a $x \in \mathcal{X}$ and add $x\widetilde{\mathbf{G}}$ to the last entry.

IBHTDF Security. We now show that the IBHTDF function $\mathcal{F}$ constructed above is selective-identity secure assuming the SIS assumption.

Theorem 3.1. *The function $\mathcal{F}$ constructed above is a selective-identity secure* IBHTDF *assuming the* $\mathrm{SIS}_{n,m_0,q,\beta_{SIS}}$ *assumption.*

Proof. Assume there exists a PPT adversary $\mathcal{A}$ that wins the security experiment $\mathbf{Exp}^{\mathrm{sID}}_{\mathcal{A},\mathsf{IBHTDF}}(1^\lambda)$ for $\mathcal{F}$ with non-negligible probability δ. We construct a PPT simulater $\mathcal{S}$ that breaks the $\mathrm{SIS}_{n,m_0,q,\beta_{SIS}}$ problem for $\mathbf{A}_0 \xleftarrow{\$} \mathbb{Z}_q^{n\times m_0}$.

Let id^* be the identity that $\mathcal{A}$ intends to attack. $\mathcal{S}$ will run the simulated algorithms ($\mathsf{IBHTDF.Setup}^*$, $\mathsf{IBHTDF.Extract}^*$).

– $\mathsf{IBHTDF.Setup}^*(1^\lambda)$: On input the same parameters as $\mathsf{IBHTDF.Setup}(1^\lambda)$, $\mathcal{S}$ does:

1. After receiving target identity $id^* \in \mathcal{I}$ and challenge matrix $\mathbf{A}_0 \in \mathbb{Z}_q^{n\times m_0}$, $\mathcal{S}$ runs $\mathsf{TrapGen}(\mathbf{A}_0, -\mathbf{H}(id^*))$ to produce a matrix $\mathbf{A} = [\mathbf{A}_0||\mathbf{A}_1] = [\mathbf{A}_0||-\mathbf{H}(id^*)\mathbf{G} - \mathbf{A}_0\mathbf{R}] \in \mathbb{Z}_q^{n\times(m_0+m_1)}$ and a trapdoor $\mathbf{R}$ such that $\mathbf{R} \leftarrow \mathcal{D}$ and $\mathbf{A}$ is negl(λ)-far from uniform. Set $msk = \mathbf{R}$.
2. $\mathcal{S}$ samples $\mathbf{S} \leftarrow \mathcal{D}$ and computes $\mathbf{A}_2 = \mathbf{A}_0\mathbf{S}$. Set $mpk = \{\mathbf{A}, \mathbf{A}_2\}$.

– $\mathsf{IBHTDF.Extract}^*(mpk, \mathbf{R}, id)$: On input a master public key mpk, a master secret key $\mathbf{R}$ and an identity $id \in \mathcal{I}$, do:
  1. Compute $\mathbf{H}(id)$ for $id \in \mathcal{I}$ and let $\mathbf{A}'_{id} = [\mathbf{A}_0||\mathbf{H}(id)\cdot\mathbf{G} + \mathbf{A}_1] = [\mathbf{A}_0||(\mathbf{H}(id) - \mathbf{H}(id^*))\mathbf{G} - \mathbf{A}_0\mathbf{R}]$ (Note that $\mathbf{R}$ is a $\mathbf{G}$-trapdoor of $\mathbf{A}'_{id}$ with tag $\mathbf{H}(id) - \mathbf{H}(id^*)$. Set $\mathbf{A}_{id} = [\mathbf{A}'_{id}||\mathbf{A}_2]$.
  2. Recall that $(\mathbf{H}(id) - \mathbf{H}(id^*))$ is invertible (by the property of $\mathbf{H}$) if $id \neq id^*$. Therefore, to respond to an identity-key query for $id \neq id^*$, $\mathcal{S}$ can run $\mathsf{SamplePre}(\mathbf{A}_0, \mathbf{R}, \mathbf{H}(id) - \mathbf{H}(id^*), \mathbf{G} - \mathbf{A_2}, O(\sqrt{n\log q}))$ and output $\mathbf{R}_{id} \in \mathbb{Z}^{(m_0+m_1)\times m_1}$ such that $\mathbf{A}'_{id}\cdot\mathbf{R}_{id} = \mathbf{G} - \mathbf{A}_2$ (Note that $\mathbf{R}_{id}$ is a $\mathbf{G}$-trapdoor of $\mathbf{A}_{id}$ with tag $\mathbf{I}_n$). Set $pk_{id} = \mathbf{A}_{id}$ and $sk_{id} = \mathbf{R}_{id}$.
  3. However, if $id = id^*$, then $\mathbf{A}_{id^*} = [\mathbf{A}_0|| - \mathbf{A}_0\mathbf{R}||\mathbf{A}_0\mathbf{S}]$ and the trapdoor disappears. Thus, the simulator $\mathcal{S}$ can not generate identity key for id^*.

The views of adversary $\mathcal{A}$ between the original experiment and the simulated experiment are indistinguishable by Lemma 2.2. Particularly, the winning probability of $\mathcal{A}$ attacking the simulated experiment is at least δ – negl(λ).

Now, we show that an adversary $\mathcal{A}$ who wins the simulated experiment $\mathbf{Exp}^{\mathrm{sID}}_{\mathcal{A},\mathsf{IBHTDF}}(1^\lambda)$ can be used to solve the SIS problem. Assume the winning adversary $\mathcal{A}$ outputs values $\mathbf{U} \neq \mathbf{U}' \in \mathcal{U}$, $x, x' \in \mathcal{X}$ such that $f_{pk_{id^*},x}(\mathbf{U}) = f_{pk_{id^*},x'}(\mathbf{U}')$. Let $\mathbf{U}^* = \mathbf{U} - \mathbf{U}'$ and $x^* = x' - x$. Then,

$$f_{pk_{id^*},x}(\mathbf{U}) = \mathbf{A}_{id^*}\mathbf{U} + x\widetilde{\mathbf{G}} = \mathbf{A}_{id^*}\mathbf{U}' + x'\widetilde{\mathbf{G}} = f_{pk_{id^*},x'}(\mathbf{U}') \;\Rightarrow\; \mathbf{A}_{id^*}\mathbf{U}^* = x^*\widetilde{\mathbf{G}}. \tag{3}$$

Recall that $\mathbf{A}_{id^*} = [\mathbf{A}_0|| - \mathbf{A}_0\mathbf{R}||\mathbf{A}_0\mathbf{S}]$. By the right hand side of Eq. (3), it holds that

$$\mathbf{A}_0\cdot\mathbf{U}^\diamond \triangleq \mathbf{A}_0\cdot([\mathbf{I}_{m_0}|| - \mathbf{R}||\mathbf{S}]\mathbf{U}^*) = x^*\widetilde{\mathbf{G}}. \tag{4}$$

Moreover, since $\mathbf{U}, \mathbf{U}' \in \mathcal{U}$, we have $||\mathbf{U}||_\infty, ||\mathbf{U}'||_\infty \leq \beta_{max}$ and thus $||\mathbf{U}^*||_\infty \leq 2\beta_{max}$. Moveover, since $\mathbf{R}, \mathbf{S}$ are sampled from $\mathcal{D}$, we also have $||\mathbf{R}||_\infty$, $||\mathbf{S}||_\infty \leq O(\sqrt{n\log q})$ and thus $||\mathbf{U}^\diamond||_\infty \leq 2\beta_{max}(2m_1\cdot O(\sqrt{n\log q}) + 1) \leq \beta_{SIS}$.

To solve the SIS problem defined by $\mathbf{A}_0 \in \mathbb{Z}_q^{n\times m_0}$, we discuss the following two cases:

– $x = x'$ (collision): In this case, it is sufficed to show that $\mathbf{U}^\diamond \neq 0$ except with negligible probability, since $\mathbf{A}_0\mathbf{U}^\diamond = x^*\widetilde{\mathbf{G}} = (x - x')\widetilde{\mathbf{G}} = 0$ and $||\mathbf{U}^\diamond||_\infty$ is small. Let $\mathbf{U}^* = (\mathbf{U}_0^*, \mathbf{U}_1^*, \mathbf{U}_2^*)$. Then, we have $\mathbf{U}^\diamond = \mathbf{U}_0^* - \mathbf{R}\mathbf{U}_1^* + \mathbf{S}\mathbf{U}_2^*$. We split to 2 distinct cases to analyze it.
  1. $\mathbf{U}_1^* = \mathbf{U}_2^* = 0$: In this case, we have $\mathbf{U}_0^* \neq 0$ since $\mathbf{U}^* \neq 0$. So, $\mathbf{U}^\diamond \neq 0$.
  2. $\mathbf{U}_1^* \neq 0$ or $\mathbf{U}_2^* \neq 0$: Without loss of generalization, we assume $\mathbf{U}_2^* \neq 0$. By Lemma 2.2, we then have that, even revealing $\mathbf{R}$, the min-entropy of $\mathbf{S}\mathbf{U}_2^*$ conditioned on the knowledge of $\mathbf{A}_0$ and $\mathbf{A}_0\mathbf{S}$ is at least $\Omega(n)$. Particularly, the probability that $\mathbf{U}^\diamond = \mathbf{0}$ is less than $2^{-\Omega(n)} = \mathrm{negl}(\lambda)$.

– $x \neq x'$ (claw): In this case, we show that the simulater $\mathcal{S}$ can use the knowledge of a small $\mathbf{U}^{\diamond} \neq \mathbf{0}$ and some $x^* \neq 0$ satisfying the Eq. (4) to find a solution of the SIS problem (similarly as [18]).

Choose $\mathbf{t} \xleftarrow{\$} \{0,1\}^{m_0}$ and set $\mathbf{r} \triangleq \mathbf{A}_0\mathbf{t}$. Compute $\mathbf{t}' = \widetilde{\mathbf{G}}^{-1}(\mathbf{r}/x^*) \in \{0,1\}^m$ such that $x^*\widetilde{\mathbf{G}}\mathbf{t}' = \mathbf{r}$. so,

$$\mathbf{A}_0(\mathbf{U}^{\diamond}\mathbf{t}' - \mathbf{t}) = (\mathbf{A}_0\mathbf{U}^{\diamond})\mathbf{t}' - \mathbf{A}_0\mathbf{t} = x^*\widetilde{\mathbf{G}}\mathbf{t}' - \mathbf{A}_0\mathbf{t} = \mathbf{r} - \mathbf{r} = 0.$$

Setting $\mathbf{u} \triangleq \mathbf{U}^{\diamond}\mathbf{t}' - \mathbf{t}$, we then have $\mathbf{A}_0\mathbf{u} = 0$ and $\|\mathbf{u}\|_\infty \leq (2m+1)\beta_{max} \leq \beta_{SIS}$. It remains to prove that $\mathbf{u} \neq 0$, i.e., $\mathbf{t} \neq \mathbf{U}^{\diamond}\mathbf{t}'$. We prove that it holds with overwhelming probability over the random $\mathbf{t}$, even given $\mathbf{A}_0, \mathbf{U}^{\diamond}, x^*$. In fact, we have

$$\widetilde{\mathbf{H}}_\infty(\mathbf{t}|\mathbf{t}') \geq \widetilde{\mathbf{H}}_\infty(\mathbf{t}|\mathbf{A}_0\mathbf{t}) \geq m_0 - n\log q = O(n).$$

where the first inequality follows from the fact that $\mathbf{t}'$ is deterministic by $\mathbf{r} = \mathbf{A}_0\mathbf{t}$, and the second inequality follows from Lemma 2.1. So, $\Pr[\mathbf{t} = \mathbf{U}^{\diamond}\mathbf{t}'] \leq 2^{-O(n)} = \text{negl}(\lambda)$.

Therefore, if the adversary $\mathcal{A}$ wins the simulated experiment $\mathbf{Exp}^{\text{sID}}_{\mathcal{A},\mathsf{IBHTDF}}(1^\lambda)$ with non-negligible probability $\delta/2 - \text{negl}(\lambda)$ in either case, the simulater $\mathcal{S}$ then will produce a valid solution for SIS problem with probability $\delta/2 - \text{negl}(\lambda)$. This finishes the proof. □

4 Homomorphic Evaluation and Noise Analysis

Although we can homomorphically compute arithmetic circuit or boolean circuit similarly as that in [18] with same-level parameters, we show how to do better in both works in this section based on the fact that the noise growth is asymmetric.

We define deterministic homomorphic addition and multiplication algorithms in Sect. 4.1. In Sect. 4.2, we show that these algorithms are not used by a naive combination of addition and multiplication, as in the work [9], but by an elaborate combination form to considerably slowing down the noise growth. The main difference between this work and [9] is that, to homomorphic evaluate, it requires us to design correspondingly two deterministic homomorphic algorithms: one for *input* and the other for *output* in this work, while it only requires to design one randomized homomorphic algorithm over ciphertexts in [9].

4.1 Basic Homomorphic Evaluation

We now define basic homomorphic addition and multiplication algorithms that will be used in IBHTDFs. These algorithms for IBHTDFs are same as that for HTDFs in [18] because of the same external structure with or without identity. Therefore, we can improve the parameters of HTDFs in [18] using asymmetric homomorphic multiplication demonstrated in this section and simplify the notations (e.g., Add^{in} instead of $\mathsf{IBHTDF.Add}^{in}$). Recall that $\mathbf{V}_i = \mathbf{A}\mathbf{U}_i + x_i\mathbf{G}$ ($i = 1, 2$), where we set $\mathbf{A} = \mathbf{A}_{id}, \mathbf{G} = \widetilde{\mathbf{G}}$ for simplicity throughout Sect. 4. Let $\|\mathbf{U}_i\|_\infty \leq \beta_i$ and $x_i \in \{0,1\}$.

Homomorphic Addition Algorithms. They are simple modulo-q addition of the *input* or *output* matrices respectively.

- $\mathsf{Add}^{in}((x_1, \mathbf{U}_1, \mathbf{V}_1), (x_2, \mathbf{U}_2, \mathbf{V}_2)) \triangleq \mathbf{U}_1 + \mathbf{U}_2 \mod q$
- $\mathsf{Add}^{out}(\mathbf{V}_1, \mathbf{V}_2) \triangleq \mathbf{V}_1 + \mathbf{V}_2 \mod q$

The addition-noise is bounded by $\beta_1 + \beta_2$. The correctness follows by $(\mathbf{V}_1 + \mathbf{V}_2) = \mathbf{A}(\mathbf{U}_1 + \mathbf{U}_2) + (x_1 + x_2)\mathbf{G}$.

Homomorphic Multiplication Algorithms. The homomorphic *input* multiplication algorithm is asymmetric and involved in whole input, partial output and index, and the homomorphic *output* multiplication algorithm is essentially a multiplicaiton of the output matrices.

- $\mathsf{Multi}^{in}((x_1, \mathbf{U}_1, \mathbf{V}_1), (x_2, \mathbf{U}_2, \mathbf{V}_2)) \triangleq x_2 \cdot \mathbf{U}_1 + \mathbf{U}_2 \cdot \widehat{\mathbf{V}}_1 \mod q$
- $\mathsf{Multi}^{out}(\mathbf{V}_1, \mathbf{V}_2) \triangleq \mathbf{V}_2 \cdot \widehat{\mathbf{V}}_1 \mod q$

The multiplication-noise is bounded by $|x_2|\beta_1 + m\beta_2 = \beta_1 + m\beta_2$. The correctness also follows by a simple computation assuming $\mathbf{V}_i = \mathbf{A}\mathbf{U}_i + x_i\mathbf{G}$.

4.2 The Homomorphic *Output* and *Input* Evaluation

Homomorphic *Output* Evaluation. We define the homomorphic *output* evaluation algorithm

$$\mathsf{Eval}^{out}(\Pi, \mathbf{V}_0, \{\mathbf{V}_{0,i}\}_{i\in[w]}, \{\mathbf{V}_j\}_{j\in[t]}) \rightarrow \mathbf{V}_\Pi$$

for a length-L permutation branching program Π, where $\mathbf{V}_0, \{\mathbf{V}_{0,i}\}_{i\in[w]}$ will be assigned in the initialization stage below and $\mathbf{V}_j$ is such that $\mathbf{V}_j = \mathbf{A}\mathbf{U}_j + x_j\mathbf{G}$. Recall that $\{(h(k), \gamma_{k,i,0}, \gamma_{k,i,1})\}_{k\in[L],i\in[w]}$ is a valid description of Π, and that the initial state vector is set to be the first w-dimensional unit vector $\mathbf{v}_0 = (1, 0, 0, \ldots, 0)$, and that for $k \in [L]$ and $i \in [w]$,

$$\mathbf{v}_k[i] = \mathbf{v}_{k-1}[\gamma_{k,i,1}] \cdot x_{h(k)} + \mathbf{v}_{k-1}[\gamma_{k,i,0}] \cdot (1 - x_{h(k)}).$$

The homomorphic *output* evaluation algorithm Eval^{out} proceeds as follows.

- **Initialization:** For $k \in [L], i \in [w]$, let $\mathbf{V}_k[i]$ be an *output* corresponding to the state $\mathbf{v}_k[i]$.
  1. Choose $\mathbf{V}_{0,i} \xleftarrow{\$} \mathbb{Z}_q^{n\times m}$ uniformly at random and set it be an initial *output* corresponding to the initial state $\mathbf{v}_0[i]$.
  2. Choose $\mathbf{V}_0 \xleftarrow{\$} \mathbb{Z}_q^{n\times m}$ uniformly at random and see it be an *output* corresponding to a constant state 1.
  3. Set $\bar{\mathbf{V}}_j \triangleq \mathbf{V}_0 - \mathbf{V}_j$ and see it be an *output* corresponding to $(1 - x_j)$, where $\mathbf{V}_j$ (so that $\mathbf{V}_j = \mathbf{A}\mathbf{U}_j + x_j\mathbf{G}$) is an *output* corresponding to x_j.

- **Computation:** For $k = 1, 2, \ldots, L$, the computation process proceeds inductively as follows. Assume that at step $t-1$, we have $\{\mathbf{V}_{k-1,i}\}_{i\in[w]}$. We compute

$$\mathbf{V}_{k,i} = \mathbf{V}_{h(k)} \cdot \widehat{\mathbf{V}}_{k-1,\gamma_{k,i,1}} + \bar{\mathbf{V}}_{h(k)} \cdot \widehat{\mathbf{V}}_{k-1,\gamma_{k,i,0}}. \tag{5}$$

- **Final *Output***: Finally, we have $\{\mathbf{V}_{L,i}\}_{i\in[w]}$ after finishing the computation process. Output $\mathbf{V}_{L,1}$ as the final *output* corresponding to $\mathbf{v}_L[1]$, i.e., $\mathbf{V}_\Pi = \mathbf{V}_{L,1}$.

Homomorphic *Input* Evaluation. We define the homomorphic *input* evaluation algorithm

$$\mathsf{Eval}^{in}(\Pi, (1, \mathbf{U}_0, \mathbf{V}_0), \{(\mathbf{v}_0[i], \mathbf{U}_{0,i}, \mathbf{V}_{0,i})\}_{i\in[w]}, \{(x_j, \mathbf{U}_j, \mathbf{V}_j)\}_{j\in[t]}) \rightarrow \mathbf{U}_\Pi$$

for a permutation branching program Π which proceeds as follows.

- **Initialization:** For $k \in [L], i \in [w]$, let $\mathbf{U}_k[i]$ be an *input* corresponding to the state $\mathbf{v}_k[i]$.
  1. Sample $\mathbf{U}_{0,i} \leftarrow \mathcal{D}_\mathcal{U}$ (such that $\mathbf{V}_{0,i} = \mathbf{A}\mathbf{U}_{0,i} + \mathbf{v}_0[i]\mathbf{G}$) and see it be an initial *input* corresponding to the initial state $\mathbf{v}_0[i]$.
  2. Sample $\mathbf{U}_0 \leftarrow \mathcal{D}_\mathcal{U}$ (such that $\mathbf{V}_0 = \mathbf{A}\mathbf{U}_0 + 1 \cdot \mathbf{G}$) and see it be an *input* corresponding to a constant state 1.
  3. Set $\bar{\mathbf{U}}_j \triangleq \mathbf{U}_0 - \mathbf{U}_j$, where $\mathbf{U}_j$ (such that $\mathbf{V}_j = \mathbf{A}\mathbf{U}_j + x_j\mathbf{G}$) is an *input* corresponding to x_j and see it be an *input* corresponding to $(1 - x_j)$.
- **Computation:** For $k = 1, 2, \ldots, L$, the computation process proceeds inductively as follows. Assume that at step $t-1$, we have $\{\mathbf{U}_{k-1,i}\}_{i\in[w]}$. We compute

$$\begin{aligned}\mathbf{U}_{k,i} = &(x_{h(k)} \cdot \mathbf{U}_{k-1,\gamma_{k,i,1}} + \mathbf{U}_{h(k)} \cdot \widehat{\mathbf{V}}_{k-1,\gamma_{k,i,1}}) \\ &+ ((1 - x_{h(k)}) \cdot \mathbf{U}_{k-1,\gamma_{k,i,0}} + \bar{\mathbf{U}}_{h(k)} \cdot \widehat{\mathbf{V}}_{k-1,\gamma_{k,i,0}}).\end{aligned} \tag{6}$$

- **Final *Input***: Finally, we have $\{\mathbf{U}_{L,i}\}_{i\in[w]}$ after finishing the computation process. Output $\mathbf{U}_{L,1}$ as the final *input* corresponding to $\mathbf{v}_L[1]$, i.e., $\mathbf{U}_\Pi = \mathbf{U}_{L,1}$.

4.3 Correctness of Homomorphic Evaluation and Noise Analysis

We will prove the correctness of above homomorphic *input-output* evaluation algorithms and analyze the noise growth under homomorphic evaluation.

Lemma 4.1. *Assuming that* $\mathsf{Eval}^{out}(\Pi, \mathbf{V}_0, \{\mathbf{V}_{0,i}\}_{i\in[w]}, \{\mathbf{V}_j\}_{j\in[t]}) \rightarrow \mathbf{V}_\Pi$ *and* $\mathsf{Eval}^{in}(\Pi, (1, \mathbf{U}_0, \mathbf{V}_0), \{(\mathbf{v}_0[i], \mathbf{U}_{0,i}, \mathbf{V}_{0,i})\}_{i\in[w]}, \{(x_j, \mathbf{U}_j, \mathbf{V}_j)\}_{j\in[t]}) \rightarrow \mathbf{U}_\Pi$ *are such that* $\mathbf{V}_0 = \mathbf{A}\mathbf{U}_0 + 1 \cdot \mathbf{G}$, $\mathbf{V}_{0,i} = \mathbf{A}\mathbf{U}_{0,i} + \mathbf{v}_0[i]\mathbf{G}$ *and* $\mathbf{V}_j = \mathbf{A}\mathbf{U}_j + x_j\mathbf{G}$ *for* $i \in [w], j \in [t]$. *For all* $k \in [L], i \in [w]$, *we then have*

$$\mathbf{V}_{k,i} = \mathbf{A}\mathbf{U}_{k,i} + \mathbf{v}_k[i]\mathbf{G}.$$

In particular, we have $\mathbf{V}_{L,1} = \mathbf{A}\mathbf{U}_{L,1} + \mathbf{v}_L[1]\mathbf{G}$.

Proof. Given the conditions in this lemma, by formulas (1), (5) and (6), we have

$$\begin{aligned}
\mathbf{A}\mathbf{U}_{k,i} + \mathbf{v}_k[i]\mathbf{G} =& \mathbf{A} \cdot \Big[\big(x_{h(k)} \cdot \mathbf{U}_{k-1,\gamma_{k,i,1}} + \mathbf{U}_{h(k)} \cdot \widehat{\mathbf{V}}_{k-1,\gamma_{k,i,1}}\big) \\
&+ \big((1 - x_{h(k)}) \cdot \mathbf{U}_{k-1,\gamma_{k,i,0}} + \bar{\mathbf{U}}_{h(k)} \cdot \widehat{\mathbf{V}}_{k-1,\gamma_{k,i,0}}\big)\Big] \\
&+ \Big(\mathbf{v}_{k-1}[\gamma_{k,i,1}] \cdot x_{h(k)} + \mathbf{v}_{k-1}[\gamma_{k,i,0}] \cdot (1 - x_{h(k)})\Big) \cdot \mathbf{G} \\
=& \Big(x_{h(k)} \cdot \mathbf{V}_{k-1,\gamma_{k,i,1}} - x_{h(k)} \cdot \mathbf{v}_{k-1}[\gamma_{k,i,1}] \cdot \mathbf{G}\Big) \\
&+ \Big(\mathbf{V}_{h(k)} \cdot \widehat{\mathbf{V}}_{k-1,\gamma_{k,i,1}} - x_{h(k)} \cdot \mathbf{V}_{k-1,\gamma_{k,i,1}}\Big) \\
&+ \Big((1 - x_{h(k)}) \cdot \mathbf{V}_{k-1,\gamma_{k,i,0}} - (1 - x_{h(k)}) \cdot \mathbf{v}_{k-1}[\gamma_{k,i,0}] \cdot \mathbf{G}\Big) \\
&+ \Big(\bar{\mathbf{V}}_{h(k)} \cdot \widehat{\mathbf{V}}_{k-1,\gamma_{k,i,0}} - (1 - x_{h(k)}) \cdot \mathbf{V}_{k-1,\gamma_{k,i,0}}\Big) \\
&+ \Big(x_{h(k)} \cdot \mathbf{v}_{k-1}[\gamma_{k,i,1}] \cdot \mathbf{G} + (1 - x_{h(k)}) \cdot \mathbf{v}_{k-1}[\gamma_{k,i,0}] \cdot \mathbf{G}\Big) \\
=& \mathbf{V}_{h(k)} \cdot \widehat{\mathbf{V}}_{k-1,\gamma_{k,i,1}} + \bar{\mathbf{V}}_{h(k)} \cdot \widehat{\mathbf{V}}_{k-1,\gamma_{k,i,0}} \\
=& \mathbf{V}_{k,i}
\end{aligned}$$

for all $k \in [L], i \in [w]$. This finishes the proof. □

Lemma 4.2. *Assuming that* $\mathsf{Eval}^{in}(\Pi, (1, \mathbf{U}_0, \mathbf{V}_0), \{(\mathbf{v}_0[i], \mathbf{U}_{0,i}, \mathbf{V}_{0,i})\}_{i\in[w]}, \{(x_j, \mathbf{U}_j, \mathbf{V}_j)\}_{j\in[t]}) \to \mathbf{U}_\Pi$ *is such that all the input-noises are bounded by* β, *i.e.,* $\|\mathbf{U}_0\|_\infty, \|\mathbf{U}_{0,i}\|_\infty, \|\mathbf{U}_j\|_\infty \leq \beta$, *it then holds that* $\|\mathbf{U}_\Pi\|_\infty \leq 3mL\beta + \beta$.

Proof. We will simply show the lemma by inductive method. Namely, we will show that $\|\mathbf{U}_{k,i}\|_\infty \leq 3km\beta + \beta$ for any step $k = 0, 1, 2, \ldots, L$ and $i \in [w]$.

If $k = 0$, there is no computation and by initialization it is very easy to see that all the initial noises are such that $\|\mathbf{U}_{0,i}\|_\infty \leq \beta, i \in [w]$.

Assume that at step $k - 1$, we have $\|\mathbf{U}_{k,i}\|_\infty \leq 3m(k-1)\beta + \beta$. By formula (6), we obtain that

$$\begin{aligned}
\|\mathbf{U}_{k,i}\|_\infty =& \|(x_{h(k)} \cdot \mathbf{U}_{k-1,\gamma_{k,i,1}} + \mathbf{U}_{h(k)} \cdot \widehat{\mathbf{V}}_{k-1,\gamma_{k,i,1}}) \\
&+ ((1 - x_{h(k)}) \cdot \mathbf{U}_{k-1,\gamma_{k,i,0}} + \bar{\mathbf{U}}_{h(k)} \cdot \widehat{\mathbf{V}}_{k-1,\gamma_{k,i,0}})\|_\infty \\
\leq& \|x_{h(k)} \cdot \mathbf{U}_{k-1,\gamma_{k,i,1}}\|_\infty + \|\mathbf{U}_{h(k)} \cdot \widehat{\mathbf{V}}_{k-1,\gamma_{k,i,1}}\|_\infty \\
&+ \|(1 - x_{h(k)}) \cdot \mathbf{U}_{k-1,\gamma_{k,i,0}}\|_\infty + \|\bar{\mathbf{U}}_{h(k)} \cdot \widehat{\mathbf{V}}_{k-1,\gamma_{k,i,0}}\|_\infty \\
\leq& x_{h(k)} \cdot (3m(k-1)\beta + \beta) + m\beta + (1 - x_{h(k)}) \cdot (3m(k-1)\beta + \beta) + 2m\beta \\
=& 3mk\beta + \beta
\end{aligned}$$

where $\|\bar{\mathbf{U}}_{h(k)}\|_\infty = \|\mathbf{U}_0 - \mathbf{U}_{h(k)}\|_\infty \leq \|\mathbf{U}_0\|_\infty + \|\mathbf{U}_{h(k)}\|_\infty \leq \beta + \beta = 2\beta$.

By induction, we get $\|\mathbf{U}_\Pi\|_\infty = \|\mathbf{U}_{L,1}\|_\infty \leq 3mL\beta + \beta$. This finishes the proof. □

Remark. By Barrington's theorem [3], a depth-d circuit can be transformed to a length $L = 4^d$ permutation branching program. Therefore, whenever

$d \le \text{poly}(\lambda)$, the maximum noise comparing to Gorbunov-Vaikuntanathan-Wichs' HTDF reduces roughly from $O(m^d\beta)$ to $O(4^d m\beta)$. In particular, we can set polynomial modulus $q = \text{poly}(\lambda) > O(4^d m\beta)$ when $d = O(\log \lambda)$ which will result in better security based on GapSVP with polynomial approximation factors.

5 Strongly-Unforgeable Identity-Based Fully Homomorphic Signatures

5.1 Definition

A single data-set identity-based homomorphic signature scheme consists of the following poly-time algorithms $(\mathsf{PrmsGen}, \mathsf{Setup}, \mathsf{Extract}, \mathsf{Sign}, \mathsf{SignEval}, \mathsf{Process}, \mathsf{Verify})$ with syntax:

- $prms \leftarrow \mathsf{PrmsGen}(1^\lambda, 1^N)$: Take the security parameter λ and the maximum data-size N. Output public parameters $prms$. The security parameter also defines the message space $\mathcal{X}$.
- $(mpk, msk) \leftarrow \mathsf{Setup}(1^\lambda)$: Take the security parameter λ. Output a master key pair (mpk, msk).
- $(pk_{id}, sk_{id}) \leftarrow \mathsf{Extract}(mpk, msk, id)$: An identity-key extraction procedure.
- $(\sigma_1, \ldots, \sigma_N) \leftarrow \mathsf{Sign}_{sk_{id}}(prms, x_1, \ldots, x_N)$: Sign message data $(x_1, \ldots, x_N) \in \mathcal{X}^N$ to id.
- $\sigma_g = \mathsf{SignEval}_{prms}(g, (x_1, \sigma_1), \ldots, (x_t, \sigma_t))$: Deterministically and homomorphically evaluate a signature σ_g for some function g over $(x_1, \ldots, x_t) \in \mathcal{X}^t$.
- $v_g = \mathsf{Process}_{prms}(g)$: Deterministically and homomorphically evaluate a *certificate* v_g for the function g from the public parameters $prms$.
- $\mathsf{Verify}_{pk_{id}}(v_g, y, \sigma_g)$: Verify that y is the correct output of g by proving σ_g corresponding to v_g.

Correctness. For $prms \leftarrow \mathsf{PrmsGen}(1^\lambda, 1^N)$, $(pk_{id}, sk_{id}) \leftarrow \mathsf{Extract}(mpk, msk, id)$, $(x_1, \ldots, x_N) \in \mathcal{X}^N$, $(\sigma_1, \ldots, \sigma_N) \leftarrow \mathsf{Sign}_{sk_{id}}(prms, x_1, \ldots, x_N)$, and $g : \mathcal{X}^N \to \mathcal{X}$, we require that the following equation

$$\mathsf{Verify}_{pk_{id}}(v_g, y = g(x_1, \ldots, x_N), \sigma_g) = \text{accept}$$

holds, where $v_g = \mathsf{Process}_{prms}(g)$ and $\sigma_g = \mathsf{SignEval}_{prms}(g, (x_1, \sigma_1), \ldots, (x_t, \sigma_t))$.

Relaxation Correctness of Leveled IBFHS. Here, the relaxation correctness of leveled IBFHS follows from that of leveled IBHTDF and hence is omitted.

Security Experiment. The experiment $\mathbf{Exp}^{\text{SU-sID-sCMA}}_{\mathcal{A},\mathsf{IBFHS}}(1^\lambda)$ defined in Fig. 2 describes the *strongly-unforgeable selective-identity static chosen-message-attack* security game, where the adversary has to fix a target identity id^* to attack and message data to sign before obtaining the master public-key and public parameters. Moreover, the adversary can query identity-keys for all identities except id^*. He is then forced to find (g, y', σ') such that the winning conditions (described in the experiment) hold. Remark that we do not require either $y = y'$ or not. So, if $y = y'$, then σ' is a strongly-forgeable signature, otherwise a existentially-forgeable signature.

$$\mathbf{Exp}^{\text{SU-sID-sCMA}}_{\mathcal{A},\mathsf{IBFHS}}(1^\lambda)$$

- $(id^*, \{x_i\}_{i\in[N]}, state) \leftarrow \mathcal{A}(1^\lambda)$
- $prms \leftarrow \mathsf{PrmsGen}(1^\lambda, 1^N)$, $(mpk, msk) \leftarrow \mathsf{Setup}(1^\lambda)$
- $(g, y', \sigma') \leftarrow \mathcal{A}^{\mathsf{Extract}(mpk,msk,\cdot)\setminus\{id^*\},\mathsf{Sign}(id^*,\{x_i\}_{i\in[N]})}(prms, mpk, state)$
- $\mathcal{A}$ wins if all of the following hold:
  1. g is admissible on the messages $x_1, \ldots, x_N$;
  2. $\sigma' \neq \sigma_g$, where $\sigma_g = \mathsf{SignEval}_{prms}(g, (x_1, \sigma_1), \ldots, (x_N, \sigma_N))$;
  3. $\mathsf{Verify}_{pk_{id^*}}(v_g, y', \sigma')$ accept, where $v_g = \mathsf{Process}_{prms}(g)$.

Fig. 2. Definition of security for IBFHS with single data-set

We say an IBFHS is *strongly-unforgeable selective-identity static chosen-message-attack* (SU-sID-sCMA) secure if $\Pr[\mathbf{Exp}^{\text{SU-sID-sCMA}}_{\mathcal{A},\mathsf{IBFHS}}(1^\lambda)] \leq \text{negl}(\lambda)$.

5.2 Construction

Let $\mathcal{F} = (\mathsf{IBHTDF.Setup}, \mathsf{IBHTDF.Extract}, f, \mathsf{Invert}, \mathsf{IBHTDF.Eval}^{in}, \mathsf{IBHTDF.Eval}^{out})$ be an IBHTDF with identity space $\mathcal{I}$, index space $\mathcal{X}$, input space $\mathcal{U}$, output space $\mathcal{V}$ and some efficiently samplable input distribution $\mathcal{D}_\mathcal{U}$ over $\mathcal{U}$. We construct an IBFHS scheme $\mathcal{S} = (\mathsf{PrmsGen}, \mathsf{Setup}, \mathsf{Extract}, \mathsf{Sign}, \mathsf{SignEval}, \mathsf{Process}, \mathsf{Verify})$ with message space $\mathcal{X}$ as follows.

- $prms \leftarrow \mathsf{PrmsGen}(1^\lambda, 1^N)$: Sample $v_i \xleftarrow{\$} \mathcal{V}, i \in [N]$ and set public parameters $prms = (v_1, \ldots, v_N)$.
- $(mpk, msk) \leftarrow \mathsf{Setup}(1^\lambda)$: Select $(mpk', msk') \leftarrow \mathsf{IBHTDF.Setup}(1^\lambda)$ and set master-key pair $(mpk = mpk', msk = msk')$.
- $(pk_{id}, sk_{id}) \leftarrow \mathsf{Extract}(mpk, msk, id)$: Run $\mathsf{IBHTDF.Extract}(mpk', msk', id)$ to get (pk'_{id}, sk'_{id}) and set $pk_{id} = pk'_{id}, sk_{id} = sk'_{id}$ for $id \in \mathcal{I}$.
- $(\sigma_1, \ldots, \sigma_N) \leftarrow \mathsf{Sign}_{sk_{id}}(prms, x_1, \ldots, x_N)$: Sample $u_i \leftarrow \mathsf{Invert}_{sk'_{id}, x_i}(v_i)$ and set $\sigma_i = u_i, i \in [N]$.
- $\sigma_g = \mathsf{SignEval}_{prms}(g, (x_1, \sigma_1), \ldots, (x_t, \sigma_t))$: Perform deterministic algorithm $\mathsf{IBHTDF.Eval}^{in}(g, (x_1, u_1, v_1), \ldots, (x_t, u_t, v_t))$ to get u_g and set $\sigma_g = u_g$.
- $v_g = \mathsf{Process}_{prms}(g)$: Perform $\mathsf{IBHTDF.Eval}^{out}(g, v_1, \ldots, v_t)$ and output the result v_g.
- $\mathsf{Verify}_{pk_{id}}(v_g, y, \sigma_g)$: If $f_{pk'_{id}, y}(\sigma_g) = v_g$ accept, else reject.

Correctness. Here, the discussion of the relaxation correctness of the leveled IBFHS constructed above follows from that of the underlying leveled IBHTDF in Sect. 3 and hence is omitted.

Security. We now show the SU-sID-sCMA security of the leveled IBFHS above.

Theorem 5.1. *The leveled* IBFHS *scheme* $\mathcal{S}$ *constructed above is* SU-sID-sCMA *secure assuming that* $\mathcal{F}$ *is a leveled selective-identity secure* IBHTDF.

Proof. Assume there exists a PPT adversary $\mathcal{A}$ that wins the security experiment $\mathbf{Exp}^{\text{SU-sID-sCMA}}_{\mathcal{A},\mathsf{IBFHS}}(1^\lambda)$ of IBFHS with non-negligible probability δ. We construct a PPT reduction $\mathcal{B}$ that breaks the selective-identity security of $\mathcal{F}$.

Let id^* be the identity that $\mathcal{A}$ intends to attack. $\mathcal{B}$ will run the changed algorithms ($\mathsf{PrmsGen}^*$, Setup^*, $\mathsf{Extract}^*$, Sign^*).

- $\mathsf{Setup}^*(1^\lambda)$: Run $(mpk', msk') \leftarrow \mathsf{IBHTDF.Setup}^*(1^\lambda)$ and set $mpk = mpk'$, $msk = msk'$.
- $\mathsf{Extract}^*(mpk, msk, id)$: Run $(pk'_{id}, sk'_{id}) \leftarrow \mathsf{IBHTDF.Extract}^*(mpk, \mathbf{R}, id)$ when $id \neq id^*$ and set $pk_{id} = pk'_{id}$, $sk_{id} = sk'_{id}$. However, if $id = id^*$, then the trapdoor disappears and $\mathcal{B}$ can not generate identity key for id^*.
- $\mathsf{PrmsGen}^*(1^\lambda, 1^N)$: Choose $u_i \leftarrow \mathcal{D}_\mathcal{U}$ and compute $v_i = f_{pk_{id^*}, x_i}(u_i)$. Output $prms = (v_1, \ldots, v_N)$.
- $\mathsf{Sign}^*(x_1, \ldots, x_N)$: Set $\sigma_i = u_i$ and output $(\sigma_1, \ldots, \sigma_N)$.

The views of adversary $\mathcal{A}$ between the original experiment and the changed experiment are indistinguishable by *Distributional Equivalence of Inversion* property of the underlying IBHTDF. In particular, the winning probability of $\mathcal{A}$ attacking the changed experiment is at least $\delta - \text{negl}(\lambda)$.

We now show that there exists a PPT reduction $\mathcal{B}$ that takes any PPT adversary $\mathcal{A}$ winning the changed experiment with non-negligible advantage $\delta - \text{negl}(\lambda)$, and that breaks the $\mathbf{Exp}^{\text{sID}}_{\mathcal{A},\mathsf{IBHTDF}}(1^\lambda)$ security of the underlying $\mathcal{F}$ with probability $\delta - \text{negl}(\lambda)$.

The reduction $\mathcal{B}$ receives the challenge identity id^* and message data-set $(x_1, \ldots, x_N)$, generates $(mpk, msk, \{\sigma_i = u_i, v_i\}_{i \in [N]})$ as in the changed experiment and sends $(mpk, \{\sigma_i, v_i\}_{i \in [N]})$ to $\mathcal{A}$. Note that $\mathcal{B}$ can respond to the identity-key query for $id \neq id^*$ using msk. But, $\mathcal{B}$ has no valid trapdoor to generate the identity key for id^*.

Assume the adversary $\mathcal{A}$ (winning the changed experiment) outputs values (g, y', σ'), where $g : \mathcal{X}^N \to \mathcal{X}$ on $(x_1, \ldots, x_N)$ is an admissible function and $\sigma' = u'$. Let $y = g(x_1, \ldots, x_N)$, $u_g = \sigma_g = \mathsf{SignEval}_{prms}(g, (x_1, \sigma_1), \ldots, (x_t, \sigma_t))$, $v_g = \mathsf{Process}_{prms}(g)$. Thus, on one hand, since the forged signature σ' verifies, $f_{pk_{id^*}, y'}(u') = v_g$ holds. On the other hand, since g is admissible, $f_{pk_{id^*}, y}(u_g) = v_g$ also holds by the correctness of homomorphic computation. Therefore, we have values $u_g \neq u' \in \mathcal{U}$ and $y, y' \in \mathcal{X}$ satisfying $f_{pk_{id^*}, y}(u_g) = f_{pk_{id^*}, y'}(u')$, which allows $\mathcal{B}$ to break $\mathbf{Exp}^{\text{sID}}_{\mathcal{A},\mathsf{IBHTDF}}(1^\lambda)$ security of $\mathcal{F}$ with probability $\delta - \text{negl}(\lambda)$ whenever $\mathcal{A}$ wins the changed experiment with probability $\delta - \text{negl}(\lambda)$. □

6 Conclusions

In this work, we defined and constructed the first leveled strongly-unforgeable IBFHS schemes. To this end, we extended Gorbunov-Vaikuntanathan-Wichs'

HTDF, the underlying primitive of FHS, to IBHTDF with stronger security and better parameters, the underlying primitive of IBFHS. The drawback is that our scheme is only a leveled IBFHS with large public parameters. It remains open to Construct a non-leveled IBFHS or a leveled IBFHS with short public parameters. One way to achieve this would be to draw on the ideas in constructing non-leveled (IB)FHEs from indistinguishability obfuscation [13,14].

Acknowledgement. We are very grateful to the anonymous ISC reviewers for valuable comments and constructive suggestions that helped to improve the presentation of this work.

References

1. Agrawal, S., Boneh, D., Boyen, X.: Efficient lattice (H)IBE in the standard model. In: Gilbert, H. (ed.) EUROCRYPT 2010. LNCS, vol. 6110, pp. 553–572. Springer, Heidelberg (2010)
2. Attrapadung, N., Libert, B.: Homomorphic network coding signatures in the standard model. In: Catalano, D., Fazio, N., Gennaro, R., Nicolosi, A. (eds.) PKC 2011. LNCS, vol. 6571, pp. 17–34. Springer, Heidelberg (2011)
3. Barrington, D.A.M.: Bounded-width polynomial-size branching programs recognize exactly those languages in NC^1. In: STOC 1986, pp. 1–5. ACM (1986)
4. Boneh, D., Freeman, D.M.: Homomorphic signatures for polynomial functions. In: Paterson, K.G. (ed.) EUROCRYPT 2011. LNCS, vol. 6632, pp. 149–168. Springer, Heidelberg (2011)
5. Boneh, D., Freeman, D.M.: Linearly homomorphic signatures over binary fields and new tools for lattice-based signatures. In: Catalano, D., Fazio, N., Gennaro, R., Nicolosi, A. (eds.) PKC 2011. LNCS, vol. 6571, pp. 1–16. Springer, Heidelberg (2011)
6. Boyen, X., Fan, X., Shi, E.: Adaptively secure fully gomomorphic signatures based on lattices. http://eprint.iacr.org/2014/916
7. Boneh, D., Gentry, C., Gorbunov, S., Halevi, S., Nikolaenko, V., Segev, G., Vaikuntanathan, V., Vinayagamurthy, D.: Fully key-homomorphic encryption, arithmetic circuit ABE and compact garbled circuits. In: Nguyen, P.Q., Oswald, E. (eds.) EUROCRYPT 2014. LNCS, vol. 8441, pp. 533–556. Springer, Heidelberg (2014)
8. Boyen, X.: Lattice mixing and vanishing trapdoors: a framework for fully secure short signatures and more. In: Nguyen, P.Q., Pointcheval, D. (eds.) PKC 2010. LNCS, vol. 6056, pp. 499–517. Springer, Heidelberg (2010)
9. Brakerski Z., Vaikuntanathan, V.: Lattice-based FHE as secure as PKE. In: ITCS, pp. 1–12 (2014)
10. Catalano, D., Fiore, D., Warinschi, B.: Efficient network coding signatures in the standard model. In: Fischlin, M., Buchmann, J., Manulis, M. (eds.) PKC 2012. LNCS, vol. 7293, pp. 680–696. Springer, Heidelberg (2012)
11. Catalano, D., Fiore, D., Warinschi, B.: Homomorphic signatures with efficient verification for polynomial functions. In: Garay, J.A., Gennaro, R. (eds.) CRYPTO 2014, Part I. LNCS, vol. 8616, pp. 371–389. Springer, Heidelberg (2014)
12. Cash, D., Hofheinz, D., Kiltz, E., Peikert, C.: Bonsai trees, or how to delegate a lattice basis. In: Gilbert, H. (ed.) EUROCRYPT 2010. LNCS, vol. 6110, pp. 523–552. Springer, Heidelberg (2010)

13. Canetti, R., Lin, H., Tessaro, S., Vaikuntanathan, V.: Obfuscation of probabilistic circuits and applications. In: Dodis, Y., Nielsen, J.B. (eds.) TCC 2015, Part II. LNCS, vol. 9015, pp. 468–497. Springer, Heidelberg (2015)
14. Clear, M., McGoldrick, C.: Bootstrappable identity-based fully homomorphic encryption. In: Gritzalis, D., Kiayias, A., Askoxylakis, I. (eds.) CANS 2014. LNCS, vol. 8813, pp. 1–19. Springer, Heidelberg (2014)
15. Dodis, Y., Ostrovsky, R., Reyzin, L., Smith, A.: Fuzzy extractors: how to generate strong keys from biometrics and other noisy data. SIAM J. Comput. **38**(1), 97–139 (2008)
16. Freeman, D.M.: Improved security for linearly homomorphic signatures: a generic framework. In: Fischlin, M., Buchmann, J., Manulis, M. (eds.) PKC 2012. LNCS, vol. 7293, pp. 697–714. Springer, Heidelberg (2012)
17. Gentry, C.: Fully homomorphic encryption using ideal lattices. In: STOC 2009, pp. 169–178. ACM (2009)
18. Gorbunov, S., Vaikuntanathan, V., Wichs, D.: Leveled fully homomorphic signatures from standard lattices. In: STOC 2015, pp. 469–477. ACM (2015)
19. Micciancio, D., Peikert, C.: Trapdoors for lattices: simpler, tighter, faster, smaller. In: Pointcheval, D., Johansson, T. (eds.) EUROCRYPT 2012. LNCS, vol. 7237, pp. 700–718. Springer, Heidelberg (2012)
20. Micciancio, D., Peikert, C.: Hardness of SIS and LWE with small parameters. In: Canetti, R., Garay, J.A. (eds.) CRYPTO 2013, Part I. LNCS, vol. 8042, pp. 21–39. Springer, Heidelberg (2013)
21. Micciancio, D., Regev, O.: Worst-case to average-case reductions based on gaussian measures. SIAM J. Comput. **37**(1), 267–302 (2007)
22. Shamir, A.: Identity-based cryptosystems and signature schemes. In: Blakely, G.R., Chaum, D. (eds.) CRYPTO 1984. LNCS, vol. 196, pp. 47–53. Springer, Heidelberg (1985)
23. Xie, X., Xue, R.: Bounded fully homomorphic signature schemes. http://eprint.iacr.org/2014/420

Graded Signatures

Aggelos Kiayias[2], Murat Osmanoglu[1](✉), and Qiang Tang[1]

[1] University of Connecticut, Mansfield, USA
{murat,qiang}@cse.uconn.edu
[2] National and Kapodistrian University of Athens, Athens, Greece
aggelos@cse.uconn.edu

Abstract. Motivated by the application of anonymous petitions, we formalize a new primitive called "graded signatures", which enables a user to consolidate a set of signatures on a message m originating from l different signers that are members of a PKI. We call the value $l \in \mathbb{N}$, the *grade* of the consolidated signature. The resulting consolidated signature object on m reveals nothing more than the grade and the validity of the original signatures without leaking the identity of the signers. Further, we require that the signature consolidation is taken place in an unlinkable fashion so that neither the signer nor the CA of the PKI can tell whether a signature is used in a consolidation action. Beyond petitions, we demonstrate the usefulness of the new primitive by providing several other applications including delegation of signing rights adhering to dynamic threshold policies and issuing graded certificates in a multi-CA PKI setting.

We present an efficient construction for graded signatures that relies on Groth-Sahai proofs and efficient arguments for showing that an integer belongs to a specified range. We achieve a linear in the grade signature size and verification time in this setting. Besides, we propose some extension that can support the certificate revocation by utilizing efficient non-membership proofs.

1 Introduction

In a petition system, a group of participants would like to send a formal request to an organization via a representative (petitioner) that helps them to express their opinions about an issue. There are several important criteria that a petition system has to satisfy: (1) the number of participants supporting the petition should be indicated; (2) the participants may prefer to remain anonymous in many scenarios, e.g., when these relate to political or religious issues; (3) the petitioner should not be able to make a false claim that the claimed number of participants is more than their actual number, e.g., duplicate participants should be removed without revealing any identities etc.

To address the above problem, we introduce a new primitive, which we call graded signature[1], that is applicable to an efficient privacy-preserving digital

[1] It is actually quite surprising that many seemingly related notions exist, however none of them satisfy all the natural requirements; we elaborate more on this below.

J. Lopez and C.J. Mitchell (Eds.): ISC 2015, LNCS 9290, pp. 61–80, 2015.
DOI: 10.1007/978-3-319-23318-5_4

petition system. In a graded signature scheme a user collects signatures from registered signers in a PKI. The primitive enables the consolidation of an arbitrary number of signatures (say l) originating from a subset of l distinct signers on the same message m. The resulting signature object, $\sigma^{(l)}$, convinces the verifier that at least l signers indeed signed on m without revealing the identities of signers. We call l, the grade of the signature. Note that l can range from 1 to n, where n is the total number of currently registered signers in the PKI. There is no need to pre-determine the value of l before the signature collecting procedure.

Applications of Graded Signatures. The new primitive can be useful in a number of applications that we discuss below.

Anonymous Petitions: In an anonymous petition, the petitioner aims at convincing an organization that a certain number of people have a consensus on one issue, and it is desired that the identity of each participant remains hidden. Our graded signature immediately solves this problem. Suppose every valid voter has a registered public key, and the one who initiates a petition on a message m, tries to get as much support (signatures) on m as she can. At the end, she consolidates all the signatures into one, and presents to the organization the message, the graded signature and the corresponding grade l. The privacy of all signers will be preserved, and the grade precisely reflects how many signatures the consolidator collected. The organization can verify that indeed l different signers are needed to produce the l-grade signature using the PKI parameters.

Anonymous Delegation of Signing Rights adhering to Threshold Policies: Consider an organization whose members are in a PKI and wish to authorize in anonymous fashion a certain individual to execute certain tasks without necessarily revealing their names. The authorization requires a certain quorum that, if reached, it should be universally accepted. For instance, suppose that the members of the board of trustees of a listed company would like to authorize the CEO to take certain decisions on behalf of them. Such authorization may require the agreement of the majority (or other suitable percentage) of the trustees. Using graded-signatures the CEO can obtain the signature of a suitable number of trustees on her public-key and then consolidate those to demonstrate the fact that a suitable number of trustees endorse her actions.

Graded Certificates for multi-CA PKI's: As a number of incidents have shown, certification authorities (CA) can be corrupted (e.g., see [21]) and in this way the security of critical Internet protocols such as TLS can be jeopardized. In a multi-CA setting a user may obtain certificates from multiple PKI's tying her identity to her public-key. Assuming the CA's themselves can be certified by an acceptable top-level CA, a user can form a "graded certificate" by consolidating her distinct certificates coming from different CA's into a single graded signature. The grade will reflect the number of certificates that the user has collected on her identity. Using graded signatures it is thus possible to enable a certificate negotiation step between two communicating parties that (1) provides sufficient assurance on their identities (by requiring a minimum signature grade for both sides) and (2) maintains their privacy in terms of their CA choices as the anonymity of the graded signature reveals only the grade but not the individual entities that have provided certificates.

Related Works. There exist many variants of PKI oriented signatures that provide anonymity, e.g., ring signatures [26] is a prominent example. Moreover there are aggregate signatures [9] and threshold signatures [15,28] which provide a form of a consolidation operation aimed at combining signatures into a single object. In some sense, a graded signature is a new primitive that brings together these lines of work. We will carefully compare our graded signature with existing related primitives below.

In a ring signature [26], the signer can anonymously sign a message on behalf of a group formed in an ad-hoc manner. Using the terminology of our paper, every ring signature will have a *fixed grade* 1, i.e., one of the signers signed the message. To form a graded signature with the grade k, for instance, the combiner can collect k ring signature on the message from k different signers. However, a regular ring signature scheme does not enable the receiver to check if there are two signatures produces by the same signer, which we need in graded signatures. To this aim, the notion of linkable ring signature [13,23], that enables one to detect whether two signatures were generated by the same signer, seems sufficient to get a graded signature scheme at first glance. However, even if we use short ring signatures [2,29], it still results in quadratic verification time since the verifier should check every pair of signatures. On the other hand, a (t,n)-threshold ring signature scheme [11] will convince the verifier that t signers agree on the message without leaking their identities. Similar to ring signatures, any (t,n)-threshold ring signature will have a *fixed grade* t. While we also require anonymity in a graded signature in a similar sense, in contrast to these previous primitives, our graded signature should enable one to produce a signature, with an arbitrary grade, solely depending on how many signers agree to sign on the message; furthermore our constructions can even allow the grade to be upgraded if the user can get more signatures from additional signers on the message.

Regarding our second application of delegation of signing rights, one may think of proxy signature [7,24], in which a proxy can sign documents on behalf of the delegator if it is granted the signing rights from the owner by running a delegation protocol. Also, other variants of proxy signatures exist, e.g., anonymous proxy signature [16] provides anonymity for the intermediate proxies if there is a chain of delegatees; and threshold proxy signature [30] in which one key owner delegates his signing rights to a bunch of proxies, but only when the total number of proxies is above the threshold, a valid proxy signature can be produced. The notion of functional signatures was also studied in [10]. It enables the key owner to delegate the signing rights according to a fine-grained policy f, such that the delegatee can only sign messages in the range of f. Our notion of graded signature is different than the above in the sense that there are multiple key owners to delegate their signing rights to one "proxy", so that the "proxy signature" can be verified according to the number of delegators (its grade) without leaking the delegators' identities.

In a threshold signature scheme [15,28] and its distributive variants [14,25], when the number of signers is below the threshold, they can not jointly produce a signature that convinces the verifier. However, if the signers are above the threshold, the signatures will look the same to the verifier. Although the signers may be allowed to change [17], normally the value of the threshold needs to be fixed during the system setup. Furthermore, they either require a fully trusted dealer to distribute signing keys, or when the number of signers is bigger than the threshold, they can recover all the secret key data. In contrast to that, in a graded signature, each consolidated signature is assigned a grade – the number of signers, and this number is not pre-determined, and it can vary from 1 to the total number of the signers which is n. Also, it can be deployed in a standard PKI setting without a trusted setup. No collusion of signers is able to produce a signature with grade larger than the size of the collusion.

The closest to our work is the notion of signature of reputation [5], which focuses only on the application of reputation systems and allows a user, as the combiner in our scheme, to consolidate all the upvotes for him as his reputation. Their construction is built on a general framework of NIZK proof systems that the user commits to each upvote and prove in zero-knowledge that each of the commitment contains a valid upvote. Note that a straigtforward application of such general framework would yield a signature of reputation with size that grows at least quadratically to the number of votes (even with the most efficient NIZK proof technique). The user has to provide a NIZK proof for each pair of commitments that they are from different identities. They resolve this problem via a clever use of the "linkability" of each commitment that the same randomness is used across all commitments of the votes, and each vote is essentially a unique signature. The verifier thus can check that each pair of commitments contain different votes which must come from different identities. However, this trick inherently incurs a quadratic (to the number of votes–grade in our terminology) verification time. Instead of only focusing on the application of reputation systems, our graded signature schemes aim at broader applications and we consider the notion as a more fundamental cryptographic primitive. Furthermore, since a signature might be verified many times, verification time is considered to be one of the most important efficiency metrics. Moreover, we want to remove the restriction that only unique signature schemes can be used for graded signatures. We propose a new way of using the general commit and prove framework in our construction that brings down the verification time to linear while still keeps the signature size linear to the grade, for a broader class of signature schemes.

Another closely related work is graded encryption [22], which is a generalization of identity based encryption (IBE). The primitive enables the user to sequentially upgrade the level of his key so that the secret key of an identity with level k can decrypt all the ciphertexts sent for the identity with level $k' \leq k$. Since IBE implies a signature scheme a graded encryption scheme also implies a graded signature scheme in the sense that the consolidation has to happen in a sequential fashion. While our graded signature scheme does not have this restriction, the signatures can be collected in an arbitrary order from signers.

It would be an interesting open question to consider graded encryption in the setting that the upgrading procedure is flexible like in graded signatures, i.e., only depending on how many secret keys received.

Besides those privacy preserving signature schemes, multi-signatures [3], and more generally, aggregate signatures [9] provide mechanisms for one to compactly represent signatures from different parties, some recent work [20] even shows that one may aggregate any type of signature from obfuscation techniques [20, 27]. However, the identities (public keys) of the signers will have to be explicitly given out for the verification. Contrary to that, a graded signature will keep the identities hidden, while reveal to the verifier only the grade of the signature. One may wonder whether adding some kind of anonymity to the signer to aggregate signatures will give us a graded signature. Specifically, if we have a trusted registration authority to issue certificates for public keys in a way that the identities are not revealed, the anonymity of the signers will be achieved. However, graded signature schemes also require the distinctness of the signers to be validated. Besides, according to the application scenarios e.g., that of an anonymous petition, the definition of anonymity has to be very strong so that even the registration authority (which might be *the* adversary in some settings) is allowed to be corrupted. Actually this anonymity requirement is a crucial difference between aggregate signatures and graded signatures that makes them incomparable. On one hand, there is no clear mechanism from aggregate signatures that can provide us strong anonymity together with the proof of distinctness of signers; on the other hand, our strong anonymity precludes the possibility for the verifier to identify the exact source of the signature.

Our Results. We first introduce formal definitions for graded signatures, including their correctness and security properties: unforgeability and anonymity. Every signer has his own key pair that is certified by the certificate authority. For correctness, when a signature is consolidated from ℓ different signatures, the consolidator should be able to convince the verifier that the signature is of grade ℓ' as long as $\ell \geq \ell'$. This allows us to define unforgeability focusing only at the attack scenario when the adversary produces a signature with grade one more than she is supposed to be able to produce. Regarding anonymity, we define it in a very strong sense: even if all parties, including the signers and the certification authority, are corrupted the consolidated signature should not leak the set of signers whose signatures were included in the consolidation process.

We provide an efficient construction for graded signatures. which (Sect. 3) achieves a constant verification and secret key size while both the graded signature size and the verification time are linear in the grade of the signature. This construction follows a "commit and prove" approach. Note that simply committing the signatures and showing that they originate from certified signers is insufficient: this is subject to a trivial attack where the consolidator uses the same signature over and over to increase the grade. In order to prevent this attack, an assurance of signature *distinctness* should be included in the proof that, if straightforwardly implemented, leads to a quadratic size or verification overhead. We go around this by introducing an order among signer public keys, and design the protocol in a way that it is compatible with the recent results of very efficient range proofs that were developed in [12]. Note that since each signing key is independent, the verification key of each signer who contributes to the graded signature should be somehow involved in the signature object generation in order for the verification of the consolidated signature to take place correctly. Thus, if we view graded signature with grade l as a "proof of knowledge" of l signatures it follows that the length of the underlying consolidated signature must be at least linear in l since it is supposed to carry information for l independent originating signers. Besides, this proof should also include an argument which shows that l *distinct* certified signers were involved in its construction.

2 Definitions and Security Modeling

In a graded signature scheme, there is a set of signers who register their public key with a certification authority, as in a traditional PKI setting, and there is a procedure which enables a privacy preserving signature combining functionality. Specifically, from several signatures on a message m originating from different signers, one can produce a "signature object" which convinces any verifier that at least "l distinct signers" signed on the message m without leaking the identity of any of them (beyond that they are members of the PKI of course). The grade l can vary from 1 to n where n is the total number of the registered signers in the system. For the ease of presentation, we differentiate the real grade ℓ which is the actual number of signatures used to consolidate the graded signature and the claimed grade ℓ' which is sent together with the graded signature for verification. Verification algorithm will accept if $\ell' \leq \ell$. The detailed definition of a graded signature is as follows:

- **Setup**: This algorithm takes the security parameter as an input, and outputs a master key pair (gsk, gpk).
- **Register**: This algorithm takes the master secret key gsk, a signer verification key vk_i as inputs and outputs a certificate $cert_i = Sign(gsk, vk_i||i)$ for the registered signer. The index $i \in \{1, \ldots, n\}$ corresponds to a unique signer.
- **Sign**: This algorithm takes a key pair (sk_i, vk_i) and a message m as inputs, and outputs a signature σ_i on m.
- **Combine**: This algorithm takes as inputs the global public parameters gpk, a message m and a set of signatures $\{\sigma_{i_1}, \ldots, \sigma_{i_l}\}$ on m from different signers and a set of verification keys $\{vk_{i_1}, \ldots, vk_{i_l}\}$ and the corresponding certifications $\{cert_{i_1}, ..., cert_{i_l}\}$. It outputs a "consolidated" signature $\sigma^{(\ell)}$ and its real grade ℓ.
- **Verify**: This algorithm inputs the global public gpk, a message signature pair $(m, \sigma^{(\ell)})$ and the claimed grade ℓ' of the signatur, and outputs 0 or 1.

Security Model for Graded Signatures. The *Correctness* of a graded signature scheme requires that if ℓ valid signatures under ℓ different certified verification keys are used to produce the graded signature, then as long as the claimed grade is no bigger than ℓ, the verification should always output 1, i.e., if $(m, \sigma^{(\ell)}) = \textbf{Combine}(m, \{(vk_i, cert_i, \sigma_i)\}_{i=i_1,\ldots,i_l})$, and for each i, (m, σ_i) is a valid message-signature pair under vk_i, and $(vk_i, cert_i)$ is valid under gpk, then $\textbf{Verify}(\ell', m, \sigma, gpk)$=1 as long as $\ell' \leq \ell$.

Next, we will define the security requirements of a graded signature. There are two major security concerns in a graded signature, unforgeability and anonymity. Unforgeability in this setting means one can not produce a graded signature with a higher grade ($\geq \ell$) than that she is supposed to be capable of, i.e., she may register new users, corrupt existing users, and receive some signatures on a target message, but the numbers add up to at most $\ell - 1$. For anonymity, we require it in a very strong sense that any two graded signatures with a same grade will look indistinguishable (even to the CA and the signers who contribute one of the signatures).

Unforgeability of Graded Signatures: In order to capture all the possible attacks that the adversary $\mathcal{A}$ may try, we make explicit all kinds queries[2] including registration queries which ask the CA to certify some public keys provided by $\mathcal{A}$, the corruption queries which enables $\mathcal{A}$ to learn the secret key of known, certified public keys, and the signature query for uncorrupted public keys. Consider the following game between an adversary $\mathcal{A}$ and a challenger C.

- $\mathcal{A}$ receives the master public key gpk.
- $\mathcal{A}$ is allowed to make registration queries, and gets certifications for the public keys that are generated by $\mathcal{A}$.
- $\mathcal{A}$ is also allowed to make corrupt queries, and gets secret keys for some existing certifications. (Note that all existing certifications and public keys together with the corresponding indices are available to the adversary.)
- $\mathcal{A}$ also adaptively chooses messages to ask C for signing queries from signers that are not queried for the secret key or the certification, and receives the corresponding signatures on those messages.
- $\mathcal{A}$ outputs a message m^* and signature with grade l.

Definition 1. *Let $Adv_{GS}^{\mathcal{A}}$ be the advantage of $\mathcal{A}$ in the game under the condition that $\mathcal{A}$ has asked at most $l - 1 = q_1 + q_2 + q_3$ queries where q_1 is the total number of the secret key queries, q_2 is the total number of the certification queries, and q_3 is the total number of signature queries for m^*. We say the graded signature is existentially unforgeable under adaptive corruption attack if $Adv_{GS}^{\mathcal{A}} \leq negl(\lambda)$.*

Remark that in our definition of unforgeability, we did not explicitly consider the attack that the adversary outputs a graded signature with $\ell + t$ for $t > 0$. However, from our definition of correctness, it is straightforward that if adversary is able to do so, she will be also capable of amounting an effective attack on our definition directly, as a forged signature with grade $\ell + t$ is also a forged signature

[2] It is possible that we may simplify the model by categorizing some of the queries into one, and argue the equivalence. Due to lack of space, we do not discuss this improvement and refer to the full version.

with grade ℓ. We may also consider weaker models such as selective corruption, and we omit the discussion of details of these weaker variants.

Anonymity of Graded Signatures: We require a strong type of anonymity for a graded signature: two graded signatures can not be distinguished with respect to any characteristic except their grade. In the anonymity definition, even the certification authority and signers will not be able to link two graded signatures for an adversarially chosen message with a same grade. Consider the following game between an adversary $\mathcal{A}$ and a challenger C.

- $\mathcal{A}$ receives the master public key gpk.
- $\mathcal{A}$ makes queries for the secret keys of signers. Note that the adversary here is allowed to corrupt all signers, even the certification authority.
- $\mathcal{A}$ also selects a grade l, a message m, and two sets of signers S_0, S_1 with size l such that $S_0 \neq S_1$. The adversary then produces two sets of tuples $D_0 = \{cert_i, \sigma_i, vk_i\}$ and $D_1 = \{cert_j, \sigma_j, vk_j\}$ where $i \in S_0$ and $j \in S_1$. Thus, $\mathcal{A}$ sends all sets S_0, S_1, D_0, D_1 and message m together with l to C.[3]
- The challenger C randomly flips a coin $b \in \{0, 1\}$, and sends $\mathcal{A}$ a graded signature $\sigma^{(\ell)}$ with grade l which is produced from l signatures on m from the set D_b.
- Finally, $\mathcal{A}$ output a guess b'.

Definition 2. *We say a graded signature is fully anonymous if the probability of guessing the bit correctly is negligibly close to $\frac{1}{2}$, i.e., $|\Pr[b = b'] - \frac{1}{2}| \leq \epsilon$, where ϵ is a negligible function.*

3 Graded Signatures with Linear Signature Size and Verification Time

In this section, we present an existentially unforgeable graded signature scheme with both linear in the grade verification time and signature size. The construction relies on involved mechanisms that are compatible with a constant size NIZK range proof together with a constant size NIZK proof of consistency of committed verification and signatures.

In order to motivate our construction recall the following generic solution for a graded signature: the user runs the aggregation algorithm of an aggregate signature scheme (or multi-signature with non-interactive signing).

On input $vk_{i_1}, \ldots, vk_{i_t}, m, \sigma$, and commits to all the verification keys, and produces a non-interactive zero-knowledge proof for the following statements: 1. σ is an aggregate signature on m under the committed verification keys; 2. all committed verification keys are certified; 3. each of the committed verification keys are different. The straightforward way of proving the third condition in zero-knowledge would be to prove that the verification-keys are pairwise different. Even with the most efficient NIZK proof of inequality, this step brings a cost at

[3] In order to simplify the game definition, we assume the sets S_0, S_1 differ only by one index, i.e., $S_0 \setminus S_1 = i_0$ and $S_1 \setminus S_0 = i_1$.

least quadratic in the grade of the signature (the number of signer public keys) that we want to avoid. We may use SNARK [4] to construct efficient graded signature schemes as the final proof size could be as short as $poly(\lambda)$ where λ is the security parameter. However, we aim to get an efficient graded signature without applying knowledge assumptions, thus we will focus on using standard building blocks as Groth-Sahai proofs [18] below.

Besides, designing a linear size signature from standard assumptions was also the main technical work of signature of reputation [5]. Unfortunately their technique inherently relies on certain kind of "linkability" among commitments and "uniqueness" of the signature scheme. They incur quadratic verification time and restrict the class of signature schemes that can be used to produce a graded signature.

We go around these problems by introducing a new technique that we assign an index from $\{1, ..., n\}$ as a part of public key of the signer where n is the maximum level. We then utilize an efficient non-interactive range proof so that we can sort the indices and sequentially prove a "larger than" statement to show that indices from which the graded signature is produced are different. In this way we can bring down the complexity from (at least) quadratic to linear. Specifically, when a signer registers his verification key, the CA will choose an index for him and sign the index together with his verification key to produce the certificate for that signer. After collecting signatures $(m, vk_{i_1}, \sigma_{i_1}, cert_{i_1}), \ldots, (m, vk_{i_t}, \sigma_{i_t}, cert_{i_t})$, the **Combine** algorithm commits to all the verification keys, all the certificates, and all the corresponding signatures. Then, the **Combine** algorithm will produce a proof that each committed signature is valid under the corresponding committed verification key; second, a proof that each certificate is valid under the public key of the certification authority; third, the algorithm will sort the indices of the verification-keys in a decreasing order and establish that each index belongs to range $[1, n]$. Due to the additive homomorphic property of the commitment scheme we use, the **Combine** algorithm will be also capable to produce a proof that $Com(i_j - i_{j-1})$ is a commitment to an integer which also falls in range $[1, n]$. So it follows that this value is bigger than 0, and hence the difference of any two neighboring indices is strict. In this fashion the algorithm will establish a proof showing that there are l valid signatures from l different certified signers on m. This completes the high level overview of the construction. What remains is how to get a constant size NIZK proof for each of the above statements. Thanks to the flexibility of this construction methodology we can choose any appropriate signature scheme as long as it can be paired with efficient NIZK proofs. We instantiate the scheme using automorphic signatures [1] together with a Groth-Sahai proof of validity of committed signatures [19], and also an efficient range proof of committed values. In this way we can see that verification only has to do a sequential scanning instead of pair-wise comparison as in [5]; furthermore, the signature size is still linear in the grade as each component only cost a constant number of group elements.

Suppose we have two signature schemes Sig, Sig', an additively homomorphic commitment scheme Com. The scheme is formally presented as follows:

- **Setup**: The algorithm runs the key generation of Sig, and generates a key pair (msk, mpk). It also generates global parameters $param$ including the CRS string for the commitment scheme and the NIZK proof system, and the total number n of allowed signers in the system. It outputs the global key pair (gsk, gpk) where $gsk = msk$, and $gpk = (mpk, param)$.
- **Register**: This is a protocol between signer and CA. Signer first runs the key generation of sig' to get his signing key pair (vk, sk), and submits vk to the CA. The CA first checks whether this signer is already registered, if not, he chooses an index i, runs the signing algorithm of Sig on (vk, i), and returns the signer $cert_i$, where $cert_i = Sig(msk, (vk, i))$.
- **Sign**: This algorithm receives as input a signer's secret key sk_i, a message m and runs the Sig' algorithm to get a signature σ_i on m, and it outputs σ_i, signer's index i and $cert_i$.
- **Combine**: This algorithm takes as inputs a message m, a sequence of signatures $(\sigma_{i_1}, \ldots, \sigma_{i_l})$ for the message m under $vk_{i_1}, \ldots, vk_{i_l}$ with the corresponding certificates $(cert_{i_1}, \ldots, cert_{i_l})$, from l different signers. It first checks the validity of the signatures and the certificates, and determines the grade l. Suppose the sequence is in a decreasing order according to the indices, i.e., $i_1 > i_2 \ldots > i_l$. It computes the commitments to all those values and gets $c^1_{i_j} = Com(\sigma_{i_j})$ for the signatures, $c^2_{i_j} = Com(vk_{i_j})$ for the signers' verification keys, $c^3_{i_j} = Com(cert_{i_j})$ for the certificates, and $c^4_{i_j} = Com(i_j)$ for the signers' indices. Using the signatures as witnesses, it constructs $4l - 1$ NIZK proofs. For each $j \in \{1, \ldots, l\}$, the proof $\pi^1_{i_j}$ establishes that $c^1_{i_j}$ commits to a valid signature on m under the verification key contained in $c^2_{i_j}$; $\pi^2_{i_j}$ proves that $c^3_{i_j}$ commits to a valid signature under mpk on the message pair contained in $c^2_{i_j}$ and $c^4_{i_j}$; $\pi^3_{i_j}$ proves $c^4_{i_j}$ commits to a value which belongs to $\{1, \ldots, n\}$; $\pi^4_{i_j}$ proves that $c^4_{i_{j+1}}/c^4_{i_j} = Com(i_{j+1} - i_j)$ also commits to a value ranging in $\{1, \ldots, n\}$. It outputs the message m and signature object as $\{c^1_{i,j}, c^2_{i_j}, c^3_{i_j}, c^4_{i_j}, \pi^1_{i,j}, \pi^2_{i_j}, \pi^3_{i_j}, \pi^4_{i_j}\}_{j=1,\ldots,l}$, together with its grade l.
- **Verify**: The verifier takes global public key gpk, a message m, and a graded signature $\{c^1_{i,j}, c^2_{i_j}, c^3_{i_j}, c^4_{i_j}, \pi^1_{i,j}, \pi^2_{i_j}, \pi^3_{i_j}, \pi^4_{i_j}\}_{j=1,\ldots,l}$ with grade l as inputs, it first parses the signature, and for $j = 1, \ldots, l$, it checks the validity of the proofs $\pi^1_{i,j}, \pi^2_{i_j}, \pi^3_{i_j}$, and for $j = 1, \ldots, l-1$, it checks the validity of $\pi^4_{i,j}$; if all checks pass, it outputs 1, otherwise 0.

Correctness: The correctness of our scheme trivially follows the correctness of the signature schemes Sig and Sig', and the completeness of the NIZK proof systems. Briefly, if the user has ℓ signatures $(\sigma_{i_j}, vk_{i_j}, cert_{i,j})$ for a message m collected from different signers such that each σ_{i_j} is a valid signature on m under vk_{i_j}, and each $cert_{i_j}$ is valid signature on (pk_{i_j}, i_j) under mpk, and if the ℓ tuples $(\sigma_{i_j}, vk_{i_j}, cert_{i,j})$ are sorted in decreasing order, and all indices i_j and all $i_{j+1} - i_j$ are in the range $[1, n]$. Then from the completeness of NIZK proof systems, the Verify algorithm accepts the signature $\sigma^{(\ell)}$ on m constructed as

$$\sigma^{(\ell)} = \{c^1_{i,j}, c^2_{i_j}, c^3_{i_j}, c^4_{i_j}, \pi^1_{i,j}, \pi^2_{i_j}, \pi^3_{i_j}, \pi^4_{i_j}\}_{j\in[\ell]} = Combine(m, (\sigma_{i_j}, vk_{i_j}, cert_{i_j})_{j\in[l]}).$$

Security Analysis: Security follows quite easily from the properties of the zero-knowledge proofs and the commitment schemes. For *unforgeability*, suppose the adversary only gets t signatures on a message m by corrupting signers or asking signing queries, and he is able to produce a signature on m with grade $t+1$. According to the soundness of the NIZK proof system, there must be $t+1$ valid signatures under $t+1$ different verification keys committed by the adversary. Note that because of the extractability property of the commitment scheme, at the beginning, the simulator can produce a simulated crs which contains an opening trapdoor for the commitment scheme, and thus the simulator can open these commitments to retrieve the $t+1$ tuple of signatures, verification keys, and certificates. If the verification keys are all certified by the CA, then the adversary must have forged one new signature against an honest signer; alternatively, the adversary could have forged a certificate for an unregistered verification key. The simulator can examine these cases and break the unforgeability of either Sig' or Sig.

Regarding *anonymity*, suppose the adversary submits m, l, S_0, S_1, D_0, D_1 as the challenge. Suppose, for simplicity, that $S_0 \setminus S_1$ contains only one index i_0 and similarly $S_1 \setminus S_0$ contains only one index i_1. The simulator can use signatures $\sigma_{i_0}, \sigma_{i_1}$ on m under pk_{i_0}, pk_{i_1} to ask as a challenge in a plaintext indistinguishability game of the underlying commitment scheme; after receiving $Com(\sigma_{i_b})$, the simulator will create a graded signature by computing the commitments to all other signatures on m and simulate all the proofs (the latter part following from the zero-knowledge property). In this way, the simulator can use the adversary's ability in breaking anonymity to break the hiding property of the commitment scheme in a straightforward fashion.

Theorem 1. *The scheme is existentially unforgeable under adaptive corruption attacks if Sig, Sig' are unforgeable digital signatures, Com is a binding (extractable) commitment scheme, and the proof system is sound.*

Proof: We show the security by a sequence of games. We start with the original game Game_0, and prove that a polynomial time attacker's advantage of distinguishing any successive games is negligible.

Game_0 :

- The simulator runs the key generation of Sig, and generates a key pair (msk, mpk). It also runs the key generation algorithm of Sig' to generate the signing key-verification pairs. Then for each verification key vk, it picks a random integer $i \in [n]$, generates the certification of the corresponding verification key using $MS.Sign$ algorithm on (vk, i), and forms a set S that contains all certifications and corresponding indices. Besides, it generates the global parameters $param$ including the crs strings for the commitment scheme and the NIZK proof system, and the total number n of allowed signers in the system. The simulator keeps gsk, and gives $gpk = (mpk, param)$ and S to the adversary.

- For each register query; the adversary $\mathcal{A}$ generates a fresh key pair $(sk, vk) \leftarrow$ *S.Setup*, and gives vk to the simulator. The simulator selects a random integer i from $[n] - S$ as the index of vk (The challenger keeps a list T for registered indices. If $i \in T \cup S$, the simulator reselects it) and computes $cert_i =$ MS.Sign$(msk, (vk_i, i))$. $\mathcal{C}$ sends $cert_i$ to $\mathcal{A}$ and writes $(i, vk, cert_i)$ to T.
- For each signing key query of an index $j \in S$; the simulator gives the corresponding signing key sk_j to the adversary. The simulator also keeps a list C for corrupted indices.
- For each signature query on the message m with index $k \in [n] - T$; the simulator computes $\sigma =$ S.Sign(sk_k, m), and gives it to $\mathcal{A}$.
- $\mathcal{A}$ submits a forgery $\sigma^{(\ell)}$ with grade $\ell > |C| + |T| + q$ for m^* where q is the number of signature queries on m^*. If Verify$(\sigma^{(\ell)}, \ell, m^*, gpk) = 1$, the adversary wins the game.

Game$_1$: Same as Game$_0$, except we substitute Setup algorithm of the commitment scheme with Extractable Setup algorithm which generates the crs string of the commitment scheme together with the extraction key ek.

Game$_2$: Same as Game$_1$, except we require that for each commitment $c^u_{i_j} = Com(crd, (X^u_{i_j}, \alpha))$ and associated proof $\pi^u_{i,j} \leftarrow Prove(crs, Ver^u_{i_j}, (X^u_{i_j}, \alpha))$ generated by the adversary in the challenge phase, $Ver(crd, E^u_{i_j}, c^u_{i_j}, \pi^u_{i,j}) = 1$ where $E^u_{i_j}$ is the corresponding verification equation.

Claim: Assuming the NIZK proof systems has two types of common reference strings (hiding and binding) which are computationally indistinguishable, for any PPT adversary $\mathcal{A}$,

$$|Adv^{(0)}_{\mathcal{A}} - Adv^{(1)}_{\mathcal{A}}| \leq negl(\lambda).$$

Proof. Suppose there exists a PPT adversary $\mathcal{A}$ such that the difference the advantages of the adversary between both games is non negligible, then we can construct a PPT algorithm $\mathcal{B}$ that use $\mathcal{A}$ to distinguish two types of CRS with non negligible advantage. □

Claim: Assuming the NIZK proof systems are sound, for any PPT $\mathcal{A}$,

$$|Adv^{(1)}_{\mathcal{A}} - Adv^{(2)}_{\mathcal{A}}| \leq negl(\lambda).$$

Proof. Suppose there exists a PPT adversary $\mathcal{A}$ such that the difference the advantages of $\mathcal{A}$ between both games is non negligible, then we will construct a PPT algorithm $\mathcal{B}$ that uses $\mathcal{A}$ to break the soundness of the proof systems.

$\mathcal{B}$ gets the crs of the commitments from the challenger of the NIZK proof system. It then computes $(msk, mpk) \leftarrow MS.Setup(\lambda)$ and gives the gsk, gpk to the adversary. $\mathcal{B}$ can simulate the corrupt queries, the registration queries, and the signature queries as in Game$_1$ and Game$_2$. The only difference between two games is that, the adversary can prove a false statement with non-negligible probability in Game$_1$. If the algorithm $\mathcal{B}$ is dealing with the proofs of false statements, then it corresponds to Game$_1$; otherwise it corresponds to Game$_2$. Thus, $\mathcal{B}$ can break the soundness of the underlying proof systems with non-negligible probability. □

Claim: Assuming Sig and Sig' are existentially unforgeable, and the commitment scheme is perfectly binding, for any PPT $\mathcal{A}$,

$$Adv_{\mathcal{A}}^{(2)} \leq negl(\lambda).$$

Proof. Suppose there exists a PPT adversary $\mathcal{A}$ such that the difference the advantages of $\mathcal{A}$ between both games is non negligible, then we will construct a PPT algorithm $\mathcal{B}$ that uses $\mathcal{A}$ to break the unforgeability of Sig or Sig'.

$\mathcal{B}$ gets mpk from the challenger of Sig and vk from the challenger of Sig', and requests a certification for vk. $\mathcal{B}$ also generates some signing key-verification key pairs, and requests the certifications for those verification keys. It then forms the set S that contains all certifications and corresponding indices. Besides, $\mathcal{B}$ generates the global parameters $param$ that includes the extractable crs strings for the commitment scheme with the extraction key ek and the NIZK proofs system, and the total number n of the allowed signers in the system. $\mathcal{B}$ keeps ek, and gives $(mpk, params)$ to the adversary.

For each register query that the adversary makes, $\mathcal{B}$ gets the corresponding certification from the challenger of Sig; for each signing key query that the adversary makes, if the corresponding verification key is vk, $\mathcal{B}$ aborts, otherwise gives the corresponding signing to the adversary; for each signature query on a message m, if the adversary requests a signature for vk, $\mathcal{B}$ asks a signature on m from the challenger of Sig', otherwise produces the signature using the corresponding signing key.

When the adversary submits a valid forgery $\sigma^{(k)}$ on a message m^* with the grade ℓ, $\mathcal{B}$ extracts all tuples $\{(vk_i, cert_i, \sigma_i)\}_{i\in[\ell]}$ uniquely from $\sigma^{(k)}$ using ek since the commitment scheme is perfectly binding (extractable). Since the number of registration queries and the corruption queries add up to be less than ℓ, there should be one tuple $(vk_i, cert_i, \sigma_i)$ such that either σ_i is a valid forgery on m^* under vk_i, or $cert_i$ is a valid forgery on (vk_i, i) under mpk. If σ_i is a valid forgery on m^*, since the probability of $vk_i = vk$ is $1/|S|$, $\mathcal{B}$ can use this forgery to break the unforgeability of Sig. If $cert_i$ is a valid forgery on (vk_i, i), then $\mathcal{B}$ can use the pair to break the unforgeability of Sig'. This concludes the proof. □

Theorem 2. *The scheme satisfies full anonymity, if Com is computationally hiding and the proof system is zero-knowledge.*

Proof: We show the security by a sequence of hybrid experiments. We start with the original experiment Game$_0$, and prove that any polynomial time attacker's advantage of distinguishing any successive experiments is negligible.

Game$_0$:

- The challenger runs the key generation of Sig, and generates a key pair (msk, mpk). It also generates global parameters $param$ including the crs

strings for the commitment scheme and the NIZK proof system, and the total number n of allowed signers in the system. The challenger gives $gpk = (mpk, param)$ and gsk to the adversary.
- The adversary selects two sets S_1 and S_0 of indexes such that $|S_0| = |S_1| = k$, $S_1 \setminus S_0 = \{i_0\}$, and $S_0 \setminus S_1 = \{j_0\}$. It first runs $S.Setup$ algorithm to generate signing key-public key pair (sk, pk) for each index, then computes a certification $cert_i$ for each public key vk_i of the index i using gsk. The adversary also produces signatures σ_i on same message m under each public key pk_i. It finally gives the index k as the level, the message m, two sets of indexes S_0, S_1, and two sets of tuples $D_0 = \{cert_i, \sigma_i, pk_i\}$, $D_1 = \{cert_j.\sigma_j, pk_j\}$ where $i \in S_0$ and $j \in S_1$.
- The challenger sets $b = 0$, produces a graded signature $\sigma^{(k)}$ on m using the tuples D_b, and gives $\sigma^{(k)}$ to the adversary.
- The adversary gives a guess b' to the challenger, and wins the game if $b = b'$.

Game$_1$: Same as Game$_0$, except we substitute Setup algorithm of the commitment scheme with SimSetup algorithm which generates the simulable crs string of the commitment schemes and proofs.

Game$_2$: Same as Game$_1$, except the challenger changes the proofs $(\pi^1_{i,0}, \pi^2_{i_0}, \pi^3_{i_0}, \pi^4_{i_0})$ of index i_0 from $\sigma^{(k)}$ with the simulated proofs $(\pi'^1_{i_0}, \pi'^2_{i_0}, \pi'^3_{i_0}, \pi'^4_{i_0})$.

Game$_3$: Same as Game$_2$, except the challenger changes the commitments $(c^1_{i,0}, c^2_{i_0}, c^3_{i_0}, c^4_{i_0})$ of index $i_0 \in S_0$ from $\sigma^{(k)}$ with $(c^1_{j,0}, c^2_{j_0}, c^3_{j_0}, c^4_{j_0})$ of the index $j_0 \in S_1$.

Game$_4$: Same as Game$_3$, except the challenger changes the simulated proofs $(\pi'^1_{i_0}, \pi'^2_{i_0}, \pi'^3_{i_0}, \pi'^4_{i_0})$ from $\sigma^{(k)}$ with the proofs $(\pi^1_{j_0}, \pi^2_{j_0}, \pi^3_{j_0}, \pi^4_{j_0})$. Thus, in the final game, the challenger generates the graded signature $\sigma^{(k)}$ using the tuples from the set D_1.

Claim: Assuming the proof systems are zero-knowledge, for any PPT $\mathcal{A}$,

$$|Adv^{(0)}_{\mathcal{A}} - Adv^{(1)}_{\mathcal{A}}| \leq negl(\lambda).$$

Proof. Suppose there exists a PPT adversary $\mathcal{A}$ such that the difference of the advantages of the adversary between both games is non negligible, then we can construct a PPT algorithm $\mathcal{B}$ that use $\mathcal{A}$ to break the zero knowledge property of the proof systems. □

Claim: Assuming the proof systems are zero-knowledge, for any PPT $\mathcal{A}$,

$$|Adv^{(1)}_{\mathcal{A}} - Adv^{(2)}_{\mathcal{A}}| \leq negl(\lambda).$$

Proof. Suppose there exists a PPT adversary $\mathcal{A}$ such that the difference of the advantages of the adversary between both games is non negligible, then we will construct a PPT algorithm $\mathcal{B}$ that use $\mathcal{A}$ to break the zero knowledge property of the proof systems.

$\mathcal{B}$ generates (gsk, gpk) and gives them to the adversary as in Game$_1$ and Game$_2$. After getting it, the simulator gives the challenge tuple $(vk_{i_0}, cert_{i_0}, \sigma_{i_0})$ to the challenger of the proof system, and gets the corresponding commitments com_{i_0} and proofs $\pi_{i_0}^{(b)}$. $\mathcal{B}$ then simulates all other commitments and proofs and gives the final signature to the adversary. If $b = 0$, then it corresponds to Game$_1$, otherwise it corresponds to Game$_2$. Thus, if the difference of the advantages of the adversary between both games is non negligible, then $\mathcal{B}$ can use $\mathcal{A}$ to break the zero knowledge of the proof system. □

Claim: Assuming the commitment scheme is computationally hiding, for any PPT adversary $\mathcal{A}$,

$$|Adv_{\mathcal{A}}^{(2)} - Adv_{\mathcal{A}}^{(3)}| \leq negl(\lambda).$$

Proof. Suppose there exists a PPT adversary $\mathcal{A}$ such that the difference of the advantages of the adversary between both games is non negligible, then we will construct a PPT algorithm $\mathcal{B}$ that use $\mathcal{A}$ to break hiding property of the commitment scheme.

$\mathcal{B}$ generates (gsk, gpk) and gives them to the adversary as in Game$_2$ and Game$_3$. After getting it, the simulator gives the challenge tuples $(vk_{i_0}, cert_{i_0}, \sigma_{i_0})$ and $(vk_{j_0}, cert_{j_0}, \sigma_{j_0})$ to the challenger of the commitment scheme, and gets the challenge commitments $(c_b^1, c_b^2, c_b^3, c_b^4)$. $\mathcal{B}$ also simulates all other commitments and corresponding proofs, and gives the final signature to the adversary. If $b = i_0$, then it corresponds to Game$_2$, otherwise it corresponds to Game$_3$. Thus, if the difference of the advantages of the adversary between both games is non negligible, then $\mathcal{B}$ can use $\mathcal{A}$ to break the hiding property of the commitment scheme. □

Claim: Assuming the proof systems are zero-knowledge, for any PPT $\mathcal{A}$,

$$|Adv_{\mathcal{A}}^{(3)} - Adv_{\mathcal{A}}^{(4)}| \leq negl(\lambda).$$

Proof. Suppose there exists a PPT adversary $\mathcal{A}$ such that the difference of the advantages of the adversary between both games is non negligible, then we will construct a PPT algorithm $\mathcal{B}$ that use $\mathcal{A}$ to break the zero knowledge property of the proof systems.

$\mathcal{B}$ generates (gsk, gpk) and gives them to the adversary as in Game$_3$ and Game$_4$. After getting it, the simulator gives the challenge tuple $(vk_{j_0}, cert_{j_0}, \sigma_{j_0})$ to the challenger of the proof system, and gets the corresponding commitments com_{j_0} and proofs $\pi_{j_0}^{(b)}$. $\mathcal{B}$ then simulates all other commitments and proofs and gives the final signature to the adversary. If $b = 0$, then it corresponds to Game$_3$, otherwise it corresponds to Game$_4$. Thus, if the difference of the advantages of the adversary between both games is non negligible, then $\mathcal{B}$ can use $\mathcal{A}$ to break the zero knowledge of the proof system.

In conclusion, since any PPT attacker's advantage of distinguishing any successive games is negligible, the adversary cannot distinguish two graded signatures with the same grade. Hence, the scheme is fully anonymous. □

An Efficient Instantiation. In order to get a graded signature with size linear in the grade, we need to make all the NIZK proofs $\pi_{i_j}^1, \pi_{i_j}^2, \pi_{i_j}^3, \pi_{i_j}^4$ to be constant size. One natural approach, that also yields a standard model construction, is to instantiate the scheme with signature schemes which are compatible with the Groth-Sahai proof system [19]. Note that a structure preserving signature or automorphic signature [1] satisfies exactly our needs – both the verification key and the signature belong to the same group, and the verification are conjunctions of pairing product equations; furthermore, this signature scheme allows signing on a pair of messages as well. For Sig, the CA needs to sign on pk and index i; we instantiate this with an automorphic signature on (g^x, g^i), where $pk = g^x$; for Sig', in order to sign a message $m \in Z_p$, we instantiate the algorithm via the same signature scheme operating on a single group element equal to g^m. It is straightforward to obtain constant size proofs realizing π_i^1, π_i^2 by applying the Groth-Sahai framework.

For π_i^3, π_i^4, we use the constant size non-interactive range proof for range $[0, H]$ proposed in [12][4]. First, we apply the range proof for the range $[1, n]$ in order to establish the "larger than" statement. Relying on the additive homomorphic property of the commitment scheme, we can do a straightforward "shift" in the protocol of [12], in order to prove $x \in [1, n]$, where x is committed in $Com(x)$. Specifically the prover executes the proof with respect to the commitment $\psi = Com(x)/Com(1;0)$ where $Com(x;r)$ denotes the commitment on x with randomness r, thus establishing that $x - 1 \in [0, H]$. With this construction at hand, it follows that the proofs π_i^3, π_i^4 are also constant size.

Now the only problem left is to show the index committed for π_i^1, π_i^2 is consistent with the value committed for π_i^3, π_i^4. We observe that the commitment schemes used in the range proof include a BBS encryption type of commitment, which is compatible with Groth-Sahai proof system and this proof can be constructed easily. Specifically, the NIZK proof establishes that the two commitments c_1, c_2 belong to the language:

$$L = \{(c_1, c_2) | \exists x, r_1, r_2, s_1, s_2, s.t, c_1 = (g^{x+r_1+r_2}, f^{r_1}, h^{r_2}) \wedge c_2 = (g^x u_1^{s_1} u_2^{s_2})\},$$

where g, f, h, u_1, u_2 are all contained in CRS.

Graded Signatures Supporting Revocation. Since our notion of graded signature is directly built upon the PKI, it would be nice if we can support certificate revocation as well due to the same reasons as in the regular PKI setting, e.g., some signing key might get compromised. A common method for revocation in the PKI setting is that the CA publishes a revocation list that maintains all the revoked certificates, and every user can check it.

In our construction of the graded signature scheme, in order to guarantee that the signatures are all from the valid signers and their privacy is preserved,

[4] Using different instantiations of parameters, they obtain suitable communication and verification complexity for different scenarios. In our case, adding CRS with $O(log^{1+\epsilon} n)$-length to the public parameters will be enough to achieve constant size range proof and verification time.

we have one important step that the user commits to the certificates and the public keys and proves that the public keys are certified, i.e., the certificates contained in the commitments are valid signatures under the master public key of the CA. We can see that in principle, it would not be very difficult to extend our construction to support revocation as we can simply let the user to add one more proof that the certificates committed are *not* in the public revocation list.[5] The challenging task is that how we can maintain the signature size still to be linear in the grade, which means we need to keep each non-membership proof to be constant![6]

Fortunately, Blazy et al. [6] propose an efficient NIZK proof system to prove an exclusive statement, i.e., the statement does not belong to a language L. We can instantiate their proof system to prove that a committed value does not belong to a given set S. The main idea of their technique is that the user first generates a "proof" $\tilde{\pi}$ showing that the statement belongs to L, and it can not pass the verification (as he does not have the witness), then he proves using another $\tilde{\pi}'$ that $\tilde{\pi}$ is generated honestly, i.e., it is indeed computed following the regular prover algorithm. In this way, $\tilde{\pi}, \tilde{\pi}'$ together convince the verifier about the negation, as if not, the prover can not generate $\tilde{\pi}, \tilde{\pi}'$ simultaneously. For details of the technique, we refer to [6]. Now to instantiate the non-membership proof, we can start with the membership proof we use [6] to generate $\tilde{\pi}$ which is constant size and we then prove each component of $\tilde{\pi}$ is generated honestly. Since [6] is compatible with Groth-Sahai [18], the validity of the components can be again proven efficiently using the Groth-Sahai proof. Thus, we can conclude that we can extend our graded signature to support certificate revocation by adding the above non-membership proof for each committed certificates. Furthermore, each pair of such non-membership proof is with constant size, thus the total signature size is still linear in the number of grade.

Acknowledgment. The first author was supported by the ERC project CODAMODA and the project FINER of the Greek Secretariat of Research and Technology.

A Preliminaries

Non-Interactive Zero-Knowledge (NIZK) Proof: Let $R = \{(x, w)\}$ be an efficiently computational binary relation, where we call x the statement and w the witness. Let L be the language which consists of the statements from R. A non-interactive argument for a relation R consists of a key generation algorithm G, which creates a common reference string crs, a prover P and a verifier V. The prover generates a non-interactive argument π for an input (crs, x, w). The verifier outputs 1 if the proof is valid; otherwise, outputs 0. Suppose ϵ_1, ϵ_2 are negligible functions,

[5] Instead of certifications, it would be enough to keep only the indices of the revoked signers in the revocation list.

[6] A straightforward way to show that the committed value does not equal to any of the set element is highly inefficient due to the inequality proof and the AND proof.

- A non-interactive argument (G, P, V) is perfectly complete if:

$$\Pr[crs \leftarrow G, \forall(x, w) \in R, V(crs, x, P(crs, x, w))] = 1.$$

- We say (G, P, V) is sound, if $\forall \mathcal{A}$,

$$Pr[crs \leftarrow G; (x, \pi) \leftarrow \mathcal{A}(crs), x \notin L \wedge V(crs, x, \pi) = 1] \leq \epsilon.$$

- (G, P, V) is zero knowledge, if there exists a simulator (S_1, S_2) such that for all non-uniform ppt adversaries $\mathcal{A}$, $\forall(x, w) \in R$

$$|\Pr[crs \leftarrow G, \mathcal{A}^{P(crs,x,w)}(crs) = 1] - Pr[(crs, t) \leftarrow S_1, \mathcal{A}^{S_2(crs,t,x)}(crs) = 1]| < \epsilon$$

Extractable Commitments: An extractable commitment scheme consists of five algorithms: *Setup*, *Com*, *ExtGen*, *Ext*. *Gen* algorithm outputs a commitment key ck, and *ExtGen* outputs (ck', td), where ck' is indistinguishable with ck, and td is an extraction key. *Com* outputs a commitment c on ck, a message m, and randomness r.

- It is perfectly binding if for any commitment c there exists exactly one m satisfying $c = Com(ck, m, r)$ for some r, further, $Ext(td, c) = m$.
- It is computationally hiding if for any messages m, m', $Com(ck, m, r)$ is indistinguishable with $Com(ck, m', r')$.

Automorphic Signatures: An automorphic signature over a bilinear group is an existentially unforgeable signature scheme whose verification keys lie in the same space with message, and the verification predicate is conjunction of pairing-product equations over the verification key, the message and the signature [1]. We can apply Groth-Sahai proof to such signature scheme to instantiate efficient NIZK proofs. Furthermore, their construction enables signing on message vectors as well which we will use for the **Register** algorithm to sign on (pk, i).

Constant Size Range Proof: A prover with the range proof given by [12] convinces a verifier that a number in a commitment belongs to the interval $[0, k]$. Setup algorithm just outputs a common reference string crs for the commitment and the public parameters for BBS encryption [8]. The common input for the range proof consists of a BBS encryption $(A_g, A_f, A_h) = (g_1^{r+i}, f^{r_1}, h^{r_2})$ and a commitment $(A_c, \hat{A}_c) = (g_1^r g_{11}^a, \hat{g_1}^r \hat{g_{11}}^a)$ where $r = r_1 + r_2$. They propose an efficient NIZK argument which convinces a verifier that the key committed in $(A_c, \hat{A}_c)$ and encrypted as A_g belongs to [0,H]. We leave the details to the paper [12]. Also, note that BBS encryption type of commitment is compatible with Groth-Sahai proof.

References

1. Abe, M., Fuchsbauer, G., Groth, J., Haralambiev, K., Ohkubo, M.: Structure-preserving signatures and commitments to group elements. In: Rabin, T. (ed.) CRYPTO 2010. LNCS, vol. 6223, pp. 209–236. Springer, Heidelberg (2010)

2. Au, M.H., Chow, S.S.M., Susilo, W., Tsang, P.P.: Short linkable ring signatures revisited. In: Atzeni, A.S., Lioy, A. (eds.) EuroPKI 2006. LNCS, vol. 4043, pp. 101–115. Springer, Heidelberg (2006)
3. Bellare, M., Neven, G.: Multi-signatures in the plain public-key model and a general forking lemma. In: ACM CCS, pp. 390–399 (2006)
4. Ben-Sasson, E., Chiesa, A., Tromer, E., Virza, M.: Succinct non-interactive zero knowledge for a von neumann architecture. In: USENIX 2014, pp. 781–796 (2014)
5. Bethencourt, J., Shi, E., Song, D.: Signatures of reputation. In: Financial Cryptography, pp. 400–407 (2010)
6. Blazy, O., Chevalier, C., Vergnaud, D.: Non-interactive zero-knowledge proofs of non-membership. In: Nyberg, K. (ed.) CT-RSA 2015. LNCS, vol. 9048, pp. 145–164. Springer, Heidelberg (2015)
7. Boldyreva, A., Palacio, A., Warinschi, B.: Secure proxy signature schemes for delegation of signing rights. J. Cryptology **25**(1), 57–115 (2012)
8. Boneh, D., Boyen, X., Shacham, H.: Short group signatures. In: Franklin, M. (ed.) CRYPTO 2004. LNCS, vol. 3152, pp. 41–55. Springer, Heidelberg (2004)
9. Boneh, D., Gentry, C., Lynn, B., Shacham, H.: Aggregate and verifiably encrypted signatures from bilinear maps. In: Biham, E. (ed.) EUROCRYPT 2003. LNCS, vol. 2656, pp. 416–432. Springer, Heidelberg (2003)
10. Boyle, E., Goldwasser, S., Ivan, I.: Functional signatures and pseudorandom functions. In: Krawczyk, H. (ed.) PKC 2014. LNCS, vol. 8383, pp. 501–519. Springer, Heidelberg (2014)
11. Bresson, E., Stern, J., Szydlo, M.: Threshold ring signatures and applications to ad-hoc groups. In: Yung, M. (ed.) CRYPTO 2002. LNCS, vol. 2442, pp. 465–480. Springer, Heidelberg (2002)
12. Chaabouni, R., Lipmaa, H., Zhang, B.: A non-interactive range proof with constant communication. In: Keromytis, A.D. (ed.) FC 2012. LNCS, vol. 7397, pp. 179–199. Springer, Heidelberg (2012)
13. Chow, S.S.M., Susilo, W., Yuen, T.H.: Escrowed linkability of ring signatures and its applications. In: Nguyên, P.Q. (ed.) VIETCRYPT 2006. LNCS, vol. 4341, pp. 175–192. Springer, Heidelberg (2006)
14. Damgård, I.B., Koprowski, M.: Practical threshold RSA signatures without a trusted dealer. In: Pfitzmann, B. (ed.) EUROCRYPT 2001. LNCS, vol. 2045, pp. 152–165. Springer, Heidelberg (2001)
15. Desmedt, Y.G.: Society and group oriented cryptography: A new concept. In: Pomerance, C. (ed.) CRYPTO 1987. LNCS, vol. 293, pp. 120–127. Springer, Heidelberg (1988)
16. Fuchsbauer, G., Pointcheval, D.: Anonymous proxy signatures. In: Ostrovsky, R., De Prisco, R., Visconti, I. (eds.) SCN 2008. LNCS, vol. 5229, pp. 201–217. Springer, Heidelberg (2008)
17. Gennaro, R., Halevi, S., Krawczyk, H., Rabin, T.: Threshold RSA for dynamic and ad-hoc groups. In: Smart, N.P. (ed.) EUROCRYPT 2008. LNCS, vol. 4965, pp. 88–107. Springer, Heidelberg (2008)
18. Groth, J., Sahai, A.: Efficient non-interactive proof systems for bilinear groups. In: Smart, N.P. (ed.) EUROCRYPT 2008. LNCS, vol. 4965, pp. 415–432. Springer, Heidelberg (2008)
19. Groth, J., Sahai, A.: Efficient noninteractive proof systems for bilinear groups. SIAM J. Comput. **41**(5), 1193–1232 (2012)
20. Hohenberger, S., Koppula, V., Waters, B.: Universal signature aggregators. In: Oswald, E., Fischlin, M. (eds.) EUROCRYPT 2015. LNCS, vol. 9057, pp. 3–34. Springer, Heidelberg (2015)

21. Kaminsky, D., Patterson, M.L., Sassaman, L.: PKI layer cake: new collision attacks against the global x.509 infrastructure. In: Sion, R. (ed.) FC 2010. LNCS, vol. 6052, pp. 289–303. Springer, Heidelberg (2010)
22. Kiayias, A., Osmanoglu, M., Tang, Q.: Graded Encryption, or how to play "Who wants to be a millionaire?" distributively. In: Chow, S.S.M., Camenisch, J., Hui, L.C.K., Yiu, S.M. (eds.) ISC 2014. LNCS, vol. 8783, pp. 377–387. Springer, Heidelberg (2014)
23. Liu, J.K., Wei, V.K., Wong, D.S.: Linkable spontaneous anonymous group signature for ad hoc groups. In: Wang, H., Pieprzyk, J., Varadharajan, V. (eds.) ACISP 2004. LNCS, vol. 3108, pp. 325–335. Springer, Heidelberg (2004)
24. Mambo, M., Usuda, K., Okamoto, E.: Proxy signatures for delegating signing operation. In: CCS 1996, pp. 48–57 (1996)
25. Pedersen, T.P.: A threshold cryptosystem without a trusted party. In: Davies, D.W. (ed.) EUROCRYPT 1991. LNCS, vol. 547, pp. 522–526. Springer, Heidelberg (1991)
26. Rivest, R.L., Shamir, A., Tauman, Y.: How to leak a secret. In: Boyd, C. (ed.) ASIACRYPT 2001. LNCS, vol. 2248, pp. 552–565. Springer, Heidelberg (2001)
27. Sahai, A., Waters, B.: How to use indistinguishability obfuscation: deniable encryption, and more. In: STOC 2014, pp. 475–484 (2014)
28. Shoup, V.: Practical threshold signatures. In: Preneel, B. (ed.) EUROCRYPT 2000. LNCS, vol. 1807, pp. 207–220. Springer, Heidelberg (2000)
29. Tsang, P.P., Wei, V.K.: Short linkable ring signatures for e-voting, e-cash and attestation. In: Deng, R.H., Bao, F., Pang, H.H., Zhou, J. (eds.) ISPEC 2005. LNCS, vol. 3439, pp. 48–60. Springer, Heidelberg (2005)
30. Zhang, K.: Threshold proxy signature schemes. In: Okamoto, E. (ed.) ISW 1997. LNCS, vol. 1396, pp. 282–290. Springer, Heidelberg (1998)

System and Software Security

Dynamically Provisioning Isolation in Hierarchical Architectures

Kevin Falzon(✉) and Eric Bodden

European Centre for Security and Privacy by Design (EC-SPRIDE), Darmstadt, Germany
{kevin.falzon,eric.bodden}@ec-spride.de

Abstract. Physical isolation provides tenants in a cloud with strong security guarantees, yet dedicating entire machines to tenants would go against cloud computing's tenet of consolidation. A fine-grained isolation model allowing tenants to request fractions of dedicated hardware can provide similar guarantees at a lower cost.

In this work, we investigate the dynamic provisioning of isolation at various levels of a system's architecture, primarily at the core, cache, and machine level, as well as their virtualised equivalents. We evaluate recent technological developments, including post-copy VM migration and OS containers, and show how they assist in improving reconfiguration times and utilisation. We incorporate these concepts into a unified framework, dubbed SafeHaven, and apply it to two case studies, showing its efficacy both in a reactive, as well as an anticipatory role. Specifically, we describe its use in detecting and foiling a *system-wide covert channel* in a matter of seconds, and in implementing a *multi-level moving target defence policy.*

Keywords: Side channels · Covert channels · Migration · Isolation

1 Introduction

The growing use of shared public computational infrastructures, most notably in the form of *cloud computing*, has raised concerns over side channel and covert channel attacks (collectively termed *illicit channels*). These are formed using unconventional and often discreet means that circumvent current security measures. This gives an attacker an edge over conventional attacks, which, while often effective, are well-characterised, conspicuous, and actively guarded against. To date, demonstrations of illicit channels have remained largely academic, with occasional influences on mainstream security practices. Nevertheless, the threat of such channels continues to grow as knowledge on the subject increases.

Hardware illicit channels are fundamentally the product of the unregulated sharing of locality, be it spatial or temporal. Side channels occur when a process

This work was supported by the BMBF within EC SPRIDE, by the Hessian LOEWE excellence initiative within CASED, and by the DFG Collaborative Research Center CROSSING.

J. Lopez and C.J. Mitchell (Eds.): ISC 2015, LNCS 9290, pp. 83–101, 2015.
DOI: 10.1007/978-3-319-23318-5_5

inadvertently leaks its internal state, whereas covert channels are built by conspiring processes that actively leak state in an effort to transmit information. To break hardware locality, processes must be confined through what has been termed *soft* or *hard* isolation [38]. Hard isolation involves giving a process exclusive access to hardware, preventing illicit channels by removing their prerequisite of *co-location*. This approach is limited by the physical hardware available, yet it offers the strongest level of isolation. In contrast, soft isolation allows hardware to be shared but attempts to mask its characteristics.

Soft isolation often incurs an ongoing performance overhead, with some fraction of the machine's capacity committed to maintaining the isolation. Hard isolation does not typically incur a maintenance cost, but it can lead to underutilised hardware [26]. Nevertheless, underused capacity is not truly lost, and can potentially be used to perform functionally useful computations. Conversely, the maintenance costs of soft isolation consume resources.

The viability of hard isolation as a general mitigation technique depends on three factors, namely the availability of hardware, the degree of utilisation supported and the cost of reconfiguration. Modern architectures are hierarchical and vast, with different regions of their hierarchy offering varying granularities of isolation. Isolated resources can thus be provisioned at a finer granularity than dedicating machines to each tenant, which enables higher rates of utilisation. The cost of reconfiguration depends on the type of isolation being provisioned. Cheap reconfiguration allows isolation to be procured temporarily and on-demand, further improving utilisation rates by minimising the duration for which resources are reserved, which translates into lowered operating costs for tenants requesting isolation

This work presents the following contributions:

- an investigation into the types of hard isolations present within modern hierarchical computer architectures, and the types of migration mechanisms available at each level, namely at the core, cache, and machine level, and their virtualised equivalents,
- the creation of a framework, dubbed SafeHaven, to orchestrate migration and distributed monitoring,
- an evaluation of the use of a series of maturing technologies, namely *post-copy live VM migration*, *OS-level containers* and *hardware counters*, and their application in improving a mitigation's agility and utilisation, and finally,
- an application of SafeHaven in mitigating a system-wide covert channel, in implementing a multi-level moving target defence, and in measuring the cost of migration at each level of the hierarchy.

2 Background and Related Work

The issue of isolating processes has been historically described as the *confinement problem* [25]. The following is an overview of the various ways in which confinements can be broken and upheld.

Attacks. Confinements can be broken at different levels of a system architecture, such as the cache level (L1 [32], L2 [41] and L3 [42]), virtual machine level [33], system level [4,40], or network level [10], through various forms of attack. Attacks are characterised by type (side or covert), scope (socket, system or network-wide), bandwidth and feasibility. Illicit channels can be broadly categorised as being *time-driven*, *trace-driven* or *access-driven* [38]. Time-driven attacks rely on measuring variations in the aggregate execution time of operations. Trace-driven cache attacks are based on analysing an operation's evolution over time. Access-driven attacks allow an attacker to correlate effects of the underlying system's internal state to that of a co-located victim.

Covert channels are generally simpler to construct due to the involved parties cooperating. Fast channels have been shown at the L2 cache level [41], which in a virtualised environment would require VCPUs to share related cores, as well as across virtual machines [40]. Scheduling algorithms can also be leveraged to form a channel by modulating the time for which a VM [30] or process [20] is scheduled.

Defences. Mitigations can broadly be categorised as being *passive*, *reactive* or *architectural*. Passive countermeasures attempt to preserve isolations through an indiscriminate process. For example, disabling hardware threads will eliminate a class of attacks [32] at the cost of performance. Alternatively, one can use a scheduling policy that only co-schedules entities belonging to the same process [24,39] or *coalition* of virtual machines [34]. Policies can also be altered to limit their preemption rate, restricting the granularity of cache-level attacks [38]. Other countermeasures include periodically flushing caches [45], changing event release rates [6], and intercepting potentially dangerous operations [35].

Reactive countermeasures attempt to detect and mitigate attacks at runtime. Frameworks for distributed event monitoring [28] can be fed events generated via introspection [14], or can enforce a defined information flow policy [34].

Architectural mitigations are changes in hardware or to the way in which it is used. One example is Intel's introduction of specialised AES instructions, which insulate the operations' internal state from external caches [18]. Other solutions include randomly permuting memory placement [39], rewriting programs to remove timing variations [5,13], reducing the precision of system clocks [19,32] or normalising timings [26], cache colouring [24] and managing virtual machines entirely in hardware [23].

3 Isolation and Co-Location

We briefly introduce the fundamental notions of co-location and migration using a simple graph model, with which the relationship between different forms of isolation can be represented.

3.1 Locality

A *confinement* delineates a boundary within which entities can potentially share state. Entities are themselves confinements, leading to a hierarchy.

Definition 1 (Locality). *A confinement (or locality) with a name N, a type $\boldsymbol{\Gamma}$, a set of capabilities* C, *and a set of sub-localities* SB *is denoted by* $\boldsymbol{\Gamma}{:}N(\mathtt{C})\mathrm{S}\textsc{b}$.

Capabilities regulate how confinements can modify each other, with operations on confinements only being allowed when they share a capability. We denote a locality X as being a sub-locality of D using $\mathrm{X} \in \mathrm{D}$. This is extended to the notion of transitive containment $\mathrm{X} \in^{+} \mathrm{D}$, where $\mathrm{X} \in^{+} \mathrm{D} \stackrel{\text{def}}{=} \mathrm{X} \in \mathrm{D} \vee \exists \mathrm{X'} \in \mathrm{D}.\ \mathrm{X} \in^{+} \mathrm{X'}$.

Example 1 (Cache Hierarchy). Intel CPUs often implement *simultaneous multithreading*, with two hardware threads (**C**) sharing an **L1** cache. A dual-core system with per-core **L2** caches and a common **L3** cache can be described as:

$$\mathbf{L3}{:}0()\,[\mathbf{L2}{:}0()\,[\mathbf{L1}{:}0()\,[\mathbf{C}{:}0()\,[]\,,\mathbf{C}{:}1()\,[]]]\,,\mathbf{L2}{:}1()\,[\mathbf{L1}{:}1()\,[\mathbf{C}{:}2()\,[]\,,\mathbf{C}{:}3()\,[]]]]$$

Definition 2 (Co-Location). *Two localities* X *and* Y *are co-located within* D *(denoted by* $\mathrm{X} \stackrel{\mathrm{D}}{\longleftrightarrow} \mathrm{Y}$*) if* $\mathrm{X} \in \mathrm{D} \wedge \mathrm{Y} \in \mathrm{D}$. *The localities are transitively co-located in* D *(denoted by* $\mathrm{X} \stackrel{\mathrm{D}}{\Leftrightarrow} \mathrm{Y}$*) if* $\mathrm{X} \in^{+} \mathrm{D} \wedge \mathrm{Y} \in^{+} \mathrm{D}$.

We denote the movement of a locality X to a parent confinement D as $\mathrm{X} \curvearrowright \mathrm{D}$.

Example 2 (Cache Co-Location). For the hierarchy defined in Example 1, given that a process $\mathrm{P_i}$ executes on a hardware thread **C**:i, process $\mathrm{P_0}$ is transitively co-located with (i) $\mathrm{P_1}$ via **L1**:0, **L2**:0 and **L3**:0, and (ii) $\mathrm{P_2}$ via **L3**:0.

3.2 Confinements

Figure 1a lists the primary types of isolations with which this work is concerned, which are broadly categorised as being static or dynamic. The former are architectural elements such as caches and networks, which, while offering some degree of configuration, exist at fixed locations in relation to each other. The latter are isolations that can be created, destroyed or otherwise moved around. Figure 1b is an example of a containment graph, with possible migration paths depicted through arrows 1–7, where paths denote how an isolation's parent can be changed. The mechanisms implementing each path will be detailed in Sect. 4.2.

An additional form of confinement is that produced by soft isolation [38], which attempts to decrease the amount of information that can be inferred from shared state, simulating a plurality of disjoint isolations. This often incurs an ongoing overhead, the severity of which varies depending on the technique being used [38]. For example, the `clflush` instruction, which flushes all cached versions of a given cache line, has been shown as an effective enabler of side-channel attacks [42,44]. Disabling the instruction would impede attacks. While `clflush` is an unprivileged instruction that does not generate a hardware trap [44], closer inspection of its semantics shows that its execution depends upon a `clflush` flag within the machine's `cpuid` register being asserted [22]. This register is generally immutable, yet virtualisation can mask it [3]. Unfortunately, hardware-assisted virtualisation, such as that used by `KVM`, bypasses the virtualised `cpuid` register,

Static Confinements		
Type	Description	Can Contain
Net	Network	**Net**, **M**
M	Machine	**L3**, **OS**
L3	L3 Cache	**L2**
L2	L2 Cache	**L1**
L1	L1 Cache	**C**
C	Physical core	**VC**, $\mathbf{P_E}$, **Con**, **VM**
OS	Operating Sys.	$\mathbf{P_E}$, **Con**, **VM**

Dynamic Confinements		
Type	Description	Can Contain
VC	Virt. CPU	**VC**, $\mathbf{P_E}$, **Con**, **VM**
VM	Virt. machine	**VC**, **OS**
$\mathbf{P_E}$	Control group	**Con**, **P**
Con	Container	**P**
P	Process	-

(a) Confinement types

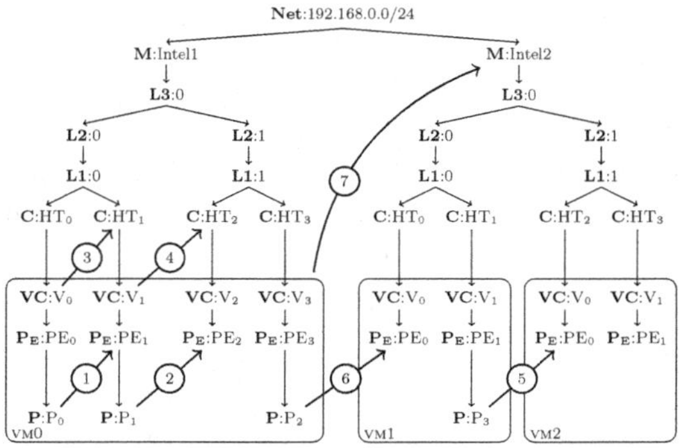

(b) Graph of 2 × **M**, 3 × **VM**. Edges denote containment. 1-7 denote migration paths.

Fig. 1. Example of a containment hierarchy, and various confinement types.

limiting one to using an emulated VCPU such as `QEMU`. While we found this to be effective in disabling `clflush` (an invalid opcode exception was thrown on its invocation), a `QEMU` VCPU is substantially slower than its `KVM` equivalent, leading to a continuous overhead.

4 SafeHaven

With the basic terminology and notation required to model locality and co-location introduced, we now describe SafeHaven, a framework designed to facilitate the creation, deployment and evaluation of isolation properties.

4.1 Overview

SafeHaven is a framework that assists in creating and deploying a network of communicating *probe* and *agent* processes. Sophisticated system-wide detectors can be built by cascading events from various probes at different system levels. A crucial aspect of this model is that detectors can be both *anticipatory* as well as *reactive*, meaning that they can either trigger isolations as a precaution or as a countermeasure to a detected attack.

SafeHaven is implemented in Erlang [16] due to its language-level support for many of the framework's requirements, with probes and agents as long-lived distributed actor processes communicating their stimuli through message passing. Other innate language features include robust process discovery and communication mechanisms and extensive support for node monitoring and error reporting. SafeHaven was developed in lieu of adapting existing cloud-management suites such as OpenStack [31] so as to focus on the event signalling and migration aspects of the approach. Erlang's functional nature, defined communication semantics and use of *generic process behaviours* help to simplify the automatic generation and verification of policy enforcement code, paving the way for future formal analysis.

Probes and Agents. A *probe* is an abstraction for an event source, typically implemented in SafeHaven as an Erlang server process. *Agents* are management probes that can modify one or more confinements.

```
1 Procs = process:recon(),                    % Get system processes
2 [CR, CA|Cs] = cpu:recon(),                  % Get available CPUs
3 lists:foreach(
4   fun(P = #locality{type = process, owner = User}) ->
5     Dest = case User of                     % Choose destination CPU
6              "root"    -> CR;
7              "apache"  -> CA;
8              _         -> Cs
9            end,
10    mig_process:migrate(P, Dest)           % Pin process
11  end, Procs).
```

Algorithm 1. Agent partitioning processes between CPUs by owner via SafeHaven.

Capabilities. An agent can create, destroy or migrate a locality if it owns its associated *capability*. Capabilities serve to describe the extent of an agent's influence. To exert influence on locations outside its scope, an agent must proxy its requests through an external agent that controls the target scope. For example, a probe within a tenant's virtual machine may ask an agent within the underlying cloud provider for an isolated **VC**, which then changes the **VC** to **C** mappings.

Communication. Communication within SafeHaven is carried out using Erlang's message passing facilities. Processes can only message others that share a token (a *magic cookie* [16]) that serves as a communication capability.

Confinement Discovery. The view of an arbitrary agent within a cloud is generally limited to its immediate environment and that of other agents with which it is co-operating. For example, a tenant's agents will be restricted to the processes and structures of their **OS** environment. Similarly, the cloud provider views **VMs** as black boxes. Knowledge of their internal structures is limited to what is exposed by the tenants' agents, bar the use of introspection or disassembly mechanisms.

To facilitate the creation of dynamic policies, SafeHaven provides a series of *reconnaissance* (or `recon`) functions that query the underlying system at runtime and build a partial model of the infrastructure, translating it into a graph of first-class Erlang objects. Algorithm 1 demonstrates an agent's use of SafeHaven's `recon` functions. Handles to the system's running processes (Line 1) and available CPU cores (Line 2) are loaded into lists of `locality` structures that can be manipulated programmatically. This example describes a simple property that partitions processes to different **C**s based on their user ID (Lines 5–9). The procedure for pinning (or migrating) processes (Line 10) will be described in the next section.

4.2 Migrating Confinements

An agent's core isolation operator is *migration.* Agents perform both *objective* and *subjective moves* [11], as they can migrate confinements to which they belong as well as external confinements. The following section describes methods with which one can migrate system structures, namely VCPUs, process groups, processes, containers and virtual machines.

Virtual CPUs (VC). Virtual CPUs in `KVM` [2] can be pinned to different sets of CPUs by means of a mask, set through libvirt [3]. **VC**s can only be migrated to cores to which the parent **VM** has been assigned.

Process/Control Groups ($\mathbf{P_E}$). Pinning processes to CPUs via affinities has a drawback in that unprivileged processes can change their own mappings at will, subverting their confinement. Instead, *control groups* (managed via `cpusets`) [27] are used to define a hierarchy of **C** partitions. Assigning processes to a partition confines their execution to that **C** group, which cannot be exited through `sched_setaffinity`. All processes are initially placed within a default *root* control group. Control groups can be created, remapped or destroyed dynamically. Destroying a group will not automatically kill its constituent processes, rather they will revert to that group's parent.

Processes and Containers (P, Con). Process migration moves a process from one $\mathbf{P_E}$ to another, using mechanisms that vary based on the level at which the control groups are co-located. Arbitrary processes can be moved directly amongst $\mathbf{P_E}$ groups within the same **OS** using `cpusets`, which is fast and can be performed in bulk. Conversely, if the target $\mathbf{P_E}$ exists within a different **OS**, additional mechanisms must be used to translate the process' data structures across system boundaries. In SafeHaven, this is handled using `criu` [1], which enables process checkpoint and restore from within user-space. Recent versions of the Linux kernel (3.11 onwards) have built-in support for the constructs required by `criu`. Migration preserves a process' $\mathbf{P_E}$ containment structure.

Cross-**OS** process migration comes with some limitations. Trivially, processes that are critical to their parent **OS** cannot be migrated away. Other restrictions stem from a process' use of shared resources. For instance, the use of interprocess communication may result in unsafe migrations, as the process will be disconnected from its endpoints. Similarly, a process cannot be migrated if it would cause a conflict at the destination, such as in the case of overlapping process IDs or changing directory structures. This problem is addressed by launching a process with its own *namespaces*, or more generally, by using a container such as LXC or Docker [1]. Live migration for LXC containers is still under active development. An alternative stop-gap measure is to perform checkpoint and restore, transferring the frozen image in a separate step [37].

Virtual Machines (VM). SafeHaven uses `KVM` for virtualisation, managed via libvirt. In the case of a cloud infrastructure, the provider's agents exist within the base **OS**, running alongside a tenant's **VM**. The framework can easily be retargeted to `Xen`-like architectures, with hypervisor-level agents residing within dom0. The choice of hypervisor largely determines what type of instrumentation can be made available to probes.

Similarly to process migration, **VM**s can be migrated locally (changing **C** pinnings) using $\mathbf{P_E}$ groups, or at the global level (changing **OS**). The latter is performed using *live migration* , backed by a *Network File System* (NFS) server storing **VM** images. Recently, experimental patches have been released that enable *post-copy* migration through libvirt, which also requires patching the kernel and QEMU[1]. Using post-copy migration, a virtual machine is immediately migrated to its destination, and pages are retrieved from the original machine on demand. The drawback of post-copy migration is that a network failure can corrupt the **VM**, as its state is split across machines. *Hybrid migration* reduces this risk by initially using standard pre-copy and switching to post-copy migration if the system determines that the transfer will not converge, which would happen when memory pages are being modified faster than they can be transferred.

Other Operations. In addition to being migrated, **VM**, **P** and **Con** isolations can be paused in memory, which can serve as a temporary compromise in cases where an imminent threat cannot be mitigated quickly enough through migration.

4.3 Allocation

To determine a destination for a confinement that must be migrated, an agent broadcasts an isolation request to its known agents. If one of these agents finds that it can serve the request whilst maintaining its existent isolation commitments, it authorises the migration. The problem of placement is equivalent to the *bin-packing problem* [7], and a greedy allocation policy will not produce an optimal allocation. Nevertheless, our scheme is sufficiently general so as to allow different allocation strategies. For example, targets can be prioritised based on their physical distance. Prioritisation can also be used in hybrid infrastructures, where certain targets may be more effective at breaking specific types of co-locations than others. For example, a cloud provider can opt to mix in a number of machines with various hardware confinements and lease them on demand.

5 Case Studies

The previous section detailed the architecture of SafeHaven and the migration techniques it employs. The following section describes the application and evaluation of these methods in the context of illicit-channel mitigation. All experiments were carried out on two Intel i7-4790 machines (4 cores × 2 hardware threads) with 8 GB RAM. **VM**s were allocated 2 **VC**s and 2 GB of RAM, and had 40 GB images. A third computer acted as an NFS server hosting the virtual

[1] https://git.cs.umu.se/cklein/libvirt.

machines' images (average measured sequential speeds: 54 MB/s read, 70 MB/s write), and all machines were connected together via a consumer-grade gigabit switch. **VMs** were connected to the network through a bridged interface. All systems were running Ubuntu 14.04 LTS with the 3.19.0-rc2+ kernel and libvirtd version 1.2.11, patched to enable post-copy support (Sect. 4.2).

5.1 Case 1: System-Wide Covert Channel

The following section describes the use of SafeHaven as an active countermeasure to thwart a system-wide covert-channel.

Overview. Wu et al. [40] demonstrated that performing an atomic operation spanning across a misaligned memory boundary will lock the memory bus of certain architectures, inducing a system-wide slowdown in memory access times. This effect was then used to implement a cross-VM covert channel.

Detection. Detecting the channel's reader process is difficult, as it mostly performs low-key memory and timing operations, and would execute in a co-located **VM**, placing it outside the victim tenant's scope. Conversely, writer processes are relatively conspicuous, in that they perform memory operations that are *atomic* and *misaligned*. Atomic instructions are used in very restricted contexts, and compilers generally align a program's memory locations to the architecture's native width. Having both simultaneously can thus be taken as a strong indication that a program is misbehaving.

Although an attack can be detected by replicating a reader process, a much more direct, precise and efficient method is to use *hardware event counters* [21] to measure the occurrence of misaligned atomic accesses. Recent versions of `KVM` virtualise a system's performance monitoring unit, allowing **VMs** to count events within their domain [15]. One limitation of hardware counters is that their implementation is not uniform across vendors, complicating their use in heterogeneous systems. In addition, while event counters are confined to their **VM** and can only be used by privileged users, one must ensure that they do not themselves enable attacks (for instance, by exposing a high resolution timer).

Policy. Algorithm 2 outlines the behaviour of the agents participating in the mitigation. Each agent takes two arguments, namely the isolation that they are monitoring and a list of additional cooperating agents. When a probe detects that a process P is emitting events at a rate exceeding a threshold ϵ, it notifies its local agent. If the environment is not already isolated, then the agent attempts to locate an isolated resource amongst its own existing tenants. Failing this, the cloud provider is co-opted into finding an isolated machine and resolving the request at the virtual machine level. If a process is mobile, then the cloud provider can opt to create a new isolated **VM** to which the process can be migrated, rather than migrating the source machine.

Require: An event rate threshold ϵ
Require: A_T set of tenant-owned agents, A_C set of cloud-owned agents

```
agent TENANT(OS:X, A_T)
  for all P:P ∈⁺ X do
    if evs(P) ≥ ε ∧ ¬isol_X(P) then
      D ← ⊥
      if mobile(P) then
        D ∈ {D' | TENANT(D', *) ∈ A_T
                  ∧isol_D'(P)}
      if D ≠ ⊥ then
        P ↷ D
      else if X ∈ VM:V then
        request CLOUD(Y, *). V ∈⁺ Y
  TENANT(X, A_T)
end agent
```

(a) Tenant agent

```
agent CLOUD(M:Y, A_C)
  receive isol request for VM:X ∈⁺ Y
  if ¬isol_Y(X) then
    D ∈ {D' | CLOUD(D', *) ∈ A_C
              ∧isol_D'(X)}
    if D ≠ ⊥ then
      X ↷ D
    else
      fallback strategies
  CLOUD(Y, A_C)
end agent
```

(b) Cloud agent

Algorithm 2. Agents for mitigating a system-wide channel.

The degree of isolation required is regulated by the $\mathtt{isol}_D(X)$ predicate, which checks whether X is isolated within D. Evaluating this accurately from within the tenant's scope requires additional information from the cloud agent regarding its neighbours. The strictest interpretation of isolation would be to allocate a physical machine to each **VM** requesting isolation. Another approach is to stratify isolation into different classes determined by user access lists [12], or to only allow a tenant's isolated **VMs** to be co-located with each other.

If an isolated destination cannot be found immediately, then soft isolation must be used as a fallback strategy. Note that soft isolation only has to disrupt the channel until hard isolation is achieved. For example, rather than migrating the locality requesting isolation, one can evict its co-residents, applying soft isolation during their eviction. A simple, general but intrusive method would be to pause the process until isolation is obtained. This should be reserved for creating temporary isolations during fast migration operations. A more targeted mitigation may attempt to degrade the attacker's signal-to-noise ratio by flooding the memory bus with its own misaligned atomic memory accesses. Finally, one may deploy a system such as BusMonitor [35] on a number of machines and migrate **VMs** requesting isolation to them. The problem with the latter solutions is that they must be changed with each discovered attack, whereas a migration-based approach would only require a change in the detector.

Implementation and Evaluation. The policy was implemented in SafeHaven as a network of Erlang server processes, with the detector running as a separate process and taking two parameters, namely (i) a set of system processes $\vec{\mathbf{P}}$ to be scanned, and (ii) a duration τ within which the scan must be performed. Hardware counters were accessed using the *Performance Application Programming Interface* (PAPI) [29] library, with calls proxied through an Erlang module using *Native Implemented Functions* (NIF) [16]. The test machines exposed

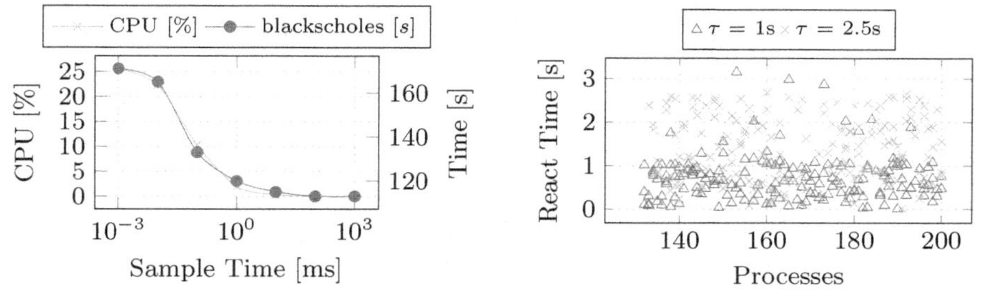

(a) Detector overhead against ϕ (b) React time on varying $\vec{\mathbf{P}}$, $\tau = 1\text{s}, 2.5\text{s}$

Fig. 2. Detector overhead and reaction times.

a native event type that counts misaligned atomic accesses (LOCK_CYCLES: SPLIT_LOCK_UC_LOCK_DURATION [21]). Conversely, another machine to which we had access, namely an AMD Phenom II X6, was found to lack such a combined event type. In this case, one would have to measure misaligned accesses and atomic operations independently, which can lead to more false positives.

The procedure for measuring a process' event emission rate is to attach a counter to it, sleep for a sample time ϕ, and read the number of events generated over that period of time. This is repeated for each process in $\vec{\mathbf{P}}$. The choice of ϕ will affect the detector's duty cycle. Setting $\phi = \tau/|\vec{\mathbf{P}}|$ guarantees that each process will have been sampled once within each τ period, but the sampling window will become narrower as the number of processes increases, raising the frequency of library calls and consequently CPU usage. Setting a fixed ϕ produces an even CPU usage, but leads to an unbounded reaction time.

We tested our hypothesis regarding the infrequency of misaligned atomic accesses by sampling each process in a virtualised and non-virtualised environment over a minute during normal execution. Most processes produced no events of the type under consideration, with the exception of certain graphical applications such as VNC, which produced spikes on the order of a few hundreds per second during use. We then measured the emission rate of the attack's sender process using the reference implementation of Wu et al. [40], compiled with its defaults. This was found to emit $\approx 1.4 \times 10^6$ events per second in both environments, with attacks for 64-byte transmissions lasting $6 \pm 2\,\text{s}$.

Figure 2a shows the detector's CPU usage (measured directly using `top`) against varying ϕ on shifting the detector's logic into a compiled C probe and enumerating processes directly from `/proc/`. To fully encompass the detector's overhead, we pinned the virtual machine to a single VCPU. At $\phi = 10\,\text{ms}$, overhead peaked at a measured $0.3\,\%$. This was confirmed by executing the CPU-intensive `blackscholes` computation from the PARSEC benchmark suite [8] in parallel with the detector, and observing a speed-up proportional to ϕ. Figure 2b describes how reaction time varied against the number of processes being monitored, where reaction time was measured as the time elapsed between the start of an attack and its detection. The reaction time was measured for $133 \leq |\vec{\mathbf{P}}| \leq 200$.

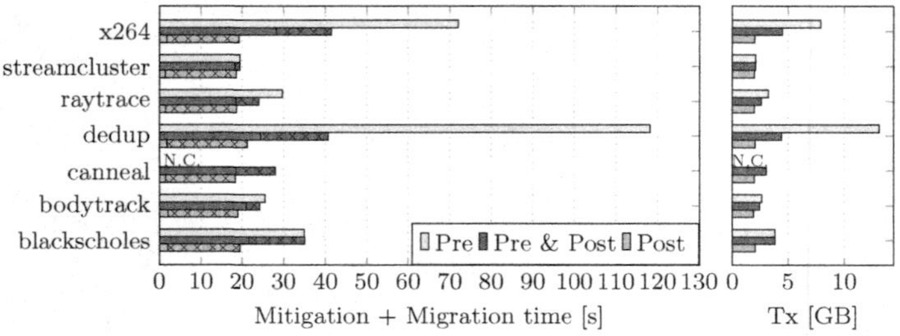

Fig. 3. Comparison of pre-copy, hybrid and post-copy migration.

The size of $\vec{\mathbf{P}}$ was raised by spawning additional processes that periodically wrote to an array. The attack was started at random points in time.

Mitigation. Once a potential attack is detected, it must be isolated. The performance of process migration will be discussed in further detail in Sect. 5.2. For now, we will focus on the different modes of **VM** migration.

Table 1. Summary of detection and mitigation times (s).

Phase	Parameters	Min	Max	Geometric mean	Arithmetic mean
Detect	$\tau = 1\,\mathrm{s}$	0.0148	3.16	0.54	0.72
	$\tau = 2.5\,\mathrm{s}$	0.0272	2.69	1.20	1.46
Migrate	Post-copy	1.2813	2.13	1.47	1.48
Detect &	Post-copy & $\tau = 1\,\mathrm{s}$	1.296	5.29	2.01	2.20
Migrate	Post-copy & $\tau = 2.5$	1.309	4.82	2.67	2.93

Figure 3 illustrates the worst case times taken to perform a single **VM** live migration using pre-copy, hybrid and post-copy while it executed various workloads from the PARSEC suite. Migrations were triggered at random points during the benchmark's execution, with 6 readings per benchmark and migration mode. The host machines were left idle to reduce additional noise. Solid bars represent the time taken for the **VM** to resume execution at the target machine, and the shaded area denotes the time spent copying over the remainder of the **VM**'s memory pages after it has been moved.

Pre-copy's performance was significantly affected by the workload being executed, with `canneal` never converging. Hybrid migration fared better as it always converged and generated less traffic. Post-copy exhibited the most consistent behaviour, both in terms of migration time as well as generated traffic. During the course of our experiments, we found that attempting to start a migration immediately in post-copy mode would occasionally trigger a race condition. This was remedied by adding a one second delay before switching to post-copy. Nevertheless, **VMs** migrated using post-copy resumed execution at the target in at most

2.13 s, and 1.51 s on average, which includes the delay. Total migration time and data transferred were also consistently low, averaging 20 s and 2 GB, respectively.

Table 1 summarises the results. Based on the detector's reaction times and post-copy's switching time, and assuming that a target machine has already been identified, a channel can be mitigated in around 1.3 s under ideal conditions, 5.3 s in the worst case, and in just under 3 s on average.

Conclusion. We have shown how hardware event counters can be used to detect an attack efficiently, quickly and precisely, and how post-copy migration considerably narrows an attack's time window. Additional improvements can be obtained by integrating event counting with the scheduling policy, where the event monitor's targets are changed on context switching. This would eliminate the need to sweep through processes and avoids missing events.

5.2 Case 2: Moving Target Defence

The following describes the use of SafeHaven in implementing a passive and preventive mitigation, specifically, a *moving target defence.*

Overview. The moving target defence [46] is based on the premise that an attacker co-located with a victim within a confinement D requires a minimum amount of time $\alpha(\mathrm{D})$ to set up and perform its attack. Attacks can thus be foiled by limiting continuous co-location with every other process to at most $\alpha(\mathrm{D})$. The defence is notable in that it does not attempt to identify a specific attacker, being driven entirely on the basis of co-location.

Policy. Algorithm 3 describes the moving target defence as a generalisation of the formulation given by Zhang et al. [46]. The policy assumes the existence of three predicates, namely: (i) $H(\mathbf{T})$, the time required to migrate a locality of type $\mathbf{T}$, (ii) $\alpha(\mathrm{D})$, the time required to attack a process through D, and (iii) $\tau(\mathrm{P})$, the duration for which a supplied predicate P holds. The following section attempts to establish practical approximations for the aforementioned predicates.

Require: A root locality R
for all $\mathbf{T}{:}\mathrm{L_0}, \mathbf{T}{:}\mathrm{L_1} \in^{+} \mathrm{R}.\ \mathrm{L_0} \neq \mathrm{L_1}$ **do**
  if $\exists \mathrm{D} \in^{+} \mathrm{R}.\ \tau(\mathrm{L_0} \overset{\mathrm{D}}{\Leftrightarrow} \mathrm{L_1}) + H(\mathbf{T}) \geq \alpha(\mathrm{D})$ **then**
    $\mathrm{L}_{i \in \{0,1\}} \curvearrowright \mathrm{S}.\ \mathrm{S} \in^{+} \mathrm{R} \wedge \neg \mathrm{L_0} \overset{\mathrm{D}}{\Leftrightarrow} \mathrm{L_1}$

Algorithm 3. General form of the moving target defence.

Defining $H()$. $H()$ must be able to predict the cost of a future migration. In addition, $H()$ varies based on the destination of a migration, thus requiring that the predicate be refined. We estimate the next value of $H()$ using an *exponential average* [36], expressed as the following recurrence relation:

$$H_{n+1}(\mathrm{T} \curvearrowright \mathrm{D}) = h\eta_n(\mathrm{T} \curvearrowright \mathrm{D}) + (1-h)H_n(\mathrm{T} \curvearrowright \mathrm{D})$$

where $\eta_n()$ is the measured duration of a migration, and $0 \leq h \leq 1$ biases predictions towards historical or current migration times. We take $h = 0.5$.

Defining $\alpha()$. A precise predicate for $\alpha()$ is difficult to define, as it would require a complete characterisation of the potential attacks that a system can face, with knowledge of the state of the art at most bounding the predicate. In the absence of a perfect model, we adopt a pragmatic approach, whereby the duration of co-locations (and, by association, the migration rate) is determined by the overhead that a tenant will bear, as this is ultimately the limiting factor.

Defining $\tau(\Leftrightarrow)$. A tenant can determine the co-location times for processes within its domain, but is otherwise oblivious to other tenants' processes. In the absence of additional isolation guarantees from the cloud provider, $\tau(\Leftrightarrow)$ must be taken as the total time spent at a location, timed from the point of entry.

Propagating Resets. The hierarchical nature of confinements can be leveraged to improve the moving target defence. Migrations at higher levels will break co-locations in their constituents. Thus, following a migration, an agent can propagate a directive to its sub-localities, resetting their $\tau(\Leftrightarrow)$ predicates. Propagation must be selective. For example, while process migration to another machine will break locality at the **OS** and **C** level, **VM** migration only breaks cache and machine-wide locality, and leaves the **OS** hierarchy intact. Similarly, a lower locality can request isolation from a higher-level parent to trigger a bulk migration action, which can resolve multiple lower-level migration deadlines.

Implementation and Evaluation. Similarly to the previous case study, a two-tiered system of agents is used. Agents are given a set of distinct locations which are guaranteed to be disjoint, which is necessary for the mitigation to work, as otherwise migrations would not break co-location.

Table 2. Migration times for different isolation types and paths (ms).

		Con $\curvearrowright$ VC		**VC $\curvearrowright$ C**		**Con $\curvearrowright$ OS**			**Con $\curvearrowright$ OS**			**VM $\curvearrowright$ OS**
	Mig. Path	1	2	3	4	5			6			7
						`rsync`	Check	Rest	`rsync`	Check	Rest	
Benchmark	blackscholes	24.14	24.07	26.84	26.93	32,508	13,695	2,027	31,235	13,636	1,876	18,781
	bodytrack	23.99	24.61	26.80	26.99	15,442	4,895	1,018	18,596	4,539	899	19,069
	canneal	25.03	25.20	27.00	27.29	68,972	24,562	7,950	55,831	21,936	6,399	18,748
	dedup	26.81	26.79	26.99	26.98	71,563	10,888	3,396	56,422	11,021	2,712	19,469
	streamcluster	24.70	24.79	26.79	26.96	19,215	5,048	842	13,016	5,104	797	18,654
	raytrace	24.30	24.85	26.92	26.96	66,881	18,668	4,804	53,223	17,057	4,255	18,841
	x264	25.65	25.56	26.99	27.04	56,224	4,262	1,095	47,580	4,392	1,228	19,410
	$H_0()$ (Geo.)	24.93	25.11	26.90	27.02	40,510	9,542	2,197	34,678	9,233	1,986	18,994

Table 2 lists the migration times measured when migrating containers and **VMs** through each migration path (paths 1–7 in Fig. 1b) whilst executing various benchmarks from PARSEC, with the hosts being otherwise idle. Given its consistent behaviour, we only considered post-copy migration when moving **VMs**. The timings for **Con** migration were broken down into its phases. To keep **Con** migration independent from the cloud provider, container images were transferred to their target using `rsync`. This was by far the dominant factor in **Con** migration times, and can largely be eliminated through shared storage. The initial value of $H_0()$ for each path was derived from the geometric mean of the migration times.

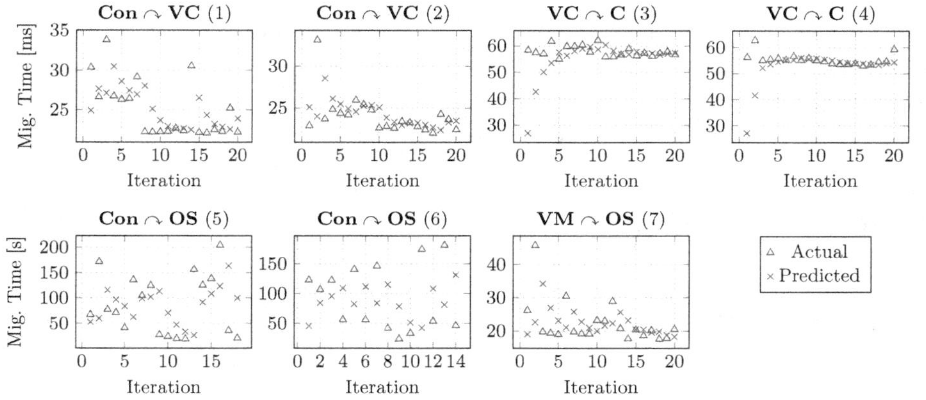

Fig. 4. Predictions of $H()$ against measured migration times.

We evaluated the relationship between performance and migration frequency on the system running at capacity. On the first machine, three **VMs** were assigned benchmarks to execute. A fourth was set as a migrating tenant, running each benchmark listed in Table 2. A fifth **VMs** for cross-**VM** process migration, and was kept idle. The second machine ws configured with three tenants running benchmarks and two idle **VMs**. Table 3 in Appendix A lists the geometric means of the benchmarks' running times, with the *All* column denoting the time required for all of the migrating tenant's benchmarks to complete. Figure 4 shows the predicted and actual migration times for the first migration operations, using the $H_0()$ values derived previously. Network effects and thrashing on the NFS server introduced a significant degree of variability. In summary, we found that migration operations generally had no discernible effect on the neighbouring tenants, although we posit that this would not hold for oversubscribed systems. Migrations at the **C** and **VC** level had no significant effect on performance. **Con** and **VM** migration did not appear to affect neighbouring tenants, but clearly affected their own execution. Migrating the **VM** every 30 s more than doubled its benchmark's running time (note that at this migration frequency, the **VM** was involved in a migration operation for two-thirds of its running time).

Conclusion. We have investigated the core components of a multi-level moving target defence, and examined the cost of migration at each level. Lower-level migrations can be performed at high frequency, but break the fewest co-locations, whereas the opposite holds at higher levels. Restricting the moving target defence to a single level limits its ability to break co-location. For example, while **VM** migration will break co-locations with other tenants, it cannot break the **OS**-level co-locations formed within it. Process and container migration can break co-location through every level, yet offline migration results in a significant downtime, rendering its application to a moving target defence limited. The advent of live process migration will thus help in making this mitigation pathway more viable.

5.3 Other Policies

HomeAlone. *HomeAlone* [43] uses a PRIME-PROBE attack to monitor cache utilisation, and a trained classifier to recognize patterns indicative of shared locality. This can be used to implement a hypervisor-independent version of the `isol`() predicate described in Sect. 5.1, or to detect adversarial behaviour.

Network Isolation. Networks can harbour illicit channels [9,10]. Isolation at this level can be achieved via a combination of soft and hard isolation, with trusted machines sharing network segments and traffic normalisers [17] monitoring communication at the edges.

6 Conclusion

In this work, we examined the use of migration, in its many forms, to dynamically reconfigure a system at runtime. Through the SAFEHAVEN framework, we described and evaluated the use of migration to implement an efficient and timely mitigation against a system-wide covert-channel attack. We also demonstrated how a moving target defence can be enhanced by considering multiple levels and granularities of isolation, examining the costs associated with migrating entities at each level, and showing how performance and granularity are correlated.

A Appendix: Migration Frequency and Performance

Table 3. Effect of migration frequency on performance when running at capacity.

			Local				Remote		
	Dispatch (ms)	Migrations	All	blackscholes	canneal	streamcluster	blackscholes	canneal	streamcluster
No migration	-	0	1,612	124	184	397	118	169	367
Con ↷ VC (1)	500	2,930	1,475	122	168	385	117	153	368
	400	3,601	1,463	120	168	385	117	154	369
	300	5,230	1,567	121	166	383	117	153	369
	200	7,889	1,590	122	170	384	118	153	369
Con ↷ VC (2)	500	3,110	1,560	124	169	391	118	153	369
	400	4,062	1,656	126	167	388	118	152	371
	300	5,034	1,521	127	171	388	117	154	367
	200	7,824	1,573	126	171	390	117	152	368
VC ↷ C (3)	500	3,117	1,562	123	172	404	118	159	373
	400	4,020	1,609	124	173	387	118	158	374
	300	5,379	1,614	124	174	388	118	160	372
	200	7,628	1,534	126	177	394	118	158	372
VC ↷ C (4)	500	3,154	1,576	125	171	395	118	157	372
	400	3,995	1,598	127	170	393	118	157	372
	300	5,413	1,630	128	173	394	119	159	372
	200	8,514	1,705	128	175	398	118	154	369
Con ↷ OS (5)	210000	14	2,886	124	167	380	119	153	369
	180000	18	3,565	124	165	380	118	152	369
Con ↷ OS (6)	210000	14	2,780	122	164	375	119	155	373
VM ↷ OS (7)	120000	17	2,028	120	179	392	121	176	375
	90000	23	2,025	122	170	384	120	162	392
	60000	39	2,282	121	162	389	122	173	390
	30000	125	3,770	121	169	384	124	177	394

References

1. CRIU project page, April 2015. http://criu.org/Main_Page
2. KVM project page, April 2015. http://www.linux-kvm.org/
3. Libvirt project page, April 2015. http://www.libvirt.org/
4. Aciiçmez, O., Koç, c.K., Seifert, J.P.: On the power of simple branch prediction analysis. In: ASIACCS 2007, pp. 312–320. ACM, New York (2007)
5. Agat, J.: Transforming out timing leaks. In: Proceedings of the 27th ACM SIGPLAN-SIGACT Symposium on Principles of Programming Languages, POPL 2000, pp. 40–53. ACM, New York (2000)
6. Askarov, A., Zhang, D., Myers, A.C.: Predictive black-box mitigation of timing channels. In: CCS 2010, pp. 297–307. ACM, New York (2010)
7. Azar, Y., Kamara, S., Menache, I., Raykova, M., Shepard, B.: Co-location-resistant clouds. In: CCSW 2014, pp. 9–20. ACM, New York (2014)
8. Bienia, C., Kumar, S., Singh, J.P., Li, K.: The parsec benchmark suite: characterization and architectural implications. In: Proceedings of the 17th International Conference on Parallel Architectures and Compilation Techniques, October 2008
9. Brumley, B.B., Tuveri, N.: Remote timing attacks are still practical. In: Atluri, V., Diaz, C. (eds.) ESORICS 2011. LNCS, vol. 6879, pp. 355–371. Springer, Heidelberg (2011)
10. Cabuk, S., Brodley, C.E., Shields, C.: IP covert timing channels: design and detection. In: CCS 2004. ACM, New York (2004)
11. Cardelli, L., Gordon, A.D.: Mobile ambients. In: POPL 1998. ACM Press (1998)
12. Caron, E., Desprez, F., Rouzaud-Cornabas, J.: Smart resource allocation to improve cloud security. In: Nepal, S., Pathan, M. (eds.) Security, Privacy and Trust in Cloud Systems. Springer, Heidelberg (2014)
13. Coppens, B., Verbauwhede, I., Bosschere, K.D., Sutter, B.D.: Practical mitigations for timing-based side-channel attacks on modern x86 processors. In: S&P 2009, pp. 45–60. IEEE Computer Society, Washington, DC (2009)
14. Dolan-Gavitt, B., Leek, T., Hodosh, J., Lee, W.: Tappan zee (north) bridge: mining memory accesses for introspection. In: CCS 2013. ACM, New York (2013)
15. Du, J., Sehrawat, N., Zwaenepoel, W.: Performance profiling in a virtualized environment. In: 2nd USENIX Workshop on Hot Topics in Cloud Computing (2010)
16. Ericsson AB: Erlang reference manual user's guide, 6.2 edn., September 2014. http://www.erlang.org/doc/reference_manual/users_guide.html
17. Gorantla, S., Kadloor, S., Kiyavash, N., Coleman, T., Moskowitz, I., Kang, M.: Characterizing the efficacy of the NRL network pump in mitigating covert timing channels. IEEE Trans. Inf. Forensics Secur. **7**(1), 64–75 (2012)
18. Gueron, S.: Intel advanced encryption standard (AES) new instructions set, May 2010. http://www.intel.com/content/dam/doc/white-paper/advanced-encryption-standard-new-instructions-set-paper.pdf
19. Hu, W.M.: Reducing timing channels with fuzzy time. In: S&P 1991, pp. 8–20. IEEE Computer Society, May 1991
20. Hu, W.M.: Lattice scheduling and covert channels. In: S&P 1992, p. 52. IEEE Computer Society, Washington, DC (1992)
21. Intel: system programming guide, Intel® 64 & IA-32 architectures software developers manual, vol. 3B. Intel, May 2011
22. Intel: instruction set reference, intel® 64 & IA-32 architectures software developers manual, vol. 2. Intel, January 2015

23. Keller, E., Szefer, J., Rexford, J., Lee, R.B.: Nohype: virtualized cloud infrastructure without the virtualization. In: 37th Annual International Symposium on Computer Architecture, ISCA 2010, pp. 350–361. ACM, New York (2010)
24. Kim, T., Peinado, M., Mainar-Ruiz, G.: Stealthmem: system-level protection against cache-based side channel attacks in the cloud. In: Security 2012. USENIX Association, Berkeley (2012)
25. Lampson, B.W.: A note on the confinement problem. CACM **16**(10), 613–615 (1973)
26. Li, P., Gao, D., Reiter, M.: Mitigating access-driven timing channels in clouds using stopwatch. In: 43rd Annual IEEE/IFIP International Conference on Dependable Systems and Networks (DSN), pp. 1–12, June 2013
27. Linux: cpuset(7) - Linux manual page, August 2014. http://www.man7.org/linux/man-pages/man7/cpuset.7.html
28. Mdhaffar, A., Ben Halima, R., Jmaiel, M., Freisleben, B.: A dynamic complex event processing architecture for cloud monitoring and analysis. In: 2013 IEEE 5th International Conference on Cloud Computing Technology and Science, CloudCom, vol. 2, pp. 270–275, December 2013
29. Mucci, P.J., Browne, S., Deane, C., Ho, G.: Papi: a portable interface to hardware performance counters. In: Proceedings of the DoD HPCMP Users Group Conference (1999)
30. Okamura, K., Oyama, Y.: Load-based covert channels between Xen virtual machines. In: 2010 ACM Symposium on Applied Computing, SAC 2010, pp. 173–180. ACM, New York (2010)
31. OpenStack foundation: OpenStack documentation, February 2015. http://www.docs.openstack.org/
32. Osvik, D.A., Shamir, A., Tromer, E.: Cache attacks and countermeasures: the case of AES. In: Pointcheval, D. (ed.) CT-RSA 2006. LNCS, vol. 3860, pp. 1–20. Springer, Heidelberg (2006)
33. Ristenpart, T., Tromer, E., Shacham, H., Savage, S.: Hey, you, get off of my cloud: exploring information leakage in third-party compute clouds. In: CCS 2009, pp. 199–212. ACM, New York (2009)
34. Sailer, R., Jaeger, T., Valdez, E., Cáceres, R., Perez, R., Berger, S., Linwood, J., Doorn, G.L.: Building a MAC-based security architecture for the Xen opensource hypervisor. In: 21st Annual Competition Section Applications Conference, ACSAC 2005 (2005)
35. Saltaformaggio, B., Xu, D., Zhang, X.: Busmonitor: a hypervisor-based solution for memory bus covert channels. In: EuroSec 2013. ACM (2013)
36. Silberschatz, A., Galvin, P.B., Gagne, G.: Operating System Concepts, Chap. 5, 7th edn, p. 161. Wiley Publishing, New York (2005)
37. Tycho: live migration of linux containers, October 2014. http://tycho.ws/blog/2014/09/container-migration.html
38. Varadarajan, V., Ristenpart, T., Swift, M.: Scheduler-based defenses against Cross-VM side-channels. In: Security 2014. USENIX Association, San Diego, August 2014
39. Wang, Z., Lee, R.B.: Covert and side channels due to processor architecture. In: 22nd Annual Computer Security Applications Conference, ACSAC 2006, pp. 473–482. IEEE Computer Society, Washington, DC (2006)
40. Wu, Z., Xu, Z., Wang, H.: Whispers in the hyper-space: high-speed covert channel attacks in the cloud. In: Security 2012. USENIX Association, Berkeley (2012)
41. Xu, Y., Bailey, M., Jahanian, F., Joshi, K., Hiltunen, M., Schlichting, R.: An exploration of L2 cache covert channels in virtualized environments. In: CCSW 2011, pp. 29–40. ACM, New York (2011)

42. Yarom, Y., Falkner, K.E.: Flush+reload: a high resolution, low noise, L3 cache side-channel attack. IACR Crypt. ePrint Arch. **2013**, 448 (2013)
43. Zhang, Y., Juels, A., Oprea, A., Reiter, M.K.: Homealone: co-residency detection in the cloud via side-channel analysis. In: S&P 2011, pp. 313–328. IEEE Computer Society, Washington, DC (2011)
44. Zhang, Y., Juels, A., Reiter, M.K., Ristenpart, T.: Cross-tenant side-channel attacks in paas clouds. In: CCS 2014, pp. 990–1003. ACM, New York (2014)
45. Zhang, Y., Reiter, M.K.: Düppel: retrofitting commodity operating systems to mitigate cache side channels in the cloud. In: CCS 2013, pp. 827–838. ACM, New York (2013)
46. Yu, M., Zang, W., Zhang, Y., Li, M., Bai, K.: Incentive compatible moving target defense against VM-colocation attacks in clouds. In: Gritzalis, D., Furnell, S., Theoharidou, M. (eds.) SEC 2012. IFIP AICT, vol. 376, pp. 388–399. Springer, Heidelberg (2012)

Factors Impacting the Effort Required to Fix Security Vulnerabilities

An Industrial Case Study

Lotfi ben Othmane[1(✉)], Golriz Chehrazi[1], Eric Bodden[1], Petar Tsalovski[2], Achim D. Brucker[2], and Philip Miseldine[2]

[1] Fraunhofer Institute for Secure Information Technology, Darmstadt, Germany
{lotfi.ben.othmane,golriz.chehrazi,eric.bodden}@sit.fraunhofer.de
[2] SAP SE, Walldorf, Germany
{petar.tsalovski,achim.brucker,philip.miseldine}@sap.com

Abstract. To what extent do investments in secure software engineering pay off? Right now, many development companies are trying to answer this important question. A change to a secure development lifecycle can pay off if it decreases significantly the time, and therefore the cost required to find, fix and address security vulnerabilities. But what are the factors involved and what influence do they have? This paper reports about a qualitative study conducted at SAP to identify the factors that impact the vulnerability fix time. The study involves interviews with 12 security experts. Through these interviews, we identified 65 factors that fall into classes which include, beside the vulnerabilities characteristics, the structure of the software involved, the diversity of the used technologies, the smoothness of the communication and collaboration, the availability and quality of information and documentation, the expertise and knowledge of developers, and the quality of the code analysis tools. These results will be an input to a planned quantitative study to evaluate and predict how changes to the secure software development lifecycle will likely impact the effort to fix security vulnerabilities.

Keywords: Human factors · Secure software · Vulnerability fix time

1 Introduction

Despite heavy investments into software security [1], security experts and attackers continue to discover code vulnerabilities in software systems on a regular basis, including buffer overflows, SQL injections, and unauthorized procedure calls. While some attack vectors relate to mis-designed software architectures, many exploit code-level vulnerabilities in the application code [2]. Major software-development companies, including SAP, embed in their development process activities (e.g., dynamic and static security testing [3]) to identify vulnerabilities early during the development of their software system. Nevertheless,

J. Lopez and C.J. Mitchell (Eds.): ISC 2015, LNCS 9290, pp. 102–119, 2015.
DOI: 10.1007/978-3-319-23318-5_6

their security development lifecycle (see, e.g., [4] for Microsoft's security development lifecycle) includes also a process for addressing vulnerabilities identified after the software is released.

Analyzing and fixing security vulnerabilities is a costly undertaking. Surely it impacts a software's time to market and increases its overall development and maintenance cost. But by how much? To answer this question directly, one would need to trace all the effort of the different actions that the developers undertake to address a security issue: initial triage, communication, implementation, verification, porting, deployment and validation of a fix. Unfortunately, such a *direct* accountability of the individual efforts associated with these action items is impossible to achieve, last but not least due to legal constraints that forbid any monitoring of the workforce. One must therefore opt for *indirect* means to relate quantitative, measurable data, such as the vulnerability type, the channel through which it was reported, or the component in which it resides, to soft human factors that correlate with the time it takes to fix the related vulnerabilities. But, which factors impact this fixing effort positively or negatively?

This paper aims to identify the factors that impact the vulnerability fix time in SAP software. (We use vulnerability fix time and vulnerability fix effort interchangeably.) For this work we interviewed 12 experts who contribute to addressing security vulnerabilities at SAP, one of the largest software vendors worldwide, and the largest in Germany. The study comprises teams located in different countries, developing diversified products. The work led to the discovery of 65 factors impacting the vulnerabilities fix time, which we classified into 8 categories. The factors could be used to estimate the required effort to fix vulnerabilities and to improve the secure development activities.

This paper is organized as follows. First, we give an overview of related work (Sect. 2) and discuss secure software development at SAP (Sect. 3). Next, we describe the research approach that we use in this work (Sect. 4), report about our findings (Sect. 5) and discuss the impact and the limitations of the study (Sect. 6). Subsequently, we discuss some of the lessons we learned from the study (Sect. 7) and conclude in Sect. 8.

2 Related Work

Several pieces of research investigate the time it takes to fix software defects [5, 6]. For instance, Hewett and Kijsanayothin applied machine-learning algorithms to defect data collected from the development of a large medical-record system to predict the duration between the time of identification of the defect and the validation of the appropriate fix [6].[1] Opposed to this previous work, we (1) focus on security vulnerabilities, not functionality errors, and (2) include in our model "human factors" such as organizational issues that cannot directly be derived from automatically collected data. In this work, as a first step, we determine the relevant factors.

[1] Among other things, the duration includes the time the defect is in the repair queue after being assigned to a developer.

Table 1. Examples of time required for fixing vulnerabilities [7].

Vulnerability type	Average fix time (min)
Dead code (unused methods)	2.6
Lack of authorization check	6.9
Unsafe threading	8.5
XSS (stored)	9.6
SQL injection	97.5

Software defects have been found to be correlated with software complexity [8], which is measured, e.g., using the size of the code and the density of its control instructions. There is a general hypothesis that software complexity is also correlated with the existence of vulnerabilities e.g., [2]. This hypothesis is often false. For example, Shin et al. [9] and Chowdhury et al. [10] found that the complexity metrics of open-source software such as Firefox only weakly correlate with the existence of vulnerabilities in those systems. Thus, the factors (e.g., code complexity) that apply to software-defects based models do not necessarily apply to vulnerabilities based models.

The only work we know that evaluates vulnerability fix time was performed by Cornell, who measured the time the developers spent fixing security vulnerabilities in 14 applications [7]. Table 1 shows the average time the developers take to fix vulnerabilities for several vulnerability types. The measured time comprises only the fix-execution phase, which includes the environment setup, implementation, validation, and deployment of the fix. Cornell found that the percentage of this time spent on the implementation of the fix is only between 29 % and 37 % of the time spent in the execution phase. The author was unable to measure the time spent on the inception (including risk assessment) and planning phases because the collected data were too inconclusive. Cornell found also that there are vulnerability types that are easy to fix, such as dead code, vulnerability types that require applying prepared solutions, such as lack of authorization, and vulnerability types that, although simple conceptually, may require a long time to fix for complex cases, such as SQL injection.

The vulnerability type is thus one of the factors that indicate the vulnerability fix time but is certainly not the only one. This paper aims to identify as many factors as possible that will likely impact the vulnerability fix time, factors that could be collected automatically but also factors that can only be inferred indirectly by observing how human analysts and developers go about fixing vulnerabilities.

3 Secure Software Development at SAP

SAP has a very diverse product portfolio: for example, a SAP product might be a small mobile application or an enterprise resource planning (ERP) system. Similarly, a large number of different programming languages and frameworks

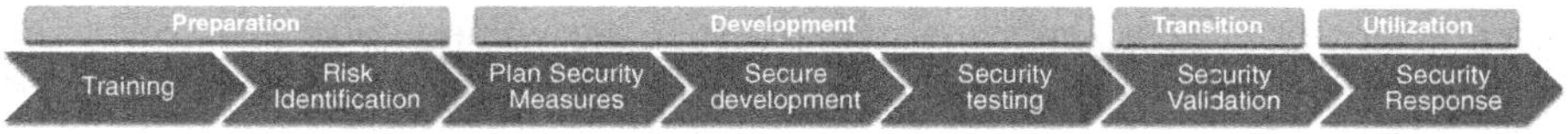

Fig. 1. High-level overview of the SAP security development lifecycle (S^2DL)

are used during their development and many different environments (e.g., web browsers, operating systems) are supported. Moreover, SAP develops also frameworks, such as SAP Netweaver, that are both offered to customers and used to build other SAP products. Finally, SAP product portfolio ranges from on-premise products to cloud offerings (including private clouds, public clouds, and hybrid clouds).

To ensure a secure software development, SAP follows the SAP Security Development Lifecycle (S^2DL). Figure 1 illustrates the main steps in this process which is split into four phases: preparation, development, transition, and utilization. For our work, the second half of the S^2DL is important:

- during the actual software development (in the steps *secure development* and *security testing*) vulnerabilities are detected, e.g., by using static and dynamic application security testing tools [3, 11];
- *security validation* is an independent quality control that acts as "first customer" during the transition from software development to release, i.e., security validation finds vulnerabilities after the code freeze, (called correction close) and the actual release;
- *security response* handles vulnerabilities reported after the release of the product, e.g., by external security researchers or customers.

To allow the necessary flexibility to adapt this process to the various application types developed by SAP as well as the different software development styles and cultural differences in a worldwide distributed organisation, SAP follows a two-staged security expert model:

1. a central security team defines the security global processes (such as the S^2DL), provides security trainings, risk identification methods, offers security testing tools, or defines and implements the security response process;
2. local security experts in each development area/team are supporting the developers, architects, and product owners in implementing the S^2DL and its supporting processes.

If a vulnerability is detected, developers and their local security experts follow a four step process: (1) analyze the vulnerability, (2) design or select a recommended solution, (3) implement and test a fix, and (4) validate and release this fix. In the security testing process, a security expert is expected to inspect the analysis results of any utilized testing tool and determine for each of the reported findings whether it is exploitable, and consequently requires fixing. The vulnerability then gets assigned to a developer who implements the suggested solution. The fix is verified by a retest of the code with the same testing rules. The fix is considered to be successful when the test passes.

While this process is the same, regardless if the vulnerability is in released code or current development code, certain administrative steps exist prior to the first step but the steps necessary to release a fix and the involved parties differ. For vulnerabilities in not yet released code, the process is locally defined by the development team and, usually, very lightweight. For vulnerabilities in *released* software, the security response team, developers and security experts are mainly involved in the first three fixing phases and the maintenance team (called IMS) is mainly involved in the last phase. Fixes of released code are reviewed and validated by the central security team. These fixes are shipped in security notes or support packages for customers to download. Security notes are patches included in support packages. Support packages are functional updates that also contain the latest security notes.

4 Research Approach

We conducted a qualitative case study to identify the factors that impact the vulnerability fix time for vulnerabilities reported to or within SAP. A case study is an empirical inquiry that investigates a phenomenon in its real-life context [12]. This study uses expert interviews as data source; that is, interview of security experts, coordinators, and developers who contribute to fixing vulnerabilities. The aim of the interviews is to use the experiences of the interviewees to identify the factors that impact the time they spend in contributing to fixing vulnerabilities.

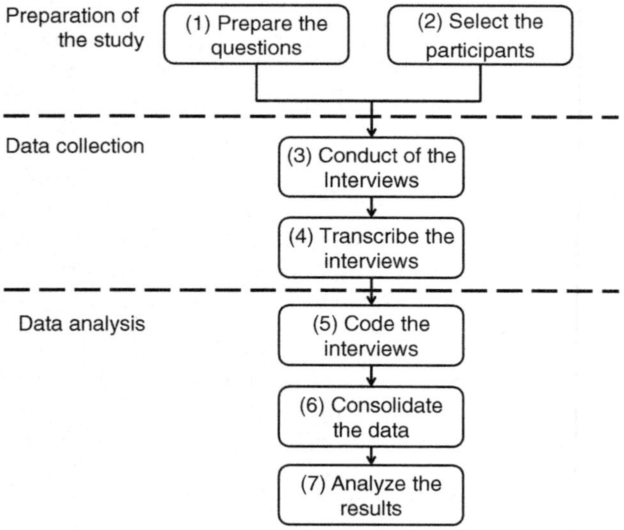

Fig. 2. The steps of the case study.

Figure 2 depicts the study process. It has three phases: study preparation, data collection, and data analysis. The description of the three phases follows.

Preparation of the Study. Initially we reviewed a set of documents that describe the processes of fixing vulnerabilities and discussed these processes with SAP security experts. We used the knowledge to develop a set of interview questions. Then, we met three security experts in pre-interviews to learn the fixing process further and to check the questions that we developed. We summarized the questions (Step 1 of Fig. 2) in an interview protocol. An interview protocol describes the procedural and the main questions that should be used in the interviews [13]. We choose semi-structured questions, which allowed us to capture similar type of data (e.g., roles, pinpoints, and recommendations) across all the interviews while being flexible to explore reported content.

Fixing vulnerabilities at SAP requires collaboration of people having different roles, who could be located in different cities and countries. We considered this contextual factor and invited representatives (Step 2 of Fig. 2) of the different roles located in several offices to participate in the study. Twelve participants accepted: nine were from Germany and three were from India. The participants were NetWeaver experts, application-layer experts, and experts in developing customer specific applications. Their roles were security experts, developers, coordinators, and project leads. This method of selecting participants complies with the maximum-variation-sampling approach [14]–a preferred participants sampling method.

We scheduled one hour for each interview. We sent the participants the interview protocol so they could prepare for the interview; e.g., prepare examples.

Table 2. Interview questions.

Question	Targeted information
1. What is your role in fixing vulnerabilities?	The role of the participant
2. How do you get the information about security vulnerabilities? And what do you do with the information?	The source of information about vulnerabilities, the steps performed by the interviewee in fixing vulnerabilities, and the tools used by the interviewee to fix the vulnerabilities
3. What are the worst and best cases of vulnerabilities you worked on? And how did do you address them?	The factors that impact the vulnerability fix time
4. How much time did you spend on [..]? Why did [it] take that long?	The factors that impact the vulnerability-fix time
5. How would we improve the way you work?	The factors that impact the vulnerability-fix time

Data Collection. We conducted the 11 interviews (Step 3 of Fig. 2) within one week (One of the interviews was conducted for 2 h with 2 interviewees as

Table 3. Used coding schema.

Code class	Description
Meta	The role of the interviewee and their experience with SAP products and with fixing vulnerabilities
Used tools	The tools used in fixing the security vulnerabilities
Process participants	The roles and teams that the interviewee collaborated with in fixing the vulnerabilities they worked on
Process activities	The activities that the interviewee performs when fixing security vulnerabilities, which are classified into pre-analysis or risk assessment, analysis, design, implementation, test, release activities
Information for activities	The information used for analysis including risk assessment analysis, design, implementation, test, and release activities
Factors	The factors affecting vulnerability fix time for the case of generic solutions (a generic solution is a way to address all the instances of a specific vulnerability type, e.g., XML code injection) and also the case of specific solutions
Complementary information	This includes generic comments, comments related to vulnerability fix time, pain points (issues), and improvement recommendations

they requested.) and initially used the questions that we prepared in the interview protocol. We let the interviewee lead and we probe issues in depth, when needed, using questions such as "Could you provide an example?" Nevertheless, we realized shortly that it was difficult for the interviewee to provide us with the maximum information related to our research goal. We adapted the questions of the interview protocol to the ones provided in Table 2. The adaptation is discussed in Sect. 7. Also, some interviewees provided us with tool demonstrations since they were aware about the interview protocol.

Next, we transcribed the interviews (Step 4 of Fig. 2) using the tool F4.[2]

Data Analysis. Subsequently, we proceeded to coding the interviews (Step 5 of Fig. 2); that is, identifying from each transcript the codes, i.e., themes and abstract concepts (e.g., code the text "I have been fixing these issues for 5 years" as "experience in fixing vulnerabilities").[3] In this step, two of the authors coded successively 3 sample interviews using the Atlas.ti tool,[4] discussed the code patterns they found, and agreed on a coding schema for the study, which is shown in Table 3. (The coding schema allows grouping the codes extracted from the interview in classes that together answer the main research question [14]). Both researchers coded each of the 11 interviews using the selected coding schema and

[2] https://www.audiotranskription.de/english/f4.htm.

[3] A code is a short phrase that assigns a summative, essence-capturing, and/or evocative attribute for a portion of text [15].

[4] http://atlasti.com/.

merged their reports in summary reports.[5] Then, we sent to each interviewee the summary report of their interview and asked them to verify the report and answer some clarification questions. The validation helps in obtaining objective results but was also important to allow the interviewee to remove any information they did not want to be processed. We ensured the anonymity of the interviewees to promote free and open discussions.

Afterwards, we merged the codes of the verified coded reports (Step 6 of Fig. 2) considering the semantic similarities between the codes extracted from different transcripts. In addition, we computed the frequency of each code in the class "Factors," that is, the number of interviewees mentioning the code. We were reluctant to generalize the factors because we did not want to bias the results with the researcher' opinions.

Thereafter, we presented the findings to the experts and the interviewees in a public meeting (Step 7 of Fig. 2). We used the frequencies of the codes as indicators (but not assertive) of the factors' importance.[6]

5 Study Results

This section presents the results of the interviews. It discusses the vulnerability-fixing process identified at SAP and the factors that impact vulnerability-fixing time along with their classification.

5.1 Vulnerability-Fixing Process

Each interviewee described a set of activities that they perform to fix vulnerabilities. The activities described by the different interviewees were sometimes incoherent–Sect. 7 discusses the challenges. However, in many ways the interviewees follow a high-level vulnerability-fixing process, which is depicted by Fig. 3. The process starts when a security expert gets notified about vulnerabilities, e.g., from customers and researchers, or when a developer identifies a vulnerability using, e.g., a code-analysis tool. The vulnerability is initially pre-analyzed, e.g., to assess its exploitability, its risk and the availability of knowledge and information to fix it. This results in three cases.

Case 1. If the type of the vulnerability is known and documented, the developer proceeds to analyze the code related to the vulnerability, to design and implement a solution, and then to test it—using the technique that was used to identify it.

Case 2. If the vulnerability type is known and documented by the central security team but the development teams (e.g., cloud applications, mobile applications, etc.) did not encounter such vulnerability before, this team collaborates

[5] The merge involves also discussing coding mismatches related to the difference in understanding the interviewee.

[6] Recall that data extracted from interviews could not be used to derive statistical assurance of the conclusions since the collected information is descriptive.

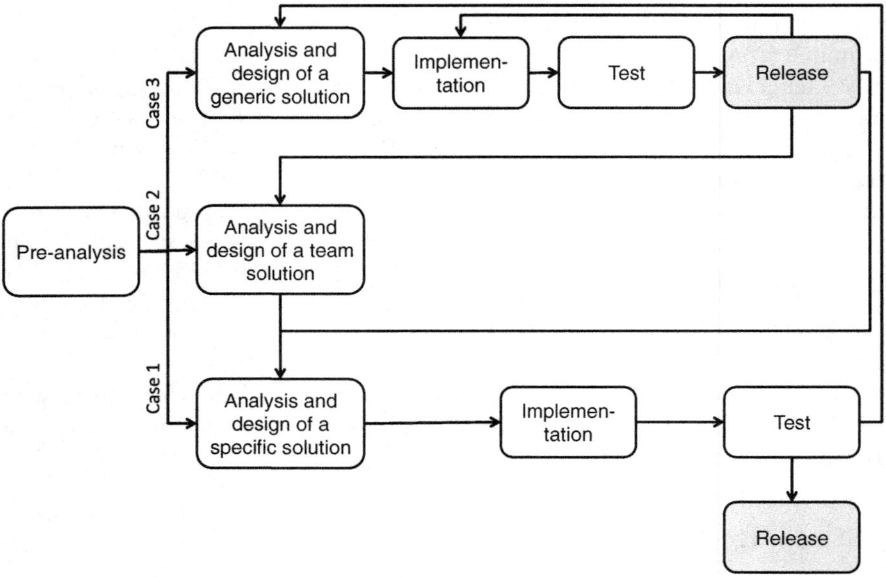

Fig. 3. High-level vulnerability-fixing process (simplified).

with the central security team to analyze the identified vulnerability and design a solution that applies to the product area as such.

Case 3. If the vulnerability type is not known before, the central security team collaborates with the experts and developers from the different areas to develop a generic solution for the vulnerability. A *generic solution* considers the different product areas, the different used technologies, and the different applicable programming languages. In addition, the security experts collaborate with the framework experts to implement libraries that the developers can use to avoid the vulnerability in the future, e.g., by using data-validation methods to avoid SQL injections; and also develop guidelines that the developers can use to address vulnerabilities of such type.

5.2 Factors that Impact the Vulnerability-Fix Time

We identified 65 factors that impact the vulnerability-fix time, each was reported by at least one interviewee. We categorized the factors based on common themes, and those that did not belong to these themes into the category "other." These categories may be generalized or consolidated further, however, we expect that such activity may be influenced by the researchers' opinions. Table 4 lists the categories, along with the number of factors that belong to each category and the number of interviewees who mentioned one or many of these factors. Table 6 of Appendix A provides the complete list of the factors that we identified.

Table 4. Classification of the factors that impact the vulnerability fix time.

Factor Categories	Number of factors	Frequency
Vulnerabilities characteristics	6	9
Software structure	19	10
Technology diversification	3	5
Communication and collaboration	7	8
Availability and quality of information and documentation	9	9
Experience and knowledge	12	11
Code analysis tool	4	4
Other	4	4

We next discuss the categories in details. To preserve the anonymity of the interviewees we identify them using identifiers, without descriptive information.

Vulnerabilities Characteristics. This category includes 6 factors that are exclusively related to the type of vulnerability. These factors are reported in 9 (about 82 % of the) interviews. For example, P01 believes that vulnerability types do not indicate the fixing time but later in the interview they find that "code injection" vulnerabilities are difficult to fix. Thus, vulnerability characteristics are commonly considered when discussing vulnerability fix time (e.g.,in [7]) and our results enforce the position.

Software Structure. This category includes 19 factors that are related to the code associated with the given vulnerability. These factors are reported in 10 (about 91 % of the) interviews. For example, P02 finds that "if the function module is the same in all these 12 or 20 releases then [..] I just have to do one correction." Generally, the interviewees find that software structure impacts the easiness to address vulnerabilities. This can also be observed from Table 1, where SQL injection vulnerability takes the most time to fix while it is conceptually easy to fix. The reason is that the complexity of the code that generates the query makes it difficult to identify the cause of the vulnerability and to fix the issue while not breaking the functional tests.

Technology Diversification. This category includes 3 factors that are related to the technologies and libraries supported by the components associated with the given vulnerability. These factors are reported in 5 (about 45 % of the) interviews. For example, P03 had to develop several solutions for a vulnerability related to random-number generation, "one for Java, one for ABAP, one for C, C++." Thus, since SAP products support different browsers, languages, and use diverse libraries, such as XML parsers, vulnerability fixes need to support these technologies as well, increasing the overall time and effort required.

Communication and Collaboration. This category includes 7 factors that are related to the communication and collaboration in fixing vulnerabilities.

These factors are reported in 8 (about 73 % of the) interviews. For example, P04 finds that "even for one team there are multiple locations and multiple responsibilities and the team at [..] is not aware" and finds that "the local teams are very smooth." Developing software at SAP involves teams located in different locations. Thus, the smoothness of the communication and collaboration between the stakeholders impacts the time spent to fix the vulnerabilities.

Availability and Quality of Information and Documentation. This category includes 9 factors that are related to the availability and the quality of information (e.g., contact information of the security experts, and uses of components) and guidelines to address vulnerabilities. These factors are reported in 9 (in 82 % of the) interviews. For example, P05 claims that a lot of time is spent on collecting information. They state: "it was taking a long time because we need to find out what are the different frameworks, what are the different applications [..] and once we had this information, we were able to use it."

Experience and Knowledge. This category includes 12 factors that are related to the experience and knowledge about the given vulnerability and the related code. It is reported in all the interviews. For example, P01 finds that "colleagues who have some background in security are able to fix them faster than the developer who is fixing the security issues for maybe the first or second time." This category of factors is often ignored in existing studies because those studies rely on data archives which do not include such human factors.

Code Analysis Tool. This category includes 4 factors that are related to the use of code analysis tools. This category of factors is reported in 4 (36 % of the) interviews. P13 for example says "what you are doing is to find out if the tool was not able to find out where is the source, where is the data coming from so find out if there is external input in this where clause or in parts of external inputs." This implies that developers spend less time to fix a given vulnerability if the tool is accurate in providing the information related to the causes of the vulnerabilities. This category of factors is often neglected because it is not common that organizations use several code-analysis tools, so their impact on the fixing time cannot be compared.

Other. This category includes 4 factors that we were not able to classify in the above 7 categories and which do not belong to a common theme. These factors are reported in 4 (36 % of the) interviews.

We note that the number of factors that belong to a given category does not assert the importance of this category but could be informative. The reason is that the descriptive nature of the interviews does not support such assertion.

5.3 Discussion

When developing the factor categories, we did not differentiate the factors based on whether the given vulnerability is generic or specific. It is true that generic vulnerabilities, i.e., vulnerabilities that are identified for the first time but apply to several products, are often addressed with generic solutions designed by the central security team. Vulnerability analysis, solution design, documentations and the provisioning of guidelines are time consuming and addressing such vulnerabilities may take years. In contrast, specific vulnerabilities, i.e., vulnerabilities that are known and apply to one product, are mostly addressed by developers in collaboration with the development team's security expert. The reason for not considering this is that most of the factors apply to both generic and specific vulnerabilities—with some exceptions.

As a second note, the positive or negative influence of each of the identified 65 factors that impact the vulnerability-fix time were not identified in the interviews and may depend, e.g., on the product. In fact, the developers gave contradicting perceptions on the influence of some factors, such as the factor risk exposure level. Also, the number of interviewees who mentioned each of the factors does not indicate the influence of the factor. We will evaluate the concrete influence of each factor in the next stage of the project.

6 Impacts and Limitations of the Study

This section discusses the impacts of the study w.r.t the state of the art and the study validity; that is, to what extent the results are objective and sound [16].

6.1 Impacts of the Study

Previous work used the attributes of collected data as factors for analyzing facts about vulnerabilities. The results of this work show that in practice there are numerous factors that impact the vulnerability fix time that should be considered. A comprehensive model for predicting the vulnerability-fix time should consider these potential factors and not only rely on the ones that are readily available in mass-collected data.

The identified vulnerabilities fixing process (Fig. 3) and the 8 factor categories may be "expected," especially from a big company like SAP. The study confirms this expectation; it makes these expectations facts that could be used for further work. The 8 factor categories indicate areas for improvement to reduce the vulnerability fix time. For example, the software structure (e.g., the dynamic construction of code and data, the cross-stack interdependency) and the use of external technologies, could be partly monitored and controlled to predict and/or reduce the vulnerability fix time, and thus, the cost of fixing vulnerabilities. In addition, experience and knowledge can be addressed with specific trainings; effectiveness of code analysis tools could be improved by enhancing the vulnerability checks, in particular their precision, and by enhancing the tool' functionalities; issues related to the availability and quality of information and documentation can also be improved by using easily accessible documentation.

In addition, we found in this study that the developers cannot identify the vulnerability fixing factors by themselves easily, except e.g., vulnerability type, which we discuss in Sect. 7. Though, they recognize the factors if extracted from their interviews. Thus, the results are not "explicit" knowledge to developers. This paper makes the information common knowledge.

6.2 Limitations of the Study

We discuss now the limitations of the study according to the commonly used validity aspects [16].

Construct Validity. We took several measures to ensure a valid relation between the performed study and the goal of the study. First, we performed three interview tests to test the interview questions. In addition, we adjusted the interview questions after the initial interviews to be more efficient in getting information. Second, we collected the information from twelve interviewees, who are located in different cities/countries and have different roles. Third, we avoided to use the researcher' opinions in working with the data, e.g., we avoided generalizing the factors extracted from the interviews.

The study has two limitations w.r.t. construct validity. First, we provided the participants with the main interview questions so they could prepare for their interviews. Thus, some participants may have prepared replies that may influence the study results. (We believe that the advantages of the measure are higher than the risk it created to the study.) Second, we used only one method to collect the data, that is, interview domain experts. Other methods that could be used to cross-validate the results include to use of data collected from the development process. Nevertheless, we observed that the attributes of collected data are among the identified factors.

Internal Validity. We took two measures to ensure a causal relationship between the study and the results of the analysis. First, we tell the interviewees at the opening of the interviews that the goal is to identify the factors that impact the vulnerability fix time.[7] Second, we did not offer any compensation to the participants, which, if done, may affect the results.

The study has two limitations w.r.t. internal validity. First, we were able to only interview two developers who currently fix vulnerabilities. The other participants have other roles in fixing vulnerabilities but most of them have developed vulnerability fixes previously.[8] Second, we did not take measures to prevent the participants from imitating each others in the response, though we believe that the participants did not talk to each other about the interviews.

Conclusion Validity. We took several measures to ensure the ability to draw correct conclusions about the relationship between the study and the results of the analysis. First, we sent each interviewee a short report about the data we extracted from the interview we conducted with them to ensure that we have a common understanding; that is, we performed member checking [17]. Second, two researchers coded each interview and we merged the collected data [14]. The measure should reduce the subjectivity of the results.

External Validity. This validity concerns the conditions to the generalization of the results. The study was conducted in the same company and was related to one vulnerability fixing process. However, given the diversity of the products

[7] This mitigates the threat ambiguity of the direction of the causality relationship.

[8] The limitation is related to the selection of participants.

Table 5. Summary of the interview protocol.

Opening	Thank the interviewee for accepting to participate and request for permission to record the interview
Questions	1. What are the steps that, in general, you follow to fix security vulnerabilities?
	2. What is the distribution of your time among planning, meeting, and implementing vulnerability fixes for the last week? Did you have any other major activity in fixing vulnerabilities?
	3. What are the major characteristics of complex vulnerabilities that you fixed last week? Are there other challenges that make simple vulnerabilities time consuming?
	4. What are the factors that quantify these characteristics?
	5. How can we improve and ease the fixing process?
Closing	Thank the interviewee for sharing his/her experience and knowledge and inform him/her about the next steps for the study

(and their respective domains, e.g., mobile, cloud.) being developed at SAP and the diversity of the developers' cultures and countries of residences. We believe that the results of the study could be generalized, especially within SAP, without high risk.

7 Lessons Learned

This section describes the lessons we learned from the case study with respect to formulating interview questions, conducting interviews, and analyzing software development processes of a big software organization.

Interview Protocol and Questions. We produced an interview protocol, summarized in Table 5. The first interviews showed that the interviewees had, in general, difficulties in answering questions 3 and 4, initially developed as main questions to achieve the study goal. This was due to the "what" type of asked questions (i.e., "what are the factors?") that require enumeration of elements while we should be limited to "how" and "why" type of questions.[9] To enhance the communication we transformed the questions accordingly (see question 3 and 4 of Table 2) and encouraged the participants to tell us *their own stories* [13] about complicated and easy vulnerabilities they addressed and the challenges they faced. Thus, with indirect questions we derive the factors that impact the fix development time from the reasons that make fixing a given vulnerability complicated, the challenges that the interviewees faced herein, and their improvement recommendations (question 5 in both tables).

Interview Conduct. We learned that some participants in interviews conducted in organizations mainly participate to deliver a message or impact the

[9] "What" type of questions are easy to answer when the purpose is to describe a concept/object.

study results. We learned to encourage the participants to talk freely for some time to build up a trust relationship, since the information they provide when getting into a flow may be important. The risk herein was the limited interview time and thereby the challenge to change the discussion smoothly such that they answer the interview questions and do not use the interview to talk about subjects not related to the research goal.

Analysis of the Software Development Processes. The intuitive approach to identify the vulnerability fix time is to define the fixing process and its phases. Following this approach, our initial attempt was to identify the different roles of the participants in the process, the activities performed in the phases and the information created and consumed by each role. We derived inconsistent models with a big variety of cases. This is due to the different perspectives of the interviewees; they work with different programming languages and tooling, have different (and multiple) roles, have expertise in different product areas, are members of different teams, and use their own internal social network to simplify the work. In addition, there were process improvements and each of the interviewees reported about the process versions that they worked with. The open structure of the interviews and the participation of long time employees made it difficult to identify a consistent fixing process. Therefore, we focused on the identification of the factors independent of the phases.

8 Conclusions

This paper reports about a case study we conducted at SAP SE to identify the factors that impact the vulnerability fix time. The study found that, for big development organizations such as SAP, there are numerous factors that impact the vulnerability fix time. We identified 65 factors, which we grouped into 8 categories: vulnerabilities characteristics, software structure, diversity of the used technology, communication and collaboration smoothness, availability and quality of the information and documentation, expertise and knowledge of developers, efficiency of the static analysis tool, and other.

The study was conducted at one organization, SAP SE, which may limit the generalization of the results. We believe that the limitation is weak because SAP development groups simulate different organizations, each has independence and specificities such as location, used programming language, and products area.

The common approach in investigating vulnerability fix time (and other facts related to vulnerabilities) is to apply machine-learning techniques on historical data related to open-source software. This work shows the limitation of this approach since it is constrained by a limited number of data attributes while the factors that potentially influence these facts are numerous.

The results of this work are being used to improve the vulnerability fixing process and to develop a model for predicting the cost of fixing vulnerabilities.

Acknowledgments. This work was supported by SAP SE, the BMBF within EC SPRIDE, and a Fraunhofer Attract grant. The authors thank the participants in the study.

Appendix A: Factors that Impact the Vulnerability Fix Time

Table 6. Factors that impact the vulnerability fix time. (The column "Freq." indicates the number of interviews–out of 11–where the given factor is mentioned.)

Factors	Freq.
Vulnerabilities characteristics	
Vulnerability type	5
Risk exposure level (CVSS)	3
Priority of the vulnerability fix	2
Availability of a generic solution	2
Simplicity of the solution type, e.g., exception catching	2
Source of the vulnerability, e.g., input and remote function call	1
Software structure	
Number of affected applications and framework releases	7
Need of the solution to change the framework, e.g., as an API	5
Number of software layers that should be analyzed	5
Number of the code changes and the distribution of their locations	4
Dependency on external software and/or technology	3
Code related to the vulnerability includes generated code	3
Complexity of the related code and reliance on configurations	2
Level of code smells (e.g., high, low)	2
The issue is related to the dynamic generation of queries	2
Dynamic generation of the input data that cause the vulnerability	2
The code related to the vulnerability involves dynamic function calls	2
Existence of the issue in shipped products	1
Need for customer intervention to apply the fix	1
Need to divide the fix in several support packages	1
Similarity of code related to the issue in the releases of the software	1
Code related to the vulnerability is in new development	1
Components related to the issue	1
Length of the data field related to the issue (for SQL injection)	1
The vulnerability is in an unused code	1
Technology diversification	
Diversity and complexity of supported technologies, e.g., browsers	5
Number of supported programming languages	2
Number of customers that have specific functionalities that should be preserved but may be affected by the vulnerability fix	1
Communication and collaboration	
Number of people involved in the fixing process	5
Dispersion of the locations of the communicating parties	4
Availability of the responsible developer	1
The software is developed by an external developer	1

Continued on next page

Table 6. *(Continued)*

Factors	Freq.
Existence of concurrent changes to code of dependent components	1
Easiness to convince the developers to apply vulnerability fixes	1
The code related to the issue is owned by other teams	1
Availability and quality of information and documentation	
Availability of information about the developer responsible for the fix	5
Availability and usability of the guidelines explaining the vulnerability	5
Availability of information about contact persons for support	2
Availability of information about the framework changes that cause the vulnerability in the applications	1
Quality of the documentation for technology types (e.g., Hana)	1
Availability of information about the work progress on the issue	1
Availability of information about contacts in other development areas	1
Difficulty to get the information about affected software	1
Difficulty to identify the use of affected code by other components	1
Experience and knowledge	
Developer expertise with fixing security vulnerabilities	7
Developer knowledge and experience about the code	7
Experience with SAP software, processes and tools	2
Developer knowledge about the vulnerability	2
Security expert knowledge of the language and affected technologies	1
Experience with the application technology	1
Knowledge about the software architecture	1
Knowledge about the development setup of affected software	1
Knowledge about the development processes for NetWeaver products	1
Availability of security expert in the area responsible for the issue	1
The coding and testing strategy of developers	1
Experience in the impacts of changes on the related products	1
Code analysis tool	
The issue is among a set of issues of the same vulnerability type and in the same code component	2
Lack of motivation due to high rate of false positives	2
The issue is for projects addressing results identified by external tools	1
The failure of the security checks to identify true positives	1
Other	
Commitment of the stakeholders	1
Source of the notification about the vulnerability, e.g., customer	1
Number of externally reported incidents	1
The issue is a false positive that does not require fixing	1

References

1. Katzeff, P.: Hacking epidemic spurs security software stocks, February 2015. Investor's business daily of 02/19/2015. http://news.investors.com/investing-mutual-funds/021915-740082-revenues-are-up-for-security-software-firms.htm
2. McGraw, G.: Software Security: Building Security In. Addison-Wesley Software Security Series. Pearson Education Inc., Boston (2006)
3. Bachmann, R., Brucker, A.D.: Developing secure software: a holistic approach to security testing. Datenschutz und Datensicherheit (DuD) **38**(4), 257–261 (2014)
4. Howard, M., Lipner, S.: The Security Development Lifecycle: SDL: A Process for Developing Demonstrably More Secure Software. Microsoft Press, CA (2006)
5. Hamill, M., Goseva-Popstojanova, K.: Software faults fixing effort: Analysis and prediction. Technical report 20150001332, NASA Goddard Space Flight Center, Greenbelt, MD United States, January 2014
6. Hewett, R., Kijsanayothin, P.: On modeling software defect repair time. Empirical Softw. Eng. **14**(2), 165–186 (2009)
7. Cornell, D.: Remediation statistics: what does fixing application vulnerabilities cost? In: Proceedings of the RSAConference, San Fransisco, CA, USA, February 2012
8. Khoshgoftaar, T.M., Allen, E.B., Kalaichelvan, K.S., Goel, N.: Early quality prediction: a case study in telecommunications. IEEE Softw. **13**(1), 65–71 (1996)
9. Shin, Y., Williams, L.: Is complexity really the enemy of software security? In: Proceedings of the 4th ACM Workshop on Quality of Protection. QoP 2008, Alexandria, VA, USA, pp. 47–50, October 2008
10. Chowdhury, I., Zulkernine, M.: Using complexity, coupling, and cohesion metrics as early indicators of vulnerabilities. J. Syst. Archit. **57**(3), 294–313 (2011). Special Issue on Security and Dependability Assurance of Software Architectures
11. Brucker, A.D., Sodan, U.: Deploying static application security testing on a large scale. In: GI Sicherheit 2014. Lecture Notes in Informatics, vol. 228, pp. 91–101, March 2014
12. Yin, R.K.: Case Study Research: Design and Methods. Sage Publications, Beverly Hills (1984)
13. Jacob, S.A., Furgerson, S.P.: Writing interview protocols and conducting interviews: tips for students new to the field of qualitative research. Qual. Rep. **17**(42), Article no. 6, 1–10, October 2012
14. Brikci, N., Green, J.: A guide to using qualitative research methodology, February 2007. http://www.alnap.org/resource/13024
15. Saldana, J.: The Coding Manual for Qualitative Researchers. SAGE Publications Ltd, London (2009)
16. Wohlin, C., Runeson, P., Host, M., Ohlsson, M., Regnell, B., Wesslen, A.: Experimentation in Software Engineering. Springer, Berlin (2012)
17. Seaman, C.: Qualitative methods in empirical studies of software engineering. IEEE Trans. Softw. Eng. **25**(4), 557–572 (1999)

Software Security Maturity in Public Organisations

Martin Gilje Jaatun[1(✉)], Daniela S. Cruzes[1], Karin Bernsmed[1], Inger Anne Tøndel[1], and Lillian Røstad[2]

[1] Department of Software Engineering, Safety and Security, SINTEF ICT, 7465 Trondheim, Norway
{martin.g.jaatun,danielac,karin.bernsmed,inger.a.tondel}@sintef.no
http://infosec.sintef.no

[2] Norwegian Agency for Public Management and eGovernment (Difi), Oslo, Norway
lillian.rostad@difi.no
http://www.difi.no/

Abstract. Software security is about building software that will be secure even when it is attacked. This paper presents results from a survey evaluating software security practices in software development lifecycles in 20 public organisations in Norway using the practices and activities of the Building Security In Maturity Model (BSIMM). The findings suggest that public organisations in Norway excel at Compliance and Policy activities when developing their own code, but that there is a large potential for improvement with respect to Metrics, Penetration testing, and Training of developers in secure software development.

Keywords: Software security · Secure software engineering · Maturity · BSIMM

1 Introduction

Society is increasingly dependent on information and communication technology (ICT). Traditionally, ICT security has primarily been about implementing security mechanisms on the system or network level. In recent times, it has become clear that it is equally important to ensure that all mechanisms of the software is secure including the code itself, i.e., develop software from scratch so that it is secure against attack [1]. This is what we call software security.

Numerous guidelines and best-practices exist, which outline processes and methodologies that can be adopted to achieve better software security. However, in practice these are only used to a limited extent and the problem of insecure software is bigger than ever [2]. We argue that organisations learn best by comparing themselves to other organisations that tackle similar challenges, rather than comparing themselves to abstract theoretical models of ideal practices for software security.

The Building Security In Maturity Model (BSIMM) [3] is a study of real-world software security initiatives that is organised so that an organisation can

J. Lopez and C.J. Mitchell (Eds.): ISC 2015, LNCS 9290, pp. 120–138, 2015.
DOI: 10.1007/978-3-319-23318-5_7

use it to determine where they stand with their software security initiative. BSIMM provides an overview over the security of software by mapping how it was built, what kind of activities that were carried out while it was built and by measuring a number of artefacts that were created when it was developed. BSIMM can also be used to measure how an organisation's software security efforts evolve over time.

The BSIMM study is dominated by large American companies, and the average number of developers in the studied organisations exceeds 4000. We were therefore curious to see how applicable the BSIMM activities were to smaller, non-commercial organisations in Europe, and if we could identify any discrepancies and obvious areas for improvement of the current practices in these organisations.

This paper reports on a software security maturity study in 20 Norwegian public organisations, based on the BSIMM framework. The organisations that we have studied are all part of or owned by the government or municipalities in Norway.

Both on a European and a national level there is a push towards a more efficient public sector through use of eGovernment services. eGovernment consists mainly of the digital interactions between a citizen and their government (C2G), between governments and government agencies (G2G) and between government and citizens (G2C) and between government and businesses/commerce (G2B). The move towards eGovernment, also means a move towards a more digitalized society; a society relying heavily on ICT and software-based services to function. Thus, security becomes a major concern.

The Cyber Security Strategy for Norway [4] describes current and future security challenges and points to where efforts should be focused in order to meet those challenges. The strategy, and it's accompanying action plan, suggests that a center of competence for information security in the public sector is needed. This study is part of the work done to collect information and knowledge, needed to be able to focus the work of this competence center. The study provides valuable insight, and acts as a benchmark study. The intention is to repeat the study, at intervals yet to be determined, to assess the effect of efforts to improve software security in the public sector.

The remainder of this paper is organised as follows: In Sect. 2 we present the theoretical background for the study, and in Sect. 3 we elaborate on the method employed. The results are described in Sect. 4, and discussed in Sect. 5. Section 6 concludes the paper and outlines further work. The questionnaire that was used in the study is presented in Appendix A.

2 Background

There are two well-known maturity models for software security; BSIMM [3] and OpenSAMM [5]. Both have a common origin, and have many similarities.

BSIMM and OpenSAMM are organised in a similar fashion, and contain many similar topics and activities. Both divide the software security practices into three main areas with twelve practices in total, and place the practices into three maturity levels. However, their content and fundamental idea differ. BSIMM is based on real-world data and only includes activities that are done by real companies. As such it does not aim to tell what activities should be performed, but rather what activities are actually performed by companies today. OpenSAMM is not based on real-world data directly, but rather on experience on what activities will improve software security, and that thus should be performed. But where BSIMM is based on practices of relatively large organisations that are in the forefront regarding software security, and may thus be most relevant for that type of organisations, the description of OpenSAMM clearly states that it was designed with flexibility in mind, and should thus be useful for any type of organisation, big or small.

We chose to base our study on BSIMM rather than OpenSAMM for two reasons: we were more familiar with BSIMM; and BSIMM is a more descriptive methodology that basically is designed to measure, while OpenSAMM has a stronger prescriptive focus, i.e., to define "the right way to do it".

2.1 OpenSAMM

The Software Assurance Maturity Model (SAMM or OpenSAMM) is an open software security framework divided into four business functions: Governance, Construction, Verification and Deployment. Each business function is composed of three security practices, as shown below:

Governance: Strategy & Metrics; Policy & Compliance; Education & Guidance.
Construction: Threat Assessment; Security Requirements; Secure Architecture.
Verification: Design Review; Code Review; Security Testing.
Deployment: Vulnerability Management; Environment Hardening; Operational Enablement.

Each practice is assessed at a maturity level from 1 to 3 (plus 0 for "no maturity"), and for each maturity level there is an objective and two activities that have to be fulfilled to achieve that level.

2.2 BSIMM

The Building Security In Maturity Model (BSIMM) measures which software security activities are included in an organisation's overall Secure Software Development Lifecycle (SSDL). A central concept in BSIMM is the Software Security Group (SSG), which is the person (or persons) responsible for software security in an organisation. The SSG can be as small as a single person, it need not be a formal role, and need not be a full-time position. In addition, there is the concept of "the satellite"; a more or less well-defined group of developers who are not part of the SSG, but still have a special interest in and knowledge of software security, and thus can operate as the extended arm of the SSG in many contexts.

The purpose of BSIMM is to quantify the software security activities performed in real software development projects in real organisations. As these projects and organisations use different methodologies and different terminology, it is necessary to use a framework that allows describing all initiatives in a unified manner. The BSIMM framework consists of twelve practices organised into four domains; Governance, Intelligence, SSDL Touchpoints and Deployment (see Table 1). Each practice has a number of activities on three levels, with level 1 being the lowest maturity and level 3 is the highest. For example, for practice Strategy and Metrics, SM1.4 is an activity on level 1, SM 2.5 is an activity on level 2, and SM 3.2 is an activity on level 3.

Table 1. The BSIMM software security framework

Governance	Intelligence	SSDL touchpoints	Deployment
Strategy and metrics	Attack models	Architecture analysis	Penetration testing
Compliance and policy	Security features and design	Code review	Software environment
Training	Standards and requirements	Security testing	Configuration management and vulnerability management

The starting point for the first BSIMM survey in 2008 [3] was to study the software security activities performed by nine selected companies. The nine companies were presumably far ahead in software security, and the activities that were observed here formed the basis of the framework in Table 1. Representatives from Cigital[1] physically visited each company, and these first surveys were done by Gary McGraw and Sammy Migues personally, using a whole day for each company.

3 Method

In this work we have performed a survey using a questionnaire with individual follow-up interviews [6]. The questionnaire (see Appendix A) is based on the BSIMM software security framework as documented in the BSIMM V report [3]. The main function of BSIMM is to serve as a yardstick to determine where an organisation stands compared with other organisations [3]. The questionnaire tells us what activities the organisation has in place, and based on how well they cover the various practices, we can determine the maturity level of each

[1] http://www.cigital.com.

organisation. BSIMM contends that a radar chart according to the high watermark method (based on three levels per practice) is sufficient to give a rough, overall picture of maturity. We have chosen to also develop two complementary maturity measures that can provide a more balanced view of maturity. The three maturity measures we use are thus:

Conservative maturity (Scale 0–3): Here an organisation is approved at a maturity level only if all the activities in the level are met ("Yes"), provided all the activities on the lower level are also fulfilled. If the organisation performs some (but not all) activities at a level this is indicated with a "+", i.e., if you have 3 of 5 activities on the first level, the result is 0+; if all the activities at level 1 are fulfilled, and 2 out of 4 activities at level 2 are fulfilled, the result is 1+, etc. In connection with calculating the average value, a "+" is counted as 0.5. As will be seen in Sect. 4, since few organisations in our study do all the activities at level 1, many end up in the category 0+.

Weighted maturity (Scale 0–6): This value gives a greater emphasis for activities at a high level, even if the lower level activities are not fully implemented. The value is calculated using the following formula:

$$\sum_{i=1}^{3} \frac{\text{Observed activities at level } i}{\text{Total number of activities at level } i} \times i$$

High Watermark Maturity (Scale 0–3): This value is calculated in the same manner as in BSIMM [3]; if the organisation has at least one activity at level 3, it gets the maturity level 3. The high watermark maturity level therefore only says something about what is the level of the highest rated activity they perform. In contrast to the conservative maturity level, it is therefore easier to reach a level 2 or 3 high watermark maturity level.

In our study it will be of most interest to compare the two first maturity measures (conservative and weighted) from a given organisation with the average values of all the studied organisations to see how they compare, as we have done in the radar diagrams in Fig. 1.

We distributed the questionnaire in Appendix A in January 2015 via email to 32 Norwegian public organisations which we had reason to believe had ongoing software development activities. 20 of these organisations returned fully filled-out questionnaires. For seven of the responses, the questionnaire had been filled out in cooperation by representatives involved in software development and in general IT security work. In the other cases, the response was made either by people working on information security or on IT in general (six responses), by people working on software development (five responses), or the main responsibility of the respondent was unclear based on the job title (two responses). In most cases, at least one of the respondents had a managing role in the organisation, e.g., information security manager, IT manager, group leader or architect. In order to verify the answers and to clarify possible misunderstandings, we organised follow-up interviews with all the involved organisations during which their answers were scrutinised and corrected whenever needed. The results were then

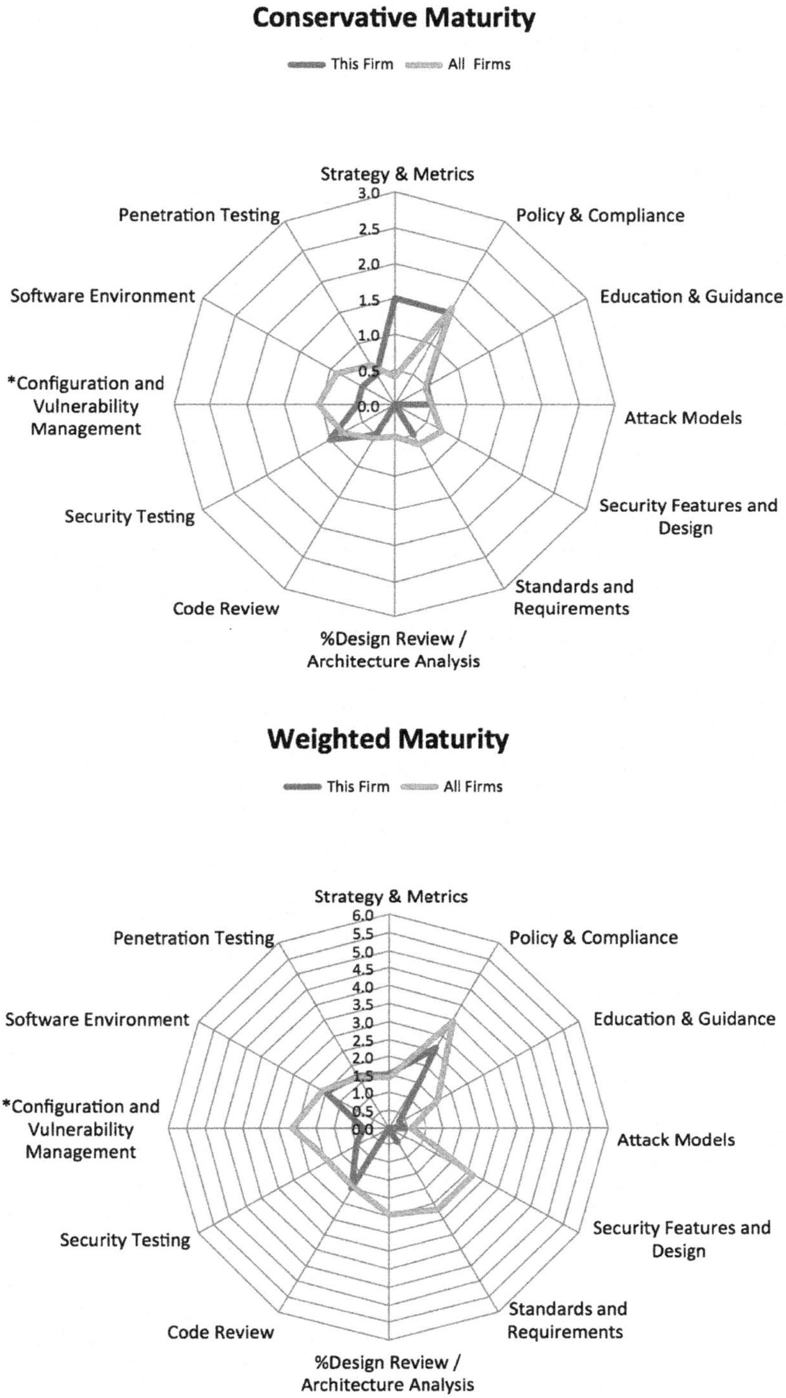

Fig. 1. Comparing an imaginary organisation with average of all organisations

compiled and the conservative, weighted and high watermark maturity measures were computed and analysed.

4 Results

In this section we present a selected set of the results from the study. For the full results, the reader is referred to the report [7].

The organisations with the lowest maturity level declared that it performed 9 activities of 112, while the organisation with the highest level of maturity performed 87 activities. Based on the boxplot chart in Fig. 2, we see that most of the organisations come halfway up the scale; they perform on average 39 % of the activities.

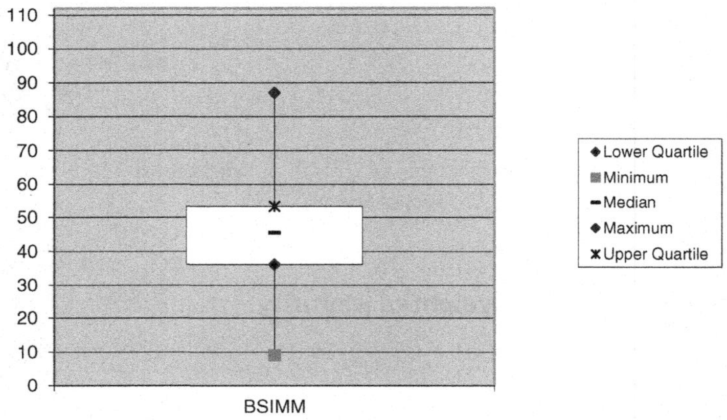

Fig. 2. The distribution of the total number of activities for all the organisations

The total number of activities for each of the organisations is shown in Fig. 3. This figure shows the "raw score" for each organisation that participated in the study.

4.1 Practices with a High Degree of Maturity

As can be seen in Fig. 4 and Table 2, we found the highest degree of maturity among the surveyed organisations ("Observed Difi") within the practice Compliance and Policy ("Guidelines and compliance with laws and regulations"); more than 80 % of the respondents answered yes to most of the activities in this area. The result is not surprising, since it concerns public organisations in Norway, which usually are accustomed to adhere to standards and government requirements. It is apparent that there is better adherence to these practices among Norwegian public organisations than the average from the official BSIMM

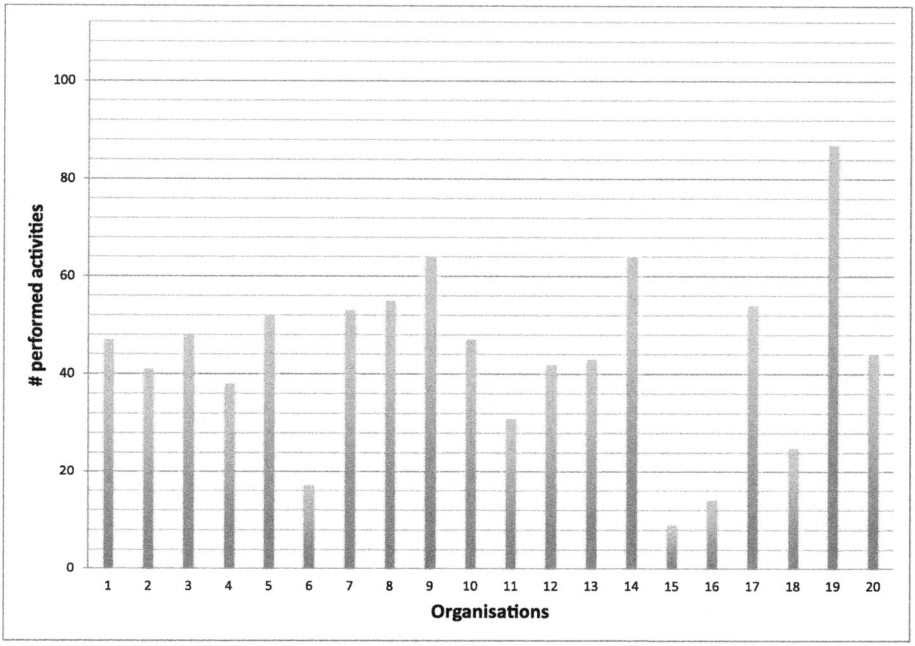

Fig. 3. The total number of activities ("raw score") for each of the organisations

Table 2. High-level results

Areas	Observed Difi	Observed BSIMM
Strategy and metrics	33 %	45 %
Compliance and policy	66 %	43 %
Training	27 %	24 %
Attack models	25 %	29 %
Security features and design	45 %	41 %
Standards and requirements	48 %	40 %
Architecture analysis	39 %	35 %
Code review	35 %	28 %
Security testing	32 %	30 %
Configuration management and vulnerability management	46 %	43 %
Software environment	41 %	42 %
Penetration testing	36 %	49 %

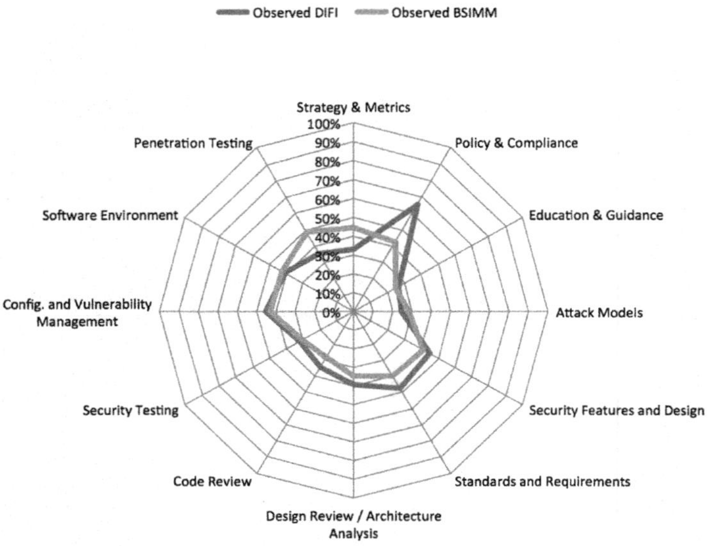

Fig. 4. High-level results illustrated

study [3] ("Observed BSIMM"). It is also important to see the positive maturity values in this area in the context of how businesses are organised. Many public organisations have their own lawyers who handle compliance, and have a good overview of the requirements of laws and regulations. However, it is not thus given that this expertise is applicable in software development projects. In many cases there might be quite a long distance (physically and organisationally) from the internal expertise related to laws and regulations to the developers or the hired consultants that are central to software development.

Three other practices that received high maturity are Construction and intelligence, Security features and design, and Standards and Requirements). 80 % of the organisations say they do SFD1.2 ("Security is a regular part of our organisation's software architecture discussion") and 80 % say that they do SR 2.3 ("We use a limited number of standard technology stacks"). Regarding the latter, in most cases this meant that the organisation uses only Microsoft products (Microsoft Active Directory, Microsoft Internet Information Services, etc.) in their software development and production processes.

In the practice Configuration Management and Vulnerability Management, 85 % of the organisations said that they satisfy CMVM1.1 ("The software security group has procedures for incident response, in collaboration with the incident response team (if it exists)"). They also claim that they do CMVM 2.2 ("We track software DEFECTS found during operations until they are closed"), but it seems as though many people equate this with their internal bug tracking system that often does not take particular account of security flaws. In such cases it is necessary that the security flaws are prioritized high enough that they *must* be handled; if not, there is no guarantee that they actually closed within a reasonable time. It is also unclear to what extent there are procedures for cooperation during security incidents, but respondents says that developers will be able to get involved when needed.

Finally, 90 % of the organisations said that they meet SE1.2 ("We use accepted good practice mechanisms for host/network security"), but this is not so surprising since this is strictly not about software security. Security seems to have a relatively large focus amongst the people involved with the network, operations and infrastructure. There are indications, however, that there is a distinction between the developers and the "security people" in the organisations. One respondent stated that security can be perceived as an obstacle among their developers, since those who work with security might restrict too much traffic through firewalls etc. There is a different culture among those who work with infrastructure than among those who are involved in the software development.

4.2 Practices with a Low Degree of Maturity

According to our results, the area with the lowest maturity is Attack models, followed closely by Strategy and Metrics. Regarding attack models, 80 % stated that they do AM1.5 ("The software security group keeps up to date by learning about new types of attacks/vulnerabilities"), and 55 % said they do AM 1.6 ("Build an internal forum to discuss attacks") but all the other activities in the AM practice are performed by fewer than 25 % of the organisations.

One reason that these activities are performed only to a small extent may be that these are activities that are very specifically related to security and therefore come in addition to, or on top of, all the other activities that are being done in the development process. During the follow-up interviews, several of the organisations said they are fully aware that these are areas where they have a potential for improvement. Regarding the two activities mentioned above that relatively many do, several of the respondents indicated that these are largely done outside the development environments. Those who work with operations and infrastructure often get alerts or information on new attacks, and these are discussed as needed. Respondents assume then that developers will be notified of things that are relevant to them, but there seems to be little systematic effort on monitoring attacks in the development environment. Several respondents said that individual developers are adept at keeping up to date also in the security field, for example, to gain knowledge about issues related to components they use themselves, but this work seems to be relatively unstructured and largely depends on the individual developer.

When it comes to work with strategy, there are some organisations that have started some activities related to this, but few organisations currently have a strategic and systematic approach to software security, where they clearly assign responsibility, make plans and strategies, and follow up the implementation and effectiveness. This could relate to the fact that few companies have a clear answer to the question of who are the SSG. All the organisations do some activities related to software security, and some do many, but since this is not done systematically, it is difficult to say something about the effect of the work done in organisations today related to this. On the basis of this investigation, we can not say anything about the reasons for this, but some statements in the interviews indicate that there is little awareness of the importance of software security within the management. As an example, in one interview it was stated that risk at the enterprise level was not relevant to software development.

4.3 Result Summary

To summarize, there is considerable variation in the maturity of the various organisations in the study. As can be seen in Fig. 2, the most mature organisation has implemented 87 of the 112 activities, and the least mature has only implemented 9. The average of the 20 businesses is about 44 of 112. If we look at the three most mature businesses as illustrated in Fig. 5, we notice that, even though the conservative maturity level for the practice Strategy and Metrics is consistently low amongst all the top three organisations, there is an extreme variation in the practice Code Review (ranging from 0.5 to 3).

5 Discussion

As became apparent during the follow-up interviews, the organisations that participated in the study vary as to how much development they perform themselves, and to what extent they use external consultants. A few rely solely on external consultants, while many operate with a mix of internal and external developers. Some largely purchase solutions from system vendors, and make adjustments internally, while others develop the bulk of the solutions themselves. This affects how they work with security, and also the extent to which they have an overview of and control over their software security activities. Some of the respondents had acquired input from their vendors on the questionnaire. Others had little overview of what the vendors did in terms of activities related to information security. This uncertainty applies both to training related to contractor developers as well as to what activities are performed in the development process. The answers to the survey did not indicate that there are any differences in the maturity levels that are due to how the software development teams are organised. However, it is clear that some of the organisations who participated in the study have relatively high levels of expertise, resources and experience related to software development in general, while others have less experience with this. At the same time, our results also show that there are many good practices also among those who do not have a large number of developers. The two organisations that received the highest maturity scores (weighted maturity) have 10 to 20 developers in total (internal and contracted).

The BSIMM framework is based on the idea that there is a formally defined software security group (SSG), and the activities are centered around this group. Few of the surveyed organisations had such a formally defined group. Several organisations have a manager with more or less explicit responsibility for software

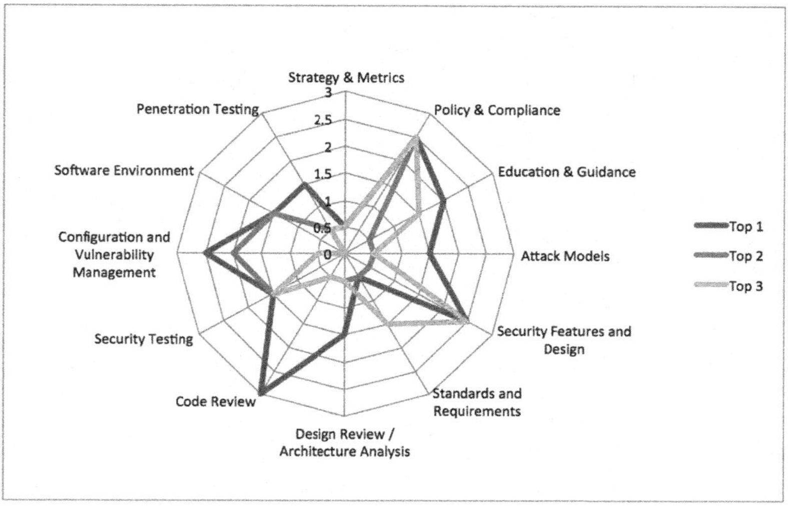

Fig. 5. Conservative and weighted maturity for the three most mature organisations

security, but then usually as part of an overall security responsibility in the organisation.

The method used in our study can be characterized as "assisted self-evaluation"; the respondents from the various organisations indicated in a questionnaire which software security activities that they do, and then they participated in a follow-up interview with the purpose of clarifying uncertainties and correcting possible errors in the questionnaire. During the interviews, many of the respondents mentioned that they found it difficult to answer some of the questions in the questionnaire. In some cases this was because they did not understand what certain activities entailed, for example because they were not familiar with the concepts and terminology that were used. In other cases, they lacked knowledge about the practices in their own organisation or among their consultants and vendors. In several cases, however, the uncertainty was linked to the challenge of responding a simple "yes" or "no" to whether they perform a certain activity. These cases were discussed in depth in the follow-up interviews, aiming to reach a most equal assessment of the various organisations. In most cases the follow-up interviews resulted in some changes to the original answers in the questionnaire. It is also important to point out that some of the respondents seemed to have different attitudes to the study. Some appeared keen to put forward as much as possible of what they do in order to get a good score, while others were more modest on their own behalf; feeling uncomfortable about the possibility that they might claim to do an activity that they did not implement to the full extent.

Another uncertainty factor in our study is that the follow-up interviews were conducted by three different researchers. However, they did synchronize their assessment criteria, both before and during the interview phase, in order to ensure that they had an as similar as possible perception of what is required to receive a "yes" for the various activities in the questionnaire. However, it is still possible that researchers may have made different assessments related to what should be approved as an activity.

Since the study is based largely on self-evaluation, there is reason to believe that the resulting "BSIMM-score" in our study is higher than it would be with a review in line with the one made by Cigital in the original BSIMM study [8], since we have not been in a position to verify the claims made by each organisation. In concrete terms, this implies that we must assume that it has been easier for the organisations to get an activity "approved" in our study than it would be if Cigital had done the survey in accordance with its usual practice. This means that although our results provide some indications of the maturity level of the evaluated organisations, none of the organisations in our study can claim that they have established their "BSIMM Score". It would also be misleading to compare their results directly with the official BSIMM reports. On the other hand, the validity of the answers in our study were increased because of the follow-up interviews, compared with the results from a pure survey. Furthermore, using a questionnaire approach significantly lowers the threshold for initiating a software security maturity study, and we maintain that it is a useful exercise for determining a baseline.

BSIMM claims to be descriptive rather than normative, but by ranking activities in maturity levels, there is an implicit statement that some activities are "better" (or more mature) than others. However, a given organisation may have good reasons for not doing a certain activity, but this will not be reflected in the results from our study. Sometimes checklists have an option to specify "Not relevant" to a given question, and it could be worth considering adding this to the BSIMM yardstick as well.

6 Conclusion and Further Work

This study shows that the public organisations that we have studied are doing a number of activities that contribute to security in the software they are developing. However, it is clear that few are working strategically with software security, where they have a comprehensive and systematic approach and follow up using metrics to evaluate the effectiveness of the various activities. Many of the organisations are very dependent on the interest, competence and initiative of individual developers when it comes to keeping up to date on software security and ensuring that security is not forgotten in the development lifecycle. This stands in contrast to the operations or network side of organisations, where security seems to have a clear priority.

Most of the organisations that participated in our study seemed to be very interested in the topic of software security, but many point out that they have limited resources and this is also reflected in the results. Some stated clearly that they have prioritized other areas in their effort to improve security, however, they all seem to realise the importance of addressing software security as well.

In order to put public organisations in a position to work strategically and systematically with software security, it is important to implement training activities on this topic. Thus far, software security seems to be a very small part of efforts to increase knowledge and awareness of information security in the various organisations.

Several respondents commented that they found many of the governance-related activities more difficult to perform when using an agile development method. This would indicate that there is a need for further study on how to ensure software security in an agile environment. It would also be interesting to compare our results with a similar study on private sector organisations in Norway, which is something we hope to be able to initiate later this year.

Acknowledgment. The research reported in this paper was commissioned by the Norwegian Agency for Public Management and eGovernment (Difi), and partially funded by the EU FP7 project OPTET, grant number 317631.

A Questionnaire

The questionnaire is taken from the BSIMM activity descriptions [3], with only minor textual modifications. Note that the official BSIMM study does not rely on questionnaires. As mentioned before, BSIMM ranks activities on three levels, but we decided not to show this to the respondents; we also re-arranged some activities, placing variations on the same activity together even though they are on different levels in the BSIMM description.

A.1 Governance

Strategy and Metrics

- We publish our process for addressing software security; containing goals, roles, responsibilities and activities.
- We have a secure software evangelist role to promote software security internally.
- We educate our executives about the consequences of inadequate software security.
- We have identified gate locations in our secure software development process where we make go/no go decisions with respect to software security.
- We enforce the identified gate locations in our secure software development process where we make go/no go decisions with respect to software security, and track exceptions.
- We have a process of accepting security risk and documenting accountability. In this process we assign a responsible manager for signing off on the state of all software prior to release.
- The software security group publishes data internally on the state of software security within the organisation.

- In addition to the software security group, we have also identified members of the development teams that have a special interest in software security, and have a process for involving them in the software security work.
- We have identified metrics that measure software security initiative progress and success.
- The software security group has a centralized tracking application to chart the progress of all software.
- The software security group advertises the software security initiative outside the organization (for example by writing articles, holding talks in conferences, etc.).

Policy and Compliance

- The software security group has an overview of the regulations that our software has to comply with.
- We have a software security policy to meet regulatory needs and customer demands.
- The software security group is responsible for identifying all legislation related to personally identifiable information (for example personopplysningsloven).
- We have identified all the personally identifiable information stored by each of our systems and data repositories.
- All identified risks have to be mitigated or accepted by a responsible manager.
- We can demonstrate compliance with regulations that we have to comply with.
- We make sure that all vendor contracts are compatible with our software security policy.
- We promote executive awareness of compliance and privacy obligations.
- We have all the documentation necessary for demonstrating the organisation's compliance with regulations we have to comply with (for ex. written policy, lists of controls, artifacts from software development).
- When managing our third party vendors, we impose our software security policies on them.
- Information from the secure software development process is routinely fed back into the policy creation process.

Education and Guidance

- We have a security awareness training program.
- We offer role-specific security courses (for example on specific tools, technology stacks, bug parade).
- The security awareness training content/material is tailored to our history of security incidents.
- We deliver on-demand individual security training.
- We encourage security learning outside of the software security group by offering specific training and events.

- We provide security training for new employees to enhance the security culture.
- We use the security training to identify individuals that have a particular interest in security.
- We have a reward system for encouraging learning about security.
- We provide security training for vendors and/or outsourced workers.
- We host external software security events.
- We require an annual software security refresher course.
- The software security group has defined office hours for helping the rest of the organization.

A.2 Construction/Intelligence

Attack Models

- We build and maintain a top N possible attacks list.
- We have a data classification scheme and an inventory of attacks so we can prioritize applications by the data handled by them.
- We maintain a list of likely attacker profiles.
- We collect and publish attack stories.
- The software security group keeps up to date by learning about new types of attacks/vulnerabilities.
- We have an internal forum to discuss attacks.
- We link abuse cases to each attacker profile.
- We have a list of technology-specific abuse cases.
- We have an engineering team that develops new attack methods.
- We have automated the attack methods developed by our engineers.

Security Features and Design

- Our software security group builds and publishes a library of security features.
- Security is a regular part of our organization's software architecture discussion.
- The software security group facilitates the use of secure-by-design middleware frameworks/common libraries.
- The software security group is directly involved in the design of security solutions.
- We have a review board to approve and maintain secure design patterns.
- We require the use of approved security features and frameworks.
- We find and publish mature design patterns from the organization.

Standards and Requirements

- The software security group create standards that explain the accepted way to carry out specific security centric operations.
- We have a portal where all security related documents are easily accessible.

- The software security group assists the software development team in translating compliance constraints (for instance from legislation) into application specific security requirements.
- We use secure coding standards in our software development.
- We have a standards review board to formalize the process used to develop security standards.
- We use a limited number of standard technology stacks.
- We have a template SLA text for use in contracts with vendors and providers, to help prevent compliance and privacy problems.
- We have procedures to communicate and promote our security standards to vendors.
- We have a list of all open source components used in our software.
- We manage the risks related to using open source components.

A.3 Verification/Touchpoints

Design Review/Architecture Analysis

- We perform security feature review.
- We perform design review for high-risk applications.
- We have a software security group that leads review efforts.
- We use a risk questionnaire to rank applications in terms of the risk they are exposed to.
- We have a defined process to do architecture analysis.
- We have a standardized format for describring architecture that also covers data flow.
- The software security group is available to support architecture analysis when needed.
- The software architects lead design review efforts to detect and correct security flaws.
- Failures identified during architecture analysis are used to update the standard architecture patterns.

Code Review

- We create a list with top N software security defects list.
- The software security group does ad-hoc code reviews.
- We use automated tools (such as static analysis) along with manual review to detect software security defects.
- We make code review mandatory for all projects before release.
- The software security defects found during code review are tracked in a centralized repository.
- We enforce coding standards to improve software security.
- We have mentors for code review tools for making most efficient use of the tools.
- We use automated tools with tailored rules to improve efficiency and reduce false positives.

- We combine assessment results so that multiple analysis techniques feed into one reporting and remediation process.
- When a software defect is found we have tools to search for that defect also in the whole codebase.
- We perform automated code review on all code to detect malicious code.

Security Testing

- We perform adversarial tests with edge and boundary values.
- We create our tests based on existing security requirements and security features.
- We integrate black box security tools into the testing process (including protocol fuzzing).
- We share security test results with QA.
- We include security tests in QA automation.
- We perform fuzz testing customized to application APIs.
- We base the security tests on the security risks analysis.
- We use code coverage tools to ensure that security tests cover all parts of the code.
- We write tests cases based on abuse cases provided by the software security group.

A.4 Deployment

Configuration and Vulnerability Management

- The software security group has procedures for incident response, in collaboration with the incident response team (if it exists).
- We are able to make quick changes in the software when under attack.
- We perform drills to ensure that incident response capabilities minimize the impact of an attack.
- We identify software defects found in operations (for ex. by intrusion detection systems) and feed back to development.
- We track software defects found during operations until they are closed.
- We maintain a matrix of all installed applications in order to identify all places that need to be updated when a piece of code needs to be changed.
- When a software defect is found in a piece of code during operations we have a process to search for that defect also in the whole codebase.
- We do software security process improvement based on the analysis of cause of software defects found in operations.
- We have a system for paying rewards to individuals who report security flaws in our software.

Software Environment

- We monitor the input to software we run in order to spot attacks on our software.
- We use accepted good practice mechanisms for host/network security.
- The software security group creates and publishes installation guides to ensure that our software is configured securely.
- We create digital signatures for all binaries that we deliver.
- We use code protection such as obfuscation to make reverse engineering harder.
- We monitor the behavior of our software looking for misbehavior and signs of attacks.

Penetration Testing

- We use external penetration testers on our software.
- Defects found in penetration testing are inserted in our bug tracking system and flagged as security defects.
- We use penetration testing tools internally.
- The penetration testers have access to all available information about our software (for example: the source code, design documents, architecture analysis results and code review results).
- We periodically perform penetration tests on all our software.
- We use external penetration testers to do deep-dive analysis for critical projects to complement internal competence.
- The software security group has created customized penetration testing tools and scripts for our organization.

References

1. McGraw, G.: Software security. IEEE Secur. Priv. **2**(2), 80–83 (2004)
2. Hope, P.: Why measurement is key to driving improvement in software security. Developer Tech (2014). http://www.developer-tech.com/news/2014/mar/06/why-measurement-key-driving-improvement-software-security/
3. McGraw, G., Migues, S., West, J.: Building Security In Maturity Model (BSIMM V) (2013). http://bsimm.com
4. Norwegian Ministries: Cyber Security Strategy for Norway (2012). https://www.regjeringen.no/globalassets/upload/fad/vedlegg/ikt-politikk/cyber_security_strategy_norway.pdf
5. OpenSAMM: Software Assurance Maturity Model (SAMM): A guide to building security into software development. http://www.opensamm.org/
6. Robson, C.: Real World Research, 3rd edn. Wiley, Chichester (2011)
7. Jaatun, M.G., Tøndel, I.A., Cruzes, D.S.: Modenhetskartlegging av programvaresikkerhet i offentlige virksomheter. Technical report A26860, SINTEF ICT (2015). http://sintef.no/publikasjoner/publikasjon/?pubid=SINTEF+A26860
8. Jaatun, M.G.: Hunting for aardvarks: can software security be measured? In: Quirchmayr, G., Basl, J., You, I., Xu, L., Weippl, E. (eds.) CD-ARES 2012. LNCS, vol. 7465, pp. 85–92. Springer, Heidelberg (2012)

Cryptanalysis I: Block Ciphers

Extending the Applicability of the Mixed-Integer Programming Technique in Automatic Differential Cryptanalysis

Siwei Sun[1,2], Lei Hu[1,2](✉), Meiqin Wang[3], Qianqian Yang[1,2], Kexin Qiao[1,2], Xiaoshuang Ma[1,2], Ling Song[1,2], and Jinyong Shan[1,2]

[1] State Key Laboratory of Information Security, Institute of Information Engineering, Chinese Academy of Sciences, Beijing 100093, China
{sunsiwei,hulei,yangqianqian,qiaokexin,maxiaoshuang, songling,shanjinyong}@iie.ac.cn

[2] Data Assurance and Communication Security Research Center, Chinese Academy of Sciences, Beijing 100093, China

[3] Key Laboratory of Cryptologic Technology and Information Security, Ministry of Education, Shandong University, Jinan 250100, China
mqwang@sdu.edu.cn

Abstract. We focus on extending the applicability of the mixed-integer programming (MIP) based method in differential cryptanalysis such that more work can be done automatically. Firstly, we show how to use the MIP-based technique to obtain almost all high probability 2-round iterative related-key differential characteristics of PRIDE (a block cipher proposed in CRYPTO 2014) automatically by treating the $g_i^{(j)}(\cdot)$ function with a special kind of modulo addition operations in the key schedule algorithm of PRIDE as an 8×8 S-box and *partially* modelling its differential behavior with linear inequalities. Note that some of the characteristics presented in this paper has not been found before, and all the characteristics we found can be used to attack the full-round PRIDE in the related-key model. Secondly, we show how to construct MIP models whose feasible regions are exactly the sets of all possible differential characteristics of SIMON (a family of lightweight block ciphers designed by the U.S. National Security Agency). With this method, there is no need to filter out invalid characteristics due to the dependent inputs of the AND operations. Finally, we present an MIP-based method which can be used to automatically analyze how the differences at the beginning and end of a differential distinguisher propagate upwards and downward. Note that how the differences at the ends of a differential distinguisher propagate, together with the probability of the differential distinguisher, determine how many outer rounds can be added to the distinguisher, which key bits can be recovered without exhaustive search, and how to identify wrong pairs in the filtering process. We think this work serves to further strengthens the position of the MIP as a promising tool in automatic differential cryptanalysis.

Keywords: Automatic cryptanalysis · Related-key differential attack · Mixed-integer programming · PRIDE

J. Lopez and C.J. Mitchell (Eds.): ISC 2015, LNCS 9290, pp. 141–157, 2015.
DOI: 10.1007/978-3-319-23318-5_8

1 Introduction

Block ciphers are probably the most widely used cryptographic algorithms today for data encryption. With the ever increasing cryptographic complexity and demand for new block ciphers which can be deployed in a diverse computing environment, it is imperative to develop tools for automatic block cipher cryptanalysis.

Differential cryptanalysis [2], introduced by Eli Biham and Adi Shamir in the late 1980s, is one of the most effective and well understood attacks on modern block ciphers. Providing a convincing security argument with respect to the differential attack have become one of the most import aspects for the design of block ciphers. Accordingly, tools for (partially) automatic differential analysis have attracted a lot of attention from the cryptographic community.

Matsui's algorithm [8] is probably the most widely used (partially) automatic method for finding good differential characteristics. Several papers were devoted to improving the efficiency of Matsui's algorithm. In [6], the concept of *search pattern* was introduced to reduce the search complexity of Matsui's algorithm by detecting unnecessary search candidates. Further improvements were obtained in [5] and [21]. In [1], Biryukov *et al.* extended Matsui's algorithm by using the partial (rather than the full) difference distribution table (pDDT) to prevent the number of examined candidates from exploding and at the same time keep the total probability of the resulting characteristic high. These methods have been employed to evaluate the security of many block ciphers with respect to the differential attack.

Automatic differential analysis is also studied under the framework of constraint programming. In recent years, there has been an increasing interest in this line of research. Compared with other methods, these methods are easier to implement and more flexible. In [9,18,19], SAT or SMT solvers are employed to find differential characteristics of Salsa and other ciphers. Mouha *et al.* [10], Wu *et al.* [14], and Sun *et al.* [15] converted the problem of counting the minimum number of differentially active S-boxes into an MIP problem which can be solved automatically with open source or commercially available optimizers. These methods have been applied in evaluating the security against (related-key) differential attacks of many block ciphers. However, these tools cannot be used to find the actual differential characteristics directly. In Asiacrypt 2014, two systematic methods for generating linear inequalities describing the differential properties of an arbitrary S-box were given in [17]. With these inequalities, the authors of [17] were able to construct an MIP model whose feasible region is a more accurate description of the differential behavior of a given cipher. Based on such MIP models, the authors of [16] proposed a heuristic algorithm for finding actual (related-key) differential characteristics, which is applicable to a wide range of block ciphers. In [16], Sun *et al.* get rid of the heuristic argument in [17] by constructing MIP models whose feasible regions are exactly the sets of all (related-key) differential characteristics. However, the method presented in [16] still has some important limitations. Firstly, it can not be applied to ciphers with modulo additions. Secondly, for the case of SIMON (a lightweight block cipher designed by the U.S. National Security Agency), the method proposed in [16] is not *exact* anymore. That is, the feasible region of the MIP model constructed for SIMON contains invalid differential characteristics due to the dependent input bits of the AND operations, and these invalid characteristics must be filtered out by other methods. This is a very inconvenient process and reduces the level of automation of the framework of MIP based automatic differential analysis. Finally, it does not support any subsequent analysis of the differential attack after finding good differentials.

Our Contribution. In this work, we mainly focus on the MIP-based method for automatic differential analysis. We do not try to improve specific results in cryptanalysis, but attempt to use the MIP-based method in a clever way such that more work in differential analysis can be done *automatically*.

So far, the MIP-based method has not been applied to ciphers with modulo addition operations, which is one of the limitations of the MIP-based method discussed in [16]. In this work, we show how a *special case* of modulo addition can be dealt with MIP method. To be more specific, we treat the modulo addition in the key schedule algorithm of PRIDE as an 8×8 S-box and *partially* model its differential behavior with MIP method. With this approach, we show how to enumerate almost all high probability 2-round iterative related-key differential characteristics of PRIDE automatically. We present some iterative related-key characteristics for the full PRIDE which have never been found before, and all the characteristics we found are of high probability such that they can be used to attack the full-round PRIDE.

Moreover, by using constraints from the H-representation of a specific convex hull, we give a method for constructing MIP models whose feasible regions are exactly the sets of all possible differential characteristics for SIMON (a family of lightweight block ciphers designed by the U.S. National Security Agency). Note that the feasible region of the MIP model constructed by the method presented in [16] for SIMON contains invalid differential characteristics due to the dependent input bits of the AND operations, and these invalid characteristics must be filtered out by other methods. This is a very inconvenient process and reduces the level of automation of the framework of MIP based automatic differential analysis.

In addition, currently the MIP-based method does not support any subsequent analysis in differential attack except finding good (related-key) differentials. After finding a good differential and building a distinguisher, the cryptanalyst needs to add several rounds at the end (or the beginning) of the distinguisher and try to mount a key recovery attack. How many rounds can be appended, which key bits can be recovered without exhaustive search, and how to filter wrong pairs are all determined by how the output difference of the distinguisher propagate through the added rounds. Typically, this analysis is done manually or by another computer program in a trial-and-error style. In this paper, we show how this error-prone process can be done automatically. This method can be integrated into the MIP framework for finding high probability differentials.

Thus the cryptanalyst using this tool can quickly examine how a differential propagate upward and downward at the beginning and end of a differential distinguisher. Although this approach is of no theoretical interest, it has been proved in our daily work that such tool is very convenient and more reliable than other methods.

Organization. In Sect. 2 we give a brief introduction of the MIP-based method for automatic differential analysis. In Sect. 3, We show how to enumerate high probability 2-round iterative related-key differential characteristics of PRIDE whose key schedule algorithm containing modulo addition operations. In Sect. 4, we show how to construct MIP models whose feasible regions are exactly the sets of all possible differential characteristics of SIMON. An MIP-based method for automating the analysis of the propagation of the differences at the ends of a differential distinguisher is presented in Sects. 5 and 6 is the conclusion.

2 MIP-based Automatic Differential Analysis

The MIP-based method for automatic differential analysis [16] can be applied to ciphers involving the following three operations:

- bitwise XOR;
- bitwise permutation L which permutes the bit positions of an n dimensional vector in $\mathbb{F}_2^n$;
- S-box, $\mathcal{S} : \mathbb{F}_2^{\omega} \to \mathbb{F}_2^{\nu}$.

Note that a general linear transformation $T : \mathbb{F}_2^n \to \mathbb{F}_2^m$ can be treated as some XOR summations and bitwise permutations of the input bits. In [16], a new variable x_i is introduced for every input and output bit-level differences, where $x_i = 1$ means the XOR difference at this position is 1 and $x_i = 0$ if there is no difference. Also, for every S-box involved in the cipher, introduce a new 0–1 variable A_j such that

$$A_j = \begin{cases} 1, & \text{if the input word of the Sbox is nonzero,} \\ 0, & \text{otherwise.} \end{cases}$$

Now, we can describe the MIP-based method [16] by clarifying the objective function and constraints in the MIP model. Note that we assume all variables involved are 0–1 variables.

Objective Function. The objective function is to minimize the sum of all variables A_j indicating the activities of the S-boxes: $\sum_j A_j$.

Constraints. Firstly, for every XOR operation $a \oplus b = c \in \{0, 1\}$, include the following constraints

$$\begin{cases} a + b + c \geq 2d_{\oplus} \\ a + b + c \leq 2 \\ d_{\oplus} \geq a, \ d_{\oplus} \geq b, \ d_{\oplus} \geq c \end{cases} \tag{1}$$

where $d_{\oplus}$ is a dummy variable.

Assuming $(x_{i_0}, \ldots, x_{i_{\omega-1}})$ and $(y_{i_0}, \ldots, y_{i_{\nu-1}})$ are the input and output differences of an $\omega \times \nu$ S-box marked by A_t, we have

$$\begin{cases} A_t - x_{i_k} \geq 0, \ k \in \{0, \ldots, \omega - 1\} \\ -A_t + \sum_{j=0}^{\omega-1} x_{i_j} \geq 0 \end{cases} \tag{2}$$

and

$$\begin{cases} \sum_{k=0}^{\omega-1} x_{i_k} + \sum_{k=0}^{\nu-1} y_{j_k} \geq \mathcal{B}_{\mathcal{S}} d_{\mathcal{S}} \\ d_{\mathcal{S}} \geq x_{i_k}, \quad 0 \leq k \leq \omega - 1 \\ d_{\mathcal{S}} \geq y_{j_k}, \quad 0 \leq k \leq \nu - 1 \end{cases} \tag{3}$$

where $d_{\mathcal{S}}$ is a dummy variable, and the branch number $\mathcal{B}_{\mathcal{S}}$ of an S-box $\mathcal{S}$, is defined as $\min_{a \neq b}\{\mathrm{wt}((a \oplus b) || (\mathcal{S}(a) \oplus \mathcal{S}(b)) : a, b \in \mathbb{F}_2^{\omega}\}$. For an bijective S-box we have

$$\begin{cases} \omega \sum_{k=0}^{\nu-1} y_{j_k} - \sum_{k=0}^{\omega-1} x_{i_k} \geq 0 \\ \nu \sum_{k=0}^{\omega-1} x_{i_k} - \sum_{k=0}^{\nu-1} y_{j_k} \geq 0 \end{cases} \tag{4}$$

Then, treat every possible input-output differential pattern $(x_0, \ldots, x_{\omega-1}) \rightarrow (y_0, \ldots, y_{\nu-1})$ of an $\omega \times \nu$ S-box as an $(\omega + \nu)$-dimensional vector $(x_0, \ldots, x_{\omega-1}, y_0, \ldots, y_{\nu-1}) \in \{0, 1\}^{\omega+\nu} \subseteq \mathbb{R}^{\omega+\nu}$, and compute the H-representation of the convex hull of all possible input-output differential patterns of the S-box. From the H-representation we can extract the critical set with the method presented in [16]. The critical set contains a small number of linear inequalities which can be used to exactly describe the differential behavior of the S-box. Finally, relate the input and output variables of the S-box using the inequalities in the critical set. Now, if we require that all the variables involved are 0–1 variables, then the feasible region of the resulting MIP model is exactly the set of all differential characteristics. By solving this kind of MIP models, we can find good differential characteristics or enumerate all characteristics with some predefined properties.

3 Automatic Search for Related-Key Differential Characteristics of PRIDE

The block cipher PRIDE [7] is one of the vast number of newly designed lightweight block ciphers. One of the novelties of PRIDE is that its linear layer is constructed according to a general methodology for constructing good, sometimes optimal linear layers allowing for a large variety of trade-offs between security and efficiency. Although PRIDE is optimized for software implementation on 8-bit micro-controllers, it is also efficient in hardware. After its publication, PRIDE has receive several cryptanalysis [3,4,12]. Very recently, Dai *et al.* presented several related-key differential attacks on the full PRIDE with the help of 16 2-round iterative related-key differential characteristics obtained by some *ad-hoc* method [20]. In this section, we show how to enumerate almost all high probability 2-round iterative related-key differential characteristics of PRIDE automatically by using the MIP technique.

3.1 Description of PRIDE

PRIDE [7] is a block cipher based on FX-construction whose block size and key size are 64-bit and 128-bit respectively. It consists of 20 rounds of iterations of which the first 19 rounds are identical, and the overall structure of PRIDE is depicted in Fig. 1.

The round function $\mathcal{R}$ of PRIDE is an SPN structure: the state is XORed with the round key $f_i(k_1)$ permuted with a bit permutation $\mathcal{P}^{-1}$, fed into 16 parallel 4-bit S-boxes and then processed by the linear layer involving bit permutations and linear transformations (see Fig. 2).

The 128-bit master key of PRIDE is divided into to 64-bit words k_0 and k_1, and k_0 is used as the pre- and post-whitening keys. The subkey $f_i(k_1)$ of the ith round of PRIDE is defined as follows

$$f_i(k_1) = k_{1,0}||g_i^{(0)}(k_{1,1})||k_{1,2}||g_i^{(1)}(k_{1,3})||k_{1,4}||g_i^{(2)}(k_{1,5})||k_{1,6}||g_i^{(3)}(k_{1,7})$$

where $k_{1,i}$ is the ith nibble of k_1 and $g_i^{(j)}(\cdot)$'s are defined as follows

$$g_i^{(0)}(x) = (x + 193i) \bmod 256, \qquad g_i^{(1)}(x) = (x + 165i) \bmod 256$$

$$g_i^{(2)}(x) = (x + 81i) \bmod 256, \qquad g_i^{(3)}(x) = (x + 197i) \bmod 256$$

For a more detailed description of PRIDE, we refer the reader to [7].

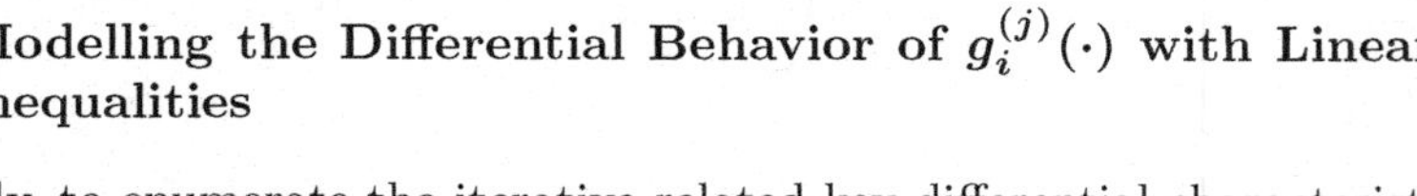

Fig. 1. The overall structure of the PRIDE

3.2 Modelling the Differential Behavior of $g_i^{(j)}(\cdot)$ with Linear Inequalities

Typically, to enumerate the iterative related-key differential characteristics of a block cipher with the MIP-based method, we need to construct an MIP model for the cipher in the related-key setting whose feasible region is exactly the set of all related-key differential characteristics of the target cipher. However, the key schedule algorithm (KSA) of PRIDE contains modulo addition operations, and it seems that there is no existing technique which can be used to generate practically solvable MIP models describing the differential behavior of ciphers involving modulo additions.

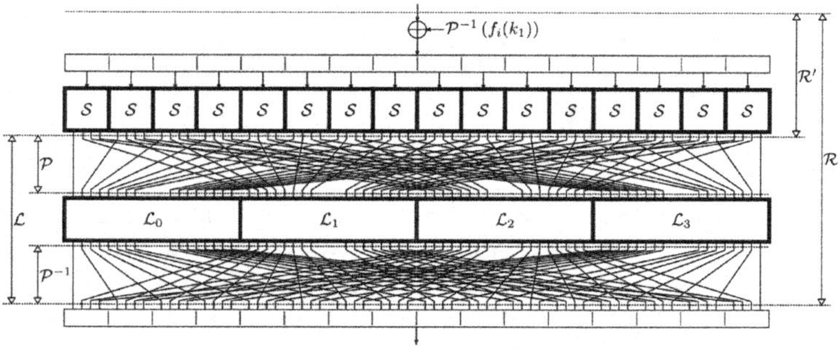

Fig. 2. The round function $\mathcal{R}$ of PRIDE

Whereas the modulo additions appearing in the KSA of PRIDE is very special. Taking the function $g_i^{(0)}(\cdot)$ for example, one of its operands is $193i$ which is a constant for a given i. Hence, $g_i^{(0)} : x \mapsto (x + 193i) \bmod 256$ can be treated as an 8×8 S-box whose differential behavior can be modeled by the convex hull computation method presented in [16]. But, it is still very difficult to compute the convex hull of all differential patterns of an 8×8 S-box. In fact, in the original papers [16,17] proposing the MIP-based method for finding good characteristics, no cipher with an 8×8 S-box has been analyzed since computing the convex hull of a subset of $\mathbb{R}^{8+8}$ with a reasonable size can be a difficult task. In particular, for the AES S-box, we are failed to get the convex hull of all the differential patterns of the AES S-box after 3 h of computation on a PC. Therefore, we take the following strategy to deal with the $g_i^{(j)}(\cdot)$ function in the KSA of PRIDE. For the sake of simplicity, we demonstrate our method on $g_1^{(0)}(\cdot)$. For other functions it can be analyzed in the same way.

Firstly, compute the differential distribution table (DDT) of $g_1^{(0)}(\cdot)$. Secondly, from the DDT, select a set H of differential patterns $(x_0, \cdots, x_7, y_0, \cdots, y_7) \in \{0,1\}^{16} \subseteq \mathbb{R}^{16}$ such that the probability of the differential $(x_0, \cdots, x_7) \to (y_0, \cdots, y_7)$ is greater than $\frac{240}{256}$. The differential patterns contained in the set H are listed below

```
00000000 --> 00000000 :        (0,0,0,0,0,0,0,0,0,0,0,0,0,0,0,0)
00100000 --> 00100000 :        (0,0,1,0,0,0,0,0,0,0,1,0,0,0,0,0)
01000000 --> 11000000 :        (0,1,0,0,0,0,0,0,1,1,0,0,0,0,0,0)
01100000 --> 11100000 :        (0,1,1,0,0,0,0,0,1,1,1,0,0,0,0,0)
10000000 --> 10000000 :        (1,0,0,0,0,0,0,0,1,0,0,0,0,0,0,0)
10100000 --> 10100000 :        (1,0,1,0,0,0,0,0,1,0,1,0,0,0,0,0)
11000000 --> 01000000 :        (1,1,0,0,0,0,0,0,0,1,0,0,0,0,0,0)
11100000 --> 01100000 :        (1,1,1,0,0,0,0,0,0,1,1,0,0,0,0,0)
```

Thirdly, using the method presented in [16], we compute the H-representation of the convex hull of H, from which we can derive the critical set $\mathcal{O}_H$ which is an exact linear inequality description of the differential patterns contained in H. This is, the set of 0–1 solutions of the system of linear inequalities $\mathcal{O}_H$ is exactly H, where $\mathcal{O}_H$ is given below

$$\begin{cases} -x_0 + x_1 + y_0 \geq 0 \\ -x_0 - x_1 - y_0 \geq -2 \\ x_0 - x_1 + y_0 \geq 0 \\ x_0 + x_1 - y_0 \geq 0 \end{cases} \quad (5)$$

By using the above constraints to describe the differential behavior of $g_1^{(0)}(\cdot)$ in an MIP model, the feasible region of this MIP model will not contain any differential pattern $(x_0, \cdots, x_7, y_0, \cdots, y_7)$ of $g_1^{(0)}(\cdot)$ such that the probability of the differential $(x_0, \cdots, x_7) \rightarrow (y_0, \cdots, y_7)$ is less than or equal to $\frac{240}{256}$.

At this point, we can construct an MIP model *partially* describing the differential behavior of PRIDE in the related-key model by using the constraints for $g_i^{(j)}(\cdot)$ generated in the above method. Note that for different i and j, a different set of constraints is generated. With this approach, we construct an MIP model for PRIDE in the related-key setting and enumerate its 2-round iterative related-key differential characteristics with probability 2^{-4} by using the method presented in [16]. We obtain 42 characteristics and they are listed in Tables 1, 2, 3 and 4 of Appendix A, where we use ΔI to denote the input difference of one round $\mathcal{R}$ (see Fig. 1) and Δk to denote the output difference of $P^{-1}(f(k_1))$.

Since for every characteristic we found, we can construct an 18-round related-key differential of the PRIDE (19-round in total) with probability at least $(2^{-4})^{18/2} = 2^{-36}$, any one of the characteristics we found can be used to attack the full PRIDE. Also note that the set of all characteristics found in [20] is only a subset of the characteristics we found. Hence, using the characteristics we found will produce attacks on full PRIDE at least as good as that presented in [20]. However, we do not take the effort to improve the attack since this is not the focus of this paper.

4 Constructing MIP Models Whose Feasible Regions are Exactly the Sets of All Differential Characteristics of SIMON

In [16], a method for constructing mixed-integer programming models whose feasible regions are exactly the sets of all possible differential (or linear) characteristics for a wide range of block ciphers is presented. These models can be used to search for or enumerate differential and linear characteristics of a block cipher automatically. However, for the case of SIMON (a lightweight block cipher designed by the U.S. National Security Agency), the method proposed in [16] is not *exact* anymore. That is, the feasible region of the MIP model constructed for SIMON contains invalid differential characteristics due to the dependent input bits of the AND operations, and these invalid characteristics must be filtered out by other methods. This is a very inconvenient process and reduces the level of automation of the framework of MIP-based automatic differential analysis. In the following, by using constraints from the H-representation of a specific convex hull, we give a method for constructing MIP models whose feasible regions are exactly the sets of all possible differential characteristics for SIMON.

We will focus on the case of SIMON32 [13] with block size 32 bits, for other cases the method is similar. The nonlinear layer of SIMON32 can be described by a non-linear function $F : \mathbb{F}_2^{16} \to \mathbb{F}_2^{16}$, such that

$$F(x) = (x <<< 1) \cdot (x <<< 8), \ x = (x_0, \cdots, x_{15}) \in \mathbb{F}_2^{16}.$$

where $\cdot$ is the bitwise AND operation.

Let $\Delta = (\Delta_0, \cdots, \Delta_{15}) \in \mathbb{F}_2^{16}$, and $\delta = (\delta_0, \cdots, \delta_{15}) \in \mathbb{F}_2^{16}$, then the differential $\Delta \to \delta$ is valid for F if and only if there exists $x \in \mathbb{F}_2^{16}$ such that $F(x) + F(x + \Delta) = \delta$, that is $((x + \Delta) <<< 1) \cdot ((x + \Delta) <<< 8) = \delta$.

Writing it bitwisely, the differential $\Delta \to \delta$ is valid if and only if the following system of equations of x_i has a solution

$$\begin{cases} \delta_0 = \Delta_1 \cdot x_8 + \Delta_8 \cdot x_1 \\ \delta_1 = \Delta_2 \cdot x_9 + \Delta_9 \cdot x_2 \\ \delta_2 = \Delta_3 \cdot x_{10} + \Delta_{10} \cdot x_3 \\ \delta_3 = \Delta_4 \cdot x_{11} + \Delta_{11} \cdot x_4 \\ \delta_4 = \Delta_5 \cdot x_{12} + \Delta_{12} \cdot x_5 \\ \delta_5 = \Delta_6 \cdot x_{13} + \Delta_{13} \cdot x_6 \\ \delta_6 = \Delta_7 \cdot x_{14} + \Delta_{14} \cdot x_7 \\ \delta_7 = \Delta_8 \cdot x_{15} + \Delta_{15} \cdot x_8 \\ \delta_8 = \Delta_9 \cdot x_0 + \Delta_0 \cdot x_9 \\ \delta_9 = \Delta_{10} \cdot x_1 + \Delta_1 \cdot x_{10} \\ \delta_{10} = \Delta_{11} \cdot x_2 + \Delta_2 \cdot x_{11} \\ \delta_{11} = \Delta_{12} \cdot x_3 + \Delta_3 \cdot x_{12} \\ \delta_{12} = \Delta_{13} \cdot x_4 + \Delta_4 \cdot x_{13} \\ \delta_{13} = \Delta_{14} \cdot x_5 + \Delta_5 \cdot x_{14} \\ \delta_{14} = \Delta_{15} \cdot x_6 + \Delta_6 \cdot x_{15} \\ \delta_{15} = \Delta_0 \cdot x_7 + \Delta_7 \cdot x_0 \end{cases} \tag{6}$$

In the work of [16], the MILP models generated for SIMON only have variables for the differences (δ_i and Δ_i) and the variables marking the activities of the AND operations. To generate exact models for SIMON, we need to introduce a new set of variables ($x_i, 0 \leq i \leq 15$) for every round of SIMON, and include the constraints which dictating that the system of equations listed in (6) has a solution. In the following, we show how to convert the constraints presented in (6) into a set of linear (in)equalities by the convex hull computation technique presented in [16].

Taking the first equation $\delta_0 = \Delta_1 \cdot x_8 + \Delta_8 \cdot x_1$ in (6) for example, let $Sol(\delta_0 = \Delta_1 \cdot x_8 + \Delta_8 \cdot x_1)$ be the set of all 0–1 solutions for this equation. Then

$$Sol(\delta_0 = \Delta_1 \cdot x_8 + \Delta_8 \cdot x_1)$$

can be treated as a subset of $\{0, 1\}^5 \subseteq \mathbb{R}^5$. The vectors $(\delta_0, \Delta_1, x_8, \Delta_8, x_1)$ in $Sol(\delta_0 = \Delta_1 \cdot x_8 + \Delta_8 \cdot x_1)$ are given below

```
(0, 0, 0, 0, 0) (0, 0, 0, 0, 1) (0, 0, 0, 1, 0) (0, 0, 1, 0, 0)
(0, 0, 1, 0, 1) (0, 0, 1, 1, 0) (0, 1, 0, 0, 0) (0, 1, 0, 0, 1)
```

```
(0, 1, 0, 1, 0) (0, 1, 1, 1, 1) (1, 0, 0, 1, 1) (1, 0, 1, 1, 1)
(1, 1, 0, 1, 1) (1, 1, 1, 0, 0) (1, 1, 1, 0, 1) (1, 1, 1, 1, 0)
```

Now, we can compute the critical set $\mathcal{O}$ of the H-representation of the convex hull of $Sol(\delta_0 = \Delta_1 \cdot x_8 + \Delta_8 \cdot x_1)$. We refer the reader to [16] for more information about how to compute the critical set and H-representation. The H-representation of the convex hull of $Sol(\delta_0 = \Delta_1 \cdot x_8 + \Delta_8 \cdot x_1)$ is given below

```
( 0, -1,  0,  0,  0, 1 ) ( 0,  0, -1,  0,  0, 1 ) ( 0,  0,  0, -1,  0, 1 )
(-1,  1,  0,  0,  1, 0 ) (-1,  0,  1,  0,  1, 0 ) (-1,  0,  0,  0,  0, 1 )
( 0,  0,  0,  0,  1, 0 ) ( 1, -1, -1,  0,  1, 1 ) ( 0,  1,  0,  0,  0, 0 )
( 0,  0,  0,  0, -1, 1 ) (-1,  1,  0,  1,  0, 0 ) ( 1,  0,  0,  0,  0, 0 )
(-1,  0,  1,  1,  0, 0 ) ( 0,  0,  0,  1,  0, 0 ) ( 1,  0,  1, -1, -1, 1 )
( 0,  0,  1,  0,  0, 0 ) ( 1, -1, -1,  1,  0, 1 ) ( 1,  1,  0, -1, -1, 1 )
(-1, -1, -1, -1, -1, 4 )
```

where a 6-dimensional vector $(\lambda_0, \cdots, \lambda_4, \gamma)$ denotes the linear inequality

$$\lambda_0\delta_0 + \lambda_1\Delta_1 + \lambda_2 x_8 + \lambda_3\Delta_8 + \lambda_4 x_1 + \gamma \geq 0.$$

From the H-representation we can derive the critical set $\mathcal{O}$, which is listed in the following

```
(-1,  0,  1,  0,  1, 0) (-1,  1,  0,  1,  0, 0) ( 1, -1, -1,  1,  0, 1)
( 1,  1,  0, -1, -1, 1) ( 1,  0,  1, -1, -1, 1) (-1, -1, -1, -1, -1, 4)
(-1,  1,  0,  0,  1, 0) (-1,  0,  1,  1,  0, 0) ( 1, -1, -1,  0,  1, 1)
```

$\mathcal{O}$ is a set of 9 linear inequalities involving the 5 variables: $\delta_0, \Delta_1, x_8, \Delta_8, x_1$. For every equation in (6), we can derive a corresponding critical set. Then we can add all these sets of linear constraints into the overall MIP model. Now, we come to an MIP model for SIMON whose feasible region is exactly the set of all differential characteristics for SIMON since every feasible solution of the new model will make the system of equations (6) have at least one solution.

Compared with the models generated in [16], the new models contain more variables and constraints which will make them more difficult to solve. So, we suggest that we should first try to find a good differential characteristic by the method presented in [16]. According to our experimental experience, we will get a valid characteristic with a very high chance. Then when we want to enumerate the characteristics in the differential $\alpha \to \beta$, we fix the variables in the MIP model according to the input and output differences and limit the number of active AND operations, and add the new constraints described in this paper to the MIP model. Now, we can enumerate all differential characteristics of this differential with the predefined properties by finding all solutions of the MIP model.

Using the above method, we enumerate all single-key differential characteristics with N_A ($50 \leq N_A \leq 300$) active AND operations for 16-round SIMON with input difference $(800000, 220082)$ and output difference $(800000, 220000)$. Note that this differential whose probability is $2^{-44.65}$ has been used in [11] to attack 23-round SIMON48. Finally, we obtain 877231 differential characteristics in no more than 3 days on a PC, and the differential probability of the differential is at least $2^{-44.26}$. While in [16], only 3822 characteristics were found. This result can be used directly to improve the currently known best differential attack on 23-round SIMON48 presented in [11].

5 Automatic Analysis of the Propagation of Differences

In a typical differential attack, after a good differential has been identified and therefore a distinguisher is built, the attacker then attempts to recover some secret key bits from the outer rounds of the distinguisher. To accomplish this, the attacker must analyze how the differences at the two ends of the distinguisher evolve through the outer rounds of the cipher under consideration (for simplicity, here we only focus on the last rounds). This is a trivial yet tedious, error-prone, and important process.

Firstly, if the difference propagates in such a way that some specific bits of the ciphertext difference must be 0 or 1 for any key, the attacker can determine in some situation that a pair could not possibly be a right pair just by looking at the ciphertext pair and discard it immediately. That is, how the difference propagate affects the filtering process of differential analysis which is essential for the success of a differential attack, since a good filtering technique will increase the signal to noise ratio.

Secondly, how the difference evolve in the outer rounds of the differential distinguisher affects which key bits can be recovered without a brute force search (we refer such key bits as the target key bits here after) in the attack and how many rounds can be added to the two ends of the distinguisher. To a large extend, the number of key bits can be recovered without a brute force search together with the probability of the distinguisher determine the overall complexity of the differential attack.

All in all, we stress once again that analysis of how the differences evolve in the outer rounds is of great importance to differential attack even though it is trivial. In this section we propose an MIP-based method for analyzing the propagation of difference automatically. For the sake of simplicity, we will describe our method by a simple example depicted in Fig. 3. Assume an attacker has built a differential distinguisher whose input and output differences are $\alpha \in \{0,1\}^n$ and $\beta \in \{0,1\}^n$ respectively. Then the attacker appends 3 more rounds at the end of the distinguisher and tries to recover key material involved in these rounds. To identify the target key bits and the filtering strategy in the differential attack, the attacker must determine first what kind of difference patterns will β_1, β_2, and β_3 take. The steps to accomplish this task under the MIP framework are listed as follows.

Step 1. Construct an MIP model describing the differential behavior of the round function R_1 using the method given in [16], and set its objective function to be any constant.

Step 2. Add the bit-level constraints (a set of equalities) which dictating that the input difference pattern to the round function R_1 is β.

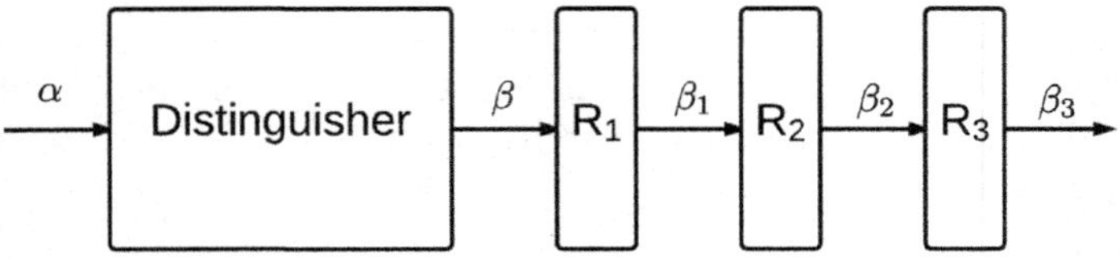

Fig. 3. Append 3 more rounds at the end of the distinguisher

Step 3. Based on the MIP model built by Step 1 and Step 2, generate n different MIP models $\mathcal{M}_0, \ldots, \mathcal{M}_{n-1}$ such that in the ith model we add one more constraint which dictates that the ith bit of the output difference β_1 of R_1 is 1.

Step 4. For $i \in \{0, \ldots, n-1\}$, solve $\mathcal{M}_i$. If $\mathcal{M}_i$ is infeasible, we claim that the ith bit of β_1 must be 0; If $\mathcal{M}_i$ is feasible, we change its constraint which dictates that the ith bit of the output difference of R_1 is 1 to the constraint which dictates that the ith bit of the output difference of R_1 is 1 and solve the updated model $\mathcal{M}_i$ again. If it is infeasible, we claim that the ith bit of β_1 must be 1; If it is feasible, we claim that the ith bit of β_1 can be any value which is denoted by a "*" or "?" typically.

Using a similar method, we can also automatically deduce the difference patterns of β_2 and β_3. For example, after the above steps, we know that β_1 is of the form $\beta_{1,j_0} = 0, \cdots, \beta_{1,j_{s_r}} = 0$; $\beta_{1,j_{s_r+1}} = 1, \cdots, \beta_{1,j_{s_t}} = 1$; $\beta_{1,j_{s_t+1}} = *, \cdots, \beta_{1,j_{n-1}} = *$, where $\beta_{1,k}$ is the kth bit of β_1. Then, to get the difference pattern of β_2, we can proceed in the same manner described in the above steps, except in step 2 we only need to add these constraints: $\beta_{1,j_0} = 0, \cdots, \beta_{1,j_{s_r}} = 0$; $\beta_{1,j_{s_r+1}} = 1, \cdots, \beta_{1,j_{s_t}} = 1$.

By using the above method, we produce the same results of Table 4 in [12], Tables 7 and 8 in [20] automatically. At first glance, our approach seems to be a overkill since it converts a simple task which can be done manually into a task of solving many small MIP instances. But as has long been recognized by the programming and computer engineering community, we support the *Rule of Economy* which states that *programmer time is expensive; conserve it in preference to machine time*. The advantage of the method presented in this section is that it can be integrated into the MIP framework for automatic differential analysis [16], therefore reduce the burden of cryptanalyst significantly (Table 5).

6 Conclusion and Discussion

This work makes some contribution to the MIP-based method for automatic differential analysis. Firstly, the MIP-based method is applied to the cipher PRIDE in the related-key model by partially modelling the differential behavior of the modulo additions involved in the key schedule algorithm of PRIDE. To the best of our knowledge, it is the first time that the MIP-based method is used to analyze a cryptographic algorithm involving the modulo addition operations. Secondly, by using constraints from the H-representation of a specific convex hull, we give a method for constructing MIP models whose feasible regions are exactly the sets of all possible differential characteristics for SIMON (a family of lightweight block ciphers designed by the U.S. National Security Agency). Thirdly, we show how to use the MIP-based method in a clever way to automatically analyze the propagation of the differences, which is an important step in the differential cryptanalysis. We think this work further strengthens the position of the MIP as a promising tool in automatic differential cryptanalysis.

Finally, we note that a future study investigating how to use the MIP-based method to analyze ciphers involving ordinary additions mod 2^n would be very interesting. Unlike the case of PRIDE where one operand of the modulo addition operation is a constant, an ordinary addition modulo 2^n can not be treated as an $n \times n$ S-box, and modelling its differential behavior will be much more difficult.

Acknowledgements. The authors would like to thank the anonymous reviewers for their helpful comments and suggestions. The work of this paper was supported by the National Key Basic Research Program of China (2013CB834203), the National Natural Science Foundation of China (Grants 61472417, 61402469 and 61472415), the Strategic Priority Research Program of Chinese Academy of Sciences under Grant XDA06010702, and the State Key Laboratory of Information Security, Chinese Academy of Sciences.

A 2-round Iterative Related-key Differential Characteristics with Probability 2^{-4} for PRIDE

Table 1. 6 characteristics with $\Delta I = 0$, two active S-boxes in the first round and zero active S-box in the second round, and the differential pattern used by the active S-box is $1000 \rightarrow 1000$

1	ΔI	0000	0000	0000	0000	0000	0000	0000	0000	0000	0000	0000	0000	0000	0000	0000	0000
	Δk	1000	0000	0000	0000	1000	0000	0000	0000	0000	0000	0000	0000	0000	0000	0000	0000
2	ΔI	0000	0000	0000	0000	0000	0000	0000	0000	0000	0000	0000	0000	0000	0000	0000	0000
	Δk	0000	1000	0000	0000	0000	1000	0000	0000	0000	0000	0000	0000	0000	0000	0000	0000
3	ΔI	0000	0000	0000	0000	0000	0000	0000	0000	0000	0000	0000	0000	0000	0000	0000	0000
	Δk	0000	0000	1000	0000	0000	0000	1000	0000	0000	0000	0000	0000	0000	0000	0000	0000
4	ΔI	0000	0000	0000	0000	0000	0000	0000	0000	0000	0000	0000	0000	0000	0000	0000	0000
	Δk	0000	0000	0000	1000	0000	0000	0000	1000	0000	0000	0000	0000	0000	0000	0000	0000
5	ΔI	0000	0000	0000	0000	0000	0000	0000	0000	0000	0000	0000	0000	0000	0000	0000	0000
	Δk	0000	0000	0000	0000	1000	0000	0000	0000	1000	0000	0000	0000	0000	0000	0000	0000
6	ΔI	0000	0000	0000	0000	0000	0000	0000	0000	0000	0000	0000	0000	0000	0000	0000	0000
	Δk	1000	0000	0000	0000	0000	0000	0000	0000	1000	0000	0000	0000	0000	0000	0000	0000

Table 2. 6 characteristics with zero active S-box in the first round and two active S-box in the second round, and the differential pattern used by the active S-boxes is 1000 → 1000

1	ΔI	1000	0000	0000	0000	1000	0000	0000	0000	0000	0000	0000	0000	0000	0000	0000	0000
	Δk	1000	0000	0000	0000	1000	0000	0000	0000	0000	0000	0000	0000	0000	0000	0000	0000
2	ΔI	0000	1000	0000	0000	0000	1000	0000	0000	0000	0000	0000	0000	0000	0000	0000	0000
	Δk	0000	1000	0000	0000	0000	1000	0000	0000	0000	0000	0000	0000	0000	0000	0000	0000
3	ΔI	0000	0000	1000	0000	0000	0000	1000	0000	0000	0000	0000	0000	0000	0000	0000	0000
	Δk	0000	0000	1000	0000	0000	0000	1000	0000	0000	0000	0000	0000	0000	0000	0000	0000
4	ΔI	0000	0000	0000	1000	0000	0000	0000	1000	0000	0000	0000	0000	0000	0000	0000	0000
	Δk	0000	0000	0000	1000	0000	0000	0000	1000	0000	0000	0000	0000	0000	0000	0000	0000
5	ΔI	0000	0000	0000	0000	1000	0000	0000	0000	1000	0000	0000	0000	0000	0000	0000	0000
	Δk	0000	0000	0000	0000	1000	0000	0000	0000	1000	0000	0000	0000	0000	0000	0000	0000
6	ΔI	1000	0000	0000	0000	0000	0000	0000	0000	1000	0000	0000	0000	0000	0000	0000	0000
	Δk	1000	0000	0000	0000	0000	0000	0000	0000	1000	0000	0000	0000	0000	0000	0000	0000

Table 3. 12 characteristics with one active S-box in the first round and one active S-box in the second round, and the differential pattern used by the active S-boxes is 1000 → 1000

1	ΔI	1000	0000	0000	0000	1000	0000	0000	0000	1000	0000	0000	0000	0000	0000	0000	0000
	Δk	1000	0000	0000	0000	1000	0000	0000	0000	0000	0000	0000	0000	0000	0000	0000	0000
2	ΔI	1000	0000	0000	0000	1000	0000	0000	0000	0000	0000	0000	0000	1000	0000	0000	0000
	Δk	1000	0000	0000	0000	1000	0000	0000	0000	0000	0000	0000	0000	0000	0000	0000	0000
3	ΔI	0000	1000	0000	0000	0000	1000	0000	0000	0000	1000	0000	0000	0000	0000	0000	0000
	Δk	0000	1000	0000	0000	0000	1000	0000	0000	0000	0000	0000	0000	0000	0000	0000	0000
4	ΔI	0000	1000	0000	0000	0000	1000	0000	0000	0000	0000	0000	0000	0000	1000	0000	0000
	Δk	0000	1000	0000	0000	0000	1000	0000	0000	0000	0000	0000	0000	0000	0000	0000	0000
5	ΔI	0000	0000	1000	0000	0000	0000	1000	0000	0000	0000	1000	0000	0000	0000	0000	0000
	Δk	0000	0000	1000	0000	0000	0000	1000	0000	0000	0000	0000	0000	0000	0000	0000	0000
6	ΔI	0000	0000	1000	0000	0000	0000	1000	0000	0000	0000	0000	0000	0000	0000	1000	0000
	Δk	0000	0000	1000	0000	0000	0000	1000	0000	0000	0000	0000	0000	0000	0000	0000	0000
7	ΔI	0000	0000	0000	1000	0000	0000	0000	1000	0000	0000	0000	1000	0000	0000	0000	0000
	Δk	0000	0000	0000	1000	0000	0000	0000	1000	0000	0000	0000	0000	0000	0000	0000	0000
8	ΔI	0000	0000	0000	1000	0000	0000	0000	1000	0000	0000	0000	0000	0000	0000	0000	1000
	Δk	0000	0000	0000	1000	0000	0000	0000	1000	0000	0000	0000	0000	0000	0000	0000	0000
9	ΔI	1000	0000	0000	0000	1000	0000	0000	0000	1000	0000	0000	0000	0000	0000	0000	0000
	Δk	0000	0000	0000	0000	1000	0000	0000	0000	1000	0000	0000	0000	0000	0000	0000	0000
10	ΔI	0000	0000	0000	0000	1000	0000	0000	0000	1000	0000	0000	0000	1000	0000	0000	0000
	Δk	0000	0000	0000	0000	1000	0000	0000	0000	1000	0000	0000	0000	0000	0000	0000	0000
11	ΔI	1000	0000	0000	0000	1000	0000	0000	0000	1000	0000	0000	0000	0000	0000	0000	0000
	Δk	1000	0000	0000	0000	0000	0000	0000	0000	1000	0000	0000	0000	0000	0000	0000	0000
12	ΔI	1000	0000	0000	0000	0000	0000	0000	0000	1000	0000	0000	0000	1000	0000	0000	0000
	Δk	1000	0000	0000	0000	0000	0000	0000	0000	1000	0000	0000	0000	0000	0000	0000	0000

Table 4. 10 characteristics with one active S-box in the first round and two active S-box in the second round, and the differential patterns used by the active S-boxes are 1011 → 0010, 1000 → 0010, 1011 → 0011, 1000 → 0011, 0110 → 0001, 0111 → 0001, 0100 → 0001, 0101 → 0001, 0110 → 0100, and 0001 → 0100 respectively; the characteristics marked by a "*" are also 1-round iterative characteristics

1	ΔI	0010	0000	0000	0000	0000	0000	0000	0010	0010	0000	0000	0000	0000	0000	0000	0000
*	Δk	0010	0000	0000	0000	1011	0000	0000	0010	0010	0000	0000	0000	0000	0000	0000	0000
2	ΔI	0010	0000	0000	0000	0000	0000	0000	0010	0010	0000	0000	0000	0000	0000	0000	0000
*	Δk	0010	0000	0000	0000	1000	0000	0000	0010	0010	0000	0000	0000	0000	0000	0000	0000
3	ΔI	0011	0000	0000	0000	0001	0000	0000	0010	0011	0000	0000	0000	0000	0000	0000	0000
	Δk	0011	0000	0000	0000	1010	0000	0000	0010	0011	0000	0000	0000	0000	0000	0000	0000
4	ΔI	0011	0000	0000	0000	0001	0000	0000	0010	0011	0000	0000	0000	0000	0000	0000	0000
	Δk	0011	0000	0000	0000	1001	0000	0000	0010	0011	0000	0000	0000	0000	0000	0000	0000
5	ΔI	0001	0000	0000	0000	0001	0000	0000	0000	0001	0000	0000	0000	0000	0000	0000	0000
	Δk	0001	0000	0000	0000	0111	0000	0000	0000	0001	0000	0000	0000	0000	0000	0000	0000
6	ΔI	0001	0000	0000	0000	0001	0000	0000	0000	0001	0000	0000	0000	0000	0000	0000	0000
	Δk	0001	0000	0000	0000	0110	0000	0000	0000	0001	0000	0000	0000	0000	0000	0000	0000
7	ΔI	0001	0000	0000	0000	0001	0000	0000	0000	0001	0000	0000	0000	0000	0000	0000	0000
	Δk	0001	0000	0000	0000	0101	0000	0000	0000	0001	0000	0000	0000	0000	0000	0000	0000
8	ΔI	0001	0000	0000	0000	0001	0000	0000	0000	0001	0000	0000	0000	0000	0000	0000	0000
	Δk	0001	0000	0000	0000	0100	0000	0000	0000	0001	0000	0000	0000	0000	0000	0000	0000
9	ΔI	0100	0000	0000	0000	0000	0000	0000	0100	0100	0000	0000	0000	0000	0000	0000	0000
	Δk	0010	0000	0000	0000	0000	0000	0000	0100	0100	0000	0000	0000	0000	0000	0000	0000
10	ΔI	0100	0000	0000	0000	0000	0000	0000	0100	0100	0000	0000	0000	0000	0000	0000	0000
	Δk	0101	0000	0000	0000	0000	0000	0000	0100	0100	0000	0000	0000	0000	0000	0000	0000

Table 5. 8 characteristics which require the output difference of $g_i^{(1)}(\cdot)$ is $0x20$

1	ΔI	0000	0000	1000	0000	0000	0000	0000	0000	0000	0000	1000	0000	0000	1000	0000	0000
	Δk	0000	0000	1000	0000	0000	0000	0000	0000	0000	0000	1000	0000	0000	0000	0000	0000
2	ΔI	0000	0000	1000	0000	0000	0000	0000	0000	0000	0000	1000	0000	0000	0000	0000	0000
	Δk	0000	0000	1000	0000	0000	0000	0000	0000	0000	0000	1000	0000	0000	0000	0000	0000
3	ΔI	0000	0000	1000	0000	0000	0000	1000	0000	0000	0000	1000	0000	0000	0000	0000	0000
	Δk	0000	0000	1000	0000	0000	0000	0000	0000	0000	0000	1000	0000	0000	0000	0000	0000
4	ΔI	0000	0000	0000	0000	0000	0000	0000	0000	0000	0000	0000	0000	0000	0000	0000	0000
	Δk	0000	0000	1000	0000	0000	0000	0000	0000	0000	0000	1000	0000	0000	0000	0000	0000
5	ΔI	0000	0000	0000	0000	0000	0000	1000	0000	0000	0000	1000	0000	0000	0000	1000	0000
	Δk	0000	0000	0000	0000	0000	0000	1000	0000	0000	0000	1000	0000	0000	0000	0000	0000
6	ΔI	0000	0000	0000	0000	0000	0000	1000	0000	0000	0000	1000	0000	0000	0000	0000	0000
	Δk	0000	0000	0000	0000	0000	0000	1000	0000	0000	0000	1000	0000	0000	0000	0000	0000
7	ΔI	0000	0000	0000	0000	0000	0000	0000	0000	0000	0000	0000	0000	0000	0000	0000	0000
	Δk	0000	0000	0000	0000	0000	0000	1000	0000	0000	0000	1000	0000	0000	0000	0000	0000
8	ΔI	0000	0000	1000	0000	0000	1000	0000	0000	0000	0000	1000	0000	0000	0000	0000	0000
	Δk	0000	0000	0000	0000	0000	0000	1000	0000	0000	0000	1000	0000	0000	0000	0000	0000

References

1. Biryukov, A., Velichkov, V.: Automatic search for differential trails in ARX ciphers. In: Benaloh, J. (ed.) CT-RSA 2014. LNCS, vol. 8366, pp. 227–250. Springer, Heidelberg (2014)
2. Biham, E., Shamir, A.: Differential cryptanalysis of DES-like cryptosystems. J. Cryptology **4**(1), 3–72 (1991)
3. Dinur, I.: Cryptanalytic Time-Memory-Data Tradeoffs for FX-Constructions with Applications to PRINCE and PRIDE. Cryptology ePrint Archive, Report 2014/656 (2014). http://eprint.iacr.org/2014/656
4. Zhao, J., Wang, X., Wang, M., Dong, X.: Differential Analysis on Block Cipher PRIDE. IACR Cryptology ePrint Archive, Report 2014/525 (2014). http://eprint.iacr.org/2014/525
5. Aoki, K., Kobayashi, K., Moriai, S.: Best differential characteristic search of FEAL. In: Biham, E. (ed.) FSE 1997. LNCS, vol. 1267, pp. 41–53. Springer, Heidelberg (1997)
6. Ohta, K., Moriai, S., Aoki, K.: Improving the search algorithm for the best linear expression. In: Coppersmith, D. (ed.) CRYPTO 1995. LNCS, vol. 963, pp. 157–170. Springer, Heidelberg (1995)
7. Albrecht, M.R., Driessen, B., Kavun, E.B., Leander, G., Paar, C., Yalçın, T.: Block ciphers – focus on the linear layer (feat. PRIDE). In: Garay, J.A., Gennaro, R. (eds.) CRYPTO 2014, Part I. LNCS, vol. 8616, pp. 57–76. Springer, Heidelberg (2014)
8. Matsui, M.: On correlation between the order of S-boxes and the strength of DES. In: Santis, A. (ed.) EUROCRYPT 1994. LNCS, vol. 950, pp. 366–375. Springer, Heidelberg (1995)
9. Mouha, N., Preneel, B.: Towards finding optimal differential characteristics for ARX: Application to Salsa20. IACR Cryptology ePrint Archive, Report 2013/328 (2013). http://eprint.iacr.org/2013/328
10. Mouha, N., Wang, Q., Gu, D., Preneel, B.: Differential and linear cryptanalysis using mixed-integer linear programming. In: Wu, C.-K., Yung, M., Lin, D. (eds.) Inscrypt 2011. LNCS, vol. 7537, pp. 57–76. Springer, Heidelberg (2012)
11. Wang, N., Wang, X., Jia, K., Zhao, J.: Differential Attacks on Reduced SIMON Versions with Dynamic Key-guessing Techniques. Cryptology ePrint Archive, Report 2014/448 (2014). http://eprint.iacr.org/2014/448
12. Yang, Q., Hu, L., Sun, S., Qiao, K., Song, L., Shan, J., Ma, X.: Improved Differential Analysis of Block Cipher PRIDE. IACR Cryptology ePrint Archive, Report 2014/978 (2014). https://eprint.iacr.org/2014/978
13. Beaulieu, R., Shors, D., Smith, J., Treatman-Clark, S., Weeks, B., Wingers, L.: The SIMON and SPECK families of lightweight block ciphers. IACR Cryptology ePrint Archive, Report 2013/404 (2013). http://eprint.iacr.org/2013/404
14. Wu, S., Wang, M.: Security evaluation against differential cryptanalysis for block cipher structures. IACR Cryptology ePrint Archive, Report 2011/551 (2011). https://eprint.iacr.org/2011/551
15. Sun, S., Hu, L., Song, L., Xie, Y., Wang, P.: Automatic security evaluation of block ciphers with S-bP structures against related-key differential attacks. In: Lin, D., Xu, S., Yung, M. (eds.) Inscrypt 2013. LNCS, vol. 8567, pp. 39–51. Springer, Heidelberg (2014)

16. Sun, S., Hu, L., Wang, M., Wang, P., Qiao, K., Ma, X., Shi, D., Song, L., Fu, K.: Towards Finding the Best Characteristics of Some Bit-oriented Block Ciphers and Automatic Enumeration of (Related-key) Differential and Linear Characteristics with Predefined Properties. Cryptology ePrint Archive, Report 2014/747 (2014). http://eprint.iacr.org/2014/747
17. Sun, S., Hu, L., Wang, P., Qiao, K., Ma, X., Song, L.: Automatic security evaluation and (related-key) differential characteristic search: application to SIMON, PRESENT, LBlock, DES(L) and other bit-oriented block ciphers. In: Sarkar, P., Iwata, T. (eds.) ASIACRYPT 2014. LNCS, vol. 8873, pp. 158–178. Springer, Heidelberg (2014)
18. Kölbl, S.: CryptoSMT - an easy to use tool for cryptanalysis of symmetric primitives likes block ciphers or hash functions. https://github.com/kste/cryptosmt
19. Kölbl, S., Leander, G., Tiessen, T.: Observations on the SIMON block cipher family. Cryptology ePrint Archive, Report 2015/145 (2015). http://eprint.iacr.org/2015/145
20. Dai, Y., Chen, S.: Cryptanalysis of Full PRIDE Block Cipher. Cryptology ePrint Archive, Report 2014/987 (2014). http://eprint.iacr.org/2014/987
21. Bao, Z., Zhang, W., Lin, D.: Speeding up the search algorithm for the best differential and best linear trails. In: Lin, D., Yung, M., Zhou, J. (eds.) Inscrypt 2014. LNCS, vol. 8957, pp. 259–285. Springer, Heidelberg (2015)

Automatic Search for Linear Trails of the SPECK Family

Yuan Yao[1,2](✉), Bin Zhang[1], and Wenling Wu[1]

[1] TCA Laboratory, Institute of Software, Chinese Academy of Sciences, Beijing, China
{yaoyuan,zhangbin,wwl}@tca.iscas.ac.cn
[2] University of Chinese Academy of Sciences, Beijing, China

Abstract. SPECK is a lightweight block cipher family designed by the U.S. National Security Agency and published in 2013. Although several cryptanalyses have been applied since then, no linear results have been proposed. In this paper, we apply Wallén's enumeration algorithm to Matsui's branch-and-bound framework and find the best correlations of SPECK reduced to various rounds, i.e. full rounds of SPECK-32 and 7/ 5/ 4/ 4 rounds of SPECK-48/ 64/ 96/ 128. Since the best 10-round correlation of SPECK-32 is as small as 2^{-17} already, SPECK-32 is immune to the 1-dimensional linear cryptanalysis. Moreover, we present several distinguishers and key recovery attacks as an application of the linear trails. Besides the search for linear trails, we also discuss possible implementations of the Wallén's algorithm and provide an implementation which is faster than the straightforward implementations.

Keywords: Automatic search · Linear cryptanalysis · SPECK · Modulo addition

1 Introduction

The SPECK family [1] is based on a Feistel-like structure and belongs to the ARX ciphers, i.e. primitives composed of modulo addition, bitwise rotation and bitwise XOR only. It is designed to provide optimal software performance on resource constrained devices and is comprised of five variants according to the block size. Despite of its simple structure, no cryptanalysis has threatened its security and particularly no linear cryptanalysis has been proposed due to the intrinsic property of modulo addition. The best previously published attacks are the improved differential cryptanalysis provided by Dinur at SAC 2014 [4].

Generally, good linear trails/approximations should be found in advance in order to launch linear attacks. A widely used approach to search for linear trails of block ciphers is the general framework proposed by Matsui at EUROCRYPT 1994 [8] and it is straightforward to apply as long as the linear approximation

This work is supported by the National Basic Research Program of China (No. 2013CB338002).

J. Lopez and C.J. Mitchell (Eds.): ISC 2015, LNCS 9290, pp. 158–176, 2015.
DOI: 10.1007/978-3-319-23318-5_9

table (LAT) of sub-components is obtained. However, the complexity to compute the LAT varies greatly from cipher to cipher. In particular, the time/memory complexity of addition modulo 2^n is $O(2^{3n})$ for a plain enumeration which is nearly impractical even with $n = 16$. Whereas the problem exists in the search for differential trails as well, Biryukov [2] has recently proposed a technique using partial differential distribution tables, called the *threshold search*, and successfully conquered this problem. Fortunately, Wallén has already provided an efficient algorithm to enumerate the LAT of modulo addition at FSE 2003 [10], thus linear approximations could be generated on the fly until it is necessary. The algorithm is further rediscovered in [9] using another approach and its efficiency has been proved by the application to SNOW 2.0 [9] and SOSEMANUK [3]. In case of possible confusions, it should be noted that another algorithm which determines the correlation of a given linear approximation with $O(\log(n))$ time was presented by Wallén in [10] as well. As the latter algorithm is never used in this paper, the Wallén's algorithm in this paper always refers to the algorithm to enumerate the LAT.

By combining Wallén's algorithm and Matsui's branch-and-bound framework, we are able to find the best linear trail of SPECK-32 of full rounds and the best linear trail of SPECK-48/ 64/ 96/ 128 reduced to 7/ 5/ 4/ 4 rounds respectively, shown in Tables 1 and 2 where "≥" denotes a lower bound of the best correlation. Since the data complexity of a 1-dimensional linear cryptanalysis is inversely proportional to the square of the correlation, the best 10-round correlation in Table 1 suggests that SPECK-32 is secure under this method. Indeed, the data complexity of a 1-dimensional linear cryptanalysis against SPECK-32 using the 10-round linear trail is 2^{34}, greater than the size of the code book which is 2^{32}. Moreover, we provide several distinguishers and key recovery attacks as an application of the linear trails. Yet, they do not pose a threat to SPECK and are worse than the differential cryptanalyses of Dinur. After all, this is the first linear cryptanalysis against the SPECK family, evaluating the security in a different perspective.

We additionally find a set of necessary conditions for correlations to be non-zero which allows us to develop an alternative implementation of Wallén's algorithm. According to experiments, this implementation is faster than straightforward implementations derived from the Wallén's theorem and thus useful when called for a tremendous number of times.

The rest of this paper is organized as follows. Section 2 introduces SPECK, Matsui's branch-and-bound framework and the previous Wallén's results on lin-

Table 1. Best correlations for SPECK-32

Rounds(r)	1	2	3	4	5	6	7	8	9	10	11
$\lvert B[r]\rvert$	1	1	2^{-1}	2^{-3}	2^{-5}	2^{-7}	2^{-9}	2^{-12}	2^{-14}	2^{-17}	2^{-19}
Rounds(r)	12	13	14	15	16	17	18	19	20	21	22
$\lvert B[r]\rvert$	2^{-20}	2^{-22}	2^{-24}	2^{-26}	2^{-28}	2^{-30}	2^{-34}	2^{-36}	2^{-38}	2^{-40}	2^{-42}

Table 2. Best correlations for SPECK48/ 64/ 96/ 128 ("≥" indicates a lower bound)

$\|B[r]\|$		Rounds(r)											
		1	2	3	4	5	6	7	8	9	10	11	12
Block length	48	1	1	2^{-1}	2^{-3}	2^{-6}	2^{-8}	2^{-12}	$\geq 2^{-17}$	$\geq 2^{-20}$	$\geq 2^{-25}$		
	64	1	1	2^{-1}	2^{-3}	2^{-6}	$\geq 2^{-10}$	$\geq 2^{-14}$	$\geq 2^{-17}$	$\geq 2^{-19}$	$\geq 2^{-21}$	$\geq 2^{-25}$	$\geq 2^{-31}$
	96	1	1	2^{-1}	2^{-3}	$\geq 2^{-6}$	$\geq 2^{-11}$						
	128	1	1	2^{-1}	2^{-3}	$\geq 2^{-6}$	$\geq 2^{-11}$						

ear approximation of modulo addition. Section 3 describes the search for linear trails on SPECK and the cryptanalytic results. Section 4 provides the alternative implementation of Wallén's algorithm. Finally, Sect. 5 draws conclusions.

2 Preliminaries

2.1 Notions

a_i	the i-th least-significant bit of word $\boldsymbol{a}$, i.e. $\boldsymbol{a} = (a_{n-1}, \cdots, a_0)$
$\mathbf{1}$	the word $(1, \ldots, 1)$
$null$	a special word of length zero, i.e. ( )
$\|\|$	the concatenation operation
$\boldsymbol{a} \cdot \boldsymbol{b}$	the inner product of $\boldsymbol{a}, \boldsymbol{b}$
$\boldsymbol{ab}$	the bitwise AND of $\boldsymbol{a}, \boldsymbol{b}$
$\underset{n}{\boxplus}$	the addition modulo 2^n and n is omitted if it is clear from the context
$\underset{n}{\boxminus}$	the subtraction modulo 2^n and n is omitted if it is clear from the context
$\Pr_D(\boldsymbol{y})$	the probability to be $\boldsymbol{y}$ given the probability distribution function D

2.2 Description of SPECK

SPECK is a family of block ciphers containing five variants according to the block size which can be further divided into ten variants regarding the key size. Each variant has two constants ς, τ depending on the block size, i.e. $\varsigma = 7, \tau = 2$ for SPECK-32 and $\varsigma = 8, \tau = 3$ otherwise. The i-th round function (Fig. 1) is defined by

$$\boldsymbol{x}[i+1] \leftarrow ((\boldsymbol{x}[i] \ggg \varsigma) \boxplus \boldsymbol{y}[i]) \oplus \boldsymbol{k}[i]$$
$$\boldsymbol{y}[i+1] \leftarrow (\boldsymbol{y}[i] \lll \tau) \oplus \boldsymbol{x}[i+1]$$

where $\boldsymbol{x}[i]$ and $\boldsymbol{y}[i]$ denote the left and right block of the input respectively, and $\boldsymbol{k}[i]$ is the round-key. The key schedule algorithm is omitted since it is irrelevant to the search, but it should be noted that the master key can be recovered with $2 \times \text{key length}/\text{block size}$ successive round-keys. For more details, please refer to [1].

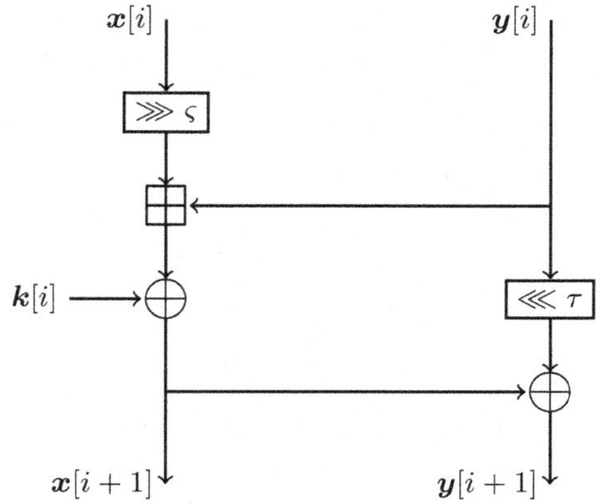

Fig. 1. The round function of SPECK

2.3 Automatic Search Framework

The following is an introduction to the general branch-and-bound search framework proposed by Matsui at EUROCRYPT 1994 [8] in the language of linear cryptanalysis.

To find the best correlation of r successive rounds $B[r]$, the framework performs a recursive search from the knowledge of shorter rounds $B[1], \ldots, B[r-1]$ and an initial estimate $\hat{B}[r]$ such that $|\hat{B}[r]| < |B[r]|$. In the search phase, an s-round trail is kept only if

$$|B[r-s]\prod_{i=1}^{s} c[i]| > |\hat{B}[r]|, 1 \leq s \leq r$$

where $c[i]$ denotes the correlation of the i-th round and $B[0]$ is defined to be 1. $\hat{B}[r]$ is updated once the correlation of a r-round trail is better than $\hat{B}[r]$. Therefore, $B[r] = \hat{B}[r]$ when the search completes. Algorithm 1 is an overview where *Get_Mask* is a cipher dependent function to extend linear trails.

2.4 Linear Approximation of Modulo Addition

In this subsection, we briefly introduce Wallén's results on linear approximations of addition modulo 2^n in [10,11].

Definition 1 (Correlation). *Let* $\boldsymbol{u}$ *be the output mask of the modulo addition and* $\boldsymbol{v}, \boldsymbol{w}$ *be the input masks. Then the correlation is defined by*

$$c(\boldsymbol{u}, \boldsymbol{v}, \boldsymbol{w}) \triangleq 2\Pr(\boldsymbol{u} \cdot (\boldsymbol{Z}_1 \boxplus \boldsymbol{Z}_2) \oplus \boldsymbol{v} \cdot \boldsymbol{Z}_1 \oplus \boldsymbol{w} \cdot \boldsymbol{Z}_2 = 0) - 1$$

where $\boldsymbol{Z}_1, \boldsymbol{Z}_2$ *are independent uniform distributed random variables.*

Algorithm 1. Matsui Search for The Best Linear Trail

function SEARCH(B, $T = \{\}$) ▷ $T = \{T[1], \ldots, T[s]\}$ denotes the linear trail
  $r \leftarrow$ SIZEOF$(B) - 1$, $s \leftarrow$ SIZEOF(T)
  if $s = r$ **then**
    $\hat{B}[r] \leftarrow \prod_{i=1}^{r} c[i]$
  else
    for T' **in** GET_MASK(T) **do** ▷ Extend T to $(s+1)$-round linear trails
      if $|B[r-(s+1)] \prod_{i=1}^{s+1} c'[i]| > |\hat{B}[r]|$ **then** ▷ $c'[i]$ is the correlation of $T'[i]$
        SEARCH(B, T')
      else
        return ▷ Pruning, supposing that T's are enumerated in decreasing order
      end if
    end for
  end if
end function

The Enumeration Algorithm.

Theorem 1. *[9, 11] Let* $S^0(0,0) \triangleq \{null\}$*,* $S^0(n,k) = S^1(n,k) \triangleq \emptyset$ *when* $k < 0$ *or* $k \geq n > 0$*, and*

$$S^0(n,k) \triangleq \left(S^0(n-1,k) \parallel \{0\}\right) \cup \left(S^1(n-1,k-1) \parallel \{1,2,4,7\}\right) \quad (1)$$

$$S^1(n,k) \triangleq \left(S^0(n-1,k) \parallel \{7\}\right) \cup \left(S^1(n-1,k-1) \parallel \{0,3,5,6\}\right) \quad (2)$$

otherwise, where $S^\star \parallel \Omega \triangleq \{\boldsymbol{a} \parallel \boldsymbol{b} \mid \boldsymbol{a} \in S^\star, \boldsymbol{b} \in \Omega\}$*. Then*

$$S(n,k) \triangleq \left\{(\boldsymbol{u}, \boldsymbol{v}, \boldsymbol{w}) \mid 4u_i + 2v_i + w_i = s_i, i = 0, \ldots, n-1, \right.$$
$$\left. \boldsymbol{s} \in S^0(n,k) \cup S^1(n,k)\right\}$$

is the set of all masks such that $c(\boldsymbol{u}, \boldsymbol{v}, \boldsymbol{w}) = \pm 2^{-k}$*.*

Example 1. $S^0(n,0) = \{(0 \cdots 0)\}, S^1(n,0) = \{(0 \cdots 07)\}$, thus $S(n,0) = \{((0 \cdots 0), (0 \cdots 0), (0 \cdots 0)), ((0 \cdots 01), (0 \cdots 01), (0 \cdots 01))\}$ is the set of all masks such that $c(\boldsymbol{u}, \boldsymbol{v}, \boldsymbol{w}) = \pm 1$.

As was pointed out by Wallén, the LAT of addition modulo 2^n can be enumerated using $O(n)$ space via Theorem 1. A trivial implementation, called the *top-down* method in this paper, can be deduced as shown in Fig. 2(a) and Appendix A.1. However, it is inefficient in the sense that the same subtree will be generated for multiple times. Another possible implementation is the *bottom-up* method which is shown in Fig. 2(b) and Appendix A.2, i.e. starting from $S^0(0,0)$ and then computing $S(1,0)$ etc. While it also generates duplicate subtrees, surprisingly it is faster than the top-down method. (See Fig. 5 for the comparison)

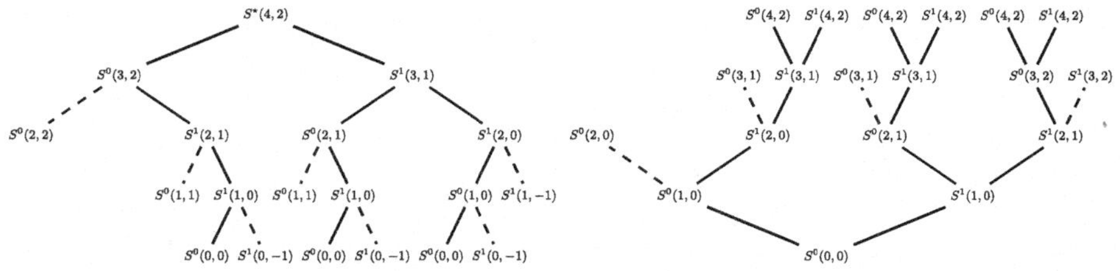

(a) The Top-down Method (b) The Bottom-up Method

Fig. 2. The computational process of $S(4,2)$

Common Prefix Mask vs. Correlation. This subsection serves for the alternative implementation of the above algorithm and may be skipped safely to understand the search.

Definition 2 (CPM). *Let $\boldsymbol{a}, \boldsymbol{b} \in \mathbb{F}_2^n$. If $n = 2$, the common prefix mask of $\boldsymbol{a}, \boldsymbol{b}$ is defined by*

$$cpm_2(\boldsymbol{a}, \boldsymbol{b}) = a_1$$

If $n > 2$, the common prefix mask of $\boldsymbol{a}, \boldsymbol{b}$ is defined by

$$cpm_n(\boldsymbol{a}, \boldsymbol{b}) = a_{n-1} \parallel cpm_{n-1}\left((a_{n-2} \oplus a_{n-1} \cdot b_{n-2}) \parallel \boldsymbol{a}', 1 \parallel \boldsymbol{b}'\right)$$

where $\boldsymbol{a}' = (a_{n-3}, \ldots, a_0)$ and $\boldsymbol{b}' = (b_{n-3}, \ldots, b_0)$.[1]

Lemma 1. *[10] Let $\boldsymbol{u}, \boldsymbol{v}, \boldsymbol{w} \in \mathbb{F}_2^n$ be defined as in Definition 1, $\boldsymbol{\phi} = \boldsymbol{v} \oplus \boldsymbol{u}, \boldsymbol{\varphi} = \boldsymbol{w} \oplus \boldsymbol{u}$ be the input masks of the carry function, $\boldsymbol{\gamma} = \boldsymbol{v} \oplus \boldsymbol{w}$ and $\boldsymbol{\delta} = cpm_{n+1}(0 \parallel \boldsymbol{u}, (0 \parallel \boldsymbol{\gamma}) \oplus \mathbf{1})$. Then*

$$c(\boldsymbol{u}, \boldsymbol{v}, \boldsymbol{w}) = \begin{cases} (-1)^{wt(\boldsymbol{\delta}\boldsymbol{\phi}\boldsymbol{\varphi})} 2^{-wt(\boldsymbol{\delta})}, & \text{if } \boldsymbol{\phi} = \boldsymbol{\phi}\boldsymbol{\delta} \text{ and } \boldsymbol{\varphi} = \boldsymbol{\varphi}\boldsymbol{\delta} \\ 0, & \text{otherwise} \end{cases}$$

where wt is the hamming weight.

Example 2. Suppose $\boldsymbol{u} = (1100), \boldsymbol{v} = \boldsymbol{w} = (1000)$, then $\boldsymbol{\phi} = \boldsymbol{\varphi} = (0100)$, $\boldsymbol{\gamma} = (0000)$ and $\boldsymbol{\delta} = (0100)$. Thus, $c(\boldsymbol{u}, \boldsymbol{v}, \boldsymbol{w}) = -2^{-1}$.

3 Linear Results on SPECK

3.1 Details of the Search

In this section, we will concentrate on the design of *Get_Mask* which will be used to extend linear trails by Algorithm 1. Firstly, we recall the linear properties of branch, bitwise XOR and bitwise rotation.

[1] This definition is the method proposed by Wallén to calculate the CPM.

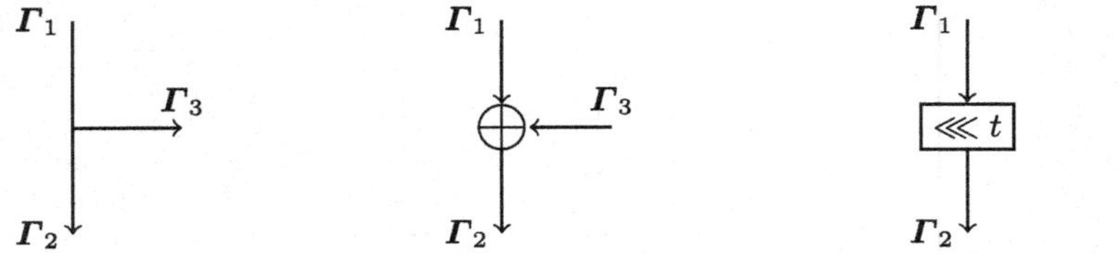

(a) Branch (b) Bitwise XOR (c) Bitwise Rotation

Fig. 3. The spread of linear masks

Property 1. Let $\boldsymbol{\Gamma}_1, \boldsymbol{\Gamma}_2, \boldsymbol{\Gamma}_3$ be linear masks defined by Fig. 3(a), then the correlation is nonzero if and only if $\boldsymbol{\Gamma}_1 \oplus \boldsymbol{\Gamma}_2 \oplus \boldsymbol{\Gamma}_3 = \mathbf{0}$.

Property 2. Let $\boldsymbol{\Gamma}_1, \boldsymbol{\Gamma}_2, \boldsymbol{\Gamma}_3$ be linear masks defined by Fig. 3(b) then the correlation is nonzero if and only if $\boldsymbol{\Gamma}_1 = \boldsymbol{\Gamma}_2 = \boldsymbol{\Gamma}_3$.

Property 3. Let $\boldsymbol{\Gamma}_1, \boldsymbol{\Gamma}_2$ be linear masks defined by Fig. 3(c) then the correlation is nonzero if and only if $\boldsymbol{\Gamma}_2 = \boldsymbol{\Gamma}_1 \lll t$.

Let the linear masks of the i-th round be defined in Fig. 4. Accordingly,

$$\begin{aligned} \boldsymbol{u}[i] &= \boldsymbol{X}[i+1] \oplus \boldsymbol{Y}[i+1] \\ \boldsymbol{v}[i] &= \boldsymbol{X}[i] \ggg \varsigma \\ \boldsymbol{w}[i] &= \boldsymbol{Y}[i] \oplus (\boldsymbol{Y}[i+1] \ggg \tau) \end{aligned}$$

Thereupon,

$$\begin{aligned} \boldsymbol{u}[r] &= \boldsymbol{X}[r+1] \oplus \boldsymbol{Y}[r+1] \\ \boldsymbol{u}[r-1] &= (\boldsymbol{v}[r] \lll \varsigma) \oplus \boldsymbol{w}[r] \oplus (\boldsymbol{Y}[r+1] \ggg \tau) \end{aligned}$$

and

$$\begin{aligned} \boldsymbol{u}[i] = (\boldsymbol{v}[i+1] \lll \varsigma) \oplus \boldsymbol{w}[i+1] \oplus \\ ((\boldsymbol{u}[i+1] \oplus (\boldsymbol{v}[i+2] \lll \varsigma)) \ggg \tau), 1 \le i \le r-2 \end{aligned}$$

If we enumerate $\boldsymbol{X}[r+1]$ and $\boldsymbol{Y}[r+1]$ directly, the complexity is at least 2^{2n} and it is a waste of efforts on masks with insignificant correlations at the initial stage. Since $\boldsymbol{X}[r+1], \boldsymbol{Y}[r+1]$ are uniquely determined by $\boldsymbol{u}[r], \boldsymbol{v}[r], \boldsymbol{w}[r], \boldsymbol{u}[r-1]$, it is equivalently and more efficiently to enumerate $\boldsymbol{u}[r], \boldsymbol{v}[r], \boldsymbol{w}[r], \boldsymbol{u}[r-1]$ using the Wallén's algorithm. On the other hand, when $1 \le i \le r-2$, $\boldsymbol{u}[i]$ can be deduced from the two following rounds. As a result, we have presented a method to extend linear trails by appending one round to the front and Algorithm 2 is the corresponding implementation of *Get_Mask*.

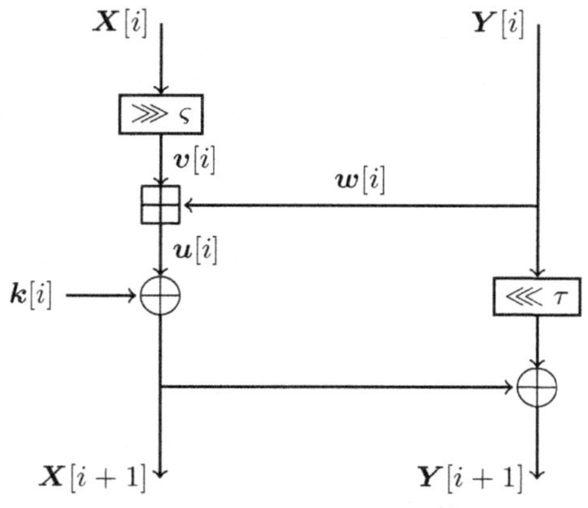

Fig. 4. Masks of the i-th round

Algorithm 2. The Implementation of *Get_Mask*

---

function GENERATE()
    for k **from** 0 **to** $n - 1$ **do**
        BU_GENERATE($n, 000, k,$ *not used, not used, not used*) ▷ bottom-up generation of $S(n, k)$ (see Appendix A.2) and 000 means totally free
    end for
end function

function GENERATE($\boldsymbol{u}$)
    for k **from** 0 **to** $n - 1$ **do**
        BU_GENERATE($n, 100, k, \boldsymbol{u},$ *not used, not used*) ▷ bottom-up generation of $S(n, k)$ (see Appendix A.2) and 100 means $\boldsymbol{u}$ is supplied and fixed
    end for
end function

function GET_MASK(T) ▷ $T = \{T[1], \ldots, T[s]\}$ and $T[r + 1 - i] = \{\boldsymbol{u}[i], \boldsymbol{v}[i], \boldsymbol{w}[i]\}$
    $s \leftarrow$ SIZEOF(T), $curr \leftarrow r - s, last \leftarrow curr + 1$
    if $s < 2$ **then**
        for *tuple* **in** GENERATE() **do**
            if $\boldsymbol{u}[r], \boldsymbol{u}[r - 1]$ don't equal $\mathbf{0}$ simultaneously **then**
                $T[s + 1] \leftarrow$ *tuple* and **yield** T
            end if
        end for
    else
        $\boldsymbol{u}[curr] \leftarrow (\boldsymbol{v}[last] \lll \varsigma) \oplus \boldsymbol{w}[last] \oplus ((\boldsymbol{u}[last] \oplus (\boldsymbol{v}[last + 1] \lll \varsigma)) \ggg \tau)$
        for *tuple* **in** GENERATE($\boldsymbol{u}[curr]$) **do**
            $T[s + 1] \leftarrow$ *tuple* and **yield** T
        end for
    end if
end function

3.2 Search Results

The automatic search is applied to variants of all block sizes and the best correlations are presented in Tables 1 and 2. Since the quotient of the best correlations of successive rounds is quite regular, $\hat{B}[r]$ is set to $2^{-3}B[r-1]$ for most of the cases. However, not all searches can finish in a reasonable time period due to the huge size of the search space, even with a tight threshold (e.g. $\hat{B}[r] \approx B[r]$). Thus, the "$\geq$" in the Table 2 denotes the best correlation that has been found in this case, i.e. a lower bound.

3.3 Linear Distinguishers

A Linear Distinguisher identifies the nonuniformity of a cipher and generally converts to a hypothesis testing problem using statistical tools. In this and subsequent sections, we make the common assumption that the correlation of a linear approximation can be estimated by the correlation of a significant linear trail. Moreover, the data complexity to distinguish two probability distributions D and D_0 is estimated by $C\left(D, D_0\right)^{-1}$ (see [5] for example) with the capacity

$$C\left(D, D_0\right) \triangleq \Sigma_{\boldsymbol{y}\in\mathcal{Y}}\left(\Pr_D(\boldsymbol{y})-\Pr_{D_0}(\boldsymbol{y})\right)^2/\Pr_{D_0}(\boldsymbol{y})$$

Since

$$\begin{aligned}\boldsymbol{x}[i]\cdot\boldsymbol{X}[i]\oplus\boldsymbol{y}[i]\cdot\boldsymbol{Y}[i] = (((\boldsymbol{x}[i-1]\ggg\varsigma)\boxplus\boldsymbol{y}[i-1])\oplus\boldsymbol{k}[i-1])\cdot \\ (\boldsymbol{X}[i]\oplus\boldsymbol{Y}[i])\oplus\boldsymbol{y}[i-1]\cdot(\boldsymbol{Y}[i]\ggg\tau)\end{aligned}$$

and $\boldsymbol{k}[i-1]\cdot(\boldsymbol{X}[i]\oplus\boldsymbol{Y}[i])$ is constant, the *absolute value* of the correlation of

$$\boldsymbol{x}[2]\cdot\boldsymbol{X}[2]\oplus\boldsymbol{y}[2]\cdot\boldsymbol{Y}[2]\oplus\boldsymbol{x}[2+r]\cdot\boldsymbol{X}[2+r]\oplus\boldsymbol{y}[2+r]\cdot\boldsymbol{Y}[2+r]$$

can be calculated from $\boldsymbol{x}[1], \boldsymbol{y}[1], \boldsymbol{x}[2+r], \boldsymbol{y}[2+r]$ without $\boldsymbol{k}[1]$. In other words, a r-round linear trail can be transformed into a $(r+1)$-round linear distinguisher by appending one round to the front. Thus, we immediately obtain the results in Table 3.

Table 3. Linear distinguishers against the SPECK family

Block length	Trail length	Correlation	Rounds	Data
32	9	2^{-14}	10	2^{28}
48	9	2^{-20}	10	2^{40}
64	11	2^{-25}	12	2^{50}
64	12	2^{-31}	13	2^{62}
96	6	2^{-11}	7	2^{22}
128	6	2^{-11}	7	2^{22}

3.4 Key Recovery Attacks

For key recovery attacks, we adopt the χ^2 extension of Matsui's Algorithm 2 which was presented by Hermelin et al. in [6] and does not require the distribution of the linear approximation for the correct key.

Let $h(\boldsymbol{a})$ denote one plus the position of the most-significant one of $\boldsymbol{a}$ and $\boldsymbol{a}^{\boxplus} \triangleq 2^{h(\boldsymbol{a})} - 1$. Because

$$\boldsymbol{a} \underset{n}{\boxplus} \boldsymbol{b} = \boldsymbol{c} \Rightarrow \boldsymbol{a} \underset{h(\boldsymbol{a})}{\boxplus} \boldsymbol{b} = \boldsymbol{c}\boldsymbol{a}^{\boxplus} \Rightarrow \boldsymbol{a}\boldsymbol{a}^{\boxplus} = \boldsymbol{c}\boldsymbol{a}^{\boxplus} \underset{h(\boldsymbol{a})}{\boxminus} \boldsymbol{b}$$

$$\boldsymbol{y}[i] = (\boldsymbol{x}[i+1] \oplus \boldsymbol{y}[i+1]) \ggg \tau$$

guessing $\boldsymbol{k}[i]\boldsymbol{v}[i]^{\boxplus}$ is enough to calculate

$$\begin{aligned}\boldsymbol{x}[i] \cdot \boldsymbol{X}[i] \oplus \boldsymbol{y}[i] \cdot \boldsymbol{Y}[i] &= (\boldsymbol{x}[i] \ggg \varsigma) \cdot \boldsymbol{v}[i] \oplus \boldsymbol{y}[i] \cdot \boldsymbol{Y}[i] \\ &= \left((\boldsymbol{x}[i] \ggg \varsigma)\boldsymbol{v}[i]^{\boxplus}\right) \cdot \boldsymbol{v}[i] \oplus \boldsymbol{y}[i] \cdot \boldsymbol{Y}[i] \\ &= \left(\left(\boldsymbol{x}[i+1]\boldsymbol{v}[i]^{\boxplus} \oplus \boldsymbol{k}[i]\boldsymbol{v}[i]^{\boxplus}\right) \underset{h(\boldsymbol{v}[i])}{\boxminus} \boldsymbol{y}[i]\right) \cdot \boldsymbol{v}[i] \oplus \boldsymbol{y}[i] \cdot \\ &\quad \boldsymbol{Y}[i]\end{aligned}$$

from $\boldsymbol{x}[i+1], \boldsymbol{y}[i+1]$. Therefore, if m rounds are appended to the back of a r-round distinguisher, then only

$$\begin{aligned}h(\boldsymbol{v}[r+1]) + (m-1)n &= h((\boldsymbol{u}[r] \oplus \boldsymbol{Y}[r+1]) \ggg \varsigma) + (m-1)n \\ &= h((\boldsymbol{u}[r] \oplus ((\boldsymbol{w}[r] \oplus \boldsymbol{Y}[r]) \lll \tau)) \ggg \varsigma) + (m-1)n \\ &= h((\boldsymbol{u}[r] \oplus ((\boldsymbol{w}[r] \oplus (\boldsymbol{u}[r-1] \oplus (\boldsymbol{v}[r] \lll \varsigma))) \lll \tau)) \ggg \\ &\quad \varsigma) + (m-1)n\end{aligned}$$

bits of key need to be guessed, i.e. $\boldsymbol{k}[r+1]\boldsymbol{v}[r+1]^{\boxplus}, \boldsymbol{k}[r+2], \ldots, \boldsymbol{k}[r+m]$. Consequently, we have Table 4 where

$$\begin{aligned}\text{Time} &= \text{Data} \times 2^{\text{guessed bits}} + 2^{\text{key length}} \times \beta \\ \text{Average Time} &= {}^{\text{Time}}/_{1-\alpha}\end{aligned}$$

and α, β are missing detection and false alarm probabilities respectively. Moreover, the results may be improved by trails of smaller $h(\boldsymbol{v}[r+1])$ or vectorial linear approximations. But it seems unable to be improved by the similar technique of [4] since the size of the equation derived from a sub-cipher is one bit instead of $2n$ bits in the case of 1-dimensional linear cryptanalysis.

4 Another Implementation of Wallén's Algorithm

In this section, we present another implementation of Wallén's algorithm, called the *CPM* method, and compare the performance of different implementations. Firstly, a set of necessary conditions for correlations to be non-zero needs to be proved.

Table 4. Key recovery attacks on the SPECK family

Block/ key length	Trail length (this paper/ [4])	Rounds (this paper/ [4]/Total)	Guessed bits	α	β	Data (this paper/ [4])	Time	Average time (this paper/ [4])
32/ 64	9/ 10	12/ 14/ 22	$13+16$	2^{-1}	2^{-6}	$2^{30.8668}/2^{31}$	$2^{60.2164}$	$2^{61.2164}/2^{63}$
48/ 72	9/ 11	11/ 14/ 22	24	2^{-2}	2^{-7}	$2^{43.727}/2^{41}$	$2^{67.93}$	$2^{68.345}/2^{65}$
48/ 96	9/ 11	12/ 15/ 23	$24+24$	2^{-2}	2^{-7}	$2^{43.727}/2^{41}$	$2^{91.93}$	$2^{92.345}/2^{89}$
64/ 96	11/ 15	13/ 18/ 26	31	2^{-2}	2^{-14}	$2^{54.6279}/2^{61}$	$2^{85.7401}$	$2^{86.1551}/2^{93}$
64/ 96	12/ 15	14/ 18/ 26	31	2^{-1}	2^{-2}	$2^{62.7302}/2^{61}$	$2^{94.8714}$	$2^{95.8714}/2^{93}$
64/ 128	11/ 15	14/ 19/ 27	$31+32$	2^{-2}	2^{-14}	$2^{54.8029}/2^{61}$	$2^{117.74}$	$2^{118.155}/2^{125}$
64/ 128	12/ 15	15/ 19/ 27	$31+32$	2^{-1}	2^{-2}	$2^{62.7302}/2^{61}$	$2^{126.871}$	$2^{127.871}/2^{125}$
96/ 96	6/ 14	8/ 16/ 28	47	2^{-3}	2^{-26}	$2^{27.6463}/2^{85}$	$2^{74.7028}$	$2^{74.8954}/2^{85}$
96/ 144	6/ 14	9/ 17/ 29	$47+48$	2^{-3}	2^{-26}	$2^{27.6463}/2^{85}$	$2^{122.703}$	$2^{122.895}/2^{133}$
128/ 128	6/ 15	8/ 17/ 32	63	2^{-5}	2^{-36}	$2^{28.2959}/2^{113}$	$2^{92.6905}$	$2^{92.7363}/2^{113}$
128/ 192	6/ 15	9/ 18/ 33	$63+64$	2^{-5}	2^{-36}	$2^{28.2959}/2^{113}$	$2^{156.69}$	$2^{156.736}/2^{177}$
128/ 256	6/ 15	7/ 19/ 34	$63+2\times 64$	2^{-5}	2^{-36}	$2^{28.2959}/2^{113}$	$2^{220.69}$	$2^{220.736}/2^{241}$

Lemma 2. *Let $\boldsymbol{u}, \gamma, \boldsymbol{\delta} \in \mathbb{F}_2^n$. Then*

$$\boldsymbol{\delta} = cpm_{n+1}\left(0 \parallel \boldsymbol{u}, (0 \parallel \gamma) \oplus \mathbf{1}\right) \Longleftrightarrow \boldsymbol{\delta} = \left(\boldsymbol{u} \oplus (\gamma \oplus \mathbf{1})\,\boldsymbol{\delta}\right) \gg 1$$

Proof. "$\Longrightarrow$". From Definition 2, it is clear that $\delta_{n-1} = 0$ and $\delta_i = u_{i+1} \oplus (\gamma_{i+1} \oplus 1)\delta_{i+1}, i = 0, \ldots, n-2$.

"$\Longleftarrow$". *Suppose* $\boldsymbol{\delta}' = cpm_{n+1}(0 \parallel \boldsymbol{u}, (0 \parallel \gamma) \oplus \mathbf{1})$, then $\delta'_i = u_{i+1} \oplus (\gamma_{i+1} \oplus 1)\delta'_{i+1}, i = 0, \ldots, n-2$. Thus $\delta_i \oplus \delta'_i = (\gamma_{i+1} \oplus 1)(\delta_{i+1} \oplus \delta'_{i+1}), i = 0, \ldots, n-2$. Finally, $\delta_{n-2} = \delta'_{n-2}, \ldots, \delta_0 = \delta'_0$ following from $\delta_{n-1} = \delta'_{n-1} = 0$. □

Theorem 2. *Let $\boldsymbol{u}, \boldsymbol{v}, \boldsymbol{w}, \boldsymbol{\phi}, \boldsymbol{\varphi}, \boldsymbol{\delta} \in \mathbb{F}_2^n$ and $\boldsymbol{\phi} = \boldsymbol{v} \oplus \boldsymbol{u}, \boldsymbol{\varphi} = \boldsymbol{w} \oplus \boldsymbol{u}, \gamma = \boldsymbol{v} \oplus \boldsymbol{w}$. Then*

$$\boldsymbol{\delta} = cpm_{n+1}(0 \parallel \boldsymbol{u}, (0 \parallel \gamma) \oplus \mathbf{1}), c(\boldsymbol{u}, \boldsymbol{v}, \boldsymbol{w}) \neq 0$$

if and only if

$$\boldsymbol{\phi} = \boldsymbol{\phi}\boldsymbol{\delta} \tag{3}$$

$$\boldsymbol{\varphi} = \boldsymbol{\varphi}\boldsymbol{\delta} \tag{4}$$

$$\gamma \gg 1 = ((\boldsymbol{u} \oplus \boldsymbol{\delta}) \gg 1) \oplus \boldsymbol{\delta} \tag{5}$$

$$\mathbf{0} = ((\boldsymbol{u} \gg 1) \oplus \boldsymbol{\delta})\,((\boldsymbol{\delta} \oplus \mathbf{1}) \gg 1) \tag{6}$$

$$\mathbf{0} = ((\boldsymbol{v} \gg 1) \oplus \boldsymbol{\delta})\,((\boldsymbol{\delta} \oplus \mathbf{1}) \gg 1) \tag{7}$$

$$\mathbf{0} = ((\boldsymbol{w} \gg 1) \oplus \boldsymbol{\delta})\,((\boldsymbol{\delta} \oplus \mathbf{1}) \gg 1) \tag{8}$$

Proof. Proof of the *only-if-part*. Since $c(\boldsymbol{u}, \boldsymbol{v}, \boldsymbol{w}) \neq 0$, (3) and (4) follow from Lemma 1 directly. According to Lemma 2, $\boldsymbol{\delta} = (\boldsymbol{u} \oplus (\boldsymbol{\gamma} \oplus \mathbf{1})\boldsymbol{\delta}) \gg 1$. Hence,

$$\begin{aligned}((\boldsymbol{u} \oplus \boldsymbol{\delta}) \gg 1) \oplus \boldsymbol{\delta} &= ((\boldsymbol{u} \oplus \boldsymbol{\delta}) \oplus (\boldsymbol{u} \oplus (\boldsymbol{\gamma} \oplus \mathbf{1})\,\boldsymbol{\delta})) \gg 1 = (\boldsymbol{\gamma}\boldsymbol{\delta}) \gg 1 \\ &= ((\boldsymbol{\phi} \oplus \boldsymbol{\varphi})\,\boldsymbol{\delta}) \gg 1 = (\boldsymbol{\phi} \oplus \boldsymbol{\varphi}) \gg 1 = \boldsymbol{\gamma} \gg 1\end{aligned}$$

Accordingly,

$$\begin{aligned}\mathbf{0} = \boldsymbol{\gamma}\,(\boldsymbol{\delta} \oplus \mathbf{1}) &= (\boldsymbol{\gamma}\,(\boldsymbol{\delta} \oplus \mathbf{1})) \gg 1 = (\boldsymbol{\gamma} \gg 1)\,((\boldsymbol{\delta} \oplus \mathbf{1}) \gg 1) \\ &= (((\boldsymbol{u} \oplus \boldsymbol{\delta}) \gg 1) \oplus \boldsymbol{\delta})\,((\boldsymbol{\delta} \oplus \mathbf{1}) \gg 1) \\ &= ((\boldsymbol{u} \gg 1) \oplus \boldsymbol{\delta})\,((\boldsymbol{\delta} \oplus \mathbf{1}) \gg 1) \oplus ((\boldsymbol{\delta}(\boldsymbol{\delta} \oplus \mathbf{1})) \gg 1) \\ &= ((\boldsymbol{u} \gg 1) \oplus \boldsymbol{\delta})\,((\boldsymbol{\delta} \oplus \mathbf{1}) \gg 1)\end{aligned}$$

(3) implies $(\boldsymbol{\phi}\,(\boldsymbol{\delta} \oplus \mathbf{1})) \gg 1 = \mathbf{0}$, thus

$$\begin{aligned}&((\boldsymbol{v} \gg 1) \oplus \boldsymbol{\delta})\,((\boldsymbol{\delta} \oplus \mathbf{1}) \gg 1) = \\ &\qquad (\boldsymbol{\phi}\,(\boldsymbol{\delta} \oplus \mathbf{1})) \gg 1 \oplus ((\boldsymbol{u} \gg 1) \oplus \boldsymbol{\delta})\,((\boldsymbol{\delta} \oplus \mathbf{1}) \gg 1) = \mathbf{0}\end{aligned}$$

(8) holds similarly.

Proof of the *if-part*. From (5),

$$\begin{aligned}((\boldsymbol{u} \oplus \boldsymbol{\delta}) \gg 1) \oplus \boldsymbol{\delta} &= \boldsymbol{\gamma} \gg 1 = (\boldsymbol{\gamma}\boldsymbol{\delta}) \gg 1 = (\boldsymbol{u} \oplus \boldsymbol{\gamma}\boldsymbol{\delta} \oplus \boldsymbol{\delta} \oplus \boldsymbol{\delta} \oplus \boldsymbol{u}) \gg 1 \\ &= ((\boldsymbol{u} \oplus (\boldsymbol{\gamma} \oplus \mathbf{1})\,\boldsymbol{\delta}) \gg 1) \oplus ((\boldsymbol{\delta} \oplus \boldsymbol{u}) \gg 1)\end{aligned}$$

Therefore,

$$\boldsymbol{\delta} = (\boldsymbol{u} \oplus (\boldsymbol{\gamma} \oplus \mathbf{1})\,\boldsymbol{\delta}) \gg 1 = cpm_{n+1}\,(0 \parallel \boldsymbol{u}, (0 \parallel \boldsymbol{\gamma}) \oplus \mathbf{1})$$

and the conclusion is derived from Lemma 1. □

We next discusses details of the CPM method under different scenarios.

Case 1: $\boldsymbol{u}$ is known and fixed. Therefore, $\boldsymbol{\delta}$ should satisfy (6) and δ_i is determined by δ_{i+1} for $0 \leq i < n-1$, i.e.

$$\delta_i = \begin{cases} 0, 1 & \text{if } \delta_{i+1} = 1 \\ u_{i+1} & \text{otherwise} \end{cases}$$

Recall that $\delta_{n-1} = 0$, thus $\boldsymbol{\delta}$ can be resolved bit by bit from left to right. But it should be noted that $\boldsymbol{\delta}$ needs to be enumerated in the order of hamming weight according to Lemma 1. We adopt a deque (i.e. a data structure supporting push and pop in both front and back directions) for this purpose, and $\boldsymbol{\delta}$ is pushed to the front whenever $\delta_{i-1} = 0$ and is pushed to the back otherwise. Details are presented in Algorithm 3.

Given $\boldsymbol{u}$ and $\boldsymbol{\delta}$, the approximation is determined by two of $\boldsymbol{v}, \boldsymbol{w}$ and $\boldsymbol{\gamma}$. Obviously, $\boldsymbol{\gamma}$ can be obtained from (5) except γ_0. Thus, the input masks are

Algorithm 3. Generate $\boldsymbol{\delta}$ given $\boldsymbol{u}$

```
function CPM_GENERATE_DELTA(u)
    deque ← {(0, n − 1)}
    while deque is not empty do
        (δ, i) ← POP_FRONT(deque)
        if i ≠ 0 then
            if δ_i = 1 then
                PUSH_FRONT(deque, (δ, i − 1))
                PUSH_BACK(deque, (δ ⊕ (1 ≪ (i − 1)), (i − 1)))
            else if u_i = 1 then
                PUSH_BACK(deque, (δ ⊕ (1 ≪ (i − 1)), (i − 1)))
            else
                PUSH_FRONT(deque, (δ, i − 1))
            end if
        else
            yield δ
        end if
    end while
end function
```

known once $\boldsymbol{v}$ or $\boldsymbol{w}$ is generated. Without loss of generality, we choose to generate $\boldsymbol{v}$ and then calculate $\boldsymbol{w}$ as $\boldsymbol{w} = \boldsymbol{v} \oplus \boldsymbol{\gamma}$. According to (3),

$$v_i = \phi_i \oplus u_i = \begin{cases} 0, 1 & \text{if } \delta_i = 1 \\ u_i & \text{otherwise} \end{cases}$$

for $0 \leq i < n$. Hence, the bits of $\boldsymbol{v}$ where $\boldsymbol{\delta}$ equals one need to be traversed to generate all valid masks. As far as we know, the most efficient method to generate all tuples is the Gray code strategy [7] which flips one bit only in each iteration as shown in Appendix B. Also, this step may be customized for special purpose, e.g. generating the tuples by hamming weight. See Algorithm 4 for details.

Case 2: $\boldsymbol{v}$ or $\boldsymbol{w}$ is known and fixed. Suppose $\boldsymbol{v}$ is known, then $\boldsymbol{\delta}$ should satisfy (7). Thus, $\boldsymbol{\delta}$ can be generated using the procedure *CPM_Generate_Delta* with the parameter $\boldsymbol{v}$ and thereupon $\boldsymbol{u}$ can be determined by (6), i.e.

$$u_i = \begin{cases} 0, 1 & \text{if } \delta_i = 1 \\ \delta_{i-1} & \text{otherwise} \end{cases}$$

for $1 \leq i \leq n-1$. Since $\phi_0\delta_0 = \delta_0 = \phi_0 = v_0 \oplus u_0$ according to (3), $u_0 = v_0$ if $\delta_0 = 0$ and $u_0 \in \{0, 1\}$ otherwise. Finally, $\boldsymbol{\gamma}$ and $\boldsymbol{w}$ are determined by (5) and (4) as in Case 1.

Case 3: $\boldsymbol{u}, \boldsymbol{v}$ or $\boldsymbol{u}, \boldsymbol{w}$ are known and fixed. Suppose $\boldsymbol{u}, \boldsymbol{v}$ are known, so $\boldsymbol{\phi} = \boldsymbol{v} \oplus \boldsymbol{u}$ is known as well. And $\boldsymbol{\delta}$ should satisfy (3), (6) and (7). Notice that the conditions may be incompatible and result in zero correlation. Indeed, since $\delta_{n-1} = 0$, $\boldsymbol{\delta}$

Algorithm 4. The case that $\boldsymbol{u}$ is known and fixed

function CPM_GENERATE_MASK($\boldsymbol{u}$, $\boldsymbol{\delta}$)
    $\boldsymbol{\gamma} \leftarrow ((\boldsymbol{u} \oplus (\boldsymbol{\delta} \ll 1)) \oplus \boldsymbol{\delta})\boldsymbol{\delta}$ ▷ $\gamma_0 \in \{0, 1\}$ if $\delta_0 = 1$ and $\gamma_0 \in \{0\}$ otherwise
    $ones \leftarrow \{0 \leq i < n : \delta_i = 1\}$
    for $\boldsymbol{v}$ **in** GRAY_VISIT($\boldsymbol{\delta} \oplus \boldsymbol{u}, ones$) **do**
        $\boldsymbol{w} \leftarrow \boldsymbol{v} \oplus \boldsymbol{\gamma}$
        yield $(\boldsymbol{u}, \boldsymbol{v}, \boldsymbol{w})$
        if $\delta_0 = 1$ **then**
            yield $(\boldsymbol{u}, \boldsymbol{v}, \boldsymbol{w} \oplus 1)$ ▷ Equivalent to flipping γ_0
        end if
    end for
end function

function GENERATE'($\boldsymbol{u}$)
    for $\boldsymbol{\delta}$ **in** CPM_GENERATE_DELTA($\boldsymbol{u}$) **do**
        CPM_GENERATE_MASK($\boldsymbol{u}$, $\boldsymbol{\delta}$)
    end for
end function

exists only if $\phi_{n-1} = 0$. By (3) and (6), we have

$$\delta_i = \begin{cases} 0, 1 & \text{if } \delta_{i+1} = 1 \text{ and } \phi_i = 0 \\ 1 & \text{if } \delta_{i+1} = 1 \text{ and } \phi_i = 1 \\ 1 & \text{if } \delta_{i+1} = 0 \text{ and } u_{i+1} = 1 \\ 0 & \text{if } \delta_{i+1} = 0 \text{ and } u_{i+1} = \phi_i = 0 \\ \perp & \text{otherwise} \end{cases}$$

for $0 \leq i < n - 1$ where $\perp$ means no solution. Consequently, $\boldsymbol{\delta}$ can be solved using procedure similar to *CPM_Generate_Delta*. At last, $\boldsymbol{\gamma}$ is resolved by (5) and $\boldsymbol{w} = \boldsymbol{v} \oplus \boldsymbol{\gamma}$.

Case 4: $\boldsymbol{v}, \boldsymbol{w}$ are known. Thus, $\boldsymbol{\gamma}$ is fixed and $\boldsymbol{\delta}$ should satisfy $\boldsymbol{\gamma\delta} = \boldsymbol{\gamma}$, (7) and (8). Similar to Case 3, $\boldsymbol{\delta}$ exists only if $\gamma_{n-1} = 0$, and

$$\delta_i = \begin{cases} 0, 1 & \text{if } \delta_{i+1} = 1 \text{ and } \gamma_i = 0 \\ 1 & \text{if } \delta_{i+1} = 1 \text{ and } \gamma_i = 1 \\ 1 & \text{if } \delta_{i+1} = 0 \text{ and } v_{i+1} = 1 \\ 0 & \text{if } \delta_{i+1} = 0 \text{ and } v_{i+1} = \gamma_i = 0 \\ \perp & \text{otherwise} \end{cases}$$

for $0 \leq i < n - 1$. Then, $\boldsymbol{u}$ is calculated by (5) except that u_0 needs to satisfy

$$(u_0 \oplus v_0)\delta_0 = u_0 \oplus v_0$$
$$(u_0 \oplus v_0)\delta_0 = u_0 \oplus v_0$$

Since $\gamma_0 = v_0 \oplus w_0 = 1 \Rightarrow \delta_0 = 1$, then $u_0 \in \{0, 1\}$ if $\delta_0 = 1$ and $u_0 = v_0 = w_0$ otherwise.

Case 5: All Masks are Free. In this case, $\boldsymbol{\delta}$ is generated first according to its hamming weight to ensure the order of approximations. Then, $\boldsymbol{u}$ is obtained as in Case 2 without the constraint on u_0. At last, the procedure *CPM_Generate_Mask* takes over. Refer to Algorithm 5 for details.

Algorithm 5. The case that all masks are free

```
function GENERATE'()
    for k from 0 to n − 1 do
        for δ of weight k do
            ones ← {0 < i < n : δ_i = 1} ∪ {0}
            for u in GRAY_VISIT(δ ≪ 1, ones) do
                CPM_GENERATE_MASK(u, δ)
            end for
        end for
    end for
end function
```

Obviously, the CPM method is not as elegant as the top-down/bottom-up method, but surprisingly it is faster for $n \geq 11$ according to Fig. 5 (note that the labels on y-axis increase *exponentially*). We believe better direct techniques to instantiate Theorem 1 exists, but *Generate′* is the most effective implementation we can think of at present and is used to replace *Generate* in Algorithm 2.

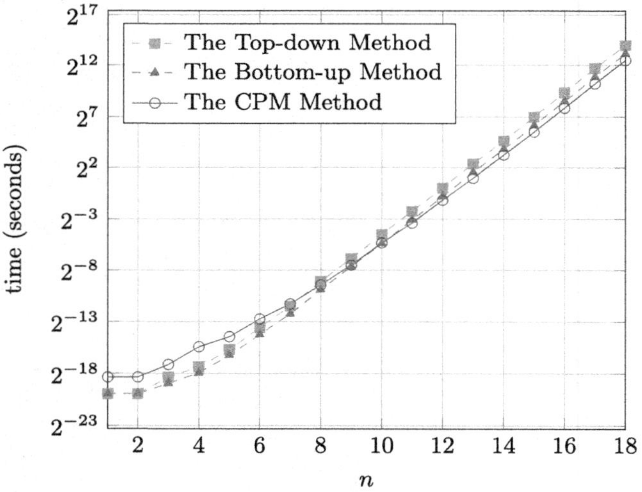

Fig. 5. The performance of generating $\bigcup_{k=0}^{n-1} S(n,k)$ Platform: 32-bit Win7 with Visual C++ 2015 CTP optimized by /Ox

5 Conclusions

In this paper, we presented a search for linear trails on the SPECK family via Wallén's enumeration algorithm and Matsui's branch-and-bound framework. The best correlation of full rounds of SPECK-32 was found as well as reduced rounds of other variants. According to the best 10-round correlation of SPECK-32 which is 2^{-17}, SPECK-32 is immune to the 1-dimensional linear cryptanalysis. We further proposed the first linear distinguishers and key recovery attacks on the SPECK family which do not threaten the security of SPECK. Finally, a CPM implementation of the Wallén's algorithm was presented which seems faster than the straightforward instantiations, i.e. the top-down and the bottom-up approaches.

Additional future work items include applying the threshold search [2] on SPECK, mounting vectorial linear cryptanalyses and implementing the search on other ARX ciphers.

A Straightforward Implementations of Wallén's Algorithm

The *mode* argument indicates whether $\boldsymbol{u}, \boldsymbol{v}, \boldsymbol{w}$ are fixed and used hereafter.

A.1 The Top-Down Method

```
function TDV(mode, t_{i+1}, t_i, i, u, v, w)
    if u, v, w, t_{i+1}, t_i and mode are compatible then
        modify the i-th bit of u, v, w and yield (u, v, w)
    end if
end function

function TDD_GENERATE(N, mode, t, n, rk, u, v, w)
    if n = 0 then
        if t = S0 then                                        ▷ S^0(0,0)
            yield (u, v, w)
        end if
        return
    end if
    i ← N − n
    if n ≠ rk or rk = 0 then
        if n = N or t = S0 then              ▷ S^0(n − 1, rk) ← S^0(n, rk)
            for u', v', w' in TDV(mode, S0, S0, i, u, v, w) do
                TDD_GENERATE(N, mode, S0, n − 1, rk, u', v', w')
            end for
        end if
        if n = N or t = S1 then              ▷ S^0(n − 1, rk) ← S^1(n, rk)
            for u', v', w' in TDV(mode, S0, S1, i, u, v, w) do
```

TDD_GENERATE($N, mode, S0, n-1, rk, \boldsymbol{u}', \boldsymbol{v}', \boldsymbol{w}'$)
end for
end if
end if
if $rk \neq 0$ **then**
if $n = N$ or $t = S0$ **then** ▷ $S^1(n-1, rk-1) \leftarrow S^0(n, rk)$
for $\boldsymbol{u}', \boldsymbol{v}', \boldsymbol{w}'$ **in** TDV($mode, S1, S0, i, \boldsymbol{u}, \boldsymbol{v}, \boldsymbol{w}$) **do**
TDD_GENERATE($N, mode, S1, n-1, rk-1, \boldsymbol{u}', \boldsymbol{v}', \boldsymbol{w}'$)
end for
end if
if $n = N$ or $t = S1$ **then** ▷ $S^1(n-1, rk-1) \leftarrow S^1(n, rk)$
for $\boldsymbol{u}', \boldsymbol{v}', \boldsymbol{w}'$ **in** TDV($mode, S1, S1, i, \boldsymbol{u}, \boldsymbol{v}, \boldsymbol{w}$) **do**
TDD_GENERATE($N, mode, S1, n-1, rk-1, \boldsymbol{u}', \boldsymbol{v}', \boldsymbol{w}'$)
end for
end if
end if
end function

function TD_GENERATE($n, mode, k, \boldsymbol{u}, \boldsymbol{v}, \boldsymbol{w}$) ▷ generate $S(n, k)$
TDD_GENERATE($n, mode, not\_used, n, k, \boldsymbol{u}, \boldsymbol{v}, \boldsymbol{w}$) ▷ start from $S(n, k)$
end function

A.2 The Bottom-Up Method

function BUV($mode, t_i, t_{i-1}, i, \boldsymbol{u}, \boldsymbol{v}, \boldsymbol{w}$)
if $\boldsymbol{u}, \boldsymbol{v}, \boldsymbol{w}, t_i, t_{i-1}$ and $mode$ are compatible **then**
modify the i-th bit of $\boldsymbol{u}, \boldsymbol{v}, \boldsymbol{w}$ and **yield** $(\boldsymbol{u}, \boldsymbol{v}, \boldsymbol{w})$
end if
end function

function BUD_GENERATE($N, mode, t, n, rk, \boldsymbol{u}, \boldsymbol{v}, \boldsymbol{w}$)
if $n = N$ **then** ▷ $S(N, k)$
yield $(\boldsymbol{u}, \boldsymbol{v}, \boldsymbol{w})$ and **return**
end if
$i \leftarrow N - 1 - n$
if $t = S0$ **then**
if $i = 0$ or $rk < i$ **then** ▷ $S^0(n, k-rk) \rightarrow S^0(n+1, k-rk)$
for $\boldsymbol{u}', \boldsymbol{v}', \boldsymbol{w}'$ **in** BUV($mode, S0, S0, i, \boldsymbol{u}, \boldsymbol{v}, \boldsymbol{w}$) **do**
BUD_GENERATE($N, mode, S0, n+1, rk, \boldsymbol{u}', \boldsymbol{v}', \boldsymbol{w}'$)
end for
end if
if $i = 0$ or $rk \neq 0$ **then** ▷ $S^0(n, k-rk) \rightarrow S^1(n+1, k-rk)$
for $\boldsymbol{u}', \boldsymbol{v}', \boldsymbol{w}'$ **in** BUV($mode, S0, S1, i, \boldsymbol{u}, \boldsymbol{v}, \boldsymbol{w}$) **do**
BUD_GENERATE($N, mode, S1, n+1, rk, \boldsymbol{u}', \boldsymbol{v}', \boldsymbol{w}'$)
end for
end if

```
    else
        if i = 0 or rk ≤ i then            ▷ S^1(n, k − rk) → S^0(n + 1, k − rk + 1)
            for u′, v′, w′ in BUV(mode, S1, S0, i, u, v, w) do
                BUD_Generate(N, mode, S0, n + 1, rk − 1, u′, v′, w′)
            end for
        end if
        if i = 0 or rk ≠ 1 then            ▷ S^1(n, k − rk) → S^1(n + 1, k − rk + 1)
            for u′, v′, w′ in BUV(mode, S1, S1, i, u, v, w) do
                BUD_Generate(N, mode, S1, n + 1, rk − 1, u′, v′, w′)
            end for
        end if
    end if
end function

function BU_Generate(n, mode, k, u, v, w)            ▷ generate S(n, k)
    BUD_Generate(n, mode, S0, 0, k, u, v, w)        ▷ start from S^0(0, k − k)
end function
```

B The Gray_Visit Procedure

```
function Gray_Visit(a, set)
    s ← sizeof(set), buf ← {1, 2, ..., s + 1}
    while true do
        yield a
        j ← buf[1], buf[1] ← 1
        if j = s + 1 then
            return
        end if
        i ← j + 1, buf[j] ← buf[i], buf[i] ← i
        flip a[set[j]]
    end while
end function
```

References

1. Beaulieu, R., Shors, D., Smith, J., Treatman-Clark, S., Weeks, B., Wingers, L.: The SIMON and SPECK families of lightweight block ciphers. Cryptology ePrint Archive, Report 2013/404 (2013). http://eprint.iacr.org/
2. Biryukov, A., Velichkov, V.: Automatic search for differential trails in ARX ciphers. In: Benaloh, J. (ed.) CT-RSA 2014. LNCS, vol. 8366, pp. 227–250. Springer, Heidelberg (2014). http://dx.doi.org/10.1007/978-3-319-04852-9_12
3. Cho, J.Y., Hermelin, M.: Improved linear cryptanalysis of SOSEMANUK. In: Lee, D., Hong, S. (eds.) ICISC 2009. LNCS, vol. 5984, pp. 101–117. Springer, Heidelberg (2010). http://dx.doi.org/10.1007/978-3-642-14423-3_8

4. Dinur, I.: Improved differential cryptanalysis of round-reduced SPECK. Cryptology ePrint Archive, Report 2014/320 (2014). http://eprint.iacr.org/. Accepted by SAC 2014
5. Hermelin, M.: Multidimensional Linear Cryptanalysis. Ph.D. thesis, Aalto University School of Science and Technology, Faculty of Information and Natural Sciences, Department of Information and Computer Science (2003). http://lib.tkk.fi/Diss/2010/isbn9789526031903/isbn9789526031903.pdf
6. Hermelin, M., Cho, J.Y., Nyberg, K.: Multidimensional extension of matsui's algorithm 2. In: Dunkelman, O. (ed.) FSE 2009. LNCS, vol. 5665, pp. 209–227. Springer, Heidelberg (2009). http://dx.doi.org/10.1007/978-3-642-03317-9_13
7. Knuth, D.: The Art of Computer Programming: Generating All Tuples and Permutations. Addison-Wesley Series in Computer Science and Information Proceedings, vol. 4. Addison Wesley Publishing Company Incorporated, Upper Saddle River (2005)
8. Matsui, M.: On correlation between the order of S-Boxes and the strength of DES. In: De Santis, A. (ed.) EUROCRYPT 1994. LNCS, vol. 950, pp. 366–375. Springer, Heidelberg (1995). http://dx.doi.org/10.1007/BFb0053451
9. Nyberg, K., Wallén, J.: Improved linear distinguishers for SNOW 2.0. In: Robshaw, M. (ed.) FSE 2006. LNCS, vol. 4047, pp. 144–162. Springer, Heidelberg (2006). http://dx.doi.org/10.1007/11799313_10
10. Wallén, J.: Linear approximations of addition modulo 2^n. In: Johansson, T. (ed.) FSE 2003. LNCS, vol. 2887, pp. 261–273. Springer, Heidelberg (2003). http://dx.doi.org/10.1007/978-3-540-39887-5_20
11. Wallén, J.: On the differential and linear properties of addition. Master's thesis, Helsinki University of Technology, Department of Computer Science and Engineering, Laboratory for Theoretical Computer Science (2003). http://www.tcs.hut.fi/Publications/bibdb/HUT-TCS-A84.pdf

From Distinguishers to Key Recovery: Improved Related-Key Attacks on Even-Mansour

Pierre Karpman[1,2(✉)]

[1] Inria, Saclay, France
pierre.karpman@inria.fr
[2] Nanyang Technological University, Singapore, Singapore

Abstract. We show that a distinguishing attack in the related key model on an Even-Mansour block cipher can readily be converted into an extremely efficient key recovery attack. Concerned ciphers include in particular all iterated Even-Mansour schemes with independent keys. We apply this observation to the CAESAR candidate PRØST-OTR and are able to recover the whole key with a number of requests linear in its size. This improves on recent forgery attacks in a similar setting.

Keywords: Even-Mansour · Related-key attacks · PRØST-OTR

1 Introduction

The Even-Mansour scheme is arguably the simplest way to construct a block cipher from publicly available components. It defines the encryption $\mathcal{E}((k_1, k_0), p)$ of the plaintext p under the (possibly equal) keys k_0 and k_1 as $\mathcal{P}(p \oplus k_0) \oplus k_1$, where $\mathcal{P}$ is a public permutation. Even and Mansour proved in 1991 that for a permutation of size n, the probability of recovering the keys is upper-bounded by $\mathcal{O}(DT \cdot 2^{-n})$ when the attacker considers the permutation as a black box, where D is the data complexity and T is the time complexity of the attack [7]. Although of considerable interest, this bound also shows at the same time that the construction is not ideal, as one gets security only up to $\mathcal{O}(2^{\frac{n}{2}})$ queries, which is less than the $\mathcal{O}(2^n)$ one would expect for an n-bit block cipher. For this reason, much later work investigated the security of variants of the Even-Mansour cipher. A simple one is the iterated Even-Mansour scheme with independent keys and independent permutations, with its r-round version defined as $\mathsf{IEM}^r((k_r, k_{r-1}, \ldots, k_0), p) := \mathcal{P}_{r-1}(\mathcal{P}_{r-2}(\ldots \mathcal{P}_0(p \oplus k_0) \oplus k_1) \ldots) \oplus k_r$, and it has been established that this construction is secure up to $\mathcal{O}(2^{\frac{rn}{r+1}})$ queries [3]. On the other hand, in a related-key model, the same construction lends itself to trivial distinguishing attacks, and one must consider alternatives if security in this model is necessary. Yet until the recent work of Cogliati and Seurin [4] and Farshim and Procter [8], no variant of the Even-Mansour construction was

Partially supported by the Direction Générale de l'Armement and by the Singapore National Research Foundation Fellowship 2012 (NRF-NRFF2012-06).

J. Lopez and C.J. Mitchell (Eds.): ISC 2015, LNCS 9290, pp. 177–188, 2015.
DOI: 10.1007/978-3-319-23318-5_10

proved to be secure in the related-key model. This is not the case anymore and it has now been proven that one can reach a non-trivial level of related-key security for IEM^r starting from $r = 3$ when using keys linearly derived from a single master key (instead of using independent keys), or even when $r = 1$ when this derivation is non-linear and meets some conditions. While related-key analysis obviously gives much more power to the attacker than the single-key setting, it is a widely accepted model that may provide useful results on primitives studied in a general context, especially as related keys may naturally arise in some protocols.

Our Contribution. We show that the distinguishing attacks on Even-Mansour ciphers in a related-key model can be extended to much more powerful key-recovery attacks by considering modular additive differences instead of XOR differences. This applies both to the trivial distinguishers on iterated Even-Mansour with independent keys and to the more complex distinguisher of Cogliati and Seurin for 2-round Even-Mansour with a linear key-schedule. While these observations are somewhat elementary, they eventually lead to a key-recovery attack on the authenticated-encryption scheme and CAESAR candidate PRØST-OTR in a related-key model. This improves on the recent work from FSE 2015 of Dobraunig, Eichlseder and Mendel who use similar methods but only produce forgeries [5].

2 Notation

We use $||$ to denote string concatenation, α^i with i an integer to denote the string made of the concatenation of i copies of the character α, and α^* to denote any string of the set $\{\alpha^i, i \in \mathbb{N}\}$, α^0 denoting the empty string ε. For any string s, we use $s[i]$ to denote its ith element (starting from zero).

We also use Δ_i^n to denote the string $0^{n-i-1}||1||0^{i-1}$. The superscript n will always be clear from the context and therefore omitted.

Finally, we identify strings of length n over the binary alphabet $\{0, 1\}$ with elements of the vector-space $\mathbb{F}_2^n$ and with the binary representation of elements of the group $\mathbb{Z}/2^n\mathbb{Z}$. The addition operation on these structures are respectively denoted by $\oplus$ (bitwise exclusive or (XOR)) and $+$ (modular addition).

3 Generic Related-Key Key-Recovery Attacks on Even-Mansour Ciphers

Since the work of Bellare and Kohno [1], it is well known that no block cipher can resist related-key attacks (RKA) when an attacker may request encryptions under related keys using two relation classes. A simple example showing why this cannot be the case is to consider the classes $\phi^{\oplus}(k)$ and $\phi^{+}(k)$ of keys related to k by the XOR and the modular addition of any constant chosen by the attacker respectively. If we have access to (related-key) encryption oracles $\mathcal{E}(k, \cdot)$,

$\mathcal{E}(\phi^{\oplus}(k),\cdot)$ and $\mathcal{E}(\phi^{+}(k),\cdot)$ for the block cipher $\mathcal{E}$ with κ-bit keys, we can easily learn the value of the bit $k[i]$ of k by comparing the results of the queries $\mathcal{E}(k + \Delta_i, p)$ and $\mathcal{E}(k \oplus \Delta_i, p)$. For $i < \kappa - 1$, the plaintext p is encrypted under the same key if $k[i] = 0$, then resulting in the same ciphertext, and is encrypted under different keys if $k[i] = 1$, then resulting in different ciphertexts with an overwhelming probability. Doing this test for every bit of k thus allows to recover the whole key with a complexity linear in κ, except its most significant bit. Indeed, the carry of a modular addition on this bit never propagates and thus there will never be a difference between the related keys. This key bit can of course easily be recovered once all the others have been determined.

In the same paper, Bellare and Kohno also show that no such trivial generic attack exists when the attacker is restricted to using only one of the two classes $\phi^{\oplus}$ or ϕ^{+}, and they prove that an ideal cipher is in this case resistant to RKA. Taken together, these results mean in essence that a related-key attack on a block cipher $\mathcal{E}$ using both classes $\phi^{\oplus}(k)$ and $\phi^{+}(k)$ does not say much on $\mathcal{E}$, as nearly all ciphers fall to an attack in the same model. On the other hand, an attack using either of $\phi^{\oplus}$ or ϕ^{+} *is* meaningful, because an ideal cipher is secure in that case.

3.1 Key-Recovery Attacks on r-round IEM with Independent Keys

Going back to Even-Mansour ciphers, we explicit the trivial related-key distinguishers mentioned in the introduction. These distinguishers exist for r-round iterated Even-Mansour block ciphers with independent keys, for any value of r. As they only use keys related with, say, the $\phi^{\oplus}$ class, they are therefore meaningful when considering the related-key security of IEM.

From the very definition of IEM^r, it is obvious to see that the two values $\mathcal{E}((k_{r-1}, k_{r-2}, \ldots, k_0), p)$ and $\mathcal{E}((k_{r-1}, k_{r-2}, \ldots, k_0 \oplus \delta), p \oplus \delta)$ are equal for any difference δ when $\mathcal{E} = \mathsf{IEM}^r$ and that this equality does not hold in general, thence allowing to distinguish IEM^r from an ideal cipher.

We now show how these distinguishers can be combined with the two-class attack of Bellare and Kohno in order to extend it to a very efficient key-recovery attack. We give a description in the case of one-round Even-Mansour, but it can easily be extended to an arbitrary r. The attack is very simple and works as follows: consider again $\mathcal{E}((k_1, k_0), p) = \mathcal{P}(p \oplus k_0) \oplus k_1$; one can learn the value of the bit $k_0[i]$ by querying $\mathcal{E}((k_1, k_0), p)$ and $\mathcal{E}((k_1, k_0 + \Delta_i), p \oplus \Delta_i)$ and by comparing their values. These differ with overwhelming probability if $k_0[i] = 1$ and are equal otherwise.

A similar attack works on the variant of the (iterated) Even-Mansour cipher that uses modular addition instead of XOR for the combination of the key with the plaintext. This variant was first analyzed by Dunkelman, Keller and Shamir [6] and offers the same security bounds as the original Even-Mansour cipher. An attack in that case works similarly by querying *e.g.* $\mathcal{E}((k_1, k_0), \Delta_i)$ and $\mathcal{E}((k_1, k_0 \oplus \Delta_i), 0^{\kappa})$.

Both attacks use a single difference class for the related keys (either $\phi^{\oplus}$ or ϕ^{+}), and they are therefore meaningful as related-key attacks. They simply emulate the attack that uses both classes simultaneously by taking advantage of the fact that the usage of key material is very simple in Even-Mansour ciphers. Finally, we can see that in the particular case of a one-round construction, the attack still works if one chooses the keys k_1 and k_0 to be equal.

3.2 Extension to 2-Round Even-Mansour with a Linear Key Schedule

As has been shown by Cogliati and Seurin [4], it is also possible to very efficiently distinguish the 2-round Even-Mansour with related keys, even when the keys are equal or derived from a master key by a linear key schedule. Similarly as for IEM, we can adapt the distinguisher and transform it into a key-recovery attack. The idea remains the same: one replaces the $\phi^{\oplus}$ class of the original distinguisher with ϕ^{+}, which makes its success conditioned on the value of a few key bits, hence allowing their recovery. We give the description of our modified distinguisher for $\mathcal{E}(k,p) := \mathcal{P}(\mathcal{P}(k \oplus p) \oplus k) \oplus k$:

1. Query $y_1 := \mathcal{E}(k + \Delta_1, x_1)$
2. Set x_2 to $x_1 \oplus \Delta_1 \oplus \Delta_2$ and query $y_2 := \mathcal{E}(k + \Delta_2, x_2)$
3. Set y_3 to $y_1 \oplus \Delta_1 \oplus \Delta_3$ and query $x_3 := \mathcal{E}^{-1}(k + \Delta_3, y_3)$
4. Set y_4 to $y_2 \oplus \Delta_1 \oplus \Delta_3$ and query $x_4 := \mathcal{E}^{-1}(k + (\Delta_1 \oplus \Delta_2 \oplus \Delta_3), y_4)$
5. Test if $x_4 = x_3 \oplus \Delta_1 \oplus \Delta_2$

If the test is successful, it means that with overwhelming probability the key bits at the positions of the differences Δ_1, Δ_2, Δ_3 are all zero, as in that case $k + \Delta_i = k \oplus \Delta_i$ and the distinguisher works "as intended", and as otherwise at least one uncontrolled difference goes through $\mathcal{P}$ or $\mathcal{P}^{-1}$. It is possible to restrict oneself to using differences in only two bits for the Δ_is, and as soon as two such zero bits have been found (which happens after an expected four trials for random keys), the rest of the key bits can be tested one by one.

We conclude this short section by showing why the test of line 5 is successful when $k + \Delta_i = k \oplus \Delta_i$, but refer to Cogliati and Seurin for a complete description of their distinguisher, including the general case of distinct permutations and keys linearly derived from a master key (this only requires slight modifications to our simplified formulation).

For the sake of clarity, we write $k \oplus \Delta_i$ for $k + \Delta_i$, as they are equal by hypothesis. By definition, $y_1 = \mathcal{P}(\mathcal{P}(x_1 \oplus k \oplus \Delta_1) \oplus k \oplus \Delta_1) \oplus k \oplus \Delta_1$ and $y_3 = \mathcal{P}(\mathcal{P}(x_1 \oplus k \oplus \Delta_1) \oplus k \oplus \Delta_1) \oplus k \oplus \Delta_1 \oplus \Delta_1 \oplus \Delta_3$ which simplifies to $\mathcal{P}(\mathcal{P}(x_1 \oplus k \oplus \Delta_1) \oplus k \oplus \Delta_1) \oplus k \oplus \Delta_3$. This yields the following expression for x_3:

$$\begin{aligned} x_3 &= \mathcal{P}^{-1}(\mathcal{P}^{-1}(\mathcal{P}(\mathcal{P}(x_1 \oplus k \oplus \Delta_1) \oplus k \oplus \Delta_1) \oplus k \oplus \Delta_3 \oplus k \oplus \Delta_3) \\ &\quad \oplus k \oplus \Delta_3) \oplus k \oplus \Delta_3 \\ &= \mathcal{P}^{-1}(\mathcal{P}^{-1}(\mathcal{P}(\mathcal{P}(x_1 \oplus k \oplus \Delta_1) \oplus k \oplus \Delta_1)) \oplus k \oplus \Delta_3) \oplus k \oplus \Delta_3 \\ &= \mathcal{P}^{-1}(\mathcal{P}(x_1 \oplus k \oplus \Delta_1) \oplus k \oplus \Delta_1 \oplus k \oplus \Delta_3) \oplus k \oplus \Delta_3 \\ &= \mathcal{P}^{-1}(\mathcal{P}(x_1 \oplus k \oplus \Delta_1) \oplus \Delta_1 \oplus \Delta_3) \oplus k \oplus \Delta_3 \end{aligned}$$

Similarly, $y_2 = \mathcal{P}(\mathcal{P}(x_1 \oplus k \oplus \Delta_1) \oplus k \oplus \Delta_2) \oplus k \oplus \Delta_2$ and $y_4 = \mathcal{P}(\mathcal{P}(x_1 \oplus k \oplus \Delta_1) \oplus k \oplus \Delta_2) \oplus k \oplus \Delta_2 \oplus \Delta_1 \oplus \Delta_3$, which yields the following expression for x_4:

$$\begin{aligned} x_4 &= \mathcal{P}^{-1}(\mathcal{P}^{-1}(\mathcal{P}(\mathcal{P}(x_1 \oplus k \oplus \Delta_1) \oplus k \oplus \Delta_2) \oplus k \oplus \Delta_2 \oplus \Delta_1 \oplus \Delta_3 \\ &\quad \oplus k \oplus \Delta_1 \oplus \Delta_2 \oplus \Delta_3) \oplus k \oplus \Delta_1 \oplus \Delta_2 \oplus \Delta_3) \oplus k \oplus \Delta_1 \oplus \Delta_2 \oplus \Delta_3 \\ &= \mathcal{P}^{-1}(\mathcal{P}^{-1}(\mathcal{P}(\mathcal{P}(x_1 \oplus k \oplus \Delta_1) \oplus k \oplus \Delta_2)) \oplus k \oplus \Delta_1 \oplus \Delta_2 \oplus \Delta_3) \\ &\quad \oplus k \oplus \Delta_1 \oplus \Delta_2 \oplus \Delta_3 \\ &= \mathcal{P}^{-1}(\mathcal{P}(x_1 \oplus k \oplus \Delta_1) \oplus k \oplus \Delta_2 \oplus k \oplus \Delta_1 \oplus \Delta_2 \oplus \Delta_3) \\ &\quad \oplus k \oplus \Delta_1 \oplus \Delta_2 \oplus \Delta_3 \\ &= \mathcal{P}^{-1}(\mathcal{P}(x_1 \oplus k \oplus \Delta_1) \oplus \Delta_1 \oplus \Delta_3) \oplus k \oplus \Delta_1 \oplus \Delta_2 \oplus \Delta_3 \end{aligned}$$

From the final expressions of x_3 and x_4, we see that their XOR difference is indeed $\Delta_1 \oplus \Delta_2$.

4 Application to Prøst-OTR

We apply the simple generic key-recovery attack to the CAESAR candidate PRØST-OTR, which is an authenticated-encryption scheme member of the PRØST family [10]. This family is based on the PRØST permutation and defines three schemes instantiating as many modes of operation, namely COPA, OTR and APE. Only the latter can readily be instantiated with a permutation, and both COPA and OTR rely on a keyed primitive. For that purpose they use a block cipher defined as a one-round Even-Mansour cipher with identical keys $\mathcal{E}(k, p) := \mathcal{P}(p \oplus k) \oplus k$ with the PRØST permutation as $\mathcal{P}$. We will denote this cipher as PRØST/SEM.

Although the attack of Sect. 3 could readily be applied to PRØST/SEM, this cipher is only meant to be embedded into a specific instantiation of a mode such as OTR, and attacking it out of context may not be relevant to its intended use. Hence we must be able to mount an attack on PRØST-COPA or PRØST-OTR as a whole for it to be really significant, which is precisely what we describe now for the latter.

Because our attack solely relies on the Even-Mansour structure of the cipher, we refer the interested reader to the submission document of PRØST for the definition of its permutation. The same goes for the OTR mode [12], as we only need to focus on a small part to describe the attack. Consequently, we just describe how the encryption of the first block of plaintext is performed in PRØST-OTR.

The mode of operation OTR is nonce-based; it takes as input a key k, a nonce n, a message m, (possibly empty) associated data a, and produces a ciphertext c corresponding to the encryption of the message with k, and a tag t authenticating m and a together with the key k. It is important for the security of the mode to ensure that one cannot encrypt twice using the same nonce. However, there are no specific restriction as to their value, and we consider throughout that one can freely choose them.

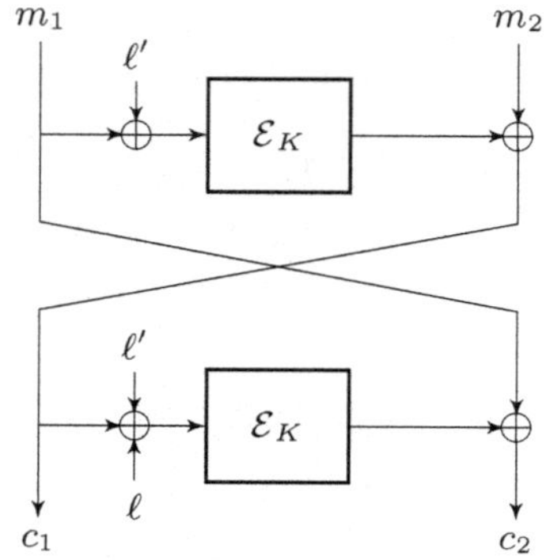

Fig. 1. The encryption of the first two blocks of message in PRØST-OTR.

The encryption of the first block of ciphertext c_1 by PRØST-OTR is defined as a function $\mathcal{F}(k, n, m_1, m_2)$ of k, n, and the first two blocks of plaintext m_1 and m_2: let $\ell := \mathcal{E}(k, n||10^*)$ be the encryption of the padded nonce and $\ell' := \pi(\ell)$, with π a linear permutation (the multiplication by 4 in some finite field), then c_1 is simply equal to $\mathcal{E}(k, \ell' \oplus m_1) \oplus m_2$. We show this schematically along with the encryption of the second block in Fig. 1. Let us now apply the attack from Sect. 3.

Step 1: Recovering the Most Significant Half of the Key. It is straightforward to see that one can recover the value of the bit $k[i]$ by performing only two queries with related keys and different nonces and messages. One just has to compare $c_1 = \mathcal{F}(k, n, m_1, m_2)$ and $\hat{c}_1 = \mathcal{F}(k + \Delta_i, n \oplus \Delta_i, m_1 \oplus \Delta_i \oplus \pi(\Delta_i), m_2)$. Indeed, if $k[i] = 0$, then the value $\hat{\ell}$ obtained in the computation of $\hat{c}_1$ is equal to $\ell \oplus \Delta_i$ and $\hat{\ell}' = \ell' \oplus \pi(\Delta_i)$, hence $\hat{c}_1 = c_1 \oplus \Delta_i$. If $k[i] = 1$, the latter equality does not hold with overwhelming probability.

Yet this does not allow to recover the whole key because the nonce in PRØST-OTR is restricted to a length half of the width of the block cipher $\mathcal{E}$ (or equivalently of the underlying PRØST permutation), *i.e.* $\frac{\kappa}{2}$. It is then possible to recover only half of the bits of k using this procedure, as one cannot introduce appropriate differences in the computation of ℓ for the other half. The targeted security of the whole primitive being precisely $\frac{\kappa}{2}$ because of the generic single key attacks on Even-Mansour, one does not make a significant gain by recovering only half of the key. Even though, it should still be noted that this yields an attack with very little data requirements and with the same time complexity as the best point on the tradeoff curve of generic attacks, which in that case has a much higher data complexity of $2^{\frac{\kappa}{2}}$.

Step 2: Recovering the Least Significant Half of the Key. Even though the generic attack in its most simple form does not allow to recover the full key of PRØST-OTR, we can use the fact that the padding of the nonce is done on the least significant bits to our advantage, and by slightly adapting the procedure we can iteratively recover the value of the least significant half of the key with no more effort than for the most significant half.

Let us first show how we can recover the most significant bit of the least significant half of the key $k[\kappa/2-1]$ (*i.e.* the first bit for which we cannot use the procedure of Step 1) after a single encryption by $\mathcal{E}$.

We note k^{MSB} the (by now known) most significant half of the key k. To mount the attack, one queries $\mathcal{E}(k - k^{\mathsf{MSB}} + \Delta_{\kappa/2-1}, p \oplus \Delta_{\kappa/2})$ and $\mathcal{E}(k - k^{\mathsf{MSB}} - \Delta_{\kappa/2-1}, p)$. We can see that the inputs to $\mathcal{P}$ in these two cases are equal iff $k[\kappa/2-1] = 1$. Indeed, in that case, the carry in the addition $(k - k^{\mathsf{MSB}}) + \Delta_{\kappa/2-1}$ propagates by exactly one position and is "cancelled" by the difference in p, and there is no carry propagation in $(k - k^{\mathsf{MSB}}) - \Delta_{\kappa/2-1}$. The result of the two queries are therefore equal to $\mathsf{C} \oplus (k - k^{\mathsf{MSB}} + \Delta_{\kappa/2-1}) = \mathsf{C} \oplus (k \oplus k^{\mathsf{MSB}} \oplus \Delta_{\kappa/2-1} \oplus \Delta_{\kappa/2})$ and $\mathsf{C} \oplus (k - k^{\mathsf{MSB}} - \Delta_{\kappa/2-1}) = \mathsf{C} \oplus (k \oplus k^{\mathsf{MSB}} \oplus \Delta_{\kappa/2-1})$ with $\mathsf{C} = \mathcal{P}(p \oplus (k - k^{\mathsf{MSB}} - \Delta_{\kappa/2-1}))$. Consequently, the XOR difference between the two results is known and equal to $\Delta_{\kappa/2}$. If on the other hand $k[\kappa/2-1] = 0$, the carry in $(k - k^{\mathsf{MSB}}) - \Delta_{\kappa/2-1})$ propagates all the way to the most significant bit of k, whereas only two differences are introduced in the input to $\mathcal{P}$ in the first query. This allows to distinguish between the two cases and thus to recover the value of this key bit.

Once the value of $k[\kappa/2-1]$ has been learned, one can iterate the process to recover the remaining bits of k. The only subtlety is that we want to ensure that if there is a carry propagation in $(k - k^{\mathsf{MSB}}) + \Delta_{\kappa/2-1-i}$ (resp. $(k - k^{\mathsf{MSB}}) - \Delta_{\kappa/2-1-i}$), it should propagate up to $k_{\kappa/2}$, the position where we cancel it with an XOR difference (resp. up to the most significant bit); this can easily be achieved by adding two terms to both keys. Let us define γ_i as the value of the key k only on positions $\kappa/2-1 \dots \kappa/2-i$, completed with zeros left and right; that is $\gamma_i[j] = k[j]$ if $\kappa/2-1 \geq j \geq \kappa/2-i$, and $\gamma_i[j] = 0$ otherwise. Let us also define $\widetilde{\gamma_i}$ as the binary complement of γ_i on its non-zero support; that is $\widetilde{\gamma_i}[j] = \widetilde{k[j]}$ if $\kappa/2-1 \geq j \geq \kappa/2-i$, and $\widetilde{\gamma_i}[j] = 0$ otherwise. The modified queries then become $\mathcal{E}(k - k^{\mathsf{MSB}} + \Delta_{\kappa/2-1-i} + \widetilde{\gamma_i}, p \oplus \Delta_{\kappa/2})$ and $\mathcal{E}(k - k^{\mathsf{MSB}} - \Delta_{\kappa/2-1-i} - \gamma_i, p)$, for which the propagation of the carries is ensured. Note that the difference between the results of these two queries when $k[\kappa/2-1-i] = 1$ is independent of i and always equal to $\Delta_{\kappa/2}$.

We conclude by showing how to apply this procedure to PRØST-OTR. For the sake of readability, let us denote by Δ_i^+ and Δ_i^- the complete modular differences used to recover one less significant bit $k[i]$. We then simply perform the two queries $\mathcal{F}(k+\Delta_i^+, n \oplus \Delta_{\kappa/2}, m_1 \oplus \Delta_{\kappa/2}, m_2)$ and $\mathcal{F}(k+\Delta_i^-, n, m_1 \oplus \pi(\Delta_{\kappa/2}), m_2)$, which differ by $\Delta_{\kappa/2}$ iff k_i is one, with overwhelming probability.

All in all, one can retrieve the whole key of size κ using only 2κ related-key encryption requests, ignoring everything in the output (including the tag) apart from the value of the first block of ciphertext. We give the entire procedure to do so in Algorithm 1. Note that it makes use of a procedure REFRESH which picks fresh values for two message words and (most importantly) for the nonce.

Because the attack is entirely practical, it can easily be tested. We give an implementation of the attack for a 64-bit toy cipher in the Appendix A[1].

Algorithm 1. Related-key key recovery for PRØST-OTR

Input: Oracle access to $\mathcal{F}(k,\cdot,\cdot,\cdot)$ and $\mathcal{F}(\phi^+(k),\cdot,\cdot,\cdot)$ for a fixed (unknown) key k of even length κ

Output: Two candidates for the key k

$k' := 0^\kappa$
for $i := \kappa - 2$ *to* $\kappa/2$ **do**
  REFRESH(n, m_1, m_2)
  $x := \mathcal{F}(k, n, m_1, m_2)$
  $y := \mathcal{F}(k + \Delta_i, n \oplus \Delta_i, m_1 \oplus \Delta_i \oplus \pi(\Delta_i), m_2)$
  if $x = y \oplus \Delta_i$ **then**
    $k'[i] := 0$
  else
    $k'[i] := 1$
for $i := \kappa/2 - 1$ *to* 0 **do**
  REFRESH(n, m_1, m_2)
  $x := \mathcal{F}(k + \Delta_i^+, n \oplus \Delta_{\kappa/2}, m_1 \oplus \Delta_{\kappa/2}, m_2)$
  $y := \mathcal{F}(k + \Delta_i^-, n, m_1 \oplus \pi(\Delta_{\kappa/2}), m_2)$
  if $x = y \oplus \Delta_{\kappa/2}$ **then**
    $k'[i] := 1$
  else
    $k'[i] := 0$
$k'' := k'$
$k''[\kappa - 1] := 1$
return (k', k'')

REMARK. If the padding of the nonce in PRØST-OTR were done on the most significant bits, no attack similar to Step 2 could recover the corresponding key bits: the modular addition is a triangular function (meaning that the result of $a + b$ on a bit i only depends on the value of bits of position less than i in a and b), and therefore no XOR in the nonce in the less significant bits could control modular differences introduced in the padding in the more significant bits. An attack in that case would thus most likely be applicable to general ciphers when using only the ϕ^+ class, and it is proven that no such attack is efficient. However, one could always imagine using a related-key class using an addition operation reading the bits in reverse. While admittedly unorthodox, this would not result in a stronger model than using ϕ^+, strictly speaking.

[1] The code is also available at https://github.com/P1K/EMRKA.

Discussion. In a recent independent work, Dobraunig, Eichlseder and Mendel use similar methods to produce forgeries for PRØST-OTR by considering related keys with XOR differences [5]. On the one hand, one could argue that the class $\phi^{\oplus}$ is more natural than ϕ^{+} and more likely to arise in actual protocols, which would make their attack more applicable than ours. On the other hand, an ideal cipher is expected to give a similar security against RKA using either class, which means that our model is not theoretically stronger than the one of Dobraunig et al., while resulting in a much more powerful key recovery attack.

5 Conclusion

We made a simple observation that allows to convert related-key distinguishing attacks on some Even-Mansour ciphers into much more powerful key-recovery attacks, and we used this observation to derive an extremely efficient key-recovery attack on the PRØST-OTR CAESAR candidate.

Primitives based on the Even-Mansour construction are quite common, and it is natural to wonder if we could mount similar attacks on other ciphers. A natural first target would be PRØST-COPA which is also based on the PRØST/SEM cipher. However, in this mode, encryption and tag generation depend on the encryption of a fixed plaintext $\ell = \mathcal{E}(k, 0)$ which is different for different keys with overwhelming probability and makes our attack fail. The forgery attacks of Dobraunig et al. seem to fail in that case for the same reason. Keeping with CAESAR candidates, another good target would be Minalpher [14], which also uses a one-round Even-Mansour block cipher as one of its components. The attack also fails in this case, though, because the masking key used in the Even-Mansour cipher is derived from the master key in a highly non-linear way. In fact, Mennink recently proved that both ciphers are resistant to related-key attacks [11]. Finally, leaving aside authentication and going back to traditional block ciphers, we could consider designs such as LED [9]. The attack also fails in that case, however, because the cipher uses an iterated construction with at least 8 rounds and only one (or two) keys.

This lack of other results is not very surprising, as we only improve existing distinguishing attacks, and this improvement cannot be used without a distinguisher as its basis. Therefore, any primitive for which resistance to related-key attacks is important should already be resistant to the distinguishing attacks and thus to ours. Yet it would be reasonable to allow the presence of a simple related-key distinguisher when designing a primitive, as this a very weak type of attack (in fact, this is for instance the approach taken by PRINCE, among others [2], which admits a trivial distinguisher due to its FX construction). What we have shown is that one must be extremely careful when contemplating such a decision for Even-Mansour ciphers, as in that case it is actually equivalent to allowing key recovery, the most powerful of all attacks.

Acknowledgements. I am grateful to Jérémy Jean, Brice Minaud and the anonymous reviewers for their comments on this work.

A Proof-of-concept Implementation for a 64-Bit Permutation

We give the source of a C program that recovers a 64-bit key from a design similar to PRØST-OTR where the permutation has been replaced by a small ARX, for compactness. For the sake of simplicity, we do not ensure that the nonce does not repeat in the queries.
This code is also available at https://github.com/P1K/EMRKA.

```
#include <stdio.h>
#include <stdint.h>
#include <stdlib.h>

#define ROL32(x,r) (((x) << (r)) ^ ((x) >> (32 - (r))))
#define MIX(hi,lo,r) { (hi) += (lo); (lo) = ROL32((lo),(r)) ; (lo) ^= (hi)
    ; }

#define TIMES2(x) ((x & 0x8000000000000000ULL) ? ((x) << 1ULL) ^
                   0x000000000000001BULL : (x << 1ULL))
#define TIMES4(x) TIMES2(TIMES2((x)))

#define DELTA(x) (1ULL << (x))
#define MSB(x) ((x) & 0xFFFFFFFF00000000ULL)
#define LSB(x) ((x) & 0x00000000FFFFFFFFULL)

/* Replace arc4random() by your favourite PRNG */

/* 64-bit permutation using Skein's MIX */
uint64_t p64(uint64_t x)
{
    uint32_t hi = x >> 32;
    uint32_t lo = LSB(x);
    unsigned rcon[8] = {1, 29, 4, 8, 17, 12, 3, 14};

    for (int i = 0; i < 32; i++)
    {
       MIX(hi, lo, rcon[i
       lo += i;
    }

    return (((uint64_t)hi) << 32) ^ lo);
}

uint64_t em64(uint64_t k, uint64_t p)
{
    return p64(k ^ p) ^ k;
}

uint64_t potr_1(uint64_t k, uint64_t n, uint64_t m1, uint64_t m2)
{
    uint64_t l, c;

    l = TIMES4(em64(k, n));
    c = em64(k, l ^ m1) ^ m2;

    return c;
}

uint64_t recover_hi(uint64_t secret_key)
{
    uint64_t kk = 0;

    for (int i = 62; i >= 32; i--)
    {
        uint64_t m1, m2, c11, c12, n;
```

```
        m1 = (((uint64_t)arc4random()) << 32) ^ arc4random();
        m2 = (((uint64_t)arc4random()) << 32) ^ arc4random();
        n  = (((uint64_t)arc4random()) << 32) ^ 0x80000000ULL;
        c11 = potr_1(secret_key, n, m1, m2);
        c12 = potr_1(secret_key + DELTA(i), n ^ DELTA(i), m1 ^ DELTA(i) ^
             TIMES4(DELTA(i)), m2);

        if (c11 != (c12 ^ DELTA(i)))
            kk |= DELTA(i);
    }

    return kk;
}

uint64_t recover_lo(uint64_t secret_key, uint64_t hi_key)
{
    uint64_t kk = hi_key;

    for (int i = 31; i >= 0; i--)
    {
        uint64_t m1, m2, c11, c12, n;
        uint64_t delta_p, delta_m;

        m1 = (((uint64_t)arc4random()) << 32) ^ arc4random();
        m2 = (((uint64_t)arc4random()) << 32) ^ arc4random();
        n  = (((uint64_t)arc4random()) << 32) ^ 0x80000000ULL;

        delta_p = DELTA(i) - MSB(kk) + (((LSB(~kk)) >> (i + 1)) << (i + 1)
             );
        delta_m = DELTA(i) + MSB(kk) + LSB(kk);
        c11 = potr_1(secret_key + delta_p, n ^ DELTA(32), m1 ^ DELTA(32),
             m2);
        c12 = potr_1(secret_key - delta_m, n, m1 ^ TIMES4(DELTA(32)), m2);

        if (c11 == (c12 ^ DELTA(32)))
            kk |= DELTA(i);
    }

    return kk;
}

int main()
{
    uint64_t secret_key = (((uint64_t)arc4random()) << 32) ^ arc4random();
    uint64_t kk1 = recover_lo(secret_key, recover_hi(secret_key));
    uint64_t kk2 = kk1 ^ 0x8000000000000000ULL;

    printf("The real key is %016llx, the key candidates are %016llx, %016
         llx   ", secret_key, kk1, kk2);
    if ((kk1 == secret_key) || (kk2 == secret_key))
        printf("SUCCESS!\n");
    else
        printf("FAILURE!\n");

    return 0;
}
```

References

1. Bellare, M., Kohno, T.: A theoretical treatment of related-key attacks: RKA-PRPs, RKA-PRFs, and applications. In: Biham, E. (ed.) EUROCRYPT 2003. LNCS, vol. 2656. Springer, Heidelberg (2003). http://dx.doi.org/10.1007/3-540-39200-9_31

2. Borghoff, J., Canteaut, A., Güneysu, T., Kavun, E.B., Knezevic, M., Knudsen, L.R., Leander, G., Nikov, V., Paar, C., Rechberger, C., Rombouts, P., Thomsen, S.S., Yalçın, T.: PRINCE – a low-latency block cipher for pervasive computing applications. In: Wang, X., Sako, K. (eds.) ASIACRYPT 2012. LNCS, vol. 7658, pp. 208–225. Springer, Heidelberg (2012). http://dx.doi.org/10.1007/978-3-642-34961-4_14
3. Chen, S., Steinberger, J.P.: Tight security bounds for key-alternating ciphers. In: Nguyen and Oswald [13], pp. 327–350. http://dx.doi.org/10.1007/978-3-642-55220-5_19
4. Cogliati, B., Seurin, Y.: On the provable security of the iterated even-mansour cipher against related-key and chosen-key attacks. In: Oswald, E., Fischlin, M. (eds.) EUROCRYPT 2015. LNCS, vol. 9056, pp. 584–613. Springer, Heidelberg (2015). http://dx.doi.org/10.1007/978-3-662-46800-5_23
5. Dobraunig, C., Eichlseder, M., Mendel, F.: Related-key forgeries for Prøst-OTR. IACR Cryptology ePrint Archive 2015, 91 (2015). https://eprint.iacr.org/2015/91. To appear in the proceedings of FSE 2015
6. Dunkelman, O., Keller, N., Shamir, A.: Minimalism in cryptography: the even-mansour scheme revisited. In: Pointcheval, D., Johansson, T. (eds.) EUROCRYPT 2012. LNCS, vol. 7237, pp. 336–354. Springer, Heidelberg (2012). http://dx.doi.org/10.1007/978-3-642-29011-4_21
7. Even, S., Mansour, Y.: A construction of a cioher from a single pseudorandom permutation. In: Matsumoto, T., Imai, H., Rivest, R.L. (eds.) ASIACRYPT 1991. LNCS, vol. 739, pp. 210–224. Springer, Heidelberg (1993). http://dx.doi.org/10.1007/3-540-57332-1_17
8. Farshim, P., Procter, G.: The related-key security of iterated even-mansour ciphers. IACR Cryptology ePrint Archive 2014, 953 (2014). https://eprint.iacr.org/2014/953. To appear in the proceedings of FSE 2015
9. Guo, J., Peyrin, T., Poschmann, A., Robshaw, M.: The LED block cipher. In: Preneel, B., Takagi, T. (eds.) CHES 2011. LNCS, vol. 6917, pp. 326–341. Springer, Heidelberg (2011)
10. Kavun, E.B., Lauridsen, M.M., Leander, G., Rechberger, C., Schwabe, P., Yalçın, T.: Prøst. CAESAR Proposal (2014). http://proest.compute.dtu.dk
11. Mennink, B.: XPX: Generalized Tweakable Even-Mansour with Improved Security Guarantees. IACR Cryptology ePrint Archive 2015, 476 (2015). http://eprint.iacr.org/2015/476
12. Minematsu, K.: Parallelizable rate-1 authenticated encryption from pseudorandom functions. In: Nguyen, P.Q., Oswald, E. (eds.) EUROCRYPT 2014. LNCS, vol. 8441, pp. 275–292. Springer, Heidelberg (2014)
13. Nguyen, P.Q., Oswald, E. (eds.): Advances in Cryptology – EUROCRYPT 2014–33rd Annual International Conference on the Theory and Applications of Cryptographic Techniques. Lecture Notes in Computer Science, vol. 8441. Springer, Heidelberg (2014). http://dx.doi.org/10.1007/978-3-642-55220-5
14. Sasaki, Y., Todo, Y., Aoki, K., Naito, Y., Sugawara, T., Murakami, Y., Matsui, M., Hirose, S.: Minalpher. CAESAR Proposal (2014). http://competitions.cr.yp.to/round1/minalpherv1.pdf

Cryptography II: Protocols

Oblivious PAKE: Efficient Handling of Password Trials

Franziskus Kiefer(✉) and Mark Manulis

Surrey Centre for Cyber Security Department of Computer Science,
University of Surrey, Surrey, UK
mail@franziskuskiefer.de, mark@manulis.eu

Abstract. In this work we introduce *Oblivious Password based Authenticated Key Exchange* (O-PAKE) and show how ordinary PAKE protocols can be transformed into O-PAKE. O-PAKE allows a client that holds multiple passwords and is registered with *one* of them at some server to use any *subset* of his passwords in a PAKE session with that server. The term *oblivious* is used to emphasise that the only information leaked to the server is whether the one password used on the server side matches any of the passwords input by the client. O-PAKE protocols can be used to improve the overall efficiency of login attempts using PAKE protocols in scenarios where users are not sure (e.g. no longer remember) which of their passwords has been used at a particular web server. Using special processing techniques, our O-PAKE compiler reaches nearly constant run time on the server side, independent of the size of the client's password set; in contrast, a naive approach to run a new PAKE session for each login attempt would require linear run time for both parties. We prove security of the O-PAKE compiler under standard assumptions using the latest game-based PAKE model by Abdalla, Fouque and Pointcheval (PKC 2005), tailored to our needs. We identify the requirements that standard PAKE protocols must satisfy in order to suit our O-PAKE transformation and give two examples.

1 Introduction

Authentication with passwords is the most common (and perhaps most critical) authentication mechanism on the modern Internet. The dominating approach today is when clients send passwords (or some function thereof) to the server over a secure channel (e.g. TLS [18]). This approach requires PKI and its security relies solely on the secure channel and the client's ability to correctly verify the server's certificate. Any impersonation of the certificate leads to password exposure. Even if no impersonation takes place, any password input on the client side is revealed to the server. This creates a different problem based on statistics, indicating that many users operate with a small set of passwords but often do not remember their correct mapping to the servers. If a user types in a password that is not shared with this server but with another one then its exposure may lead to subsequent impersonation attacks on the client. The studies in [20,21]

J. Lopez and C.J. Mitchell (Eds.): ISC 2015, LNCS 9290, pp. 191–208, 2015.
DOI: 10.1007/978-3-319-23318-5_11

show that every user has 6.5 passwords on average, used on 25 different websites and that on average 2.4 password trials are required until the user types in the correct password. These numbers suggest that in case where a server limits a number of failed attempts to say 3, in the worst case roughly 2 passwords from the client's set could potentially be revealed to the server within a single TLS session — a significant threat for the client. Note that the amount of work for processing failed login attempts on the server side is negligible since all trials are performed through the same secure channel.

The notion of *Password-based Authenticated Key Exchange (PAKE)*, introduced by Bellovin and Merritt [8], initially formalised in [6,14], and later explored in numerous further works [1–5,11,16,22,23,26,27], is considered as a more secure alternative to the above approach. The standard model of PAKE does not require any PKI and assumes that only a human-memorable password is shared between both parties. PAKE protocols solve the problem of potential password leakage, inherent to the previously described approach. They aim to protect against offline dictionary attacks but require the same method of protection against online dictionary attacks as the aforementioned TLS-based approach, namely by restricting the number of failed password trials. While passwords can be retransmitted and checked by the server, using the same TLS channel, the only way for current PAKE protocols to deal with failed password trials is to repeat the entire protocol. This however implies that the computational costs on the server side, in particular for (costly) public key-operations that are inherent to all PAKE protocols, increase *linearly* with the number of attempts. This can be seen as a reason for the limited progress on the adoption of PAKE on the Internet (in addition to unrelated issues such as browser incompatibility, patent considerations, and the lack of adopted standards).

While handling multiple password trials with PAKE may seem like a pure implementation problem at first sight, the problem becomes non-trivial if we want to avoid linear increase of public key operations on the server side. This seems to be avoidable only if in a single PAKE execution the client can use several passwords, while the server would use only the one password, shared with the client. Yet this idea alone is not sufficient for breaking the linear bound on the server side: for instance, assume that one PAKE execution is built out of n independent (possibly parallelised) runs of some secure PAKE protocol, where the client uses a different password in each run but the server uses the same one in all of them. The amount of work for the server in this case would still remain $O(n)$. Therefore, something non-trivial must additionally happen in order to reduce the amount of work on the server side to $O(1)$.

However, we still need to fulfil basic PAKE requirements like addressing the persistent threat of online dictionary attacks by enforcing that the number of passwords that can be tested by the client in one session remains below some threshold, which is set by the server. For the server there is no difference whether a client is given the opportunity to perform at most c independent PAKE sessions (password trials) with one input password per session, or only one session but with at most c input passwords. Finally, we must be able to prevent a possibly malicious server from obtaining any password from the set of passwords input by the client.

1.1 Oblivious PAKE and Our Contributions

We solve the aforementioned problem of efficient handling of password trials on the server side by proposing a compiler that transforms PAKE protocols in a black-box way into what we call an *Oblivious PAKE* (O-PAKE). To describe and analyse the proposed O-PAKE notion we introduce a new algorithmic way to model PAKE protocols that also allows for easy compilation as done with O-PAKE, and real-world implementation.

The functionality of O-PAKE protocols resembles that of PAKE except that the client inputs a set **pw** of $n \in [1, c]$ passwords while the server's input is limited to one password pw. The use of **pw** does *not* increase the overall probability for online dictionary attacks in comparison to running a separate PAKE session for each tried password because the *maximum number* of passwords c that the client can try with O-PAKE is fixed by the server. The client can still input less than c passwords, i.e. if the client is confident about validity of some particular pw for a given server then pw can be used as the sole input, in which case O-PAKE is equivalent to PAKE. In general, O-PAKE protocol execution succeeds if and only if the server's password pw is part of the client's password set **pw**. We use the standard (game-based) PAKE model by Bellare, Pointcheval, and Rogaway [6] in its (stronger) Real-or-Random flavour from [4] and update it to account for the use of **pw** as client's input. In this model passwords are assumed to be distributed uniformly at random. In practice, the use of passwords with different strengths in the same O-PAKE session would lower the overall security to the probability for guessing the weakest password (irrespective of the adopted strength metric).

The crucial idea behind our O-PAKE compiler is to let each client execute n sessions of secure PAKE protocol in parallel and let the server execute only *one* PAKE session. The challenging part is to enable the server to actually identify the correct PAKE session in which the client used the correct password pw, while preserving security against offline dictionary attacks for all passwords in the client's password set **pw**. This is the trickiest part of the compiler. Intuitively, if the server can recover the messages of the correct PAKE session, it can answer them according to the specification of the PAKE protocol. By repeating this approach in each communication round of the given PAKE protocol both parties will be able to successfully accomplish the protocol. If identification of the correct PAKE session by the server requires only a constant amount of (costly) operations, then the total amount of server's work in the resulting O-PAKE protocol will also remain constant. The amount on the client side remains linear in the size n of input passwords. This stems from the obvious fact that the client has to compute messages for all PAKE sessions without knowing the correct password. We show how to apply our O-PAKE compiler to two concrete PAKE protocols: the SPAKE protocol from [5] and the PAKE protocol from [28] (for space limitations the second construction is given in the full version of this work [29]).

2 Oblivious PAKE Model

In this section we recall the PAKE security model from [4], tailored to the needs of O-PAKE. The security model for O-PAKE protocols is in the multi-user setting and utilises the Real-or-Random approach for AKE-security from [4,6]. Note that the AKE-security definition addresses the aforementioned security against malicious servers, trying to retrieve client passwords. A server learning information about the additional passwords in the client's password set **pw** can easily break AKE-security by using this password in another session with the same client.

Participants and Passwords. An O-PAKE protocol is executed between two parties P and P', chosen from the universe of participants $\Omega = \mathcal{S} \cup \mathcal{C}$, where $\mathcal{S}$ denotes the universe of servers and $\mathcal{C}$ the universe of clients, such that if $P \in \mathcal{C}$ then $P' \in \mathcal{S}$, and vice versa. We assume the scenario where every client in $\mathcal{C}$ is registered with every servers from $\mathcal{S}$. For each such pair $(P, P') \in \mathcal{C} \times \mathcal{S}$, a password $\mathrm{pw}_{P,P'}$ (shared between client P and server P') is drawn uniformly at random from the dictionary $\mathcal{D}$ of size $|\mathcal{D}|$. Execution of an oblivious PAKE protocol between P and P' uses $\mathrm{pw}_{P,P'}$ on the server and a password vector $\mathbf{pw}_P \subseteq \{\mathrm{pw}_{P,P'_{x_1}}, \ldots, \mathrm{pw}_{P,P'_{xn}}\}$ for $1 \leq n \leq c$ and client-server pairs (P, P'_{x_i}) for $i \in [2, n]$ on the client side. For the protocol to be successful it is necessary that $\mathrm{pw}_{P,P'} \in \mathbf{pw}_P$. The value c is a global parameter with $c \leq |\mathcal{S}|$. We will sometimes write **pw** and pw instead of $\mathbf{pw}_P$ and $\mathrm{pw}_{P,P'}$ when the association with the participants is clear or if it applies to every participant. We will further write PAKE for O-PAKE protocols with $n = 1$, i.e. standard class of PAKE protocols where the client uses $\mathbf{pw}_P = \mathrm{pw}_{P,P'}$.

Protocol Instances. For $i \in \mathbb{N}$, we denote by P_i the i-th instance of $P \in \Omega$. In order to model uniqueness of P_i within the model we use i as a counter. For each instance P_i we consider further a list of parameters:

- $\mathtt{pid}_P^i$ is the partner id of P_i, defined upon initialisation, subject to following restriction: if $P_i \in \mathcal{C}$ then $\mathtt{pid}_P^i \in \mathcal{S}$, and if $P_i \in \mathcal{S}$ then $\mathtt{pid}_P^i \in \mathcal{C}$.
- $\mathtt{sid}_P^i$ is the session id of P_i, modelled as ordered (partial) protocol transcript $[m_{\mathrm{in}}^1, m_{\mathrm{out}}^1, \ldots, m_{\mathrm{in}}^r, m_{\mathrm{out}}^r]$ of incoming and outgoing messages of P_i in rounds 1 to r. $\mathtt{sid}_P^i$ is thus updated on each sent or received protocol message.
- $\mathtt{k}_P^i$ is the value of the session key of instance P_i, which is initialised to `null`.
- $\mathtt{state}^{i_P}$ is the internal state of instance P_i.
- $\mathtt{used}^{i_P}$ indicates whether P_i has already been used.
- $\mathtt{role}_P^i$ indicates whether P_i acts as a `client` or a `server`.

Partnered Instances. Two instances P_i and P'_j are *partnered* if all of the following holds: (i) $(P, P') \in \mathcal{C} \times \mathcal{S}$, (ii) $\mathtt{pid}_P^i = P'$ and $\mathtt{pid}_{P'}^j = P$, and (iii) $\mathtt{match}(\mathtt{sid}_P^i, \mathtt{sid}_{P'}^j) = 1$, where Boolean algorithm `match` is defined according to the matching conversations from [7], i.e. outputs 1 if and only if round messages (in temporal order) in $\mathtt{sid}_P^i$ equal to the corresponding round messages in $\mathtt{sid}_{P'}^j$, except for the final round, in which the incoming message of one instance may differ from the outgoing message of another instance.

Oblivious PAKE. We define O-PAKE using an initialisation algorithm `init` and a stateful interactive algorithm `next`, which handles protocol messages and eventually outputs the session key.

Definition 1 (Oblivious PAKE). *An O-PAKE protocol* O-PAKE = (`init`, `next`) *over a message space* $\boldsymbol{\mathcal{M}} = (\bigcup_r \mathcal{M}_C^r) \cup (\bigcup_r \mathcal{M}_S^r)$, *where* $\mathcal{M}_C^r$ *resp.* $\mathcal{M}_S^r$ *denotes the space of outgoing server's resp. client's messages in the r-th invocation of* `next`, *a dictionary* $\mathcal{D}$, *and a key space* $\mathcal{K}$ *consists of two polynomial-time algorithms:*

$P_i \leftarrow \mathtt{init}(\mathbf{pw}, \mathtt{role}, P', \mathtt{par})$: *On input* $\mathbf{pw}$, $\mathtt{role} \in \{\mathtt{client}, \mathtt{server}\}$, $P' \in \Omega$ *and the public parameters* `par`, *the algorithm initialises a new instance* P_i *with the internal O-PAKE state information* `state`, *defines the intended partner id as* $\mathtt{pid}_P^i = P'$ *and session key* $\mathtt{k}_P^i = \mathtt{null}$, *and stores protocol parameters* `par`. *The* `role` *indicates whether the participant acts as* `client` *or* `server`.

$(m_{\mathrm{out}}, \mathtt{k}_P^i) \leftarrow \mathtt{next}(m_{\mathrm{in}})$: *On input* $m_{\mathrm{in}} \in \mathcal{M}_{[S,C]}^r \cup \emptyset$ *with implicit access to internal* `state`, *the algorithm outputs the next protocol message* $m_{\mathrm{out}} \in \mathcal{M}_{[S,C]}^{r+1} \cup \emptyset$ *and updates* $\mathtt{k}_P^i$ *with* $\mathtt{k}_P^i \in \mathcal{K} \cup \mathtt{null} \cup \perp$. *As long as the instance has not terminated the key* $\mathtt{k}_P^i$ *is* `null`. *If* m_{in} *leads to acceptance then* $\mathtt{k}_P^i$ *is from* $\mathcal{K}$, *otherwise* $\mathtt{k}_P^i = \perp$. *We also assume that* `next` *implicitly updates the internal* `state` *prior to each output and sets* `used` *to* `true`.

Note that $\boldsymbol{\mathcal{M}} = (\bigcup_r \mathcal{M}_S^r) \cup (\bigcup_r \mathcal{M}_C^r)$ is the union of outgoing client's message spaces $\mathcal{M}_C^r$ and server's message spaces $\mathcal{M}_S^r$ over all protocol rounds r. We may further view each round's message space $\mathcal{M}_C^r$ as a Cartesian product $\mathcal{M}_C^{r,1} \times \cdots \times \mathcal{M}_C^{r,l}$ for up to l different classes of message components, e.g. to model labels, identities, group elements, etc. When clear from the context, we will write $\mathcal{M}_C^r$ instead of $\mathcal{M}_C^{r,1} \times \cdots \times \mathcal{M}_C^{r,l}$.

Correctness. Let P_i be an instance initialised through $\mathtt{init}(\mathbf{pw}_P, \mathtt{client}, P', \mathtt{par})$ and P_j' be an instance initialised through $\mathtt{init}(\mathrm{pw}_{P,P'}, \mathtt{server}, P, \mathtt{par})$ where $P \in \mathcal{C}$, $P' \in \mathcal{S}$, and $\mathrm{pw}_{P,P'} \in \mathbf{pw}_P$. Assume that all outgoing messages, generated by `next` are faithfully transmitted between P_i and P_j' so that the instances become partnered. An O-PAKE = (`init`, `next`) is said to be *correct* if for all partnered P_i and P_j' it holds that $\mathtt{k}_P^i \in \mathcal{K}$ and $\mathtt{k}_P^i = \mathtt{k}_{P'}^j$.

Adversary Model. The adversary $\mathcal{A}$ is modelled as a probabilistic-polynomial time (PPT) algorithm, with access to the following oracles:

$m_{\text{out}} \leftarrow \mathsf{Send}(P, i, m_{\text{in}})$: the oracle processes the incoming message $m_{\text{in}} \in \mathcal{M}^r_{[C,S]}$ for the instance P_i and returns its outgoing message $m_{\text{out}} \in \mathcal{M}^{r+1}_{[C,S]} \cup \emptyset$. If P_i does not exist, a new session is created with P' as partner, where P' is given in m_{in}.

$\mathtt{trans} \leftarrow \mathsf{Execute}(P, P')$: if $(P, P') \in \mathcal{C} \times \mathcal{S}$ the oracle creates two new instances P_i and P'_j via appropriate calls to `init` and returns the transcript `trans` of their protocol execution, obtained through invocations of corresponding `next` algorithms and faithful transmission of generated messages amongst the two instances.

$\text{pw} \leftarrow \mathsf{Corrupt}(P, P')$: if $P \in \mathcal{C}$ and $P' \in \mathcal{S}$ then return $\text{pw}_{P,P'}$ and mark (P, P') as a corrupted pair.

AKE-Security. The following definition of AKE-security follows the *Real-Or-Random* (ROR) approach from [4], which provides the adversary multiple access to the Test oracle for which the randomly chosen bit $b \in_R \{0, 1\}$ is fixed in the beginning of the experiment:
$\mathtt{k}_{\mathcal{A}} \leftarrow \mathsf{Test}_b(P, i)$, depending on the values of bit b and $\mathtt{k}^i_P$, this oracle responds with key $\mathtt{k}_{\mathcal{A}}$ defined as follows:

- If, while $\mathtt{k}^i_P = \mathtt{null}$, either (P, P') or $(\mathtt{pid}^i_P, P)$ were queried to the Corrupt oracle for, w.l.o.g., any client-server pair (P, P') with $\text{pw}_{P,P'} \in \mathbf{pw}_P$, then abort. Note that this prevents $\mathcal{A}$ from obtaining any $\text{pw}_{P,P'} \in \mathbf{pw}$ and then testing new instances of P and P', or instances that were still in the process of establishing session keys when corruption took place.
- If some previous query $\mathsf{Test}(P', j)$ was asked for an instance P'_j, which is partnered with P_i, then return the same response as to that query. Note that this guarantees consistency of oracle responses.
- If $\mathtt{k}^i_P \in \mathcal{K}$ then if $b = 1$, return $\mathtt{k}^i_P$, else if $b = 0$, return a randomly chosen element from $\mathcal{K}$ and store it for later use.
- Else return $\mathtt{k}^i_P$. Note that in this case $\mathtt{k}^i_P$ is either $\perp$ or `null`.

According to [4] a session is an online session when $\mathcal{A}$ queried the Send oracle on one of the participants.

Definition 2 (AKE-Security). *An O-PAKE protocol Π with up to c passwords on client side is* AKE-secure *if for all dictionaries $\mathcal{D}$ with corresponding universe of participants Ω and for all PPT adversaries $\mathcal{A}$ using at most t online sessions there exists a negligible function $\varepsilon(\cdot)$ such that:*

$$\mathsf{Adv}^{\mathsf{AKE}}_{\Pi,\mathcal{A}}(\lambda) = \left|\Pr[\mathsf{Exp}^{\mathsf{AKE}}_{\Pi,\mathcal{A}}(\lambda) = 1] - \frac{1}{2}\right| \le \frac{c \cdot \mathcal{O}(t)}{|\mathcal{D}|} + \varepsilon(\lambda).$$

$\mathsf{Exp}^{\mathsf{AKE}}_{\Pi,\mathcal{A}}(\lambda)$: $c \in \mathbb{N}; b \in_R \{0, 1\}; \forall (P, P') \in \mathcal{C} \times \mathcal{S}$ *choose* $\text{pw}_{P,P'} \in_R \mathcal{D}$; $b' \leftarrow \mathcal{A}^{\mathsf{Send},\mathsf{Execute},\mathsf{Corrupt},\mathsf{Test}_b}(\lambda, c)$; *return* $b = b'$.

The above definition (without Corrupt) reverts to RoR AKE-security from [4] for $c = 1$. We have to factor in the maximal size of $|\mathbf{pw}| = n \le c$ into the original adversarial advantage bound $\mathcal{O}(t)/|\mathcal{D}|$ to account for the adversarial possibility of testing up to c passwords per session in the role of the client.

PAKE vs. O-PAKE. The actual relation between common PAKE and O-PAKE security may not be immediately evident. For clarification, we discuss the relation between O-PAKE and the simple repetition of a PAKE protocol c times, and the implication of user's password choice.

The advantage of an adversary that is allowed to query up to c passwords in one session is not greater than the advantage of an adversary that runs c online sessions using one password in each of them. The typical advantage of a PAKE adversary $\mathcal{A}$ in an AKE-security experiment, e.g. [4,6], is bounded by $\mathcal{O}(t)/|\mathcal{D}| + \varepsilon(\lambda)$. In contrast, we limit the advantage of an O-PAKE adversary to $c \cdot \mathcal{O}(t)/|\mathcal{D}| + \varepsilon'(\lambda)$. We give the following lemma to formalise the relation between the two notions.

Lemma 1. $\mathsf{Adv}^{\mathsf{AKE}}_{\Pi_c,\mathcal{A}} \leq c \cdot \mathsf{Adv}^{\mathsf{AKE}}_{\Pi,\mathcal{A}}$ *for O-PAKE protocol Π_c allowing up to c passwords in one session, built from PAKE protocol Π.*

Proof. The lemma follows directly from the following observations. O-PAKE can be realised in the naïve way by running c separate PAKE sessions. That results in an advantage of at most $c \cdot \mathsf{Adv}^{\mathsf{AKE}}_{\Pi,\mathcal{A}} = c \cdot \mathcal{O}(t)/|\mathcal{D}| + \varepsilon'(\lambda)$. Information gathered from Send and Execute oracle invocations are the same for the O-PAKE and PAKE adversary. Corrupt and Test_b queries of the O-PAKE adversary return one password, respectively key, independent from c, while the PAKE adversary gets c passwords, respectively keys. Thus, the resulting advantage of the O-PAKE adversary is at most $c \cdot \mathcal{O}(t)/|\mathcal{D}| + \varepsilon'(\lambda)$, but depending on the implementation most probably lower. □

Assuming malicious servers one may also be concerned about the client's password choice considering a client entering passwords with different levels of entropy. Similar to the standard PAKE case the weakest password from $\mathbf{pw}$ would determine the security of O-PAKE. However, the used model considers uniformly at random chosen passwords from one dictionary such that the case of varying password probabilities can not be adequately addressed in this model (as is also the case for the models in [4,6]).

3 Transforming PAKE Protocols into O-PAKE

Recall that one may realise O-PAKE in a naïve way by running the input PAKE protocol n times, which is not efficient on the server side due to the linearly increasing round complexity. The idea of the O-PAKE compiler is to mix the n PAKE messages on client side such that the server can extract the "right" message using the shared password and reply only to that. This, however, is a non-trivial problem because PAKE messages do not provide information that would allow the server to check locally whether a given password was used in their computation; as this would offer the possibility of offline dictionary attacks. Note that we assume throughout this section that $n \geq 2$ and $\mathrm{pw}_{P,P'} \in \mathbf{pw}_P$. Our solution for the identification of the "right" PAKE session is a careful composition of two encoding techniques that were introduced in a different context yet allow us to generically construct AKE-secure O-PAKE protocols from (suitable) AKE-secure PAKE protocols, preserving constant round complexity and offering nearly constant server load.

Our first building block is *Index-Hiding Message Encoding (IHME)* [30,31]. An IHME scheme assigns a different index to each given message and encodes the resulting index-message pairs into a single structure from which messages can be recovered on the receiver side using the corresponding indices. The IHME structure hides indices that were used for encoding and therefore all encoded messages must contain enough entropy to prevent dictionary attacks over the index space. An IHME scheme consists of two algorithms `iEncode` and `iDecode`. The `iEncode` algorithm takes as input a set of index-message pairs $(\mathrm{ix}_1, m_1), \ldots, (\mathrm{ix}_n, m_n)$ and outputs a structure S whereas the `iDecode` algorithm can extract m_j, $j \in [1, n]$ from S using the corresponding index ix_j. For formal definitions surrounding IHME we refer to the original work and only mention that the original IHME construction in [30] assumes $(\mathrm{ix}_j, m_j) \in \mathbb{F}$ for a prime-order finite field $\mathbb{F}$ and defines the IHME structure S through coefficients of the interpolated polynomial by treating index-message pairs as its points. There exists a more efficient IHME version from [31] for longer messages, which uses $(\mathrm{ix}_j, m_j) \in \mathbb{F} \times \mathbb{F}^\nu$ and thus splits m_j into ν components each being an element of $\mathbb{F}$. The corresponding index-hiding property demands that no information about indices ix_j is leaked to the adversary that doesn't know the corresponding messages m_j and is defined for messages that are chosen uniformly from the IHME message space. For the aforementioned IHME schemes the message space is given by $\mathbb{F}$ (or $\mathbb{F}^\nu$) and their index-hiding property is perfect (in the information-theoretic sense). Note that this approach still allows the server to learn which of the n PAKE sessions is the correct one without revealing any password to the server.

In order to enable encoding of PAKE messages using IHME with passwords as indices we apply our second building block, namely *admissible encoding* [13,15,19]. Briefly, a function $F : S \to R$ is an *ϵ-admissible encoding* for (S, R) with $|S| > |R|$ when for all uniformly distributed $r \in R$, the distribution of the inverse transformation $\mathcal{I}_F(r)$ is ϵ-statistically indistinguishable from the uniform distribution over S. We refer to [15,19] for more details. $\mathcal{I}_F$ enables us to map PAKE messages into the IHME message space where necessary. In Sect. 3.5 we will discuss suitable PAKE message spaces and their admissible encodings offering compatibility with the message space $\mathbb{F}$ of the IHME schemes from [30,31].

In the following we describe our compiler that transforms suitable AKE-secure PAKE protocols into AKE-secure O-PAKE protocols. The intuition behind the compiler is to let the client run n PAKE sessions, one session for each of the n input passwords $\mathbf{pw}$, and apply an index-hiding message encoding on each message-password pair. The server can apply the shared password pw as index to IHME to extract the "right" PAKE message. For this message the server executes the algorithm `next` of the given PAKE protocol and returns the resulting PAKE message to the client. As soon as the algorithm `next` terminates, the server generates a confirmation message, which is then used by the client to derive the final session key.

3.1 Requirements on PAKE

Our O-PAKE compiler can be used to convert any AKE-secure R-round PAKE protocol Π where in each round $r \in [1, \ldots, R]$ the client sends messages from $\mathcal{M}_C^r$ that can be processed using a compatible admissible encoding $F^r : \mathcal{M}^{\mathtt{IHME},r} \rightarrow \mathcal{M}_C^r$. In order to guarantee that client messages from $\mathcal{M}_C^r$, when mapped into $\mathcal{M}^{\mathtt{IHME},r}$ using the inverse transformation $\mathcal{I}_F$, are uniformly distributed over $\mathcal{M}^{\mathtt{IHME},r}$, the underlying Π itself must output client messages whose joint distribution over all R rounds remains indistinguishable from a distribution where for each round r the output client message is chosen uniformly at random from $\mathcal{M}_C^r$. For this purpose Π must satisfy a stronger notion of AKE security that in addition to the indistinguishability of session keys requires indistinguishability of client messages. This requirement is formalised in Definition 3 that extends the AKE-security experiment for PAKE from Definition 2, using $\mathsf{Execute}, \mathsf{Send}, \mathsf{Corrupt}$ and Test_b definitions from there. We assume that $c = 1$ and define two oracles Send_b and $\mathsf{Execute}_b$ that are parameterised with the bit b as used in the Test_b oracle. Any query $\mathsf{Send}_b(P, i, m_{\text{in}})$ for a client $P \in \mathcal{C}$ made by the adversary $\mathcal{A}$ first triggers the invocation of $m_{\text{out}} \leftarrow \mathsf{Send}(P, i, m_{\text{in}})$. If $\mathcal{A}$ queried $\mathsf{Corrupt}(P, \mathtt{pid}_P^i)$ or $\mathsf{Corrupt}(\mathtt{pid}_P^i, P)$ while $\mathtt{k}_P^i = \mathtt{null}$ or if $b = 1$ then m_{out} is returned to $\mathcal{A}$ without any modification. The additional condition on the $\mathsf{Corrupt}$ queries prevents $\mathcal{A}$ from trivially distinguishing the client messages by corrupting passwords and then communicating with client instances that were still in the process of establishing the session keys. If $b = 0$ then m_{out} is set to a random message from $\mathcal{M}_C^r$ and returned to $\mathcal{A}$. Any $\mathsf{Execute}_b(P, P')$ query first triggers the invocation of $\mathtt{trans} \leftarrow \mathsf{Execute}(P, P')$. If $b = 0$ then for each round r the corresponding client's message in $\mathtt{trans}$ is replaced with an independently at random chosen message from $\mathcal{M}_C^r$, else if $b = 1$ then $\mathtt{trans}$ is forwarded without any modification. Note that if $\mathcal{A}$ mounts an online attack with a correct password then it can easily distinguish so that the lower bound of $\frac{\mathcal{O}(t)}{|\mathcal{D}|}$ that accounts for online dictionary attacks still applies in the definition.

Definition 3 (AKE-Security with Indistinguishable Client Messages). *A PAKE protocol Π is* AKE-secure with indistinguishable client messages *if for all dictionaries $\mathcal{D}$ with corresponding universe of participants Ω and for all PPT adversaries $\mathcal{A}$ using at most t online sessions there exists a negligible function $\varepsilon(\cdot)$ such that:*

$$\mathsf{Adv}_{\Pi,\mathcal{A}}^{\mathsf{AKE\text{-}ICM}}(\lambda) = \left| \Pr[\mathsf{Exp}_{\Pi,\mathcal{A}}^{\mathsf{AKE\text{-}ICM}}(\lambda) = 1] - \frac{1}{2} \right| \leq \frac{\mathcal{O}(t)}{|\mathcal{D}|} + \varepsilon(\lambda).$$

$\mathsf{Exp}_{\Pi,\mathcal{A}}^{\mathsf{AKE}}(\lambda) : c = 1; b \in_R \{0,1\}; \forall (P, P') \in \mathcal{C} \times \mathcal{S}$ *choose* $\mathrm{pw}_{P,P'}$;
$b' \leftarrow \mathcal{A}^{\mathsf{Send}_b, \mathsf{Execute}_b, \mathsf{Corrupt}, \mathsf{Test}_b}(\lambda, c)$; *return* $b = b'$.

The above requirement is stronger than AKE-security. In particular, it cannot be satisfied by PAKE protocols where client messages depend on those of the server or where client messages sent in later rounds depend on client messages that were sent in previous rounds. Nonetheless, there exist efficient AKE-secure

PAKE protocols with indistinguishable client messages as discussed in Sect. 3.5. In particular, for AKE-secure *one-round* PAKE protocols, where the client can send its message independently of the server's message the indistinguishability property can be argued based on the uniformity of the client's message in the message space.

3.2 The O-PAKE Compiler

Our compiler takes as input a PAKE protocol Π and outputs its O-PAKE version, denoted C_Π. The compiled protocol C_Π follows Definition 1 and consists of the two algorithms C_Π.`init` and C_Π.`next`. For the passwords in **pw** used as input to C_Π.`init` we assume that each $\mathbf{pw}[i] = (\text{ix}, \pi) \in \mathbb{F} \times \mathcal{D}_\Pi$, where ix denotes an *index* and π the corresponding *password* for the underlying PAKE protocol Π, whereby the distributions of ix and π are independent and no two pairs $(\text{ix}_1, \pi_1), (\text{ix}_2, \pi_2) \in \mathbf{pw}$ have $\text{ix}_1 = \text{ix}_2$. For each PAKE round r the compiler uses a corresponding instance $\mathtt{IHME}^r$ with message space $\mathcal{M}^{\mathtt{IHME}^r}$ and a compatible admissible encoding $F^r : \mathcal{M}^{\mathtt{IHME}^r} \to \mathcal{M}_C^r$ where $\mathcal{M}_C^r$ is the space of clients messages of Π in that round. In the following we assume that the underlying Π.`next` algorithm outputs messages that can be seen as one element and thus can be processed using one instance $(F^r, \mathtt{IHME}^r)$ in each round. Note that this allows for a more comprehensible description and is not a restriction of the O-PAKE compiler. We discuss the case of multi-set messages $\mathcal{M}_C^r = \mathcal{M}_C^{r,1} \times \cdots \times \mathcal{M}_C^{r,l}$ that will require composition of up to l instances of encoding schemes per round in Sect. 3.6.

The C_Π.`next` algorithm on the client side computes corresponding PAKE round messages for all passwords in **pw** using the original Π.`next` algorithm and encodes them with $\mathcal{I}_{F^r}$ and $\mathtt{IHME}^r$.`iEncode` prior to transmission to the server. On the server side C_Π.`next` decodes the incoming PAKE message using F^r and $\mathtt{IHME}^r$.`iDecode` (using its input $\mathbf{pw}[i]$.ix as index) and replies with the message output by Π.`next`. Note that the server only decodes messages but never encodes them. If $pw \in \mathbf{pw}$ then at the end of its n PAKE sessions the client will hold n intermediate PAKE keys, whereas the server holds only one such key. The additional key confirmation and key derivation steps allow the client to determine which of its n PAKE session keys matches the one held by the server, in which case both participants will derive the same session key. In the following we describe the two algorithms C_Π.`init` and C_Π.`next` more in detail.

Algorithm C_Π.init The algorithm makes n calls to Π.`init`, one for each password $\mathbf{pw}[i].\pi$, to generate corresponding `state` for each of the n PAKE sessions that are stored in $\mathbf{state}_P^i$. An ith session of Π run by the client using the corresponding password $\mathbf{pw}[i].\pi$ is denoted by $\Pi[i]$. The partner id $\mathtt{pid}_P^i$ is set to P' and the instance P_i with the given `role` and a vector of n local states in $\mathbf{state}_P^i$ is established. We require that no two passwords in **pw** are identical, which is necessary to ensure the correctness of the IHME step. Note that if `role` = `server` then $n = 1$, i.e. servers run only one PAKE session (Fig.1).

a $C_{\Pi}.\mathtt{next}(m_{\text{in}})$ — Client

Input: m_{in}
Output: $(m_{\text{out}}, \mathtt{k})$
  $E = \emptyset$; $m_{\text{out}} = \emptyset$
  for $i = 1 \ldots n$ **do**
    if $\Pi[i]$ has not finished **then**
      $(m'_{\text{out}}, \Pi[i].\mathtt{k}) \leftarrow \Pi[i].\mathtt{next}(m_{\text{in}})$
      if $m'_{\text{out}} \neq \emptyset$ **then**
        $E = E \cup \{(\mathbf{pw}[i].\text{ix}, \mathcal{I}_{F^r}(m'_{\text{out}}))\}$
    else if $\Pi[i].\mathtt{k} \in \mathcal{K}_{\Pi}$ and
        $m_{\text{in}} = \mathtt{PRF}_{\Pi[i].\mathtt{k}}(\mathtt{sid}^i_P || P_i || \mathtt{pid}^i_P || 0)$ **then**
      $\mathtt{k} = \mathtt{PRF}_{\Pi[i].\mathtt{k}}(\mathtt{sid}^i_P || P_i || \mathtt{pid}^i_P || 1)$
    else
      $\mathtt{k} = \perp$
  if $E \neq \emptyset$ **then**
    $m_{\text{out}} = \mathtt{IHME}^r.\mathtt{iEncode}(E)$
  return $(m_{\text{out}}, \mathtt{k})$

b $C_{\Pi}.\mathtt{next}(m_{\text{in}})$ — Server

Input: m_{in}
Output: $(m_{\text{out}}, \mathtt{k})$
  $m_{\text{out}} = \emptyset$
  if Π has not finished **then**
    $m \leftarrow \mathtt{IHME}^r.\mathtt{iDecode}(\text{pw.ix}, m_{\text{in}})$
    $m' = F^r(m)$
    $(m_{\text{out}}, \Pi.\mathtt{k}) \leftarrow \Pi.\mathtt{next}(m')$
  if $\Pi.\mathtt{k} \in \mathcal{K}_{\Pi}$ **then**
    $m_{\text{out}} = \mathtt{PRF}_{\Pi.\mathtt{k}}(\mathtt{sid}^j_{P'} || \mathtt{pid}^j_{P'} || P'_j || 0)$
    $\mathtt{k} = \mathtt{PRF}_{\Pi.\mathtt{k}}(\mathtt{sid}^j_{P'} || \mathtt{pid}^j_{P'} || P'_j || 1)$
  else
    $\mathtt{k} = \perp$
  return $(m_{\text{out}}, \mathtt{k})$

Fig. 1. $C_{\Pi}.\mathtt{next}$ algorithms

Algorithm $\mathbf{C}_{\Pi}$.next We distinguish between $C_{\Pi}.\mathtt{next}$ specifications for clients (Algorithm 1a) and servers (Algorithm 1b) as they are significantly different. We write $\Pi[i].\mathtt{next}$ for the invocation of $\Pi.\mathtt{next}$ for the ith session of Π run by the client using $\mathbf{pw}[i].\pi$. On the client side $C_{\Pi}.\mathtt{next}$ computes messages m'_{out} for all running PAKE sessions and encodes them. The server decodes the incoming IHME structure and computes its response using $\Pi.\mathtt{next}$. If any PAKE session $\Pi[i]$ at the client has finished with $\Pi[i].\mathtt{k} \in \mathcal{K}_{\Pi}$ then the client expects a valid confirmation message from the server prior to derivation of the resulting session key $\mathtt{k}$ with $\mathtt{PRF}$ using $\Pi[i].\mathtt{k}$. An invalid confirmation message implies that $\mathtt{k}$ is set to $\perp$. This confirmation message is generated on the server side using $\mathtt{PRF}$ only if and immediately after $\Pi.\mathtt{next}$ outputs $\Pi[i].\mathtt{k} \in \mathcal{K}_{\Pi}$; in which case a valid resulting session key $\mathtt{k}$ is also derived. If, however, Π finishes with $\Pi[i].\mathtt{k} = \perp$ then $\mathtt{k}$ will also be set to $\perp$.

3.3 Relation to LAKE

A Language Authenticated Key Exchange (LAKE) protocol, proposed by Benhamouda et al. in [10], authenticates two parties, client C and server S holding each a word in an algebraic languages. In particular, let $R : \{0,1\}^* \times P \times W \rightarrow \{0,1\}$ denote a relation and $L_R(\mathsf{pub}, \mathsf{priv}) \subseteq W$ a language with $\mathsf{pub} \in \{0,1\}^*$ and $\mathsf{priv} \in P$. A word $w \in W$ is in the language L_R iff $R(\mathsf{pub}, \mathsf{priv}, w) = 1$. The client holds a word w_c for relation R_C and the server holds a word w_s for relation R_S. They agree on public parameters pub, exchange ephemeral public keys, and *think* of a value priv'_C, resp. priv'_S, they expect to be used by the other party. To instantiate the LAKE framework it is necessary to specify client and server languages and according commitments with associated smooth projective hash functions (SPHF) [17]. We briefly recall how to instantiate LAKE with passwords from [10, Sect. 6.2], i.e. how to build PAKE protocols in the LAKE framework. The languages are defined as $L_C = \{w_c\}$ for the client and $L_S = \{w_s\}$ for the server, such that $\mathsf{priv}'_C = \mathsf{priv}'_S = w_c = w_s$ is the password and the relations are $R_C = R_S = (\emptyset, \mathsf{priv}, w) = 1 \iff \mathsf{priv} = w$, i.e. equality test for the password.

To instantiate O-PAKE in LAKE we define client relation $R_C(\emptyset,\text{pw}', \mathbf{pw}) = 1 \iff \text{pw}' \in \mathbf{pw}$ and server relation $R_S(\emptyset, \text{pw}', \text{pw}) = 1 \iff \text{pw}' = \text{pw}$. While the server relation stays the same as in PAKE, the client language $L_{R_C}(\emptyset, \text{pw}') \subseteq \{\text{pw}_1, \dots, \text{pw}_n\} = \mathbf{pw}$ uses a relation that takes a set of passwords $\mathbf{pw}$ and an *expected* password pw′ as input, and is fulfilled iff $\text{pw}' \in \mathbf{pw}$. Following [10, Fig. 4] we realise O-PAKE in the LAKE framework as follows: First, the client (initiator) generates a multiDLCSCom′ commitment (C_C, C'_C) on word w_c, i.e. a multi-commitment to all passwords $\text{pw} \in (\text{pw}_1, \dots, \text{pw}_n)$, as well as a Pedersen commitment C''_C on C'_C, and sends (C_C, C''_C) to the server S. The server replies with $(C_S, \varepsilon, \mathtt{k}_{\mathtt{p}_S}, \sigma_S)$, computed as follows: C_S is a multi-LCS commitment on $w_s = \text{pw}_S$; ε is a challenge vector on C_C of length n; $\mathtt{k}_{\mathtt{p}_S}$ is a projection key for a suitable SPHF for C_C; and σ_S is a signature on all flows. In the final round, the client checks σ_S before returning $(C'_C, t, \mathtt{k}_{\mathtt{p}_C}, \sigma_C)$ to the server, which is computed as follows: (C'_C, t) is the decommitment to C''_C, where t is the used randomness; $\mathtt{k}_{\mathtt{p}_C}$ is a projection key for a suitable SPHF for C_S; and σ_C is a signature on all flows. After checking all signatures and commitments, session keys are computed as multiplication of projection and hash function on $\mathsf{Com}_C = C_C \cdot {C'_C}^{\varepsilon}$ and $\mathsf{Com}_S = C_S$.

So while it seems possible to instantiate O-PAKE in the LAKE framework (after specifying necessary primitives), the construction is rather inefficient. In particular, an instantiation of O-PAKE in LAKE needs four rounds, our O-PAKE compiler adds only one round to the round-complexity of the underlying PAKE, i.e. can be instantiated with three rounds. Further, server-complexity is linear in the number of client-passwords n. This stems from the observation that the projection key $\mathtt{k}_{\mathtt{p}_S}$, as well as the computation of the hash function, requires a linear number of public key operations, e.g., exponentiations, in n. Performance of O-PAKE instantiated in the LAKE framework is therefore not more efficient than the naïve construction, and in particular does not fulfil our requirement of nearly constant server performance.

3.4 Security Analysis

AKE-security of the protocol generated with the O-PAKE compiler is established in Theorem 1.

Theorem 1. *If Π is an R-round AKE-secure PAKE protocol with indistinguishable client messages in $\mathcal{M}^r_C$ for $r \in [1, \dots, R]$, $F^r : \mathcal{M}^{\mathtt{IHME}^r} \to \mathcal{M}^r_C$ is an ϵ-admissible encoding, and $\mathtt{IHME}^r$ is an index-hiding message encoding, then C_Π is an $R+1$-round AKE-secure O-PAKE protocol.*

Proof (sketch). The proof uses a sequence of experiments Exp_i, $i = 1, \dots 4$, where each Exp_i is based on a small modification of Exp_{i-1}. At a high level we first replace real client messages and session keys of underlying PAKE sessions $\Pi[i]$ with random messages and keys while ensuring the consistency against an adversary that corrupted passwords and then mounts online attacks. This modification remains unnoticeable to $\mathcal{A}$ if the underlying PAKE protocol Π is AKE-secure with indistinguishable client messages. Then, we replace the outputs of the inverse transformations $\mathcal{I}_{F^r}$ applied in each round r by choosing random elements from the corresponding round's $\texttt{IHME}^r$ message space and show that this remains unnoticeable assuming that F^r is an ϵ_{F^r}-admissible encoding F^r. Then, we replace each real password index in the $\texttt{IHME}^r$ encoding process with a random password and show that this remains unnoticeable due to the index-hiding property of each $\texttt{IHME}^r$ scheme. Finally, we modify the computation of the server's confirmation message and of the session keys that are returned in Test_b queries by using random elements from the corresponding spaces. This remains unnoticeable due to the pseudorandomness of the $\texttt{PRF}$ function that is used to derive their values. We refer to the full version [29] of this work for the full proof due to space limitations.

3.5 Oblivious PAKE Instantiation

An AKE-secure PAKE protocol Π is suitable for our O-PAKE transformation if it is also AKE-ICM-secure and there exist admissible encodings to map those messages into the message space of the IHME scheme. In the following we list four sets R with suitable admissible encodings. Thus, any AKE-secure PAKE protocol whose client messages contain components from these four sets can be transformed into an O-PAKE protocol using our compiler.

Definition 4 (Admissible Encodings for Client Messages). *An admissible encoding $F : \{0,1\}^{\ell(\lambda)} \rightarrow R$ with polynomial $\ell(\lambda)$ exists for any of the following four sets:*

(1) Set $R = \{0, \ldots, N-1\} = \mathbb{Z}_N$ of natural numbers, for arbitrary $N \in \mathbb{N}$. (cf. [19, Lemma 13])
(2) The set of quadratic residues modulo safe primes p, i.e. $R = QR(p) \subseteq \mathbb{Z}_p^\times$. (cf. [19, Lemma 13])
(3) Arbitrary subgroups $G \subseteq \mathbb{Z}_p^\times$ of prime order q. (cf. [19, Lemma 13])
(4) The set $R = E(\mathbb{F})$ of rational points on (certain) elliptic curves, defined over a finite field (cf. [15]).

Computing Indices. We require that password pw used in O-PAKE consists of two independent components ix and π. For instance, it is sufficient for the user to choose $\pi \in_R \mathcal{D}$ and compute the index $\text{pw.ix} = f(\rho, \text{pw}.\pi)$ using some fresh randomness ρ and a function f with output independent from π, i.e. the probability that π was used as input to f to produce ix must remain $1/|\mathcal{D}|$. Note that this approach requires a pre-flow to the protocol to exchange randomness ρ, which can however be easily integrated into the overall login process. Furthermore, it is crucial that randomness ρ is fresh for every execution of the protocol as any reuse of ρ would offers an attacker the possibility to distinguish between real and simulated O-PAKE messages.

Remark 1. Verifier-based PAKE (VPAKE) protocols, such as [9,12,24,25], where only some password-dependent verification information (e.g. a randomised password hash with a random salt) is stored on the server side are not formally considered in this work. Nonetheless, the techniques underlying our O-PAKE compiler seem also applicable to VPAKE protocols as long as their messages satisfy the identified AKE-ICM requirement.

3.6 Processing Multi-Component Messages

In the following we describe how the compiler can handle PAKE protocol messages consisting of multiple elements, possibly from different sets. We observe that any such PAKE message can be seen as an element of a combined message space that is formed through a Cartesian product of those sets and distinguish between two types of message components, namely components that represent constants and components that depend on passwords, including integer values and group elements. Since constants are password-independent they do not need to be processed by the compiler and can be communicated directly. All other message components have to be encoded according to the compiler specification. In order to encode those components we use ν-fold IHME introduced in [31], which allows to encode a list of ν message components from the same finite field. The compiler splits message components from different finite fields into corresponding classes and applies appropriate IHME encoding to each class separately in order to compute the corresponding IHME structure. The IHME structures for all message components are then concatenated and treated as a single compiler message. This processing of multi-component messages requires existence of admissible encodings and index-hiding message encodings for each component class m_j of m. In order to process the components, a loop over $m_1, \ldots, m_l$ adds $(\mathbf{pw}[i], \mathcal{I}_{F^{r,j}}(m_j))$ to the input set of ν−fold−IHME_j^r.`iEncode` according to their classes (e.g. finite fields). Likewise, the output message m_{out} of the `next` algorithm is the concatenation of the encoded component classes. Upon receiving a client message m_{in}, the server has to decompose it to retrieve the IHME encoded messages. After decoding the message parts with $m_j \leftarrow \nu$−fold−$\texttt{IHME}_j^r$.`iDecode`$(\text{pw}_{P,P'}, m_{\text{in}}^j)$ the original PAKE message of Π is reassembled by decoding messages $F^{r,j}(m_j)$.

Adopting this approach for multi-component messages, the AKE-security remains preserved. This is due to the following observation about the proof of Theorem 1: in the game-hopping sequence the adversary will be provided with l IHME encoded messages (one for each message element class that requires encoding). The corresponding index-hiding advantage will therefore be multiplied by l. The remaining parts of the proof remain as is.

4 Concrete Instantiation Examples

In this section we give concrete instantiation of the O-PAKE compiler, using the random-oracle based SPAKE protocol by Abdalla and Pointcheval [5]. A second instantiation using the common-reference-string-model protocol from Katz and Vaikuntanathan [28] can be found in the full version [29].

4.1 Oblivious SPAKE

We demonstrate how the compiler can be applied to PAKE protocols using the AKE-secure, random-oracle-based SPAKE protocol from [5]. The resulting O-SPAKE is specified in Fig. 2 and involves steps of the original SPAKE protocol from [5, Sect. 5], which is a secure variant of [8], whose security has been proven in the random oracle model.[1] SPAKE uses a prime-order cyclic group G for which the Computational Diffie-Hellman (CDH) problem is assumed to be hard. The shared SPAKE password pw is chosen from $\mathbb{Z}_q$. Let $M, N \in G$ denote two public group elements. The protocol proceeds in one round, where the client sends $X^* \leftarrow g^x \cdot M^{\text{pw}.\pi}, x \in_R \mathbb{Z}_q$ and the server responds with $Y^* \leftarrow g^y \cdot N^{\text{pw}.\pi}, y \in_R \mathbb{Z}_q$. The actual order of these messages does not matter since they are independent. The algorithm `next` computes an intermediate value s and derives the session key as $\Pi.\mathtt{k} \leftarrow H(P, P', X^*, Y^*, \text{pw}, s)$. We refer to the original work [5, Sect. 5] for more details on SPAKE. The SPAKE protocol is a suitable input PAKE protocol for our O-PAKE compiler since it can be instantiated using subgroups $G \subseteq \mathbb{Z}_p^\times$ of prime order q in which the CDH problem is believed to be hard. We can apply the admissible encodings (3) from [19, Lemma 13] due to the fact that client's SPAKE message $X^* = g^x \cdot M^{\text{pw}.\pi}$ is uniformly distributed in G, given the uniformity of $x \in \mathbb{Z}_q$. We formalise this by showing that SPAKE fulfils our definition of AKE-ICM, before defining suitable admissible encodings, which concludes the instantiation of O-SPAKE.

Lemma 2 (SPAKE is AKE-ICM Secure). *The SPAKE protocol from [5, Sect. 5] is AKE-ICM secure.*

Proof. The initial experiment in the proof for SPAKE security in [5] corresponds to the AKE-ICM experiment with $b = 1$. In the following we show that the proof in [5, Appendix C] can be modified without changing the adversaries advantage such that the final experiment is equal to the AKE-ICM experiment with $b = 0$, which concludes the proof. We first change experiment one by additionally simulating the Corrupt oracle and using a global bit b in simulating the Test oracle. This does not change the adversary's success probability. The second experiment, aborting on hash collisions, stays unchanged. In the following two experiments we have to make sure that the adversary does not win trivially by returning the correct key to Test queries on corrupted sessions and only modify oracle replies to uncorrupted sessions. While the original proof only changes the calculation of the session key in passive sessions to a random element in experiment three, we also change client messages produced in Execute queries to random elements. Note that this is implicitly already done in the original proof. However, we formalise it here again and change experiment three as follows: Invocations of the

[1] Note that the very similar SOKE protocol from [1] can also be used in the O-PAKE compiler following the here given description of O-SPAKE.

Client $(C, S, \mathbf{pw})$		Server (C, S, pw)
for $i = 1$ **to** $\lvert\mathbf{pw}\rvert$		$y \in_R \mathbb{Z}_q;\ Y \leftarrow g^y$
$x \in_R \mathbb{Z}_q;\ x \in_R \mathbb{Z}_q$		$X^* \leftarrow \mathtt{IHME.iDecode}(\mathrm{pw.ix}, S)$
$X \leftarrow \mathcal{I}_F(g^x \cdot M^{\mathbf{pw}[i].\pi})$	S →	$X^* \leftarrow F(X^*)$
$E = E \cup \{(\mathbf{pw}[i].\mathrm{ix}, X)\}$		$Y^* \leftarrow Y \cdot N^{\mathrm{pw}.\pi}$
$S \leftarrow \mathtt{IHME.iEncode}(E)$		$S \leftarrow \mathtt{IHME.iEncode}(E)$
for $i = 1$ **to** $\lvert\mathbf{pw}\rvert$		$s \leftarrow (X^*/M^{\mathrm{pw}.\pi})^y$
$s = (Y^*/N^{\mathbf{pw}[i].\pi})^x$	← Y^*, c	$\mathbf{k} \leftarrow H(C, S, X^*, Y^*, \mathrm{pw}, s)$
$\Pi[i].\mathbf{k} \leftarrow H(C, S, X^*, Y^*, \mathbf{pw}[i], s)$		$c \leftarrow \mathtt{PRF}_{\Pi.\mathbf{k}}(S\|\|Y^*\|\|C\|\|S\|\|0)$
$c' \leftarrow \mathtt{PRF}_{\Pi[i].\mathbf{k}}(S\|\|Y^*\|\|C\|\|S\|\|0)$		$\mathbf{k} \leftarrow \mathtt{PRF}_{\Pi.\mathbf{k}}(S\|\|Y^*\|\|C\|\|S\|\|1)$
if $c = c'$ **then**		
$\mathbf{k} \leftarrow \mathtt{PRF}_{\Pi[i].\mathbf{k}}(S\|\|Y^*\|\|C\|\|S\|\|1)$		

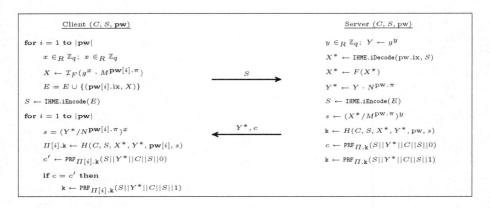

Fig. 2. Oblivious SPAKE (O-SPAKE) public input: $G, g, p, q, M, N, H, \mathtt{IHME}, F$

Execute oracle on uncorrupted parties are answered with uniformly at random chosen messages, i.e. $X^* = Ag^x$ and $Y^* = Bg^y$ with $x, y \in_R \mathbb{Z}_p$, for some DH instance (A, B). Experiment three corresponds now to the AKE-ICM experiment with Send_1, Test_b and $\mathsf{Execute}_0$. The lemma follows by noting that after our modifications of experiment three the last experiment of the AKE-security proof of SPAKE in [5, Appendix C] is equivalent to the AKE-ICM experiment with $\mathsf{Execute}_0$, Send_0 and Test_0, i.e. the adversary only wins by guessing the correct password. □

Admissible Encodings for SPAKE We use admissible encodings (1) and (3) from [19, Lemma 13] to encode SPAKE client messages. To implement the inverse encoding of (1) $:= \mathcal{I}_{F^{(1)}} : \mathbb{Z}_N \rightarrow \{0,1\}^{\ell(\lambda)}$ we use the inverse of encoding (3) $:= \mathcal{I}_{F^{(1)}} : G \rightarrow \mathbb{Z}_p^\times$. This results in a combined inverse encoding of $\mathcal{I}_{F^{(3,1)}} : G \rightarrow \mathbb{Z}_p^\times \rightarrow \{0,1\}^{\ell(\lambda)}$ with $\ell(\lambda) > 2|N|$ and $p = N$. Implementation of $F^{(3,1)} : \mathbb{Z}_{q'} \rightarrow G$ and $\mathcal{I}_{F^{(3,1)}}$ follows the specification from [19, Lemma 12] with prime $|q'| = \ell(\lambda) > 2|N|$ to meet IHME requirements.

5 Conclusion

In this paper we addressed the problem of handling multiple password trials efficiently within the execution of PAKE protocols; in particular, aiming to optimise the amount of work on the server side. The proposed O-PAKE compiler results in almost constant computational complexity for the server without significantly increasing the computation costs on the client side, yet preserving all security guarantees offered by standard PAKE protocols. It can be used with PAKE protocols that fulfil our new definition of AKE-ICM security and whose client messages can be encoded through a suitable admissible encoding scheme. The security of the compiler has been proven under standard assumptions in an extension of the widely used PAKE model from [4] and exemplified on the PAKE protocol and [28].

References

1. Abdalla, M., Bresson, E., Chevassut, O., Möller, B., Pointcheval, D.: Provably secure password-based authentication in TLS. In: ASIACCS 2006, pp. 35–45. ACM (2006)
2. Abdalla, M., Catalano, D., Chevalier, C., Pointcheval, D.: Efficient two-party password-based key exchange protocols in the UC framework. In: Malkin, T. (ed.) CT-RSA 2008. LNCS, vol. 4964, pp. 335–351. Springer, Heidelberg (2008)
3. Abdalla, M., Chevassut, O., Fouque, P.-A., Pointcheval, D.: A simple threshold authenticated key exchange from short secrets. In: Roy, B. (ed.) ASIACRYPT 2005. LNCS, vol. 3788, pp. 566–584. Springer, Heidelberg (2005)
4. Abdalla, M., Fouque, P.-A., Pointcheval, D.: Password-based authenticated key exchange in the three-party setting. In: Vaudenay, S. (ed.) PKC 2005. LNCS, vol. 3386, pp. 65–84. Springer, Heidelberg (2005)
5. Abdalla, M., Pointcheval, D.: Simple password-based encrypted key exchange protocols. In: Menezes, A. (ed.) CT-RSA 2005. LNCS, vol. 3376, pp. 191–208. Springer, Heidelberg (2005)
6. Bellare, M., Pointcheval, D., Rogaway, P.: Authenticated key exchange secure against dictionary attacks. In: Preneel, B. (ed.) EUROCRYPT 2000. LNCS, vol. 1807, pp. 139–155. Springer, Heidelberg (2000)
7. Bellare, M., Rogaway, P.: Entity authentication and key distribution. In: Stinson, D.R. (ed.) CRYPTO 1993. LNCS, vol. 773, pp. 232–249. Springer, Heidelberg (1994)
8. Bellovin, S.M., Merritt, M.: Encrypted key exchange: password-based protocols secure against dictionary attacks. In: IEEE Symposium on Research in Security and Privacy, pp. 72–84 (1992)
9. Bellovin, S.M., Merritt, M.: Augmented encrypted key exchange: a password-based protocol secure against dictionary attacks and password file compromise. In: CCS 1993, pp. 244–250. ACM (1993)
10. Ben Hamouda, F., Blazy, O., Chevalier, C., Pointcheval, D., Vergnaud, D.: Efficient UC-secure authenticated key-exchange for algebraic languages. In: Kurosawa, K., Hanaoka, G. (eds.) PKC 2013. LNCS, vol. 7778, pp. 272–291. Springer, Heidelberg (2013)
11. Benhamouda, F., Blazy, O., Chevalier, C., Pointcheval, D., Vergnaud, D.: New techniques for SPHFs and efficient one-round PAKE protocols. In: Canetti, R., Garay, J.A. (eds.) CRYPTO 2013, Part I. LNCS, vol. 8042, pp. 449–475. Springer, Heidelberg (2013)
12. Benhamouda, F., Pointcheval, D.: Verifier-based password-authenticated key exchange: new models and constructions. Cryptology ePrint Archive, report 2013/833 (2013). http://eprint.iacr.org/2013/833
13. Boneh, D., Franklin, M.: Identity-based encryption from the weil pairing. In: Kilian, J. (ed.) CRYPTO 2001. LNCS, vol. 2139, pp. 213–229. Springer, Heidelberg (2001)
14. Boyko, V., MacKenzie, P.D., Patel, S.: Provably secure password-authenticated key exchange using Diffie-Hellman. In: Preneel, B. (ed.) EUROCRYPT 2000. LNCS, vol. 1807, pp. 156–171. Springer, Heidelberg (2000)
15. Brier, E., Coron, J.-S., Icart, T., Madore, D., Randriam, H., Tibouchi, M.: Efficient indifferentiable hashing into ordinary elliptic curves. In: Rabin, T. (ed.) CRYPTO 2010. LNCS, vol. 6223, pp. 237–254. Springer, Heidelberg (2010)
16. Canetti, R., Halevi, S., Katz, J., Lindell, Y., MacKenzie, P.: Universally composable password-based key exchange. In: Cramer, R. (ed.) EUROCRYPT 2005. LNCS, vol. 3494, pp. 404–421. Springer, Heidelberg (2005)

17. Cramer, R., Shoup, V.: Universal hash proofs and a paradigm for adaptive chosen ciphertext secure public-key encryption. In: Knudsen, L.R. (ed.) EUROCRYPT 2002. LNCS, vol. 2332, pp. 45–64. Springer, Heidelberg (2002)
18. Dierks, T., Rescorla, E.: RFC 5246 - the transport layer security (TLS) protocol version 1.2, August 2008. Updated by RFCs 5746, 5878, 6176
19. Fleischhacker, N., Günther, F., Kiefer, F., Manulis, M., Poettering, B.: Pseudorandom signatures. In: ASIA CCS 2013, pp. 107–118. ACM (2013)
20. Florencio, D., Herley, C.: A large-scale study of web password habits. In: 16th International Conference on World Wide Web, WWW 2007, pp. 657–666. ACM (2007)
21. Gaw, S., Felten, E.W.: Password management strategies for online accounts. In: Symposium on Usable Privacy and Security, SOUPS 2006, pp. 44–55. ACM (2006)
22. Gennaro, R.: Faster and shorter password-authenticated key exchange. In: Canetti, R. (ed.) TCC 2008. LNCS, vol. 4948, pp. 589–606. Springer, Heidelberg (2008)
23. Gennaro, R., Lindell, Y.: A framework for password-based authenticated key exchange. ACM Trans. Inf. Syst. Secur. **9**(2), 181–234 (2006)
24. Gentry, C., MacKenzie, P.D., Ramzan, Z.: A method for making password-based key exchange resilient to server compromise. In: Dwork, C. (ed.) CRYPTO 2006. LNCS, vol. 4117, pp. 142–159. Springer, Heidelberg (2006)
25. Jablon, D.P.: Extended password key exchange protocols immune to dictionary attacks. In: WETICE, pp. 248–255. IEEE Computer Society (1997)
26. Katz, J., Ostrovsky, R., Yung, M.: Efficient and secure authenticated key exchange using weak passwords. J. ACM **57**(1), 3:1–3:39 (2009)
27. Katz, J., Vaikuntanathan, V.: Round-optimal password-based authenticated key exchange. In: Ishai, Y. (ed.) TCC 2011. LNCS, vol. 6597, pp. 293–310. Springer, Heidelberg (2011)
28. Katz, J., Vaikuntanathan, V.: Round-optimal password-based authenticated key exchange. J. Cryptology **26**(4), 714–743 (2013)
29. Kiefer, F., Manulis, M.: Oblivious pake: efficient handling of password trials. Cryptology ePrint Archive, report 2013/127 (2013). http://eprint.iacr.org/2013/127
30. Manulis, M., Pinkas, B., Poettering, B.: Privacy-preserving group discovery with linear complexity. In: Zhou, J., Yung, M. (eds.) ACNS 2010. LNCS, vol. 6123, pp. 420–437. Springer, Heidelberg (2010)
31. Manulis, M., Poettering, B.: Practical affiliation-hiding authentication from improved polynomial interpolation. In: ASIACCS 2011, pp. 286–295. ACM (2011)

Secure and Efficient Private Set Intersection Cardinality Using Bloom Filter

Sumit Kumar Debnath(✉) and Ratna Dutta

Department of Mathematics, Indian Institute of Technology Kharagpur, Kharagpur 721302, India
sd.iitkgp@gmail.com, ratna@maths.iitkgp.ernet.in

Abstract. We first present a Private Set Intersection Cardinality (PSI-CA) protocol followed by its authorized variant, APSI-CA, utilizing *Bloom filter* (BF). We further extend these to PSI and APSI protocols. All the constructions are proven to be secure in standard model with *linear* complexities. Moreover, our protocols hide the size of the client's private set which may be sensitive in application specific scenarios. The proposed PSI-CA and APSI-CA are the *first* to achieve security in *standard* model under the Quadratic Residuosity (QR) assumption with *linear complexities.*

Keywords: (A)PSI-CA · (A)PSI · Semi-honest adversary · Malicious adversary · Privacy · Bloom filter

1 Introduction

Private Set Intersection (PSI) is a two-party protocol carried out between a client (C) and a server (S), and allows them to compute *privately* the intersection of their respective private sets. Consider the following problem, suppose department of homeland security (DHS) wants to detect whether anyone on its terror watch list is on a flight's passenger list while preserving privacy for terror watch list. Besides, the innocent flight passengers may not want to reveal their privacy. A PSI between DHS and Airlines is an ideal cryptographic primitive to address the above problem. PSI protocols have been used extensively for many practical applications. Privacy preserving data mining, location-based services, social networks, testing of fully sequenced human genomes, collaborative botnet detection, on-line gaming are a few to name.

A variant of PSI, where the client's set is made authorized by a trusted third party at the beginning of PSI protocol, is known as *Authorized Private Set Intersection* (APSI).

In social networking, suppose two parties want to secretly compute the number of common connections (or interests) in order to decide whether or not to become friends. *Private Set Intersection Cardinality* (PSI-CA) is a proper choice for this scenario which yields only size of the intersection of the private sets of C

J. Lopez and C.J. Mitchell (Eds.): ISC 2015, LNCS 9290, pp. 209–226, 2015.
DOI: 10.1007/978-3-319-23318-5_12

and S. PSI-CA is applicable in various field such as genomic operations, role-based association mining, location sharing, affiliation-hiding authentication, etc.

It is essential to keep secret the size of the client's input set from the server in applications where (i) input size may represent sensitive information, (ii) fluctuations in input size may be equally (or even more) sensitive when multiple interactions between the same two parties are given or (iii) the amount of computation imposed on the server may concern input size privacy. For instance, department of homeland security (DHS) does not reveal the number of names on the terror watch list because the list is dynamic in the sense that names can be added and removed frequently to the list and disclosing its size may leak sensitive information. The concept of size-hiding PSI was introduced by Ateniese et al. [1].

Recently, several PSI protocols are proposed based on Bloom filters [2] which provide space and time-efficient storage of sets. Bloom filter is a simple randomized data structure that possess a linear relation between its size and the number of element that can be stored in it and has applications in testing membership in a set and many areas including PSI [6,10].

Our Contribution: The main contribution in this paper is the construction of efficient PSI-CA and APSI-CA protocols using Bloom filter achieving *security in standard model, linear complexity in terms of both communication and computation* and *independency of the size of client's set.* To the best of our knowledge, there is no PSI-CA and APSI-CA with the above mentioned properties. These are further modified to PSI and APSI respectively. We summarize the security features of our constructions below.

1. The proposed PSI, PSI-CA are provably secure in semi-honest model where adversaries follow the prescribed protocol but try to learn more information than allowed from the protocol transcript. The security of APSI is in fully malicious model where adversaries can run any efficient strategy to carry out attack and can deviate at will from the protocol specification. The proposed APSI-CA is secure against malicious client and semi-honest server.
2. All our constructions are secure in standard model under the Quadratic Residuosity (QR) assumption.
3. The client need not to reveal the size of his private set to the server. Thus our designs are independent of the size of the client's set. Only the upper bound of the client's set size is disclosed to the server. As the upper bound is different from the security parameter, the computational effort of the server is independent of this bound.
4. The client's set needs to be authorized by a trusted third party in APSI and APSI-CA designs which prevents the client from including arbitrary elements in its input set to steal server's element.

The performance of our designs over prior works are summarized in Tables 1, 2, 3 and 4. We point out the following facts:

1. Till now, the constructions of [4] are the most efficient to the best of our knowledge. However, the security proofs are in the random oracle model (ROM)

Table 1. Comparison of PSI-CA protocols

Protocol	Security model	Adv. model	Security assumption	Comm.	Comp.	Based on	Size hiding
Sch. 1 of [4]	ROM	SH	DDH and GOMDH	$O(w+v)$	$O(w+v)$		no
Sch. 2 of [4]	ROM	MS, SHC	GOMDH	$O(w+v)$	$O(w+v)$		no
Our PSI-CA	Std	SH	QR	$O(w+v)$	$O(w+v)$	BF	yes

Table 2. Comparison of APSI-CA protocols

Protocol	Security model	Adv. model	Security assumption	Comm.	Comp.	Based on	Size hiding
[3]	Std	Mal	Strong RSA	$O(wv)$	$O(wv)$	OPE	no
[4]	ROM	SH	GOMDH	$O(w+v)$	$O(w+v)$		no
Our APSI-CA	Std	MC, SHS	QR	$O(w+v)$	$O(w+v)$	BF	yes

Table 3. Comparison of PSI protocols based on BF

Protocol	Security model	Adv. model	Security assumption	Comm. cost	Comp. cost	Based on	Size hiding
[10]	Std	SH	QR	$O(w+v)$	$O(w+v)$	BF+SYY	yes
[9]	Std	Mal	d-strong DDH	$O(w+v)$	$O(w+v)$	OPRF	no
Sch. 1 of [6]	ROM	SH	CDH	$O(w+v)$	$O(w+v)$	BF	yes
Sch. 2 of [6]	ROM	Mal	CDH	$O(w+v)$	$O(w+v)$	BF	no
[11]	ROM	SH		$O(w+v)$	$O(w+v)$	BF	yes
our PSI	Std	SH	QR	$O(w+v)$	$O(w+v)$	BF	yes

Table 4. Comparison of APSI protocols

Protocol	Security model	Adv. model	Security assumption	Comm.	Comp.	Based on	Size hiding
[3]	Std	Mal	Strong RSA	$O(w+v)$	$O(wv)$	OPE	no
[5]	ROM	Mal	RSA	$O(w+v)$	$O(w+v)$		no
[12]	ROM	Mal	CBDH	$O(v)$	$O(v \log \log v)$		no
[10]	Std	Mal	QR	$O(w+v)$	$O(w+v)$	BF+SYY	yes
our APSI	Std	Mal	QR	$O(w+v)$	$O(w+v)$	BF	yes

VE=Verifiable Encryption, CBDH=Computational Bilinear Diffie-Hellman, Mal=Malicious, OPRF=Oblivious Pseudorandom Function, SYY= Sander, Young and Yung Technique, MC=Malicious Client, SHS=Semi-honest Server, CDH=Computational Diffie-Hellman, SD=Subgroup Decision, SC=Subgroup Computation, CE=Commutative Encryption, GOMDH=Gap-One-More-Diffie-Hellman, OPE=Oblivious Polynomial Evaluation, Std=Standard, SH=Semi-honest, MS=Malicious Server, SHC=Semi-honest Client, DDH=Decisional Diffie-Hellman, HE=Homomorphic Encryption, w= Size of Client's set, v= Size of Server's set

under non-standard cryptographic assumption and require $O(v+w)$ *modular exponentiations.* Here w, v are the sizes of input sets of the two parties. In contrast, our designs are secure in *standard* model under QR assumption. We emphasize that our PSI-CA utilize $O(v+w)$ *modular multiplications* and no modular exponentiations. On the other side, our APSI-CA requires $O(v+w)$ modular multiplications together with $O(w)$ many signature verifications. Recently, Dong et al. [6] and Pinkas et al. [11] proposed very efficient PSI protocols using the garbled Bloom filter GBF. However, extending these PSI to PSI-CA seems to be a non-trivial task.
2. The Boom filter based PSI protocols of [6,11] are secure in ROM, although they are computationally less expensive as compared to our scheme, especially when the sizes of input sets are large. Constructions of [10] also have the properties similar to our schemes. However, computationally they are more expensive and have more false positive rate than our constructions. Besides, server's elements must be fixed length bit string in the PSI of [10], whereas our PSI is not restricted to this. In the APSI of [10], the trusted party has to send an exponent secretly to the server which needs a secure communication channel between them. On the other hand, we do not require any such secret sharing. Both the client and the trusted third party are able to compute the intersection in the APSI of [10] as both of them hold the secret key of associated encryption scheme. Our APSI enables only the client to get the intersection and no other party.

2 Preliminaries

Throughout the paper the notations κ, $a \leftarrow A$, $x \leftarrowtail X$ and $\{\mathcal{X}_t\}_{t\in\mathcal{N}} \equiv^c \{\mathcal{Y}_t\}_{t\in\mathcal{N}}$ are used to represent "security parameter", "a is output of the procedure A", "variable x is chosen uniformly at random from set X" and "the distribution ensemble $\{\mathcal{X}_t\}_{t\in\mathcal{N}}$ is computationally indistinguishable from the distribution ensemble $\{\mathcal{Y}_t\}_{t\in\mathcal{N}}$" respectively. Informally, $\{\mathcal{X}_t\}_{t\in\mathcal{N}} \equiv^c \{\mathcal{Y}_t\}_{t\in\mathcal{N}}$ means for all probabilistic polynomial time (PPT) distinguisher $\mathcal{Z}$, there exists a negligible function $\epsilon(t)$ such that $|Prob_{x\leftarrow\mathcal{X}_t}[\mathcal{Z}(x)=1] - Prob_{x\leftarrow\mathcal{Y}_t}[\mathcal{Z}(x)=1]| \leq \epsilon(t)$.

Definition 1. Negligible Function: *A function $\epsilon : \mathbb{N} \to \mathbb{R}$ is said to be negligible function of κ if for each constant $c > 0$, we have $\epsilon(\kappa) = o(\kappa^{-c})$ for all sufficiently large κ.*

Definition 2. Quadratic Residuosity (QR) Assumption [8]**:** *On input 1^κ, let the algorithm $\mathcal{R}Gen$ generates an RSA modulus $n = PQ$, where P and Q are distinct primes and let X be the subgroup of $\mathbb{Z}_n^*$ of elements having Jacobi symbol equal to 1. The QR assumption states that, given an RSA modulus n (without its factorization), it is computationally infeasible to distinguish a random element u of $X \subseteq \mathbb{Z}_n^*$ from an element of the subgroup $\{x^2 | x \in \mathbb{Z}_n^*\}$ of quadratic residues modulo n i.e., for every PPT algorithm $\mathcal{A}$, $|Prob[\mathcal{A}(n, x^2) = 1] - Prob[\mathcal{A}(n, u) = 1]|$ is negligible function of κ.*

2.1 Security Model for Semi-honest Adversary [7]

A two-party protocol, Π is a random process that computes a function f from pair of inputs (one per party) to pair of outputs i.e.,

$$f = (f_1, f_2) : \{0,1\}^* \times \{0,1\}^* \rightarrow \{0,1\}^* \times \{0,1\}^*.$$

Let $x, y \in \{0,1\}^*$ be the inputs of parties P_1, P_2 respectively. Then the outputs of the parties P_1, P_2 are $f_1(x,y), f_2(x,y)$ respectively. A protocol Π is said to be secure in semi-honest model if whatever can be computed by a party after participating in the protocol, it could obtain from its input and output only. This is formalized using the simulation paradigm. On the input pair (x, y), view of the party P_i during an execution of Π is denoted by $\mathsf{View}_i^{\Pi}(x,y) = (w, r^{(i)}, m_1^{(i)}, ..., m_t^{(i)})$, where $w \in \{x, y\}$ represents the input of the party P_i, $r^{(i)}$ is the outcome of P_i's internal coin tosses, and $m_j^{(i)}$ $(j = 1, 2, ..., t)$ represents the j-th message which has received by P_i during the execution of Π.

Definition 3. *Let $f = (f_1, f_2)$ be a deterministic function. Then we say that the protocol Π securely computes f if there exists probabilistic polynomial-time adversaries, denoted by S_1 and S_2, controlling P_1 and P_2 respectively, such that*

$$\{S_1(x, f_1(x,y))\}_{x,y\in\{0,1\}^*} \equiv^c \mathsf{View}_1^{\Pi}(x,y)_{x,y\in\{0,1\}^*}$$

$$\{S_2(y, f_2(x,y))\}_{x,y\in\{0,1\}^*} \equiv^c \mathsf{View}_2^{\Pi}(x,y)_{x,y\in\{0,1\}^*}$$

2.2 Security Model for Malicious Adversary [7]

The formal security framework of a two-party protocol in malicious model is described below:

The Real World: In the real world a protocol Π is executed. An honest party follows the instructions of Π, but an adversary $\mathcal{A}_i$, controlling the party P_i, can behave arbitrarily. Let the party P_1 has the private input X, the party P_2 has the private input Y and the adversary $\mathcal{A}_i$ has auxiliary input Z. At the end of the execution an honest party outputs whatever prescribed in the protocol, a corrupted party outputs nothing and an adversary outputs its view which consists of the transcripts available to the adversary. The joint output in the real world is denoted by $\mathsf{REAL}_{\Pi,\mathcal{A}_i(Z)}(X,Y)$.

The Ideal Process: Let $\mathcal{SIM}_i$ be the ideal process adversary that corrupts a party $P_i, i \in \{1,2\}$. The ideal process involves an incorruptible trusted third party.

Input: Let X and Y be the inputs of parties P_1 and P_2 respectively, $\mathcal{SIM}_i$ gets P_i's input and an auxiliary input Z.

Sending Inputs to the Trusted Party: An honest party always sends his original input to the trusted party whereas a corrupted party may send "abort" or an arbitrary input. Let the trusted party receives $(\widetilde{X}, \widetilde{Y})$, where $\widetilde{X}, \widetilde{Y}$ may be different from X, Y respectively. If anyone of $\widetilde{X}, \widetilde{Y}$ is "abort", then the trusted party sends $\perp$ to both the parties.

The Trusted Party Answers the Adversary: The trusted party computes the functionality $\mathcal{F} : (\widetilde{X}, \widetilde{Y}) \rightarrow (F_1(\widetilde{X}, \widetilde{Y}), F_2(\widetilde{X}, \widetilde{Y}))$ and sends $F_i(\tilde{X}, \widetilde{Y})$ to $\mathcal{SIM}_i$. Then $\mathcal{SIM}_i$ sends "abort" or "continue" to the trusted party.

The trusted Party Answers the Honest Party: If the trusted party receives "continue" from $\mathcal{SIM}_i$, then the trusted party sends $F_j(\tilde{X}, \widetilde{Y})$ to the honest party $P_j, j \in \{1,2\} \setminus \{i\}$. Otherwise, the trusted party sends $\perp$ to the honest party.

Output: An honest party always outputs the output value it obtained from the trusted party. The corrupted party outputs nothing. The adversary outputs his view. The joint output of the ideal process is denoted by $\mathsf{IDEAL}_{\mathcal{F},\mathcal{SIM}_i(Z)}(X, Y)$.

Definition 4. Simulatability: *Let Π be a protocol and $\mathcal{F}$ be a functionality. Then protocol Π is said to securely compute $\mathcal{F}$ in the malicious model if for every PPT adversary $\mathcal{A}_i$ in the real world, there exists a PPT adversary $\mathcal{SIM}_i$ in the ideal model, such that for every $i \in \{1,2\}$,*
$\mathsf{IDEAL}_{\mathcal{F},\mathcal{SIM}_i(Z)}(X, Y) \equiv^c \mathsf{REAL}_{\Pi,\mathcal{A}_i(Z)}(X, Y)$.

2.3 Goldwasser-Micali (GM) Encryption [8]

The Goldwasser-Micali (GM) encryption consists of three algorithm (KGen, Enc, Dec) and works as follows:

KGen. On input 1^κ, an user generates secret key as $sk = (P, Q)$ and public key as $pk = (n, u)$, where $n = PQ$ is an RSA modulus, P, Q are distinct primes, u is a pseudo quadratic residue i.e., $L(\frac{u}{P}) = -1$ and $L(\frac{u}{Q}) = -1$ but $J(\frac{u}{n}) = 1$, where L and J denote respectively the Legendre symbol and Jacobi symbol.

Enc. Encryptor encrypts a message $m \in \{0,1\}$ using the public key $pk = (n, u)$ by picking $r \leftarrow \mathbb{Z}_n$ and outputs the corresponding ciphertext as

$$c = \mathsf{Enc}_{pk}(x) = \begin{cases} r^2 \bmod n \text{ if } m = 0 \\ ur^2 \bmod n \text{ if } m = 1 \end{cases}$$

Note that c is a quadratic residue modulo n if $m = 0$ and quadratic non-residue modulo n if $m = 1$ with $J(\frac{c}{n}) = 1$

Dec. Decryptor has the the secret key $sk = (P, Q)$. On receiving the ciphertext c, the decryptor computes $L(\frac{c}{P})$. If $L(\frac{c}{P}) = 1$, then the decryptor outputs the message m as 0 . Otherwise, the decryptor outputs the message m as 1.

This encryption scheme is semantically secure under the QR assumption. It also satisfies homomorphic property under the binary operations exclusive-or $\oplus$ and modulo multiplication on the message space and the ciphertext space i.e., $\mathsf{Enc}_{pk}(x \oplus y) = \mathsf{Enc}_{pk}(x) \cdot \mathsf{Enc}_{pk}(y)$. Encryption cost of GM encryption is maximum 2 modular multiplications and decryption cost is $O((\log n)^2)$ bit operations.

2.4 Bloom Filter [2]

Bloom filter (BF) is a data structure that represents a set $X = \{s_1, s_2, ..., s_v\}$ of v elements by an array of m bits and uses k independent hash functions $H = \{h_0, h_1, ..., h_{k-1}\}$ with $h_i : \{0,1\}^* \rightarrow \{0, 1, ..., m-1\}$ for $i = 0, 1, ..., k-1$. Let $\mathsf{BF}_X \in \{0,1\}^m$ represents a Bloom filter for the set X and $\mathsf{BF}_X[i]$ represents the bit at the index i in BF_X. We describe below a variant of Bloom filter [2] that completes in three steps- Initialization, Add, Check.

Initialization: Set 1 to all the bits of an m-bit array, which is an empty Bloom filter with no element in that array.

Add(s): To add an element $s \in X \subseteq \{0,1\}^*$ into a Bloom filter, s is hashed with the k hash functions $\{h_0, h_1, ..., h_{k-1}\}$ to get k indices $h_0(s), h_1(s), ..., h_{k-1}(s)$. Set 0 to the indices $h_0(s), h_1(s), ..., h_{k-1}(s)$ of the Bloom filter. Each $s \in X$ needs to be added to get $\mathsf{BF}_X \in \{0,1\}^m$.

Check ($\hat{s}$)*:* To check if an element $\hat{s}$ belongs to X or not, $\hat{s}$ is hashed with the k hash functions $\{h_0, h_1, ..., h_{k-1}\}$ to get k indices $h_0(\hat{s}), h_1(\hat{s}), ..., h_{k-1}(\hat{s})$. Now if at least one of $\mathsf{BF}_X[h_0(\hat{s})], ..., \mathsf{BF}_X[h_{k-1}(\hat{s})]$ is 1 then $\hat{s}$ is not in X, otherwise $\hat{s}$ is *probably* in X.

Bloom filter allows false positive whereby an element that has not been inserted in the filter can mistakenly pass the set membership test. This happens due to the fact that an element $\hat{s}$ may not belong to X but $\mathsf{BF}_X[h_i(\hat{s})] = 0$ for all $i = 0, 1, ..., k-1$. On the contrary, Bloom filter never yields false negative i.e., an element that has been inserted in the filter will always pass the test. This is because if $\hat{s}$ belongs to X, then each of $\mathsf{BF}_X[h_0(\hat{s})], ..., \mathsf{BF}_X[h_{k-1}(\hat{s})]$ is 0.

Theorem 1. *Given the number v of elements to be added and a desired maximum false positive rate $\frac{1}{2^k}$, the optimal size m of the Bloom filter is $m = \frac{vk}{\ln 2}$.*

3 Protocol

Protocol Requirements: Each of our protocols is run between a client C with private input set $Y = \{c_1, c_2, ..., c_w\}$ and a server with private input set $X = \{s_1, s_2, ..., s_v\}$, where $w \leq v$. In the rest of our discussion, k represents number of hash functions for Bloom filter, H denotes the set $\{h_0, h_1, ..., h_{k-1}\}$ of k hash functions with $h_i : \{0,1\}^* \rightarrow \{0, 1, ..., m-1\}$ for $i = 0, 1, ..., k-1$, m stands for optimal size of Bloom filter, pk_C/sk_C indicates public/secret key for GM encryption, $\mathsf{Enc}_{pk_C}/\mathsf{Dec}_{sk_C}$ stands for Encryption/Decryption function for GM under pk_C/sk_C and $\bar{s}_{i,j}$ represents the j-th bit of the element $\bar{s}_i \in \{0,1\}^k$ for $j = 0, 1, ..., k-1$. Let $E(\bar{s}_i) = \{\mathsf{Enc}_{pk_C}(\bar{s}_{i,0}), ..., \mathsf{Enc}_{pk_C}(\bar{s}_{i,k-1})\}$ and decryption of $E(\bar{s}_i)$ be $D(E(\bar{s}_i)) = \{\mathsf{Dec}_{sk_C}(\mathsf{Enc}_{pk_C}(\bar{s}_{i,0})), ..., \mathsf{Dec}_{sk_C}(\mathsf{Enc}_{pk_C}(\bar{s}_{i,k-1})\}) = \{\bar{s}_{i,0}, ..., \bar{s}_{i,k-1}\} = \bar{s}_i$. The auxiliary inputs include security parameter κ, the maximum set size v, the optimal Bloom filter parameters m, H and k.

1. The client C generates a public key $pk_C = (n, u)$ and a secret key $sk_C = (P, Q)$ for GM encryption using KGen algorithm described in section 2.3. Then C does the following:
   (a) constructs a Bloom filter $\mathsf{BF}_Y = (\mathsf{BF}_Y[0], ..., \mathsf{BF}_Y[m-1]) \in \{0,1\}^m$ of the set $Y = \{c_1, c_2, ..., c_w\} \subseteq \{0,1\}^*$ following the procedure described in section 2.4,
   (b) generates $b_i = \mathsf{Enc}_{pk_C}(\mathsf{BF}_Y[i]) \in \mathbb{Z}_n$ for $i = 0, 1, ..., m-1$,
   (c) sets $\overline{Y} = \{b_0, ..., b_{m-1}\} \in \mathbb{Z}_n^m$,
   (d) sends $\overline{Y}$ and $pk_C = (n, u)$ to S.
2. The server S with private input set $X = \{s_1, s_2, ..., s_v\} \subseteq \{0,1\}^*, pk_C = (n, u)$ does the followings on receiving $\overline{Y} = \{b_0, ..., b_{m-1}\} \in \mathbb{Z}_n^m$ from C: For $i = 1, 2, ..., v$, the server S
   (a) computes $h_0(s_i), ..., h_{k-1}(s_i) \in \{0, 1, ..., m-1\}$;
   (b) extracts $b_{h_0(s_i)}, ..., b_{h_{k-1}(s_i)} \in \mathbb{Z}_n$ from $\overline{Y}$; and
   (c) sets $E(\bar{s}_i) = \{b_{h_0(s_i)} \cdot r_{i,0}^2 \bmod n, ..., b_{h_{k-1}(s_i)} \cdot r_{i,k-1}^2 \bmod n\} \in \mathbb{Z}_n^k$, where $r_{i,0}, ..., r_{i,k-1} \leftarrow \mathbb{Z}_n$.

   Finally S sends $\overline{X} = \{E(\bar{s}_1), ..., E(\bar{s}_v)\} \subseteq \mathbb{Z}_n^k$ to C.
3. On receiving $\overline{X} = \{E(\bar{s}_1), ..., E(\bar{s}_v)\} \subseteq \mathbb{Z}_n^k$ from S, the client C first sets $card = 0$. For $i = 1, 2, ..., v$, the client C
   (a) decrypts $E(\bar{s}_i) \in \mathbb{Z}_n$ to get $\bar{s}_i \in \{0,1\}^k$; and
   (b) checks whether $\bar{s}_i \in \{0,1\}^k$ is all-zero string. If yes, then sets $card = card + 1$.

   Finally, C outputs $card$ as the cardinality of $X \cap Y$.

Fig. 1. Description of our PSI-CA

3.1 The PSI-CA

We present the description of our PSI-CA in Fig. 1, where the client C has the private set $Y = \{c_1, c_2, ..., c_w\} \subseteq \{0,1\}^*$ and the server S has the private set $X = \{s_1, s_2, ..., s_v\} \subseteq \{0,1\}^*$.

Correctness: First note that as $r_{i,j}^2$ is a quadratic residue modulo n, we have $r_{i,j}^2 = \mathsf{Enc}_{pk_C}(0)$ for all $i = 1, 2, ..., v; j = 0, 1, ..., k-1$. Since GM Encryption is homomorphic under the operation $\oplus$, we have

$$\begin{aligned} E(\bar{s}_i) &= \{b_{h_0(s_i)} \cdot r_{i,0}^2 \bmod n, ..., b_{h_{k-1}(s_i)} \cdot r_{i,k-1}^2 \bmod n\} \\ &= \{\mathsf{Enc}_{pk_C}(\mathsf{BF}_Y[h_0(s_i)]) \cdot \mathsf{Enc}_{pk_C}(0), ..., \mathsf{Enc}_{pk_C}(\mathsf{BF}_Y[h_{k-1}(s_i)]) \cdot \mathsf{Enc}_{pk_C}(0)\} \\ &= \{\mathsf{Enc}_{pk_C}(\mathsf{BF}_Y[h_0(s_i)] \oplus 0), ..., \mathsf{Enc}_{pk_C}(\mathsf{BF}_Y[h_{k-1}(s_i)] \oplus 0)\} \\ &= \{\mathsf{Enc}_{pk_C}(\mathsf{BF}_Y[h_0(s_i)]), ..., \mathsf{Enc}_{pk_C}(\mathsf{BF}_Y[h_{k-1}(s_i)])\} \end{aligned}$$

Therefore $\bar{s}_i = \{\mathsf{BF}_Y[h_0(s_i)], ..., \mathsf{BF}_Y[h_{k-1}(s_i)]\} \in \{0,1\}^k$ for all $i = 1, 2, ..., v$. Now we have the following claim:

Claim 1. An element $\bar{s}_i \in \{0,1\}^k$ is a all-zero string if and only if $s_i \in X \cap Y$.

Proof. If $s_i \in X \cap Y$, then by Add step of Bloom filter construction defined in Sect. 2.4, $\mathsf{BF}_Y[h_j(s_i)] = 0$ for all $j = 0, 1, ..., k-1$. This in turn implies $\bar{s}_i \in \{0,1\}^k$ is a all-zero string. On the other hand, if $\bar{s}_i \in \{0,1\}^k$ is a all-zero string, then $s_i \in X$ satisfies the set membership test for the Bloom filter $\mathsf{BF}_Y \in \{0,1\}^m$ with high probability by the Check step of the Bloom filter construction defined in Sect. 2.4 for the set Y. Hence $s_i \in Y$ and thereby $s_i \in X \cap Y$ except with negligible probability $\frac{1}{2^k}$. □

The variable *card* is incremented only when $\bar{s}_i$ is a all-zero string and hence it gives the cardinality of $X \cap Y$. Note that the server S first multiplies $b_{h_j(s_i)}$ with a random value $r_{i,j}^2$ for all $i = 1, 2, ..., v; j = 0, 1, ..., k-1$ and then sends $E(\bar{s}_i) = \{b_{h_j(s_i)} \cdot r_{i,j}^2\}_{j=0}^{k-1}$ for $i = 1, 2, ..., v$ to the client C. Due to this randomization C would not be able to predict the value of $b_{h_j(s_i)}$ from $E(\bar{s}_i)$ although C is having the set $\{b_0, ..., b_{k-1}\}$ i.e., s_i is not revealed to C. Otherwise, if S does not multiply $b_{h_j(s_i)}$ with a random value $r_{i,j}^2$, then from $b_{h_j(s_i)}$ the client C would be able to compute the position $h_j(s_i)$ for all $i = 1, 2, ..., v; j = 0, 1, ..., k-1$ as C is having the set $\{b_0, ..., b_{k-1}\}$. On the other hand, C would be able to compute the Bloom filter BF_X of S's set X by setting $\mathsf{BF}_X[h_j(s_i)]$ as 0 for all $i = 1, 2, ..., v; j = 0, 1, ..., k-1$ and remaining positions as 1 which in turn enables C to compute S's private set X i.e., S's privacy is not preserved in this case.

3.2 The APSI-CA

Our APSI-CA involves three parties – a client C with private set $Y = \{c_1, c_2, ..., c_w\} \subseteq \{0,1\}^*$, a server S with private set $X = \{s_1, s_2, ..., s_v\} \subseteq \{0,1\}^*$ and a certifying authority CA who is assumed to be mutually trusted to both C and S. The protocol completes in two phases: off-line phase described in Fig. 2 and online phase described Fig. 3.

Note that given $(\{b_0, ..., b_{m-1}\}, Sig(\bar{h}(b_0), ..., \bar{h}(b_{m-1})), pk_{DSig})$, if someone wants to verify the signature then he does the following steps:

1. computes $Y' = \{\bar{h}(b_0), ..., \bar{h}(b_{m-1})\}$ from $Sig(\bar{h}(b_0), ..., \bar{h}(b_{m-1}))$ using pk_{DSig}.
2. computes hash $\bar{h}$ of each elements of the set $\overline{Y} = \{b_0, ..., b_{m-1}\}$ to compute the set $\widehat{Y} = \{\bar{h}(b_0), ..., \bar{h}(b_{m-1})\}$ and checks that whether i-th member of the set $\widehat{Y}$ is same as i-th member of the set Y'.

The correctness can be shown in the same way as explained for PSI-CA in Sect. 3.1. Note that in APSI-CA the certifying authority CA generates $(\overline{Y} = \{b_0, ..., b_{m-1}\} \in \mathbb{Z}_n^m, Sig(\bar{h}(b_0), ..., \bar{h}(b_{m-1})))$ on behalf of the client C in the off-line phase to control over the malicious behavior of C, the server S checks the validity of the signature $Sig(\bar{h}(b_0), ..., \bar{h}(b_{m-1}))$ received from C in the online phase. For this S uses public key pk_{DSig} that S has received from CA in the off-line phase.

1. The client C generates a key pair $(pk_C = (n, u), sk_C = (P, Q))$ for GM encryption using KGen algorithm described in section 2.3 and sends $Y = \{c_1, c_2, ..., c_w\} \subseteq \{0,1\}^*$, $pk_C = (n, u)$ to the certifying authority CA.
2. On receiving $Y = \{c_1, c_2, ..., c_w\} \subseteq \{0,1\}^*$, $pk_C = (n, u)$ from C, the certifying authority CA constructs a Bloom filter $\mathsf{BF}_Y \in \{0,1\}^m$, generates a key pair (pk_{DSig}, sk_{DSig}) for some publicly verifiable digital signature scheme $DSig$ over composite order group $\mathbb{Z}_n$ and for each $i = 0, 1, ..., m-1$,
   (a) sets $b_i = \mathsf{Enc}_{pk_C}(\mathsf{BF}_Y[i]) \in \mathbb{Z}_n$,
   (b) computes $\bar{h}(b_i)$, where $\bar{h} : \{0,1\}^* \to \mathbb{Z}_n$ is a hash function.
   (c) makes a signature $Sig(\bar{h}(b_0), ..., \bar{h}(b_{m-1}))$ on $Y' = \{\bar{h}(b_0), ..., \bar{h}(b_{m-1})\}$ with the secret key sk_{DSig}.

   Finally, CA sends $(\overline{Y} = \{b_0, ..., b_{m-1}\} \in \mathbb{Z}_n^m, Sig(\bar{h}(b_0), ..., \bar{h}(b_{m-1})), pk_{DSig})$ to C and pk_{DSig} to S.
   Note that in digital signature scheme, hash of a message is signed, instead of the message to overcome forgery.

Fig. 2. Description of off-line phase of our APSI-CA

1. The client C forwards $(\overline{Y} = \{b_0, ..., b_{m-1}\} \in \mathbb{Z}_n^m, Sig(\bar{h}(b_0), ..., \bar{h}(b_{m-1})))$ received from CA in off-line phase together with public key $pk_C = (n, u)$ to the server S.
2. The server S, on receiving $(\overline{Y} = \{b_0, ..., b_{m-1}\} \in \mathbb{Z}_n^m, Sig(\bar{h}(b_0), ..., \bar{h}(b_{m-1}))$, $pk_C = (n, u))$ from C, verifies the validity of signature $Sig(\bar{h}(b_0), ..., \bar{h}(b_{m-1}))$ using pk_{DSig} received from CA in off-line phase. If verification fails, then S aborts the protocol. Otherwise, for each $i = 1, 2, ..., v$, the server S
   (a) computes $h_0(s_i), ..., h_{k-1}(s_i) \in \{0, 1, ..., m-1\}$;
   (b) extracts $b_{h_0(s_i)}, ..., b_{h_{k-1}(s_i)} \in \mathbb{Z}_n$ from $\overline{Y}$;
   (c) sets $E(\bar{s}_i) = \{b_{h_0(s_i)} \cdot r_{i,0}^2 \bmod n, ..., b_{h_{k-1}(s_i)} \cdot r_{i,k-1}^2 \bmod n\} \in \mathbb{Z}_n^k$, where $r_{i,0}, ..., r_{i,k-1} \xleftarrow{} \mathbb{Z}_n$.

   Finally S sends $\overline{X} = \{E(\bar{s}_1), ..., E(\bar{s}_v)\} \subseteq \mathbb{Z}_n^k$ to C.
3. On receiving $\overline{X} = \{E(\bar{s}_1), ..., E(\bar{s}_v)\} \subseteq \mathbb{Z}_n^k$ from S, the client C first sets $card = 0$. For each $i = 1, 2, ..., v$, the client C
   (a) decrypts $E(\bar{s}_i) \in \mathbb{Z}_n$ to get $\bar{s}_i \in \{0,1\}^k$,
   (b) increments $card$ by 1 if $\bar{s}_i \in \{0,1\}^k$ is all-zero string.

   Finally, C outputs $card$ as the cardinality of $X \cap Y$.

Fig. 3. Description of online phase of our APSI-CA

3.3 The PSI

Let $\phi : \{0,1\}^* \to \{0,1\}^k$ be a collision resistant hash function. Our PSI protocol is described in Fig. 4, where the client C has the private set $Y = \{c_1, c_2, ..., c_w\} \subseteq \{0,1\}^*$ and the server S has the private set $X = \{s_1, s_2, ..., s_v\} \subseteq \{0,1\}^*$.

Correctness: As GM Encryption is homomorphic under $\oplus$, we have

$$E(\bar{s}_i) = \{b_{h_0(s_i)} \cdot \mathsf{Enc}_{pk_C}(s_{i,0}), ..., b_{h_{k-1}(s_i)} \cdot \mathsf{Enc}_{pk_C}(s_{i,k-1})\}$$

$$= \{\mathsf{Enc}_{pk_C}(\mathsf{BF}_Y[h_0(s_i)] \oplus s_{i,0}), ..., \mathsf{Enc}_{pk_C}(\mathsf{BF}_Y[h_{k-1}(s_i)] \oplus s_{i,k-1})\}$$

Therefore $\bar{s}_i = \{\mathsf{BF}_Y[h_0(s_i)] \oplus s_{i,0}, ..., \mathsf{BF}_Y[h_{k-1}(s_i)] \oplus s_{i,k-1}\} \in \{0,1\}^k$ for all $i = 1, 2, ..., v$. Now we can conclude that $\{c_i \in Y | \phi(c_i) \in \widehat{X}\} = X \cap Y$ from the following claim:

Claim 2. Let $\widetilde{X} = \{\phi(s_i)\}_{i=1}^{v}$. Then an element $\phi(s_i) \in \widetilde{X}$ is same as $\bar{s}_i \in \widehat{X}$ if and only if $s_i \in X \cap Y$.

Proof. If $s_i \in X \cap Y$, then by Add step of Bloom filter construction for the set Y, $\mathsf{BF}_Y[h_j(s_i)] = 0$ for all $j = 0, 1, ..., k-1$ which in turn implies $\bar{s}_i \in \widehat{X}$ is same as $\phi(s_i) \in X$. On the other side, if $\phi(s_i) \in \widetilde{X}$ is same as $\bar{s}_i \in \widehat{X}$, then $s_i \in X$ satisfies the set membership test for the Bloom filter $\mathsf{BF}_Y \in \{0,1\}^m$ with high probability by the Check step of the Bloom filter for the set Y. Hence $s_i \in Y$ and thereby $s_i \in X \cap Y$ except with negligible probability $\frac{1}{2^k}$. □

1. The client C generates a public key $pk_C = (n, u)$ and a secret key $sk_C = (P, Q)$ for GM encryption using the KGen algorithm as in section 2.3. Then C proceeds as follows:
   (a) constructs a Bloom filter $\mathsf{BF}_Y \in \{0,1\}^m$ of the set Y,
   (b) encrypts each of $\mathsf{BF}_Y[i] \in \{0,1\}$ for $i = 0, 1, ..., m-1$ using GM encryption under public key pk_C,
   (c) sends $\overline{Y} = \{b_0 = \mathsf{Enc}_{pk_C}(\mathsf{BF}_Y[0]), ..., b_{m-1} = \mathsf{Enc}_{pk_C}(\mathsf{BF}_Y[m-1])\} \in \mathbb{Z}_n^m, pk_C = (n, u)$ to S.
2. The server S, on receiving $\overline{Y} = \{b_0, ..., b_{m-1}\} \in \mathbb{Z}_n^m, pk_C = (n, u)$ from C does the followings:
   For $i = 1, 2, ..., v$, the server S
   (a) computes $h_0(s_i), ..., h_{k-1}(s_i) \in \{0, 1, ..., m-1\}$;
   (b) extracts $b_{h_0(s_i)}, ..., b_{h_{k-1}(s_i)} \in \mathbb{Z}_n$ from $\overline{Y}$;
   (c) generates $\mathsf{Enc}_{pk_C}(s_{i,0}), ..., \mathsf{Enc}_{pk_C}(s_{i,k-1})$, where $s_{i,j}$ is j-th bit of $\phi(s_i) \in \{0,1\}^k$ for $j = 0, 1, ..., k-1$; and
   (d) sets $E(\bar{s}_i) = \{b_{h_0(s_i)} \cdot \mathsf{Enc}_{pk_C}(s_{i,0}) \bmod n, ..., b_{h_{k-1}(s_i)} \cdot \mathsf{Enc}_{pk_C}(s_{i,k-1}) \bmod n\} \in \mathbb{Z}_n^k$.
   Finally S sends $\overline{X} = \{E(\bar{s}_1), ..., E(\bar{s}_v)\} \subseteq \mathbb{Z}_n^k$ to C.
3. On receiving $\overline{X} = \{E(\bar{s}_1), ..., E(\bar{s}_v)\} \subseteq \mathbb{Z}_n^k$ from S, C does the followings:
   (a) decrypts $E(\bar{s}_i) \in \mathbb{Z}_n$ for $i = 1, 2, ..., v$ to get the set $\widehat{X} = \{\bar{s}_1, ..., \bar{s}_v\} \subseteq \{0,1\}^k$.
   (b) computes $\widetilde{Y} = \{\phi(c_i)\}_{i=1}^{w}$
   Finally, C outputs $\{c_i \in Y | \phi(c_i) \in \widehat{X}\}$ as $X \cap Y$.

Fig. 4. Description of our PSI

1. The client C generates a key pair $(pk_C = (n,u), sk_C = (P,Q))$ for GM encryption using KGen algorithm described in section 2.3 and sends $Y = \{c_1, c_2, ..., c_w\} \subseteq \{0,1\}^*, pk_C = (n,u)$ to the certifying authority CA.
2. The certifying authority CA, on receiving $Y = \{c_1, c_2, ..., c_w\} \subseteq \{0,1\}^*, pk_C = (n,u)$ from C, constructs a Bloom filter $\mathsf{BF}_Y \in \{0,1\}^m$, generates a key pair (pk_{DSig}, sk_{DSig}) for some publicly verifiable digital signature scheme $DSig$ over composite order group $\mathbb{Z}_n$ and for each $i = 0, 1, ..., m-1$,
   (a) sets $b_i = \mathsf{Enc}_{pk_C}(\mathsf{BF}_Y[i]) \in \mathbb{Z}_n$,
   (b) computes $\bar{h}(b_i)$, where $\bar{h} : \{0,1\}^* \rightarrow \mathbb{Z}_n$ is a hash function.
   (c) generates a signature $Sig(\bar{h}(b_0), ..., \bar{h}(b_{m-1}))$ on $Y' = \{\bar{h}(b_0), ..., \bar{h}(b_{m-1})\}$ with the secret key sk_{DSig}

   Finally, the certifying authority CA , sends $(\overline{Y} = \{b_0, ..., b_{m-1}\} \in \mathbb{Z}_n^m$, $Sig(\bar{h}(b_0), ..., \bar{h}(b_{m-1})), pk_{DSig})$ to C and pk_{DSig} to S.

Fig. 5. Description of off-line phase of our APSI

1. The client C sends $(\overline{Y} = \{b_0, ..., b_{m-1}\} \in \mathbb{Z}_n^m, Sig(\bar{h}(b_0), ..., \bar{h}(b_{m-1})))$ received from CA in off-line phase together with public key $pk_C = (n,u)$ to the server S.
2. The server S, on receiving $(\overline{Y} = \{b_0, ..., b_{m-1}\} \in \mathbb{Z}_n^m, Sig(\bar{h}(b_0), ..., \bar{h}(b_{m-1}))$, $pk_C = (n,u))$ from C, verifies the signature $Sig(\bar{h}(b_0), ..., \bar{h}(b_{m-1}))$ using pk_{DSig} which S has received in off-line phase from CA. If verification fails, then S aborts the protocol. Otherwise, for each $i = 1, 2, ..., v$, the server S
   (a) computes $h_0(s_i), ..., h_{k-1}(s_i) \in \{0, 1, ..., m-1\}$;
   (b) extracts $b_{h_0(s_i)}, ..., b_{h_{k-1}(s_i)} \in \mathbb{Z}_n$ from $\overline{Y}$;
   (c) generates $\mathsf{Enc}_{pk_C}(s_{i,0}), ..., \mathsf{Enc}_{pk_C}(s_{i,k-1})$, where $s_{i,j}$ is j-th bit of $s_i \in \{0,1\}^k$ for $j = 0, 1, ..., k-1$; and
   (d) sets $E(\bar{s}_i) = \{b_{h_j(s_i)} \cdot \mathsf{Enc}_{pk_C}(s_{i,j}) \bmod n\}_{j=0}^{k-1} \in \mathbb{Z}_n^k$.

   Finally, S sends $\overline{X} = \{E(\bar{s}_1), ..., E(\bar{s}_v)\} \subseteq \mathbb{Z}_n^k$ to C.
3. On receiving $\overline{X} = \{E(\bar{s}_1), ..., E(\bar{s}_v)\} \subseteq \mathbb{Z}_n^k$ from S, C decrypts $E(\bar{s}_i) \in \mathbb{Z}_n$ for $i = 1, 2, ..., v$ to get the set $\widehat{X} = \{\bar{s}_1, ..., \bar{s}_v\} \subseteq \{0,1\}^k$. Finally, C outputs the set $\widehat{X} \cap Y$ as intersection of X and Y.

Fig. 6. Description of online phase of our APSI

3.4 The APSI

Similar to APSI-CA, this protocol also involves three participants – client C with private set $Y = \{c_1, c_2, ..., c_w\} \subseteq \{0,1\}^*$, server S with private set $X = \{s_1, s_2, ..., s_v\} \subseteq \{0,1\}^k$ and a certifying authority CA that is assumed to be trusted to both C and S. The protocol completes in two phases: off-line phase and online phase described in Figs. 5 and 6 respectively. Note that in this protocol the certifying authority CA does some short of signature on the client's inputs in the off-line phase to control over the malicious behavior of C, the server S checks the validity of the messages received from C in the online phase. For this S uses a public key pk_{DSig} that S has received from CA in the off-line phase.

Note that each $s_i \in X$ is chosen as k-bit string so that simulator can extract corrupt server S's private set X in ideal world. The correctness of this protocol can be shown in the same way as explained for PSI in Sect. 3.3. The only difference is that $\phi(s_i)$ will be replaced by s_i in case of APSI.

4 Security

All our protocols are based on Bloom filter. As Bloom filter allows false positive, therefore an element of the server can be revealed to the client with negligible probability $\epsilon = \frac{1}{2^k}$. We describe below the security proofs of PSI-CA, APSI-CA, PSI and APSI.

Theorem 2. *If the quadratic residuosity assumption holds, then PSI-CA protocol presented in Sect. 3.1 is a secure computation protocol for functionality* $\mathcal{F}_{card} : (Y, X) \longrightarrow (|X \cap Y|, \perp)$ *in the security model described in the Sect. 2.1 against semi-honest server and semi-honest client except with negligible probability* ϵ, *where* $Y = \{c_1, c_2, ..., c_w\} \subseteq \{0,1\}^*$ *and* $X = \{s_1, s_2, ..., s_v\} \subseteq \{0,1\}^*$ *with* $w \leq v$.

Proof. **Case I (Server is Corrupted):** We construct a simulator $\mathcal{SIM}$ that has given access to the server's private input X and output $\perp$. $\mathcal{SIM}$ then chooses m random elements $z_0, z_1, ..., z_{m-1} \leftarrow \mathbb{Z}_n$ and outputs the simulated view as $(X, z_0, z_1, ..., z_{m-1})$. The view in the real protocol execution consists of X and the ciphertexts $\{b_i = \mathsf{Enc}_{pk_C}(\mathsf{BF}_Y[i])\}_{i=0}^{m-1} \in \mathbb{Z}_n^m$. Input set X in a real view is same as input set in simulated view and the distribution of $\{z_0, z_1, ..., z_{m-1}\} \in \mathbb{Z}_n^m$ in the ideal model is computationally indistinguishable from the distribution of $\{b_0, ..., b_{m-1}\} \in \mathbb{Z}_n^m$ as GM encryption scheme is IND-CPA secure under quadratic residuosity assumption. Hence the the simulated view $(X, z_0, z_1, ..., z_{m-1})$ is indistinguishable from the view $(X, \mathsf{Enc}_{pk_C}(\mathsf{BF}_Y[0]), ..., \mathsf{Enc}_{pk_C}(\mathsf{BF}_Y[m-1]))$ of a real protocol execution.

Case II (Client is Corrupted): Let us construct a simulator $\mathcal{SIM}$ that has given access to the client's input Y and output $|X \cap Y|$. $\mathcal{SIM}$ outputs the simulated view as $(Y, a_1, a_2, ..., a_v)$, where $|X \cap Y|$ many a_i's are all-zero strings and remaining $v - |X \cap Y|$ many a_i's are randomly chosen non-zero strings of length k each. In a real execution the view contains Y, $|X \cap Y|$ many all-zero strings and $v - |X \cap Y|$ many non-zero strings of length k each, except with a negligible probability ϵ. Input set Y in a real view and simulated view are same. Also $\{a_1, a_2, ..., a_v\}$ is a set of $|X \cap Y|$ many of all-zero strings and $v - |X \cap Y|$ many non-zero strings of length k each. Thus the simulated view is indistinguishable from the view of a real protocol execution except with negligible probability ϵ.

Theorem 3. *If the quadratic residuosity assumption holds, then APSI-CA protocol presented in Sect. 3.2 is a secure computation protocol for functionality* $\mathcal{F}_{card} : (Y, X) \longrightarrow (|X \cap Y|, \perp)$ *in the security models described in the Sects. 2.1 and 2.2 against semi-honest server and malicious client except with negligible probability* ϵ*, where* $Y = \{c_1, c_2, ..., c_w\} \subseteq \{0,1\}^*$ *and* $X = \{s_1, s_2, ..., s_v\} \subseteq \{0,1\}^*$ *with* $w \leq v$.

Proof. **Case I (Server is Corrupted):** This case is exactly same as case I in the proof of the Theorem 2.

Case II (Client is Corrupted): Let the client C be corrupted by an adversary $\mathcal{A}$ in the real world while the server S is honest. We construct the corresponding ideal world adversary $\mathcal{SIM}$ who has oracle access to $\mathcal{A}$ and simulates S in the ideal world as follows. Note that the ideal process additionally involves an incorruptible trusted third party, say T.

1. $\mathcal{SIM}$ invokes the adversary $\mathcal{A}$ with the input $Y = \{c_1, c_2, ..., c_w\} \in \{0,1\}^*$ and generates a key pair (pk_{DSig}, sk_{DSig}) for the digital signature used by CA in real world.
2. First, $\mathcal{SIM}$ plays the role of CA. On receiving $Y = \{c_1, c_2, ..., c_w\} \in \{0,1\}^*$, $pk_C = (n, u)$ from $\mathcal{A}$, $\mathcal{SIM}$ constructs a Bloom filter $\mathsf{BF}_Y \in \{0,1\}^m$, encrypts each of $\mathsf{BF}_Y[i] \in \{0,1\}$ using GM encryption to get $b_i = \mathsf{Enc}_{pk_C}(\mathsf{BF}_Y[i]) \in \mathbb{Z}_n$ and generates signature $Sig(\bar{h}(b_0), ..., \bar{h}(b_{m-1}))$ using sk_{DSig}. $\mathcal{SIM}$ then sends $(\overline{Y} = \{b_0, ..., b_{m-1}\} \in \mathbb{Z}_n^m, Sig(\bar{h}(b_0), ..., \bar{h}(b_{m-1})), pk_{DSig})$ to $\mathcal{A}$.
3. $\mathcal{SIM}$ next plays the role of real world server. $\mathcal{SIM}$, on receiving $(\overline{Y} = \{b_0, ..., b_{m-1}\} \in \mathbb{Z}_n^m, Sig(\bar{h}(b_0), ..., \bar{h}(b_{m-1})), pk_C = (n, u))$ from $\mathcal{A}$, verifies the signature $Sig(\bar{h}(b_0), ..., \bar{h}(b_{m-1}))$ using public key pk_{DSig}. If the verification fails, then $\mathcal{SIM}$ aborts. Otherwise, $\mathcal{SIM}$ plays the role of an ideal world client by sending the set Y to T, whereas the ideal world server sends the set $X = \{s_1, s_2, ..., s_v\} \subseteq \{0,1\}^*$ to T, T being an incorruptible trusted party involved in the ideal process. The ideal functionality $\mathcal{F}_{card}$ is computed in turn by T on the inputs X and Y. As the output of the ideal functionality, $\mathcal{SIM}$ receives the cardinality of $X \cap Y$ from T.
4. $\mathcal{SIM}$ constructs a set $\widehat{X} = \{\bar{s}_1, ..., \bar{s}_v\} \subseteq \{0,1\}^k$ by setting $|X \cap Y|$ many $\bar{s}_i$'s to be all-zero strings of length k and the remaining $v - |X \cap Y|$ many $\bar{s}_i$'s to be k-bit non-zero strings each. $\mathcal{SIM}$ encrypts each $\bar{s}_i \in \{0,1\}^k$ for $i = 1, 2, ..., v$ and sends $\overline{X} = \{E(\bar{s}_1), ..., E(\bar{s}_v)\} \subseteq \mathbb{Z}_n^k$ to $\mathcal{A}$, where $\bar{s}_i = \{\bar{s}_{i,0}, ..., \bar{s}_{i,k-1}\} \in \{0,1\}^k$ and $E(\bar{s}_i) = \{\mathsf{Enc}_{pk_C}(\bar{s}_{i,0}), ..., \mathsf{Enc}_{pk_C}(\bar{s}_{i,k-1})\} \in \mathbb{Z}_n^k$. The simulator $\mathcal{SIM}$ then outputs whatever $\mathcal{A}$ outputs and terminates.

As the honest party has no output, it is sufficient to show that the adversary $\mathcal{A}$'s view in the ideal process is indistinguishable from a view in the real world. The input set Y is same in simulation and in real protocol execution. Note that the simulator $\mathcal{SIM}$ sets the set $\overline{X}$ above as the encryption of v number of k-bit strings of the set $\widehat{X}$, out of which $|X \cap Y|$ many strings are all-zero k-bit strings and $v - |X \cap Y|$ many strings are randomly chosen non-zero strings of length k. Also in the real protocol execution, number of encrypted all-zero strings and non-zero strings in $\overline{X}$ are respectively $|X \cap Y|$ and $v - |X \cap Y|$ except with negligible probability ϵ. Thus the views of $\mathcal{A}$ in the real world and ideal process are indistinguishable except with negligible probability ϵ. □

Theorem 4. *If the quadratic residuosity assumption holds, then PSI protocol presented in Sect. 3.3 is a secure computation protocol for functionality* $\mathcal{F}_\cap : (Y, X) \longrightarrow (X \cap Y, \perp)$ *in the security model described in the Sect. 2.1 against semi-honest server and semi-honest client except with negligible probability* ϵ*, where* $Y = \{c_1, c_2, ..., c_w\} \subseteq \{0,1\}^*$ *and* $X = \{s_1, s_2, ..., s_v\} \subseteq \{0,1\}^*$ *with* $w \leq v$*.*

Proof. **Case I (Server is Corrupted):** This case is exactly same as case I in the proof of the Theorem 2.

Case II (Client is Corrupted): This case is obvious because the client C only receives the messages for its output, i.e. the intersection $X \cap Y$. Thus the simulated view $(Y, X \cap Y)$ is same as the view of a real protocol execution. □

The APSI is proven to be secure against malicious server and malicious client.

Theorem 5. *If the quadratic residuosity assumption holds, then APSI protocol presented in Sect. 3.4 is a secure computation protocol for functionality* $\mathcal{F}_\cap : (Y, X) \longrightarrow (X \cap Y, \perp)$ *in the security model described in the Sect. 2.2 against malicious server and malicious client except with negligible probability* ϵ*, where* $Y = \{c_1, c_2, ..., c_w\} \subseteq \{0,1\}^*$ *and* $X = \{s_1, s_2, ..., s_v\} \subseteq \{0,1\}^k$ *with* $w \leq v$*.*

Proof. **Case I (Server is Corrupted):** Let the adversary $\mathcal{A}$ corrupts the server S in the real world and the client C be honest party. We construct the corresponding ideal world adversary $\mathcal{SIM}$ who has oracle access to $\mathcal{A}$ and simulates C in the ideal world as follows:

1. $\mathcal{SIM}$ first acts as certifying authority by generating key pair (pk_{DSig}, sk_{DSig}) for the digital signature used by CA in real world and sending pk_{DSig} to $\mathcal{A}$. The adversary $\mathcal{A}$ that has access to the input $X = \{s_1, s_2, ..., s_v\} \subseteq \{0,1\}^k$ of the corrupted server S is also invoked by $\mathcal{SIM}$.
2. $\mathcal{SIM}$ next plays the role of real world client as follows:
   (a) runs the key generation algorithm KGen to generate the key pair $(pk_C = (n, u), sk_C = (P, Q))$ for GM encryption,
   (b) creates a Bloom filter $\mathsf{BF}_Y \in \{0,1\}^m$, by setting $\mathsf{BF}_Y[i] = 0$ for $i = 0, ..., m-1$,
   (c) sends $(\overline{W} = \{\hat{b}_i = \mathsf{Enc}_{pk_C}(\mathsf{BF}_Y[i])\}_{i=0}^{m-1} \in \mathbb{Z}_n^m, Sig(\bar{h}(\hat{b}_0), ..., \bar{h}(\hat{b}_{m-1})), pk_C = (n, u))$ to $\mathcal{A}$.
3. On receiving $\overline{X} = \{E(\bar{s}_1), ..., E(\bar{s}_v)\} \subseteq \mathbb{Z}_n^k$ from $\mathcal{A}$, the simulator $\mathcal{SIM}$ decrypts each element of $\overline{X}$ to get the set $\widehat{X} = \{\bar{s}_1, ..., \bar{s}_v\} \subseteq \{0,1\}^k$ using the secret key sk_C of GM encryption generated by $\mathcal{SIM}$ himself.
4. $\mathcal{SIM}$ then plays the role of ideal world server by sending the set $\widehat{X}$ to the incorruptible trusted third party T involved in the ideal process, whereas the ideal world client sends the set Y to T. The deal functionality $\mathcal{F}_\cap$ is then computed by T on the inputs $\widehat{X}$, Y and as the output of the ideal functionality, the ideal world client receives $\widehat{X} \cap Y$ from T.

Note that $\widehat{X} \cap Y$ is equal to $X \cap Y$ as the Bloom filter BF_Y set by $\mathcal{SIM}$ is a all-zero string which makes the elements of $\widehat{X} = \{\bar{s}_1, ..., \bar{s}_v\} \subseteq \{0,1\}^k$ same as the elements of $X = \{s_1, ..., s_v\} \subseteq \{0,1\}^k$. Therefore the honest party C's output is same in the real and ideal world. The following argument shows that the view of the adversary $\mathcal{A}$ in the ideal process is also indistinguishable from a view in the real world. As the input set X in simulation and $\widehat{X}$ in real protocol execution are same and GM encryption is IND-CPA secure under quadratic residuosity assumption, the simulated sets $\overline{W} = \{\hat{b}_i\}_{i=0}^{m-1} \in \mathbb{Z}_n^m, Sig(\bar{h}(\hat{b}_0), ..., \bar{h}(\hat{b}_{m-1}))$ in the ideal world are computationally indistinguishable from the sets $\overline{Y} = \{b_i\}_{i=0}^{m-1} \in \mathbb{Z}_n^m, Sig(\bar{h}(b_0), ..., \bar{h}(b_{m-1}))$ in the real world respectively. Hence the views of the adversary $\mathcal{A}$ in the ideal process and in the real world are computationally indistinguishable.

Case II (Client is Corrupted): Let the client C be corrupted by an adversary $\mathcal{A}$ in the real world and the server S be honest party. Then the corresponding ideal world adversary $\mathcal{SIM}$ who has oracle access to $\mathcal{A}$ and simulates S in the ideal world, can be constructed with similar manner as in case II in the proof of the Theorem 3. The only difference is that in this case $\mathcal{SIM}$ receives $X \cap Y$ rather than $|X \cap Y|$ from the incorruptible trusted party T and constructs $\widehat{X} \subseteq \{0,1\}^k$ by including all the elements of $X \cap Y$ and $v - |X \cap Y|$ many randomly chosen k-bit strings.

We need to show only that the adversary $\mathcal{A}$'s view in the ideal process is indistinguishable from a view in the real world as the honest party has no output. In the both simulation and real protocol execution the input set is Y. The simulator $\mathcal{SIM}$ sets the set $\overline{X}$ as the encryption of all the elements of $X \cap Y$ and $v - |X \cap Y|$ many randomly chosen k-bit strings. Also in the real protocol execution $\overline{X}$ contains the encryption of the elements of $X \cap Y$ and $v - |X \cap Y|$ number of strings except with negligible probability ϵ. Thus the views of $\mathcal{A}$ in the real world and ideal process are indistinguishable except with negligible probability ϵ. □

5 Efficiency

The communication cost in our protocol is measured by counting number of group elements transmitted publicly by an user. Also each entity in our protocols performs same operations like modular multiplication, signature, Jacobi/Legendre symbol computation, hash function evaluation. These incur computation overheads. Tables 5 and 6 exhibit the complexity of our construction. Once the digital signature, used by our constructions APSI-CA and APSI, is fixed, then the additional group elements and the computation cost due to the digital signature can be computed. In particular, the number of group elements to be transferred during the APSI-CA (or APSI) is $O(v + w)$ and the computation cost of APSI-CA (or APSI) is $O(v + w)$. Note that for PSI-CA and APSI-CA, number of modulo multiplications MUL will be $2(\frac{vk}{\ln 2} + kv)$ instead of $2\frac{vk}{\ln 2} + 3kv$ in Tables 5 and 6 respectively. Furthermore, without including any extra cost, our PSI-CA or

Table 5. Complexity of PSI-CA and PSI

Party	Hash	MUL	GE	LC
C	$(k+1)w$	$2m = 2\frac{vk}{\ln 2}$	$m+2$	kv
S	$(k+1)v$	$3kv$	kv	
Total	$(k+1)(w+v)$	$2\frac{vk}{\ln 2} + 3kv$	$\frac{vk}{\ln 2} + kv + 2$	kv

Table 6. Complexity of APSI-CA and APSI without digital signature

Party	Hash	MUL	GE	LC
CA	$kw + m$	$2m$	m	
C			$m + w + 4$	kv
S	$kv + m$	$3kv$	kv	
Total	$k(w+v) + 2\frac{vk}{\ln 2}$	$2\frac{vk}{\ln 2} + 3kv$	$2\frac{vk}{\ln 2} + kv + w + 4$	kv

Hash=Number of hash query, LC=Number of Legendre symbol computation, MUL=Number of multiplications, GE= Number of group element

APSI-CA protocol can be used to compute Private set union cardinality (PSU-CA) or Authorized PSU-CA using the formula $|X \cup Y| = |X| + |Y| - |X \cap Y|$.

Note: Recently proposed very efficient BF based PSI protocols are [6,11], where the authors used a variant of BF, called Garbled Bloom Filter (GBF). Similar to BF, GBF uses k independent hash functions $H = \{h_0, h_1, ..., h_{k-1}\}$ but instead of single bit, each of $GBF[i]$, $i = 0, 1, ..., k-1$ contains λ-bit string, where λ is a security parameter. In [6], Dong et al. represented a set X using a GBF as follows:

1. Set each $\mathsf{GBF}_X[i]$ as unoccupied for $i = 0, 1, ..., m-1$.
2. For each $x \in X$,
   (a) find a hash $h_l \in \{h_0, h_1, ..., h_{k-1}\}$ such that $\mathsf{GBF}_X[h_l(x)]$ is unoccupied;
   (b) for $j = 0, 1, ..., k-1$ $(j \neq l)$, if $\mathsf{GBF}_X[h_j(x)]$ is unoccupied then set $\mathsf{GBF}_X[h_j(x)]$ as a random λ-bit string;
   (c) finally, sets $\mathsf{GBF}_X[h_l(x)] = x \oplus (\oplus_{j=0,j\neq l}^{k-1} \mathsf{GBF}_X[h_j(x)])$ to obtain a valid XOR sharing of x.

Thus the PSI of [6] may be converted to PSI-CA by setting $\mathsf{GBF}_X[h_l(x)]$ as $(\oplus_{j=0,j\neq l}^{k-1} \mathsf{GBF}_X[h_j(x)])$ instead of $x \oplus (\oplus_{j=0,j\neq l}^{k-1} \mathsf{GBF}_X[h_j(x)])$ i.e., by XOR sharing of λ-bit all-zero string. However, in that case client also can compute the intersection by inserting the elements of his private set for which $(\oplus_{j=0}^{k-1} \mathsf{GBF}_X [h_j(x)])$ is a λ-bit all-zero string. This breaks the security of PSI-CA, thereby extension of the PSI of [6] to PSI-CA seems to be a non-trivial task. Similar argument holds for the PSI of [11].

6 Conclusion

We have presented efficient constructions for PSI-CA, APSI-CA, PSI and APSI protocols with linear complexities based on Bloom filter and homomorphic GM encryption. In our protocols, client's input set size need not be revealed to the server. Proposed PSI-CA and APSI-CA are the *first* cardinality set intersection protocols secure in standard model with linear complexity and preserving client's input set size independency.

References

1. Ateniese, G., De Cristofaro, E., Tsudik, G.: (If) size matters: size-hiding private set intersection. In: Catalano, D., Fazio, N., Gennaro, R., Nicolosi, A. (eds.) PKC 2011. LNCS, vol. 6571, pp. 156–173. Springer, Heidelberg (2011)
2. Bloom, B.H.: Space/time trade-offs in hash coding with allowable errors. Commun. ACM **13**(7), 422–426 (1970)
3. Camenisch, J., Zaverucha, G.M.: Private intersection of certified sets. In: Dingledine, R., Golle, P. (eds.) FC 2009. LNCS, vol. 5628, pp. 108–127. Springer, Heidelberg (2009)
4. De Cristofaro, E., Gasti, P., Tsudik, G.: Fast and private computation of cardinality of set intersection and union. In: Pieprzyk, J., Sadeghi, A.-R., Manulis, M. (eds.) CANS 2012. LNCS, vol. 7712, pp. 218–231. Springer, Heidelberg (2012)
5. De Cristofaro, E., Kim, J., Tsudik, G.: Linear-complexity private set intersection protocols secure in malicious model. In: Abe, M. (ed.) ASIACRYPT 2010. LNCS, vol. 6477, pp. 213–231. Springer, Heidelberg (2010)
6. Dong, C., Chen, L., Wen, Z.: When private set intersection meets big data: an efficient and scalable protocol. In: Proceedings of the 2013 ACM SIGSAC Conference on Computer and Communications Security, pp. 789–800. ACM (2013)
7. Goldreich, O.: Foundations of Cryptography: Volume 2, Basic Applications, vol. 2. Cambridge University Press, New York (2009)
8. Goldwasser, S., Micali, S.: Probabilistic encryption. J. Comput. Syst. Sci. **28**(2), 270–299 (1984)
9. Hazay, C.: Oblivious polynomial evaluation and secure set-intersection from algebraic PRFs. IACR Cryptology ePrint Archive 2015, 4 (2015)
10. Kerschbaum, F.: Outsourced private set intersection using homomorphic encryption. In: Proceedings of the 7th ACM Symposium on Information, Computer and Communications Security, pp. 85–86. ACM (2012)
11. Pinkas, B., Schneider, T., Zohner, M.: Faster private set intersection based on ot extension. USENIX Secur. **14**, 797–812 (2014)
12. Stefanov, E., Shi, E., Song, D.: Policy-enhanced private set intersection: sharing information while enforcing privacy policies. In: Fischlin, M., Buchmann, J., Manulis, M. (eds.) PKC 2012. LNCS, vol. 7293, pp. 413–430. Springer, Heidelberg (2012)

On the Efficiency of Multi-party Contract Signing Protocols

Gerard Draper-Gil[1(✉)], Josep-Lluís Ferrer-Gomila[2], M. Francisca Hinarejos[2], and Jianying Zhou[3]

[1] University of New Brunswick, Fredericton, NB E3B 5A3, Canada
gerard.draper@unb.ca
[2] Universitat de les Illes Balears, 07122 Palma, Spain
{jlferrer,xisca.hinarejos}@uib.es
[3] Institute for Infocomm Research, Singapore 138632, Singapore
jyzhou@i2r.a-star.edu.sg

Abstract. This paper presents an efficiency study of fair exchange protocols for Multi-Party Contract Signing (MPCS) from their architecture point of view, an approach that has not been previously explored. A set of common topologies is presented and defined: ring, star sequential and mesh. Some common terms and notions, such as round and message, are defined according to the topology where they are applied. The suitability of such common terms to measure the efficiency of the protocols is discussed. Finally, we present the design of optimal asynchronous optimistic MPCS protocols for different topologies and evaluate them under the unified definition/criterion of the efficiency parameters. These results are important to support secure and efficient online business which is part of our efforts for building secure and smart cyber society.

Keywords: Multi-party contract signing · Contract signing efficiency · Abuse-freeness

1 Introduction

The objective of a Multi-Party Contract Signing (MPCS) protocol is to allow a set of participants P_i $(2 < i < N)$ to exchange a valid signature on a contract C, without any of them gaining advantage over the others. We can describe the protocol as an application of fair exchange: N parties want to sign a contract C, but none of the participants is willing to give his signature away unless he has an assurance that he will receive all the other participants' signatures.

Most of the solutions we can find in the literature for MPCS protocols are based on the existence and possible involvement of a Trusted Third Party (TTP). The TTP is an external entity that assures the protocol is executed correctly, providing the participants who contact it with evidence proving the state of the execution. In fact, even for two-party contract signing protocols there is a consensus that solutions without a TTP are not practical. One step further is

J. Lopez and C.J. Mitchell (Eds.): ISC 2015, LNCS 9290, pp. 227–243, 2015.
DOI: 10.1007/978-3-319-23318-5_13

to decide if this TTP will intervene in every protocol run (inline or online TTP) or only in case of exception (offline TTP, also called optimistic solutions). The majority of scientific proposals tend to use offline TTPs, where the TTP is only involved if a dispute arises, which is expected to be an exceptional case.

We can find different proposals for MPCS in the scientific literature [3,4,8, 12–14]. Some of them claim to propose optimal solutions or define lower-bounds to design MPCS protocols [8,12,13], but the different criteria applied to define requirements like fairness, or terms like round, step, etc., makes it difficult to assert the validity of such optimal solutions. Moreover, even though we can use different topologies to design MPCS protocols, none of these solutions contemplates the influence of the topology over the efficiency of the final result.

The objective of this paper is to design asynchronous protocols in which N participants sign the same contract C. We choose to design asynchronous protocols instead of synchronous ones, to avoid the problems related to the participant's clock synchronization.

Our Contribution: The contributions of this paper are manifold. First, we discuss the parameters that are generally used to measure the efficiency of MPCS protocols, making clear definitions of each one and defining new ones when the commonly used parameters are not good for measuring efficiency. Second, we describe four of the most common architectures (ring, star, sequential and mesh) and we define them according to the efficiency parameters. Finally, we describe a method to design asynchronous optimistic MPCS protocols, and we propose one as example. We also informally prove that our proposals are optimal, improving the existent proposals of lower-bounds for asynchronous optimistic MPCS protocols.

2 MPCS Requirements

MPCS is a particular case of fair exchange protocols in which we have more than 2 participants and the items to be exchanged are signatures on a contract. Requirements for optimistic fair exchange protocols were defined by Asokan *et al.* [1]: *effectiveness, fairness (strong and weak), timeliness and non-repudiation*, and later, re-formulated by Zhou *et al.* [19]. In this section we will adapt these requirements to the asynchronous optimistic MPCS scenario.

Effectiveness. If all participants in a MPCS protocol behave correctly (and there are no network or system errors), the protocol will finish without the intervention of the TTP.

Strong Fairness. Upon finalization of a MPCS protocol, either all honest participants have the signature from the other participants, or all of them have proof that the signature has been canceled. None of the participants can receive evidence that contradicts the final state of the protocol execution.

Weak Fairness. Upon finalization of a MPCS protocol, either strong fairness is met or all honest participants can prove they have behaved correctly.

Non-repudiation. Upon finalization of a MPCS protocol, none of the participants can deny having participated. In particular, the participants cannot deny having originated (non-repudiation of origin) the signatures exchanged.

Timeliness. Any participant in a MPCS protocol can be sure that the duration of the protocol execution is finite. And once the protocol is finished, any honest participant will maintain the level of fairness obtained.

In addition to the requirements stated by Asokan *et al.* [1] and Zhou *et al.* [19], Garay *et al.* [7] introduced abuse-freeness. Its objective is to avoid dishonest participants to misuse the information acquired during the protocol execution (e.g., the commitment to sign a contract from other participants).

Abuse-Freeness. After receiving P_i a partial signature from another participant P_j, the recipient P_i cannot convince others but himself that the partial signature is from the sender P_j.

3 Efficiency

It is usually accepted that a protocol is efficient when it makes a *reasonable* use of resources to fulfil its purpose. But we do not have a reference measure to distinguish reasonable from unreasonable, therefore authors usually talk about the efficiency of their solutions compared to others. Most of the papers use the computational power as *the resource* to measure, giving their value of efficiency in terms of number of mathematical operations, but it is not always easy to grasp the real value of these measures.

Throughout the solutions found in the literature, authors use the terms 'round' and 'step' without clearly defining them, which often brings on confusion with respect to the metric to be used for its efficiency evaluation, or what they are exactly measuring. Another value typically provided is the number of messages required to complete a protocol execution, but again they fail to give a clear description of it. In our opinion, the term round should not be used for measuring the efficiency of a protocol, but to help in its description. As we will see in Sect. 4, the problem is that rounds in different topologies are not equal, e.g., in a ring topology a round requires the participants to make N transmissions of information (1 per participant). Moreover, in a ring topology the protocol execution must follow a certain order, and this information can be used by the TTP to detect malicious users (see TTP rules for ring topology, in Sect. 7.1), meanwhile in a mesh topology there is no execution order among participants. The use of message as a parameter to measure efficiency has also the same problem: a transmission of information may contain more than one message. In this paper we will take a different approach, we will focus on the participants to measure the protocol efficiency. We will measure the protocol complexity in terms of how many transmissions are required, and how many messages each user has to generate. These terms can later be translated in a time estimation, giving each participant an idea of time, how long will a protocol execution take. Following we make a definition of message and transmission, to clearly state their meaning.

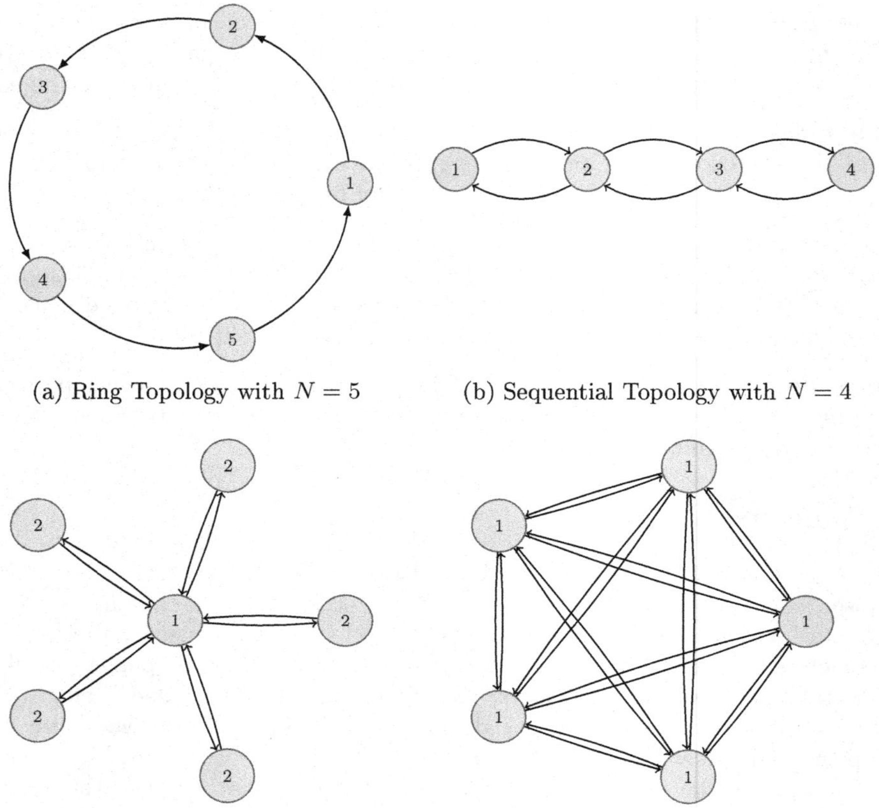

(a) Ring Topology with $N = 5$ (b) Sequential Topology with $N = 4$

(c) Star Topology with $N = 6$ (d) Full Mesh Topology with $N = 5$

Fig. 1. Multi-party contract signature topologies

Definition 1 ***(Transmission)*** *The action of transmitting one or more messages from an originator A to a recipient B.*

Definition 2 ***(Message)*** *A logical set of information sent from an originator A to B, where B can be a set of recipients $\{B_1, ..., B_N\}$.*

4 Topologies

In this section we will define four of the most used topologies when designing MPCS protocols. For each topology we will define the meaning of round, and we will calculate the number of transmissions required to complete a round, assuming N participants $\{P_1, ..., P_N\}$.

4.1 Ring

In a ring topology the transmissions occur between two adjacent nodes P_i and $P_{(i+1)}$, until the execution flow reaches P_N, whose transmission recipient is P_1, the initiator node. The ring architecture executes the transmissions on a serial basis. In Fig. 1a we have depicted a complete round of a ring topology.

Definition 3 ***(Ring-Round.)*** *A round begins when P_1 executes a transmission to P_2, then P_2 transmits to P_3, ..., and ends when P_1 receives the transmission from P_N, closing the ring.*

A complete *ring-round* requires N transmissions, and generates information on the execution order that can be used by the TTP to detect attempts of misbehaviour.

4.2 Sequential

In a sequential topology the transmissions are executed on a serial basis. The protocol execution flows from P_1 to P_N, and back to P_1, going through all the participants in between. In Fig. 1b we have a complete round execution of a sequential topology depicted.

Definition 4 ***(Sequential-Round.)*** *A round is started by the participant P_1, transmitting one or more messages to P_2. The transmissions continue through all the participants in a certain order (e.g., incrementing the subindex i: $P_i, P_{(i+1),..}$), until it reaches P_N, who reverses the order transmitting to $P_{(N-1)}$, who executes a transmission to $P_{(N-2)}$, etc. The round ends when P_1 receives a transmission from P_2.*

A complete *sequential-round* requires $2(N-1)$ transmissions. It also generates information on the execution order of the transmissions.

4.3 Star

In a star topology the transmissions between participants are routed through a central node/participant. The central node P_1 receives all the transmissions from the participants P_j ($j \in [2..N]$), and then P_1 returns to each P_j the corresponding messages. Figure 1c depicts a complete round execution of star topology depicted.

Definition 5 ***(Star-Round.)*** *A round begins when the initiator P_1 transmits some message or messages to all P_j ($j \in [2..N]$), and ends when P_1 has received the corresponding transmission from each P_j. Alternatively, the round can be started by all P_j ($j \in [2..N]$) transmitting to P_1, and finish when each P_j has received P_1's transmission.*

The star topology only generates information about who initiated the *star-round*. It requires $2(N-1)$ transmissions.

4.4 Mesh

In a mesh topology the transmissions are executed on a parallel basis. Each P_i, with $1 \leq i \leq N$ will execute a transmission to each P_j, with $j \in [1..N]$, $j \neq i$. In Fig. 1d we have a complete round execution of mesh topology.

Definition 6 ***(Mesh-Round.)*** *A round begins when P_i, with $1 \leq i \leq N$ executes a transmission to each P_j, with $j \in [1..N]$, $j \neq i$. The round will end when every participant has received a transmission from the other $N-1$ participants.*

A complete *mesh-round* requires $N(N-1)$ transmissions, and it does not generate additional information: the participants are not ordered.

5 Related Work

Baum-Waidner *et al.* propose in [3] a MPCS protocol that requires $N+1$ rounds without abuse-freeness, and $N+3$ with it (optimistic case). Their protocol uses a mesh topology, but they also describe how to transform it into a protocol with star topology. The number of rounds and messages required for each protocol are presented as a function of the number of dishonest parties t. But it is not clear the usefulness of it, because we cannot know the number of dishonest participants beforehand. In [2] Baum-Waidner presents an optimization of the previous proposal [3] where they assume a number of dishonest participants $t < (N-1)$, but it is difficult to see the utility of this enhancement because, as we just said, we cannot predict the number of dishonest parties beforehand.

Khill *et al.* [11] propose a protocol for multi-party fair exchange using a ring topology. They affirm that the ring model is more efficient than the full mesh topology. Their protocol consists on 3 rounds and $3N$ messages in the optimistic case, and $7N$ messages in the worst case. A serious drawback of the protocol is that it is supposed that the TTP broadcasts its decision to all parties. This assumption is dangerous, because the channels can be resilient, but some party can be unreachable for other reasons.

Chadha *et al.* [4] analyze formally two previous works: Garay *et al.* [8] and Baum-Waidner *et al.* [3]. They focus on three properties: fairness, timeliness and abuse-freeness. They conclude that the proposal of Baum-Waidner *et al.* [3] has no security problems. On the other hand, they prove that the proposal of Garay *et al.* [8] presents a security flaw when $N = 4$: it is not fair. Mukhamedov *et al.* [15] prove that Chadha *et al.*'s [4] proposal (a fix to [8]) is also flawed. An interesting issue discussed in that paper [15] is the abort chaining problem (or resolve impossibility): non-honest parties can group together to propagate a TTP's abort decision. In fact, this is a way to prove the necessity of more than $(N-1)$ rounds for N users (in order to avoid the abort chaining attack). The abort chaining attack is a sequence of requests made by dishonest participants trying to force the TTP to deliver cancel evidence, even though some other honest participant may have already signed the contract.

Ferrer *et al.* present in [5] an optimal solution for asynchronous optimistic MPCS with a ring topology, requiring quasi N rounds (more than $N-1$ but less than N) for N parties. Their solution meets the following requirements: effectiveness, weak fairness, timeliness, non-repudiation and verifiability of TTP. The proposal takes into account the abort chaining problem.

In [14] Mukhamedov and Ryan criticize the work of Baum-Waidner *et al.* [3] alleging that they use a non-standard notion of signed contract and they need $(N+1)N(N-1)$ messages, more than in the solution provided in [14], $N(N-1)(\lceil N/2\rceil+1)$. In [14] fairness, abuse-freeness and timeliness are considered. They use a *hybrid* topology, a mixture of sequential and mesh, where the participants are ordered. The protocol needs $(\lceil N/2\rceil + 1)$ rounds, and authors observe that it is not coherent with Garay's Theorem of [7], but they argue that the concept of round, used in different papers, is not clear.

Mauw *et al.* [13] use the concept of abort chaining of Mukhamedov *et al.* [15] to derive a lower bound on the number of messages in MPCS protocols. The authors model contract signing protocols as sequences of numbers. They consider three security requirements: fairness, timeliness and abuse-freeness (but they affirm that the latter "will not play a role in our observations on message minimality").

Zhang *et al.* propose in [18] a game-based verification of MPCS protocols. They assume that MPCS protocols have to satisfy three properties: fairness, timeliness and abuse-freeness. They analyze the protocols provided in [13,16], proving the latter to be flawed for 3 signers and proposing a fix. Authors assume that "once having contacted TTP by initiating a sub-protocol, the signers would never be allowed to proceed the main protocol any more", but we cannot forbid a dishonest party to contact the TTP and proceed with the main protocol.

Following a similar reasoning than [13], Kordy *et al.* [12] propose protocols derived from sequences of numbers. They consider the following requirements: fairness, timeliness and abuse-freeness. An example with $N = 3$ results in a protocol (sequential topology) with 18 messages, that can be converted to 12 messages. They cannot provide closed expressions for all values of N, and only provide upper bounds.

6 MPCS Protocols Overview

This section presents a simple method to design asynchronous optimistic MPCS protocols. An asynchronous optimistic MPCS protocol will be composed of two sub-protocols: the exchange sub-protocol and the resolution sub-protocol. If all participant behave correctly and there are no network errors, only the exchange sub-protocol will be executed and the TTP will not intervene. As defined in Sect. 2, the MPCS protocol will meet the security requirements for asynchronous optimistic MPCS: effectiveness, weak fairness, non-repudiation and timeliness. And since it is a contract signing protocol, it should also consider the abuse-freeness requirement.

The MPCS protocol execution will follow a simple principle: in turns, the participants will exchange series of commitments to sign the contract C, until they have enough evidence to consider the contract signed. The commitments are signatures on the contract C and an index k. What is a "turn" or what is "enough evidence" will be determined by the topology of the protocol (ring, star, etc.).

In Sect. 7 we have an example of an asynchronous optimistic MPCS protocol using a ring topology. Along the rest of the paper, we will use the following notation:

- **N** Number of participants.
- $\mathbf{P_i}$ Participant i, $1 \leq i \leq N$.
- $\mathbf{Tx_{(r,i)}}$ Transmission generated by the participant i, during round r.
- **C** Contract to be signed.
- **CID** unique Contract IDentifier. A random number used to uniquely identify a protocol execution.
- $\mathbf{h(M_i)}$ Hash Function of message M_i.
- $\mathbf{S_j[M_i] = SK_j[h(M_i)]}$ j's Digital Signature on M_i (where SK_j is j's private key).

We assume that the contract C includes the necessary information, as the identity of the participants, the TTP, the number N of participants, etc.

As regards the communications channels we make some usual assumptions ([1,6]):

- channels among participants P_i are unreliable, the messages can be delayed or lost.
- channels among participants P_i and the TTP are resilient, the messages can be delayed but not lost.

To meet the abuse-freeness requirement we can use signature schemes like Designated Verifier Signatures (DVS) presented by Jakobsson *et al.* [10], Multi DVS (MDVS) [17], Private Contract Signatures (PCS) introduced by Garay *et al.* [7] or the Ambiguous Optimistic Fair Exchange (AOFE) scheme from Huang *et al.* [9]. In essence, these signature schemes allow the participants to generate a "weak" signature as commitment (partial signature), that can only be verified by the intended recipient (or recipients in the case of MDVS). Once all commitments are exchanged, they can generate a signature that can be verified by third parties (full signature).

6.1 The TTP

The TTP is a third-party that assures the fairness of the protocol providing the participants with proof of the protocol execution state. When a participant does not receive a signature expected, either because an error occurred or because a misbehaving participant, he can send a resolution request to the TTP. The TTP will answer with a canceled or a signed token.

To solve the resolution requests (Table 1) the TTP follows a set of rules. These rules are based on a group of variables the TTP updates on every request received, indicating the state of a protocol execution. Following we have this group of variables, their definition, and some notation used along the rules definition.

- $\overline{X_N} = \{P_1, ..., P_N\}$ set of participants in a MPCS.
- $\overline{XR}$ set of participants who already requested resolution.
- $\overline{XC}$ set of participants who have received a canceled token from the TTP.
- $\overline{XS}$ set of participants who have received a signed token from the TTP.
- $\overline{TxR}$ set of transmissions $Tx_{(r',i')}$ received by the TTP.
- $\overline{PC}$ set of participants that are allowed to cancel the contract signature.
- *canceled* boolean value stating that the contracting protocol execution has been canceled if its value is true.
- *signed* boolean value stating that the contracting protocol execution has been finished (signed) if its value is true.

The rules are the same for all the protocols, but there are some particularizations depending on the topology, that we will explain in the example of MPCS protocol (see Sects. 7.1 and 7.2). Following we have the common set of rules that the TTP will follow to solve the resolution requests (the term *x-round* refers to the particular round of each topology):

RULE 0 (R0). The TTP will only accept one resolution request per participant P_i: if $P_i \in \overline{XR}$, the TTP will dismiss the request.

RULE 1 (R1). If the TTP receives a request from $P_i \in \overline{PC}$ during *x-round* $r = 1$, and the execution has not been previously finished (*signed=true*) by other party, the TTP will cancel it and send a canceled token to P_i.

RULE 2 (R2). If the TTP receives a request from P_i during *x-round* $r > 1$, and the execution has not been previously canceled by other party, the TTP will finish it (*signed=true*) and send a signed token to P_i.

RULE 3 (R3). If the TTP receives a request from P_i during *x-round* $r \geq 1$, and the execution has been previously finished (*signed=true*) by other party, the TTP will send a signed token to P_i.

RULE 4 (R4). If the TTP receives a request from P_i during *x-round* $r > 1$, and the execution has been previously canceled (*canceled=true*) by other party, the TTP will check the previously received requests. If the TTP can prove that all previous requestors cheated, it will change the protocol status from canceled to finished, and deliver the signed token to P_i. Otherwise the TTP will send a cancel token to P_i.

7 Asynchronous Optimistic MPCS Protocols

In this section we present a set of asynchronous optimistic MPCS protocols, one for each topology (ring, sequential, star and mesh) described in Sect. 4. All

Table 1. Resolution sub-protocol for all topologies

MPCS resolution sub-protocol
$P_i \rightarrow$ TTP: $CID, C, r, Tx_{(r,i)}, S_{P_i}[CID, C, r, Tx_{(r,i)}]$
if the TTP decides canceled
$P_i \leftarrow$ TTP: Cancel Token
else
$P_i \leftarrow$ TTP: Signed Token
Cancel Token: $S_{TTP}[CID, C, r, CANCELED]$; where CANCELED is a string
Signed Token: $S_{TTP}[CID, C, \overline{m_{(k,i)}}_N]$

protocols meet the requirements defined in Sect. 2: effectiveness, weak fairness, timeliness, non-repudiation, and abuse-freeness. The examples presented in this section assume the use of the Private Contract Signature (PCS) scheme ([7]) to meet the abuse-freeness requirement. We could replace the PCS signature scheme for any other mentioned in the precious section, and the protocol would still be valid (TTP rules, number of transmissions, security requirements, etc.), according we make the necessary modifications on the content of each transmission $Tx_{(r,i)}$.

7.1 An Asynchronous Optimistic MPCS Protocol Using Ring Topology

This protocol is based on the optimistic MPCS protocol from Ferrer-Gomila *et al.* [5], where the authors present a solution with *quasi* N rounds (more than $N-1$ but less than N) for N parties, meeting the security requirements (Sect. 2). Table 2 shows the exchange sub-protocol execution using the PCS signature scheme, following the nomenclature from Garay *et al.* [7] for the PCS.

In every turn each participant generates a commitment, a private signature (PCS), for each of the other participants in the protocol execution. The index of the commitments, k, is decremented by one every time a participant receives all k-commitments from the other participants (see Table 3 for values of k in a ring protocol). From then, the participants generate $(k-1)$-commitments, until again a participant receives all $(k-1)$-commitments and he decrements its value again. This process is repeated until k reaches the value -1. When the index k reaches the value 0, the commitments will be generated on the contract C, without index. The next iteration, when k is -1, the participants will start to transmit the full signature on the contract C, final evidence that the protocol has finished successfully. In Table 4 we can see an example of a complete execution for $N = 3$.

All participants except P_N can cancel the protocol, therefore we have that in TTP's rule R1 $\overline{PC} = \{P_1, ..., P_{(N-1)}\}$. When P_N receives the first transmission $Tx_{(1,(N-1))}$ he already has evidence that proves that all other participants are willing to sign the contract. If he does not want to sign the contract, he only

Table 2. Asynchronous optimistic MPCS protocol with ring topology and PCS signature scheme

MPCS protocol with ring topology

for $r = 1$

  for $i = 1$ **to** N: $P_i \rightarrow P_{(i+1)}\ Tx_{(1,i)}$

    $PCS_i((C,k), P_j, TTP) \quad \forall j \in [1..N] \setminus [i]$

    $PCS_{(i-1)}((C,k), P_j, TTP)\ \forall j \in [1..N] \setminus [i, (i-1)]$

    ⋮

    $PCS_1((C,k), P_j, TTP) \quad \forall j \in [1..N] \setminus [i, (i-1), ...1]$

for $r = 2$ **to** $(\mathbf{N-1})$

  for $i = 1$ **to** N: $P_i \rightarrow P_{(i+1)}\ Tx_{(r,i)}$

    $PCS_i((C,k), P_j, TTP) \quad \forall j \in [1..N] \setminus [i]$

    $PCS_{(i-1)}((C,k), P_j, TTP)\ \forall j \in [1..N] \setminus [i, (i-1)]$

    ⋮

    $PCS_{(i-K)}((C,k), P_j, TTP)\ \forall j \in [1..N] \setminus [i, (i-1), ...(i-K)]$

for $r = N$

  for $i = 1$ **to** $(N-1)$: $P_i \rightarrow P_{(i+1)}\ Tx_{(N,i)}$

    $S - Sign_j(C) \quad \forall j \in [1..N] \setminus [2, ..(i+1)]$

$K = N - 2$

$PCS_i(C, P_j, TTP)$, Private Contract Signature of P_i over C for P_j with Trusted Third Party TTP

$S - Sign_i(C)$ Universally verifiable signature of P_i over C

Operations are mod, e.g., $P_i \rightarrow P_{(i+1)}$ when $i = N$ is $P_N \rightarrow P_1$

needs to discontinue the protocol execution. In a protocol with ring topology, TTP's rule R4 states:

- if $\exists\ Tx_{(r',i')} \in \overline{TxR}\ /\ (r' = r)$ or $(r' = r-1$ and $i' > i)$, the TTP will send a cancel token to P_i to maintain fairness for the previous honest requestors.
- if $\forall\ Tx_{(r',i')} \in \overline{TxR}\ /\ r' < r-1$, then $P_{i'}$ cheated.
- if $\forall\ Tx_{(r',i')} \in \overline{TxR}\ /\ r' = r-1$ and $i' < i$, then $P_{i'}$ cheated.

Notice that the TTP's rule R4 for a ring topology uses the information generated by the protocol flow (when comparing the index i with i'), the execution order to detect cheating participants.

The *cancel Token* ($S_{TTP}[CID, C, r, CANCELED]$) is evidence provided by the TTP proving the contract signature has been canceled. It is the TTP's universally verifiable signature on the unique Contract Identifier CID, the contract C itself, the round number r in which the request was sent to the TTP (it can be used later to prove the validity of the assertion), and a value indicating the final state of the protocol execution.

The *signed Token* is evidence provided by the TTP proving the contract is signed. It will depend on the signature scheme used. In the case of the PCS

Table 3. Value of k according to the round r and the participant i for an asynchronous optimistic MPCS protocol with ring topology

r	i	k
1	$1 \leq i \leq (N-1)$	$k = (N-2)$
1	$i = N$	$k = (N-3)$
2	$1 \leq i \leq (N-2)$	$k = (N-3)$
2	$(N-1) \leq i \leq N$	$k = (N-4)$
3	$1 \leq i \leq (N-3)$	$k = (N-4)$
3	$(N-2) \leq i \leq N$	$k = (N-5)$
⋮		
(N-2)	$1 \leq i \leq 2$	$k = 1$
(N-2)	$3 \leq i \leq N$	$k = 0$
(N-1)	$i = 1$	$k = 0$
(N-1)	$2 \leq i \leq N$	$k = -1$
N	$1 \leq i \leq (N-1)$	$k = -1$

and ambiguous signatures, both schemes have a method to transform the partial signatures in full signatures. Therefore the signed token will be the TTP's full signature on the CID, C, the round r, and the partial signatures converted into full signatures.

Lemma 1. *All asynchronous optimistic MPCS protocols with ring topology, meeting timeliness, require at least* $(N+1)(N-1)$ *transmissions to be fair.*

Proof. We will prove it by contradiction. Assume that $(N+1)(N-1)-1$ transmissions are enough (we eliminate the last transmission: $Tx_{(N,(N-1))}$). It means that P_N has all the evidence when he receives $Tx_{(N-1),(N-1)}$. Now we will construct the abort-chaining attack.

Let us suppose $P_{(N-1)}$ sends a resolution request claiming he has sent $Tx_{(1,(N-1))}$ but he has not received $Tx_{(2,(N-2))}$. He is the first to contact the TTP, therefore the TTP will apply rule R1, canceling the protocol and delivering a canceled token to $P_{(N-1)}$.

Next, $P_{(N-2)}$ sends a resolution request claiming he has sent $Tx_{(2,(N-2))}$ but he has not received $Tx_{(3,(N-3))}$. The TTP will apply rule R4 ($r' = r-1$ and $i' > i$) and it will send a canceled token to $P_{(N-2)}$.

Following, $P_{(N-3)}$ sends a resolution request claiming he has sent $Tx_{(3,(N-3))}$ but he has not received $Tx_{(4,(N-4))}$. The TTP will apply rule R4. This time, the TTP will detect that $P_{(N-1)}$ $((r' = 1) < (r-1 = 2))$ cheated, but to maintain fairness for $P_{(N-2)}$ $(1 = 2-1$ and $(N-1) > (N-2))$, it will send a canceled token to $P_{(N-3)}$.

We can continue this abort-chaining attack, until P_2 sends a resolution request claiming he has sent $Tx_{((N-2),2)}$ but he has not received $Tx_{((N-1),1)}$. Applying

R4, the TTP detects that P_4 cheated, but to maintain fairness for P_3 it sends a canceled token to P_2.

Finally, P_1 sends a resolution request claiming he has sent $Tx_{((N-1),1)}$ but he has not received $Tx_{((N-1),N)}$. Again, the TTP will apply rule R4, and deliver a canceled token to P_1 to maintain fairness for P_2 (the TTP can prove that $\{P_3, P_4, ..., P_{(N-2)}, P_{(N-1)}\}$ have cheated). In this scenario, an honest P_N may have received all evidence, but an honest P_1 has a canceled token from the TTP ($Tx_{((N-1),N)}$ may be lost due to a network error), therefore fairness is broken.

But if we add another transmission, $Tx_{(N,(N-1))}$, we can avoid the abort chaining attack. Continuing the previous execution, with the additional transmission, we have two possibilities:

- If P_1 is honest, he will not continue with the protocol execution, therefore P_N will send a resolution request to the TTP claiming the missing evidence. The TTP will be able to prove that P_2 cheated, but again, to maintain fairness for the honest participants it will send a canceled token to P_N. Both P_1 and P_N, honest, will have a canceled token.
- If P_1 is dishonest, and all other dishonest participants continue with the protocol execution, P_N will receive $Tx_{(N,(N-1))}$, therefore he will have evidence the contract has been signed.

In both cases weak fairness is met. Therefore we can affirm that the minimum number of transmissions that an asynchronous optimistic MPCS protocol with ring topology needs to be fair is $(N+1)(N-1)$.

7.2 An Asynchronous Optimistic MPCS Protocol with Sequential, Star and Mesh Topology

Following the same method we used to design the MPCS protocol with ring topology, we can design a protocol using a sequential, star and mesh topology. Due to the lack of space, we cannot include their full description, but we briefly present their measures of efficiency and the TTP rules.

Sequential: $(N+1)(N-1)$ transmissions are necessary.
In a protocol with sequential topology we have: $\overline{PC} = \{P_1, ..., P_{(N-1)}\}$, i.e. all participants except P_N can cancel the protocol. To apply R4, the TTP follows these statements:

- if $\exists\, Tx_{(r',i')} \in \overline{MR}$ / $(r' = r)$ or $(r' = r-1$ and $i' < i)$, the TTP will send a canceled token to P_i to maintain fairness for the previous honest requesters.
- if $\forall\, Tx_{(r',i')} \in \overline{MR}$ / $r' < r-1$, then $P_{i'}$ cheated.
- if $\forall\, Tx_{(r',i')} \in \overline{MR}$ / $r' = r-1$ and $i' > i$, then $P_{i'}$ cheated.

Star: $(2N-1)(N-1)$ transmissions are necessary.
In an asynchronous optimistic MPCS protocol with star topology, all participants except P_1 can cancel the protocol: $\overline{PC} = \{P_2, ..., P_N\}$, TTP's R1. To apply R4 the TTP follows the next assertions:

Table 4. Example of asynchronous optimistic MPCS protocol, ring topology, PCS signature scheme and $N = 3$

$r = 1$	
$P_1 \rightarrow P_2$	$PCS_1((C,1), P_2, TTP)$, $PCS_1((C,1), P_3, TTP)$
$P_2 \rightarrow P_3$	$PCS_2((C,1), P_3, TTP)$, $PCS_2((C,1), P_1, TTP)$
	$PCS_1((C,1), P_3, TTP)$
$P_3 \rightarrow P_1$	$PCS_3(C, P_1, TTP)$, $PCS_3(C, P_2, TTP)$
	$PCS_2((C,1), P_1, TTP)$
$r = 2$	
$P_1 \rightarrow P_2$	$PCS_1(C, P_2, TTP)$, $PCS_1(C, P_3, TTP)$
	$PCS_3(C, P_2, TTP)$
$P_2 \rightarrow P_3$	$S - Sig_2$
	$PCS_1(C, P_3, TTP)$
$P_3 \rightarrow P_1$	$S - Sig_3$
	$S - Sig_2$
$r = 3$	
$P_1 \rightarrow P_2$	$S - Sig_1$
	$S - Sig_3$
$P_2 \rightarrow P_3$	$S - Sig_1$

- if $\exists\ Tx_{(r',i')} \in \overline{MR}\ /\ r' \geq r - 1$, the TTP will send a canceled token to P_i to maintain fairness for the previous honest requesters.
- if $\forall\ Tx_{(r',i')} \in \overline{MR}\ /\ r' < r - 1$, then $P_{i'}$ cheated.

Mesh: $N^2(N-1)$ transmissions are necessary.
In the mesh topology, all participants can cancel the protocol: $\overline{PC} = \{P_1, ..., P_N\}$, in TTP's R1. Regarding the detection of cheating users, R4 for a mesh topology states:

- if $\exists\ Tx_{(r',i')} \in \overline{MR}\ /\ r' \geq r - 1$, the TTP will send a canceled token to P_i to maintain fairness for the previous honest requesters.
- if $\forall\ Tx_{(r',i')} \in \overline{MR}\ /\ r' < r - 1$, then $P_{i'}$ cheated.

8 Protocol Comparison

In Table 5 we can compare the efficiency of our MPCS protocol proposals and some of the most relevant presented in the related work (we eluded proposals that have been proved flawed). In Sect. 3 we have given a definition of message and transmission to avoid confusions (a transmission can include several messages).

Within our solutions, the proposals with ring and sequential architecture are the most efficient, requiring only $(N+1)(N-1)$ transmissions. But comparing the

Table 5. Efficiency of asynchronous optimistic MPCS protocols

Protocol	Topology	Transmissions[(1)]	Messages/User[(2)]
★	Ring	$(N+1)(N-1)$	$(N-1)^2+1$ when $P_i = P_1$
			$(N-2)(N-1)+1$ when $P_i \neq P_1$
★	Sequential	$(N+1)(N-1)$	$\lceil (N-1)/2 \rceil (N-1)+1$ when N is odd
			$\lceil (N-1)/2 \rceil (N-1)-i+2$ when N is even
[14]	Sequential/ Mesh	$(\lceil N/2 \rceil +1)N(N-1)$	$(\lceil N/2 \rceil +1)(N-1)+1$
★	Star	$(2N-1)(N-1)$	$(N-2)(N-1)+1$
[3]	Star	$4(N+3)(N-1)$	$(N+2)$
★	Mesh	$N^2(N-1)$	$(N-1)^2+1$
[3]	Mesh	$N(N-1)(N+3)$	$(N+2)$

★ Our proposal.
N, number of participants.
(1) Optimistic case, the TTP does not intervene, and $N-1$ malicious participants assumed.
(2) Number of messages (signatures) generated by each user. $i \in [1..N]$ i^{th} participant

number of messages each user has to generate, we can see that the sequential protocol has advantage, requiring approximately half the messages, which is translated in less computational power needed.

Baum-Waidner *et al.* [3] propose a MPCS protocol that achieves abuse-freeness using *standard* public key cryptography. In their paper, the authors propose a mesh solution and they also explain how to convert it into a star solution. Comparing the efficiency of their solution, we see that our proposals require less transmissions in both cases: ring and mesh architecture. But the number of messages generated by each user is lower in their proposals, due to the use of *standard* public key cryptography.

The proposal from Mukhamedov *et al.* [14] uses an architecture that is a mix of sequential and mesh. Compared with Mukhamedov *et al.* [14], our proposals with ring and sequential architecture require less transmissions. Therefore both should be more efficient than Mukhamedov *et al.*'s. But comparing the number of messages generated by each user, we have that Mukhamedov *et al.*'s proposal requires each user to generates less messages than our solution with ring topology, but more than our sequential proposal. Therefore, requiring less transmissions and less messages, we can affirm that our solution with sequential topology is the most efficient asynchronous optimistic MPCS protocol, setting a lower-bound for the minimum number of transmissions required in MPCS protocols.

With these results we can see that with our new definition of efficiency parameters we are able to compare protocols using different topologies, but only if they use the same cryptographic techniques (signature schemes, encryption schemes, etc.). A step further into measuring the efficiency of MPCS protocols is to find other parameters that would allow us to compare protocols, even when they use different cryptographic techniques, and to have a better notion of the influence of the architecture, e.g., sending messages in a sequence (ring, sequential) is slower than sending them in parallel (star, mesh), but sending messages in parallel (independent threads) requires more resources than sending messages in a sequence. This topic is part of the future work.

9 Conclusions

Using a common approach for the design of asynchronous optimistic MPCS protocols, we have proposed an asynchronous optimistic MPCS protocol for a ring, sequential, star and mesh topology. Each proposal meets the asynchronous optimistic MPCS protocol requirements, including abuse-freeness. Moreover, the number of transmissions required by each protocol is the minimum needed to maintain fairness, therefore we have also defined a new set of lower-bounds, minimum number of transmissions, for each topology.

As future work we plan to extend this work including hybrid topologies (mesh/sequential, etc.) and a study of different efficiency parameters. Our final goal is to use these results to enhance the efficiency of existing MPCS protocols, and establish a common framework for the evaluation of MPCS topologies.

Acknowledgements. This work is the result of a visiting Ph.D. Fellowship done by Gerard Draper-Gil at the Institute for Infocomm Research (I2R). Jianying Zhou's work is partially supported by A*STAR funded project SPARK-1224104047.

References

1. Asokan, N., Shoup, V., Waidner, M.: Asynchronous protocols for optimistic fair exchange. In: IEEE Symposium on Security and Privacy, S&P 1998, pp. 86–99. IEEE Computer Society, Los Alamitos (1998)
2. Baum-Waidner, B.: Optimistic asynchronous multi-party contract signing with reduced number of rounds. In: Orejas, F., Spirakis, P.G., van Leeuwen, J. (eds.) ICALP 2001. LNCS, vol. 2076, pp. 898–911. Springer, Heidelberg (2001)
3. Baum-Waidner, B., Waidner, M.: Round-optimal and abuse-free optimistic multi-party contract signing. In: Welzl, E., Montanari, U., Rolim, J.D.P. (eds.) ICALP 2000. LNCS, vol. 1853, pp. 524–535. Springer, Heidelberg (2000)
4. Chadha, R., Kramer, S., Scedrov, A.: Formal analysis of multi-party contract signing. In: Proceedings of the 17th IEEE workshop on Computer Security Foundations, 28–30 June 2004, pp. 266–279. IEEE Computer Society, Washington, DC, Pacific Grove (2004)
5. Ferrer-Gomila, J.L., Payeras-Capellà, M.M., Huguet-Rotger, L.: Optimality in asynchronous contract signing protocols. In: Katsikas, S.K., López, J., Pernul, G. (eds.) TrustBus 2004. LNCS, vol. 3184, pp. 200–208. Springer, Heidelberg (2004)

6. Ferrer-Gomilla, J.L., Onieva, J.A., Payeras, M., Lopez, J.: Certified electronic mail: properties revisited. Elsevier Comput. Secur. **29**, 167–179 (2010)
7. Garay, J.A., Jakobsson, M., MacKenzie, P.D.: Abuse-free optimistic contract signing. In: Wiener, M. (ed.) CRYPTO 1999. LNCS, vol. 1666, pp. 449–466. Springer, Heidelberg (1999)
8. Garay, J.A., MacKenzie, P.D.: Abuse-free multi-party contract signing. In: Jayanti, P. (ed.) DISC 1999. LNCS, vol. 1693, pp. 151–166. Springer, Heidelberg (1999)
9. Huang, Q., Yang, G., Wong, D.S., Susilo, W.: Ambiguous optimistic fair exchange. In: Pieprzyk, J. (ed.) ASIACRYPT 2008. LNCS, vol. 5350, pp. 74–89. Springer, Heidelberg (2008)
10. Jakobsson, M., Sako, K., Impagliazzo, R.: Designated verifier proofs and their applications. In: Maurer, U.M. (ed.) EUROCRYPT 1996. LNCS, vol. 1070, pp. 143–154. Springer, Heidelberg (1996)
11. Khill, I., Kim, J., Han, I., Ryou, J.: Multi-party fair exchange protocol using ring architecture model. Elsevier Comput. Secur. **20**, 422–439 (2001)
12. Kordy, B., Radomirovic, S.: Constructing optimistic multi-party contract signing protocols. In: Proceedings of the 25th Computer Security Foundations Symposium, CSF 2012, pp. 215–229. IEEE Computer Society, Los Alamitos (2012)
13. Mauw, S., Radomirovic, S., Dashti, M.T.: Minimal message complexity of asynchronous multi-party contract signing. In: Proceedings of the 2009 22nd IEEE Computer Security Foundations Symposium, CSF 2009, pp. 13–25. IEEE Computer Society, Washington, DC (2009)
14. Mukhamedov, A., Ryan, M.D.: Improved multi-party contract signing. In: Dietrich, S., Dhamija, R. (eds.) FC 2007 and USEC 2007. LNCS, vol. 4886, pp. 179–191. Springer, Heidelberg (2007)
15. Mukhamedov, A., Ryan, M.D.: Resolve-impossibility for a contract-signing protocol. In: Proceedings of the 19th IEEE Workshop on Computer Security Foundations, CSFW 2006, pp. 167–176. IEEE Computer Society, Washington, DC (2006)
16. Mukhamedov, A., Ryan, M.D.: Fair multi-party contract signing using private contract signatures. Elsevier Inf. Comput. **206**, 272–290 (2008)
17. Tian, H.: A new strong multiple designated verifiers signature. Int. J. Grid Util. Comput. **3**(1), 1–11 (2012)
18. Zhang, Y., Zhang, C., Pang, J., Mauw, S.: Game-based verification of multi-party contract signing protocols. In: Degano, P., Guttman, J.D. (eds.) FAST 2009. LNCS, vol. 5983, pp. 186–200. Springer, Heidelberg (2010)
19. Zhou, J., Deng, R., Bao, F.: Some remarks on a fair exchange protocol. In: Imai, H., Zheng, Y. (eds.) PKC 2000. LNCS, vol. 1751, pp. 46–57. Springer, Heidelberg (2000)

On the Provable Security of the Dragonfly Protocol

Jean Lancrenon and Marjan Škrobot[(✉)]

Interdisciplinary Centre for Security, Reliability and Trust (SnT), University of Luxembourg, Luxembourg City, Luxembourg
{jean.lancrenon,marjan.skrobot}@uni.lu

Abstract. **Dragonfly** is a password-authenticated key exchange protocol that was proposed by Harkins [11] in 2008. It is currently a candidate for standardization by the Internet Engineering Task Force, and would greatly benefit from a security proof. In this paper, we prove the security of a very close variant of **Dragonfly** in the random oracle model. It shows in particular that **Dragonfly**'s main flows - a kind of Diffie-Hellman variation with a password-derived base - are sound. We employ the standard Bellare et al. [2] security model, which incorporates forward secrecy.

1 Introduction

Authenticated Key Exchange (AKE) is a cryptographic service run between two or more parties over a network with the aim of agreeing on a secret, high-quality, session key to use in higher-level applications (e.g., to create an efficient and secure channel.) One talks of *Password*-Authenticated Key Exchange (PAKE) if the message flows of the protocol itself are authenticated by means of a *low-entropy password* held by each user. The inherent danger in this setup is its vulnerability to *dictionary attacks*, wherein an adversary - either eavesdropping or impersonating a user - tries to correlate protocol messages with password guesses to determine the correct password being used.

PAKE research is very active. New protocol designs are regularly proposed and analyzed, and PAKE itself has been subject to standardization since at least 2002. In 2008 Harkins proposed **Dragonfly** [11]: Specifically tailored for mesh networks, it is up for IETF (Internet Engineering Task Force) standardization [12]. However, proving the security of **Dragonfly** remains open.

This paper proves secure a protocol similar to the version of **Dragonfly** up for standardization, in the random oracle (RO) model [4]. Thus we can at least assert that the scheme's main flows - a Diffie-Hellman [10] variant with a password-derived base - are sound. **Dragonfly**'s design is similar to that of Jablon's **SPEKE** [13]. MacKenzie having proved **SPEKE** secure in [16], we followed [16]'s proof to structure ours. However, unlike in [16], we incorporated forward secrecy into the analysis, and chose to work in the Bellare et al. model [2]. To our knowledge, this is the first time a protocol employing a password-derived

J. Lopez and C.J. Mitchell (Eds.): ISC 2015, LNCS 9290, pp. 244–261, 2015.
DOI: 10.1007/978-3-319-23318-5_14

Diffie-Hellman base is proven forward-secure and analyzed using [2]. As in [16], **Dragonfly**'s security is based on the Computational Diffie-Hellman (CDH) and Decisional Inverted-Additive Diffie-Hellman (DIDH) assumptions (see Sect. 2.2).

Related Work. PAKE has been heavily studied in the last decade. It began with the works of Bellovin and Merrit [5] and Jablon [13], but with no precise security analysis. Security models in the vein of [3,19] were then introduced by Bellare et al. [2] and Boyko et al. [6] respectively, and the number of provably secure schemes - with random oracles (RO) or ideal ciphers [2,7], common reference strings [8,14], universal composability [8], to name a few - has exploded. We refer to Pointcheval's survey [18] for a more complete picture. As for **Dragonfly**, it first appeared in [11]. The attention it has received as an IETF proposal has led it to being broken by Clarke and Hao [9], and subsequently fixed.

Organization. The rest of the paper is structured as follows. In Sect. 2, we recall the commonly-used security model of [2]. Section 3 contains a description of the version of **Dragonfly** we analyze. Next, Sect. 4 sketches the security proof, the details being in the appendix. Finally, the paper is concluded in Sect. 5.

2 Security Model

We use the indistinguishability-based framework of [2], designed for two-party PAKE. In what follows, we will assume some familiarity with the model in [2].

2.1 Model

Participants, Passwords and Initialization. Each principal U that can participate in a PAKE protocol P comes from either the $Client$ or $Server$ set, which are finite, disjoint, nonempty sets whose union is the set ID. We assume that each client $C \in Client$ is in possession of a password π_C, while each server $S \in Server$ holds a vector of the passwords of all clients $\pi_S = \langle \pi_S[C] \rangle_{C \in Client}$. Before the execution of a protocol, an initialization phase occurs, in which public parameters are fixed and a secret π_C, drawn uniformly (and independently) at random from a finite set **Passwords** of size N, is generated for each client and given to all servers.

Protocol Execution. The protocol P is a probabilistic algorithm that defines the way principals behave in response to messages from the environment. In the real world, each principal may run multiple executions of P with different partners, and to model this we allow each principal to have an unlimited number of *instances* executing P in parallel. We denote client instances by C^i and server instances by S^j. Each instance maintains local state (i.e. $state_U^i$, sid_U^i, pid_U^i, sk_U^i, acc_U^i, $term_U^i$) and can be used only once. To assess the security of P, we assume that an adversary $\mathcal{A}$ has complete control of the network. Thus, $\mathcal{A}$ provides the inputs to instances, via the following *queries*:

- **Send**(U^i, M): $\mathcal{A}$ sends message M to instance U^i. As a result, U^i processes M according to P, updates its local state, and outputs a reply. A **Send**$(C^i$, **Start**$)$ has client C^i output P's first message. This query models active attacks.
- **Execute**(C^i, S^j): This triggers an honest run of P between C^i and S^j, and its transcript is given to $\mathcal{A}$. It covers passive eavesdropping on protocol flows.
- **Reveal**(U^i): $\mathcal{A}$ receives the current value of the session key sk_U^i. $\mathcal{A}$ may do this only if U^i holds a session key. This captures session key leakage.
- **Test**(U^i): A bit b is flipped. If $b = 1$, $\mathcal{A}$ gets sk_U^i. Otherwise, it receives a random string from the session key space. $\mathcal{A}$ may only make one such query at any time during the execution. This query measures sk_U^i's semantic security.
- **Corrupt**(U): π_U is given to $\mathcal{A}$. This models compromise of the long-term key.[1]

Accepting and Terminating. An instance U^i accepts ($acc_U^i = 1$) if it holds a session key sk_U^i, a session ID sid_U^i and a partner ID pid_U^i. An instance U^i terminates ($term_U^i = 1$) if it will not send nor receive any more messages. U^i may accept and terminate *once*.

Partnering. Instances C^i and S^j are partnered if: (1) $acc_C^i = 1$ and $acc_S^j = 1$; (2) $sid_C^i = sid_S^j \neq \perp$; (3) $pid_C^i = S$ and $pid_S^j = C$; (4) $sk_C^i = sk_S^j$; and (5) no other instance accepts with the same sid.

Freshness. Freshness captures the idea that the adversary should not trivially know the session key being tested. An instance U^i is said to be fresh with forward secrecy if: (1) $acc_U^i = 1$; (2) no **Reveal** query was made to U^i nor to its partner U'^j (if it has one); (3) no **Corrupt**(U') query was made before the **Test** query and a **Send**(U^i, M) query was made at some point, where U' is any participant.

Advantage of the Adversary. Now that we have defined freshness and all the queries available to the adversary $\mathcal{A}$, we can formally define the authenticated key exchange (ake) advantage of $\mathcal{A}$ against P. We say that $\mathcal{A}$ wins and breaks the ake security of P, if upon making a **Test** query to a fresh instance U^i that has terminated, $\mathcal{A}$ outputs a bit b', such that $b' = b$ where b is the bit from the **Test** query. We denote the probability of this event by $\mathbf{Succ}_{\mathrm{P}}^{ake}(\mathcal{A})$. The *ake*-advantage of $\mathcal{A}$ in breaking P is

$$\mathbf{Adv}_{\mathrm{P}}^{ake}(\mathcal{A}) = 2\ \mathbf{Succ}_{\mathrm{P}}^{ake}(\mathcal{A}) - 1. \tag{1}$$

Authentication. Another of $\mathcal{A}$'s goals is violating authentication. In [2], Bellare et al. define three notions of authentication: client-to-server (**c2s**), server-to-client (**s2c**), and mutual (**ma**). We denote by $\mathbf{Succ}_{\mathrm{P}}^{c2s}(\mathcal{A})$ (respectively, $\mathbf{Succ}_{\mathrm{P}}^{s2c}(\mathcal{A})$) the probability that **c2s** (resp., **s2c**) authentication is violated, which happens if some server (resp., client) has terminated before any **Corrupt** query without being partnered with a client (resp., server). The adversary is said to violate mutual authentication if there exists some instance that terminates before any **Corrupt** query without a partner. We denote by $\mathbf{Succ}_{\mathrm{P}}^{ma}(\mathcal{A})$ the probability of this event occurring, and the *ma*-advantage of $\mathcal{A}$ in breaking P is

$$\mathbf{Adv}_{\mathrm{P}}^{ma}(\mathcal{A}) = \mathbf{Succ}_{\mathrm{P}}^{ma}(\mathcal{A}). \tag{2}$$

[1] This is the weak-corruption model of [2].

2.2 Security Assumptions

Here we state the assumptions upon which the security of **Dragonfly** rests. Let $\varepsilon \in [0,1]$, and let $\mathcal{B}$ and $\mathcal{D}$ be probabilistic algorithms running in time t. Let $\mathbb{G}$ be a finite group of prime order q, and g be a generator of $\mathbb{G}$. We say that the assumption holds if there is no (t,ε)-solver for polynomial t (in the security parameter k governing the size of $\mathbb{G}$) and non-negligible ε (also in k).

Computational Diffie-Hellman (CDH). Set $DH_g(g^x, g^y) := g^{xy}$, for any x and y in $\mathbb{Z}_q$. We say that $\mathcal{B}$ is a (t,ε)-CDH solver if

$$\mathbf{Succ}^{cdh}_{g,\mathbb{G}}(\mathcal{B}) := \Pr[\mathcal{B}(g, g^x, g^y) = DH_g(g^x, g^y)] \;\geq\; \varepsilon, \tag{3}$$

where x and y are chosen uniformly at random.

Decisional Inverted-Additive Diffie-Hellman (DIDH). For x and y in $\mathbb{Z}^*_q$, where $x + y \neq 0$, set $IDH_g(g^{1/x}, g^{1/y}) := g^{1/(x+y)}$. An algorithm $\mathcal{D}$ is a (t,ε)-DIDH solver if

$$\mathbf{Adv}^{didh}_{g,\mathbb{G}}(\mathcal{D}) := \mathbf{Succ}^{didh}_{g,\mathbb{G}}(\mathcal{D}) - \frac{1}{2} \;\geq\; \varepsilon, \tag{4}$$

where $\mathbf{Succ}^{didh}_{g,\mathbb{G}}(\mathcal{D}) := \Pr[b' = b]$ in the following game. First, x, y, and z are chosen uniformly at random and a bit b is flipped. Let $X := g^{1/x}$ and $Y := g^{1/y}$. If $b = 0$, set $Z := g^{1/z}$, and if $b = 1$, set $Z := IDH_g(X, Y)$. $\mathcal{D}$ gets as input (g, X, Y, Z), and outputs bit b'.

The DIDH assumption is less-known than the CDH one. It states that it is hard to tell apart $g^{1/(x+y)}$ and a random $g^{1/z}$ when given $g^{1/x}$ and $g^{1/y}$. [16] shows that DIDH is as hard as the Decisional Diffie-Hellman problem in generic groups. For a nice overview of the relations between the DIDH assumption and other discrete-logarithm-style assumptions we refer the reader to [1].

3 The Dragonfly Protocol

We first fix some notation and then describe the version of **Dragonfly** to analyze. Its cryptographic core is a Diffie-Hellman key exchange similar to the one used in **SPEKE** [13], where a function of the password is the base for group values.

Notation. Let $\mathbb{G}$ be a finite multiplicative group of prime order q, and k be the security parameter. When we sample elements from $\mathbb{Z}_q$, it is understood that they are viewed as integers in $[1 \ldots q]$, and all operations on these are performed mod q. Let H_0 be a full-domain hash mapping $\{0,1\}^*$ to $\mathbb{G}$. We also define a hash function H_1 from $\{0,1\}^*$ to $\{0,1\}^{3k}$. $a \leftarrow A$ denotes selecting a uniformly at random from the set A. Let the function $\text{Good}(E, s)$ be true iff: (1) $s \in [1 \ldots q]$ and (2) $E \in \mathbb{G}$. We assume the existence of an efficient algorithm to perform the latter check; this is important, as it prevents instantiation-specific attacks, like the small subgroup attack in [9].

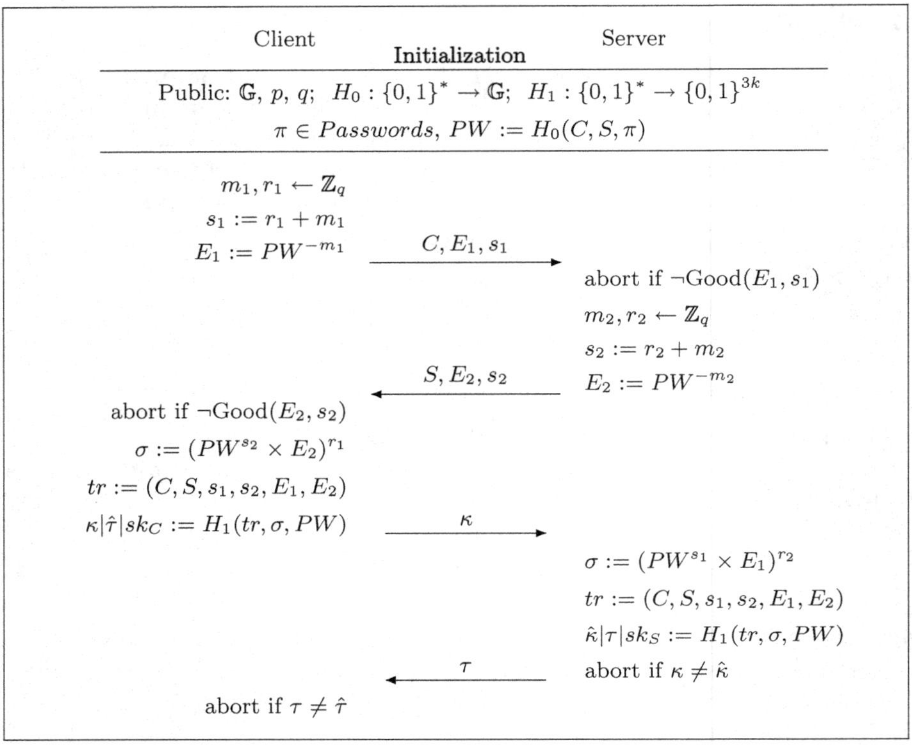

Fig. 1. The Dragonfly protocol.

Protocol Description. A high-level protocol description is shown in Fig. 1. At initialization, the password is chosen at random from **Passwords** and given to the client and server. Then, both parties compute a base $PW = H_0(C, S, \pi)$ for Diffie-Hellman values, where C and S are ID strings.

In nutshell, the protocol runs in two phases. In the first phase, each participant chooses a random exponent r_i and mask m_i, computes their sum $s_i \in \mathbb{Z}_q$ and the group element $E_i := PW^{-m_i}$, and sends the commit message (ID, E_i, s_i), where $i = 1, 2$. Upon receiving this message, Good(E_i, s_i) is called to check its validity. At this point, the session IDs sid_C and sid_S are set to $(C, S, s_1, s_2, E_1, E_2)$ for each participant. In the second phase, both participants derive the Diffie-Hellman value $\sigma = PW^{r_1 r_2}$. This is followed by a computation of a hash value (using the derived σ value), parsed into three k-bit strings: an authenticator for each participant and the session key sk. Then, the authenticators are exchanged. If the received authenticator is valid, the participant accepts and terminates the execution, saving session key sk. Otherwise, it aborts, deleting its state.

Remarks. We point out here the main areas where the presented protocol slightly differs from the IETF proposal. First of all, we do not model the "hunting-and-pecking" procedure explicitly, but this is not a problem here. As pointed out in the proposal, "hunting-and-pecking" is just one way among others to deterministically obtain a base group element from a password. Thus, simply using a random oracle taking as input the participants' identities in addition to the password is appropriate.

The procedure we use to compute the confirmation codes and the session keys is not that of the proposal. In particular, our construction makes all of these direct functions of the shared secret, both identities, the main protocol's message flows, and the password element. This is similar to the PAK and PPK protocols [6], for instance, as well as in MacKenzie's analysis of **SPEKE** [16]. In our view, it is more prudent to follow this pattern, as either removing identities - or replacing them with generic "role" strings - can lead to attacks, e.g. [17] and [1]. Thus, we recommend adding the receiver's identity in the IEFT proposal's computation of the "confirm" message.

Finally, the protocol could have been dropped to three flows, but we chose to keep it four, for two reasons. First, this way we stay close to the IETF protocol, and secondly, despite reducing communication efficiency, four-flow PAKEs - in which the first two flows commit to a shared password and the second two are proofs-of-possession of the session key - are by design secure against many-to-many guessing attacks on the server side, see [15].

4 Security Proof of Dragonfly Protocol

We now present a proof of security for **Dragonfly** in the RO model [4]. We show that **Dragonfly** distributes semantically secure session keys, provides mutual authentication, and enjoys forward secrecy. We also adopt the convenient notations of [7].

Theorem 1. *We consider **Dragonfly** as described in Sect. 3, with a password set of size N. Let $\mathcal{A}$ be an adversary that runs in time at most t, and makes at most n_{se} **Send** queries, n_{ex} **Execute** queries, and n_{h0} and n_{h1} RO queries to H_0 and H_1, respectively. Then there exist two algorithms $\mathcal{B}$ and $\mathcal{D}$ running in time t' such that $\boldsymbol{Adv}^{ake}_{\mathtt{dragonfly}}(\mathcal{A}) \leq T$ and $\boldsymbol{Adv}^{ma}_{\mathtt{dragonfly}}(\mathcal{A}) \leq T$ where*

$$T := \frac{6n_{se}}{N} + \frac{4(n_{se}+n_{ex})(2n_{se}+n_{ex}+n_{h1})}{q^2} + \frac{n_{h0}^2+2n_{h1}}{q} + \frac{n_{h1}^2+2n_{se}}{2^k} + 2n_{h1}(1+{n_{se}}^2) \times Succ^{cdh}_{PW,\mathbb{G}}(\mathcal{B}) + 4n_{h0}^3 \times \left(\boldsymbol{Adv}^{didh}_{g,\mathbb{G}}(\mathcal{D}) + \frac{n_{h1}^3+3n_{se}}{q}\right) \quad (5)$$

and where $t' = O(t + (n_{se}+n_{ex}+n_{ro})t_{exp})$ with t_{exp} being a time required for exponentiation in $\mathbb{G}$.

Proof. Our proof is given as a sequence of games $\mathbf{G}_0, ..., \mathbf{G}_4$. Our goal is to prove that **Dragonfly** resists offline dictionary attacks, i.e. that $\mathcal{A}$'s advantage is proportional to that of the easily detected "dummy" online guesser. We define events, corresponding to $\mathcal{A}$ attacking the protocol in game $\mathbf{G}_m$ and breaking semantic security, and **c2s** and **s2c** authentication, for $m = 0, ..., 4$.

Send queries made to a client instance C^i are answered as follows:

- A **Send**(C^i, **Start**) query is executed according to the following rule:
  - ⋆ **Rule C1**$^{(1)}$
    Choose an ephemeral exponent $r_1 \leftarrow \mathbb{Z}_q$ and a mask $m_1 \leftarrow \mathbb{Z}_q$, compute $s_1 := r_1 + m_1$ and $E_1 := PW^{-m_1}$.

  The client instance C^i then replies to the adversary $\mathcal{A}$ with (C, E_1, s_1) and goes to an expecting state EC_1.
- If the instance C^i is in the expecting state EC_1, a received **Send**($C^i, (S, E_2, s_2)$) query is first parsed and Good(E_2, s_2) is called. If the check passes, the instance continues processing the query according to the following rules:
  - ⋆ **Rule C2**$^{(1)}$
    Compute $\sigma := (PW^{s_2} \times E_2)^{r_1}$.
  - ⋆ **Rule C3**$^{(1)}$
    Compute $\kappa|\hat{\tau}|sk_C := H_1(C, S, s_1, s_2, E_1, E_2, \sigma, PW)$.

  The instance C^i accepts, replies to $\mathcal{A}$ with κ, and goes to an expecting state EC_2. Otherwise, it terminates (rejecting), saving no state.
- In case C^i is in the expecting state EC_2, a **Send**(C^i, τ) query is processed according to the following rule:
  - ⋆ **Rule C4**$^{(1)}$
    Check if $\tau = \hat{\tau}$. If so, the instance terminates, saving sk_C as a state.

  If the equality does not hold, the instance terminates (rejecting), saving no state.

Fig. 2. Simulation of the Send queries to the client.

- $\mathbf{S}_m$ occurs if $\mathcal{A}$ returns b' equal to the bit b chosen in the **Test** query.
- $\mathbf{Auth}_m^{c2s}$ occurs if an S^j terminates saving sk_S as a state without being partnered with some C^i.
- $\mathbf{Auth}_m^{s2c}$ occurs if a C^i terminates saving sk_C as a state without being partnered with some S^j.

Throughout the proof, we call $\mathcal{A}$'s oracle query of the form $H_1(C, S, s_1, s_2, E_1, E_2, \sigma, PW)$ **bad** if $\sigma = DH_{PW}(m, \mu)$ (where $m = PW^{s_1}E_1$ and $\mu = PW^{s_2}E_2$), and Good(E_1, s_1) and Good(E_2, s_2) are true. Also, we denote by π_C the value of the password selected for C and by $PW_{C,S}$ the value of the base derived from it with server S. The number of instances that $\mathcal{A}$ can activate and of hash queries that $\mathcal{A}$ can make are bounded by t. In addition, in case $\mathcal{A}$ does not output b' after time t, b' is chosen randomly. Let us now proceed with a detailed proof.

Game $\mathbf{G}_0$: This game is our starting point, with **Dragonfly** defined as in Fig. 1. $\mathcal{A}$ may make **Send**, **Execute**, **Reveal**, **Corrupt**, and **Test** queries and these queries are simulated as shown in Figs. 2, 3, and 4. From Definition 1 we have

$$\mathbf{Adv}^{ake}_{\mathtt{dragonfly}}(\mathcal{A}) = 2\Pr[\mathbf{S}_0] - 1. \tag{6}$$

Send queries made to a server instance S^j are answered as follows:

- A **Send**$(S^j, (C, E_1, s_1))$ query is first parsed and then Good(E_1, s_1) is called. If both values are valid, the instance continues processing the query according to the following rules:
  - ⋆ **Rule S1**$^{(1)}$
    Choose an ephemeral exponent $r_2 \leftarrow \mathbb{Z}_q$ and a mask $m_2 \leftarrow \mathbb{Z}_q$, compute $s_2 := r_2 + m_2$ and $E_2 := PW^{-m_2}$.

  The server instance S^j then replies to the adversary $\mathcal{A}$ with (S, E_2, s_2) and goes to an expecting state ES_1. Otherwise, it terminates (rejecting), saving no state.
- If the instance S^j is in the expecting state ES_1, a **Send**(S^j, κ) query is executed according to the following rules:
  - ⋆ **Rule S2**$^{(1)}$
    Compute $\sigma := (PW^{s_1} \times E_1)^{r_2}$.
  - ⋆ **Rule S3**$^{(1)}$
    $\hat{\kappa}|\tau|sk_S := H_1(C, S, s_1, s_2, E_1, E_2, \sigma, PW)$.
  - ⋆ **Rule S4**$^{(1)}$
    Check if $\kappa = \hat{\kappa}$. If so, S^j accepts, replies with τ, and terminates while saving sk_S as a state.

  If the equality does not hold, the S^j terminates (rejecting), saving no state.

Fig. 3. Simulation of the Send queries to the server.

Game G$_1$: This is our first simulation, in which hash queries[2] to H_0, H_1 and H_1' are answered by maintaining lists $\mathcal{L}_{h0}, \mathcal{L}_{h1}$ and $\mathcal{L}'_{h1}$, respectively (see Fig. 5). The simulator also maintains a separate list $\mathcal{L}_\mathcal{A}$ of all hash queries asked by $\mathcal{A}$. Note that we assume that the simulator knows the discrete logarithms of the outputs of H_0 queries. The simulator also keeps track of all honestly exchanged protocol messages in the list $\mathcal{L}_\mathcal{P}$. We say that a client instance C^i and a server instance S^j are paired if $((C, E_1, s_1), (S, E_2, s_2)) \in \mathcal{L}_P$. We can easily see that this simulation is perfectly indistinguishable from the attack in $\mathbf{G}_0$. Thus,

$$\Pr[\mathbf{S}_1] = \Pr[\mathbf{S}_0]. \tag{7}$$

Game G$_2$: In this game, collisions on the outputs of H_0 queries and collisions on the partial transcripts $((C, E_1, s_1), (S, E_2, s_2))$ are avoided. Let list $\mathcal{L}_\mathcal{R}$ keep track of the replies generated by client and server instances as answers to **Send** queries. We abort if a pair (E_1, s_1) generated by a client instance is already in the list $\mathcal{L}_\mathcal{R}$ as a result of previous **Send** or **Execute** queries, or in the list $\mathcal{L}_\mathcal{A}$ as an input to an H_1 query. Similarly, we abort in case a pair (E_2, s_2) generated by a server instance is already in $\mathcal{L}_\mathcal{R}$ or $\mathcal{L}_\mathcal{A}$.

[2] The private oracle $H_1' : \{0,1\}^* \rightarrow \{0,1\}^{3k}$ will be used in the later, starting from the $\mathbf{G}_3$.

An **Execute**(C^i, S^j) query is simulated by successively running the honest simulations of **Send** queries. After the completion, the transcript is given to the adversary.

As a result of the **Reveal**(U^i) query, the simulator returns the session key (either sk_C or sk_S) to $\mathcal{A}$, only in case the instance U^i has already computed the key and accepted.

As a result of the **Corrupt**(U) query, if $U \in Client$ the simulator returns the password π_C, and otherwise the vector of passwords $\pi_S = \langle \pi_S[C] \rangle_{C \in Client}$.

As a result of the **Test**(U^i) query, the simulator flips a bit b. If $b = 1$, it returns session key sk_U^i to $\mathcal{A}$. Otherwise, $\mathcal{A}$ receives a random string drawn from $\{0,1\}^k$.

Fig. 4. Simulation of the Execute, Reveal, Corrupt, and Test queries.

$\boldsymbol{H_0}$: For each hash query $H_0(w)$, if the same query was previously asked, the simulator retrieves the record (w, r, α) from the list $\mathcal{L}_{h0}$ and answers with r. Otherwise, the answer r is chosen according to the following rule:

⋆ **Rule** $H_0^{(1)}$
Choose $\alpha \leftarrow \mathbb{Z}_q$. Compute $r := g^\alpha$ and write the record (w, r, α) to $\mathcal{L}_{h0}$.

$\boldsymbol{H_1}$: For each hash query $H_1(w)$ (resp. $H_1'(w)$), if the same query was previously asked, the simulator retrieves the record (w, r) from the list $\mathcal{L}_{h1}$ (resp. $\mathcal{L}_{h1}'$) and answers with r. Otherwise, the answer r is chosen according to the following rule:

⋆ **Rule** $H_1^{(1)}$
Choose $r \leftarrow \{0,1\}^{3k}$, write the record (w, r) in the list $\mathcal{L}_{h1}$ (resp. $\mathcal{L}_{h1}'$), and answer with r.

Fig. 5. Simulation of the hash functions.

⋆ **Rule C1**$^{(2)}$
Choose an ephemeral exponent $r_1 \leftarrow \mathbb{Z}_q$ and a mask $m_1 \leftarrow \mathbb{Z}_q$, compute $s_1 := r_1 + m_1$ and $E_1 := PW^{-m_1}$. If $(E_1, s_1) \in \mathcal{L}_\mathcal{R} \cup \mathcal{L}_\mathcal{A}$, abort the game.

⋆ **Rule S1**$^{(2)}$
Choose an ephemeral exponent $r_2 \leftarrow \mathbb{Z}_q$ and a mask $m_2 \leftarrow \mathbb{Z}_q$, compute $s_2 := r_2 + m_2$ and $E_2 := PW^{-m_2}$. If $(E_2, s_2) \in \mathcal{L}_\mathcal{R} \cup \mathcal{L}_\mathcal{A}$, abort the game.

Additionally, we abort in case of collisions on H_0 outputs. This event's probability is bounded by the birthday paradox $n_{h0}^2/2q$.

⋆ **Rule** $H_0^{(2)}$
Choose $\alpha \leftarrow \mathbb{Z}_q$. Compute $r := g^\alpha$ and write the record (w, r, α) to $\mathcal{L}_{h0}$. If $(*, r, *) \in \mathcal{L}_\mathcal{A}$, abort the game.

The rule modifications in this game ensure the uniqueness of honest instances and that distinct passwords do not map to the same base PW. So we have:

$$|\Pr[S_2] - \Pr[S_1]| \leq \frac{2(n_{se} + n_{ex})(2n_{se} + n_{ex} + n_{h1})}{q^2} + \frac{n_{h0}^2}{2q}. \tag{8}$$

Game $\mathbf{G_3}$: In this game, we first define event **Corrupted** that occurs if the adversary makes a **Corrupt** query while the targeted client and server instance are not paired. From now on, if **Corrupted** is false, instead of using H_1 to compute session keys and authenticators, the simulator uses a private oracle H_1'. The rules change as follows:

- ⋆ **Rule C3$^{(3)}$**
  If **Corrupted** is false, compute $\kappa|\hat{\tau}|sk_C := H_1'(C, S, s_1, s_2, E_1, E_2)$. Otherwise, compute $\kappa|\hat{\tau}|sk_C := H_1(C, S, s_1, s_2, E_1, E_2, \sigma, PW)$.
- ⋆ **Rule S3$^{(3)}$**
  If **Corrupted** is false , compute $\hat{\kappa}|\tau|sk_S := H_1'(C, S, s_1, s_2, E_1, E_2)$. Otherwise, compute $\hat{\kappa}|\tau|sk_C := H_1(C, S, s_1, s_2, E_1, E_2, \sigma, PW)$.

Then, since the shared secret σ and base PW are no longer used in above computations in case the event **Corrupted** is false, we can further modify the following rules:

- ⋆ **Rule C2$^{(3)}$**
  If **Corrupted** is false, do nothing. Otherwise, compute $\sigma := (PW^{s_2} \times E_2)^{r_1}$.
- ⋆ **Rule S2$^{(3)}$**
  If **Corrupted** is false, do nothing. Otherwise, compute $\sigma := (PW^{s_1} \times E_1)^{r_2}$.
- ⋆ **Rule C1$^{(3)}$**
  Choose $\psi_1, s_1 \leftarrow \mathbb{Z}_q$ and compute $E_1 := g^{\psi_1}$. If $(E_1,\ s_1) \in \mathcal{L}_\mathcal{R} \cup \mathcal{L}_\mathcal{A}$, abort.
- ⋆ **Rule S1$^{(3)}$**
  Choose $\psi_2, s_2 \leftarrow \mathbb{Z}_q$ and compute $E_2 := g^{\psi_2}$. If $(E_2,\ s_2) \in \mathcal{L}_\mathcal{R} \cup \mathcal{L}_\mathcal{A}$, abort.

Note that after the above modification the simulator can determine correct and incorrect password guesses and answer perfectly to all queries using the ψ_1, ψ_2, and α values and the lists $\mathcal{L}_{h0}$, $\mathcal{L}_{h1}$, $\mathcal{L}'_{h1}$, $\mathcal{L}_\mathcal{A}$, $\mathcal{L}_\mathcal{P}$, and $\mathcal{L}_\mathcal{R}$. Also, the values s_1, s_2, E_1, E_2 obtained after applying rules $C1^{(3)}$ and $S1^{(3)}$ are identically distributed to those generated in game $\mathbf{G_2}$.

Now that the password-derived base is absent from protocol executions (if **Corrupted** is false[3]), we can dismiss the event that $\mathcal{A}$ has been lucky in guessing the right $PW_{C,S}$ without making the corresponding H_0 query. Hence we abort the simulation if the adversary $\mathcal{A}$ submits a $H_1(C,\ S,\ *,\ *,\ *,\ *,\ *,\ PW_{C,S})$ query without prior $H_0(C, S, \pi_C)$ query. The probability of this event occurring is n_{h1}/q.

- ⋆ **Rule $H_1^{(3)}$**
  If $w = (C, S, *, *, *, *, *, PW_{C,S})$, $((C, S, \pi_C), PW_{C,S}) \notin \mathcal{L}_\mathcal{A}$, and **Corrupted** is false, abort. Otherwise, choose $r \leftarrow \{0,1\}^{3k}$, write the record (w, r) in the list $\mathcal{L}_{h1}$, and answer with r.

[3] Notice that in case **Corrupted** is true, a password-derived base is still used in H_1 computations, hence we can not apply the same argument.

Next, we avoid the cases where the adversary $\mathcal{A}$ may have guessed one of the authenticators (κ or τ) without having made an appropriate H_1 query (when **Corrupted** is false). A "lucky guess" occurs if $\mathcal{A}$ submits a $\mathbf{Send}(S^j, \kappa)$ query with the correct authenticator κ to an unpartnered server instance S^j without previously submitting a **bad** $H_1(C, S, s_1, s_2, E_1, E_2, \sigma, PW_{C,S})$ query. In this case S^j aborts, even though it should have accepted. Similarly, if $\mathcal{A}$ submits a $\mathbf{Send}(C^i, \tau)$ query with the correct τ to an unpartnered C^i without having submitted a **bad** $H_1(C, S, s_1, s_2, E_1, E_2, \sigma, PW_{C,S})$ query, C^i aborts.

⋆ **Rule S4**$^{(3)}$
Check if $\kappa = \hat{\kappa}$. If so, check if $((C, S, s_1, s_2, E_1, E_2, \sigma, PW_{C,S}), \kappa) \notin \mathcal{L}_{\mathcal{A}}$, where $\text{Good}(E_1, s_1)$ is true, $\sigma = DH_{PW_{C,S}}(m, \mu)$, and **Corrupted** is false. If the latter check is true, the server instance S^j aborts. Otherwise, S^j accepts, replies to the adversary with τ, and terminates while saving sk_S as a state.

⋆ **Rule C4**$^{(3)}$
Check if $\tau = \hat{\tau}$. If so, check if $((C, S, s_1, s_2, E_1, E_2, \sigma, PW_{C,S}), \tau) \notin \mathcal{L}_{\mathcal{A}}$, where $\text{Good}(E_2, s_2)$ is true, $\sigma = DH_{PW_{C,S}}(m, \mu)$, and **Corrupted** is false. If the latter check is true, the client instance C^i aborts. Otherwise, C^i terminates, saving sk_C as a state.

Since the authenticators are computed using a private random oracle H_1' (when **Corrupted** is false), we can argue that the adversary can not do better than a random guess per an authentication attempt via **Send** query. Therefore, the probability of "lucky guessing" is bounded by $n_{se}/2^k$.

Without the collisions on the partial transcripts and the "lucky guesses" on the password-derived base and authenticators, one can see that $\mathcal{A}$ has to make the specific combination of H_0 and H_1 hash queries for games $\mathbf{G_2}$ and $\mathbf{G_3}$ to be distinguished. Let $\mathbf{AskH1}_3$ be the event that $\mathcal{A}$ makes the **bad** query $H_1(C, S, s_1, s_2, E_1, E_2, \sigma, PW_{C,S})$ for some transcript $((C, E_1, s_1), (S, E_2, s_2), \kappa, \tau)$, where $H_0(C, S, \pi_C)$ has been already made. Depending on how the transcript is generated, we distinguish between four disjoint sub-cases $\mathbf{AskH1}_3$:

- $\mathbf{AskH1\text{-}Passive}_3$: $((C, E_1, s_1), (S, E_2, s_2), \kappa, \tau)$ comes from an honest execution between C^i and S^j (via an $\mathbf{Execute}(C^i, S^j)$ query);
- $\mathbf{AskH1\text{-}Paired}_3$: $((C, E_1, s_1), (S, E_2, s_2))$ comes from an honest execution between C^i and S^j, while (κ, τ) may come from $\mathcal{A}$;
- $\mathbf{AskH1\text{-}withC}_3$: before any **Corrupt** query, $\mathcal{A}$ interacts with C^i, so (C, E_1, s_1) is generated by C^i, while (S, E_2, s_2) is not from S^j;
- $\mathbf{AskH1\text{-}withS}_3$: before any **Corrupt** query, $\mathcal{A}$ interacts with S^j, so (S, E_2, s_2) is generated by S^j, while (C, E_1, s_1) is not from C^i.

Since session key(s) are computed using the private oracle H_1', the only way $\mathcal{A}$ can break semantic security is via a **Reveal** query to honest instances that generated the same transcript $((C, E_1, s_1), (S, E_2, s_2), \kappa, \tau)$, a case we dismissed in $\mathbf{G_2}$. Thus,

$$\Pr[\mathbf{S}_3] = \frac{1}{2}, \quad |\Pr[\mathbf{S}_3] - \Pr[\mathbf{S}_2]| \leq \frac{n_{h1}}{q} + \frac{n_{se}}{2^k} + \Pr[\mathbf{AskH1}_3]. \tag{9}$$

Similarly - and as already previously mentioned - the authenticators are computed using H_1' as well, and due to $\mathbf{G_2}$, $\mathcal{A}$ cannot reuse authenticators from other instances. Thus,

$$\Pr[\mathbf{Auth}_3^{c2s}] \leq \frac{n_{se}}{2^k}, \quad \Pr[\mathbf{Auth}_3^{s2c}] \leq \frac{n_{se}}{2^k}. \tag{10}$$

Game $\mathbf{G_4}$: In this game, we estimate the probability of the event $\mathbf{AskH1}_3$ occurring and thus conclude the proof. Notice that the probability of **AskH1** occurring does not change between games $\mathbf{G_3}$ and $\mathbf{G_4}$. We also have that

$$\Pr[\mathbf{S}_4] = \Pr[\mathbf{S}_3], \quad \Pr[\mathbf{Auth}_4^{c2s\ (s2c)}] = \Pr[\mathbf{Auth}_3^{c2s\ (resp.,\ s2c)}]. \tag{11}$$

Since all the sub-cases of $\mathbf{AskH1}_4$ are disjoint, we will treat them independently:

The following lemma upper bounds the probability of $\mathbf{AskH1\text{-}Passive}_4$:

Lemma 1. *For any $\mathcal{A}$ running in time t that asks a bad query $H_1(C, S, s_1, s_2, E_1, E_2, \sigma, PW_{C,S})$ for some transcript $((C, E_1, s_1), (S, E_2, s_2), \kappa, \tau)$ that comes from an honest execution between C^i and S^j, there is an algorithm $\mathcal{B}$ running in time $t' = O(t + (n_{se} + n_{ex} + n_{ro})t_{exp})$ that can solve the CDH problem:*

$$\Pr[\mathbf{AskH1\text{-}Passive}_4] \leq n_{h1} \times Succ_{PW,\mathbb{G}}^{cdh}(\mathcal{B}). \tag{12}$$

Proof. Concretely, this shows that under the CDH assumption, an eavesdropping $\mathcal{A}$ has a negligible advantage in discarding a single password. The formal proof is in Appendix A.1. □

The next lemma bounds the chance of $\mathbf{AskH1\text{-}Paired}_4$ occurring:

Lemma 2. *For any $\mathcal{A}$ running in time t that asks a bad query $H_1(C, S, s_1, s_2, E_1, E_2, \sigma, PW_{C,S})$ for some partial transcript $((C, E_1, s_1), (S, E_2, s_2)) \in \mathcal{L}_{\mathcal{P}}$ that comes from an honest execution between C^i and S^j, there is an algorithm $\mathcal{B}$ running in time $t' = O(t + (n_{se} + n_{ex} + n_{ro})t_{exp})$ that can solve the CDH problem:*

$$\Pr[\mathbf{AskH1\text{-}Paired}_4] \leq n_{se}^2 n_{h1} \times Succ_{PW,\mathbb{G}}^{cdh}(\mathcal{B}). \tag{13}$$

Proof. The proof is similar to the previous one, except that the simulator needs to guess the client and server instances whose execution is going to be tested. The reason for this comes from the fact that the private exponents of all the instances would be unknown to the simulator if we applied the same reduction as in the proof of Lemma 1 (see A.1). The problem in the simulation could arise in case the adversary sends the authenticator after making the **Corrupt** query. Therefore, if the simulator makes the right guess, the given random Diffie-Hellman values will be inserted in the instances that are fresh (no **Corrupt** query). The formal proof is in Appendix A.2 . □

Before estimating the probability of $\mathbf{AskH1\text{-}withS}_4$ occurring, we evaluate that of $\mathbf{Coll}_S$, which happens if $\mathcal{A}$ makes two explicit password guesses at the same server instance. Since there are no collisions on H_0 outputs, the only way for $\mathcal{A}$ to accomplish this is if a collision occurs on the first k-bits of two H_1 queries made by $\mathcal{A}$, with $PW_1 \neq PW_2$. The probability of this occurring is bounded by the birthday paradox $n_{h1}^2/2^{k+1}$.

⋆ **Rule** $H_1^{(4)}$
If $w = (C, S, *, *, *, *, *, PW_{C,S})$, $((C, S, \pi_C), PW_{C,S}) \notin \mathcal{L}_\mathcal{A}$, and **Corrupted** is false or if $\mathbf{Coll}_S$ event occurs abort. Otherwise, choose $r \leftarrow \{0,1\}^{3k}$, write the record (w, r) in the list $\mathcal{L}_{h1}$, and answer with r.

Now, without any collision on H_0 and H_1 oracles, each authenticator κ coming from $\mathcal{A}$ via a **Send** query corresponds only to one password π. Therefore,

$$\Pr[\mathbf{AskH1\text{-}withS}_4] \leq \frac{n_{se}}{N}. \tag{14}$$

To bound the probability of $\mathbf{AskH1\text{-}withC}_4$, we first bound the probability of $\mathbf{Coll}_C$, which happens if $\mathcal{A}$ makes three implicit password guesses against the same client instance. The following lemma gives such a bound:

Lemma 3. *For any $\mathcal{A}$ running in time t that asks at least three bad H_1 queries with distinct values of PW for the same transcript $((C, E_1, s_1), (S, E_2, s_2), \kappa, \tau)$, generated in a communication between $\mathcal{A}$ and C^i, there exists an algorithm $\mathcal{D}$ running in time $t' = O(t + (n_{se} + n_{ex} + n_{ro})t_{exp})$ that can solve the DIDH problem:*

$$\Pr[\mathbf{Coll}_C] \leq 2n_{h0}^3 \times \left(\mathbf{Adv}_{g,\mathbb{G}}^{didh}(\mathcal{D}) + \frac{n_{h1}^3 + 3n_{se}}{2q} \right). \tag{15}$$

Proof. This lemma actually shows that the DIDH assumption prevents the adversary from making more than *two* password guesses per online attempt on the client. The proof is in Appendix A.3. □

Now, without any collision on the H_0 and H_1 outputs, $\mathcal{A}$ impersonating the server to an honest client instance can test at most two passwords per impersonation attempt. Therefore,

$$\Pr[\mathbf{AskH1\text{-}withC}_4] \leq \frac{2n_{se}}{N}. \tag{16}$$

Thus,

$$\Pr[\mathbf{AskH1}_4] \leq \frac{3n_{se}}{N} + \frac{n_{h1}^2}{2^{k+1}} + n_{h1}(1 + n_{se}^2) \times Succ_{PW,\mathbb{G}}^{cdh}(t') + \Pr[\mathbf{Coll}_C]. \tag{17}$$

By combining the above equations the bound for semantic security follows. The bound for the mutual authentication is derived in a similar way, by noting that from Definition 2 we have $\mathbf{Adv}_{\texttt{dragonfly}}^{ma}(\mathcal{A}) \leq \Pr[\mathbf{Auth}_0^{c2s}] + \Pr[\mathbf{Auth}_0^{s2c}]$. □

5 Conclusion

In this paper, using techniques similar to those MacKenzie used to study **SPEKE** in [16], we proved the security of a slight variant of **Dragonfly**, which gives some evidence that the IETF proposal of **Dragonfly** is sound. Furthermore, unlike the analysis of [16], we also include forward secrecy. (It is highly probable that **SPEKE** is forward secure as well.) Note that Theorem 1's statements indicate that the adversary may successfully guess up to six passwords per send query. Using a much more complex analysis, most likely we could replace the constant 6 in the non-negligible term by 2, and count per instance rather than per send query, but this would be at the cost of readability of the already intricate proof. Also, by virtue of the contents of Lemma 3, 2 is certainly the best we could do with this particular analysis.

It would also be nice to see if the proof can be made tighter. In particular, while it helps readability, the technique of using private oracles as in [7] seems less fine-grained than the systematic "backpatching" of, e.g. [17]. Finally, it would be interesting to see if the security of **Dragonfly** (and **SPEKE**) could be based on an assumption other than DIDH.

Acknowledgments. We thank the anonymous reviewers for their helpful comments. This work was partially supported by project SEQUOIA, a joint project between the *Fonds National de la Recherche, Luxembourg* and the *Agence Nationale de la Recherche* (France).

A Proofs of Lemma 1, 2 and 3

A.1 Proof of Lemma 1

Proof. We construct an algorithm $\mathcal{B}$ that, for given random Diffie-Hellman values $\langle X, Y \rangle$ such that $X \leftarrow g^x$ and $Y \leftarrow g^y$, attempts to break the CDH assumption (i.e. computes Z such that $Z = DH_g(X, Y)$) by running the adversary $\mathcal{A}$ as a subroutine. The algorithm $\mathcal{B}$ simulates the protocol for $\mathcal{A}$ with the modification of the rules **C1** and **S1** in case an **Execute**(C^i, S^j) query was made:

- ⋆ **Rule** $\mathbf{C1}^{(4)}_{exe}$
  Choose $\psi_1, s_1 \leftarrow \mathbb{Z}_q$ and compute $E_1 := Xg^{\psi_1}$. If $(E_1, s_1) \in \mathcal{L_R} \cup \mathcal{L_A}$, abort the game.
- ⋆ **Rule** $\mathbf{S1}^{(4)}_{exe}$
  Choose $\psi_2, s_2 \leftarrow \mathbb{Z}_q$ and compute $E_2 := Yg^{\psi_2}$. If $(E_2, s_2) \in \mathcal{L_R} \cup \mathcal{L_A}$, abort the game.

After the game ends, for every $H_1(C, S, s_1, s_2, E_1, E_2, \sigma, PW_{C,S})$ query the adversary $\mathcal{A}$ makes, where the values s_1, s_2, E_1 and E_2 were generated by honest client and server instances (after an **Execute**(C^i, S^j) query), the password-derived base is correct, and the corresponding $H_0(C, S, \pi_C)$ query was made,

$$(\sigma Y^{-\frac{\psi_1}{\alpha}} X^{-\frac{\psi_2}{\alpha}} E_2{}^{-s_1} E_1{}^{-s_2} g^{-\frac{\psi_1 \psi_2}{\alpha}} g^{-s_1 s_2 \alpha})^{\alpha} \tag{18}$$

is added to the list $\mathcal{L}_Z$ of possible values for $Z = DH_g(X, Y)$. Equation 18 follows from the fact that a base $PW := g^{\alpha}$ is generated in such a way that the discrete logarithm α is known. Thus, the Diffie-Hellman values X and Y can be represented as $PW^{\frac{x}{\alpha}}$ and $PW^{\frac{y}{\alpha}}$, respectively. So we have:

$$\begin{aligned}\sigma &= DH_{PW}(E_2 PW^{s_2}, E_1 PW^{s_1}) \\ &= DH_{PW}(PW^{\frac{y+\psi_2}{\alpha}} PW^{s_2}, PW^{\frac{x+\psi_1}{\alpha}} PW^{s_1}) \\ &= Z^{\frac{1}{\alpha}} Y^{\frac{\psi_1}{\alpha}} X^{\frac{\psi_2}{\alpha}} E_2{}^{s_1} E_1{}^{s_2} g^{\frac{\psi_1\psi_2}{\alpha}} g^{s_1 s_2 \alpha}.\end{aligned} \tag{19}$$

From the adversary's view, the simulation $\mathcal{B}$ runs is indistinguishable from the protocol in the game $\mathbf{G_3}$ up to the point **AskH1-Passive**$_4$ occurs, and in that case, the correct $DH_g(X, Y)$ value is added to the list $\mathcal{L}_Z$ of size at most n_{h1}. The running time of $\mathcal{B}$ is $t' = O(t + (n_{ex} + n_{ro} + n_{se})t_{exp})$. Thus, Lemma 1 follows from the fact that the probability of $\mathcal{B}$ breaking CDH assumption is at least $\Pr[\textbf{AskH1-Passive}_4]/n_{h1}$. □

A.2 Proof of Lemma 2

Proof. We construct an algorithm $\mathcal{B}$ that, for given random Diffie-Hellman values $\langle X, Y\rangle$ such that $X \leftarrow g^x$ and $Y \leftarrow g^y$, attempts to solve the CDH assumption (i.e. computes Z such that $Z = DH_g(X, Y)$) by running the adversary $\mathcal{A}$ as a subroutine. The algorithm $\mathcal{B}$ chooses distinct random indexes $b_1, b_2 \leftarrow \{1, 2, \ldots, n_{se}\}$ and simulates the protocol for $\mathcal{A}$ with the modification of the rule $\mathbf{C1}^{(3)}$ in case of a b_1th **Send**(C^i, **Start**) query and the rule $\mathbf{S1}^{(3)}$ in case of a b_2th **Send**($S^j, (C, E_1, s_1)$) query:

⋆ **Rule** $\mathbf{C1}^{(4)}$
For the b_1th query choose $s_1 \leftarrow \mathbb{Z}_q$ and set $E_1 := X$. Otherwise, choose $\psi_1, s_1 \leftarrow \mathbb{Z}_q$ and compute $E_1 := g^{\psi_1}$. In any case, if $(E_1, s_1) \in \mathcal{L}_{\mathcal{R}} \cup \mathcal{L}_{\mathcal{A}}$, abort the game.

⋆ **Rule** $\mathbf{S1}^{(4)}$
For the b_2th query choose $s_2 \leftarrow \mathbb{Z}_q$ and set $E_2 := Y$. Otherwise, choose $\psi_2, s_2 \leftarrow \mathbb{Z}_q$ and compute $E_2 := g^{\psi_2}$. In any case, if $(E_2, s_2) \in \mathcal{L}_{\mathcal{R}} \cup \mathcal{L}_{\mathcal{A}}$, abort the game.

After the game ends, for every $H_1(C, S, s_1, s_2, E_1, E_2, \sigma, PW_{C,S})$ query the adversary $\mathcal{A}$ makes, where pairs (E_1, s_1) and (E_2, s_2) were generated after b_1th and b_2th **Send** query, the password-derived base is correct, and the corresponding $H_0(C, S, \pi_C)$ query was made,

$$(\sigma Y^{-s_1} X^{-s_2} g^{-s_1 s_2 \alpha})^{\alpha} \tag{20}$$

is added to the list $\mathcal{L}_Z$ of possible values for $DH_g(X, Y)$ of size at most n_{h1}.

From the adversary's view, the simulation $\mathcal{B}$ runs is indistinguishable until the adversary triggers **AskH1-Paired**$_4$. The probability that $\mathcal{B}$ will guess the correct client instance, the correct server instance, and the correct Z value from $\mathcal{L}_Z$ is at least $1/(n_{se}^2 n_{h1})$. The running time of $\mathcal{B}$ is $t' = O(t + (n_{ex} + n_{ro} + n_{se})t_{exp})$. Thus, the Lemma 2 follows from the fact that the probability of $\mathcal{B}$ solving the CDH assumption is at least $\Pr[\textbf{AskH1-Paired}_4]/(n_{se}^2 n_{h1})$. □

A.3 Proof of Lemma 3

Proof. We construct an algorithm $\mathcal{D}$ that given a triple $\langle X, Y, Z\rangle$ as input, where $X \leftarrow g^{1/x}$, $Y \leftarrow g^{1/y}$ and $Z \in \mathbb{G}$, attempts to break the DIDH assumption (i.e. determine whether Z is random or $Z = IDH_g(X, Y) = g^{1/(x+y)}$) by running the adversary $\mathcal{A}$ as a subroutine. The algorithm $\mathcal{D}$ chooses pair-wise distinct random indexes d_1, d_2, $d_3 \leftarrow \{1, 2, \dots, n_{h0}\}$, chooses random non-zero exponents u_1, u_2, u_3 in $\mathbb{Z}_q$, and simulates the protocol for $\mathcal{A}$ as follows.

The simulation will be running as in the previous game $\mathbf{G_3}$ until the selected H_0 queries d_1, d_2, or d_3 are made. The simulator will abort the game if the inputs to three selected H_0 queries do *not* satisfy following conditions : (a) the passwords π_1, π_2, and π_3 are pair-wise distinct and different from the correct password π_C; and (b) the strings (C, S) in all three queries are the same.

If the selected H_0 queries are valid, $\langle X, Y, Z\rangle$ values will be plugged in according to the following rules:

$\star$ **Rule** $H_0^{(4)}$
For the d_1th query set $r := X^{u_1}$. For the d_2th query set $r := Y^{u_2}$. For the d_3th query set $r := Z^{u_3}$. For all three selected queries set $\alpha = \perp$. Otherwise, choose $\alpha \leftarrow \mathbb{Z}_q$ and compute $r := g^{\alpha}$. In any case, write the record (w, r, α) to $\mathcal{L}_{h0}$. If $(*, r, *) \in \mathcal{L}_{\mathcal{A}}$, abort the game.

The prerequisites for the $\mathbf{Coll}_C$ event to occur are: (1) valid d_1th, d_2th, or d_3th H_0 queries are selected by the simulator; (2) a pair (E_1, s_1) is generated by an honest client instance after a **Send**$(C^i$, **Start**$)$ query; (3) the adversary generated a pair (E_2, s_2) and made a **Send**$(C^i, (S, E_2, s_2))$ query, where Good(E_2, s_2) is true and $E_2 \notin \{X, Y, Z\}$; (4a) for each PW_i, received from the selected H_0 queries, at least one bad $H_1(C, S, s_1, s_2, E_1, E_2, \sigma_i, PW_i)$ query is made for the same transcript, where $i \in \{1, 2, 3\}$. (4b) $\sigma_i \neq 1$.

After the game ends, for every $H_1(C,\ S,\ s_1,\ s_2,\ E_1,\ E_2,\ \sigma,\ PW_i)$ query $\mathcal{A}$ made, where PW_i is equal to any of the plugged values $\{X^{u_1}, Y^{u_2}, Z^{u_3}\}$, a pair

$$\left(E_2, (\sigma^{\frac{u_i}{\psi_1}} E_2^{\frac{-u_i s_1}{\psi_1}} PW_i^{\frac{-u_i s_1 s_2}{\psi_1}} g^{-u_i s_2})\right) \tag{21}$$

is added to the list $\mathcal{L}^i_{bad}$.

So, in the case of an $H_1(C,\ S,\ s_1,\ s_2,\ E_1,\ E_2,\ \sigma,\ X^{u_1})$ query, by stripping away known values from σ, we may identify a guess at E_2^x and place it in the list $\mathcal{L}^1_{bad}$ together with the E_2 value. Remember that the client instance uses rule $\mathbf{C1}^{(4)}$ to compute E_1, which can be represented with $X^{u_1 \psi_1 x}$. In order to extract the $F_1 = E_2^x$ value we do as follows. Since

$$\begin{aligned} \sigma &= DH_{X^{u_1}}(E_2 X^{u_1 s_2}, E_1 X^{u_1 s_1}) \\ &= E_2^{s_1} X^{u_1 s_1 s_2} E_2^{\frac{\psi_1 x}{u_1}} g^{s_2 \psi_1}, \end{aligned} \tag{22}$$

we get

$$E_2^x = \sigma^{\frac{u_1}{\psi_1}} E_2^{\frac{-u_1 s_1}{\psi_1}} PW^{\frac{-u_1 s_1 s_2}{\psi_1}} g^{-u_1 s_2}. \tag{23}$$

The same goes for H_1 queries where the values Y and Z are plugged, in which case the corresponding F_2 and F_3 are computed, respectively. At the end of the simulation, $\mathcal{D}$ checks if for any E_2 value there exist pairs $(E_2, F_1) \in \mathcal{L}^1_{bad}$, $(E_2, F_2) \in \mathcal{L}^2_{bad}$ and $(E_2, F_3) \in \mathcal{L}^3_{bad}$, such that $F_1 F_2 = F_3$. If there exist three such pairs, then $\mathcal{D}$ will output $b' = 1$, and otherwise $b' = 0$.

Now let us analyze the probability that $\mathcal{D}$ returns a correct answer. Suppose first that Z is random. The algorithm $\mathcal{D}$ will return a wrong answer if by chance equation $F_1 F_2 = F_3$ holds. Since the u_i values are random, the probability of this happening is at most n_{h1}^3/q by the union bound. Now suppose that $Z = IDH_g(X, Y)$. The probability of aborting in case E_2 is equal to X, Y or Z is at most $3n_{se}/q$. If the adversary triggers $\mathbf{Coll}_C$, then $\mathcal{D}$ will correctly answer with $b' = 1$ only in case it correctly guessed d_1, d_2, and d_3 from $\{1, 2, \ldots, n_{h0}\}$, which happens with probability of $1/n_{h0}^3$. Therefore, the probability of $\mathcal{D}$ returning a correct answer is at least

$$\begin{aligned} \Pr[b' = b] &\geq \Pr[b' = 1|b = 1]\Pr[b = 1] + \Pr[b' = 0|b = 0]\Pr[b = 0] \\ &\geq \left(\frac{\Pr[\mathbf{Coll}_C]}{n_{h0}^3} - \frac{3n_{se}}{q}\right)\left(\frac{1}{2}\right) + \left(1 - \frac{n_{h1}^3}{q}\right)\left(\frac{1}{2}\right). \end{aligned} \tag{24}$$

Thus,

$$\Pr[\mathbf{Coll}_C] \leq 2n_{h0}^3 \times \left(\mathbf{Adv}_{g,\mathbb{G}}^{didh}(\mathcal{D}) + \frac{n_{h1}^3 + 3n_{se}}{2q}\right). \tag{25}$$

From the adversary's view, the simulation $\mathcal{D}$ runs is indistinguishable unless $\mathbf{Coll}_C$ event occurs. The probability of this happening is bounded by (25). $\mathcal{D}$'s running time is $t' = O(t + (n_{ex} + n_{ro} + n_{se})t_{exp})$ and thus Lemma 3 follows. □

References

1. Abdalla, M., Benhamouda, F., MacKenzie, P.: Security of the J-PAKE password-authenticated key exchange protocol. In: 2015 IEEE Symposium on Security and Privacy, pp. 6–11 (2015)
2. Bellare, M., Pointcheval, D., Rogaway, P.: Authenticated key exchange secure against dictionary attacks. In: Preneel, B. (ed.) EUROCRYPT 2000. LNCS, vol. 1807, pp. 139–155. Springer, Heidelberg (2000)
3. Bellare, M., Rogaway, P.: Entity authentication and key distribution. In: Stinson, D.R. (ed.) CRYPTO 1993. LNCS, vol. 773, pp. 232–249. Springer, Heidelberg (1994)
4. Bellare, M., Rogaway, P.: Random oracles are practical: a paradigm for designing efficient protocols. In: ACM Conference on Computer and Communications Security, pp. 62–73. ACM Press (1993)
5. Bellovin, S.M., Merritt, M.: Encrypted key exchange: password-based protocols secure against dictionary attacks. In: 1992 IEEE Computer Society Symposium on Research in Security and Privacy, 4–6 May 1992, pp. 72–84 (1992)
6. Boyko, V., MacKenzie, P.D., Patel, S.: Provably secure password-authenticated key exchange using diffie-hellman. In: Preneel, B. (ed.) EUROCRYPT 2000. LNCS, vol. 1807, pp. 156–171. Springer, Heidelberg (2000)

7. Bresson, E., Chevassut, O., Pointcheval, D.: New security results on encrypted key exchange. In: Bao, F., Deng, R., Zhou, J. (eds.) PKC 2004. LNCS, vol. 2947, pp. 145–158. Springer, Heidelberg (2004)
8. Canetti, R., Halevi, S., Katz, J., Lindell, Y., MacKenzie, P.: Universally composable password-based key exchange. In: Cramer, R. (ed.) EUROCRYPT 2005. LNCS, vol. 3494, pp. 404–421. Springer, Heidelberg (2005)
9. Clarke, D., Hao, F.: Cryptanalysis of the Dragonfly Key Exchange Protocol. Cryptology ePrint Archive, Report 2013/058 (2013). http://eprint.iacr.org/
10. Diffie, W., Hellman, M.: New directions in cryptography. IEEE Trans. Inf. Theor. **22**(6), 644–652 (1976)
11. Harkins, D.: Simultaneous authentication of equals: a secure, password-based key exchange for mesh networks. In: Second International Conference on Sensor Technologies and Applications, 2008, SENSORCOMM 2008, pp. 839–844, August 2008
12. Harkins, D.: Dragonfly Key Exchange (2015). https://datatracker.ietf.org/doc/draft-irtf-cfrg-dragonfly/
13. Jablon, D.P.: Strong password-only authenticated key exchange. ACM SIGCOMM Comput. Commun. Rev. **26**(5), 5–26 (1996)
14. Katz, J., Ostrovsky, R., Yung, M.: Efficient password-authenticated key exchange using human-memorable passwords. In: Pfitzmann, B. (ed.) EUROCRYPT 2001. LNCS, vol. 2045, pp. 475–494. Springer, Heidelberg (2001)
15. Kwon, T.: Practical authenticated key agreement using passwords. In: Zhang, K., Zheng, Y. (eds.) ISC 2004. LNCS, vol. 3225, pp. 1–12. Springer, Heidelberg (2004)
16. MacKenzie, P.: On the Security of the SPEKE Password-Authenticated Key Exchange Protocol. Cryptology ePrint Archive, Report 2001/057 (2001). http://eprint.iacr.org/2001/057
17. MacKenzie, P.: The PAK Suite: Protocols for Password-Authenticated Key Exchange. DIMACS Technical report 2002–46, p. 7 (2002)
18. Pointcheval, D.: Password-based authenticated key exchange. In: Fischlin, M., Buchmann, J., Manulis, M. (eds.) PKC 2012. LNCS, vol. 7293, pp. 390–397. Springer, Heidelberg (2012)
19. Shoup, V.: On Formal Models for Secure Key Exchange. Cryptology ePrint Archive, Report 1999/012 (1999). http://eprint.iacr.org/1999/012

Network and Cloud Security

Multipath TCP IDS Evasion and Mitigation

Zeeshan Afzal(✉) and Stefan Lindskog

Karlstad University, Karlstad, Sweden
{zeeshan.afzal,stefan.lindskog}@kau.se

Abstract. The existing network security infrastructure is not ready for future protocols such as Multipath TCP (MPTCP). The outcome is that middleboxes are configured to block such protocols. This paper studies the security risk that arises if future protocols are used over unaware infrastructures. In particular, the practicality and severity of cross-path fragmentation attacks utilizing MPTCP against the signature-matching capability of the Snort intrusion detection system (IDS) is investigated. Results reveal that the attack is realistic and opens the possibility to evade any signature-based IDS. To mitigate the attack, a solution is also proposed in the form of the *MPTCP Linker* tool. The work outlines the importance of MPTCP support in future network security middleboxes.

Keywords: IDS evasion · Multipath transfers · TCP · Snort · Middleboxes

1 Introduction

The single path nature of traditional TCP[1] is arguably its greatest weakness. Traditional TCP implements connections between two sockets (pairs of host IP addresses and port numbers) and this 4-tuple needs to remain constant during the lifetime of the connection. Today, end-hosts are equipped with multiple network interfaces, all of which can have a unique IP address. A traditional TCP connection from/to such a host will be limited to using only one path (defined by the two sockets) at a time. Thus, there is a potential and a need to implement TCP connections between end-hosts that can utilize all possible paths that the hosts provide. Such connections will provide higher *availability* and higher *throughput*, among many other advantages. They can also solve a number of existing problems in the Internet of today [11,29].

Multipath TCP (MPTCP[2]) is an extension to traditional TCP that adds the missing ability and enables the use of multiple paths between hosts. It is carefully designed to work on the Internet of today. It also has a fallback mechanism that allows it to switch to traditional TCP when MPTCP is not feasible. The developers of MPTCP have until now specially focused on the feasibility aspects of the protocol and also ensure that no residual security vulnerabilities

[1] Traditional TCP is the same TCP we know and use today.
[2] Multipath TCP is also referred to as MPTCP.

J. Lopez and C.J. Mitchell (Eds.): ISC 2015, LNCS 9290, pp. 265–282, 2015.
DOI: 10.1007/978-3-319-23318-5_15

exist in the protocol itself. However, the network security impacts of using the new MPTCP protocol on the existing infrastructure are yet to be thoroughly investigated.

1.1 Motivation and Research Questions

Internet hosts a number of middleboxes. These middleboxes are deployed either in the form of dedicated hardware or software-based solutions. Nevertheless, most of them are transparent (implicit) with end-hosts unaware of their existence. These middleboxes are deployed to do more processing than simply forwarding the packets that pass through. More and more enterprises are implementing and deploying middleboxes in the form of load balancers, Network Address Translators (NAT), firewalls and Intrusion Detection and Prevention Systems (IDPS) to optimize performance and enhance network security. The middleboxes that are used to improve security perform intrusion detection and prevention. One widely used technique in such systems is based on pre-defined signatures. The signatures used for detection have been developed over the years by making a number of assumptions about the behavior and pattern of the traffic. As revealed by [23], many of those assumptions may no longer be valid with the advent of new protocols like MPTCP. In fact, MPTCP opens the possibility of intrusion detection system (IDS) evasion, where a sender can fragment the data stream and send the fragments across multiple paths in a way such as to bypass the IDS.

This has left the enterprises with a headache as the middlebox infrastructure is unaware of MPTCP and allowing such traffic to go through might come at the cost of degraded security. An increasing number of enterprises have decided not to take that risk and instead configure their middleboxes to remove the MPTCP option and force the protocol to fall back to traditional TCP. A few years ago, Honda et al. [12] found that 14 percent of the tested paths use middleboxes to eliminate the MPTCP options. This behavior of middleboxes is a stumbling block in the universal deployment of MPTCP. There is therefore a need to investigate whether the concerns related to the use of MPTCP are actually true and how dangerous the potential attacks could be.

IDS evasion using Multipath TCP is possible, as shown in [23], but how do the current IDS solutions react under such an intrusion attempt? What is the severity or seriousness of the situation? Is there a solution to the problem? In this paper, we try to answer these questions. We use *Snort* [24] as an example of a popular software-based middlebox and investigate how cross-path data fragmentation using MPTCP affects its detection capability. We do so by generating attack traffic corresponding to the latest Snort rule set using the tools we have developed. The traffic is fragmented and sent across a varying number of paths using MPTCP to a server where Snort is running (loaded with the same rule set) as a middlebox. The number of intrusions detected by Snort is counted and compared to the benchmark results collected using the same set-up but with traditional TCP (or a single subflow). The goal is to establish the extent to which the detection suffers and security degrades as a consequence of cross-path fragmentation. Finally, to mitigate the degraded detection, a novel solution is proposed in the form of the *MPTCP Linker* tool.

1.2 Contribution

The key contributions of this paper are as follows: (1) A statistical analysis of the latest Snort rule set (snapshot 2970); (2) Development of novel tools to parse Snort rules, generate relevant payloads, fragment equally across available paths and send them to a server using MPTCP; (3) Development and implementation of a research methodology to test the effects of cross-path data fragmentation using MPTCP on Snort's detection engine; and (4) Introduction of a solution in the form of the MPTCP Linker tool that can be used as an extension to Snort to mitigate the cross-path fragmentation attack.

1.3 Paper Structure

The rest of the paper is structured as follows. Section 2 summarizes related work. Section 3 provides some background on MPTCP and Snort, which is relevant for future sections. Section 4 provides a description of the research methodology developed to carry out the work and the working of different tools. Section 5 presents a statistical analysis on the Snort rule set. Section 6 shows the results of the testing and evaluation of Snort. Section 7 describes a solution to mitigate the attack. Section 8 provides an outlook and, finally, Sect. 9 provides concluding remarks.

2 Related Work

Much effort has been put into the feasibility and functionality side of MPTCP. Honda et al. [12] studied the real world feasibility of extending TCP in their work. They used volunteers across 24 countries to test the traversal of unknown TCP options through the middleboxes deployed in different access networks. In total, 142 paths were tested between September 2010 and April 2011. The results showed that 20 of the 142 paths removed or stripped the unknown MPTCP options, while the remaining 122 paths passed the options intact. Since most paths allowed the unknown option, the authors concluded that extending TCP using new options is feasible as long as the new extension has a fallback mechanism. They also outlined that the paths with middleboxes that strip the options can easily be configured to allow the unknown options to pass through, provided that the new options do not introduce a security risk. Lanley [14] also studied the viability of TCP extensions. The work of both Honda et al. and Lanley has influenced the design of MPTCP in its current form.

Some focus has also been placed on the inherent threats in the MPTCP protocol extension. A draft [6] has been proposed to identify potential vulnerabilities in the MPTCP design. The overall goal is to ensure that MPTCP is no worse than traditional TCP in terms of security. A number of potential attacks have been identified and their possible solutions have been proposed. Such solutions, if implemented, can help push MPTCP to become a standard.

However, the network security implications of MPTCP have not been studied very much. The most significant work in this regard was that of Pearce and Thomas [23]. In their work, they investigated the effects of MPTCP on current network security and indicated that the existing security infrastructure is not MPTCP aware. To demonstrate the risk, a tool [28] was developed to show preliminary IDS evasion. Our work will further contribute to this area and highlight the network security implications of MPTCP using novel tools and will propose novel solutions.

3 Background

In this section, we discuss some key concepts and information that will be beneficial to comprehend the future sections. First, multipath transfers are discussed with a focus on MPTCP. Network security implications are then emphasized. Finally, the Snort IDS is presented.

3.1 Multipath Networking

TCP has enjoyed success for decades and will continue to do so. However, there have been an ever increasing number of situations where it falls short. The dependency of a TCP connection on the same pair of IP addresses and port numbers throughout the life of the connection is becoming an issue for a number of use cases and applications. Therefore, researchers attempted to address the issue as early as 1995 in the form of a draft [13]. The draft identified different cases where the above mentioned dependency is harmful and proposed modifying TCP and adding a new Protocol Control Block (PCB) parameter. This parameter will allow the IP addresses to change during the course of a connection. Recent efforts have seen the development of the SCTP protocol [26], which has a great deal of potential but has so far failed to achieve wide-scale deployment. There are two main reasons behind the failure of such efforts. First is their revolutionary nature, which requires changes in the software and sometimes even hardware. Second is the feasibility aspect. Any attempt to introduce an almost completely new protocol on the Internet will most likely fail. Such a protocol will not be able to traverse far across the Internet because most of the networking infrastructure on the Internet assumes that TCP and UDP are the only two transport layer protocols that exist. In addition, the proposed solutions had no fallback mechanism, which made them a failure.

In this regard, the latest efforts in multipath networking have more of an evolutionary nature. The feasibility of the solution has been a paramount consideration because, no matter how good a solution is, it is only going to succeed if it will be feasible to use it on a wide scale. The IETF established a working group called Multipath TCP in 2010. The group was tasked to develop mechanisms that can add the capability of multiple paths to the traditional TCP without requiring any significant modifications to the existing Internet infrastructure. The deployability and usability of the solution were also two key goals. The first draft was put forward by the working group in 2011 in the form of RFC 6824 [9].

Overview of MPTCP. MPTCP is an extension to traditional TCP that enables a TCP connection to operate across multiple paths simultaneously [9]. This brings the support to a number of use cases, which was not possible before. It is designed to run on top of today's Internet infrastructure and has a fallback mechanism that allows it to be backward compatible. Crudely, an MPTCP connection consists of one or more subflows. Each of these subflows is a proper TCP connection but with additional MPTCP options that allow every subflow to be linked to an MPTCP connection. A detailed technical discussion of the protocol is beyond the scope of this work. Hence, we discuss only some key concepts and the operation of MPTCP in the subsequent text.

Implementation. MPTCP is realized using the options field available in the TCP header. IANA has assigned a special TCP option (value 30) to MPTCP. Individual messages use MPTCP option subtypes. MPTCP implementations are already available on a number of operating systems. It is available for Linux [20], BSD [4] and Android [7]. Commercially, Apple has implemented it in iOS7 [5] and OS X Yosemite [22]. In our work, the Linux kernel implementation [20] of MPTCP is used.

Initiating an MPTCP Connection. An MPTCP connection uses the same three-way connection establishment handshake as the traditional TCP but with an *MP_CAPABLE* option attached to all the exchanged messages. This option serves two purposes. First, it announces to the remote host that the sender supports MPTCP. Second, it carries additional information, e.g. random keys, which can be used in forthcoming exchanges. Figure 1 shows the required interaction between a multipath capable client and server to successfully complete the MPTCP handshake. This initial handshake is also called the *MP_CAPABLE* handshake.

Addition of a New Subflow. Additional subflows can be added to an established MPTCP connection. This is achieved in the same way as initiating a new

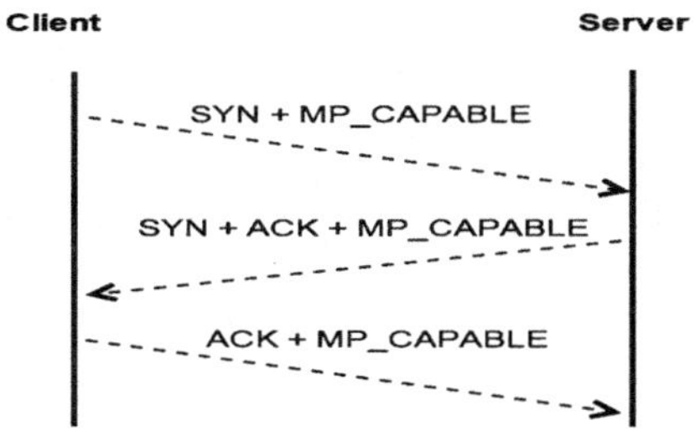

Fig. 1. MP_CAPABLE handshake.

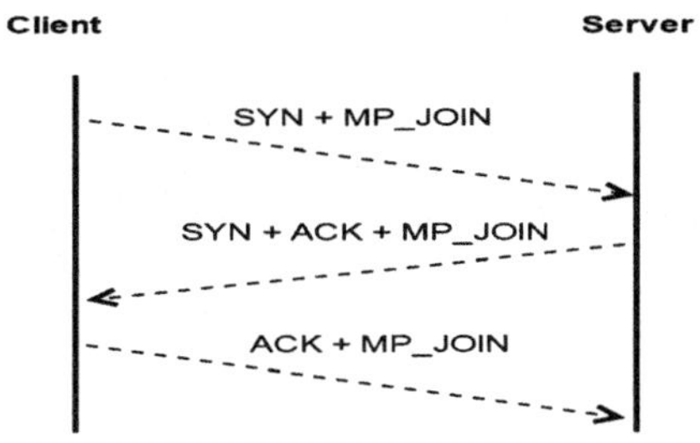

Fig. 2. MP_JOIN handshake.

MPTCP connection but instead making use of the *MP_JOIN* option. This option uses the keys exchanged in the *MP_CAPABLE* handshake to tell the remote end that the connection request is not for a new connection but relates to an existing one. Figure 2 shows the handshake involved. This handshake is also known as *MP_JOIN* handshake. New subflows can be added or removed at any time during the lifetime of a connection. For further details, see [9].

Data Transfer using MPTCP. MPTCP ensures reliable and in-order delivery of the data across all subflows of an MPTCP connection using a Data Sequence Number (DSN). Every subflow has its own transmission window (sequence number space), and the DSS option of MPTCP is used to map the subflow sequence space to the overall MPTCP connection space. This enables data to be retransmitted on different subflows in the event of failure. On the receiver side, MPTCP uses a single receive window across all subflows.

The important thing to note is that the proposed MPTCP standard leaves the exact routing or scheduling of traffic among the subflows up to the implementation. In a common use case where a higher throughput is desired, all available paths (subflows) can be used simultaneously [9]. A sender of the data can tell the receiver how the data are routed among the subflows using the DSS option. The receiver uses this information to re-order the data received over different subflows before passing them on to the application layer in the correct order. Thus, MPTCP enables the sender to choose how to split the input data among the available subflows.

3.2 Network Security Reflections

MPTCP has a number of network security implications. It affects the expectations of other entities in the environment where the protocol extension is used. Network infrastructures can not expect MPTCP to behave in ways similar to those of the traditional TCP.

In this regard, one such observation that is most relevant for this work is *cross-path data fragmentation.* As discussed earlier, MPTCP allows the use of multiple paths simultaneously. A sender can also distribute the data stream among the subflows as it wishes. This opens the possibility to perform *cross-path fragmentation attacks.* A sender can send a known malicious payload by fragmenting it across the subflows in a smart way. The activity will not be detectable by any existing (or non-MPTCP aware) network security middleboxes because, for all they know, every subflow (path) is an independent TCP connection with an unknown fragmented payload.

To further exacerbate the situation, this is just a single problem. The fact that the network paths that are part of the MPTCP connection could be controlled by different Internet Service Providers (ISPs) implies that there may not be any single point in the network that can be used to observe the traffic from all paths. This in turn implies that, even if a middlebox is intelligent enough to know that different subflows make up one MPTCP connection, it may not be able to properly aggregate the traffic and inspect, simply because some subflows may not be visible to it.

3.3 Snort

In the open source world, Snort [24] is the de facto standard IDPS. Snort is effective and is available under the GNU General Public License [25], making it the most widely deployed solution in the world [27].

Snort Operation. Snort provides protection in two ways. It can provide detailed statistics on traffic that can be used for detection of anomalies. It can also provide pattern-matching, which can be used for signature detection. Snort utilizes a rule-based detection approach to perform signature matching on the contents of traffic and detect a variety of attacks [24]. It can currently analyze packets belonging to the four protocols TCP, UDP, ICMP and IP. The detection engine of Snort is configured based on rules. Rules are used to define per packet tests and actions. Once Snort is running with a set of rules, it analyzes every packet that passes through and checks whether the specification given in any of the rules exactly matches the packet. If it detects a match, then it has the possibility to generate and send real time alerts to the *syslog* facility, a UNIX socket or a CSV formatted alert file. The paper by Roesch [24], who is the founder of Snort, provides more detailed information on it.

Rules. Snort rules are written in a simple and flexible, yet powerful, language. Basically, every Snort rule can be divided into two logical parts, a rule *header* and the rule *options*. The rule *header* defines the action of the rule, the protocol it detects, the pair of source IP address/port number, and the pair of destination IP address/port number. The rule *options* consist of the alert message to display and further information about the processing to perform on the packet as well as which parts of the packet should be inspected. The following is an example of a Snort rule.

```
alert tcp any any -> any any (msg:''Sample alert'';content:''Hello'';)
```

The above rule should trigger on an incoming TCP packet from any source IP address and port going to any destination IP address and port as long as it contains the text "Hello" in its payload.

4 Experimental Methodology

This section will provide information about the experimental methodology that was developed and was used to carry out the testing. Figure 3 shows the overall set-up. The client and server sides, both, have component modules that work together to achieve the task. The client and server sides are described in the forthcoming text.

4.1 Client Side

The client side of the methodology is made up of the latest Snort rule set and three tools, namely the *Rule Analyzer*, the *Rule Parser* and the *MPTCP tool.* We will describe all of these components in the forthcoming text.

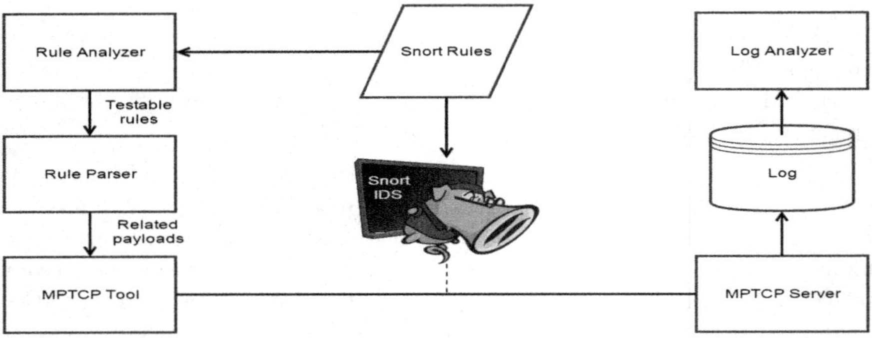

Fig. 3. Experimental setup.

Snort Rules. The latest Snort registered user rule snapshot (2970), available at [27], is used in this work. All the rule files inside the folder (except three[3]) are passed on to the Rule Analyzer.

Rule Analyzer. *Rule Analyzer* performs the following tasks. It reads all input files one by one, analyzes every single rule and then outputs statistics. The statistics provide a distribution of rules by protocol, keyword and other parameters as requested by the user. The Rule Analyzer is used to perform the statistical analysis on the rules.

Rule Parser. A mechanism to parse every Snort rule and craft a consistent payload according to the details in the rule was required. A literature review uncovered tools such as Stick [10], PCP [21] and Mucus [18]. Further investigation found two main drawbacks to these related approaches. First, they are stateless, which means that they can only handle rules relating to stateless protocols such as ICMP and UDP. TCP, on the other hand, maintains states, and any tool not supporting it will be limited in its coverage. Secondly, they are at least ten years old and are thus not compatible with the rule syntax of modern IDSs. A more recent attempt has been the rule2alert [16] tool which, although it is a great improvement over other approaches, still lacks some required functionality. For this work, a further enhancement of [16] was conducted. The tool developed has been presented at [3]. It is able to translate a large number of Snort rules into corresponding payloads. Such payloads can later be encapsulated in packets and used to test the detection accuracy of Snort. This is due to the facts that the payloads are generated directly from the details in the rule and that Snort uses the same rules for detection.

The tool developed searches for the *content* keyword in the *option* field of every rule. The keyword *content* is used to define the signature that Snort should search for in a packet payload. The signature can be a text sequence, binary

[3] Among these three files, one contains old and deleted rules, one is for local rules and the third is for obsolete X windows rules. All of these were deemed irrelevant.

data or even a combination of text and binary data. If Snort detects data that exactly match that given using the *content* keyword in a packet payload, the test is successful and the remainder of rule *options* are checked. The tool uses regular expressions to find the signatures mentioned in every rule (signatures follow the *content* keyword, e.g. *content:"Hello"*) and extracts them. A payload is crafted for each rule and sent to the MPTCP tool, which will be described later.

Snort also supports a number of additional modifier keywords, e.g. *dsize*, *offset*, *distance* and *within* in the rule *options* field. These keywords modify the semantics of the *content* keyword. As an example, the *offset* keyword tells Snort that, instead of starting pattern matching for a signature at the beginning of the payload in a packet, Snort should actually start pattern matching from the given offset value. Hence, if the modifier *keywords* are ignored when crafting the payloads, then it is implausible to expect Snort to generate alerts. The *Rule Parser* tool developed is smart enough to craft the payload by taking modifier *keywords* (if any) into account.

Limitation. It is not currently possible to test all Snort rules. There are a number of keywords that are more complex and require extra effort to build the corresponding payload for. That is why Snort rules are classified into *testable* and *untestable* categories when the statistical analysis of the rules is conducted in Sect. 5.

MPTCP Tool. This is the core tool responsible for generating MPTCP packets with the given payloads and sending them to the destination server. It runs in a virtual machine with Ubuntu as the operating system. The Linux kernel implementation [20] of MPTCP is used, which adds the MPTCP support to the kernel's networking subtree. The tool is inspired by the software [28] released by Pearce et al. [23] with a few improvements.

The tool implements an ad hoc MPTCP scheduler as in [28]. The main criterion used by the tool to fragment the data stream is that the destination should be able to correctly reassemble data in the right order. In that sense, the minimum size of a fragment has to be one byte or two hex digits. The tool attempts to fragment data equally across all available subflows as much as possible. This means that, if a way to split the data equally among all available subflows exists, the tool will do that. However if, for a given set of data, it is not possible to split it equally using all available subflows, the tool will use a subset of available subflows. The tool uses the following formula to calculate packet size for each subflow:

$$pkt\_size = ceiling\left(\frac{length\_of\_data}{available\_subflows}\right)$$

As an example, let us consider a data stream, "netmap" or 0x6e65746d6170 in hex, and two available subflows. The data stream will be fragmented in the following manner. Subflow 1 will be used to send 0x6e6574 and subflow 2 will be used to send 0x6d6170. If the available subflows are three, then subflow 1 will send 0x6e65, subflow 2 send 0x746d and finally subflow 3 send 0x6170. However, if the available subflows are four, there is no way to divide the data stream equally using all subflows. Thus, one subflow (subflow 4) will not be used at all. Regardless of how data are fragmented, it should be noted that the destination always receives data in a way such that they can be reassembled back to create the original data stream of 0x6e65746d6170 or "netmap".

The ad hoc scheduler used might not be representative of the actual intention of MPTCP developers, but there is nothing to stop an attacker from exploiting things in a way in which they were not intended to be used. In fact, splitting data equally among available paths (subflows) might also not be the most effective way from an attacker's point of view. There could be smarter ways of fragmenting data across subflows while still ensuring that the receiver gets the original data stream. We believe that the scheduler implemented is good enough to show the scope of the problem. The improved version of our tool can perform the following tasks:

1. Test the server for MPTCP support.
2. Perform a three-way MPTCP handshake with the MPTCP server.
3. Add the user defined number of subflows to the connection.
4. Send the given payload (can be text, binary or both) to the server using any specific user defined subflow.
5. Send the given payload (can be text, binary or both) to the server by fragmenting it equally across all available subflows (using the ad hoc scheduler).
6. Terminate the MPTCP connection.

4.2 Server Side

The server side of our methodology is made up of the *Snort* IDS, the *MPTCP server*, a log file generated by Snort and the *Log Analyzer* tool. Snort was described in Sect. 3, and the remaining components are discussed below.

MPTCP Server. A virtual remote machine with Ubuntu and the MPTCP kernel implementation [20] is used as the server. A simple off-the-shelf server (*http-server* [19]) is utilized to listen to incoming connections on port 80. The server accepts any incoming MPTCP connections on port 80 and receives the data. It is powerful enough to deal with a high number of simultaneously incoming requests.

Log Analyzer. Manually analyzing the log file generated by an IDS is a hectic task. Log Analyzer aims to automate this process. Snort writes its alerts to a CSV formatted log file in real time. Every value is at a fixed position on each line of the log file and can be extracted. The *Log Analyzer* reads the CSV file, parses and extracts the important features and then displays an output table similar to Table 1. We are thus able to see how many alerts per rule category exist in the log file.

Table 1. Output of the Log Analyzer tool.

Index	Category	Triggered alerts

5 Statistical Analysis of Snort Rules

This section presents the results of a statistical analysis conducted on the latest Snort rules. The motivation for this study was an investigation of how much of the Snort rule set might be affected by the cross-path data fragmentation attack using MPTCP. Thus only the rules relating to the TCP protocol were of interest. The results are based on Snort registered user rules v2.9 (snapshot 2970). The Rule Analyzer described in Sect. 4 is used to perform the analysis.

5.1 Results

Table 2 shows the results when the Rule Analyzer tool processes the Snort rules folder with all the relevant rule files in it. Table 3 shows the breakdown of TCP rules.

Table 2. Distribution of Snort rules v2.9 (excluding deleted rules).

Protocol	TCP	UDP	ICMP	IP	Total
Rules	18577 (84.17 %)	3134 (14.20 %)	156 (0.70 %)	203 (0.91 %)	22070

Table 3. Break down of TCP rules.

TCP	Rules with *content*	Rules with *offset*	Testable	Untestable
18577	18398 (99.03 %)	959 (5.16 %)	9857 (53.06 %)	8720 (46.93 %)

5.2 Trends

To investigate the evolution of Snort rules and any possible trends, the results obtained can be compared to older rule sets. However, due to licensing issues, it is not easy to acquire old Snort rules. An old study conducted on Snort rules v2.4 [17] is nevertheless relevant. In addition, two old rule set versions from October 2000 and June 2001 were found at [1]. The results of the old study on rules v2.4 and the breakdown of rules available at [1], performed by the Rule Analyzer tool that was developed, are shown in Table 4.

On the basis of these results, Fig. 4 shows a comparison chart. The following observations can be made from the comparison: (1) the total number of rules have increased significantly over the years (from 422 in October 2000 to 22070 today; (2) IP rules were non-existent at least until June 2001; (3) ICMP rules have increased but not considerably (from 47 to 156); (4) UDP rules have increased by a factor of 47 (from 66 to 3134); (5) TCP rules have increased rapidly by a factor of 60 (from 309 to 18577); and (6) TCP rules dominate other protocol rules over the years.

Table 4. Distribution of Snort rules v2.4, June'01 and Oct'00.

Rule Set	TCP	UDP	ICMP	IP	Total
v2.4	6494	356	132	39	7021
June 2001	404	86	55	0	545
October 2000	309	66	47	0	422

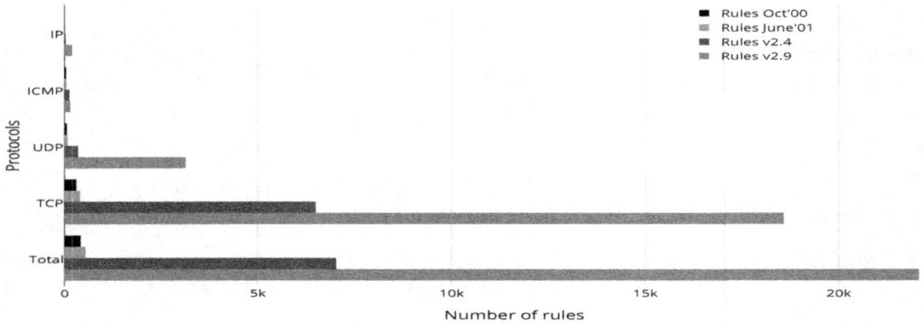

Fig. 4. Bar chart with a comparison of the number of rules in different rule sets.

6 Evaluation of Snort

This section will discuss the outcome of implementing the methodology discussed in Sect. 4. The overall operation is described next, and is followed by the results and a discussion.

6.1 Operation

The experimental methodology (see Fig. 3) discussed in Sect. 4 was put into operation in the following way. First, every rule category[4] file of Snort rules v2.9 is provided to the *Rule Analyzer* tool, which performs the analysis and classifies the rules into different protocols. Since UDP and ICMP rules will not be affected by any change at the transport layer, they are discarded. The remaining TCP rules are further classified into testable and untestable rules by the *Rule Analyzer* depending on whether the *Rule Parser* tool can create an accurate payload for them. Next, the testable TCP rules of every category are used as input to the *Rule Parser* tool to create a unique payload for each rule. These payloads are passed on to the *MPTCP tool*, which acts as a client. It establishes an MPTCP connection with the server, adds additional subflows to the connection and sends the payload across by equally fragmenting it among the available subflows. Every payload is tested using up to five subflows, starting from one subflow (which is equivalent to traditional TCP).

[4] According to the new Snort rule categories.

On the server side, a software-based middlebox Snort IDS runs with the same rules as were used to create the payloads. Every packet received on the server is logged and analyzed in detail by Snort before its being passed on to the Multipath TCP capable TCP/IP stack. Ideally, Snort should generate an alert for every rule due to the fact that all the incoming packets are specially crafted to trigger an alert. Snort logs its alerts in an alert file that is inspected by the *Log Analyzer* to extract the results discussed in the next subsection.

6.2 Results

Table 5 shows the test results for all Snort rule categories. For every category, an evaluation is conducted five times by changing the number of subflows and effectively increasing data fragmentation. The results with one subflow can be considered to be the baseline, since an MPTCP connection with one subflow is essentially a traditional TCP connection.

6.3 Discussion

The findings in Table 5 are of considerable significance. The results show that the Rule Parser tool has some inaccuracies. This leads to a lower number of triggered alerts than expected. Nevertheless, the number of alerts generated by Snort should be the same for any number of subflows, whether it be one subflow (equivalent to a traditional TCP session) or five subflows. This is not the case according to the results[5]. It can be observed from the results that, in general, the number of subflows in an MPTCP connection has an inversely proportional relationship with the detection capability of Snort. As the number of subflows increase (so does the data fragmentation), the detection of Snort suffers even more. With the fragmented data, Snort is only able to get partial matches of the signatures it is looking for in the packets. Snort is still able to detect some intrusions, even with five subflows. A deep look at those rules reveals a common characteristic. All such rules that still work search for a very small signature. Therefore, the MPTCP tool was not able to equally fragment the corresponding data stream of those rules using all available subflows.

The rationale behind this degraded detection capability of Snort is the lack of MPTCP awareness. Snort interprets all TCP subflows within an MPTCP connection as independent TCP connections. It has no awareness that multiple subflows could actually be components of the same MPTCP connection. Snort analyzes every TCP subflow in isolation and expects to see all the traffic required for matching a signature in that TCP session. That is not the case with an MPTCP connection of multiple subflows. The intrusion data are present in the overall MPTCP session (fragmented across subflows), and Snort can still not detect it. This behavior was referred to as *single-session bias* in a recent IETF draft [15]. These results confirm the concerns raised by [23]. The lack of MPTCP

[5] About one half of the Snort rule set is evaluated, but similar results are expected from the remaining rules.

Table 5. Results from all categories.

Index	Category	Subflows	Payloads sent	Triggered alerts
1	FILE	1	2064	2018
		2	2064	120
		3	2064	110
		4	2064	98
		5	2064	97
2	PROTOCOL	1	151	123
		2	151	9
		3	151	8
		4	151	8
		5	151	8
3	POLICY	1	364	340
		2	364	31
		3	364	28
		4	364	27
		5	364	25
4	SERVER	1	1753	1475
		2	1753	951
		3	1753	951
		4	1753	935
		5	1753	928
5	BROWSER	1	963	946
		2	963	8
		3	963	5
		4	963	4
		5	963	4
6	MALWARE	1	2539	1959
		2	2539	679
		3	2539	565
		4	2539	576
		5	2539	537
7	OS	1	288	271
		2	288	57
		3	288	52
		4	288	51
		5	288	50
8	INDICATOR	1	339	269
		2	339	70
		3	339	60
		4	339	57
		5	339	60
9	PUA	1	617	527
		2	617	449
		3	617	413
		4	617	351
		5	617	211
10	MISC.	1	779	680
		2	779	362
		3	779	156
		4	779	146
		5	779	154

awareness on the part of middleboxes performing traffic inspection is the primary reason why they are configured to force MPTCP into using traditional TCP instead. Essentially, these middleboxes currently see MPTCP traffic as attack traffic. If they allow it to pass, then the outcome is degraded functionality and degraded security.

7 Proposed Solution

We introduce the *MPTCP Linker*, a tool that captures and analyzes MPTCP packets to link sessions and correlate MPTCP subflows with the goal to mitigate the above discussed cross-path fragmentation attack. Using the MPTCP options, TCP flags and a few tricks, it is able to associate MPTCP subflows along with the data sent on them with the respective MPTCP connections (something Snort can not do). Figure 5 depicts the higher level concept behind the tool and the forthcoming subsections discuss the implementation and the validation of the tool.

7.1 Implementation

MPTCP Linker is implemented as a python script and is available at [2]. It sniffs the chosen network interface card for MPTCP packets and performs its processing using few commodity and open source modules. As output, it can generate TCP based pcap files. A separate pcap file is generated for each MPTCP

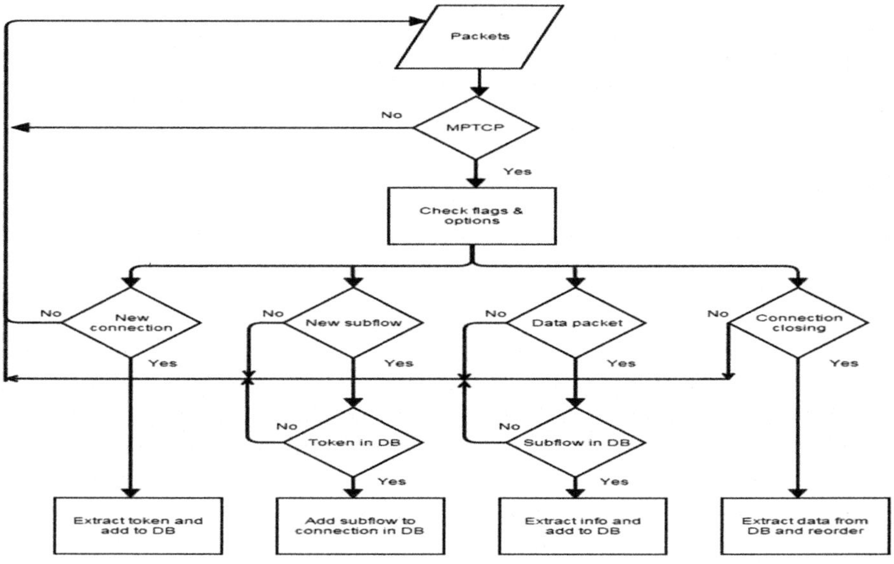

Fig. 5. Flow chart for operation of the MPTCP linker

Table 6. Validation of the MPTCP linker.

Category	Subflows	Payloads Sent	Alerts (before[6])	Alerts (after[7])
PROTOCOL	1	151	123	123
	2	151	9	123
	3	151	8	123
	4	151	8	123
	5	151	8	123

[6] Snort performed detection directly.
[7] Snort performs detection using packets generated by the MPTCP Linker.

session. Every pcap file contains in-order data from all subflows of that MPTCP session.

7.2 Validation

We validate the MPTCP Linker under the same attack traffic that was used for the evaluation of Snort in Sect. 6. Snort has an offline mode where it reads pcap files and detects intrusions from the packets. This offline mode is utilized for the validation. MPTCP Linker runs on the server side and performs its processing on all attack traffic to generate pcap files. The pcap files are then used by Snort in offline mode to perform intrusion detection. Table 6 shows the evaluation results from one category (due to space limitations) of Snort rules.

As can be seen in Table 6, the MPTCP Linker mitigates the earlier discussed cross-path fragmentation attack. The number of intrusions detected by Snort are consistent (123), irrespective of whether one subflow or up to five subflows are used to fragment the data stream. The MPTCP Linker correlates MPTCP subflows, links them to their respective MPTCP sessions and keeps track of the data on those subflows. Once it detects the termination of a session, it extracts data from all subflows of that session and reassembles them to recreate the original data stream in the correct order.

8 Outlook

The ultimate solution to this problem will be the evolution of the network security infrastructure to fully support MPTCP. However, there could be interim solutions that can be employed in the meantime to partially support the protocol and still ensure security. One such solution has been proposed and implemented in this work in the form of the MPTCP Linker [2]. There are also suggestions for developing TCP-MPTCP (and vice versa) proxies or a protocol converter [8]. Such solutions can help in the deployment of MPTCP as well as benefit network security.

9 Concluding Remarks

This paper investigated the reasons why existing network security middleboxes block MPTCP connections. With the help of a systematic experimental methodology, it was established that MPTCP could indeed be used by an attacker to degrade the functionality of existing network security middleboxes. To take a step towards adding MPTCP support and making security middleboxes MPTCP aware, a solution was also proposed and evaluated. The solution has been released under an open-access license for the benefit of the whole community. It also merits mentioning that only one of potentially many issues has been investigated in this work. Other security issues that arise with the advent of MPTCP, particularly the common scenario where only partial traffic passes through security middleboxes, also need to be explored and resolved.

Acknowledgments. The work was carried out in the High Quality Networked Services in a Mobile World (HITS) project, funded partly by the Knowledge Foundation of Sweden. The authors are grateful for the support provided by Catherine Pearce of Cisco.

References

1. Advanced Reference Archive of Current Heuristics for NIDS: Arachnids event signatures export for snort (2000–2001). http://www.autoshun.org/downloads/vision.conf, http://www.autoshun.org/downloads/vision18.conf
2. Afzal, Z.: MPTCP-Linker (2015). https://github.com/zafzal/MPTCP-Linker
3. Afzal, Z., Lindskog, S.: Automated testing of IDS rules. In: Proceedings of the 8th International Conference on Software Testing, Verification and Validation Workshops (ICSTW), pp. 1–2. IEEE, April 2015
4. Armitage, G., Williams, N., et al.: FreeBSD kernel patch to enable Multipath TCP (2014). http://caia.swin.edu.au/urp/newtcp/mptcp/tools.html
5. Bonaventure, O.: Apple seems to also believe in Multipath TCP (2013). http://perso.uclouvain.be/olivier.bonaventure/blog/html/2013/09/18/mptcp.html
6. Braun, M.B., Paasch, C., Gont, F., Bonaventure, O., Raiciu, C.: Analysis of MPTCP Residual Threats and Possible Fixes. Internet Draft draft-ietf-mptcp-attacks-02, IETF (2014). https://tools.ietf.org/id/draft-ietf-mptcp-attacks-02.txt
7. Detal, G.: MPTCP-enabled kernel for the nexus 5 (2014). https://github.com/gdetal/mptcp_nexus5
8. Detal, G., Paasch, C., Bonaventure, O.: Multipath in the middle (box). In: Proceedings of the 2013 Workshop on Hot Topics in Middleboxes and Network Function Virtualization, pp. 1–6. ACM (2013)
9. Ford, A., Raiciu, C., Handley, M., Bonaventure, O., et al.: TCP extensions for multipath operation with multiple addresses. Experimental RFC 6824, IETF (2013). https://tools.ietf.org/html/rfc6824
10. Giovanni, C.: Fun with packets: Designing a stick (2002). http://repo.hackerzvoice.net/depot_ouah/dos_ids.html
11. Han, H., Shakkottai, S., Hollot, C., Srikant, R., Towsley, D.: Multi-path TCP: a joint congestion control and routing scheme to exploit path diversity in the internet. IEEE/ACM Trans. Networking **14**(6), 1260–1271 (2006)

12. Honda, M., Nishida, Y., Raiciu, C., Greenhalgh, A., Handley, M., Tokuda, H.: Is it still possible to extend TCP? In: Proceedings of the 11th ACM SIGCOMM Internet Measurement Conference (IMC), pp. 181–194. ACM (2011)
13. Huitema, C.: Multi-homed TCP. Internet Draft draft-huitema-multi-homed-01, IETF (1995). https://tools.ietf.org/html/draft-huitema-multi-homed-01
14. Langley, A.: Probing the viability of TCP extensions (2008). http://www.imperialviolet.org/binary/ecntest.pdf
15. Lopez, E.: Multipath TCP middlebox behavior. Internet Draft draft-lopez-mptcp-middlebox-00, IETF (2014). https://tools.ietf.org/html/draft-lopez-mptcp-middlebox-00
16. Manev, P.: Rule2alert (2014). https://github.com/pevma/rule2alert
17. Münz, G., Weber, N., Carle, G.: Signature detection in sampled packets. In: Proceedings of the Workshop on Monitoring, Attack Detection and Mitigation (MonAM). IEEE (2007)
18. Mutz, D., Vigna, G., Kemmerer, R.: An experience developing an IDS stimulator for the black-box testing of network intrusion detection systems. In: Proceedings of the 19th Annual Computer Security Applications Conference (ACSAC), pp. 374–383. IEEE (2003)
19. Nodejitsu: http-server (2014). https://github.com/nodeapps/http-server
20. Paasch, C., Barré, S., et al.: Multipath TCP implementation (v0.88) in the Linux kernel (2013). http://www.multipath-tcp.org
21. Patton, S., Yurcik, W., Doss, D.: An achilles heel in signature-based IDS: squealing false positives in SNORT. In: Proceedings of 4th International Symposium on Recent Advances in Intrusion Detection (RAID) (2001)
22. Pearce, C.: MPTCP roams free (by default!) (2014). http://labs.neohapsis.com/2014/10/20/mptcp-roams-free-by-default-os-x-yosemite/
23. Pearce, C., Thomas, P.: Multipath TCP: breaking today's networks with tomorrow's protocols. In: Black Hat USA, August 2014
24. Roesch, M., et al.: Snort: lightweight intrusion detection for networks. In: Proceedings of the 13th Conference on Systems Administration, pp. 229–238 (1999)
25. Stallman, R.: GNU General Public License, version 2 (1991)
26. Stewart, R.: Stream control transmission protocol. RFC 4960, IETF (2007). https://tools.ietf.org/html/rfc4960
27. The Snort Team: Snort official website. https://www.snort.org/
28. Thomas, P.: mptcp-abuse (2014). https://github.com/Neohapsis/mptcp-abuse
29. Wischik, D., Handley, M., Braun, M.B.: The resource pooling principle. ACM SIGCOMM Comput. Commun. Rev. **38**(5), 47–52 (2008)

Provenance Based Classification Access Policy System Based on Encrypted Search for Cloud Data Storage

Xinyu Fan(✉), Vijay Varadharajan, and Michael Hitchens

Advanced Cyber Security Research Centre, Department of Computing, Macquarie University, Sydney, NSW 2109, Australia
xinyu.fan@students.mq.edu.au,
{vijay.varadharajan,michael.hitchens}@mq.edu.au

Abstract. Digital provenance, as an important type of cloud data, has aroused increasing attention on improving system performance. Currently, provenance have been employed to provide hints on access control and estimate data quality. However, provenance itself might also be sensitive information. Therefore, provenance might be encrypted to store on cloud. In this paper, we provide a mechanism to classify cloud documents by searching specific keywords from their encrypted provenance, and we proof our scheme achieves semantic security.

Keywords: Security · Provenance · Access policies · Encrypted search

1 Introduction

There has been a growing trend in recent times to store data in the cloud due to a dramatic increase in the amount of digital information. This ranges from consumers' personal data to larger enterprises wanting to back-up databases or storing archival data. Cloud data storage can be particularly attractive for users, individuals or enterprises, with unpredictable storage demands, requiring an inexpensive storage tier or a low-cost, long-term archive. By outsourcing users' data to the cloud, service providers can focus more on the design of functions to improve user experience of their services without worrying about resources to store the growing amount of data. The cloud can also provide on-demand resources for storage, which can help service providers to reduce their maintenance costs. Furthermore, cloud storage provides a flexible and convenient way for users to access their data from anywhere on any device.

An important issue when it comes to management of data, in the cloud as much as any other storage platform, is who can access the data under what circumstances. Typically a cloud service also provides an access control service to manage access to data by different users and applications. With the vast array of data stored in the cloud, and the range of possible access options, the need has grown for improved access control approaches. At base the access

J. Lopez and C.J. Mitchell (Eds.): ISC 2015, LNCS 9290, pp. 283–298, 2015.
DOI: 10.1007/978-3-319-23318-5_16

control question is which system entities can access what data? However, this simple statement does not capture the nuances of data handling. Data in the cloud may be shared between users, may be constructed from contributions from multiple users and gone through numerous processing steps before being accessed by an end user. Access decisions may not rest solely on the end data container, but also on what operations were applied by which users in the process of constructing the data. Such concepts are not new to access control, the idea of access control based on the operations carried out on the data (i.e., history-based approaches [17,22,23]) being well known. The scope and complexity of data in the cloud, however, gives renewed emphasis to the consideration of history-based approaches.

The cloud, as in many other areas of security, adds to this an extra layer of complication. By taking a service-based approach to provisioning, the cloud separates the authority over data from the authority over service provision. Historically, many access control mechanisms were controlled by system authorities that implicitly had complete system access (label-based access control being an exception). This meant that access control mechanisms could be assumed to have free access to all system information, including the data and associated meta-data. This is not true in the cloud.

Users of the cloud may wish to protect their data from the cloud storage providers. This has led to work on access control mechanisms for the cloud that allow the data to be encrypted to protect against storage provider access. Just as users may wish to protect their data from the cloud storage provider, they may wish to protect the history of operations (both which operation and who carried them out) from the cloud storage provider where those are central to the access control decision. In a service-based approach it should not be a requirement to reveal such information to the cloud storage provider, any more than it should be a requirement to reveal the data itself to the data storage provider. However, the data storage provider needs to be able to supply an access control mechanism.

Combining these requirements leads to the need for an access control mechanism that makes decisions based on historical operations on data, and can do so when the data itself and the record of operations are both encrypted.

Historical operations on data can be referred to as the data provenance. A number of proposals for access control based on the provenance of the data have been made [1,2]. Provenance is a well-known concept in the areas of art and archaeology. Provenance concerns not only the origin or creator of data but also the storage and ownership path and what operations have carried out by whom and in what context. As such, representation of data provenance provides information about the ownership as well as actions and modifications which have been performed on the data, describing the process by which it arrives at a particular point in the cloud. In terms of secure decision making, characteristics such as data accuracy, timeliness and the path of transfer of data are important. With increasing regulation, such as the Sarbanes-Oxley Act, the consequences for signing incorrect statements has significantly increased, even if the signer was not directly responsible for the invalid sections. Therefore it is important to track which entities were responsible for the process that led to the final form of the data.

In this paper, we consider the approach where the data carries with it provenance information which can be used to make access control decisions. Provenance information can be represented as a combination of what, when, where, how, who and why. *What* denotes an event that affected data during its lifetime. *When* refers to the time at which the event occurred. *Where* tells the location of the event. *How* depicts the action leading up to the event. *Who* tells about the agent(s) involved in the event, and *Why* represents the reasons for the occurrence of events. Such a generic representation is sufficiently extensible to capture the essence of the semantics of provenance across various domains. If such a representation can be captured in a secure manner, then it will be useful in tackling the issue of attribution of data as it moves around the cloud. For instance, information about the origin of data together with the conditions and the state under which it was created along with the modifications that have been made and the conditions under which these modifications have been made will allow the access control service to more robustly make security decisions. Such an approach would transform the access control service to a more stateful decision and make it more context dependent.

The issue of securing provenance information in a cloud environment poses a variety of challenges. Secure generation and transfer of provenance information and its secure management over a distributed cloud infrastructure pose several challenges. There would be a need to create provenance aware cloud infrastructure with additional trusted authorities for provenance based policy decision making and enforcement. Policy decision authorities are needed to ensure provenance validity and policy compliance of data prior to allowing it into cloud storage. Policy enforcement authorities would collect provenance from various entities in the infrastructure and act as the arbiters of whether to allow reading or writing to cloud storage. There is also a need for provenance databases to store provenance for querying by policy decision authorities. Recently there has been several research that has addressed the issue of provenance based access control (e.g. [1,9]). These research works have used the provenance information in plain format in the design of access control policies. Our focus in this paper is on *encrypted provenance* information and its use in secure access control decision making. Note as the provenance is transferred over the network infrastructure in the cloud, it needs to be secured using encryption techniques to achieve confidentiality (as provenance can contain sensitive information). Similarly the data transferred over the cloud networks is also encrypted. Encrypted provenance associated with encrypted data pose further challenges in the design of provenance based access control, as the policy decision authority checking provenance validity and compliance before storing the data may not have the key to decrypt the provenance information. This paper addresses the issue of encrypted provenance information and its use in access decision making regarding storage of encrypted data in a public cloud.

In particular, our paper makes the following contributions: (1) allowing the policy decision server to check the encrypted provenance without decrypting the provenance, while at the same time (2) providing guarantees to the policy decision server that the provenance is from a genuine source and is linked with the particular data or file. Such a solution will enable authenticated and confidential provenance information to be used in the access control service without revealing its plain content. To achieve such a solution, we introduce a new notion of Encrypted Provenance Search Scheme (EPSS). EPSS is based on the searchable encryption method proposed by Boneh et al. [12].

In cloud storage systems, to provide appropriate management and security protection, files could be stored into separate units. While each unit keeps a category of files. Classifying files as categories is an effective mechanism for organising files and management access to files. In this paper, we focus on classifying files according to provenance which records generating process of files. Specifically, our system identifies and classifies files by their own preferences on what process worked on files. For example, when medical records or governmental survey documents are anonymised, it removes sensitive personal information. Then, these files could be access by the public, students and scholars for the purpose of research. Therefore, it identifies files after anonymization as "public education". On the contrary, if files are combined with judgement or comments with sensitive agents, they might wish to keep secret from access with the public or attackers. Then these files are classifies as "sensitive information" to take higher level protection and deny access from unauthorised users. However, the third party to execute classification might not be full trusted. To prevent internal attack and keep confidentiality of data, we hope keep secret of data information by encryption as well as classifying them. We provide Provenance-based classification system to implement this goal. In this system, we propose a scheme to search keywords from encrypted provenance. When specific keywords are found, files are classified by according system policies.

The organisation of this paper is as follows. Section 2 briefly presents some research works in the areas of provenance and encrypted search that are relevant to our work. Section 3 gives a brief introduction on the representation of provenance and its characteristics. Section 4 presents our Provenance based Classification Access Policy (PBCAP) System. It gives an outline of the system architecture and describes provenance based classification policy. Section 5 propose the Provenance based Classification Scheme. Then it presents semantic-secure game for it and proofs that this scheme is semantic secure. Finally, Sect. 6 concludes the paper and states the future work.

2 Related Work

In recent years, defining provenance models [5,6,13] which are adopted by various systems is a topic of research. [4] provides Open Provenance Model (OPM) which could be presented as a form of directed acyclic graph (DAG). It consists of five main entities as nodes and dependencies as edges and records processes taken on data. The three main entities are "Artifact", "Process" and "Agent" respectively, and each edge represents a causal dependency. OPM meets the requirement allowing provenance information to be exchanged between systems, and it is possible for developers to build and share tools operating on OPM. Currently, it has been employed by some provenance-aware systems. A paper [13] provides another provenance model that presents as a set of provenance records. It defines five kinds of records which are operation records, message records, actor records, preference records, and context records and each records consists of several attributes. However, these attributes are optional in that might be null. To carry contextual information such as time, temporal aspects, user ID and so on, [2] presents an extensional model of OPM by adding attributes data of a transaction to vertex "Action" of each transaction. For instance, these attribute information supports extra access control policies of Dynamic Separation of Duties (DSOD) [14]. And there are other proposals [18–21] focusing on specific application domains such as electronic health data and scientific records.

Data provenance might be sensitive information, then the security of provenance (for example [9–11]) has increasing aroused attention. There are several attempts to encrypt provenance information to keep its confidentiality. Paper [15] proposes a provenance-aware system based on Attribute-based signature (ABS) which supports fine-grained access control policies. The users' privacy is also protected because attribute private key of users is issued with an anonymous key-issuing protocol from multiple attribute authorities. However, the whole computation is built on the assumption that could server has large computational ability. In paper [16], authors proposed a cryptographic design for cloud storage systems supporting dynamic users and provenance data.

And several papers attempt to use data provenance to make access control decisions. Paper [1,8] proposes a basic provenance-based access control model $PBAC_B$, which facilitates additional capabilities beyond those available in traditional access control models. This paper also mentions a family of PBAC models, $PBAC_B$ being the basic model in this family. It defines three criteria for the provenance-based access control family, namely (1) the kind of provenance data in the system, (2) whether policies are based on acting user dependencies and object dependencies, and (3) whether the policies are readily available or need to be retrieved. The three other models $PBAC_U$, $PBAC_A$ and $PBAC_{PR}$ extend one of these three criteria respectively. However, in this paper, the mechanism to list dependency paths manually and search dependency paths from provenance graphs are not unrealistic, because in real cloud systems, the items in dependency path lists might be huge numbers.

[13] proposes an access control language influenced by the XACML language which supports both actor preferences and organisational access control policies. In their system, applicable organisational policies and applicable preferences are evaluated together for a given query. However, there are obviously shortcomings for the evaluation, such as uncertain decisions are inevitable due to the lack of privileges.

In the area of encrypted data search, [12] presents a Public Key Encryption with keyword Search (PKES) scheme. We will be making use of this work in the design of our Provenance based Classification Access scheme. Essentially, the work in [3,12] considers the following scenario: when Alice receives emails, she would like to set a gate that helps her to check whether the incoming emails contain certain sensitive keywords such as "urgent". However, the emails are encrypted to protect privacy. As the gateway is not fully trusted, Alice does not want to grant the gateway the ability to decrypt her emails. The PKES scheme enables the gateway to conduct a test to verify if the encrypted emails contain the keywords while learning nothing else about the content of the emails themselves.

3 Provenance

Though provenance has long played a major role in the context of art and archaeology (in terms of lineage or pedigree), more recently it has become more important for data in various sectors such as finance and medicine. It is not just about the origin or creator of data but also what sort of operations have been done by whom and in what context, especially when it comes to security and privacy. As such, representation of data provenance provides information about the ownership as well as actions and modifications which have been performed on the data. In terms of secure decision making, characteristics such as data accuracy, timeliness and the path of transfer of data are important. For instance, with the Sarbanes-Oxley Act, the consequences for signing incorrect corporate financial statements became contractual. Therefore it is important to keep track of data which contributed to financial reports and to authenticate the people who worked on it.

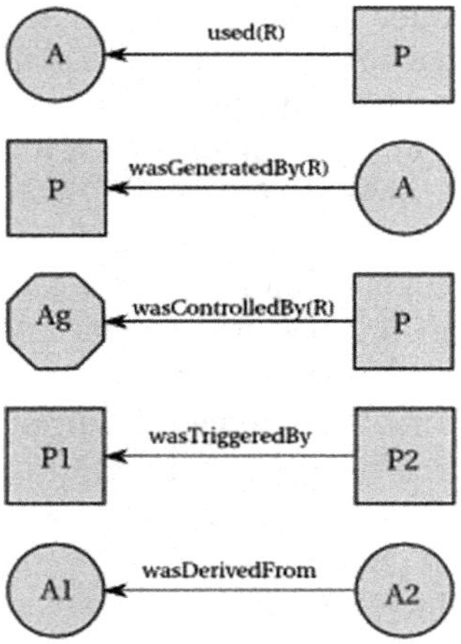

Fig. 1. Dependencies in Provenance Model [4]

In our system, we adopt the representation described in the Open Provenance Model (OPM) [4]. The OPM takes the form of a directed acyclic graph (DAG). It is based on three primary entities, namely *Artifact*, *Process* and *Agent*. Specifically, *Artifact* is a piece of data which may be a physical object or a digital representation stored in computers; *Process* is an action affecting the artifact and creating new artifacts; *Agent* is an actor enabling and controlling the process. These main entities act as nodes in the provenance directed acyclic graph. Dependencies describe the relationships between these entities and connect the nodes in the graph.

Figure 1 shows the five main dependencies. They are "used" (Process used Artifact); "wasGeneratedBy" (Artifact was generated by Process); "wasControlledBy" (Process was controlled by Agent); "wasTriggeredBy" (Process2 was triggered by Process 1); "wasDerivedFrom" (Artifact2 was derived from Artifact1).

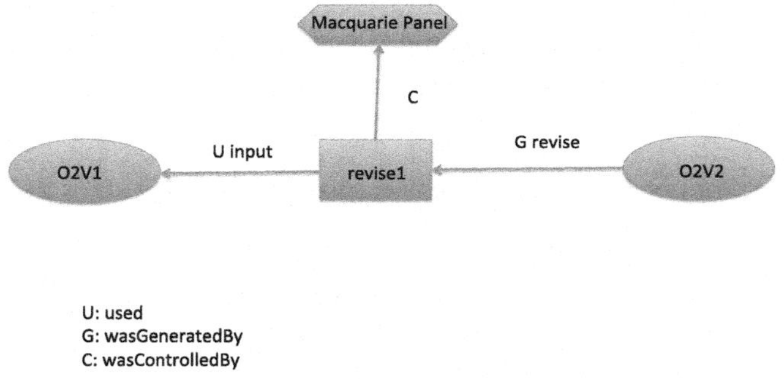

Fig. 2. Provenance example

In the example shown in Fig. 2, it records the following transaction: Macquarie Panel revised Artifact O_2V_1 and generated a new Artifact O_2V_2. Macquarie Panel is the Agent, revise1 is the Process and O_2V_1 and O_2V_2 are Artifacts. The dependency relationship between Macquarie Panel and revise1 is "c", which implies that Macquarie Panel controls the Process "revise1". The Process "revise1" uses Artifact O_2V_1 as input; Artifact O_2V_2 is the generated output from the Process "revise1".

The provenance information is normally recorded as metadata. For the above example, the metadata will be: $MacquariePanel, revise1, ArtifactO_2V_1, ArtifactO_2V_2 >:< revise1, MacquariePanel, wasControlledBy >< revise1, Artifact\, O_2V_1, U_{input} >, < Artifact\, O_2V_2, revise1, wasGeneratedBy(revise) >$. In our system, we will be focussing mainly on the first two items of each record, which are the Agent and the Process. Hence such a provenance fragment in the above example is "Macquarie Panel, revise1". In our system, the owners of files encrypt such provenance fragments and generate cipher texts which will be employed by our following scheme.

4 PBCAP System Design

4.1 System Architecture

We assume that a cloud service provider has several remote storage units available for storing data that is received from different users of the system. Users wish to store their files with provenance in the cloud, and send them in encrypted format to cloud service provider. The goal is to design an access control system that the cloud service provider can use to classify encrypted files by searching keywords from encrypted provenance. We refer to this access control system as Provenance based Classification Access Policy (PBCAP) System.

Figure 3 givens an outline of our system architecture. The remote data storage units are managed by a cloud server which classifies files and allocates them in corresponding storage units. Our PBCAP system achieves the following objectives: (1) the cloud server will classify the encrypted files that it receives from users based on the attached encrypted provenance information; (2) the encrypted provenance information is checked for policy compliance while they remain encrypted (hence the confidentiality of both the encrypted files and their provenance information are guaranteed); and (3) provides a guarantee to the cloud server that the provenance is from a genuine source.

The components of the system architecture are as follows:

- Users are the owners of files who send encrypted files to the cloud server for storage. The files along with their provenance are encrypted by the user before they are sent to the cloud server. In our scheme, users also generate a pair of public and private keys which are used for an embedded short signature verification mechanism.
- Cloud Server classifies the received encrypted files and stores them in different storage units. Each unit stores a bunch of files that shares common attributes.

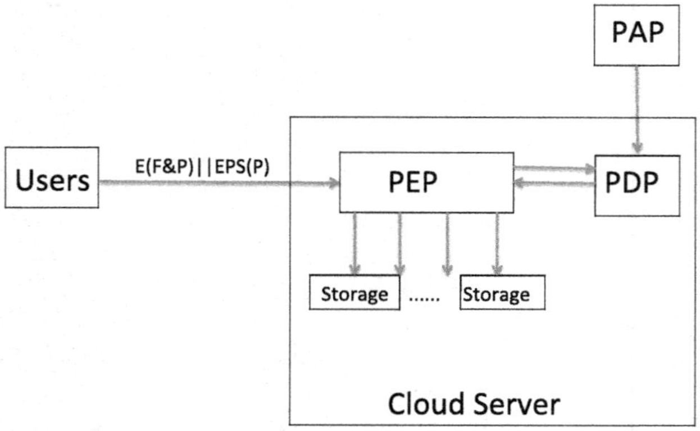

Fig. 3. PBCAP system architecture

For instance, an attribute might be undergoing a specific process (e.g. being graded by Alice, or being edited by Bob). This benefits management of files as well as providing corresponding levels security protections.
- Policy Administrator Point (PAP) generates polices and send them to Policy Decision Points (PDP) for implementation. To keep confidentiality, PAP encrypted sensitive information in policies before sending them to PDP.
- PDP Execute Encrypted Provenance Search Scheme (EPSS) as the following the steps: verifying short signatures of provenance cipher texts to make sure they are from genuine users; searching keywords from encrypted provenance, output results to Policy Enforcement Point (PEP).
- PEP receives results from PDP and allocates files to corresponding storages units.

4.2 Provenance Based Classification Policy

In this system, the rules to classify files only care about historical transactions taken on them. For example, one educational storage unit might only accept documents after anonymous processing due to the consider of comprehensive. Then, the system formulate polices to map sets of historical transactions to file categories and storage units. The example below maps a set of transactions to a category of "Medical Documents" and a storage unit "Hospital". As one transaction corresponds with a specific piece of data in provenance, these transactions was expressed as pieces of provenance data which are named provenance fragments in this paper.

However, a provenance files might contain provenance fragments which map more than one category. To avoid this conflict, we set priorities of each category. When conflict happens, files are identifies as the highest priority category.

$< PolicyID = \text{“}1'' >$
$\quad < ProvenanceFragmentSet > Set_i < /ProvenanceFragmentSet >$
$\quad < Priority > m < /Priority >$
$\quad < Category > \text{“}MedicalDocuments'' < /Category >$
$\quad < CloudStorageUnit > \text{“}Hospital'' < /CloudStorageUnit >$
$< /Policy >$

Set_1=[$ProvenanceFragment_{(1,1)}$,$ProvenanceFragment_{(1,2)}$,
... $ProvenanceFragment_{(1,n)}$],
Set_2=[$ProvenanceFragment_{(2,1)}$,$ProvenanceFragment_{(2,2)}$,
... $ProvenanceFragment_{(2,n)}$],
......
Set_n=[$ProvenanceFragment_{(n,1)}$,$ProvenanceFragment_{(n,2)}$,
... $ProvenanceFragment_{(n,n)}$],

5 Provenance Based Classification Scheme

In this section, we describe our provenance based classification scheme. First we give a brief overview of preliminaries needed for our scheme. We provide details of our scheme which consists of setup phase and verification phase. Finally, after presenting security game of a chosen-word-attack, we give the security proof of our scheme showing that is semantically secure.

5.1 Preliminaries

Let $\mathbb{G}_1, \mathbb{G}_2$ be two cyclic multiplicative groups with the same order p. The size of $\mathbb{G}_1, \mathbb{G}_2$ is determined by the security parameter. Let $\hat{e} : \mathbb{G}_1 \times \mathbb{G}_1 \rightarrow \mathbb{G}_2$ be a bilinear map with the following properties:

- Bilinearity: $\hat{e}(g_1^a, g_2^b) = \hat{e}(g_1, g_2)^{ab}$ for all $\{g_1, g_2\} \in \mathbb{G}_1, \{a, b\} \in \mathbb{Z}_q$.
- Non-degeneracy: There exists $g \in \mathbb{G}_1$ such that $\hat{e}(g, g) \neq 1$.
- Computability: There exists an efficient algorithm to compute $\hat{e}(g_1, g_2)$ for all $\{g_1, g_2\} \in \mathbb{G}_1$.

The construction of the Provenance-based Classification Scheme is based on identity-based encryption [3]. We build a non-interactive searchable encryption scheme from the Bilinear map above and hash functions $H_1 : \{0,1\}^* \rightarrow \mathbb{G}_1$ and $H_2 : \mathbb{G}_2 \rightarrow \{0,1\}^{\log p}$. In particular, H_2 is a collision resistant hash function. The functions in scheme work as follows:

- $\mathsf{KeyGen1}$: Takes a security parameter 1^λ as input; then the algorithm picks at random an $\alpha \in Z_p^*$ and a generator $g \in \mathbb{G}_1$, where p is a prime and it is the size of $\mathbb{G}_1$ and $\mathbb{G}_2$. It outputs the public key $A_{pub} = [g, h_1 = g^\alpha]$ and the private key $A_{priv} = \alpha$.
- $\mathsf{KeyGen2}$: Takes a security parameter 1^λ as input; then the algorithm picks at random a $\beta \in Z_p^*$ and a generator $g \in \mathbb{G}_1$, where p is a prime and it is the size of $\mathbb{G}_1$ and $\mathbb{G}_2$. It outputs the public key $B_{pub} = [g, h_2 = g^\beta]$ and the private key $B_{priv} = \beta$.
- $\mathsf{PBCT}(A_{pub}, B_{priv})$: Generates a Provenance based Classification Tags (PBCTs) for provenance fragments for the purpose of searching. Then Computes $t = \hat{e}(H_1(\mathcal{P})^\beta, h_1^r) \in \mathbb{G}_2$ for a random $r \in Z_p^*$ and a provenance fragment $\mathcal{P}$. Output $\mathsf{PBCT}(A_{pub}, \beta) = [h_1^\beta, h_2^r, H_2(t)] \equiv [X, Y, Z]$.
- $\mathsf{Trapdoor}(A_{priv})$: Output $T_\mathcal{P} = H_1(\mathcal{P}')^\alpha \in \mathbb{G}_1$, where $\mathcal{P}'$ is provenance fragments chosen by the administrator PAP.
- $\mathsf{Test}(A_{pub}, B_{pub}, T_\mathcal{P}, S)$: Test if $H_2(\hat{e}(T_\mathcal{P}, Y)) = Z$ and $\hat{e}(X, g) = \hat{e}(h_1, h_2)$. If both are true, output 1; otherwise 0. The test function using A_{pub}, B_{pub} checks if the encrpyted provenance matching $T_\mathcal{P}$ satisfies the policies; it also verifies if the provenance is generated by authenticated users by checking the short signature.

5.2 Policy Based Classification Scheme

Our policy-based classification scheme (Fig. 4) has two phases, namely the setup phase and the verification phase. Initially, in the setup phase, both the administrator PAP and users generate their own pair of public and private keys. Then

PAP calculates *Trapdoor* for sets of provenance fragments listed in the access control policies and sends them with policies to PDP which executes the test function. In the verification phase, before users send encrypted files and provenance to the Cloud Server, they calculate *PBCT* and attach them to the files. After receiving files, the PDP classifies them by running the *Test* function. Our scheme involves encrypted provenance search and is constructed using the technique mentioned in [12].

- Setup Phase:
  - PAP runs KeyGen1, taking an input security parameter 1^λ; the algorithm picks a random $\alpha \in Z_p^*$ and a generator $g \in \mathbb{G}_1$. It outputs the public key $A_{pub} = [g, h_1 = g^\alpha]$ and the private key $A_{priv} = \alpha$. Then PAP sends the public keys to users and PDP.
  - Users run KeyGen2 taking an input security parameter 1^λ; the algorithm picks a random $\beta \in Z_p^*$ and a generator $g \in \mathbb{G}_1$. It outputs the public key $B_{pub} = [g, h_2 = g^\beta]$ and the private key $B_{priv} = \beta$. Then, users send public keys to PDP.
  - PAP runs $\mathsf{Trapdoor}(A_{priv})$ to output $T_{\mathcal{P}'} = H_1(\mathcal{P})^\alpha \in \mathbb{G}_1$, and then sends the policies with $T_{\mathcal{P}}$ to PDP.
- Verification Phase:
  - Users run function $\mathsf{PBCT}(A_{pub}, B_{priv})$ to compute tags where t = $\hat{e}(H_1(\mathcal{P})^\beta, h_1^r) \in \mathbb{G}_2$ for a random $r \in Z_p^*$ and a provenance $\mathcal{P}$. Output $\mathsf{PBCT}(A_{pub}, \beta) = [h_1^\beta, h_2^r, H_2(t)] \equiv [X, Y, Z]$. Users then attach the tags with encrypted files and provenance.
  - When the encrypted files with tags are sent to PDP, PDP checks if the provenance has the specified keywords in the policies, by running the Test function. Test if $\hat{e}(X, g) \stackrel{?}{=} \hat{e}(h_1, h_2)$ (1) and $H_2(\hat{e}(T_{\mathcal{P}}, Y)) \stackrel{?}{=} Z$ (2). If both are true, then output 1; otherwise 0. The result will then be sent to PEP which executes further operations.

In formula (1), the left hand side $\hat{e}(X, g) = \hat{e}(h_1^\beta, g) = \hat{e}(h_1, g^\beta)$, according to the property of the Bilinear Map. By definition $h_2 = g^\beta$, and hence the left hand

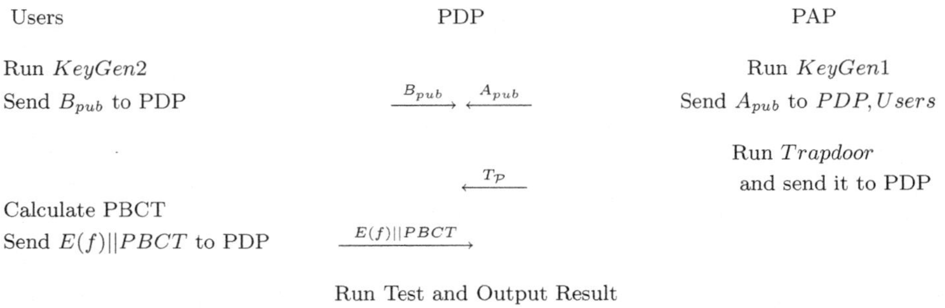

Fig. 4. Provenance based classification scheme

side equals the right hand side. This formula verifies if the users are authenticated by checking whether they have the corresponding private keys. Similarly, we also prove that the left hand side equals the right hand side in formula (2). It tests if PBCT matches the chosen provenance fragments specified by the administrator.

$$H_2(\hat{e}(T_{\mathcal{P}}, Y)) \stackrel{?}{=} Z \ (2)$$
$$H_2(\hat{e}(H_1(\mathcal{P})^{\alpha}, h_2^r)) \stackrel{?}{=} H_2(\hat{e}(H_1(\mathcal{P})^{\beta}, h_1^r))$$
$$H_2(\hat{e}(H_1(\mathcal{P}), h_2^{\alpha r})) \stackrel{?}{=} H_2(\hat{e}(H_1(\mathcal{P}), h_1^{\beta r}))$$
$$H_2(\hat{e}(H_1(\mathcal{P}), (g^{\beta})^{\alpha r})) \stackrel{?}{=} H_2(\hat{e}(H_1(\mathcal{P}), (g^{\alpha})^{\beta r}))$$

5.3 Security Proof

Let us now consider the security proof. This scheme has the property of semantic-security against a chosen word attack. That is, PBCT does not reveal any information of provenance to PDP except that $T_{\mathcal{P}}$ is available to PDP. By simulating our scheme with the game below, an active attacker can obtain $T_{\mathcal{P}}$ for any provenance fragment that s/he chooses. However the attacker could not distinguish PBCT for $\mathcal{P}_0$ and $\mathcal{P}_1$ for which it does not know the $T_{\mathcal{P}}$.

We define the security game between an attacker and the challenger as follows:

Provenance-based Classification Security Game:

1. The challenger runs $KeyGen_1$ and $KeyGen_2$ functions to obtain A_{pub}, A_{priv} and B_{pub}, B_{priv}. The challenger then sends A_{pub} and B_{pub} to the attacker.
2. The attacker sends provenance fragments $\mathcal{P} \in \{0,1\}^*$ of its choice to the challenger. Then the attacker receives trapdoor $T_{\mathcal{P}}$ calculated by the challenger.
3. Then, the attacker sends two random provenance fragments $\mathcal{P}_0$ and $\mathcal{P}_1$ for which it did not ask previously $T_{\mathcal{P}}$.
4. The challenger chooses a random $b \in \{0,1\}$, and returns C= PBCT(A_{pub}, B_{priv}, $\mathcal{P}_b$) to the attacker.
5. The attacker can continue to retrieve $T_{\mathcal{P}}$ from the challenger of any random provenance fragment as long as it is neither $\mathcal{P}_0$ nor $\mathcal{P}_1$.
6. Finally, the attacker makes a guess for $b \in \{0,1\}$ and wins if $b' = b$.

We define the attacker's advantage to break the Provenance based Classification Scheme as

$$Adv(s) = |Pr[b = b'] - 1/2|$$

To complete the security proof, we define an **External Bilinear Diffie-Hellman Problem.** We use the *Corollary A.3.* in [7] to get a new hard problem by setting $P = (1, a, b, c, d, ab, bc)$, $Q = (1)$, $f = abcd$.

Corollary A.3. in [7]. Let $P, Q \in F_p[X_1, .., X_n]^s$ be two s-tuples of n-variable polynomials over F_p and let $f \in F_p[X_1, ...X_n]$. Let d $= \max(2dp, d_Q, d_f)$. If f is independent of (P, Q) then any A that has advantage 1/2 in solving the decision

(P,Q,f)-Diffie-Hellman Problem in a generic bilinear group G must take time at least $\Omega(\sqrt{p/d} - s)$.

External Bilinear Diffie-Hellman Problem(XBDH). Let $\hat{e} : \mathbb{G}_1 \times \mathbb{G}_1 \rightarrow \mathbb{G}_2$ be a bilinear map. For a generator g of G_1, the BDH problem is as follows: given $\{g, g^a, g^b, g^c, g^d, g^{ab}, g^{bc}\} \in G_1$ as input, compute $\hat{e}(g,g)^{abcd}$.

Theorem 1. The Provenance based Classification Scheme given above is semantically secure against a chosen-word-attack in the random oracle model if the XBDH problem is hard.

Proof. Suppose the attacker makes at most q_1 hash function queries to H_2 and at most q_2 trapdoor queries. Assume the attack algorithm has an advantage ϵ in breaking the scheme. Then the challenger is able to solve the XBDH problem with an advantage $\epsilon' = \epsilon/(eq_1q_2)$. We know, in G_1, XBDH is a hard problem and ϵ' is negligible. Therefore, ϵ must be negligible and the Provenance based Classification protocol is semantic-secure.

We simulate the game between the attacker A and the challenger B. The challenger is given g, $u_1 = g^{\alpha}$, $u_2 = g^{\beta}$, $u_3 = g^{\gamma}$, $u_4 = g^{\delta}$, $u_5 = g^{\alpha\beta}$, $u_6 = g^{\beta\gamma}$. The goal of the challenger is to successfully output $v = e(g,g)^{\alpha\beta\gamma\delta}$. The attacker wins the game if it is able to distinguish $PBCT(\mathcal{P}_0)$ and $PBCT(\mathcal{P}_1)$.

KeyGen: Challenger sends public keys $[g, u_1, u_2]$ to attacker A.

H_1-queries: At anytime, attacker A could query the random oracles H_1 by sending a random $\mathcal{P}_i$, which is a provenance fragment in an item of provenance, while the challenger B keeps a H_i-list recorded as $< \mathcal{P}_i, h_i, a_j, c_j >$ to answer the queries. The list is initially empty. When the attacker A sends $\mathcal{P}_i \in \{0,1\}^*$ as a query, the challenger B calculates the following:

1. The challenger send h_i back directly as $H_i(\mathcal{P}_i) = h_i \in G_1$ if $\mathcal{P}_i$ exists in the current list.
2. Or else, the challenger B chooses a random $c_i \in \{0,1\}$, with $Pr[c_i = 0] = 1/(q_2+1)$.
3. Then, the challenger B generates a random $a_i \in Z_p$, and then computes $h_i \leftarrow u_4 * g^{a_i} \in G_1$ if c_i=0, and $h_i \leftarrow g^{a_i} \in G_1$ if c_i=1.
4. Then the challenger adds the newly generated $< \mathcal{P}_i, h_i, a_j, c_j >$ to the H_i-list and responds to the attacker h_i.

H_2-queries: The attacker A sends t as a H_2 query, and the challenger picks a random $V \in \{0,1\}^{log_p}$ as $H_2(t) = V$. Adds the set (t, V) to the H_2 list if this does not exist in the list previously. H_2 is initially empty.

Trapdoor queries: The attacker A sends random $\mathcal{P}_i$ as trapdoor queries. Then the challenger calculates the following:

1. Run H_1 query algorithm to obtain c_i. If c_i=0, and outputs failure and terminates.
2. If $c_i = 1$, outputs $T_i = u_1^{a_i}$ as the result. Note that $T_i = H(\mathcal{P}_i)^{\alpha}$ as $h_i = g^{\alpha}$. Then $T_i = H(\mathcal{P}_i)^{\alpha} = g^{a_i\alpha} = u_1^{a_i}$.

Challenge: The attacker A picks two provenance fragments $\mathcal{P}_0$ and $\mathcal{P}_1$ for the challenge. Note that both $\mathcal{P}_0$ and $\mathcal{P}_1$ should not have been challenged previously. The challenger B calculates PBCT as follows:

1. The challenger B runs H_1-query algorithm to generate c_0 and c_1: if c_0=1 and c_1=1, reports failure and terminates; if there is one between c_0 and c_1 equals 0, then sets that one to c_b; if both of them equals 0, then randomly chooses one of them to be c_b.
2. Then generates a challenge C for $\mathcal{P}_b$ as $C = [u_5, u_6, J]$, where $J \in \{0,1\}^{log_p}$ is a random number. The challenger B defines $J = H_2(\hat{e}(H_1(\mathcal{P}_b)^\beta, u_1^\gamma)) = H_2(\hat{e}(u_4 g^{a_b}, g^{\alpha\beta\gamma})) = H_2(\hat{e}(g,g)^{\alpha\beta\gamma(\delta+a_b)})$

More trapdoor queries: The attacker could continue to ask trapdoor for $\mathcal{P}_i$, where $\mathcal{P}_i \neq \mathcal{P}_0$, $\mathcal{P}_1$.

Output: Finally, the attacker A outputs a guess $b' \in \{0,1\}$ which represents whether the challenge C is calculated for $\mathcal{P}_0$ or $\mathcal{P}_1$. Then, the challenger chooses a random pair (t, V) from H_2 list and calculates $t/\hat{e}(u_5, u_3)^{a_b}$ as the output for $\hat{e}(g,g)^{\alpha\beta\gamma\delta}$, where a_b is known as a parameter to calculate the challenge C.

For the simulation process described above, the probability that a challenger B correctly outputs $\hat{e}(g,g)^{\alpha\beta\gamma\delta}$ is ϵ'. The challenger B wins the game if s/he chooses the correct H_2 pair, and does not abort during the trapdoor queries period and the challenge period.

Claim1 : The probability that challenger outputs $e(g,g)^{\alpha\beta\gamma\delta}$ is $\epsilon' = \epsilon/(eq_1q_2)$.

Proof. Briefly, the challenger's algorithm does not abort mean that it does not abort during both the trapdoor queries period and during the challenge period. The probability that a trapdoor query causes challenger to abort is $1/(q_2+1)$. Because attacker makes at most q_2 trapdoor queries, the probability that challenger does not abort at the trapdoor queries phrase is at least $(1-1/(q_2+1))^{q_T} \geq 1/e$. Similarly, it will abort at the challenge phrase when $c_0 = c_1 = 1$ with $Pr[c_0 = c_1 = 1] = (1-1/(q_2+1))^2 \leq 1 - 1/q_2$. In the opposite way, it does not abort at the challenge phrase is at least $1/q_2$. Therefore, we have the corresponding probabilities are $Pr[\xi_1] \geq 1/e$ and $Pr[\xi_2] \geq 1/q_2$ respectively. Note that these two events are independent; therefore, the probability that the challenger's algorithm does not abort is $Pr[\xi_1 \wedge \xi_2] \geq 1/(eq_2)$. Following that, the attacker A issues a query for $H_2(e(H_1(W_b)^\beta, u_1^\gamma))$ with probability at least ϵ; then the challenger chooses the right pair with probability $1/q_1$. As these processes are independent from one other, we can conclude that the probability that the challenger outputs $\hat{e}(g,g)^{\alpha\beta\gamma\delta}$ is ϵ/eq_1q_2. As this is a hard problem, the attacker can break the game with negligible probability. In other words, the attacker A cannot distinguish whether $\mathcal{P}_0$ or $\mathcal{P}_1$ is $PBCT(\mathcal{P}_b)$. Hence Then the Provenance based Classification Scheme is semantic-secure.

6 Concluding Remarks

In this paper, we have proposed a Provenance-based Access Policy System which can be used to classify encrypted files sent to the cloud by different users. The access control decisions are made according to the provenance attached to the encrypted files. The provenance information itself is in encrypted form. The cloud server is able to check whether the provenance satisfies certain policies specified by the administrator without decrypting the provenance. That is, the scheme allows searching encrypted provenance. Furthermore, the cloud server is also able to check the identity of users who sent these files as that is part of the provenance information. We have described the scheme in detail and developed a provenance-based classification security game and proof to show that the proposed scheme is semantically-secure based on a known hard problem.

However, provenance-based access control is still at its initial stage. There are still interesting work are left. That will include examination of the granularity of access control and the range of policy types that can be provided using provenance. By employing provenance, access control systems might support more kinds of policies beyond traditional scope. Provenance records could be formulated with other attributes such as time stamps, users' ID, agent *etc.* to fertilise more policies. In the meanwhile, decision uncertainties might arouse in the evaluation of provenance based access control policies, specially for a fine-grained approach. Then, conflict solutions might be required.

We also recognise that long lived and much handled data can acquire extensive provenance information. In practice, system administrators may need to limit the lifespan of provenance data if this is found to cause unacceptable performance issues. Moreover, an scheme with adaptive semantic secure will improve the security level of this system.

References

1. Jaehong, P., Dang, N., Ravi, S.S.: A provenance-based access control model. In: Tenth Annual International Conference on Privacy, Security and Trust, Paris, pp. 137–144 (2012)
2. Dang, N., Jaehong, P., Ravi, S.S.: A provenance-based access control model for dynamic separation of duties. In: Eleventh Annual International Conference on Privacy, Security and Trust, Tarragona, Catalonia, Spain, pp. 247–256 (2013)
3. Boneh, D., Franklin, M.: Identity-based encryption from the weil pairing. In: Kilian, J. (ed.) CRYPTO 2001. LNCS, vol. 2139, pp. 213–229. Springer, Heidelberg (2001)
4. Luc, M., Ben, C., Juliana, F., Joe, F., Yolanda, G., Paul, T.G., Natalia, K., Simon, K., Paolo, M., Jim, M., Beth, P., Yogesh, S., Eric, G.S., Jan, V.D.B.: The open provenance model core specification (v1.1). Future Gener. Comp. Syst. **27**, 743–756 (2011)
5. Umut, A.A., Peter, B., James, C., Jan, V.D.B., Natalia, K., Stijn, V.: A graph model of data and workflow provenance. In: TaPP (2010)
6. Martin, D., Maria, T.: CRMdig: a generic digital provenance model for scientific observation. In: TaPP (2011)

7. Boneh, D., Boyen, X., Goh, E.-J.: Hierarchical identity based encryption with constant size ciphertext. In: Cramer, R. (ed.) EUROCRYPT 2005. LNCS, vol. 3494, pp. 440–456. Springer, Heidelberg (2005)
8. Christoph, B.: How usage control and provenance tracking get together - a data protection perspective. In: Symposium on Security and Privacy Workshops, pp. 13–17. IEEE, San Francisco (2013)
9. Uri, B., Avraham, S., Margo, I.S.: Securing provenance. In: 3rd USENIX Workshop on Hot Topics in Security, CA, USA (2008)
10. Elisa, G., Gabriel, G., Murat, K., Dang, N., Jae, P., Ravi, S.S., Bhavani, M.T., Shouhuai, X.: A roadmap for privacy-enhanced secure data provenance. J. Intell. Inf. Syst. **43**, 481–501 (2014)
11. Syed, R.H., Changda, W., Salmin, S., Elisa, B.: Secure data provenance compression using arithmetic coding in wireless sensor networks. In: 33rd International Performance Computing and Communications Conference, IPCCC, TX, USA, pp. 1–10 (2014)
12. Boneh, D., Di Crescenzo, G., Ostrovsky, R., Persiano, G.: Public key encryption with keyword search. In: Cachin, C., Camenisch, J.L. (eds.) EUROCRYPT 2004. LNCS, vol. 3027, pp. 506–522. Springer, Heidelberg (2004)
13. Ni, Q., Xu, S., Bertino, E., Sandhu, R., Han, W.: An access control language for a general provenance model. In: Jonker, W., Petković, M. (eds.) SDM 2009. LNCS, vol. 5776, pp. 68–88. Springer, Heidelberg (2009)
14. Dang, N., Jaehong, P., Ravi, S.S.: A provenance-based access control model for dynamic separation of duties. In: Eleventh Annual International Conference on Privacy, Security and Trust, PST, Catalonia, Spain, pp. 247–256 (2013)
15. Jin, L., Xiaofeng, C., Qiong, H., Duncan, S.W.: Digital provenance: enabling secure data forensics in cloud computing. Future Gener. Comp. Syst. **37**, 259–266 (2014)
16. Sherman, S.M., Cheng-Kang, W., Jianying, Z., Robert, H.D.: Dynamic secure cloud storage with provenance. In: Cryptography and Security, pp. 442–464 (2012)
17. Banerjee, A., Naumann, D.A.: History-based access control and secure information flow. In: Barthe, G., Burdy, L., Huisman, M., Lanet, J.-L., Muntean, T. (eds.) CASSIS 2004. LNCS, vol. 3362, pp. 27–48. Springer, Heidelberg (2005)
18. Omar, B., Anish, D.S., Alon, Y.H., Martin, T., Jennifer, W.: Databases with uncertainty and lineage. VLDB J. **17**, 243–264 (2008)
19. Peter, B., Adriane, C., James, C.: Provenance management in curated databases. In: International Conference on Management of Data, Chicago, Illinois, USA, pp. 539–550 (2006)
20. Manish, K.A., Shawn, B., Timothy, M.M., Bertram, L.: Efficient provenance storage over nested data collections. In: 12th International Conference on Extending Database Technology, Saint Petersburg, Russia, pp. 958–969 (2009)
21. Thomas, H., Gustavo, A.: Efficient lineage tracking for scientific workflows. In: International Conference on Management of Data, Vancouver, BC, Canada, pp. 1007–1018 (2008)
22. Ali, N.R., Jafar, H.J., Morteza, A., Rasool, J.: GTHBAC: a generalized temporal history based access control model. Telecommun. Syst. **45**, 111–125 (2010)
23. Karl, K., Mogens, N., Vladimiro, S.: A logical framework for history-based access control and reputation systems. J. Comput. Secur. **16**, 63–101 (2008)

Multi-user Searchable Encryption in the Cloud

Cédric Van Rompay(✉), Refik Molva, and Melek Önen

EURECOM, Sophia Antipolis, France
{vanrompa,molva,onen}@eurecom.fr

Abstract. While Searchable Encryption (SE) has been widely studied, adapting it to the multi-user setting whereby many users can upload secret files or documents and delegate search operations to multiple other users still remains an interesting problem. In this paper we show that the adversarial models used in existing multi-user searchable encryption solutions are not realistic as they implicitly require that the cloud service provider cannot collude with some users. We then propose a stronger adversarial model, and propose a construction which is both practical and provably secure in this new model. The new solution combines the use of bilinear pairings with private information retrieval and introduces a new, non trusted entity called "proxy" to transform each user's search query into one instance per targeted file or document.

1 Introduction

Cloud computing nowadays appears to be the most prominent approach for outsourcing storage and computation. Despite well known advantages in terms of cost reduction and efficiency, cloud computing also raises various security and privacy issues. Apart from classical exposures due to third party intruders one of the new requirements akin to outsourcing is the privacy of outsourced data in the face of a potentially malicious or careless Cloud Service Provider (CSP).

While data encryption seems to be the right countermeasure to prevent privacy violations, classical encryption mechanisms fall short of meeting the privacy requirements in the cloud setting. Typical cloud storage systems also provide basic operations on stored data such as statistical data analysis, logging and searching and these operations would not be feasible if the data were encrypted using classical encryption algorithms.

Among various solutions aiming at designing operations that would be compatible with data encryption, Searchable Encryption (SE) schemes allow a potentially curious party to perform searches on encrypted data without having to decrypt it. SE seems a suitable approach to solve the data privacy problem in the cloud setting.

A further challenge is raised by SE in the multi-user setting, whereby each user may have access to a set of encrypted data segments stored by a number of different users. Multi-user searchable encryption schemes allow a user to search through several data segments based on some search rights granted by the owners of those segments. Privacy requirements in this setting are manifold, not only the

J. Lopez and C.J. Mitchell (Eds.): ISC 2015, LNCS 9290, pp. 299–316, 2015.
DOI: 10.1007/978-3-319-23318-5_17

confidentiality of the data segments but also the privacy of the queries should be assured against intruders and potentially malicious CSP. Recently, few research efforts [5,8,12,15] came up with multi-user keyword search schemes meeting these privacy requirements, either through some key sharing among users or based on a Trusted Third Party (TTP).

In this paper, we first investigate the new privacy challenges for keyword search raised by the multi-user setting beyond the basic privacy concerns about data, queries and responses by focusing on the relationship among multiple queries and responses. We realize that while as analyzed in [7], the protection of the *access pattern privacy* (privacy of the responses) is optional for single-user searchable encryption mechanisms, this requirement becomes mandatory in the multi-user setting. Unfortunately all existing Multi-User Searchable Encryption (MUSE) schemes [5,8,12,15] suffer from the lack of such protection. We further come up with a new adversary model for MUSE that takes into account new security exposures introduced by the possible collusion of some users with the CSP.

After showing that all existing MUSE schemes fail at meeting the privacy requirements in our new adversarial model, we suggest a new solution for MUSE for which it is not the case, i.e., all users who have not been explicitly authorized to search a document can collude with the adversary without threatening the privacy of that document. Our solution for MUSE inherently ensures access pattern privacy through the use of Private Information Retrieval (PIR). While the PIR protocol together with the multi-user setting may add a significant complexity overhead, this overhead is outsourced from the users to a third party our scheme introduces, the *proxy*, that is in charge of multiplexing a user query into several PIR queries. Moreover the overhead of PIR is further lowered by querying binary matrices representing the keyword indices instead of querying the bulky keyword lists themselves. As opposed to most existing solutions based on a TTP [3,5,8,15], the proxy in our scheme does not need to be trusted. With the sole assumptions that the CSP and the proxy are honest-but-curious and that they do not collude with one another, we prove that our solution meets the privacy requirements defined for MUSE.

Section 2 states the problem addressed by MUSE. Section 3 describes our solution for MUSE and Sect. 4 defines the security properties for MUSE. Section 5 proves that our solution achieves the security properties we defined and Sect. 6 studies the algorithmic complexity of our solution. Section 7 reviews the state of the art and, finally, Sect. 8 concludes the paper.

2 Multi-user Searchable Encryption (MUSE)

A MUSE mechanism extends existing keyword search solutions into a multi-writer multi-reader [6] architecture involving a large number of users, each of which having two roles:

- as a *writer*, the user uploads documents to the server and delegates keyword search rights to other users.

- as a *reader,* the user performs keyword search operations on the documents for which she received delegated rights.

As any SE solution, MUSE raises two privacy requirements:

- *index privacy:* unauthorized parties should not discover information about the content of uploaded documents.
- *query privacy:* no one should get information about the targeted word and the result of a search operation apart from the reader who sent the corresponding query.

In addition to the CSP, any user that has not been explicitly given search rights on an index should be considered as potentially colluding with the CSP in order to violate index or query privacy. This assumption leads to a model in which the adversary is composed of a coalition of the CSP and some non-delegated users. This new adversary model extends the one used in other existing MUSE schemes [5, 7, 12, 15], which although secure in their own adversary model do not achieve index and query privacy any more if non-delegated users collude with the CSP.

Figure 1 illustrates one example of the impact of a collusion between a CSP and a user on privacy by taking advantage of the lack of access pattern privacy. Assuming that R_1 is authorized to query both indices I_1 and I_2, by observing the access pattern of R_1's queries, the CSP can discover similarities between I_a and I_b. In a second phase, the CSP corrupts reader R_2 who is authorized to query I_b only. By exploiting the similarities between I_a and I_b and discovering the content of I_b through R_2, the CSP can easily discover the content of I_a. The index privacy is thus violated for I_a since the CSP partially learns the content of I_a although R_1, the only reader having delegated search rights for I_a, was not corrupted. Furthermore, once the CSP obtains information about the content of an index, observing the access pattern of the queries targeting this index enables the CSP to violate the privacy of these queries. This attack allows to violate both query and index privacy in all existing MUSE schemes since they all let the CSP discover the access pattern of the queries. The new adversary model we introduce not only prevents such an attack but also prevents any attack that would require the corruption of a non-delegated user.

3 Our Solution

3.1 Idea

Our solution introduces a third party called the *proxy* that performs an algorithm called *QueryTransform* to transform a single reader query into one query per targeted document[1]. For each of these queries, the proxy sends to the CSP a specific

[1] Note that the set of targeted document can reveal the authorized set of documents for this particular user. However, such an additional information does not have a serious impact on index or query privacy as access pattern leakage has.

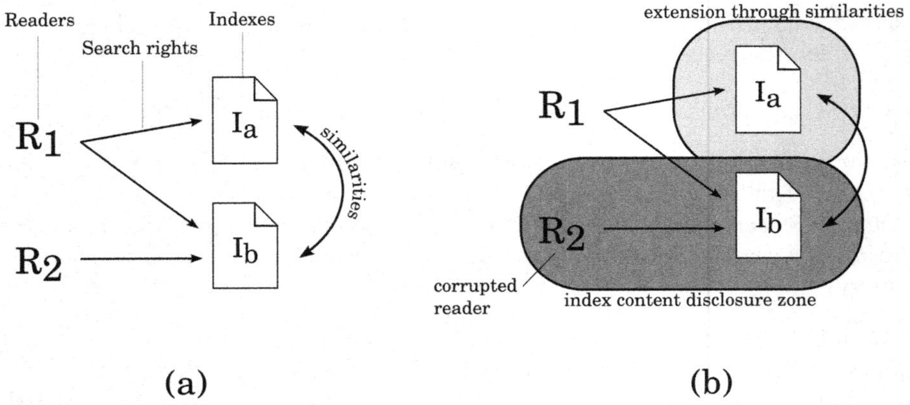

Fig. 1. In (a), discovery of similarities through the access pattern. In (b), use of the similarities to extend index privacy violation.

PIR request. Thanks to the PIR protocol the CSP does not have access neither to the content of the query nor to its result, which makes our scheme achieving query privacy (including access pattern privacy) against the CSP. While the use of PIR provides privacy against the CSP, a new privacy exposure raises with respect to the proxy. Indeed through the execution of *QueryTransform*, the proxy is able to discover the relationship between a query and the different ciphertexts in the targeted indices which are the encryption of the same keyword. However with the assumption that the proxy does not collude with the CSP, the proxy cannot realize whether these ciphertexts are present in their respective indices or not; thus, our scheme achieves index privacy against the proxy. Moreover thanks to some randomization of the queries and the encryption of the responses by the CSP with the reader's key, the proposed solution also ensures query privacy against the proxy. Consequently, while our solution does introduce a third party (the proxy), **this third party does not need to be trusted** and is considered as an adversary. Both the CSP and the proxy are then considered as potentially malicious in our scheme, and are only assumed *honest-but-curious* and non colluding with each other.

Another advantage of introducing the proxy into this new MUSE solution is scalability: Indeed, thanks to the $QueryTransform$ algorithm executed by the proxy a user does not need to generate several PIR queries (one per index) for the same keyword.

3.2 Preliminaries

Bilinear Pairings Let G_1, G_2 and G_T be three groups of prime order q and g_1, g_2 generators of G_1 and G_2 respectively. $e : G_1 \times G_2 \rightarrow G_T$ is a bilinear map if e is:

- efficiently computable
- non-degenerate: if x_1 generates G_1 and x_2 generates G_2, then $e(x_1, x_2)$ generates G_T
- bilinear: $e(g_1^a, g_2^b) = e(g_1, g_2)^{ab}\ \forall (a, b) \in \mathbb{Z}^2$

We assume that the widely used eXternal Diffie-Hellman (XDH) assumption [4] holds.

Definition 1 (External Diffie Hellman Assumption). *Given three groups G_1, G_2 and G_T and a bilinear map $e : G_1 \times G_2 \rightarrow G_T$, the Decisional Diffie-Hellman (DDH) problem is hard in G_1, i.e., given $(g_1, g_1^\alpha, g_1^\beta, g_1^\delta) \in G_1^4$, it is computationally hard to tell if $\delta = \alpha\beta$.*

Private Information Retrieval (PIR). A PIR protocol allows a user to retrieve data from a database without revealing any information about the retrieved data.

PIR consists of five algorithms:

- **PIR.Setup**() $\rightarrow PIRParams$
- **PIR.KeyGen**() $\rightarrow (PirKey)$: this algorithm outputs the keying material for PIR.
- **PIR.Query**$(PirKey, size, target) \rightarrow Query$: given PIR parameters, the size of the targeted database and a target position, this algorithm outputs a PIR query targeting the given position in a database of the given size.
- **PIR.Process**$(Query, DataBase) \rightarrow R$: this algorithm applies the query $Query$ on the database $DataBase$ and outputs a response R.
- **PIR.Retrieve**$(R, PirKey) \rightarrow Cell$: given a PIR response R and the PIR key used in corresponding query, this algorithm outputs the value of the retrieved database cell.

Single-database computational PIR has already been widely studied [1,2,10, 11], and the results presented in [1] show that solutions with practical performances already exist. Our solution uses the technique of recursive PIR which allows to reduce communication complexity as explained in [1]: The database is viewed as a matrix each row of which is considered as a sub-database. To query the whole database a single query is sent and this query is further applied on each row, resulting in the generation of many PIR responses.

3.3 Protocol Description

Figure 2 illustrates the structure and the various flows of our solution. We define two phases in the protocol, the *upload phase* and the *search phase*: During the upload phase, a writer A uploads a secure index to the CSP by encrypting each keyword with the $Index$ algorithm. A then delegates search rights to reader B using the $Delegate$ algorithm which computes an authorization token using B's public key and A's private key. The authorization token is sent to the proxy. During the search phase, B can further search all the indices for which she has

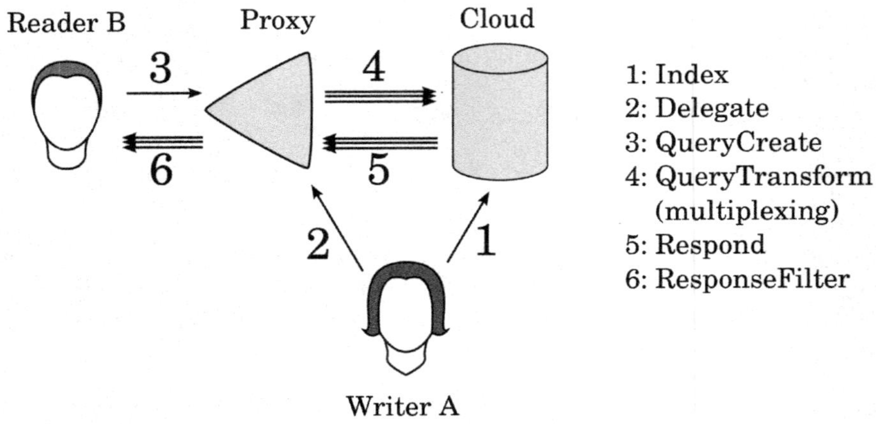

Fig. 2. Overview of our solution.

been given search rights, by creating a single query through the execution of *QueryCreate*. Whenever the proxy receives the B's query it uses the authorization tokens attributed to B to transform this query into one PIR query per authorized index through the execution of $QueryTransform$. Upon reception of a PIR query, the CSP through the execution of *Respond* builds a binary matrix using the corresponding encrypted index, applies the query to the matrix and encrypts the resulting PIR answers. The responses are then pre-processed by the proxy through the execution of $ResponseFilter$. Finally B obtains the result of her search query by executing $ResponseProcess$.

Revocation in our solution only consists in the deletion of the appropriate authorizations by the proxy upon a writer's request.

The set of users is denoted by $u_{i1 \leq i \leq N}$. For the sake of clarity, each user u_i is assumed to own only one index I_i.

- **Setup**(κ) $\rightarrow$ *params* : given the security parameter κ, this algorithm outputs the parameters *param* consisting in:
  - a description of the bilinear map that will be used: the three groups G_1, G_2, G_T of prime order q, the two generators g_1 and g_2 and the map itself e.
  - a cryptographically secure hash function $h: \{0,1\}^* \rightarrow G_1$
  - the size n of the matrices for PIR, and a hash function $H: G_T \rightarrow [\![0, n-1]\!]$ to transform encrypted keywords into positions in the matrices. Without loss of generality, n is assumed to be a perfect square.
  - the PIR parameters $PIRParams$ from the execution of $PIR.Setup$
  - a symmetric cipher *Enc* and the corresponding decipher algorithm *Dec*.

  All these parameters are considered implicit for each further algorithm.
- **KeyGen**(κ) $\rightarrow (\gamma, \rho, P, K)$: given the security parameter κ, a user u_i generates the following keys:
  - a *secret writer key* $\gamma_i \xleftarrow{\$} \mathbb{Z}_q^*$
  - a *private reader key* $\rho_i \xleftarrow{\$} \mathbb{Z}_q^*$

  - a *public reader key* $P_i = g_2^{\frac{1}{\rho}}$
  - a *transmission key* K_i used for *Enc*/*Dec*. This key is shared with the CSP.
- **Index**$(w, \gamma_i) \rightarrow \tilde{w}$: Writer u_i executes this algorithm to encrypt keyword w with his key γ_i. The algorithm outputs $\tilde{w} = e(h(w)^{\gamma_i}, g_2)$.
- **Delegate**$(\gamma_i, P_j) \rightarrow \Delta_{i,j}$: Provided with the public key P_j of reader u_j, writer u_i executes this algorithm using its secret key γ_i to generate $\Delta_{i,j} = P_j^{\gamma_i}$ the authorization token that authorizes u_j to search the index I_i. The output $\Delta_{i,j}$ is sent to the proxy which adds it to the set D_j. Note that this token can only be created by the legitimate data owner and cannot be forged by any other party including the CSP and the proxy.
- **QueryCreate**$(w, \rho_j) \rightarrow Q_j$: This algorithm is run by an authorized reader to generate a query for keyword w using its private reader key ρ_j. The algorithm draws a randomization factor $\xi \xleftarrow{\$} \mathbb{Z}_q^*$ and outputs $Q_j = h(w)^{\xi\rho_j}$.
- **QueryTransform**$(Q_j, D_j) \rightarrow <Q'_{i,j}>$: Whenever the proxy receives a reader's query Q, it calls this algorithm together with the set D_j.
  For each authorization token $\Delta_{i,j}$ in D, the algorithm creates a PIR query Q'_j as follow:
  - compute $\tilde{Q}_{i,j} \leftarrow e(Q_j, \Delta_{i,j})$
  - compute $x'||y' \leftarrow H(\tilde{Q}_{i,j})$
  - some PIR keying material is generated: $PirKey \leftarrow PIR.KeyGen()$
  - a $\sqrt{n}$-size PIR query is created that targets position y': $Q'_{i,j} \leftarrow PIR.Query(PirKey, \sqrt{n}, y')$

  The algorithm outputs $<Q'_{i,j}>$ which are forwarded to the CSP together with the corresponding identifiers i of the indices. The proxy additionally stores each generated $PIRKey$ and x' in a table in order to use them upon reception of the corresponding response.
- **Respond**$(Q', I, \xi) \rightarrow R$: Whenever the CSP receives an individual PIR query Q', it executes this algorithm using the corresponding index I and the randomization factor ξ corresponding to this query.
  The algorithm initializes a $\sqrt{n} \times \sqrt{n}$ matrix M with "0". Then for each encrypted word $\tilde{w} \in I$, the cell $M_{x,y}$ is set to "1" where $x||y \leftarrow H(\tilde{w}^{\xi})$ (recall that $\tilde{w} \in G_T$). The response is the tuple of the outputs from the application of the PIR query Q' on each row of the binary matrix M: $\tilde{R} \leftarrow (PIR.Process(Q', M_x) \mid M_x \text{ a row of } M)$. Each component of $\tilde{R}$ is then encrypted with algorithm *Enc* using the transmission key K of the querying reader to obtain R which the algorithm outputs. This layer of encryption prevents the proxy from reading the result of the query.
- **ResponseFilter**$(R, x', PirKey) \rightarrow (R', PirKey)$: Whenever the proxy receives a response R it calls this algorithm together with the x' and $PirKey$ associated to the corresponding query. The purpose of this algorithm is to reduce the communication cost for the reader. Indeed the algorithm extracts the x'-th component of R and outputs it together with the value for $PirKey$. This results in a filtered response which is much smaller than the original response.

- **ResponseProcess**$(R', PirKey, K) \rightarrow b \in \{0,1\}$: On receiving the filtered response R' with the corresponding $PirKey$, the reader executes this algorithm using her transmission key K. The algorithm further outputs the value of $PIR.Retrieve(Dec_K(R'), PirKey)$ which corresponds to the content of the retrieved matrix cell. An output of 1 means that the searched keyword is present in the index, and a 0 means that it is absent.

3.4 Correctness

We now show that a query correctly retrieves a particular cell which content corresponds to whether the queried keyword has been uploaded or not.

Let γ be the encryption key of a given index. If keyword w has been uploaded to that index, then the cell $M_{x,y}$ of the corresponding matrix is equal to 1 with $x||y = H(e(h(w), g_2^{\gamma}))$. Conversely if a given cell $M_{x,y}$ is equal to 1 then with high probability the corresponding keyword w where $x||y = H(e(h(w), g_2^{\gamma}))$ has been uploaded. A false positive implies a collision in either H or h. Thus the content of $M_{x,y}$ corresponds to the upload of w.

Secondly, a query for keyword w in that index will retrieve cell $M_{x',y'}$ with:

$$x'||y' = H(e(h(w)^{\rho}, g_2^{\frac{\gamma}{\rho}})) = H(e(h(w), g_2^{\gamma})) = x||y \ . \tag{1}$$

Thus a response to a query will decrypt to the content of the proper cell and our scheme is correct.

4 Security Model

Our security definitions are game-based definitions, where the games are represented by algorithms. Since the CSP and the proxy are considered as two non-colluding adversaries, security will be defined for each of them independently. The consequence of the non-collusion assumption is that each adversary will see the other one as an oracle. For each adversary type we define one game for index privacy and one game for query privacy. For each definition, the corresponding game consists of seven phases: a setup phase, a learning phase, a challenge phase, a restriction phase, second learning and restriction phases identical to the previous ones, and finally a response phase. The adversary is denoted by $\mathcal{A}$.

4.1 Security with the CSP as Adversary

We now formally define *Index Privacy* and *Query Privacy* considering the CSP as the adversary. In the following two definitions the setup and the learning phases are the same and are described in Algorithm 1. The challenge and restriction phases for index privacy are further described in Algorithm 2 and the ones for query privacy are described in Algorithm 3. Finally during the *response phase*, $\mathcal{A}$ outputs a bit b^* representing its guess for the challenge bit b.

```
/* Setup phase                                                        */
A ← Setup();
for i = 1 to N do
    (γ_i, ρ_i, P_i, K_i) ← KeyGen(κ) ;
    A ← (i, P_i, K_i) ;
end
/* First learning phase                                               */
for j = 1 to a polynomial number l_1 do
    A → query ;
    switch query do
        case Index for word w and user u_i
            A ← Index(w, u_i);
        case Corrupt user u_i
            A ← (ρ_i, γ_i, K_i)
        case Delegation of user u_i by user u_j
            /* A does not receive any value, but the delegation will
               modify the set D_i used in QueryTransform               */
        end
        case Queries for word w from user u_i
            /* D_i comes from the Delegations queried by A             */
            A ← QueryTransform(QueryCreate(w, ρ_i), D_i);
            /* A also receives the randomization factor ξ              */
            A ← ξ;
        case Queries for user query Q from corrupted user u_i
            A ← QueryTransform(Q, D_i);
        case Filtered response for response R from corrupted user u_i
            A ← ResponseFilter(R)
    endsw
end
```

Algorithm 1. Setup and learning phases of both index privacy and query privacy games, whereby $\mathcal{A}$ is the CSP

```
/* Challenge phase                                                    */
A → (u_chall, w*_0, w*_1);
b ←$ {0, 1};
A ← Index(w*_b, u_chall);
/* Restriction phase                                                  */
if u_chall is corrupted OR Index for w*_0 or w*_1 for user u_chall has been previously
queried OR a corrupted user has been delegated by u_chall then
    HALT;
end
```

Algorithm 2. Challenge and restriction phases of the index privacy game whereby $\mathcal{A}$ is the CSP

```
/* Challenge phase                                                    */
A → (u_chall, w*_0, w*_1);
b ←$ {0,1};
A ← QueryTransform(QueryCreate(w*_b, ρ_chall), D_chall);
/* Restriction phase                                                  */
if u_chall is corrupted then
  | HALT;
end
```

Algorithm 3. Challenge and restriction phases of the query privacy game whereby $\mathcal{A}$ is the CSP

Definition 2 (Index Privacy Against the CSP). *We say that a MUSE scheme achieves index privacy against the CSP when the following holds for the index privacy game (Algorithms 1 and 2):* $|Pr[b = b^*] - \frac{1}{2}| \leq \epsilon$, *with* ϵ *a negligible function in the security parameter* κ.

Definition 3 (Query Privacy Against the CSP). *We say that a MUSE scheme achieves query privacy against the CSP when the following holds for the query privacy game (Algorithms 1 and 3):* $|Pr[b = b^*] - \frac{1}{2}| \leq \epsilon$, *with* ϵ *a negligible function in the security parameter* κ.

4.2 Security with the Proxy as Adversary

Due to space limitations we do not provide the detailed description of index and query privacy games whereby the proxy is considered as the adversary. In a nutshell, the main differences with the previous games are the following:

- during the learning phase the proxy can query for the *Respond* algorithm executed by the CSP, but does not query for the *QueryTransform* and *ResponseFilter* algorithms. Moreover the proxy receives the output of the *Delegate* algorithm, but does not get the transmission key and the randomization factors of the users.
- during the challenge phase, the proxy does not receive the output of the *Index* algorithm for index privacy, and receives the output of *QueryCreate* for query privacy.

5 Security Analysis

Inspired by the methodology in [14], in order to prove each security property we define a sequence of games $(game_i)_{i=0..n}$, the first game being the original security definition. For each game $game_i$ a "success event" S_i is defined as the event when the adversary $\mathcal{A}_i$ correctly guesses the challenge bit b used as part of the challenge. For every two consecutive games $game_i$ and $game_{i+1}$, it is shown that $|\Pr[S_i] - \Pr[S_{i+1}]|$ is negligible. Then it is shown that the probability of success $Pr[S_n]$ of the last game is the target probability, namely 0.5. Hence the probability of success of the first game is negligibly close to the target probability, which ends the proof.

Due to space limitations we provide a detailed proof for index privacy against the CSP only.

5.1 Index Privacy with the CSP as the Adversary

Theorem 1. *Our construction achieves index privacy against the CSP.*

***game*$_0$.** Let $game_0$ be the game of Definition 2 (Algorithms 1 and 2). The success event S_0 is "$b = b^*$".
***game*$_1$.** The only difference between $game_0$ and $game_1$ is that in $game_1$, the adversary $\mathcal{A}_1$ can no longer send queries requesting the corruption of a user. Consequently $\mathcal{A}_1$ can neither send queries related to corrupted users, namely queries for $QueryTransform$ and $ResponseFilter$.

Lemma 1. *If* $\Pr[S_1]$ *is negligibly close to* 0.5*, then* $\Pr[S_1]$ *and* $\Pr[S_0]$ *are negligibly close.*

Proof. This Lemma is proved by introducing an adversary $\mathcal{A}_1$ executing the algorithm depicted in Algorithm 4.

$\mathcal{A}_1$ plays $game_1$ using adversary $\mathcal{A}_0$ playing $game_0$. To that effect, $\mathcal{A}_1$ simulates an instance of $game_0$ with respect to $\mathcal{A}_0$ and responds at $game_1$ using the response of $\mathcal{A}_0$. Since, as opposed to $\mathcal{A}_0$, $\mathcal{A}_1$ cannot corrupt any user, $\mathcal{A}_1$ has to fabricate responses to $\mathcal{A}_0$'s corruption queries as part of the simulated instance of $game_0$. To do so, $\mathcal{A}_1$ simulates corrupted users by locally generating keys which are sent to $\mathcal{A}_0$ as a response to the corruption query. These same generated keys must be used in all responses related to this corrupted user in order for $\mathcal{A}_1$ to simulate a consistent instance of $game_0$. However $\mathcal{A}_0$ may have sent queries related to this user before the corruption query. A way for $\mathcal{A}_1$ to ensure the required consistency is to choose a set of users that will be simulated from the beginning. If $\mathcal{A}_0$ sends a request to corrupt a user that $\mathcal{A}_1$ chose not to simulate, $\mathcal{A}_1$ cannot simulate a proper instance of $game_0$ any more. Simulation also fails if a user that was simulated by $\mathcal{A}_1$ is chosen by $\mathcal{A}_0$ to be the challenge user or a delegate of the challenge user. We define the event C as when none of the previous cases occur, i.e., C is "$\mathcal{A}_0$ does not corrupt any non-simulated user and $\mathcal{A}_0$ does not chose any simulated user as either the challenge user or a delegate of the challenge user". We also define the event C' as "all users but the challenge user and her delegates are simulated". Since C' implies C we have $Pr[C] \geq Pr[C']$, and actually $Pr[C]$ is expected to be much greater than $Pr[C']$. Whenever the event C occurs, $\mathcal{A}_0$ received a valid instance of $game_0$ with the challenge value from the instance of $game_1$, and thus the probability for $\mathcal{A}_1$ to succeed at $game_1$ is the probability of $\mathcal{A}_0$ to succeed at $game_0$:

$$Pr[S_1|C] = Pr[S_0]. \tag{2}$$

```
A₁ receives data from game₁ setup phase;
/* A₁ simulates some users to A₀                                        */
Sim ←$ P([1..N]);
for i ∈ Sim do
 | (γ'_i, ρ'_i, P'_i, K'_i) ← KeyGen(κ)
end
for i from 1 to N do
 |  if i ∈ Sim then
 |   | A₀ ← (i, P'_i, K'_i);
 |  else
 |   | A₀ ← (i, P_i, K_i)
 |  end
end
/* Learning phase 1                                                     */
for a polynomial number l₁ of times do
 |  A₀ → query;
 |  if A₁ knows all the input values for the corresponding algorithm then
 |   | A₁ runs the algorithm locally and sends back the answer;
 |  else
 |   |  if query was for corruption then
 |   |   |  /* exit with random guess                                   */
 |   |   |  b* ←$ 0,1;
 |   |   |  A₁ → b*;
 |   |   |  HALT;
 |   |  else
 |   |   | A₁ forwards the call to game₁ and forwards the answer to A₀;
 |   |  end
 |  end
end
/* Challenge phase                                                      */
A₁ forwards everything from A₀ to game₁ and back.
/* Learning phase 2                                                     */
Same as learning phase 1;
/* Response phase                                                       */
A₁ forwards the bit b* outputted by A₀;
```

Algorithm 4. Algorithm run by $\mathcal{A}_1$ the transition adversary from $game_0$ to $game_1$. Restrictions phases are omitted.

If the simulation of $game_0$ fails, $\mathcal{A}_1$ can still give a random answer to $game_1$ which implies:

$$Pr[S_1|\neg C] = 0.5. \tag{3}$$

Finally we define the event C'_i as "user u_i is either simulated or challenge-or-delegate, but not both". We have $Pr[C'_i] = 0.5$ and $Pr[C'] = \prod_{i=1..N} Pr[C'_i]$ thus $Pr[C'] = 2^{-N}$ and it follows that $Pr[C] \geq 2^{-N}$. It seems reasonable to assume that the number N of users grows at most polylogarithmically with the security parameter κ, which implies that $Pr[C]$ is non-negligible:

$$\exists p \text{ polynomial in } \kappa,\ \frac{1}{Pr[C]} \leq p. \tag{4}$$

Then the following holds:

$$\begin{aligned}
Pr[S_1] &= Pr[S_1|C].Pr[C] + Pr[S_1|\neg C].Pr[\neg C] \\
Pr[S_1] &= Pr[S_0].Pr[C] + 0.5(1 - Pr[C]) \\
Pr[S_1] &= Pr[C].\,(Pr[S_0] - 0.5) + 0.5 \\
Pr[S_0] &= 0.5 + \frac{1}{Pr[C]}\,(Pr[S_1] - 0.5) \\
Pr[S_0] - Pr[S_1] &= (0.5 - Pr[S_1])\left(1 - \frac{1}{Pr[C]}\right).
\end{aligned}$$

Then from (4) we have that if $(0.5 - Pr[S_1])$ is negligible then $|Pr[S_0] - Pr[S_1]|$ is negligible also. This conclude the proof of Lemma 1.

***game*$_2$**. In $game_2$, calls to *QueryCreate* are replaced by the generation of random bits.

Lemma 2. $\Pr[S_2]$ *is negligibly close to* $\Pr[S_1]$.

Proof. Distinguishing between $game_1$ and $game_2$ is equivalent to breaking the security of the encryption scheme used in the PIR construction. This holds because corruption is not allowed in $game_1$, and hence the adversary cannot obtain the required PIR parameters to open the PIR query. It follows that Lemma 2 is true.

***game*$_3$**. In $game_3$, the call to *Index* in the challenge phase is replaced by picking a random element in G_T.

Lemma 3. $\Pr[S_3]$ *is negligibly close to* $\Pr[S_2]$.

Proof. To prove this Lemma we build a distinguishing algorithm $\mathcal{D}_{DDH}$, described in Algorithm 5, which uses a $game_2$ adversary $\mathcal{A}_2$ and which advantage at the DDH game is:

$$\epsilon_{DDH} = O(\frac{1}{Nl})|\Pr[S_3] - \Pr[S_2]|\ . \tag{5}$$

Given the DDH problem instance $(g_1, g_1^{\alpha}, g_1^{\beta}, g_1^{\delta}) \in G_1^4$, the intuition behind algorithm $\mathcal{D}_{DDH}$ is to "put" β in the challenge word, α in the challenge user key, and δ in the value given to $\mathcal{A}_2$ during the challenge phase. $\mathcal{D}_{DDH}$ makes some predictions on the queries of $\mathcal{A}_2$, namely on the user $\mathcal{A}_2$ will choose as the challenge user and on the moment $\mathcal{A}_2$ will call the hash function h on the challenge word. If these predictions prove false $\mathcal{D}_{DDH}$ sends a random answer

```
D_DDH ← (g_1, g_1^α, g_1^β, g_1^δ);
A ← Setup();
predict ←$ [1..N];
I ←$ [0, .., l];
for i from 1 to N do
    (γ_i, ρ_i, P_i, K_i) ← KeyGen(κ);
    A ← (i, P_i, K_i)
end
for a polynomial number l of times do
    A → query;
    switch query do
        case hash of word w through h
            if this is the I-th call to O then
                A ← g_1^β
            else
                A ← g_1^{O[w]}
            end
        case Index for word w and user u_predict
            A ← e((g_1^α)^{O[w]}, g_2);
        otherwise
            normal handling of the query;
        end
    endsw
end
A → (u_chall, w_0^*, w_1^*);
b ←$ {0, 1};
if chall ≠ predict OR I = 0 and O has been called with input w_b^* OR I ≠ 0 and
w_b^* does not correspond to the I-th call to O then
    b_DDH ←$ 0, 1;
    D_DDH → b_DDH;
    HALT;
end
A ← e(g_1^δ, g_2);
A → b^*;
if b^* = b then
    D_DDH → 1;
else
    D_DDH → 0;
end
```

Algorithm 5. Listing for the distinguishing algorithm $\mathcal{D}_{DDH}$ from $game_2$ to $game_3$.

to the DDH problem. Otherwise if the predictions prove true $\mathcal{D}_{DDH}$ outputs 1 if $\mathcal{A}_2$ wins the game and 0 if not. If the correct answer to the DDH game was 1 then $\mathcal{A}_2$ was playing $game_2$ and $\mathcal{D}_{DDH}$ outputs 1 with probability $Pr[S_2]$.

Similarly if the answer to DDH was 0 $\mathcal{D}_{DDH}$ outputs 1 with probability $Pr[S_3]$ During the whole game $\mathcal{D}_{DDH}$ simulates the hash function h as a random oracle, using $\mathcal{O}$ which behaves the following way: if $\mathcal{O}$ has stored a value for keyword w, $\mathcal{O}[w]$ returns this value; else it returns a random value and stores it for future queries.

The following variable change shows that if the predictions prove right, the adversary received a proper game instance:

$$\alpha \leftrightarrow \gamma_{chall},\ g_1^{\beta} \leftrightarrow h(w_b^*). \tag{6}$$

The probability that the predictions were correct is clearly non-negligible: $\Pr[u_{predict} = u_{chall}] = 1/N$ and the probability that predicted I is correct is $O(1/l)$, N and l being at most polynomial in the security parameter κ.

Finally from the XDH assumption, DDH is a hard problem in G_1. Thus ϵ_{DDH} is negligible in κ. Given that N and l are at most polynomial in κ and from (5), we have that $|\Pr[S_3] - \Pr[S_2]|$ is negligible which concludes the proof of Lemma 3.

Proof of Theorem 1. In $game_3$ the adversary does not receive any value which depends on the challenge bit, so $\Pr[S_3] = 0.5$. Then Lemma 3 implies that $Pr[S_2]$ is negligibly close to 0.5, Lemma 2 implies that $Pr[S_1]$ is negligibly close to 0.5 and finally Lemma 1 implies that $Pr[S_0]$ is negligibly close to 0.5. This concludes the proof of Theorem 1.

6 Performance Analysis

During the upload phase, the cost for a user of running the *Index* algorithm over the entire index is naturally linear towards the number of keywords in the index. The most costly operation within the *Index* algorithm is one pairing computation; however since inside a same index the second argument of the pairing remains the same between two executions of *Index*, pairing becomes much more efficient than in the case with independent pairings [13].

Furthermore, the *Delegate* algorithm only consists in one exponentiation.

As the search phase involves three parties, namely the reader, the proxy and the CSP, we evaluate the computational cost for each of them.

The *QueryCreate* algorithm executed by the reader is not costly since it only consists of one hashing and one exponentiation. This algorithm outputs a unique query for a given keyword to be searched in several indices. Therefore, the cost of this algorithm does not depend on the number of searched indices. On the other hand, the reader will receive one response per targeted index and will have to execute *ResponseProcess* over each received response. The cost for one response consists in one decryption through *Dec* and one *PIR.Retrieve* operation. Note that the retrieved value for each index is a single bit, and based on [1] the computational overhead can be considered as reasonable for a lightweight user.

The cost for the proxy of multiplexing the queries with *QueryTransform* and filtering the responses with *ResponseFilter* is linear towards the number of indices the querying reader is authorized to search, and for each queried index the proxy performs a pairing and one execution of *PIR.Query*. The *ResponseFilter* algorithm can be considered negligible as it only extracts the relevant part of the response.

For a given keyword search query, the CSP builds one matrix per queried index, and executes *PIR.Process* on each matrix. The building of one matrix requires one exponentiation in G_T per keyword. The operations performed by the CSP being similar to the ones in [9], the workload of the CSP is considered affordable for a cloud server.

To conclude our scheme achieves a very low cost at the reader side, which usually is the main requirement for a cloud computing scenario, and a reasonable cost at the CSP and the proxy. Figure 3 summarizes the cost of each algorithm considering a scenario where one writer uploads several indices and one reader send one query.

Algorithm	Cost	number of executions
Index	$h + e + exp_{G_1}$	$\mathfrak{i}.\mathfrak{k}$
Delegate	exp_{G_2}	$\mathfrak{i}.\mathfrak{d}$
QueryCreate	$h + mult_{\mathbb{Z}_q} + exp_{G_1}$	
QueryTransform	$\mathfrak{a}(e + H + PIR.KeyGen + PIR.Query)$	
Respond	$\mathfrak{k}(exp_{G_T} + h) + \sqrt{n}(PIR.Process + Enc)$	$\mathfrak{a}$
ResponseFilter	negligible (data forwarding)	$\mathfrak{a}$
ResponseProcess	$Dec + PIR.Retrieve$	$\mathfrak{a}$

Key:

- exp_X: cost of an exponentiation in X
- $mult_X$: cost of a multiplication in X
- $\mathfrak{k}$: number of keyword per index
- $\mathfrak{i}$: number of index owned by a writer
- $\mathfrak{d}$: number of reader with delegated search rights per index
- $\mathfrak{a}$: number of indices the reader is authorized to search
- name of a function: execution cost of this function

Fig. 3. Computational cost of each algorithm.

7 Related Work

Our review of the related work focuses on fully multi-user SE schemes. For a detailed survey on SE in general, we refer the reader to [6].

While solutions in [7,9] seem very efficient in the case where there is a single writer authorizing multiple readers, they become unpractical for the multi writer-multi reader case. Indeed each reader should at least store one key per writer and send one query (even if the same) per writer.

Among the few existing MUSE solutions [3,5,8,12,15], all of them except the one described in [12] require the existence of a TTP, which is an unpractical assumption that our solution does not make. Finally, all the solutions share a common pitfall as they do not ensure access pattern privacy. As already discussed in this paper, this leads to a serious privacy exposure in the case where users collude with the CSP. Furthermore the execution of $PIR.Process$ in our solution is less costly compared to the search operation at the CSP in all existing MUSE schemes, since in these schemes the trapdoor must be tested with each encrypted keyword in the index either until the test shows that the keyword is present, or until all the keywords in the index have been tested.

8 Conclusion

We have presented a new multi-user searchable encryption scheme that is provably secure under the newly proposed adversarial model witch considers the case where some users can collude with the CSP. All existing schemes become insecure under this new model. The proposed solution is very efficient for the user as it introduces a new party, the proxy, which bears most of the overhead. At the same time this overhead remains reasonable for both the CSP and the proxy.

Future work on this scheme will include implementation and benchmark results of the presented scheme with realistic datasets.

Acknowledgements. The authors thank the anonymous reviewers for their suggestions for improving this paper.

This work was partially funded by the FP7-USERCENTRICNETWORKING european ICT project (grant 611001).

References

1. Aguilar-Melchor, C., Barrier, J., Fousse, L., Killijian, M.O.: Xpir: Private information retrieval for everyone. Cryptology ePrint Archive, Report 2014/1025 (2014). http://eprint.iacr.org/
2. Aguilar-Melchor, C., Gaborit, P.: A lattice-based computationally-efficient private information retrieval protocol. In: WEWORC 2007 (2007)
3. Asghar, M.R., Russello, G., Crispo, B., Ion, M.: Supporting complex queries and access policies for multi-user encrypted databases. In: Proceedings of the 2013 ACM Workshop on Cloud Computing Security Workshop, CCSW 2013, pp. 77–88. ACM, New York (2013)
4. Ballard, L., Green, M., de Medeiros, B., Monrose, F.: Correlation-resistant storage via keyword-searchable encryption. Cryptology ePrint Archive, Report 2005/417 (2005). http://eprint.iacr.org/
5. Bao, F., Deng, R.H., Ding, X., Yang, Y.: Private query on encrypted data in multi-user settings. In: Chen, L., Mu, Y., Susilo, W. (eds.) ISPEC 2008. LNCS, vol. 4991, pp. 71–85. Springer, Heidelberg (2008)
6. Bösch, C., Hartel, P., Jonker, W., Peter, A.: A survey of provably secure searchable encryption. ACM Comput. Surv. **47**(2), 1–51 (2014). http://dl.acm.org/citation.cfm?doid=2658850.2636328

7. Curtmola, R., Garay, J., Kamara, S., Ostrovsky, R.: Searchable symmetric encryption: Improved definitions and efficient constructions. Cryptology ePrint Archive, Report 2006/210 (2006). http://eprint.iacr.org/
8. Dong, C., Russello, G., Dulay, N.: Shared and searchable encrypted data for untrusted servers. In: Atluri, V. (ed.) DAS 2008. LNCS, vol. 5094, pp. 127–143. Springer, Heidelberg (2008)
9. Elkhiyaoui, K., Önen, M., Molva, R.: Privacy preserving delegated word search in the Cloud. In: SECRYPT 2014, 11th International conference on Security and Cryptography, 28–30 August 2014, Vienna, Austria (2014). http://www.eurecom.fr/publication/4345
10. Gentry, C., Ramzan, Z.: Single-database private information retrieval with constant communication rate. In: Caires, L., Italiano, G.F., Monteiro, L., Palamidessi, C., Yung, M. (eds.) ICALP 2005. LNCS, vol. 3580, pp. 803–815. Springer, Heidelberg (2005)
11. Lipmaa, H.: An oblivious transfer protocol with log-squared communication. In: Zhou, J., López, J., Deng, R.H., Bao, F. (eds.) ISC 2005. LNCS, vol. 3650, pp. 314–328. Springer, Heidelberg (2005)
12. Popa, R.A., Zeldovich, N.: Multi-Key Searchable Encryption (2013). http://people.csail.mit.edu/nickolai/papers/popa-multikey-eprint.pdf
13. Scott, M.: On the efficient implementation of pairing-based protocols. In: Chen, L. (ed.) IMACC 2011. LNCS, vol. 7089, pp. 296–308. Springer, Heidelberg (2011)
14. Shoup, V.: Sequences of games: a tool for taming complexity in security proofs. IACR Cryptology ePrint Archive 2004, 332 (2004). http://www.shoup.net/papers/games.pdf
15. Yang, Y., Lu, H., Weng, J.: Multi-user private keyword search for cloud computing. In: 2011 IEEE Third International Conference on Cloud Computing Technology and Science, pp. 264–271. IEEE, November 2011

Cryptography III: Encryption and Fundamentals

CCA Secure PKE with Auxiliary Input Security and Leakage Resiliency

Zhiwei Wang[1,2](✉) and Siu Ming Yiu[2]

[1] College of Computer, Nanjing University of Posts and Telecommunications, Nanjing 210003, Jiangsu, China
zhwwang@njupt.edu.cn

[2] University of Hong Kong, Pokfulam, Hong Kong

Abstract. Under the strengthened subgroup indistinguishability assumption, we present a new generic construction of chosen ciphertext attack (CCA) secure public key encryption scheme, achieve resilience to auxiliary input information as well as resilience to secret key leakage, from an all-but-one lossy function. In particular, under a special case of SSI assumption, we construct a scheme, if chose the proper parameters for 80-bit security, then it remains CCA secure if any 2^{-2048}-weakly uninvertible functions of secret key is given to the adversary. Furthermore, our scheme also remains CCA secure if any efficient leakage function of secret key is given to the adversary. The leakage rate is $1 - \frac{1690}{l}$, where l is the length of binary representation of secret key. If we choose a sufficiently large l, then the leakage rate is arbitrarily close to 1.

Keywords: CCA secure · PKE · Auxiliary input · Leakage resilient · All-bust-one lossy function

1 Introduction

Traditionally, the security model of cryptographic schemes assumes that an adversary has no access to its internal secret state. Unfortunately, in the real world, an adversary may often learn some partial information about secret key via *side channel attacks* [1–6]. The physical realization of a cryptographic primitive can leak additional information, such as the computation time, power-consumption, radiation/noise/heat emission etc. Recently, much progress has been made in obtaining increasing complex systems with strongly secure against side channel attacks. The emergence of *leakage resilient cryptography* has lead to constructions of many cryptographic primitives [7–15]. A variety of leakage models have been proposed, such as *Exposure-resilient* [19,20], *Only computation leaks information* [21,22], *Bounded leakage model* [8,23,24], *Continual leakage model* [17,25,26], and *Auxiliary input model* [7,16,18] etc.

Auxiliary input model is developed from the *relative leakage* model [7], which allow any uninvertible function f that no probabilistic polynomial-time (PPT) adversary can compute the actual pre-image with non-negligible probability.

J. Lopez and C.J. Mitchell (Eds.): ISC 2015, LNCS 9290, pp. 319–335, 2015.
DOI: 10.1007/978-3-319-23318-5_18

That is to say, although such a function information-theoretically reveals the entire secret key sk, it still computationally infeasible to recover sk from $f(sk)$. If an encryption scheme that is secure w.r.t. any auxiliary input, then user's secret and public key pair can be used for multiple tasks. Dodis et al. [18] firstly introduced the notion of *auxiliary input*, and proposed the public key encryption schemes in this model. Yuen et al. [16] proposed the first IBE scheme that is proved secure even when the adversary is equipped with auxiliary input. In [16], they also propose a model of *continual auxiliary leakage* which combines the concepts of auxiliary inputs with continual memory leakage. However, up to now, all the cryptographic schemes resilience to auxiliary input are only proved chosen plaintext secure (CPA) and cannot achieve chosen ciphertext security (CCA)[1].

Bounded leakage model is a simple but general model, which is formalized by allowing an adversary to adaptively and repeatedly choose functions of the secret key and gain the outputs of the functions, and the total amount of leaked information on the secret key is bounded by λ-bit (called the leakage amount). Bounded leakage model is presented by Akavia et al. [7] and further explored by Naor et al. [1]. In their definition, the leakage amount must strictly smaller than $|sk|$. Bound leakage model is simple and powerful, but a thorough understanding of this model is essential to many other models. If a cryptosystem is secure against key-leakage attacks, we call it *leakage resilient*. We call the ratio $\lambda/|sk|$ the leakage rate of a cryptosystem, which is an obvious goal of designing a leakage resilient cryptosystem. Many leakage resilient PKE schemes with CPA secure have been proposed. In particular, Naor et al. [1] proposed a generic construction of CPA secure leakage resilient PKE schemes from any hash proof systems (HPSs) [27]. The leakage rate of Naor et al. [1] is flexible range over $[0, 1)$, which is called *leakage flexible*. The open problem of leakage resilient CCA secure PKE was also solved by Naor et al. [1], which is relied on the simulation-sound Non-Interactive Zero-Knowledge proof (impractical) or HPSs with leakage rate 1/6. Later, some new variants [28,29] of Cramer and Shoup system [27] by using HPSs are show to be leakage resilient CCA secure with leakage rate 1/4. Very recently, Baodong et al. proposed a new notion called *one-time lossy filter* [14], which can be used to construct leakage resilient CCA secure PKE schemes with leakage rate $1/2 - o(1)$. In [15], Baodong et al. presented the *refined subgroup indistinguishable* (RSI) assumption and a simple case of one-time lossy filter called *all-but-one lossy function* (ABOLF). With the RSI assumption over a specific group, they obtain a CCA secure PKE scheme with leakage rate $1 - o(1)$.

Our Contributions. In this paper, we propose a generic construction of CCA secure PKE schemes, which can achieve *auxiliary input security* as well as *leakage resiliency*. In 2010, Brakerski et al. [9] proposed a generic PKE construction with circular security, leakage resiliency and auxiliary input security, however, their

[1] Indistinguishability under chosen ciphertext attack (CCA) uses a definition similar to that of CPA. However, in addition to the public key, the adversary is given access to a decryption oracle which decrypts arbitrary ciphertexts at the adversary's request, returning the plaintext.

construction only can be proved CPA secure. We define a new assumption called *strengthened subgroup indistinguishable* (SSI) assumption over a finite communicative multiplicative group G, which is a direct product of two cyclic groups $G = G_{\omega_1} \times G_{\omega_2}$ of order ω_1 and ω_2 respectively. The SSI assumption is similar to Baodong et al.'s RSI assumption [15] except that ω_1 and ω_2 are both primes, which can be used to construct simple and efficient ABOLF with large tag space. In our generic PKE construction, an ABOLF is used to verify whether the ciphertext is well-formed, which helps our construction to achieve the CCA security. We also propose an instantiation over a group of known order, which achieves auxiliary input security with any 2^{-2048}-weakly uninvertible functions for 80-bit security, and leakage resiliency with leakage rate $1 - o(1)$.

Organization. The rest of this paper is organized as follows. Basic definitions and related notations are introduced in Sect. 2. The security model of PKE with auxiliary input security and leakage resiliency are proposed in Sect. 3. The generic construction under SSI assumption, and its proofs are presented in Sect. 4. An instantiation over a group of known order is proposed in Sect. 5. Finally, we conclude our paper in Sect. 6.

2 Preliminaries

2.1 Strengthened Subgroup Indistinguishability Assumption

Brakerski et al. [9] defined a generalized class of assumptions called *subgroup indistinguishability* (SI) assumptions. A SI problem is defined by a group G which is a direct product of two groups $G = G_{\omega_1} \times G_{\omega_2}$, and their orders are ω_1 and ω_2 respectively. Here, $\gcd(G_{\omega_1}, G_{\omega_2}) = 1$ and G_{ω_1} is a cyclic group. Essentially, the SI assumption is that a random element of G is computationally indistinguishable from a random element in G_{ω_1}.

In our construction, we should proposed a new notion of SSI assumption, which is strengthened from SI assumption, and similar to Baodong et al.'s RSI assumption [15]. Let $Gen(1^\kappa)$ be a group generation algorithm that, on input a security parameter 1^κ, outputs a description of a finite commutative multiplicative group $\Xi = (G, T, g, h)$, where $G = G_{\omega_1} \times G_{\omega_2}$, and g, h are generators of $G_{\omega_1}, G_{\omega_2}$. The SSI assumption requires that:

- ω_1 and ω_2 are both prime numbers, which is the difference from RSI assumption[2]. This implies that G_{ω_1} and G_{ω_2} are all cyclic groups, and G is also a cyclic group with the order $T = \omega_1 \cdot \omega_2$.
- Elements in G are efficiently checkable.
- An upper bound $T = \omega_1 \cdot \omega_2$ given in the group description, such that for $x \leftarrow_R Z_T$, $x \mod T$ is ϵ-uniform over Z_T, where $\epsilon = \epsilon(\kappa)$ is negligible in κ. This implies that for $x \leftarrow_R Z_T$, g^x (resp. h^x) is also ϵ-uniform over G_{ω_1} (resp. G_{ω_2}).

[2] RSI assumption only requires that G_{ω_1} and G_{ω_2} are both cyclic groups.

Definition 1. *Let* $\Xi = (G, T, g, h) \leftarrow Gen(1^\kappa)$*. The SSI assumption in group* G *states that for any PPT adversary* $\mathcal{A}$*, the advantage*

$$Adv^{ssi}_{\Xi,\mathcal{A}} := |Pr[\mathcal{A}(\Xi, x) = 1 | x \leftarrow_R G_{\omega_1}] - Pr[\mathcal{A}(\Xi, x) = 1 | x \leftarrow_R G]|,$$

$$Adv^{ssi}_{\Xi,\mathcal{A}} := |Pr[\mathcal{A}(\Xi, x) = 1 | x \leftarrow_R G_{\omega_2}] - Pr[\mathcal{A}(\Xi, x) = 1 | x \leftarrow_R G]|.$$

is negligible in κ*.*

From the above definition of SSI assumption, it is easy to derive the following lemma.

Lemma 1. *Let* $\Xi = (G, T, g, h) \leftarrow Gen(1^\kappa)$*. If the SSI assumption holds in group* G*, then for any PPT adversary* $\mathcal{B}$*,*

$$|Pr[\mathcal{B}(\Xi, x) = 1 | x \leftarrow_R G_{\omega_1}] - Pr[\mathcal{B}(\Xi, x) = 1 | x \leftarrow_R G \setminus G_{\omega_1}]| \leq 2Adv^{ssi}_{\Xi,\mathcal{A}}(\kappa),$$

$$|Pr[\mathcal{B}(\Xi, x) = 1 | x \leftarrow_R G_{\omega_1}] - Pr[\mathcal{B}(\Xi, x \cdot h) = 1 | x \leftarrow_R G_{\omega_1}]| \leq 2Adv^{ssi}_{\Xi,\mathcal{A}}(\kappa).$$

Example: Let $\mathbf{P}, p, q$ be distinct prime such that $\mathbf{P} = 2pq + 1$. The lengths of p and q are both at least κ bits, where κ is a security parameter. Obviously, $Z^*_{\mathbf{p}}$ is a quadratic residues group $QR_{\mathbf{p}}$, which order is $T = pq$ [15]. $QR_{\mathbf{p}}$ can be denoted as a direct product $QR_{\mathbf{p}} = G_p \times G_q$, where G_p and G_q are cyclic groups of prime order p and q respectively. Gonzalez et al. [33] proved that the SSI assumption holds over group $QR_{\mathbf{p}}$ if the factoring problem of N is hard.

2.2 All-but-One Lossy Functions

Recently, Baodong et al. proposed a new function called *One-time Lossy Filter* (OT-LF) [14], which is a simple version of *Lossy Algebraic Filter* [30]. If the *One-time Lossy Filter* is operated at "injective mode", then the function is injective. Otherwise, if it is operated at "lossy mode", then the function is non-injective. Compared with *Lossy Algebraic Filter*, *One-time Lossy Filter* does not require efficiently invertible in "injective mode". After then, Baodong et al. proposed a simple variant of OT-LF, namely *all-but-one lossy functions* (ABOLF) [15]. ABOLF is a family of functions parameterized with a tag. In which, all tags are injective and only one lossy tag.

Definition 2. *A collection of* (Dom, ι) *- ABOLF consists of two algorithms:* ABOLF.Gen, ABOLF.Eval*, where*

ABOLF.Gen*: On input* 1^κ *and any* $c^* \in \mathfrak{C}$[3]*, this algorithm generates an evaluation key* EK*.*

ABOLF.Eval*:On input the evaluation key* EK*, a tag* $c \in \mathfrak{C}$ *and* $X \in Dom$*, this algorithm computes* $\text{ABOLF}_{EK,c}(X)$*.*

[3] $\mathfrak{C}$ is a tag space.

We require that ABOLF has the following properties:

Lossiness: If c is an injective tag, then $\text{ABOLF}_{EK,c}(\cdot)$ is also an injective function. Otherwise, If c is a lossy tag, then $\text{ABOLF}_{EK,c}(\cdot)$ is also a lossy function, which only has 2^{ι} possible outputs.

Hidden lossy tag: For any PPT adversary $\mathcal{A}$, and for any $c_0^*, c_1^* \in \mathfrak{C}$, the following advantage is negligible in κ,

$$Adv_{ABO,\mathcal{A}}(\kappa) = |Pr[\mathcal{A}(1^\kappa, EK_0) = 1] - Pr[\mathcal{A}(1^\kappa, EK_1]|$$

where $EK_0 \leftarrow \text{ABOLF.Gen}(1^\kappa, c_0^*)$, $EK_1 \leftarrow \text{ABOLF.Gen}(1^\kappa, c_1^*)$.

Here is an instantiation of ABOLF under the SSI assumption.

Let $\Xi = (G, T, g, h) \leftarrow Gen(1^\kappa)$ and $G = G_{\omega_1} \times G_{\omega_2}$ be defined as the SSI assumption. Let $\mathfrak{C} = \{0,1\}^{\lfloor \log \omega_2 \rfloor - 1}$ and $Dom = Z_T$. Algorithms ABOLF.Gen, ABOLF.Eval can be defined as follows.

ABOLF.Gen$(1^\kappa, c^*)$ On input $c^* \in \mathfrak{C}$, it randomly selects $v \leftarrow_R Z_T$. Finally, ABOLF.Gen returns the $EK = g^v h^{c^*} \in G$.

ABOLF.Eval(EK, c, x) On input EK, $c \in \mathfrak{C}$ and $x \in Z_T$, it computes $f_{EK,c}(x) = (EK \cdot h^{-c})^x = (g^v h^{c^*-c})^x$.

Lemma 2. *The above construction is a family of* $(Z_T, \log \omega_1)$*-ABOLF with tag space* $\mathfrak{C} = \{0,1\}^{\lfloor \log \omega_2 \rfloor - 1}$.

Proof: Obviously, the above construction has the *lossiness* property, since (1) for $c = c^*$, $f_{EK,c^*}(x) = g^{vx}$ has only ω_1 possible values; (2) for $c \neq c^*$, $f_{EK,c}(x) = (g^v h^{c^*-c})^x$ is an injective map. The remainder is to show its hidden lossy tag property. For any $c_0^*, c_1^* \in \mathfrak{C}$, let $\mathcal{EK}_0 = g^v h^{c_0^*}$ be the distribution output by ABOLF.Gen$(1^\kappa, c_0^*)$ and $\mathcal{EK}_1 = g^v h^{c_1^*}$ be the distribution output by ABOLF.Gen$(1^\kappa, c_1^*)$. It suffices to show that $\mathcal{EK}_0$ and $\mathcal{EK}_1$ are computationally indistinguishable under the SSI assumption. To do so, we again define two distributions $\mathcal{EK}_0' = (g \cdot h)^v h^{c_0^*}$ and $\mathcal{EK}_1' = (g \cdot h)^v h^{c_1^*}$. From Lemma 1, it follows that

$$Pr[\mathcal{A}(EK) = 1 | EK \leftarrow_R \mathcal{EK}_0] - Pr[\mathcal{A}(EK) = 1 | EK \leftarrow_R \mathcal{EK}_0'] \leq 2Adv_{\Xi,\mathcal{A}}^{ssi}(\kappa), \tag{1}$$

$$Pr[\mathcal{A}(EK) = 1 | EK \leftarrow_R \mathcal{EK}_1] - Pr[\mathcal{A}(EK) = 1 | EK \leftarrow_R \mathcal{EK}_1'] \leq 2Adv_{\Xi,\mathcal{A}}^{ssi}(\kappa). \tag{2}$$

Additionally, since v is chosen from Z_T uniformly at random, $v \mod \omega_2$ is also uniform over Z_{ω_2} even conditioned on the value of $v \mod \omega_2$ and $g^v = g^{v \mod \omega_1}$, according to the Chinese Remainder Theorem. Consequently

$$(gh)^v h^{c_0^*} = g^v h^{v \mod \omega_2 + c_0^*} \approx_s g^v h^{v \mod \omega_2 + c_1^*} = (gh)^v h^{c_1^*}.$$

So, $\mathcal{EK}_0' \approx_s \mathcal{EK}_1'$, which derives $\mathcal{EK}_0 \approx_c \mathcal{EK}_1$.

Since ω_2 is a prime number, we can set $\mathfrak{C} = \{0,1\}^{\lfloor \log \omega_2 \rfloor - 1}$. In this case, $\gcd(c^* - c, \omega_2) = 1$, hence $(c^* - c)^{-1} \mod \omega_2$ always exists. Thus, the tag space can be set $\mathfrak{C} = \{0,1\}^{\lfloor \log \omega_2 \rfloor - 1}$. This completes the proof of lemma. □

2.3 Chameleon Hash Function

A chameleon hash function **Ch** is essentially a keyed and randomized hash function, which consists of three PPT algorithms (**Ch.Gen, Ch.Eval, Ch.Equiv**). The key generation algorithm $Ch.Gen(1^\lambda)$ takes as input a security parameter 1^λ, and returns a key pair (ek_{Ch}, td_{Ch}). The evaluation algorithm $Ch.Eval(ek_{Ch}, x; r)$ takes as input a preimage $x \in \{0,1\}^*$ and a randomness $r \in \mathcal{R}$, and computes a hash value y. If r is uniformly distributed over $\mathcal{R}$, so is y over its range. **Ch** is collision resistant on the condition that for any PPT adversary $\mathcal{A}$, the following probability

$$Adv^{cr}_{Ch,\mathcal{A}}(1^\lambda) := Pr[(x', r') \neq (x, r) \wedge Ch.Eval(ek_{Ch}, x'; r') = Ch.Eval(ek_{Ch}, x; r) | \\ (ek_{Ch}, td_{Ch}) \leftarrow Ch.Gen(1^\lambda), (x', r', x, r) \leftarrow \mathcal{A}(ek_{Ch})]$$

is negligible in λ. However, given x, r, x' and the trapdoor key td_{Ch}, $Ch.Equiv(td_{Ch}, x, r, x')$ can compute r' such that $Ch.Eval(ek_{Ch}, x; r) = Ch.Eval(ek_{Ch}, x'; r')$ and the distribution of r' is also uniform over $\mathcal{R}$ given only ek_{Ch} and x.

According to [15], an ABOLF can be transformed into a OT-LF by using a chameleon hash function. The Fig. 1 shows such a construction.

2.4 Goldreich-Levin Theorem for Large Fields

We recall the Goldreich-Levin Theorem for Large Fields [18] over any field $GF(q)$ for a prime q.

Theorem 1 ***(GL Theorem for Large Fields)****. Let q be a big prime, and let H be a subset of $GF(q)$. Let f mapping from H^m to $\{0,1\}^*$ be any function. Randomly chooses a vector s from H^m, and computes $y = f(s)$. Then, randomly selects a vector r for $GF(q)^m$. If a PPT distinguisher $\mathfrak{D}$ runs in time t, and there exists a negligible probability ϵ such that*

$$|Pr[\mathfrak{D}(y, r, < r, s >) = 1] - Pr[u \leftarrow GF(q) : \mathfrak{D}(y, r, u) = 1]| = \epsilon,$$

then given $y \leftarrow f(s)$, there exists an inverter $\mathfrak{A}$ can compute s from y in time $t' = t \cdot poly(m, |H|, 1/\epsilon)$ with the probability

$$Pr[s \leftarrow H^m, y \leftarrow f(s) : \mathfrak{A}(y) = s] \geq \frac{\epsilon^3}{512 \cdot m \cdot q^2}.$$

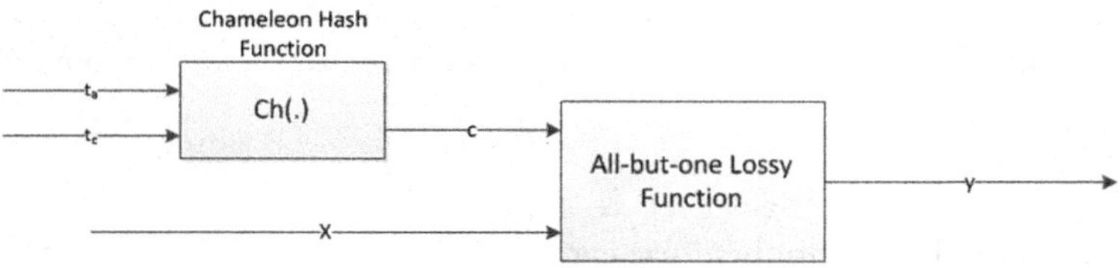

Fig. 1. Construction of a one-time lossy filter

2.5 DDH Assumption

Definition 3 ***(The Decisional Diffie Hellman Assumption*** *[18]). Let G be a cyclic group with prime order q and g is the generator of G. The Decisional Diffie Hellman (DDH) Assumption holds on G iff*

$$Adv^{ddh}_{G,\mathcal{A}}(\kappa) = |Pr[\mathcal{A}(g_1, g_2, g_1^r, g_2^r) = 1] - Pr[\mathcal{A}(g_1, g_2, g_1^r, g_2^{r'}) = 1]|$$

is negligible in κ *for any PPT adversary* $\mathcal{A}$*, where* $g_1, g_2 \leftarrow G$*,* $r \leftarrow Z_q$ *and* $r' \leftarrow Z_q \setminus \{r\}$.

Naor et al. proposed a Lemma [31] which states that a natural generalization of DDH assumption which considers $l > 2$ generators is actually equivalent to DDH.

Lemma 3 *Under the DDH assumption on G, for any positive integer l,*

$$Adv^{ddh}_{G,\mathcal{A}}(\kappa) = |Pr[\mathcal{A}(g_1, \cdots, g_l, g_1^r, \cdots, g_l^r) = 1] - Pr[\mathcal{A}(g_1, \cdots, g_l, g_1^{r_1}, \cdots, g_l^{r_l}) = 1]|$$

is negligible in κ *for any PPT adversary* $\mathcal{A}$*, where* $g_1, \cdots, g_l \leftarrow G$*,* $r \leftarrow Z_q$ *and* $r_1, \cdots, r_l \leftarrow Z_q \setminus \{r\}$.

2.6 Min-entropy

We define $SD(X, Y) = 1/2 \sum_{\omega \in \Omega} |Pr[X = \omega] - Pr[Y = \omega]|$ as the statistical distance of random variables X and Y over domain Ω. The *min-entropy* of X is $H_\infty(X) = -\log(max_{w in \Omega} Pr[X = \omega])$. The average min-entropy of X conditioned on Y can be defined as $\tilde{H}_\infty(X|Y) = -\log(E_{y \leftarrow Y}[2^{-H_\infty(X|Y=y)}])$. Dodis et al. [32] proposed the following property of average min-entropy.

Lemma 4 *Let X, Y and Z be random variables. If Z has at most* 2^r *possible values, then* $\tilde{H}_\infty(X|(Y, Z)) \geq \tilde{H}_\infty(X|Y) - r$.

3 Auxiliary Input and Leakage Resilient Public Key Encryption

A public key encryption (PKE) scheme with secret key space $\mathcal{SK}$, public key space $\mathcal{PK}$ and message space $\mathcal{M}$ consists of a triple of PPT algorithms (PKE.G, PKE.E, PKE.D). The key generation algorithm PKE.G takes as input 1^κ, and outputs a pair of public/secret keys (pk, sk). The encryption algorithm PKE.G takes as input a public key pk and a message $m \in \mathcal{M}$, and outputs a ciphertext $CT = PKE.E(pk, m)$. The decryption algorithm PKE.D takes as input a secret key sk and a ciphertext CT, and returns a message m or $\perp$. For consistency, we require that $PKE.D(sk, PKE.E(pk, m)) = m$ holds for all $(pk, sk) \leftarrow$ PKE.G(1^κ) and all messages $m \in \mathcal{M}$.

3.1 Auxiliary Input CCA Security of PKE

In the scenario of auxiliary input, the attacker can access additional information about the secret key. The auxiliary input model defines a class of computationally uninvertible function families $\mathcal{F}$ to simulate a large class of leakage. However, although such a function $f \in \mathcal{F}$ can information-theoretically reveal the entire secret key sk, but it still computationally infeasible to recover sk. For any family of functions $\mathcal{F} = \{f : \mathcal{SK} \rightarrow \{0,1\}^*\}$, we define the inverting advantage and weakly inverting advantage of an adversary $\mathcal{A}$ as follows:

$$Adv_f = Pr[\mathcal{A}(1^\kappa, f(sk)) = sk|(pk, sk) \leftarrow PKE.G(1^\kappa)]$$
$$Adv_f^{weak} = Pr[\mathcal{A}(1^\kappa, pk, f(sk)) = sk|(pk, sk) \leftarrow PKE.G(1^\kappa)]$$

Let l denote the length of binary representation of sk. A polynomial time computable function f is $\epsilon = \epsilon(l)$-weakly hard to invert(resp. ϵ-hard to invert) if for any PPT $\mathcal{A}$ it holds that $Adv_f^{weak} \leq \epsilon$ (resp. $Adv_f \leq \epsilon$). Then, for any efficiently computable function family $\mathcal{F}$, we provide the definition of auxiliary input game as follows:

Initialize: The simulator selects $b \leftarrow_R \{0,1\}$ and generates a key pair $(pk, sk) \leftarrow PKE.G(1^\kappa)$. The simulator sends pk to the adversary.

Auxiliary Input: The challenger computes $z = f(sk)$ and sends z to the adversary.

Query1: The adversary can make decryption queries (polynomial times) on any ciphertext CT, and the challenger returns with the messages $m =$ PKE.D(sk, CT).

Challenge: The adversary sends $m_0, m_1 \in \mathcal{M}$ to the challenger. The challenger computes $CT^* =$ PKE.E(pk, m_b) and sends CT^* to the adversary.

Query2: In this phase, the adversary also can make decryption queries (polynomial times) on any ciphertext CT except for CT^*.

Output: The adversary outputs a guess b' for b.

The PKE scheme is called ε-weak auxiliary input CCA secure if for any ε-weakly uninvertible f and for any PPT adversary $\mathcal{A}$, $Adv_{f,\mathcal{A}}^{weak} = Pr[b' = b]$ is negligible. The PKE scheme is called ε-auxiliary input CCA secure if the above holds for any ε-uninvertible fs.

3.2 Leakage Resilient CCA Security of PKE

The scenario of leakage resilient is quite similar to that of auxiliary input. The attacker also can get some bounded information about the secret key. However, the restriction on the amount of information is *information theoretic* rather than *computational.* We define the λ-leakage game as follows:

Initialize: The simulator selects $b \leftarrow_R \{0,1\}$ and generates a key pair $(pk, sk) \leftarrow PKE.G(1^\kappa)$. The simulator sends pk to the adversary.

Leakage: The adversary sends an efficient computable function $f : \mathcal{SK} \rightarrow \{0,1\}^\lambda$ to the challenger. The challenger computes $f(sk)$ and returns it to the adversary.

Query1: The adversary can make decryption queries (polynomial times) on any ciphertext CT, and the challenger returns with the messages $m =$ PKE.D(sk, CT).

Challenge: The adversary sends $m_0, m_1 \in \mathcal{M}$ to the challenger. The challenger computes $CT^* =$ PKE.E(pk, m_b) and sends CT^* to the adversary.

Query2: In this phase, the adversary also can make decryption queries (polynomial times) on any ciphertext CT except for CT^*.

Output: The adversary outputs a guess b' for b.

The PKE scheme is called λ-bit leakage resilient CCA secure if for any PPT adversary $\mathcal{A}$, $Adv_{\lambda,\mathcal{A}} = Pr[b' = b]$ is negligible. In the case of $\lambda = 0$, the PKE scheme is just the standard CCA secure scheme.

4 The Generic Construction

The construction is parameterized by a group G as described in Sect. 2.1. Namely, the probabilistic algorithm $Gen(1^\kappa)$ produce an instance $\Xi = \{G, T, g, h\}$, where $G = G_{\omega_1} \times G_{\omega_2}$, $T = \omega_1 \cdot \omega_2$. An additional parameter is the value l that is polynomial in the security parameter κ but its exact value is determined based on the specific application. Let $\mathcal{H}$ be a 2-universal hash family[4], consisting of (deterministic) functions $H : G_{\omega_1}^l \times G \rightarrow \{0,1\}^{\lfloor \log \omega_2 \rfloor - 1}$.[5] We assume that there exists an efficient injective mapping $inj : G_{\omega_1} \rightarrow Z_T$. The encryption scheme PKE = (KeyGen, Encrypt, Decrypt) can be defined as follows.

KeyGen: This algorithm first samples an instance $\Xi = \{G, T, g, h\}$ of SSI assumption. Then, it selects a uniformly random $\mathbf{s} = (s_1, \cdots, s_l) \in \{0,1\}^l$, and $\mathbf{g} = (g_1, \cdots, g_l) \in G_{\omega_1}^l$. It sets $g_0 = \prod_{i \in [l]} g_i^{s_i}$, and then randomly chooses $c^* \in \{0,1\}^{\lfloor \log \omega_2 \rfloor - 1}$ and $v \leftarrow_R Z_T$. It computes $EK = g^v h^{c^*} \in G$, and selects H uniformly at random from $\mathcal{H}$. Then, it chooses a chameleon hash function **Ch**=$(Ch.Gen, Ch.Eval, Ch.Equiv)$ with key pair (ek_{Ch}, td_{Ch}) and randomness space $\mathcal{R}$. Finally, outputs the public key $pk = (\mathbf{g}, g_0, ek_{Ch}, EK, H)$, and the secret key $sk = \mathbf{s}$. The instance Ξ is an additional implicit public parameter.

Encrypt: This algorithm takes as input a public key pk and a message $M \in G$. It randomly selects $r \in Z_T, t_c \in \mathcal{R}$ and computes $C = (g_1^r, \cdots, g_l^r)$. It then computes

$$\psi = g_0^r \cdot M, H(C, \psi) = c, \Pi = (EK \cdot h^{-c})^{inj(g_0^r)}$$

The ciphertext is $CT = (C, \psi, \Pi, c)$.

---

[4] For any $x \neq y$, we have $Pr_{H \leftarrow_R \mathcal{H}}[H(x) = H(y)] \leq \frac{1}{2^{\lfloor \log \omega_2 \rfloor - 1}}$.

[5] We assume that G is q-order elliptic curve group over finite field $\mathbb{F}_p$. For 80-bit security, p and q can be chosen to be 160-bit prime. Thus, in such a group, each element can be denoted as a 160-bit strings. So, in this case, the hash function H maps a $(l+1) \times 160$-bit string to a $\lfloor \log \omega_2 \rfloor - 1$-bit string.

Decrypt: This algorithm takes as input a secret key $sk = \mathbf{s}$ and a ciphertext $CT = (C, \psi, \Pi, c)$ where $C = (g_1^r, \cdots, g_l^r)$, computes $K = \prod_{i\in[l]}(g_i^r)^{s_i}$ and $\Pi' = (EK \cdot h^{-c})^{inj(K)}$. It checks whether $\Pi = \Pi'$. If not, it rejects with $\perp$. Otherwise, it outputs the message $M = \psi/K$.

Note: *1) In our construction, both the* ABOLF *and the whole PKE scheme are over the same group* $G = G_{\omega_1} \times G_{\omega_2}$*, which will greatly reduce the number of system parameters. 2) Since there is only one challenge ciphertext in the CCA secure proof, and the* ABOLF *has only one lossy tag, we only use the* ABOLF *to authenticate the encapsulated key.*

Theorem 2. *Let* $l = (4 \log \omega_1)^{1/\epsilon}$*, given an instance* $\Xi = \{G, T, g, h\}$ *of SSI assumption, if DDH assumption holds in group* G_{ω_1}*, then the encryption scheme* PKE *in this section is auxiliary input CCA secure with any* 2^{-l^ϵ}*-weakly uninvertible function under the Goldreich-Levin Theorem for Large Fields. Particularly,*

$$Adv_{PKE,\mathcal{A}}^{AI-CCA} \leq 4Adv_{\Xi,\mathcal{B}_1}^{ssi}(\kappa) + Adv_{G_{\omega_1},\mathcal{B}_4}^{ddh} + \frac{1}{2^{\lfloor \log \omega_2 \rfloor}} + negl(\kappa).$$

Proof: We prove by a game argument by using a sequence of games, $\text{Game}_0, \cdots, \text{Game}_6$, played between a simulator $\mathcal{S}$ and a PPT adversary $\mathcal{A}$. In each game, the adversary $\mathcal{A}$ outputs a bit b' as a guess for the random bit b selected by $\mathcal{S}$. Let E_i denote the event that $b' = b$ in G_i.

Game_0: This is the original auxiliary input CCA secure game. The simulator $\mathcal{S}$ generates the public/secret key pair (pk, sk) by using the KeyGen algorithm, and sends pk to $\mathcal{A}$. when $\mathcal{A}$ makes decryption query CT or leakage query, $\mathcal{S}$ responses with Decrypt(sk, CT) or $f(sk)$ using the secret key sk and public key pk. Then, $\mathcal{A}$ provides two messages M_0 and M_1 of equal length to $\mathcal{S}$. $\mathcal{S}$ randomly selects a bit b and sends the challenge ciphertext $CT^* = \text{Encrypt}(pk, M_b)$ to $\mathcal{A}$. Following that, $\mathcal{S}$ continues to answer $\mathcal{A}$'s decryption queries except that $CT = CT^*$. Finally, $\mathcal{A}$ outputs a bit b' as a guess of b. We have $Adv_{PKE,\mathcal{A}}^{AI-CCA} = |Pr[E_0] - 1/2|$.

Game_1: The game is exactly like Game_0, except for the generation of the tag c^* in the challenge ciphertext. When $\mathcal{S}$ uses the KeyGen algorithm, it keeps the lossy tag c^* of EK as well as sk. Instead of randomly choosing $c^* \in \mathfrak{C}$, $\mathcal{S}$ uses the lossy tag c^* which it keeps from the KeyGen algorithm. Due to ABOLF's property of hidden lossy tag (hlt), we have $|Pr[E_1] - Pr[E_0]| \leq Adv_{\text{ABOLF},\mathcal{B}_1}^{hlt}(\kappa)$ for a PPT adversary $\mathcal{B}_1$ on ABOLF's hidden lossy tag. Furthermore, from (1) and (2) in the proof of Lemma 2, we have

$$|Pr[E_1] - Pr[E_0]| \leq Adv_{\text{ABOLF},\mathcal{B}_1}^{hlt}(\kappa) \leq 4Adv_{\Xi,\mathcal{B}_1}^{ssi}(\kappa).$$

Game_2: This game is exactly like the Game_1, except for a copied lossy tag. That is to say, when the adversary $\mathcal{A}$ queries on a ciphertext $CT = (C, \psi, \Pi, c)$ such that $c = c^*$, the decryption oracle should halt immediately, and outputs $\perp$. If $\Pi = \Pi^*$ and H is a 2-universal hash function, then it implies $CT = CT^*$ with high probability ($\geq 1 - \frac{1}{2^{\lfloor \log \omega_2 \rfloor - 1}}$). In this case, $\mathcal{A}$ is not allowed to ask the decryption oracle for the challenge ciphertext in both Game$_1$ and Game$_2$. If $\Pi \neq \Pi^*$, since $c = H(C, \psi) = H(C^*, \psi^*) = c^*$ and H is a 2-universal hash function, it follows that $(C, \psi) = (C^*, \psi^*)$ with high probability ($\geq 1 - \frac{1}{2^{\lfloor \log \omega_2 \rfloor - 1}}$), which implies $K = K^*$ with the same probability. Then, from ABOLF.Eval$_{EK,c}(inj(K))$ = ABOLF.Eval$_{EK,c^*}(inj(K^*)) = \Pi^*$, such decryption queries still should be rejected in Game$_1$. From the discussion above, we have $Pr[E_2] - Pr[E_1] \leq \frac{1}{2^{\lfloor \log \omega_2 \rfloor - 1}}$.

Game$_3$: The game is exactly like Game$_2$, except for the generation of $g_0^{r^*}$ used in the challenge ciphertext. In this game, $\mathcal{S}$ computes $g_0^{r^*}$ using the secret key sk and C^* instead of using the public key pk. Since $g_0^{r^*} = \prod_{i\in[l]} (g_i^{r^*})^{s_i}$, this change is purely conceptual, and thus $Pr[E_3] = Pr[E_2]$.

Game$_4$: The game is exactly like Game$_3$, except for the generation of vector C^* in the challenge ciphertext $(C^*, \psi^*, \Pi^*, c^*)$. In this game, $\mathcal{S}$ selects the vector $C^* = (f_1^*, \cdots, f_l^*)$ randomly from $G_{\omega_1}^l$, and $\psi^* = \prod_{i\in[l]} f_i^{s_i} \cdot M_b$ as before. Since the DDH problem is hard for G_{ω_1}, we have $Pr[E_4] - Pr[E_3] \leq Adv^{ddh}_{G_{\omega_1}, \mathcal{B}_4}$

Game$_5$: The game is exactly like Game$_4$, except that a special rejection rule is applied to the decryption oracle. If the adversary queries a ciphertext $CT = (C, \psi, \Pi, t_c)$ for decryption such that C is invalid ($d\log_{g_i} C_i \neq d\log_{g_0} \prod_{i\in[l]} C_i^{s_i}$[6] where C_i is the ith element of C), then the decryption oracle immediately halts and outputs $\perp$. Let Γ denote the event that a ciphertext is rejected by Game$_5$, while it would not be rejected by Game$_4$. Thus, we have $Pr[E_5] - Pr[E_4] \leq Pr[\Gamma]$. We prove the following lemma which guarantees that Γ occurs with a negligible probability.

Lemma 5. *The event Γ occurs with a negligible probability*

$$Pr[\Gamma] \leq \frac{1}{2^{\lfloor \log \omega_2 \rfloor - 1}} + negl(\kappa).$$

Proof: We define Δ to be a event that in Game$_4$ there exists a decryption query $CT = (C, \psi, \Pi, c)$ such that c is a copied lossy tag, then we have

$$Pr[\Gamma] = Pr[\Gamma \wedge \Delta] + Pr[\Gamma \wedge \bar{\Delta}] \leq Pr[\Gamma|\Delta] + Pr[\Gamma|\bar{\Delta}].$$

- Γ cannot occur given Δ. Since $c = H(C, \psi) = c^*$ is a copied lossy tag, however, since H is 2-universal hash function, we have $(C, \psi) = (C^*, \psi^*)$ with high probability. In this case, $\mathcal{A}$ is not allowed to ask the decryption oracle for the challenge ciphertext in Game$_2$ – Game$_4$. Thus, $Pr[\Gamma|\Delta] \leq \frac{1}{2^{\lfloor \log \omega_2 \rfloor - 1}}$.
- Suppose that Γ happens given $\bar{\Delta}$. In this case, $d\log_{g_i} C_i \neq d\log_{g_0} \prod_{i\in[l]} C_i^{s_i}$ but

$$\Pi = \Pi' = ABOLF.Eval(EK, c, inj(\prod_{i\in[l]} C_i^{s_i})) \tag{3}$$

while c is an injective ABOLF tag. We assume that $d\log_g g_i = k_i$ and $\mathbf{k} = (k_1, \cdots, k_l)$, then $\prod_{i\in[l]} C_i^{s_i}$ can be denoted as $r < \mathbf{k}, \mathbf{s} >$. Since C is

[6] Here, $d\log$ denote the discrete logarithm.

invalid ($d\log_{g_i} C_i \neq d\log_{g_0} \prod_{i\in[l]} C_i^{s_i}$), we have $r'(\neq r)$ is uniformly random. As the definition of ABOLF, when c is an injective tag, *ABOLF.Eval* is also an injective function. Thus, if $\mathcal{A}$ want to make (3) hold, then it should make $< r\cdot \mathbf{k}, \mathbf{s} >= r'u$. When the adversary $\mathcal{A}$ gets the right value u (corresponding to a random r') according to $< r\cdot \mathbf{k}, \mathbf{s} >$, it can make the event $\varGamma$ occur. Since r' is uniformly random, $u =< r\mathbf{k}, \mathbf{s} > /r'$ is also uniformly random. If the adversary $\mathcal{A}$ want to determine u (corresponding to a random r'), then it should firstly know the exactly value of $< r\cdot \mathbf{k}, \mathbf{s} >$, where $r\cdot \mathbf{k}$ is public, and thus $\mathcal{A}$ should know $\mathbf{s}$. However, from the view of adversary $\mathcal{A}$, it only get the uninvertible auxiliary input $f(\mathbf{s})$ and the public key pk. Assume towards contradiction that there exists a non-negligible probability δ such that $Pr[\varGamma|\bar{\varDelta}] = \delta$.

We reduce the task of inverting $f(\mathbf{s})$ (with suitable probability) to the task of gaining some non-negligible $\delta = \delta(\kappa)$ advantage of event $\varGamma$. Let $l = (4\log\omega_1)^{1/\epsilon}$, and let f be any 2^{-l^ϵ}-weakly uninvertible function (more precisely, family of functions). It follows that for any adversary $\mathcal{C}$, $Pr[\mathcal{C}(f(\mathbf{s}))] < 2^{-l^\epsilon}$. Our assumption implies that when the event $\varGamma$ occurs with the probability δ, the distribution

$$(pk, f(\mathbf{s}), r\mathbf{k}, < r\mathbf{k}, \mathbf{s} >) \text{ and } (pk, f(\mathbf{s}), r\mathbf{k}, u)$$

are distinguishable with the advantage greater than δ. In this case, it follows from the Goldreich-Levin Theorem for Large Fields that there exists a $\mathcal{C}$ whose running time is at most $poly(l, 2, 1/\delta) = poly(1/\delta)$, such that

$$Pr[\mathcal{C}(f(\mathbf{s})) = \mathbf{s}] \geq \omega_1 \cdot \frac{\delta^3}{512l\cdot\omega_1^3} > \omega_1 \cdot \frac{1}{512l\cdot 2^{3l^\epsilon/4}\cdot poly(l)} > 2^{-l^\epsilon}.$$

We reach a contradiction and therefore, $Pr[\varGamma|\bar{\varDelta}] \leq negl(\kappa)$.

Thus, $Pr[\varGamma] \leq Pr[\varGamma|\varDelta] + Pr[\varGamma|\bar{\varDelta}] \leq \frac{1}{2^{\lfloor\log\omega_2\rfloor-1}} + negl(\kappa)$. □

From the Lemma 5, we have $Pr[E_5] - Pr[E_4] \leq Pr[\varGamma] \leq \frac{1}{2^{\lfloor\log\omega_2\rfloor-1}} + negl(\kappa)$.

Game$_6$: The game is exactly like Game$_5$, except for the generation of ψ^* in CT^*. In this game, $\mathcal{S}$ selects ψ^* randomly in G instead of computing $\psi^* = \prod_{i\in[l]} f_i^{s_i} \cdot M_b$. In Game$_4$, $f_1, \cdots, f_l$ are randomly chosen from G_{ω_1}. Thus, if l is large enough, then the left over hash lemma [9] implies that $\prod_{i\in[l]} f_i^{s_i}$ is almost uniform. Therefore, if $\prod_{i\in[l]} f_i^{s_i}$ is uniform, then it is distributed identically to $\prod_{i\in[l]} f_i^{s_i} \cdot M_b$. Hence, $Pr[E_6] = Pr[E_5]$. Observe that in Game$_6$, the challenge ciphertext is complete independent of the random coin b picked by $\mathcal{S}$. Thus, $Pr[E_6] = 1/2$. □

Theorem 3. *The encryption scheme* PKE *in this section is leakage resilient CCA secure with leakage bits* $\lambda \leq l - \log\omega_1 - \log T - \omega(\log\kappa)$. *Particularly,*

$$Adv_{PKE,\mathcal{A}}^{LR-CCA} \leq 4Adv_{\varXi,\mathcal{A}}^{ssi}(\kappa) + Adv_{G_{\omega_1},\mathcal{B}_4}^{ddh} + \frac{1}{2^{\lfloor\log\omega_2\rfloor}} + \frac{Q(\kappa)2^{\lambda+\log\omega_1+\log T}}{2^l - Q(\kappa)}.$$

The proof (below) follows the same steps as the proof of Theorem 2. The only change is the Game_5.

Proof: We use the exact same games as in the proof of Theorem 2 (with the exception that f is a length bounded leakage function rather that a $2^{-l^{\epsilon}}$-weakly uninvertible function). The exact same arguments imply that

$$\begin{aligned} Adv_{PKE,\mathcal{A}}^{LR-CCA} &= |Pr[E_0] - 1/2| \\ |Pr[E_1] - Pr[E_0]| &\leq 4Adv_{\Xi,\mathcal{B}_1}^{ssi}(\kappa) \\ Pr[E_2] - Pr[E_1] &\leq \frac{1}{2^{\lfloor \log \omega_2 \rfloor - 1}} \\ Pr[E_3] &= Pr[E_2] \\ Pr[E_4] - Pr[E_3] &\leq Adv_{G_{\omega_1},\mathcal{B}_4}^{ddh} \\ Pr[E_6] &= Pr[E_5] \\ Pr[E_6] &= 1/2 \end{aligned}$$

Therefore, it remains to prove a bound on $Pr[E_5] - Pr[E_4]$. Game_5 is exactly like Game_4, except that a special rejection rule is applied to the decryption oracle. If the adversary queries a ciphertext $CT = (C, \psi, \Pi, t_c)$ for decryption such that C is invalid ($d\log_{g_i} C_i \neq d\log_{g_0} \prod_{i\in[l]} C_i^{s_i}$). We also assume that Γ denote the event that a ciphertext is rejected by Game_5, while it would not be rejected by Game_4. Thus, we have $Pr[E_5] - Pr[E_4] \leq Pr[\Gamma]$.

Lemma 6. *Suppose that the adversary $\mathcal{A}$ makes at most $Q(\kappa)$ decryption queries, then*

$$Pr[\Gamma] \leq \frac{1}{2^{\lfloor \log \omega_2 \rfloor - 1}} + \frac{Q(\kappa)2^{\lambda + \log \omega_1 + \log T}}{2^l - Q(\kappa)}$$

where l is the bit length of the secret key.

Proof: We also define Δ to be a event that in Game_4 there exists a decryption query $CT = (C, \psi, \Pi, c)$ such that c is a copied lossy tag, then we have

$$Pr[\Gamma] = Pr[\Gamma \wedge \Delta] + Pr[\Gamma \wedge \bar{\Delta}] \leq Pr[\Gamma|\Delta] + Pr[\Gamma|\bar{\Delta}].$$

- $Pr[\Gamma|\Delta] \leq \frac{1}{2^{\lfloor \log \omega_2 \rfloor - 1}}$, and the proof is the same as Lemma 5.
- Suppose that $CT = (C, \psi, \Pi, c)$ is the first ciphertext that makes Γ happen given $\bar{\Delta}$. As we analyzed in the proof of Lemma 5, if the adversary $\mathcal{A}$ want to make the event Γ occur given $\bar{\Delta}$, then it should know the secret key $\mathbf{s}$. However, observe that only pk, the challenge ciphertext CT^*, and the leakage of at most λ bits of the secret key can be used to reconstruct $\mathbf{s}$. In the challenge ciphertext CT^*, only ψ^* and Π^* are related to the secret key $\mathbf{s}$, and ψ^* has at most $2^{\log T}$ possible values and Π^* has at most $2^{\log \omega_1}$ possible values. Since C is invalid, C and $\mathbf{s}$ are independent. Thus, $\tilde{H}_\infty(\mathbf{s}|(pk, C)) = H_\infty(\mathbf{s}) = l$. According to Lemma 4, we have

$$\begin{aligned} \tilde{H}_\infty(\mathbf{s}|pk, C, CT^*, \lambda - leakage) &\geq \tilde{H}_\infty(\mathbf{s}|pk, C, CT^*) - \lambda \\ &\geq \tilde{H}_\infty(\mathbf{s}|pk, C) - \lambda - \log \omega_1 - \log T \\ &= l - \lambda - \log \omega_1 - \log T \end{aligned}$$

Thus, the probability of recovering $\mathbf{s}$ is at most $2^{\lambda+\log\omega_1+\log T}/2^l$. Hence, in Game$_5$, the decryption algorithm accepts the first invalid ciphertext with the same probability. Since $\mathcal{A}$ makes at most $Q(\kappa)$ decryption queries, it follows that

$$Pr[\Gamma|\bar{\Delta}] \leq \frac{Q(\kappa)2^{\lambda+\log\omega_1+\log T}}{2^l - Q(\kappa)},$$

which is negligible in κ if $\lambda \leq l - \log\omega_1 - \log T - \omega(\log\kappa)$.

Thus, $Pr[\Gamma] \leq Pr[\Gamma|\Delta] + Pr[\Gamma|\bar{\Delta}] \leq \frac{1}{2^{\lfloor\log\omega_2\rfloor-1}} + \frac{Q(\kappa)2^{\lambda+\log\omega_1+\log T}}{2^l-Q(\kappa)}$. □

From the Lemma 6, we have $Pr[E_5] - Pr[E_4] \leq Pr[\Gamma] \leq \frac{1}{2^{\lfloor\log\omega_2\rfloor-1}} + \frac{Q(\kappa)2^{\lambda+\log\omega_1+\log T}}{2^l-Q(\kappa)}$. □

5 Instantiation over a Group of Known Order

Let $\mathbf{P}, p, q$ be distinct prime such that $\mathbf{P} = 2pq + 1$. The lengths of p and q are both at least κ bits, where κ is a security parameter. Obviously, $Z_{\mathbf{p}}^*$ is a quadratic residues group $QR_{\mathbf{p}}$, which order is $T = pq$ [15]. $QR_{\mathbf{p}}$ can be denoted as a direct product $QR_{\mathbf{p}} = G_p \times G_q$, where G_p and G_q are cyclic groups of prime order p and q respectively. If we randomly chose $x, y \leftarrow_R Z_{\mathbf{p}}^*$, then $g = x^q$ and $h = y^p$ are the generators of G_p and G_q respectively. If integer factorization of T is hard, the SSI assumption holds over the group $QR_{\mathbf{p}}$. Thus, let $G = QR_{\mathbf{p}}$, $G_{\omega_1} = G_p$, $G_{\omega_2} = G_q$, $T = pq$, $g = x^q$ (for $x \leftarrow_R Z_{\mathbf{p}}^*$), and $h = y^p$ (for $y \leftarrow_R Z_{\mathbf{p}}^*$), we can obtain an instantiation (over a group of known order) of SSI assumption by setting $I_G = (G, T, g, h)$.

Firstly, we can use I_G to realize an instantiation of ABOLF in Sect. 2.3, which is a $(Z_T, \log p)$-ABOLF with tag space $\mathfrak{C} = \{0,1\}^{\lfloor\log q\rfloor-1}$. Secondly, we can use I_G to initiate our generic PKE construction under the SSI assumption. For a 80-bit security level, we choose $|p| = |q| = 512$ bits, then $|T| = 1024$ bits, which suffices to guarantee that T is hard to be factored, and thus the SSI assumption holds over $QR_{\mathbf{p}}$. In this case, this instantiation can be proved auxiliary input CCA secure with any 2^{-2048}-weakly uninvertible functions. For the bounded leakage model, we choose $\omega(\log\kappa) = 160$ bits, then the leakage bits of secret key $\lambda \leq l - 512 - 1024 - 160 = l - 1690$, where $l = |sk|$. Therefore, the leakage rate of $\frac{\lambda}{|sk|} = \frac{l-1690}{l} = 1 - \frac{1690}{l}$. Thus, if we choose a sufficiently large l, then the leakage rate is arbitrarily close to 1. So, this instantiation is also a leakage-flexible scheme.

6 Conclusions

We present a new generic construction of PKE scheme secure against auxiliary input and leakage resilient, based on the SSI assumption. In order to achieve CCA secure, our construction use an ABOLF to verify whether the ciphertext is well-formed. Instantiation over a group of known order, which is a special case of

SSI assumption, shows that our construction can achieve auxiliary input CCA secure with any 2^{-2048}-weakly uninvertible functions, for the 80-bit security. Moreover, this instantiation is also a leakage-flexible scheme, since the leakage rate of $\frac{\lambda}{|sk|} = \frac{l-1690}{l} = 1 - \frac{1690}{l}$, where l is the bit length of secret key. If we choose a sufficiently large l, then the leakage rate is arbitrarily close to 1.

Acknowledgments. This research is partially supported by the National Natural Science Foundation of China under Grant No.61373006.

References

1. Naor, M., Segev, G.: Public-key cryptosystems resilient to key leakage. In: Halevi, S. (ed.) CRYPTO 2009. LNCS, vol. 5677, pp. 18–35. Springer, Heidelberg (2009)
2. Boneh, D., Brumley, D.: Remote timing attacks are practical. Comput. Netw. **48**(5), 701–716 (2005)
3. Biham, E., Carmeli, Y., Shamir, A.: Bug attacks. In: Wagner, D. (ed.) CRYPTO 2008. LNCS, vol. 5157, pp. 221–240. Springer, Heidelberg (2008)
4. Nguyen, P.Q., Shparlinski, I.: The insecurity of the digital signature algorithm with partially known nonces. J. Cryptology **15**(3), 151–176 (2002)
5. Halderman, A., Schoen, S., Heninger, N., Clarkson, W., Paul, W., Calandrino, J., Feldman, A., Applebaum, J., Felten, E.: Lest we remember: Cold boot attacks on encryption keys. In: USENIX Security Symposium, pp. 45–60 (2008)
6. Quisquater, J.-J., Samyde, D.: Electromagnetic analysis (EMA): measures and counter-measures for smart cards. In: Attali, S., Jensen, T. (eds.) E-smart 2001. LNCS, vol. 2140, pp. 200–210. Springer, Heidelberg (2001)
7. Akavia, A., Goldwasser, S., Vaikuntanathan, V.: Simultaneous hardcore bits and cryptography against memory attacks. In: Reingold, O. (ed.) TCC 2009. LNCS, vol. 5444, pp. 474–495. Springer, Heidelberg (2009)
8. Alwen, J., Dodis, Y., Naor, M., Segev, G., Walfish, S., Wichs, D.: Public-key encryption in the bounded-retrieval model. In: Gilbert, H. (ed.) EUROCRYPT 2010. LNCS, vol. 6110, pp. 113–134. Springer, Heidelberg (2010)
9. Brakerski, Z., Goldwasser, S.: Circular and leakage resilient public-key encryption under subgroup indistinguishability. In: Rabin, T. (ed.) CRYPTO 2010. LNCS, vol. 6223, pp. 1–20. Springer, Heidelberg (2010)
10. Brakerski, Z., Kalai, Y.T., Katz, J., Vaikuntanathan, V.: Overcoming the hole in the bucket: Public-key cryptography resilient to continual memory leakage. In: FOCS 2010, pp. 501–510. IEEE Computer Society (2010)
11. Chow, S.S.M., Dodis, Y., Rouselakis, Y., Waters, B.: Practical leakage-resilient identity-based encryption from simple assumptions. In: Al-Shaer, E., Keromytis, A.D., Shmatikov, V. (eds.) ACM CCS 2010, pp. 152–161. ACM (2010)
12. Galindo, D., Herranz, J., Villar, J.: Identity-based encryption with master key-dependent message security and leakage-resilience. In: Foresti, S., Yung, M., Martinelli, F. (eds.) ESORICS 2012. LNCS, vol. 7459, pp. 627–642. Springer, Heidelberg (2012)
13. Lewko, A., Rouselakis, Y., Waters, B.: Achieving leakage resilience through dual system encryption. In: Ishai, Y. (ed.) TCC 2011. LNCS, vol. 6597, pp. 70–88. Springer, Heidelberg (2011)

14. Qin, B., Liu, S.: Leakage-resilient chosen-ciphertext secure public-key encryption from hash proof system and one-time lossy filter. In: Sako, K., Sarkar, P. (eds.) ASIACRYPT 2013, Part II. LNCS, vol. 8270, pp. 381–400. Springer, Heidelberg (2013)
15. Qin, B., Liu, S.: Leakage-flexible CCA-secure public-key encryption: simple construction and free of pairing. In: Krawczyk, H. (ed.) PKC 2014. LNCS, vol. 8383, pp. 19–36. Springer, Heidelberg (2014)
16. Yuen, T.H., Chow, S.S.M., Zhang, Y., Yiu, S.M.: Identity-based encryption resilient to continual auxiliary leakage. In: Pointcheval, D., Johansson, T. (eds.) EUROCRYPT 2012. LNCS, vol. 7237, pp. 117–134. Springer, Heidelberg (2012)
17. Zhang, M., Shi, W., Wang, C., Chen, Z., Mu, Y.: Leakage-resilient attribute-based encryption with fast decryption: models, analysis and constructions. In: Deng, R.H., Feng, T. (eds.) ISPEC 2013. LNCS, vol. 7863, pp. 75–90. Springer, Heidelberg (2013)
18. Dodis, Y., Goldwasser, S., Tauman Kalai, Y., Peikert, C., Vaikuntanathan, V.: Public-key encryption schemes with auxiliary inputs. In: Micciancio, D. (ed.) TCC 2010. LNCS, vol. 5978, pp. 361–381. Springer, Heidelberg (2010)
19. Canetti, R., Dodis, Y., Halevi, S., Kushilevitz, E., Sahai, A.: Exposure-resilient functions and all-or-nothing transforms. In: Preneel, B. (ed.) EUROCRYPT 2000. LNCS, vol. 1807, pp. 453–469. Springer, Heidelberg (2000)
20. Kamp, J., Zuckerman, D.: Deterministic extractors for bit-xing sources and exposure-resilient cryptography. In: FOCS, pp. 92–101 (2003)
21. Dziembowski, S., Pietrzak, K.: Leakage-resilient cryptography. In: FOCS, pp. 293–302 (2008)
22. Faust, S., Kiltz, E., Pietrzak, K., Rothblum, G.N.: Leakage-resilient signatures. In: Micciancio, D. (ed.) TCC 2010. LNCS, vol. 5978, pp. 343–360. Springer, Heidelberg (2010)
23. Alwen, J., Dodis, Y., Wichs, D.: Leakage-resilient public-key cryptography in the bounded-retrieval model. In: Halevi, S. (ed.) CRYPTO 2009. LNCS, vol. 5677, pp. 36–54. Springer, Heidelberg (2009)
24. Di Crescenzo, G., Lipton, R.J., Walfish, S.: Perfectly secure password protocols in the bounded retrieval model. In: Halevi, S., Rabin, T. (eds.) TCC 2006. LNCS, vol. 3876, pp. 225–244. Springer, Heidelberg (2006)
25. Dodis, Y., Haralambiev, K., Lopez-Alt, A., Wichs, D.: Cryptography against continuous memory attacks. In: FOCS, pp. 511–520 (2010)
26. Lewko, A., Rouselakis, Y., Waters, B.: Achieving leakage resilience through dual system encryption. In: Ishai, Y. (ed.) TCC 2011. LNCS, vol. 6597, pp. 70–88. Springer, Heidelberg (2011)
27. Cramer, R., Shoup, V.: Universal hash proofs and a paradigm for adaptive chosen ciphertext secure public-key encryption. In: Knudsen, L.R. (ed.) EUROCRYPT 2002. LNCS, vol. 2332, pp. 45–64. Springer, Heidelberg (2002)
28. Li, S., Zhang, F., Sun, Y., Shen, L.: A new variant of the Cramer-Shoup leakage resilient public key encryption. In: Xhafa, F., Barolli, L., Pop, F., Chen, X., Cristea, V. (eds.) INCoS 2012, pp. 342–346. IEEE (2012)
29. Liu, S., Weng, J., Zhao, Y.: Efficient public key cryptosystem resilient to key leakage chosen ciphertext attacks. In: Dawson, E. (ed.) CT-RSA 2013. LNCS, vol. 7779, pp. 84–100. Springer, Heidelberg (2013)
30. Hofheinz, D.: All-but-many lossy trapdoor functions. In: Pointcheval, D., Johansson, T. (eds.) EUROCRYPT 2012. LNCS, vol. 7237, pp. 209–227. Springer, Heidelberg (2012)

31. Naor, M., Reingold, O.: Number-theoretic constructions of efficient pseudo-random functions. J. ACM **51**(2), 231–262 (2004)
32. Dodis, Y., Ostrovsky, R., Reyzin, L., Smith, A.: Fuzzy extractors: How to generate strong keys from biometrics and other noisy data. SIAM J. Comput. **38**(1), 97–139 (2008)
33. Gonzalez, N., Colin, B., Ed, D.: A public key cryptosystem based on a subgroup membership problem. Des. Codes Crypt. **36**, 301–316 (2005)

General Circuit Realizing Compact Revocable Attribute-Based Encryption from Multilinear Maps

Pratish Datta, Ratna Dutta, and Sourav Mukhopadhyay(✉)

Department of Mathematics, Indian Institute of Technology Kharagpur, Kharagpur 721302, India
{pratishdatta,ratna,sourav}@maths.iitkgp.ernet.in

Abstract. This paper demonstrates *new technique* for managing *revocation* in the context of *attribute-based encryption* (ABE) and presents two *selectively* secure *directly revocable* ABE (RABE) constructions

- supporting decryption policies realizable by *polynomial size* Boolean circuits of *arbitrary fan-out* and
- featuring *compactness* in the sense that the number of revocation controlling components in ciphertexts and decryption keys are *constant*.

In fact, our RABE schemes are the *first* to achieve these parameters. Both our constructions utilize *multilinear maps*. The size of public parameter in our first construction is *linear* to the maximum number of users supported by the system while in the second construction we reduce it to *logarithmic*.

Keywords: RABE for circuits · Polynomial size circuits · Multilinear map

1 Introduction

In recent times, the cost effectiveness and greater flexibility of cloud technology has triggered an emerging trend among individuals and organizations to outsource potentially sensitive private data to the "cloud", an external large and powerful server. Attribute-based encryption (ABE), a noble paradigm for public key encryption in which ciphertexts are encrypted for entities possessing specific decryption credentials, has been extensively deployed to realize complex access control functionalities in cloud environment. ABE comes in two flavors, namely, key-policy and ciphertext-policy. However, in spite of its promising properties, the adoption of ABE in cloud management requires further refinements.

A crucial feature of ABE systems is the *expressiveness* of the supported decryption policies. Recently, few independent seminal works [4,10] have extended the class of admissible policies for ABE to arbitrary polynomial size Boolean circuits of unbounded fan-out in contrast to circuits of fan-out one realized by all ABE constructions prior to their works.

J. Lopez and C.J. Mitchell (Eds.): ISC 2015, LNCS 9290, pp. 336–354, 2015.
DOI: 10.1007/978-3-319-23318-5_19

The other significant requirement in the context of ABE is user *revocation*, a tool for *changing* the users' decryption rights. Over time many users' private keys might get compromised, users might leave or be dismissed due to the revealing of malicious activities. In the literature several revocation mechanisms have been proposed in ABE setting [1–3, 11, 14–16]. The *direct* revocation technique [1, 2, 14, 15], that controls revocation by specifying a revocation list directly during encryption, does not involve any additional proxy server [16] or key update phase [1, 3, 11]. Consequently, the non-revoked users remain unaffected and revocation can take effect instantly without requiring to wait for the expiration of the current time period.

However, in all the above *revocable* ABE (RABE) systems the decryption policies were restricted to circuits of fan-out one, paving the way for a "back-tracking" attack [10] on the policy circuits by unauthorized users, thereby completely breaking the confidentiality of ciphertexts. Further, all currently available standard model RABE constructions supporting direct revocation mode [1, 2, 14] essentially follow the tree-based revocation mechanism of Naor et al. [12], as a result of which the number of components for managing user revocation contained in the ciphertexts and decryption keys are respectively $O(\widehat{r} \log \frac{N_{\max}}{\widehat{r}})$ and $O(\log N_{\max})$, where $N_{\max}$ is the maximum number of users supported by the system and $\widehat{r}$ is the number of revoked users.

Our Contribution: In this paper, we apply the revocation technique introduced in [5] and its improved variant [6] in the ABE setting and propose two RABE schemes for *general circuit* realizable decryption policies supporting *direct* revocation and featuring *constant* number of components for enforcing revocation in the ciphertexts and decryption keys.

More precisely, we integrate the revocation strategy of [5, 6] with the ABE scheme of [10]. As an outcome, we develop the *first* RABE constructions that support the *most expressive* form of decryption policies achieved so far for ABE, namely, arbitrary polynomial size circuits having unbounded fan-out with bounded depth and input length. Although the basic conception may sound simple, its exact realization involves many subtleties that we address with innovative ideas. Our schemes employ multilinear map for which some approximate candidates have recently been proposed [7, 8, 10]. Both our schemes support direct revocation and are proven secure in the *selective revocation list model* under the *Multilinear Diffie-Hellman Exponent* [4] and the *Compressed Multilinear Diffie-Hellman Exponent* assumptions [13], which are multilinear equivalents of the *Bilinear Diffie-Hellman Exponent* assumption. Our security analyses do not use random oracles or generic multilinear group framework. We emphasize that selective security can be a reasonable trade-off for performance in some circumstances. Moreover, applying a standard *complexity leveraging* argument, as in [4], our selectively secure constructions can be made *adaptively* secure.

Our first RABE scheme, which is a blend of the revocation technique of [5] and an improved version of the ABE construction proposed in [10], has ciphertext consisting of only 3 group elements (or encodings). The decryption keys comprise of $\ell + 4q + 1$ group elements in the worst case, ℓ and q being the input length and number of gates in the policy circuits. This is the same as all currently available vanilla ABE constructions for general circuits based on multilinear maps [4,10]. Consequently, we achieve very short ciphertext size without imposing any extra overhead on the decryption key for the added revocation functionality. To the best of our knowledge, our work is the *first* to achieve this property.

However, the number of group elements in the public parameters in our first RABE construction is linear to $N_{\max}$. In order to overcome this bottleneck, we modify our first construction by replacing the revocation method with that of [6] taking advantage of a multilinear map of (possibly) slightly higher multilinearity level compared to the one used in the first scheme. We reduce the number of group elements in the public parameters to $\log N_{\max}$ in our second RABE scheme. This is comparable with the previous standard model RABE constructions supporting direct revocation [1,2,14]. However, we retain the same property for ciphertext and decryption keys, i.e., the number of ciphertext and decryption key components do not grow with $N_{\max}$.

Finally, while both our RABE schemes are of *key-policy* variety, using the notion of *universal circuits*, as in [10], both our constructions can be extended to realize *ciphertext-policy* style RABE for arbitrary bounded size circuits achieving the same parameters.

2 Preliminaries

▪ **Circuit Notation**: We adopt the same notations for circuits as in [10]. First note that without loss of generality we can consider only those circuits which are *monotone*, where gates are either OR or AND having fan-in two, and *layered* (see [10] for details). Our circuits will have a single output gate. A circuit will be represented as a six-tuple $f = (\ell, q, d, \mathbb{A}, \mathbb{B}, \mathsf{GateType})$. Here, ℓ, q respectively denote the length of the input, the number of gates, and d represents the depth of the circuit which is one plus the length of the shortest path from the output wire to any input wire. We designate the set of input wires as $\mathsf{Input} = \{1, \ldots, \ell\}$, the set of gates as $\mathsf{Gates} = \{\ell + 1, \ldots, \ell + q\}$, the total set of wires in the circuit as $W = \mathsf{Input} \cup \mathsf{Gates} = \{1, \ldots, \ell + q\}$, and the wire $\ell + q$ as the output wire. Let $\mathbb{A}, \mathbb{B} : \mathsf{Gates} \to W \backslash \{\ell + q\}$ be functions. For all $w \in \mathsf{Gates}$, $\mathbb{A}(w)$ and $\mathbb{B}(w)$ respectively identify w's first and second incoming wires. Finally, $\mathsf{GateType} : \mathsf{Gates} \to \{\mathsf{AND}, \mathsf{OR}\}$ defines a functions that identifies a gate as either an AND or an OR gate. We follow the convention that $w > \mathbb{B}(w) > \mathbb{A}(w)$ for any $w \in \mathsf{Gates}$.

We also define a function $\mathsf{depth} : W \to \{1, \ldots, d\}$ such that if $w \in \mathsf{Input}$, $\mathsf{depth}(w) = 1$, and in general $\mathsf{depth}(w)$ of wire w is equal to one plus the length of the shortest path from w to an input wire. Since our circuit is layered, we have, for all $w \in \mathsf{Gates}$, if $\mathsf{depth}(w) = t$ then $\mathsf{depth}(\mathbb{A}(w)) = \mathsf{depth}(\mathbb{B}(w)) = t - 1$.

We will abuse notation and let $f(x)$ be the evaluation of the circuit f on input $x \in \{0, 1\}^\ell$, and $f_w(x)$ be the value of wire w of the circuit f on input x.

2.1 The Notion of RABE for General Circuits

■ **Syntax of RABE for circuits**: Consider a circuit family $\mathbb{F}_{\ell,d}$ that consists of all circuits f with input length ℓ and depth d characterizing decryption rights. A (key-policy) revocable attribute-based encryption (RABE) scheme for circuits in $\mathbb{F}_{\ell,d}$ with message space $\mathbb{M}$ consists of the following algorithms:

RABE.Setup$(1^\lambda, \ell, d, N_{\max})$: The trusted key generation center takes as input a security parameter 1^λ, the length ℓ of Boolean inputs to decryption circuits, the allowed depth d of the decryption circuits, and the maximum number $N_{\max}$ of users supported by the system. It publishes the public parameters PP along with the empty user list $\mathsf{UL} = \varnothing$, while keeps the master secret key MK to itself.

RABE.KeyGen$(\mathsf{PP}, \mathsf{MK}, \mathsf{UL}, \mathsf{ID}, f)$: On input the public parameters PP, the master secret key MK, the current user list UL, and a user identity ID together with the decryption policy circuit description $f \in \mathbb{F}_{\ell,d}$ of that user, the key generation center provides a decryption key $\mathsf{SK}_{f,\mathsf{ID}}$ to the user and publishes the user list UL updated with the information of the newly joined user.

RABE.Encrypt$(\mathsf{PP}, \mathsf{UL}, x, \mathsf{RL}, M)$: Taking in the public parameters PP, the current user list UL, a descriptor input string $x \in \{0,1\}^\ell$, a set of revoked user identities RL, and a message $M \in \mathbb{M}$, the encrypter prepares a ciphertext $\mathsf{CT}_{x,\mathsf{RL}}$.

RABE.Decrypt$(\mathsf{PP}, \mathsf{UL}, \mathsf{CT}_{x,\mathsf{RL}}, \mathsf{SK}_{f,\mathsf{ID}})$: A user takes as input the public parameters PP, the current user list UL, a ciphertext $\mathsf{CT}_{x,\mathsf{RL}}$ encrypted for x along with a list RL of revoked user identities, and its decryption key $\mathsf{SK}_{f,\mathsf{ID}}$ corresponding to its decryption policy circuit $f \in \mathbb{F}_{\ell,d}$ as well as user identity ID. It attempts to decrypt the ciphertext and outputs the message $M \in \mathbb{M}$ if successful; otherwise, it outputs the distinguished symbol $\perp$.

■ **Correctness:** The correctness of RABE for general circuits is defined as follows: For all $(\mathsf{PP}, \mathsf{UL}, \mathsf{MK}) \leftarrow \mathsf{RABE.Setup}(1^\lambda, \ell, d, N_{\max})$, $\mathsf{SK}_{f,\mathsf{ID}} \leftarrow \mathsf{RABE.KeyGen}(\mathsf{PP}, \mathsf{MK}, \mathsf{UL}, \mathsf{ID}, f)$ for any ID and $f \in \mathbb{F}_{\ell,d}$, $\mathsf{CT}_{x,\mathsf{RL}} \leftarrow \mathsf{RABE.Encrypt}(\mathsf{PP}, \mathsf{UL}, x, \mathsf{RL}, M)$ for any $x \in \{0,1\}^\ell, \mathsf{RL}$ and $M \in \mathbb{M}$, $\big([f(x) = 1] \wedge [\mathsf{ID} \notin \mathsf{RL}] \implies \mathsf{RABE.Decrypt}(\mathsf{PP}, \mathsf{UL}, \mathsf{CT}_{x,\mathsf{RL}}, \mathsf{SK}_{f,\mathsf{ID}}) = M\big)$.

■ **Security Model:** The security of RABE under *selective revocation list model* against *chosen plaintext attacks* (CPA) is defined in terms of the following experiment between a probabilistic challenger $\mathcal{B}$ and a probabilistic polynomial-time adversary $\mathcal{A}$:

Init: $\mathcal{A}$ commits to a challenge descriptor input string $x^* \in \{0,1\}^\ell$ along with a challenge revoked user identity list RL^*.

Setup: $\mathcal{B}$ creates a user list UL including all users with identities in RL^* in it; generates a master secret key MK together with the public parameters PP by running $\mathsf{RABE.Setup}(1^\lambda, \ell, d, N_{\max})$; keeps MK to itself; and gives PP, UL to $\mathcal{A}$.

Phase 1: $\mathcal{A}$ adaptively requests a polynomial number of decryption keys for circuit description $f \in \mathbb{F}_{\ell,d}$ along with user identity ID of its choice subject to the restriction that $[f(x^*) = 0] \vee [\mathsf{ID} \in \mathsf{RL}^*]$. $\mathcal{B}$ returns the corresponding decryption keys $\mathsf{SK}_{f,\mathsf{ID}}$ along with the updated user list UL to $\mathcal{A}$ by executing $\mathsf{RABE.KeyGen}(\mathsf{PP}, \mathsf{MK}, \mathsf{UL}, ID, f)$.

Challenge: $\mathcal{A}$ submits two equal length messages $M_0^*, M_1^* \in \mathbb{M}$. $\mathcal{B}$ flips a random coin $b \in \{0,1\}$ and hands the challenge ciphertext CT^* to $\mathcal{A}$ by performing $\mathsf{RABE.Encrypt}(\mathsf{PP}, \mathsf{UL}, x^*, \mathsf{RL}^*, M_b^*)$.

Phase 2: $\mathcal{A}$ may continue adaptively to make a polynomial number of decryption key queries as in **Phase 1** with the same constraint as above.

Guess: Finally, $\mathcal{A}$ outputs a guess $b' \in \{0,1\}$ and wins the game if $b = b'$.

Definition 1. *An* RABE *scheme for circuits is said to be secure under selective revocation list model against* CPA *if the advantage of all probabilistic polynomial time adversaries* $\mathcal{A}$ *in the above game,* $\mathsf{Adv}_{\mathcal{A}}^{\mathsf{RABE,SRL\text{-}CPA}}(\lambda) = |\mathsf{Pr}[b' = b] - 1/2|$, *is at most negligible.*

2.2 Multilinear Maps and Complexity Assumptions

A (leveled) multilinear map [7–9] consists of the following two algorithms:

(I) $\mathcal{G}^{\mathsf{MLM}}(1^\lambda, \kappa)$: It takes as input a security parameter 1^λ and a positive integer κ indicating the number of allowed pairing operations. It outputs a sequence of groups $\overrightarrow{\mathbb{G}} = (\mathbb{G}_1, \ldots, \mathbb{G}_\kappa)$ each of large prime order $p > 2^\lambda$ together with the canonical generators g_i of $\mathbb{G}_i$. We call $\mathbb{G}_1$ the source group, $\mathbb{G}_\kappa$ the target group, and $\mathbb{G}_2, \ldots, \mathbb{G}_{\kappa-1}$ intermediate groups. Let $\mathsf{PP}_{\mathsf{MLM}} = (\overrightarrow{\mathbb{G}}, g_1, \ldots, g_\kappa)$ be the description of the multilinear group with canonical generators.

(II) $e_{i,j}(g, h)$ (for $i, j \in \{1, \ldots, \kappa\}$ with $i + j \leq \kappa$): On input two elements $g \in \mathbb{G}_i$ and $h \in \mathbb{G}_j$ with $i + j \leq \kappa$, it outputs an element of $\mathbb{G}_{i+j}$ such that $e_{i,j}(g_i^a, g_j^b) = g_{i+j}^{ab}$ for $a, b \in \mathbb{Z}_p$. We often omit the subscripts and just write e. We can also generalize e to multiple inputs as $e(\chi^{(1)}, \ldots, \chi^{(t)}) = e(\chi^{(1)}, e(\chi^{(2)}, \ldots, \chi^{(t)}))$.

We refer g_i^a as a level-i encoding of $a \in \mathbb{Z}_p$. The scalar a itself is referred to as a level-0 encoding of a. Then the map e combines a level-i encoding of an element $a \in \mathbb{Z}_p$ and a level-j encoding of another element $b \in \mathbb{Z}_p$, and produces level-$(i + j)$ encoding of the product ab.

Assumption 1 [**(κ, N)-Multilinear Diffie-Hellman Exponent: (κ, N)-MDHE** [4]]. *The (κ, N)-Multilinear Diffie-Hellman Exponent ((κ, N)-MDHE) problem is to guess $\widetilde{b} \in \{0,1\}$ given $\varrho_{\widetilde{b}} = (\mathsf{PP}_{\mathsf{MLM}}, \vartheta_1, \ldots, \vartheta_N, \vartheta_{N+2}, \ldots, \vartheta_{2N}, \Upsilon, \tau_1, \ldots, \tau_{\kappa-2}, \Re_{\widetilde{b}})$ generated by $\mathcal{G}_{\widetilde{b}}^{(\kappa,N)\text{-}\mathsf{MDHE}}(1^\lambda)$, where $\mathcal{G}_{\widetilde{b}}^{(\kappa,N)\text{-}\mathsf{MDHE}}(1^\lambda)$ operates as follows: It runs $\mathcal{G}^{\mathsf{MLM}}(1^\lambda, \kappa)$ to generate $\mathsf{PP}_{\mathsf{MLM}}$ of order p; picks random $\alpha, \varsigma, \psi_1, \ldots, \psi_{\kappa-2} \in \mathbb{Z}_p$; computes $\vartheta_j = g_1^{\alpha^{(j)}}$ for $j = 1, \ldots, N, N+2, \ldots, 2N$, $\Upsilon = g_1^\varsigma$, $\tau_i = g_1^{\psi_i}$ for $i = 1, \ldots, \kappa - 2$; sets $\Re_0 = g_\kappa^{\alpha^{(N+1)}\varsigma \prod_{i=1}^{\kappa-2}\psi_i}$ while $\Re_1 =$ some random element in $\mathbb{G}_\kappa$; and finally returns $\varrho_{\widetilde{b}} = (\mathsf{PP}_{\mathsf{MLM}}, \vartheta_1, \ldots, \vartheta_N, \vartheta_{N+2}, \ldots, \vartheta_{2N}, \Upsilon, \tau_1, \ldots, \tau_{\kappa-2}, \Re_{\widetilde{b}})$.*

The (κ, N)-MDHE *assumption is that the advantage of all probabilistic polynomial time algorithms* $\mathcal{B}$ *in solving the above problem,* $\mathsf{Adv}_{\mathcal{B}}^{(\kappa,N)\text{-}\mathsf{MDHE}}(\lambda) = |\mathsf{Pr}[\mathcal{B}(1^\lambda, \varrho_0) \to 1] - \mathsf{Pr}[\mathcal{B}(1^\lambda, \varrho_1) \to 1]|$ *is at most negligible.*

Assumption 2 [(n, k, l) **-Compressed Multilinear Diffie-Hellman Exponent:** (n, k, l)-cMDHE [13]]. *The* (n, k, l)*-Compressed Multilinear Diffie-Hellman Exponent* $((n, k, l)$-$\mathsf{cMDHE})$ *problem is to guess* $\widetilde{b} \in \{0, 1\}$ *given* $\varrho_{\widetilde{b}} = (\mathsf{PP}_{\mathsf{MLM}}, \xi_0, \ldots, \xi_n, \tau_1, \ldots, \tau_k, \Upsilon, \Re_{\widetilde{b}})$ *generated by* $\mathcal{G}_{\widetilde{b}}^{(n,k,l)\text{-}\mathsf{cMDHE}}(1^\lambda)$, *where* $\mathcal{G}_{\widetilde{b}}^{(n,k,l)\text{-}\mathsf{cMDHE}}(1^\lambda)$ *operates as follows: It runs* $\mathcal{G}^{\mathsf{MLM}}(1^\lambda, \kappa = n + k + l - 1)$ *to generate* $\mathsf{PP}_{\mathsf{MLM}}$ *of order* p*; picks random* $\alpha, \varsigma, \psi_1, \ldots, \psi_k \in \mathbb{Z}_p$*; computes* $\xi_\iota = g_1^{\alpha^{(2^\iota)}}$ *for* $\iota = 0, \ldots, n$, $\tau_h = g_1^{\psi_h}$ *for* $h = 1, \ldots, k$, $\Upsilon = g_l^{\varsigma}$*; sets* $\Re_0 = g_\kappa^{\alpha^{(2^n-1)}\varsigma \prod_{h=1}^{k} \psi_h}$ *while* $\Re_1 =$ *some random element of* $\mathbb{G}_\kappa$*; and finally returns* $\varrho_{\widetilde{b}} = (\mathsf{PP}_{\mathsf{MLM}}, \xi_0, \ldots, \xi_n, \tau_1, \ldots, \tau_k, \Upsilon, \Re_{\widetilde{b}})$.

The (n, k, l)-cMDHE *assumption is that the advantage of all probabilistic polynomial time algorithms* $\mathcal{B}$ *in solving the above problem,* $\mathsf{Adv}_{\mathcal{B}}^{(n,k,l)\text{-}\mathsf{cMDHE}}(\lambda) = |\mathsf{Pr}[\mathcal{B}(1^\lambda, \varrho_0) \to 1] - \mathsf{Pr}[\mathcal{B}(1^\lambda, \varrho_1) \to 1]|$ *is at most negligible.*

3 RABE-I

▪ **The Construction:**

RABE.Setup$(1^\lambda, \ell, d, N_{\max})$: The trusted key generation center takes as input a security parameter 1^λ, the length ℓ of Boolean inputs to the decryption circuits, the allowed depth d of decryption circuits, and the maximum number $N_{\max}$ of users supported by the system. Let $\mathcal{N} = \{1, \ldots, N_{\max}\}$ be the set of user key indices. It proceeds as follows:

1. It runs $\mathcal{G}^{\mathsf{MLM}}(1^\lambda, \kappa = \ell + d + 1)$ to obtain $\mathsf{PP}_{\mathsf{MLM}} = \big(\vec{\mathbb{G}} = (\mathbb{G}_1, \ldots, \mathbb{G}_\kappa), g_1, \ldots, g_\kappa\big)$ of prime order $p > 2^\lambda$.
2. It selects random $(a_{1,0}, a_{1,1}), \ldots, (a_{\ell,0}, a_{\ell,1}) \in \mathbb{Z}_p^2$, and computes $A_{i,\beta} = g_1^{a_{i,\beta}}$ for $i = 1, \ldots, \ell$; $\beta \in \{0, 1\}$.
3. It selects random $\alpha, \gamma, \theta \in \mathbb{Z}_p$ and computes $\vartheta_j = g_1^{\alpha^{(j)}}$ for $j = 1, \ldots, N_{\max}, N_{\max} + 2, \ldots, 2N_{\max}$, $Y = g_1^{\gamma}$, $Z = g_{d-1}^{\theta}$, $\Omega = g_{d+1}^{\alpha^{(N_{\max}+1)}\theta}$.
4. It initializes the user list UL, which would consist of ordered pairs (ID, u) such that ID is the identity of an user who has participated in the system and $u \in \mathcal{N}$ is the unique index assigned to ID by the key generation center at the time of subscription, as an empty set, i.e., it sets $\mathsf{UL} = \varnothing$.
5. Finally it publishes the public parameters $\mathsf{PP} = \big(\mathsf{PP}_{\mathsf{MLM}}, \{A_{i,\beta}\}_{i=1,\ldots,\ell;\beta\in\{0,1\}}, \{\vartheta_j\}_{j=1,\ldots,N_{\max},N_{\max}+2,\ldots,2N_{\max}}, Y, Z, \Omega\big)$ along with the empty user list UL, while keeps the master secret key $\mathsf{MK} = (\alpha, \gamma, \theta)$ to itself.

RABE.KeyGen$(\mathsf{PP}, \mathsf{MK}, \mathsf{UL}, \mathsf{ID}, f)$: The key generation center takes the public parameters PP, the master secret key MK, the current user list UL, and the user identity ID together with the description $f = (\ell, q, d, \mathbb{A}, \mathbb{B}, \mathsf{GateType})$ of the decryption circuit from a user as input. Our circuit has $\ell+q$ wires $\{1,\ldots,\ell+q\}$ where $\{1,\ldots,\ell\}$ are ℓ input wires, $\{\ell+1,\ldots,\ell+q\}$ are q gates (OR or AND gates), and the wire $\ell+q$ is designated as the output wire. It proceeds as follows:

1. It first assigns an index $u \in \mathcal{N}$ such that $(\cdot, u) \notin \mathsf{UL}$ to ID and updates UL by adding the pair (ID, u).
2. It chooses random $r_1,\ldots,r_{\ell+q} \in \mathbb{Z}_p$ where we think of randomness r_w as being associated with wire $w \in \{1,\ldots,\ell+q\}$. It produces the "header" component $\mathcal{K} = g_d^{\alpha^{(u)}\theta\gamma - r_{\ell+q}}$.
3. It generates key components for every wire w. The structure of the key component depends upon the category of w, i.e., whether w is an Input wire, OR gate, or AND gate. We describe below how it generates the key components in each case.
   - *Input wire*: If $w \in \{1,\ldots,\ell\}$ then it corresponds to the w-th input. It computes the key component $\mathcal{K}_w = e(A_{w,1}, g_1)^{r_w} = g_2^{r_w a_{w,1}}$.
   - *OR gate*: Suppose that wire $w \in \mathsf{Gates}$, $\mathsf{GateType}(w) = \mathsf{OR}$, and $t = \mathsf{depth}(w)$. It picks random $\mu_w, \nu_w \in \mathbb{Z}_p$ and creates the key component

$$\mathcal{K}_w = \big(K_{w,1} = g_1^{\mu_w}, K_{w,2} = g_1^{\nu_w}, K_{w,3} = g_t^{r_w - \mu_w r_{\mathbb{A}(w)}}, K_{w,4} = g_t^{r_w - \nu_w r_{\mathbb{B}(w)}}\big).$$

   - *AND gate*: Let wire $w \in \mathsf{Gates}$, $\mathsf{GateType}(w) = \mathsf{AND}$, and $t = \mathsf{depth}(w)$. It selects random $\mu_w, \nu_w \in \mathbb{Z}_p$ and forms the key component

$$\mathcal{K}_w = \big(K_{w,1} = g_1^{\mu_w}, K_{w,2} = g_1^{\nu_w}, K_{w,3} = g_t^{r_w - \mu_w r_{\mathbb{A}(w)} - \nu_w r_{\mathbb{B}(w)}}\big).$$

4. It provides the decryption key $\mathsf{SK}_{f,\mathsf{ID}} = \big(f, \mathsf{ID}, \mathcal{K}, \{\mathcal{K}_w\}_{w\in\{1,\ldots,\ell+q\}}\big)$ to the user and publishes the updated user list UL.

RABE.Encrypt(PP, UL, x, RL, M): Taking as input the public parameters PP, the current user list UL, a descriptor input string $x = x_1 \ldots x_\ell \in \{0,1\}^\ell$, a list RL of revoked user identities, and a message $M \in \mathbb{G}_\kappa$, the encrypter forms the ciphertext as follows:

1. It first defines the revoked user key index set $\mathsf{RI} \subseteq \mathcal{N}$ corresponding to RL using UL, i.e., if $\mathsf{ID} \in \mathsf{RL}$ and $(\mathsf{ID}, j) \in \mathsf{UL}$ it includes j in RI. It then determines $\mathsf{SI} = \mathcal{N} \backslash \mathsf{RI}$.
2. It picks random $s \in \mathbb{Z}_p$ and computes

$$C_M = e(\Omega, A_{1,x_1}, \ldots, A_{\ell,x_\ell})^s M = g_\kappa^{\alpha^{(N_{\max}+1)}\theta s \delta(x)} M,$$
$$C = g_1^s,\ C' = \big(Y \prod_{j\in\mathsf{SI}} \vartheta_{N_{\max}+1-j}\big)^s = \big(g_1^\gamma \prod_{j\in\mathsf{SI}} g_1^{\alpha^{(N_{\max}+1-j)}}\big)^s,$$

   where we define $\delta(x) = \prod_{i=1}^{\ell} a_{i,x_i}$ for the ease of exposition.
3. It outputs the ciphertext $\mathsf{CT}_{x,\mathsf{RL}} = (x, \mathsf{RL}, C_M, C, C')$.

RABE.Decrypt(PP, UL, $\mathsf{CT}_{x,\mathsf{RL}}$, $\mathsf{SK}_{f,\mathsf{ID}}$): A user, on input the public parameters PP, the current user list UL, a ciphertext $\mathsf{CT}_{x,\mathsf{RL}} = (x, \mathsf{RL}, C_M, C, C')$ encrypted for descriptor input string $x = x_1 \dots x_\ell \in \{0,1\}^\ell$ and a list of revoked user identities RL, along with its decryption key $\mathsf{SK}_{f,\mathsf{ID}} = \big(f, \mathsf{ID}, \mathcal{K}, \{\mathcal{K}_w\}_{w \in \{1,\dots,\ell+q\}}\big)$ for its decryption circuit $f = (\ell, q, d, \mathbb{A}, \mathbb{B}, \mathsf{GateType})$ as well as its user identity ID, where $u \in \mathcal{N}$ is the index assigned to ID (say), outputs $\perp$, if $[f(x) = 0] \vee [\mathsf{ID} \in \mathsf{RL}]$. Otherwise, (if $[f(x) = 1] \ \wedge \ [\mathsf{ID} \notin \mathsf{RL}]$) it proceeds as follows:

1. First, as a "header" computation it computes

$$D = e(A_{1,x_1}, \dots, A_{\ell,x_\ell}) = g_\ell^{\delta(x)}, \ \widehat{E} = e(\mathcal{K}, D, C) = g_\mathcal{K}^{(\alpha^{(u)}\theta\gamma - r_{\ell+q})s\delta(x)},$$

   extracting $\{A_{i,x_i}\}_{i=1,\dots,\ell}$ from PP.
2. Next, it performs the bottom-up evaluation of the circuit. For every wire w with corresponding $\mathsf{depth}(w) = t$, if $f_w(x) = 0$, nothing is computed for that wire, otherwise (if $f_w(x) = 1$), it attempts to compute $E_w = g_{\ell+t+1}^{r_w s\delta(x)}$ as follows. The user proceeds iteratively starting with computing E_1 and moves forward *in order* to finally compute $E_{\ell+q}$. Note that computing these values in order ensures that the computation on a wire w with $\mathsf{depth}(w) = t-1$ that evaluates to 1 will be defined before the computation on a wire w with $\mathsf{depth}(w) = t$. The computation procedure depends on whether the wire is an Input wire, OR gate, or AND gate.
   - *Input wire*: If $w \in \{1,\dots,\ell\}$ then it corresponds to the w-th input and $t = \mathsf{depth}(w) = 1$. Suppose that $x_w = f_w(x) = 1$. Extracting $\mathcal{K}_w$ from its decryption key $\mathsf{SK}_{f,\mathsf{ID}}$, the user computes

$$E_w = e(\mathcal{K}_w, A_{1,x_1}, \dots, A_{w-1,x_{w-1}}, A_{w+1,x_{w+1}}, \dots, A_{\ell,x_\ell}, C) = g_{\ell+1+1}^{r_w s\delta(x)}.$$

   - *OR gate*: Consider a wire $w \in \mathsf{Gates}$ with $\mathsf{GateType}(w) = \mathsf{OR}$ and $t = \mathsf{depth}(w)$. Assume that $f_w(x) = 1$. Then the user checks whether $f_{\mathbb{A}(w)}(x) = 1$, i.e., the first input of gate w evaluated to 1, and if so, then the user extracts $K_{w,1}, K_{w,3}$ from $\mathcal{K}_w$ included in $\mathsf{SK}_{f,\mathsf{ID}}$ and computes

$$E_w = e(E_{\mathbb{A}(w)}, K_{w,1})e(K_{w,3}, D, C) = g_{\ell+t+1}^{r_w s\delta(x)}.$$

     Note that $E_{\mathbb{A}(w)}$ is already computed at this stage in the bottom-up circuit evaluation as $\mathsf{depth}(\mathbb{A}(w)) = t-1$.
     Alternatively, if $f_{\mathbb{A}(w)}(x) = 0$ then it must be the case that $f_{\mathbb{B}(w)}(x) = 1$ as $f_w(x) = 1$, and it computes

$$E_w = e(E_{\mathbb{B}(w)}, K_{w,2})e(K_{w,4}, D, C) = g_{\ell+t+1}^{r_w s\delta(x)}$$

     extracting $K_{w,2}, K_{w,4}$ from $\mathcal{K}_w$ contained in $\mathsf{SK}_{f,\mathsf{ID}}$.
   - *AND gate*: Consider a wire $w \in \mathsf{Gates}$ with $\mathsf{GateType}(w) = \mathsf{AND}$ and $t = \mathsf{depth}(w)$. Suppose that $f_w(x) = 1$. Then $f_{\mathbb{A}(w)}(x) = f_{\mathbb{B}(w)}(x) = 1$. The user computes

$$E_w = e(E_{\mathbb{A}(w)}, K_{w,1})e(E_{\mathbb{B}(w)}, K_{w,2})e(K_{w,3}, D, C) = g_{\ell+t+1}^{r_w s\delta(x)}$$

     extracting $K_{w,1}, K_{w,2}, K_{w,3}$ from $\mathcal{K}_w$ in $\mathsf{SK}_{f,\mathsf{ID}}$.
     The user finally computes $E_{\ell+q} = g_\mathcal{K}^{r_{\ell+q}s\delta(x)}$, as $f(x) = f_{\ell+q}(x) = 1$.

3. It determines the revoked user key index set $\mathsf{RI} \subseteq \mathcal{N}$ corresponding to RL using UL and obtains $\mathsf{SI} = \mathcal{N} \backslash \mathsf{RI}$ which contains all the non-revoked user key indices. Note that since $\mathsf{ID} \notin \mathsf{RL}$, $u \in \mathsf{SI}$. The user retrieves the message by the following computation:

$$C_M \widehat{E} E_{\ell+q} e\Big(\prod_{j \in \mathsf{SI} \backslash \{u\}} \vartheta_{N_{\max}+1-j+u}, Z, D, C \Big) e\big(\vartheta_u, Z, D, C'\big)^{-1} = M.$$

■ **Security Analysis**:

Theorem 1. RABE-I *is secure in the selective revocation list model against* CPA *as per the security model of Sect. 2.1 if the* $(\ell + d + 1, N_{\max})$-MDHE *assumption holds for the underlying multilinear group generator* $\mathcal{G}^{\mathsf{MLM}}$, *described in Sect. 2.2, where* ℓ, d, *and* $N_{\max}$ *denote respectively the input length of decryption circuits, depth of the decryption circuits, and the maximum number of users supported by the system.*

Proof. Suppose that there exists a probabilistic polynomial-time adversary $\mathcal{A}$ that attacks RABE-I as per the selective revocation list model under CPA with a non-negligible advantage. We construct a probabilistic algorithm $\mathcal{B}$ that attempts to solve an instance of the $(\ell + d + 1, N_{\max})$-MDHE problem using $\mathcal{A}$ as a subroutine. $\mathcal{B}$ is given a challenge instance

$$\varrho_{\widetilde{b}} = (\mathsf{PP}_{\mathsf{MLM}}, \vartheta_1, \ldots, \vartheta_{N_{\max}}, \vartheta_{N_{\max}+2}, \ldots, \vartheta_{2N_{\max}}, \Upsilon, \tau_1, \ldots, \tau_{\ell+d-1}, \Re_{\widetilde{b}})$$
$$\text{where } \{\vartheta_j = g_1^{\alpha^{(j)}}\}_{j=1,\ldots,N_{\max},N_{\max}+2,\ldots,2N_{\max}}, \{\tau_i = g_1^{\psi_i}\}_{i=1,\ldots,\ell+d-1}, \Upsilon = g_1^{\varsigma}$$

such that $\alpha, \varsigma, \psi_i$ are random elements of $\mathbb{Z}_p$, and $\Re_{\widetilde{b}}$ is $g_{\ell+d+1}^{\alpha^{(N_{\max}+1)} \varsigma \prod_{i=1}^{\ell+d-1} \psi_i}$ or some random element in $\mathbb{G}_{\ell+d+1}$ according as $\widetilde{b}$ is 0 or 1. $\mathcal{B}$ plays the role of the challenger in the CPA security game as per the selective revocation list model of Sect. 2.1 and interacts with $\mathcal{A}$ as follows:

Init: $\mathcal{A}$ declares the challenge input string $x^* = x_1^* \ldots x_\ell^* \in \{0,1\}^\ell$ along with the challenge revocation list RL^* to $\mathcal{B}$. Let $\mathcal{N} = \{1, \ldots, N_{\max}\}$ be the set of user key indices. $\mathcal{B}$ first initializes the user list $\mathsf{UL} = \varnothing$. Next for each $\mathsf{ID} \in \mathsf{RL}^*$ it selects an index $j \in \mathcal{N}$ such that $(\cdot, j) \notin \mathsf{UL}$ and adds (ID, j) to UL. Let $\mathsf{RI}^* \subseteq \mathcal{N}$ be the revoked set of user key indices corresponding to RL^* and $\mathsf{SI}^* = \mathcal{N} \backslash \mathsf{RI}^*$.

Setup: $\mathcal{B}$ chooses random $z_1, \ldots, z_\ell, \varphi \in \mathbb{Z}_p$ and sets

$$A_{i,\beta} = \tau_i = g_1^{\psi_i}, \text{ if } \beta = x_i^*, \; A_{i,\beta} = g_1^{z_i}, \text{ if } \beta \neq x_i^*, \text{ for } i = 1, \ldots, \ell; \; \beta \in \{0,1\},$$
$$Y = g_1^{\varphi} \Big(\prod_{j \in \mathsf{SI}^*} \vartheta_{N_{\max}+1-j} \Big)^{-1} = g_1^{\gamma}, \; Z = e(\tau_{\ell+1}, \ldots, \tau_{\ell+d-1}) = g_{d-1}^{\theta},$$
$$\Omega = e(\vartheta_{N_{\max}}, \vartheta_1, \tau_{\ell+1}, \ldots, \tau_{\ell+d-1}) = g_{d+1}^{\alpha^{(N_{\max}+1)} \theta}.$$

Note that the above setting corresponds to (possibly) *implicitly* letting

$$a_{i,\beta} = \psi_i, \text{ if } \beta = x_i^*, \; a_{i,\beta} = z_i, \text{ if } \beta \neq x_i^*, \text{ for } i = 1, \ldots, \ell; \; \beta \in \{0,1\},$$
$$\gamma = \varphi - \sum_{j \in \mathsf{SI}^*} \alpha^{(N_{\max}+1-j)}, \; \theta = \prod_{h=\ell+1}^{\ell+d-1} \psi_h = \Gamma(\ell+1, \ell+d-1),$$

where we define $\Gamma(v_1, v_2) = \prod_{h=v_1}^{v_2} \psi_h$ for positive integers v_1, v_2, with the convention that $\Gamma(v_1, v_2) = 1$ if $v_1 > v_2$, for the purpose of enhancing readability in subsequent discussion. $\mathcal{B}$ hands the public parameters

$$\mathsf{PP} = \big(\mathsf{PP}_{\mathsf{MLM}}, \{A_{i,\beta}\}_{i=1,\ldots,\ell;\beta\in\{0,1\}}, \{\vartheta_j\}_{j=1,\ldots,N_{\max},N_{\max}+2,\ldots,2N_{\max}}, Y, Z, \Omega\big)$$

along with the user list UL to $\mathcal{A}$.

Phase 1 and **Phase 2**: Both the key query phases are executed in the same manner by $\mathcal{B}$. So, we describe them once here. $\mathcal{A}$ adaptively queries a decryption key for a circuit $f = (\ell, q, d, \mathbb{A}, \mathbb{B}, \mathsf{GateType})$ and user identity ID to $\mathcal{B}$ subject to the restriction that $[f(x^*) = 0] \lor [\mathsf{ID} \in \mathsf{RL}^*]$. $\mathcal{B}$ answers the query as follows:

Case (I) ($\mathsf{ID} \in \mathsf{RL}^*$): $\mathcal{B}$ retrieves the index $u \in \mathcal{N}$ already assigned to ID in the initialization phase from UL. $\mathcal{B}$ forms the decryption key components $\mathcal{K}_w$ corresponding to all the wires $w \in \{1, \ldots, \ell+q\}$ of the circuit f exactly as in the real scheme. Next $\mathcal{B}$ sets the "header" component $\mathcal{K}$ of the decryption key as

$$\mathcal{K} = e(\vartheta_u, Z)^{\varphi}\Big[\prod_{j\in\mathsf{SI}^*} e(\vartheta_{N_{\max}+1-j+u}, Z)\Big]^{-1} g_d^{-r_{\ell+q}},$$

where $r_{\ell+q} \in \mathbb{Z}_p$ is the randomness associated with the wire $\ell+q$ already selected by $\mathcal{B}$ at the time of computing the decryption key component $\mathcal{K}_{\ell+q}$. The above simulation of $\mathcal{K} = g_d^{\alpha^{(u)}\theta\gamma - r_{\ell+q}}$ is valid since

$$\begin{aligned}\alpha^{(u)}\theta\gamma - r_{\ell+q} &= \alpha^{(u)}\Gamma(\ell+1, \ell+d-1)\big[\varphi - \sum_{j\in\mathsf{SI}^*} \alpha^{(N_{\max}+1-j)}\big] - r_{\ell+q} \\ &= \alpha^{(u)}\Gamma(\ell+1, \ell+d-1)\varphi - \Gamma(\ell+1, \ell+d-1)\sum_{j\in\mathsf{SI}^*} \alpha^{(N_{\max}+1-j+u)} - r_{\ell+q}.\end{aligned}$$

Further, notice that since $\mathsf{ID} \in \mathsf{RL}^*$, $u \notin \mathsf{SI}^*$. Hence, none of the $\alpha^{(N_{\max}+1-j+u)}$ in the preceding equation matches $\alpha^{(N_{\max}+1)}$, enabling $\mathcal{B}$ to simulate $\mathcal{K}$ as above using the available information.

Case (II) ($\mathsf{ID} \notin \mathsf{RL}^*$): In this case $\mathcal{B}$ assigns an index $u \in \mathcal{N}$ such that $(\cdot, u) \notin \mathsf{UL}$ to ID and adds (ID, u) to UL. Now observe that due to the restriction on $\mathcal{A}$'s decryption key queries we must have $f(x^*) = 0$ in this case. As in [10], we will think of the simulation as having some invariant property on the depth of the wire we are looking at. Consider a wire w with $\mathsf{depth}(w) = t$. $\mathcal{B}$ views r_w, the randomness associated with the wire w, as follows: If $f_w(x^*) = 0$, then $\mathcal{B}$ will *implicitly* view r_w as the term $-\alpha^{(N_{\max}+1)}\Gamma(\ell+1, \ell+t-1)$ plus some additional known randomization term. Otherwise (if $f_w(x^*) = 1$), $\mathcal{B}$ will view r_w as 0 plus some additional known randomization term. We keep this property intact for the bottom-up key simulation of the circuit. This makes $\mathcal{B}$ to view $r_{\ell+q}$ as $-\alpha^{(N_{\max}+1)}\Gamma(\ell+1, \ell+d-1)$ plus some additional known randomization term since $f_{\ell+q}(x^*) = f(x^*) = 0$. Then $\mathcal{B}$ can simulate the "header" component $\mathcal{K}$ by cancelation as will be explained shortly.

The bottom-up simulation of the key component for each wire w by $\mathcal{B}$ varies depending on whether w is an Input wire, OR gate, or AND gate as follows:

- *Input wire*: Consider $w \in \{1, \ldots, \ell\}$, i.e., an input wire. Hence, $\mathsf{depth}(w) = 1$.
  - If $x^*_w = 1$ then $\mathcal{B}$ picks random $r_w \in \mathbb{Z}_p$ (as is done honestly) and sets the key component $\mathcal{K}_w = e(\tau_w, g_1)^{r_w} = g_2^{r_w a_{w,1}}$.
  - Otherwise, if $x^*_w = 0$ then $\mathcal{B}$ *implicitly* lets $r_w = -\alpha^{(N_{\max}+1)} + \eta_w = -\alpha^{(N_{\max}+1)}\Gamma(\ell+1, \ell+1-1) + \eta_w$, where $\eta_w \in \mathbb{Z}_p$ is randomly selected by $\mathcal{B}$, and sets the key component $\mathcal{K}_w = e(\vartheta_{N_{\max}}, \vartheta_1)^{-z_w} g_2^{\eta_w z_w} = g_2^{r_w a_{w,1}}$.
- *OR gate*: Consider a wire $w \in \mathsf{Gates}$ with $\mathsf{GateType}(w) = \mathsf{OR}$ and $t = \mathsf{depth}(w)$. Then, $\mathsf{depth}(\mathbb{A}(w)) = \mathsf{depth}(\mathbb{B}(w)) = t - 1$ as our circuit is layered.
  - If $f_w(x^*) = 1$ then $\mathcal{B}$ chooses random $\mu_w, \nu_w, r_w \in \mathbb{Z}_p$ as in the real scheme, and forms the key component as

$$\mathcal{K}_w = \big(K_{w,1} = g_1^{\mu_w}, K_{w,2} = g_1^{\nu_w}, K_{w,3} = g_t^{r_w - \mu_w r_{\mathbb{A}(w)}}, K_{w,4} = g_t^{r_w - \nu_w r_{\mathbb{B}(w)}}\big).$$

    Lets have a closer look to the simulation of $K_{w,3}$ and $K_{w,4}$ in $\mathcal{K}_w$ by $\mathcal{B}$ above. Since $f_w(x^*) = 1$, the $\mathbb{A}(w)$ and $\mathbb{B}(w)$ gates might evaluate to 1 or 0 upon input x^* with the only restriction that both of them cannot be 0 at the same time. Consider the case of $K_{w,3}$. Observe that if $f_{\mathbb{A}(w)}(x^*) = 1$, then $r_{\mathbb{A}(w)}$ is a random element in $\mathbb{Z}_p$ already selected by $\mathcal{B}$ at this stage due to the bottom-up key simulation. Thus, in this case $\mathcal{B}$ can simulate $K_{w,3}$ exactly as in the real scheme. Now, let $f_{\mathbb{A}(w)}(x^*) = 0$. Therefore, $r_{\mathbb{A}(w)}$ has been implicitly set as $-\alpha^{(N_{\max}+1)}\Gamma(\ell+1, \ell+t-2) + \eta_{\mathbb{A}(w)}$ by $\mathcal{B}$ in the course of its bottom-up key simulation, where $\eta_{\mathbb{A}(w)} \in \mathbb{Z}_p$ is randomly chosen by $\mathcal{B}$. Thus, in this case $\mathcal{B}$ can create $K_{w,3}$ as

$$K_{w,3} = e(\vartheta_{N_{\max}}, \vartheta_1, \tau_{\ell+1}, \ldots, \tau_{\ell+t-2})^{\mu_w} g_t^{r_w - \mu_w \eta_{\mathbb{A}(w)}} = g_t^{r_w - \mu_w r_{\mathbb{A}(w)}}.$$

    A similar argument holds for $K_{w,4}$.
  - On the other hand, if $f_w(x^*) = 0$ then $\mathcal{B}$ picks random $\sigma_w, \zeta_w, \eta_w \in \mathbb{Z}_p$, *implicitly* sets $\mu_w = \psi_{\ell+t-1} + \sigma_w, \nu_w = \psi_{\ell+t-1} + \zeta_w$, along with $r_w = -\alpha^{(N_{\max}+1)}\Gamma(\ell+1, \ell+t-1) + \eta_w$, and creates the key component $\mathcal{K}_w = (K_{w,1}, K_{w,2}, K_{w,3}, K_{w,4})$ as follows:

$$\begin{aligned}
K_{w,1} &= \tau_{\ell+t-1} g_1^{\sigma_w} = g_1^{\mu_w}, K_{w,2} = \tau_{\ell+t-1} g_1^{\zeta_w} = g_1^{\nu_w},\\
K_{w,3} &= e(\tau_{\ell+t-1}, g_{t-1})^{-\eta_{\mathbb{A}(w)}} e(\vartheta_{N_{\max}}, \vartheta_1, \tau_{\ell+1}, \ldots, \tau_{\ell+t-2})^{\sigma_w} g_t^{\eta_w - \sigma_w \eta_{\mathbb{A}(w)}}\\
&= g_t^{\eta_w - \psi_{\ell+t-1}\eta_{\mathbb{A}(w)} - \sigma_w(-\alpha^{(N_{\max}+1)}\Gamma(\ell+1,\ell+t-2) + \eta_{\mathbb{A}(w)})} = g_t^{r_w - \mu_w r_{\mathbb{A}(w)}},\\
K_{w,4} &= e(\tau_{\ell+t-1}, g_{t-1})^{-\eta_{\mathbb{B}(w)}} e(\vartheta_{N_{\max}}, \vartheta_1, \tau_{\ell+1}, \ldots, \tau_{\ell+t-2})^{\zeta_w} g_t^{\eta_w - \zeta_w \eta_{\mathbb{B}(w)}}\\
&= g_t^{\eta_w - \psi_{\ell+t-1}\eta_{\mathbb{B}(w)} - \zeta_w(-\alpha^{(N_{\max}+1)}\Gamma(\ell+1,\ell+t-2) + \eta_{\mathbb{B}(w)})} = g_t^{r_w - \nu_w r_{\mathbb{B}(w)}}.
\end{aligned}$$

Note that since $f_w(x^*) = 0$, $f_{\mathbb{A}(w)}(x^*) = f_{\mathbb{B}(w)}(x^*) = 0$. Therefore, $\mathcal{B}$'s bottom-up key simulation has implicitly set $r_{\mathbb{A}(w)} = -\alpha^{(N_{\max}+1)}\Gamma(\ell+1, \ell+t-2) + \eta_{\mathbb{A}(w)}$, where $\eta_{\mathbb{A}(w)} \in \mathbb{Z}_p$ is randomly picked by $\mathcal{B}$. Hence,

$$\begin{aligned} & r_w - \mu_w r_{\mathbb{A}(w)} \\ & = \eta_w - \psi_{\ell+t-1}\eta_{\mathbb{A}(w)} - \sigma_w\big(-\alpha^{(N_{\max}+1)}\Gamma(\ell+1, \ell+t-2) + \eta_{\mathbb{A}(w)}\big) \qquad (1) \end{aligned}$$

establishing that the distribution of simulated $K_{w,3}$ by $\mathcal{B}$ is identical to that in the actual construction. Analogous argument holds for $K_{w,4}$.

- *AND gate*: Consider wire $w \in \mathsf{Gates}$ with $\mathsf{GateType}(w) = \mathsf{AND}$ and $t = \mathsf{depth}(w)$. Then $\mathsf{depth}(\mathbb{A}(w)) = \mathsf{depth}(\mathbb{B}(w)) = t-1$ for the reason that our circuit is layered.

  - Let $f_w(x^*) = 1$. Then $f_{\mathbb{A}(w)}(x^*) = f_{\mathbb{B}(w)}(x^*) = 1$. $\mathcal{B}$ picks random $\mu_w, \nu_w, r_w \in \mathbb{Z}_p$ and forms the key component

  $$\mathcal{K}_w = \big(K_{w,1} = g_1^{\mu_w}, K_{w,2} = g_1^{\nu_w}, K_{w,3} = g_t^{r_w - \mu_w r_{\mathbb{A}(w)} - \nu_w r_{\mathbb{B}(w)}}\big)$$

  exactly as in the real scheme. Observe that, since $f_{\mathbb{A}(w)}(x^*) = f_{\mathbb{B}(w)}(x^*) = 1$, $r_{\mathbb{A}(w)}$ and $r_{\mathbb{B}(w)}$ are random elements of $\mathbb{Z}_p$ already chosen by $\mathcal{B}$ in the course of the bottom-up simulation.
  - Alternatively, let $f_w(x^*) = 0$. Then, $f_{\mathbb{A}(w)}(x^*) = 0$ or $f_{\mathbb{B}(w)}(x^*) = 0$. If $f_{\mathbb{A}(w)}(x^*) = 0$, then $\mathcal{B}$ selects $\sigma_w, \zeta_w, \eta_w \in \mathbb{Z}_p$, *implicitly* defines $\mu_w = \psi_{\ell+t-1} + \sigma_w, \nu_w = \zeta_w$, and $r_w = -\alpha^{(N_{\max}+1)}\Gamma(\ell+1, \ell+t-1) + \eta_w$, and determines the decryption key component $\mathcal{K}_w = (K_{w,1}, K_{w,2}, K_{w,3})$ by setting

  $$\begin{aligned} K_{w,1} &= \tau_{\ell+t-1} g_1^{\sigma_w} = g_1^{\mu_w}, K_{w,2} = g_1^{\zeta_w} = g_1^{\nu_w}, \\ K_{w,3} &= e(\tau_{\ell+t-1}, g_{t-1})^{-\eta_{\mathbb{A}(w)}} e(\vartheta_{N_{\max}}, \vartheta_1, \tau_{\ell+1}, \ldots, \tau_{\ell+t-2})^{\sigma_w} g_t^{\eta_w - \sigma_w \eta_{\mathbb{A}(w)}}. \\ & \quad (g_t^{r_{\mathbb{B}(w)}})^{-\zeta_w} = g_t^{\eta_w - \psi_{\ell+t-1}\eta_{\mathbb{A}(w)} - \sigma_w(-\alpha^{(N_{\max}+1)}\Gamma(\ell+1, \ell+t-2) + \eta_{\mathbb{A}(w)}) - \zeta_w r_{\mathbb{B}(w)}} \\ &= g_t^{r_w - \mu_w r_{\mathbb{A}(w)} - \nu_w r_{\mathbb{B}(w)}}. \end{aligned}$$

  The simulated $K_{w,3}$ by $\mathcal{B}$ above is identically distributed as that in the original construction. This follows from the fact that, the $\mathbb{A}(w)$ gate being evaluated to 0, $r_{\mathbb{A}(w)}$ has already been implicitly set as $r_{\mathbb{A}(w)} = -\alpha^{(N_{\max}+1)}\Gamma(\ell+1, \ell+t-2) + \eta_{\mathbb{A}(w)}$ by $\mathcal{B}$ upon selecting random $\eta_{\mathbb{A}(w)} \in \mathbb{Z}_p$ in the course of the bottom-up key simulation. Therefore, as in Eq. (1), we have

  $$\begin{aligned} & r_w - \mu_w r_{\mathbb{A}(w)} \\ & = \eta_w - \psi_{\ell+t-1}\eta_{\mathbb{A}(w)} - \sigma_w\big(-\alpha^{(N_{\max}+1)}\Gamma(\ell+1, \ell+t-2) + \eta_{\mathbb{A}(w)}\big). \end{aligned}$$

  Notice that $g_t^{r_{\mathbb{B}(w)}}$ is always computable by $\mathcal{B}$ from the available information regardless of whether the $\mathbb{B}(w)$ gate evaluates to 1 or 0 upon input x^*. If $f_{\mathbb{B}(w)}(x^*) = 1$, then $r_{\mathbb{B}(w)}$ is a random element of $\mathbb{Z}_p$ chosen by $\mathcal{B}$

itself during the bottom-up simulation process. Hence, the computation of $g_t^{r_{\mathbb{B}(w)}}$ is straightforward in this case. Otherwise, if $f_{\mathbb{B}(w)}(x^*) = 0$, then $\mathcal{B}$ has already set $r_{\mathbb{B}(w)}$ as $r_{\mathbb{B}(w)} = -\alpha^{(N_{\max}+1)}\Gamma(\ell+1, \ell+d-2) + \eta_{\mathbb{B}(w)}$ at this stage by selecting random $\eta_{\mathbb{B}(w)} \in \mathbb{Z}_p$. Therefore, in this case $\mathcal{B}$ can compute $g_t^{r_{\mathbb{B}(w)}}$ as $g_t^{r_{\mathbb{B}(w)}} = e(\vartheta_{N_{\max}}, \vartheta_1, \tau_{\ell+1}, \ldots, \tau_{\ell+t-2})^{-1} g_t^{\eta_{\mathbb{B}(w)}}$.
The case where $f_{\mathbb{B}(w)}(x^*) = 0$ and $f_{\mathbb{A}(w)}(x^*) = 1$ can be argued analogously, with the roles of μ_w and ν_w reversed.

Since $f(x^*) = f_{\ell+q}(x^*) = 0$, $r_{\ell+q} = -\alpha^{(N_{\max}+1)}\Gamma(\ell+1, \ell+d-1) + \eta_{\ell+q}$, where $\eta_{\ell+q} \in \mathbb{Z}_p$ is randomly selected by $\mathcal{B}$. Also, since $\mathsf{ID} \notin \mathsf{RL}^*$, $u \in \mathsf{SI}^*$. These two facts allow $\mathcal{B}$ to compute the "header" component of the key as

$$\begin{aligned}
\mathcal{K} &= e(\vartheta_u, Z)^{\varphi}\Big[\prod_{j\in\mathsf{SI}^*\setminus\{u\}} e(\vartheta_{N_{\max}+1-j+u}, Z)\Big]^{-1} g_d^{-\eta_{\ell+q}} \\
&= g_d^{\alpha^{(u)}\Gamma(\ell+1,\ell+d-1)\varphi - \Gamma(\ell+1,\ell+d-1)\sum_{j\in\mathsf{SI}^*\setminus\{u\}}\alpha^{(N_{\max}+1-j+u)} - \eta_{\ell+q}} \\
&= g_d^{\alpha^{(u)}\Gamma(\ell+1,\ell+d-1)\left[\varphi - \sum_{j\in\mathsf{SI}^*}\alpha^{(N_{\max}+1-j)}\right] - r_{\ell+q}} = g_d^{\alpha^{(u)}\theta\gamma - r_{\ell+q}}.
\end{aligned}$$

$\mathcal{B}$ provides $\mathcal{A}$ the decryption key $\mathsf{SK}_{f,\mathsf{ID}} = \big(f, \mathsf{ID}, \mathcal{K}, \{\mathcal{K}_w\}_{w\in\{1,\ldots,\ell+q\}}\big)$ along with the updated user list UL.

Challenge: $\mathcal{A}$ submits two challenge messages $M_0^*, M_1^* \in \mathbb{G}_{\ell+d+1}$ to $\mathcal{B}$. $\mathcal{B}$ flips a random coin $b \in \{0,1\}$, sets the challenge ciphertext

$$\mathsf{CT}^* = \big(x^*, \mathsf{RL}^*, C_M^* = \Re_{\widetilde{b}} M_b^*, C^* = \Upsilon = g_1^{\varsigma}, C'^* = \Upsilon^{\varphi} = (Y \prod_{j\in\mathsf{SI}^*} \vartheta_{N_{\max}+1-j})^{\varsigma}\big),$$

and gives it to $\mathcal{A}$.

Guess: $\mathcal{B}$ eventually receives back the guess $b' \in \{0,1\}$ from $\mathcal{A}$. If $b = b'$, $\mathcal{B}$ outputs $\widetilde{b}' = 1$; otherwise, it outputs $\widetilde{b}' = 0$.

Note that if $\widetilde{b} = 0$, then

$$C_m^* = \Re_{\widetilde{b}} M_b^* = g_{\ell+d+1}^{\alpha^{(N_{\max}+1)}\varsigma\Gamma(1,\ell+d-1)} M_b^* = g_{\mathcal{K}}^{\alpha^{(N_{\max}+1)}\theta\varsigma\delta(x^*)} M_b^*,$$

where $\delta(x^*) = \prod_{i=1}^{\ell} a_{i,x_i^*}$. Thus, we can see that the challenge ciphertext CT^* is properly generated by $\mathcal{B}$ in this case by *implicitly* letting s, the randomness used to prepare the ciphertext, as ς. On the other hand, if $\widetilde{b} = 1$, then $\Re_{\widetilde{b}}$ is a random element of $\mathbb{G}_{\ell+d+1}$, so that, the challenge ciphertext is completely random. Hence the result. □

4 RABE-II

▪ **The Construction**:

RABE.Setup$(1^\lambda, \ell, d, N_{\max})$: Taking as input a security parameter 1^λ, the length ℓ of Boolean inputs to the decryption circuits, the allowed depth d of decryption circuits, and the maximum number $N_{\max}$ of users supported by the system, the trusted key generation center proceeds as follows:

1. It chooses two positive integers n, m suitably such that $N_{\max} \leq \binom{n}{m}$. Let $\mathcal{N}$ denotes the set of all integers $j \in \{1, \ldots, 2^n - 2\}$ of Hamming weight $\mathsf{HW}(j) = m$ when expressed as a bit string of length n. $\mathcal{N}$ is considered as the set of possible user key indices.
2. It executes $\mathcal{G}^{\mathsf{MLM}}(1^\lambda, \kappa = n + d + m - 1)$ to generate $\mathsf{PP}_{\mathsf{MLM}} = (\vec{\mathbb{G}} = (\mathbb{G}_1, \ldots, \mathbb{G}_\kappa), g_1, \ldots, g_\kappa)$ of prime order $p > 2^\lambda$.
3. It picks random $a_1, \ldots, a_\ell \in \mathbb{Z}_p$ and computes $A_i = g_m^{a_i}$ for $i = 1, \ldots, \ell$.
4. It chooses random $\alpha, \gamma, \theta \in \mathbb{Z}_p$ and computes $\xi_\iota = g_1^{\alpha^{(2^\iota)}}$ for $\iota = 0, \ldots, n$, $Y = g_{n-1}^{\gamma}$, $Z = g_d^{\theta}$, $\Omega = g_\kappa^{\alpha^{(2^n-1)}\theta}$.
5. It initializes the user list UL, which would consist of ordered pairs (ID, u) such that ID is the identity of an user who has participated in the system and $u \in \mathcal{N}$ is the unique index assigned to ID by the key generation center at the time of subscription, as an empty set, i.e., it sets $\mathsf{UL} = \varnothing$.
6. It keeps the master secret key $\mathsf{MK} = (\alpha, \gamma, \theta)$ to itself while publishes the public parameters $\mathsf{PP} = (\mathsf{PP}_{\mathsf{MLM}}, n, m, \{A_i\}_{i=1,\ldots,\ell}, \{\xi_\iota\}_{\iota=0,\ldots,n}, Y, Z, \Omega)$ along with the empty user list UL.

RABE.KeyGen($\mathsf{PP}, \mathsf{MK}, \mathsf{UL}, \mathsf{ID}, f$): The key generation center intakes the public parameters PP, the master secret key MK, the current user list UL, and the user identity ID together with the description $f = (\ell, q, d, \mathbb{A}, \mathbb{B}, \mathsf{GateType})$ of the decryption circuit from a user. Our circuit has $\ell + q$ wires $\{1, \ldots, \ell + q\}$ where $\{1, \ldots, \ell\}$ are ℓ input wires, $\{\ell + 1, \ldots, \ell + q\}$ are q gates (OR or AND gates), and the wire $\ell + q$ is distinguished as the output wire. It proceeds as follows:

1. It first assigns to ID an index $u \in \mathcal{N}$ such that $(\cdot, u) \notin \mathsf{UL}$ and updates UL by adding the pair (ID, u).
2. It chooses random $r_1, \ldots, r_{\ell+q} \in \mathbb{Z}_p$ where we think of randomness r_w as being associated with wire $w \in \{1, \ldots, \ell + q\}$. It produces the "header" component $\mathcal{K} = g_{n+d-1}^{\alpha^{(u)}\theta\gamma - r_{\ell+q}}$.
3. It forms key components for every wire w. The structure of the key component depends upon the category of w, i.e., whether w is an Input wire, OR gate, or AND gate. We describe below how it generates the key components in each case.
   - *Input wire*: If $w \in \{1, \ldots, \ell\}$ then it corresponds to the w-th input. It chooses random $z_w \in \mathbb{Z}_p$ and computes the key component

$$\mathcal{K}_w = \left(K_{w,1} = g_n^{r_w} e(A_w, g_{n-m})^{z_w} = g_n^{r_w} g_n^{a_w z_w},\ K_{w,2} = g_n^{-z_w}\right).$$

   - *OR gate*: Suppose that wire $w \in \mathsf{Gates}$, $\mathsf{GateType}(w) = \mathsf{OR}$, and $t = \mathsf{depth}(w)$. It picks random $\mu_w, \nu_w \in \mathbb{Z}_p$ and creates the key component

$$\mathcal{K}_w = \left(K_{w,1} = g_1^{\mu_w}, K_{w,2} = g_1^{\nu_w}, K_{w,3} = g_{n+t-1}^{r_w - \mu_w r_{\mathbb{A}(w)}}, K_{w,4} = g_{n+t-1}^{r_w - \nu_w r_{\mathbb{B}(w)}}\right).$$

   - *AND gate*: Let wire $w \in \mathsf{Gates}$, $\mathsf{GateType}(w) = \mathsf{AND}$, and $t = \mathsf{depth}(w)$. It selects random $\mu_w, \nu_w \in \mathbb{Z}_p$ and forms the key component

$$\mathcal{K}_w = \left(K_{w,1} = g_1^{\mu_w}, K_{w,2} = g_1^{\nu_w}, K_{w,3} = g_{n+t-1}^{r_w - \mu_w r_{\mathbb{A}(w)} - \nu_w r_{\mathbb{B}(w)}}\right).$$

4. It provides the decryption key $\mathsf{SK}_{f,\mathsf{ID}} = \big(f, \mathsf{ID}, \mathcal{K}, \{\mathcal{K}_w\}_{w\in\{1,\ldots,\ell+q\}}\big)$ to the user and publishes the updated user list UL.

RABE.Encrypt($\mathsf{PP}, \mathsf{UL}, x, \mathsf{RL}, M$): On input the public parameters PP, the current user list UL, a descriptor input string $x = x_1 \ldots x_\ell \in \{0,1\}^\ell$, a revoked user identity list RL, and a message $M \in \mathbb{G}_\kappa$, the encrypter proceeds as follows:

1. It defines the revoked user key index set $\mathsf{RI} \subseteq \mathcal{N}$ corresponding to RL using UL, i.e., if $\mathsf{ID} \in \mathsf{RL}$ and $(\mathsf{ID}, j) \in \mathsf{UL}$ it puts j in RI, and sets $\mathsf{SI} = \mathcal{N} \backslash \mathsf{RI}$.
2. It computes ϑ_{2^n-1-j} for all $j \in \mathsf{SI}$ utilizing the ξ_ι values included in PP and multilinear map as follows, where we define $\vartheta_\varpi = g_{n-1}^{\alpha^{(\varpi)}}$ for positive integer ϖ. Observe that any $j \in \mathsf{SI} \subseteq \mathcal{N}$ can be expressed as a bit string of length n with $\mathsf{HW}(j) = m$. Hence, j can be written as $j = \sum_{\iota\in J} 2^\iota$ where $J \subseteq \{0, \ldots, n-1\}$ of size m. Now $2^n - 1 = \sum_{\iota=0}^{n-1} 2^\iota$. Thus, $2^n - 1 - j = \sum_{\iota\in\overline{J}} 2^\iota$ where $\overline{J} = \{0, \ldots, n-1\}\backslash J = \{\iota_1, \ldots, \iota_{n-m}\}$. It computes ϑ_{2^n-1-j} as
$$\vartheta_{2^n-1-j} = e(\xi_{\iota_1}, \ldots, \xi_{\iota_{n-m}}, g_{m-1}) = g_{n-1}^{\alpha^{(2^n-1-j)}}.$$
3. It picks random $s \in \mathbb{Z}_p$ and computes
$$C_M = \Omega^s M = g_\kappa^{\alpha^{(2^n-1)}\theta s} M,\ C = g_m^s,$$
$$C_i' = A_i^s = g_m^{a_i s} \text{ for } i \in \mathcal{S}_x = \{i | i \in \{1, \ldots, \ell\} \wedge x_i = 1\},$$
$$C'' = \big(Y \prod_{j\in\mathsf{SI}} \vartheta_{2^n-1-j}\big)^s = \big(g_{n-1}^\gamma \prod_{j\in\mathsf{SI}} g_{n-1}^{\alpha^{(2^n-1-j)}}\big)^s.$$
4. It outputs the ciphertext $\mathsf{CT}_{x,\mathsf{RL}} = (x, \mathsf{RL}, C_M, C, \{C_i'\}_{i\in\mathcal{S}_x}, C'')$.

Remark 1. We would like to mention that the number of ciphertext components could be made constant (precisely only 4), as in RABE-I, rather than scaling with the size of $\mathcal{S}_x$ using a $(\ell+n+d+m-2)$-leveled multilinear map. However, since in current approximate multilinear map candidates the multilinearity is expensive, we opt for a construction that requires lower multilinearity level.

RABE.Decrypt($\mathsf{PP}, \mathsf{UL}, \mathsf{CT}_{x,\mathsf{RL}}, \mathsf{SK}_{f,\mathsf{ID}}$): A user intakes the public parameters PP, current user list UL, a ciphertext $\mathsf{CT}_{x,\mathsf{RL}} = (x, \mathsf{RL}, C_M, C, \{C_i'\}_{i\in\mathcal{S}_x}, C'')$ encrypted for descriptor input string $x = x_1 \ldots x_\ell \in \{0,1\}^\ell$ along with a revoked user identity list RL, and its decryption key $\mathsf{SK}_{f,\mathsf{ID}} = \big(f, \mathsf{ID}, \mathcal{K}, \{\mathcal{K}_w\}_{w\in\{1,\ldots,\ell+q\}}\big)$ for its decryption policy circuit $f = (\ell, q, d, \mathbb{A}, \mathbb{B}, \mathsf{GateType})$ as well as its user identity ID, where $u \in \mathcal{N}$ is the index assigned to ID (say). It outputs $\perp$, if $[f(x) = 0] \vee [\mathsf{ID} \in \mathsf{RL}]$; otherwise, (if $[f(x) = 1] \wedge [\mathsf{ID} \notin \mathsf{RL}]$) proceeds as follows:

1. First, as a "header" computation, it computes
$$\widehat{E} = e(\mathcal{K}, C) = e\big(g_{n+d-1}^{\alpha^{(u)}\theta\gamma - r_{\ell+q}}, g_m^s\big) = g_\kappa^{(\alpha^{(u)}\theta\gamma - r_{\ell+q})s}.$$
2. Next, it performs the bottom-up evaluation of the circuit. For every wire w with corresponding $\mathsf{depth}(w) = t$, if $f_w(x) = 0$, nothing is computed for that wire, otherwise (if $f_w(x) = 1$), it attempts to compute $E_w = g_{n+t+m-1}^{r_w s}$

as described below. The user proceeds iteratively starting with computing E_1 and moves forward in order to finally compute $E_{\ell+q}$. Note that computing these values *in order* ensures that the computation on a wire w with $\mathsf{depth}(w) = t-1$ that evaluates to 1 will be defined before the computation on a wire w with $\mathsf{depth}(w) = t$. The computation procedure depends on whether the wire is an Input wire, OR gate, or AND gate.

- *Input wire*: If $w \in \{1, \ldots, \ell\}$ then it corresponds to the w-the input and $t = \mathsf{depth}(w) = 1$. Suppose that $x_w = f_w(x) = 1$. The user extracts $K_{w,1}, K_{w,2}$ from $\mathcal{K}_w$ included in its decryption key $\mathsf{SK}_{f,\mathsf{ID}}$ and computes
$$E_w = e(K_{w,1}, C)e(K_{w,2}, C'_w) = g_{n+m}^{r_w s} = g_{n+1+m-1}^{r_w s}.$$
- *OR gate*: Consider a wire $w \in \mathsf{Gates}$ with $\mathsf{GateType}(w) = \mathsf{OR}$ and $t = \mathsf{depth}(w)$. Let $f_w(x) = 1$. Then the user checks whether $f_{\mathbb{A}(w)}(x) = 1$. If so, then it extracts $K_{w,1}, K_{w,3}$ from $\mathcal{K}_w$ contained in $\mathsf{SK}_{f,\mathsf{ID}}$ and computes
$$E_w = e(E_{\mathbb{A}(w)}, K_{w,1})e(K_{w,3}, C) = g_{n+t+m-1}^{r_w s}$$
Alternatively, if $f_{\mathbb{A}(w)}(x) = 0$, then it must hold that $f_{\mathbb{B}(w)}(x) = 1$ as $f_w(x) = 1$. In this case, it extracts $K_{w,2}, K_{w,4}$ from $\mathcal{K}_w$ in $\mathsf{SK}_{f,\mathsf{ID}}$ and computes
$$E_w = e(E_{\mathbb{B}(w)}, K_{w,2})e(K_{w,4}, C) = g_{n+t+m-1}^{r_w s}$$
- *AND gate*: Consider a wire $w \in \mathsf{Gates}$ with $\mathsf{GateType}(w) = \mathsf{AND}$ and $t = \mathsf{depth}(w)$. Suppose that $f_w(x) = 1$. Then $f_{\mathbb{A}(w)}(x) = f_{\mathbb{B}(w)}(x) = 1$. The user extracts $K_{w,1}, K_{w,2}, K_{w,3}$ from $\mathcal{K}_w$ included in $\mathsf{SK}_{f,\mathsf{ID}}$ and computes
$$E_w = e(E_{\mathbb{A}(w)}, K_{w,1})e(E_{\mathbb{B}(w)}, K_{w,2})e(K_{w,3}, C) = g_{n+t+m-1}^{r_w s}.$$
Note that both $E_{\mathbb{A}(w)}$ and $E_{\mathbb{B}(w)}$ are already computed at this stage in the course of the bottom-up evaluation of the circuit as $\mathsf{depth}(\mathbb{A}(w)) = \mathsf{depth}(\mathbb{B}(w)) = t - 1$.

  At the end, the user computes $E_{\ell+q} = g_{\kappa}^{r_{\ell+q} s}$, as $f(x) = f_{\ell+q}(x) = 1$.

3. It determines the revoked user key index set $\mathsf{RI} \subseteq \mathcal{N}$ corresponding to RL using UL and obtains $\mathsf{SI} = \mathcal{N} \backslash \mathsf{RI}$. Note that since $\mathsf{ID} \notin \mathsf{RL}$, $u \in \mathsf{SI}$.
4. It computes $\vartheta'_u = g_m^{\alpha^{(u)}}$ and $\vartheta_{2^n-1-j+u} = g_{n-1}^{\alpha^{(2^n-1-j+u)}}$ for all $j \in \mathsf{SI} \backslash \{u\}$ using the ξ_ι values included in PP and multilinear map as follows:
   (a) (Computing ϑ'_u) Note that u can be expressed as a bit string of length n with $\mathsf{HW}(u) = m$ as $u \in \mathsf{SI} \subseteq \mathcal{N}$. Let $u = \sum_{\iota \in U} 2^\iota$ where $U = \{\iota'_1, \ldots, \iota'_m\} \subseteq \{0, \ldots, n-1\}$. It computes $\vartheta'_u = e(\xi_{\iota'_1}, \ldots, \xi_{\iota'_m}) = g_m^{\alpha^{(u)}}$.

   (b) (Computing $\vartheta_{2^n-1-j+u}$ for $j \in \mathsf{SI} \backslash \{u\}$) Let $2^n - 1 - j = \sum_{\iota \in \overline{J}} 2^\iota$ where $\overline{J} = \{\iota_1, \ldots, \iota_{n-m}\} \subseteq \{0, \ldots, n-1\}$ as earlier. Now U and $\overline{J}$ are disjoined only if $\overline{J} \cup U = \{0, \ldots, n-1\}$, i.e., $2^n - 1 - j + u = \sum_{\iota \in \overline{J}} 2^\iota + \sum_{\iota \in U} 2^\iota = \sum_{\iota=0}^{n-1} 2^\iota = 2^n - 1$, i.e., $j = u$. Since $j \neq u$, there must exist at least one $\widehat{\iota} \in \{0, \ldots, n-1\}$ such that $\widehat{\iota} \in \overline{J} \cap U$. Without loss of generality, let $\widehat{\iota} = \iota_{n-m} = \iota'_m$. Then $2^n - 1 - j + u = \sum_{\iota \in \overline{J} \backslash \{\iota_{n-m}\}} 2^\iota + \sum_{\iota \in U \backslash \{\iota'_m\}} 2^\iota +$

$2 \cdot 2^{\widehat{\iota}} = \sum_{\iota \in \overline{J} \setminus \{\widehat{\iota}\}} 2^{\iota} + \sum_{\iota \in U \setminus \{\widehat{\iota}\}} 2^{\iota} + 2^{\widehat{\iota}+1}$ where $[\overline{J} \setminus \{\widehat{\iota}\}] \cap [U \setminus \{\widehat{\iota}\}]$ may not be empty. It computes $\vartheta_{2^n-1-j+u}$ as

$$\vartheta_{2^n-1-j+u} = e(\xi_{\iota_1}, \ldots, \xi_{\iota_{n-m-1}}, \xi_{\iota'_1}, \ldots, \xi_{\iota'_{m-1}}, \xi_{\widehat{\iota}+1}) = g_{n-1}^{\alpha^{(2^n-1-j+u)}}.$$

Note that $\xi_{\widehat{\iota}+1}$ is extractable from PP since $\widehat{\iota} \in \{0, \ldots, n-1\}$.

5. Finally, utilizing the fact that $u \in \mathsf{SI}$, the user retrieves the message by the following computation:

$$C_M \widehat{E} E_{\ell+q} e\Big(\prod_{j \in \mathsf{SI} \setminus \{u\}} \vartheta_{2^n-1-j+u}, Z, C\Big) e\big(\vartheta'_u, Z, C''\big)^{-1} = M.$$

▪ **Security:**

Theorem 2. $\mathsf{RABE\text{-}II}$ *is secure in the selective revocation list model against* CPA *as per the security model of Sect. 2.1 if the* (n, d, m)*-*cMDHE *assumption holds for the underlying multilinear group generator* $\mathcal{G}^{\mathsf{MLM}}$ *described in Sect. 2.2, such that* d *denotes the allowed depth of the decryption circuits, and* n, m *are two integers for which* $N_{\max} \leq \binom{n}{m}$*, where* $N_{\max}$ *is the maximum number of users supported by the system.*

The proof of Theorem 2 closely resembles that of Theorem 1 and is omitted here due to page restriction.

5 Efficiency

Both our RABE schemes permit general Boolean circuits of arbitrary polynomial size and unbounded fan-out with bounded depth and input length. This is the most expressive form of decryption policies accomplished for ABE till date [4,10]. We utilize the power of the multilinear map framework. All previous RABE constructions in the standard model [1–3,11,14], could support at most polynomial size monotone Boolean formulae because of the inherent limitation [10] of the traditional bilinear map setting underlying those schemes.

Another drawback of the previous standard model RABE schemes supporting direct revocation mode [1,2,14] is that they essentially utilize the tree-based revocation mechanism of Naor et al. [12]. As a result, the number of group elements comprising the revocation controlling segments of the ciphertexts and decryption keys are $O(\widehat{r} \log \frac{N_{\max}}{\widehat{r}})$ and $O(\log N_{\max})$ respectively, where $N_{\max}$ and $\widehat{r}$ denote respectively the maximum number of users supported by the system and the number of revoked users. Moreover, the number of group elements in the ABE realizing portion of the ciphertexts scales with the size of the attribute set or the complexity of the decryption policy associated to it. Our first RABE construction, RABE-I, which is designed by carefully integrating the revocation strategy introduced in [5] with an improved variant of [10], features only 3 group elements in the ciphertexts. Furthermore, the number of decryption key components is $\ell + 4q + 1$ in the worst case, ℓ and q being respectively the input length and number of gates in the policy circuits. This is the same in all currently available multilinear map-based vanilla ABE constructions for general circuits [4,10].

This the added revocation functionality is attained without any extra overhead on the decryption keys.

One problem in RABE-I is that the number of PP elements is linear to $N_{\max}$ and, hence, the construction can accommodate only a small number of users. In our second RABE scheme, RABE-II, we attempt to reduce it by applying a more advanced revocation technique [6] so that we can support potentially large number of users. For RABE-II we consider $\kappa = n + d + m - 1$ such that $N_{\max} \leq \binom{n}{m}$ and the number of PP components becomes linear to n. As discussed in [6], a judicious choice of n and m would require $n \approx \log N_{\max}$. Therefore, the number of PP components reduces approximately to $\log N_{\max}$ in RABE-II. This is comparable to the best PP size attained by previous RABE constructions with direct revocation mode secure in the standard model [1,2,14]. Also, in RABE-II we need to provide only one component in PP in place of two in case of RABE-I corresponding to each input of the decryption policy circuits. However, observe that in this scheme also we could maintain the property that the number of ciphertext and decryption key components meant for revocation do not grow with $N_{\max}$. To the best of our knowledge, no previous RABE scheme with direct revocation could achieve such parameters.

Regarding computational complexity, note that the (worst case) number of multilinear operations involved in the setup, key generation, encryption, and decryption algorithms are respectively $2\ell+2N_{\max}+2$, $2\ell+4q+2$, 4, and $\ell+3q+4$ for RABE-I while $\ell + 2n + 5$, $4\ell + 4q + 3$, $\ell + 3$, and $2\ell + 3q + 3$ for RABE-II. Thus, we can see that RABE-II involves slightly more computation in the key generation, encryption and decryption procedures compared to RABE-I.

Remark 2. Note that a recent work [13] has applied the same revocation technique as ours [5,6] in the context of *identity-based encryption* (IBE). We emphasize that although IBE and ABE are related concepts, the richer functionality offered by the latter, especially when the access structures are highly expressive such as general polynomial-size circuits, poses significantly more challenges in enforcing revocation and necessitates more elegant techniques which we have developed in this work.

6 Conclusion

In this work, employing multilinear map [7–9], we have adopted a new technique [5,6] for enforcing direct user revocation in the context of ABE. Following that method, we have developed two selectively secure RABE schemes, both of which support decryption policies representable as general polynomial-size circuits as well as features very short ciphertexts without imposing any extra overhead in the decryption keys for the added revocation functionality. In our first construction, the size of the public parameters is linear to the maximum number of users supported by the system, while we have shrunk it to logarithmic in our second construction.

To the best of our knowledge, our work is the first in the literature which attained these features. Both our RABE constructions are proven secure in the selective revocation list model under reasonable assumptions.

References

1. Attrapadung, N., Imai, H.: Attribute-based encryption supporting direct/indirect revocation modes. In: Parker, M.G. (ed.) Cryptography and Coding 2009. LNCS, vol. 5921, pp. 278–300. Springer, Heidelberg (2009)
2. Attrapadung, N., Imai, H.: Conjunctive broadcast and attribute-based encryption. In: Shacham, H., Waters, B. (eds.) Pairing 2009. LNCS, vol. 5671, pp. 248–265. Springer, Heidelberg (2009)
3. Boldyreva, A., Goyal, V., Kumar, V.: Identity-based encryption with efficient revocation. In: Proceedings of the 15th ACM Conference on Computer and Communications Security, pp. 417–426. ACM (2008)
4. Boneh, D., Gentry, C., Gorbunov, S., Halevi, S., Nikolaenko, V., Segev, G., Vaikuntanathan, V., Vinayagamurthy, D.: Fully key-homomorphic encryption, arithmetic circuit abe and compact garbled circuits. In: Nguyen, P.Q., Oswald, E. (eds.) EUROCRYPT 2014. LNCS, vol. 8441, pp. 533–556. Springer, Heidelberg (2014)
5. Boneh, D., Gentry, C., Waters, B.: Collusion resistant broadcast encryption with short ciphertexts and private keys. In: Shoup, V. (ed.) CRYPTO 2005. LNCS, vol. 3621, pp. 258–275. Springer, Heidelberg (2005)
6. Boneh, D., Waters, B., Zhandry, M.: Low overhead broadcast encryption from multilinear maps. In: Garay, J.A., Gennaro, R. (eds.) CRYPTO 2014, Part I. LNCS, vol. 8616, pp. 206–223. Springer, Heidelberg (2014)
7. Coron, J.-S., Lepoint, T., Tibouchi, M.: Practical multilinear maps over the integers. In: Canetti, R., Garay, J.A. (eds.) CRYPTO 2013, Part I. LNCS, vol. 8042, pp. 476–493. Springer, Heidelberg (2013)
8. Coron, J.S., Lepoint, T., Tibouchi, M.: New multilinear maps over the integers (2015)
9. Garg, S., Gentry, C., Halevi, S.: Candidate multilinear maps from ideal lattices. In: Johansson, T., Nguyen, P.Q. (eds.) EUROCRYPT 2013. LNCS, vol. 7881, pp. 1–17. Springer, Heidelberg (2013)
10. Garg, S., Gentry, C., Halevi, S., Sahai, A., Waters, B.: Attribute-based encryption for circuits from multilinear maps. In: Canetti, R., Garay, J.A. (eds.) CRYPTO 2013, Part II. LNCS, vol. 8043, pp. 479–499. Springer, Heidelberg (2013)
11. Liang, X., Lu, R., Lin, X., Shen, X.S.: Ciphertext policy attribute based encryption with efficient revocation. Technical report, University of Waterloo (2010)
12. Naor, D., Naor, M., Lotspiech, J.: Revocation and tracing schemes for stateless receivers. In: Kilian, J. (ed.) CRYPTO 2001. LNCS, vol. 2139, pp. 41–62. Springer, Heidelberg (2001)
13. Park, S., Lee, K., Lee, D.H.: New constructions of revocable identity-based encryption from multilinear maps. IACR Cryptology ePrint Archive 2013, 880 (2013)
14. Qian, J., Dong, X.: Fully secure revocable attribute-based encryption. J. Shanghai Jiaotong Univ. (Sci.) **16**, 490–496 (2011)
15. Shi, Y., Zheng, Q., Liu, J., Han, Z.: Directly revocable key-policy attribute-based encryption with verifiable ciphertext delegation. Inf. Sci. **295**, 221–231 (2015)
16. Yu, S., Wang, C., Ren, K., Lou, W.: Attribute based data sharing with attribute revocation. In: Proceedings of the 5th ACM Symposium on Information, Computer and Communications Security, pp. 261–270. ACM (2010)

Hashing into Jacobi Quartic Curves

Wei Yu[1,2](✉), Kunpeng Wang[1], Bao Li[1], Xiaoyang He[1], and Song Tian[1]

[1] Institute of Information Engineering, Chinese Academy of Sciences, Beijing 100093, China
yuwei_1_yw@163.com

[2] Data Assurance and Communication Security Research Center, Chinese Academy of Sciences, Beijing 100093, China

Abstract. Jacobi quartic curves are well known for efficient arithmetics in regard to their group law and immunity to timing attacks. Two deterministic encodings from a finite field $\mathbb{F}_q$ to Jacobi quartic curves are constructed. When $q \equiv 3 \pmod 4$, the first deterministic encoding based on Skalba's equality saves two field squarings compared with birational equivalence composed with Fouque and Tibouchi's brief version of Ulas' function. When $q \equiv 2 \pmod 3$, the second deterministic encoding based on computing cube root costs one field inversion less than birational equivalence composed with Icart's function at the cost of four field multiplications and one field squaring. It costs one field inversion less than Alasha's encoding at the cost of one field multiplication and two field squarings. With these two deterministic encodings, two hash functions from messages directly into Jacobi quartic curves are constructed. Additionally, we construct two types of new efficient functions indifferentiable from a random oracle.

Keywords: Deterministic encoding · Hash function · Random oracle · Jacobi quartic curves · Timing attacks

1 Introduction

Many algebraic curve cryptosystems require hashing into an algebraic curve. Boneh–Franklin's identity-based encryption scheme [1] proposes a one-to-one mapping f from a finite field $\mathbb{F}_q$ to a particular supersingular elliptic curve. Based on this mapping f, they constructed a hash function $f(h(m))$ where message $m \in \{0,1\}^*$ and h is a classical hash function. Many other identity-based schemes need messages to be hashed into an algebraic curve, such as encryption schemes [2,3], signature schemes [4,5], signcryption schemes [6,7], and Lindell's universally-composable commitment scheme [8]. Password-based authentication protocols also require messages to be hashed into algebraic curves. The simple password exponential key exchange [9] and the password authenticated key exchange [10]

This research is supported in part by National Research Foundation of China under Grant No. 61379137, No. 61272040, and in part by National Basic Research Program of China(973) under Grant No.2013CB338001.

J. Lopez and C.J. Mitchell (Eds.): ISC 2015, LNCS 9290, pp. 355–375, 2015.
DOI: 10.1007/978-3-319-23318-5_20

protocols both require a hash algorithm to map the password into a point of the algebraic curve.

Boneh and Franklin [1] proposed an algorithm mapping an element of $\mathbb{F}_{p^n}$ to a rational point of an ordinary elliptic curve, which is probabilistic and fails to return a point for a fraction 2^{-k} of the inputs, where k is a predetermined bound. One drawback of the algorithm is that the number of steps in the algorithm depends on the input u where u is an element of a finite field $\mathbb{F}_q$. Thus, the number of operations is not constant. If the input u has to be secret in practice, this algorithm may be threatened by timing attacks [11] for not being run in constant time. Therefore, hashing into an algebraic curve in a constant number of operations is significant.

There are various algorithms mapping elements of $\mathbb{F}_q$ to an elliptic curve in deterministic polynomial time. Shallue and Woestijne's algorithm [12] is based on Skalba's equality [14] and uses a modification of Tonelli–Shanks algorithm for computing square roots efficiently as $x^{1/2} = x^{(q+1)/4}$ when $q \equiv 3 \pmod 4$. Icart [13] in Crypto 2009 proposed an algorithm based on finding cube roots efficiently as $x^{1/3} = x^{(2q-1)/3}$ when $q \equiv 2 \pmod 3$. Both algorithms encode an element of a finite field into Weierstrass-form elliptic curves. Later, Fouque, Joux and Tibouchi introduced an injective encoding to elliptic curves [15].

There are some methods of hashing into hyperelliptic curves [16,17]. Their main idea is to compute square roots. Fouque and Tibouchi [18] gave the brief version of Ulas' function [16]. Later, hashing into Hessian curves [19] and Montgomery curves [20] were proposed. Alasha [21] proposed deterministic encodings into Jacobi quartic curves, Edwards curves and Huff elliptic curves based on calculating a cube root. A hash function from plaintext to C_{34}-curves is constructed in [22] by finding a cube root.

Jacobi quartic curves [23], one type of elliptic curves, are widely used for efficient arithmetics and immunity to timing attacks. The order of group of rational points on Jacobi quartic curves is divisible by 2 [24,25]. Jacobi quartic curves can provide a larger group than Huff elliptic curves, Montgomery-form elliptic curves, Hessian curves and Edwards curves. The pairing computation of Jacobi quartic curves were well studied in [26,27].

We propose two deterministic encodings directly from finite fields to Jacobi quartic curves: Shallue-Woestijne-Ulas (SWU) encoding and cube root encoding. SWU encoding is based on Skalba's equality [14]. Its main operation is finding a square root. Since SWU encoding has two variables, we analyze brief SWU encoding having only one variable. It costs two field squarings less than birational equivalence from Weierstrass curve to Jacobi quartic curves composed with Fouque and Tibouchi's brief version [18] of Ulas' function [16]. We estimate the character sum of any non-trivial character defined over Jacobi Quartic curves through brief SWU encoding. Cube root encoding is based on finding a cube root. It saves one field inversion compared with birational equivalence from Weierstrass curves to Jacobi quartic curves composed with Icart's function at the cost of four field multiplications and one field squaring. It saves one field inversion compared with Alasha's encoding at the cost of one field multiplication and two field squarings. We give the size of images of this hash function and also estimate the relevant character sum through cube root encoding.

We do experiments over 192-bit prime field $\mathbb{F}_{P192}$ and 384-bit prime field $\mathbb{F}_{P384}$ recommended by NIST in the elliptic curve standard [28]. All elements in both fields have a unique square root and a unique cube root. On $\mathbb{F}_{P192}$, our cube root encoding saves 23.8 % running time compared with birational equivalence composed with Ulas' function, 21.7 % with birational equivalence composed with Fouque and Tibouchi's brief version of Ulas' function, 20.2 % with our brief SWU encoding, 5 % with birational equivalence composed with Icart's function and 5.8 % with Alasha's work. On $\mathbb{F}_{P384}$, Cube root encoding is also fastest among these deterministic encodings, but saves less than on $\mathbb{F}_{P192}$.

Based on these two encodings, we construct two hash functions from messages into Jacobi quartic curves. Moreover, we provide new efficient functions indifferentiable from a random oracle based on these encodings.

This paper is organized as follows. In Sect. 2, we recall some basics about Jacobi quartic curves. In Sect. 3, we propose SWU encoding, brief SWU encoding and construct a hash function based on brief SWU encoding, and estimate its character sum. In Sect. 4, we give a cube root encoding, construct a hash function based on cube root encoding, estimate its character sum, and give the size of images of this hash function. In Sect. 5, two efficient functions indifferentiable from a random oracle are constructed. Section 6 gives the time complexity of these algorithms and the experimental results. Section 7 concludes the paper.

2 Jacobi Quartic Curves

Jacobi quartic curves, which are Jacobi quartic form elliptic curves, were first introduced to cryptography by Billet and Joye [23] for their efficient group law.

Let $\mathbb{F}_q$ be a finite field of odd characteristic. Jacobi quartic curves, defined over $\mathbb{F}_q$, can be written as

$$E_{a,d} : y^2 = dx^4 + 2ax^2 + 1,$$

where $a, d \in \mathbb{F}_q$ with d a non-quadratic residue. The discriminant of $E_{a,d}$ is $\Delta = 256(a^2 - d)^2 \neq 0$, and the j-invariant is $\frac{64(a^2+3d)^3}{d(a^2-d)^2}$. ECC-standards [28] recommend that for cryptographic applications the cofactor of the elliptic curve should be no greater than four, which Jacobi quartic curves can supply.

Next we present the group law of Jacobi quartic curves.

Group Law

1. Identity: $\mathcal{O} = (0, 1)$: $\forall\ P \in E(\mathbb{F}_q)$, $P + \mathcal{O} = \mathcal{O} + P = P$.
2. Negative point: if $P = (x, y) \in E(\mathbb{F}_q)$, then $(x, y) + (-x, y) = \mathcal{O}$, denote the negative of P is $-P = (-x, y)$, and $-\mathcal{O} = \mathcal{O}$.
3. Unified point-addition formulae: let $P_1 = (x_1, y_1)$, $P_2(x_2, y_2) \in E(\mathbb{F}_q)$, $P_1 \neq -P_2$, then $P_1 + P_2 = (x_3, y_3)$, where

$$\begin{aligned} x_3 &= \tfrac{x_1y_2+y_1x_2}{1-dx_1^2x_2^2}, \\ y_3 &= \tfrac{(y_1y_2+2ax_1x_2)(1+dx_1^2x_2^2)+2dx_1x_2(x_1^2+x_2^2)}{(1-dx_1^2x_2^2)^2}. \end{aligned}$$

Its group law is efficient and resistant to timing attacks. $(0, -1)$ is the point of order 2.

Wang et al. [26] provided a nice geometric explanation of this law and comes to a way to compute the pairing on Jacobi quartic curves. Duquesne and Fouotsa [27] gave a more efficient computation of pairing.

Next, our deterministic encodings will be given.

3 SWU Encoding

Ulas' achievement [16], constructing rational points on $y^2 = x^n + ax^2 + bx$, is used to construct our deterministic encoding.

3.1 SWU Encoding

We construct the deterministic encoding from $\mathbb{F}_q$ to $g(s) = s(s^2 - 4as + 4a^2 - 4d)$. $g(s)$ is an intermediate variable for the convenience of constructing encoding to Jacobi quartic curves.

Let u be an element of a finite field $\mathbb{F}_q$ and

$$\begin{aligned} X_1(u,r) &= r, \\ X_2(u,r) &= \frac{(a^2-d)[u^2 g(r) - 1]}{au^2 g(r)}, \\ X_3(u,r) &= u^2 g(r) X_2(u,r), \\ U(u,r) &= u^3 g(r)^2 g(X_2(u,r)), \end{aligned}$$

then

$$U(u,r)^2 = g(X_1(u,r))g(X_2(u,r))g(X_3(u,r)). \tag{1}$$

From Eq. (1), at least one of $g(X_1(u,r)), g(X_2(u,r)), g(X_3(u,r))$ is a quadratic residue. Thus, one of $X_1(u,r), X_2(u,r), X_3(u,r)$ is the abscissa of a point on the curve $t^2 = g(s)$. If $g(X_1(u,r))$ is a quadratic residue, then $s = X_1(u,r), t = -\sqrt{g(X_1(u,r))}$, else if $g(X_2(u,r))$ is a quadratic residue, then $s = X_2(u,r), t = \sqrt{g(X_2(u,r))}$, else $s = X_3(u,r), t = -\sqrt{g(X_3(u,r))}$.

Let

$$\begin{aligned} x &= \tfrac{2t}{(s-2a)^2 - 4d}, \\ y &= \tfrac{s^2 - 4(a^2-d)}{(s-2a)^2 - 4d}. \end{aligned}$$

If $q \equiv 3 \pmod 4$, $\sqrt{x}$ is simply an exponentiation $x^{\frac{1}{2}} = x^{\frac{q+1}{4}}$. This map, $u \mapsto (x, y)$, is called an SWU map and denoted by f_0, whose main operation is to compute square roots (see [12,16]). Substitute x, y of f_0 into the Jacobi quartic equation $y^2 = dx^4 + 2ax^2 + 1$, then $t^2 = s(s^2 - 4as + 4a^2 - 4d)$. Thus, x and y satisfy the Jacobi quartic equation. We simplify f_0 in the following.

3.2 Brief SWU Encoding

Note that the value of r is not required to be known in computing X_2, X_3 and U; indeed, these only depend on $g(r)$. For this reason, r does not have to be explicitly computed and we can take $g(r) = -1$. -1 is a quadratic non-residue when $q \equiv 3 \pmod 4$. Even if the value of r does not necessarily exist in $\mathbb{F}_q$, it exists in $\mathbb{F}_{q^3}$. With this value of r, Eq. (1) is still correct. Rewriting the SWU maps as a single variable with $g(r) = -1$ gives the following maps.

$$\begin{aligned} X_2(u) &= \frac{(a^2-d)(u^2+1)}{au^2}, \\ X_3(u) &= \frac{-(a^2-d)(u^2+1)}{a} = -u^2 X_2(u), \\ U(u) &= u^3 g(X_2(u)). \end{aligned}$$

Thus

$$U(u)^2 = -g(X_2(u))g(X_3(u)). \tag{2}$$

Therefore either $g(X_2(u))$ or $g(X_3(u))$ must be a quadratic residue. This leads to the simplified SWU encoding.

If $g(X_2(u))$ is a quadratic residue, then $s = X_2(u), t = \sqrt{g(X_2(u))}$, else $s = X_3(u), t = -\sqrt{g(X_3(u))}$. Then

$$\begin{aligned} x &= \tfrac{2t}{(s-2a)^2-4d}, \\ y &= \tfrac{s^2-4(a^2-d)}{(s-2a)^2-4d}. \end{aligned}$$

Denote the map $u \mapsto (s,t)$ by ρ, and denote $(s,t) \mapsto (x,y)$ by ψ, we call the composition $f_1 = \psi \circ r$ brief SWU encoding. $X_2(t)$ has at most two solutions, and $X_3(t)$ has at most two solutions ($\deg X_3(t) = 2$). Therefore a point has at most four pre-images.

We estimate the character sum of any non-trivial character defined over Jacobi Quartic curves through brief SWU encoding.

3.3 Character Sum

Definition 1 (Character Sum). *Suppose f is an encoding from $\mathbb{F}_q$ into an elliptic curve E, and $J(\mathbb{F}_q)$ denotes the Jacobian group of E, χ is a character of $J(\mathbb{F}_q)$. We define the character sum*

$$S_f(\chi) = \sum_{s \in \mathbb{F}_q} \chi(f(s)).$$

And we say f is B-well-distributed if for any nontrivial character χ of $J(\mathbb{F}_q)$, the inequality $|S_f(\chi)| \leqslant B\sqrt{q}$ holds.

Lemma 1 (Corollary 2, [30]). *If $f : \mathbb{F}_q \to E(\mathbb{F}_q)$ is a B-well-distributed encoding into a curve E, then the statistical distance between the distribution defined by $f^{\otimes s}$ on $J(\mathbb{F}_q)$ and the uniform distribution is bounded as:*

$$\sum_{D \in J(\mathbb{F}_q)} \left| \frac{N_s(D)}{q^s} - \frac{1}{\#J(\mathbb{F}_q)} \right| \leqslant \frac{B^s}{q^{s/2}} \sqrt{\#J(\mathbb{F}_q)},$$

where

$$f^{\otimes s}(u_1, \ldots, u_s) = f(u_1) + \ldots + f(u_s),$$
$$N_s(D) = \#\{(u_1, \ldots, u_s) \in (\mathbb{F}_q)^s | D = f(u_1) + \ldots + f(u_s)\},$$

i.e., $N_s(D)$ is the size of preimage of D under $f^{\otimes s}$. In particular, when s is greater than the genus of E, the distribution defined by $f^{\otimes s}$ on $J(\mathbb{F}_q)$ is statistically indistinguishable from the uniform distribution. Especially, the hash function construction

$$m \mapsto f^{\otimes s}(h_1(m), \ldots, h_s(m)) \qquad (s = g_E + 1)$$

is indifferentiable from a random oracle if $h_1, \ldots, h_s$ are seen as independent random oracles into $\mathbb{F}_q$(See [30]).

Hence, it is of great importance to calculate the encoding into a curve, and we will study the case of Jacobi Quartic curves.

Definition 2 (Artin Character). *Let E be an elliptic curve, $E(\mathbb{F}_q)$ be Jacobian group of E. Let χ be a character of $E(\mathbb{F}_q)$. Its extension is a multiplicative map $\bar{\chi} : Div_{\mathbb{F}_q}(E) \to \mathbb{C}$,*

$$\bar{\chi}(n(P)) = \begin{cases} \chi(P)^n, & P \in S, \\ 0, & P \notin S. \end{cases}$$

Here P is a point on $E(\mathbb{F}_q)$, S is a finite subset of $E(\mathbb{F}_q)$, usually denotes the ramification locus of a morphism $Y \to X$. Then we call $\bar{\chi}$ an Artin character of X.

Theorem 1 *[Theorem 3 of [30]]. Let $h : \tilde{X} \to X$ be a nonconstant morphism of projective curves, and χ be an Artin character of X. Suppose that $h^*\chi$ is unramified and nontrivial, φ is a nonconstant rational function on $\tilde{X}$. Then*

$$\left| \sum_{P \in \tilde{X}(\mathbb{F}_q)} \chi(h(P)) \right| \leqslant (2\tilde{g} - 2)\sqrt{q}$$

and

$$\left| \sum_{P \in \tilde{X}(\mathbb{F}_q)} \chi(h(P)) \left(\frac{\varphi(P)}{q} \right) \right| \leqslant (2\tilde{g} - 2 + 2 \deg \varphi)\sqrt{q},$$

where $\left(\frac{\cdot}{q}\right)$ denotes Legendre symbol, and $\tilde{g}$ is the genus of $\tilde{X}$.

Let $S = \{0\} \bigcup \{\text{roots of } g(X_j(u)) = 0, j = 2, 3\}$ where $X_j(u) = 0, j = 2, 3$ are defined as in Sect. 3.2. For any $u \in \mathbb{F}_q \backslash S$, $X_2(u)$ and $X_3(u)$ are both well defined and nonzero. Let $C_j = \{(u, s, t) \in \bar{\mathbb{F}}_q^3 | s = X_j(u), t = (-1)^j \sqrt{g(X_j(u))}\}, j = 2, 3$ be the smooth projective curves. There exists a one-to-one map $P_j : u \mapsto (u, s, t)$ from $\mathbb{F}_q$ to $C_j(\mathbb{F}_q)$ where $\rho(u) = (s, t)$. Let h_j be the projective map on C_j satisfying $\rho(u) = h_j \circ P_j(u)$, $S_j = P_j(S \cup \{\infty\})$.

Theorem 2. *Let f_1 be a map from $\mathbb{F}_q$ to $E_{a,d}$ defined as Eq. (2). For any nontrivial character χ of $E_{a,d}(\mathbb{F}_q)$, the character sum $S_{f_1}(\chi)$ satisfies:*

$$S_{f_1}(\chi) \leqslant 52\sqrt{q} + 65. \tag{3}$$

Proof. $S_{f_1}(\chi)$ can be estimated as

$$\begin{aligned} S_{f_1}(\chi) &= \left| \sum_{u \in \mathbb{F}_q \backslash S} (f_1^* \chi)(u) + \sum_{u \in S} (f_1^* \chi)(u) \right| \\ &\leqslant \left| \sum_{u \in \mathbb{F}_q \backslash S} (f_1^* \chi)(u) \right| + \#S, \end{aligned}$$

we observe that

$$\begin{aligned} \left| \sum_{u \in \mathbb{F}_q \backslash S} (f_1^* \chi)(u) \right| &= \left| \sum_{\substack{P \in C_2(\mathbb{F}_q) \backslash S_2 \\ \left(\frac{t(P)}{q}\right) = 1}} (h_2^* \psi^* \chi)(P) + \sum_{\substack{P \in C_3(\mathbb{F}_q) \backslash S_3 \\ \left(\frac{t(P)}{q}\right) = -1}} (h_3^* \psi^* \chi)(P) \right| \\ &\leqslant \#S_2 + \#S_3 + \left| \sum_{\substack{P \in C_2(\mathbb{F}_q) \\ \left(\frac{t(P)}{q}\right) = 1}} (h_2^* \psi^* \chi)(P) \right| \\ &\quad + \left| \sum_{\substack{P \in C_3(\mathbb{F}_q) \\ \left(\frac{t(P)}{q}\right) = -1}} (h_3^* \psi^* \chi)(P) \right|, \end{aligned}$$

and

$$\begin{aligned} &2 \left| \sum_{\substack{P \in C_2(\mathbb{F}_q) \\ \left(\frac{t(P)}{q}\right) = 1}} (h_2^* \psi^* \chi)(P) \right| \\ &= \left| \sum_{P \in C_2(\mathbb{F}_q)} (h_2^* \psi^* \chi)(P) + \sum_{P \in C_2(\mathbb{F}_q)} (h_2^* \psi^* \chi)(P) \cdot \left(\frac{t(P)}{q}\right) \right. \end{aligned}$$

$$- \sum_{\left(\frac{t(P)}{q}\right)=0} (h_2^*\psi^*\chi)(P) \Bigg|$$

$$\leqslant \left| \sum_{P\in C_2(\mathbb{F}_q)} (h_2^*\psi^*\chi)(P) \right| + \left| \sum_{P\in C_2(\mathbb{F}_q)} (h_2^*\psi^*\chi)(P) \cdot \left(\frac{t(P)}{q}\right) \right|$$
$$+ \#\{\text{roots of } g(X_2(u)) = 0\}.$$

From the covering $\psi \circ h_2 : C_2 \to E_{a,d}$, $X_2(u) = s \circ \psi^{-1}(x, y)$, which implies

$$\begin{aligned} T(u) &= \left(ya^2 + d + 3\,a^2 - yd\right) u^4 + \left(-2\,a^2 + 2\,d - 2\,ya^2 - 2\,yd\right) u^2 \\ &\quad + (y-1)\left(a^2 - d\right) \\ &= 0. \end{aligned}$$

Ramifications occur when $disc(T(u)) = 0$. Then $y \in \left\{\pm\frac{\sqrt{-d(a^2-d)}}{d}, -\frac{d+3\,a^2}{a^2-d}, 1\right\}$. When $y = 1$, $\psi \circ h_2$ has ramification index 2; when $y \in \left\{\pm\frac{\sqrt{-d(a^2-d)}}{d}, -\frac{d+3\,a^2}{a^2-d}\right\}$, $\psi \circ h_2$ is of ramification type $(2, 2)$. By Riemann-Hurwitz formula,

$$2g_{C_2} - 2 = 0 + 2\cdot 1 + 2\cdot 2 + 2\cdot 2 + 2\cdot 2 = 14.$$

Hence curve C_2 is of genus 8. Similarly, C_3 is also of genus 8.

Observe that $\deg t = [\mathbb{F}_q(s,t,u) : \mathbb{F}_q(t)] = [\mathbb{F}_q(s,t,u) : \mathbb{F}_q(s,t)][\mathbb{F}_q(s,t) : \mathbb{F}_q(t)] = 4\cdot 3 = 12$. Furthermore, $\left|\sum_{P\in C_2(\mathbb{F}_q)}(h_2^*\psi^*\chi)(P)\right| \leqslant (2g_{C_2}-2)\sqrt{q} = 14\sqrt{q}$, $\left|\sum_{P\in C_2(\mathbb{F}_q)}(h_2^*\psi^*\chi)(P)\cdot\left(\frac{t(P)}{q}\right)\right| \leqslant (2g_{C_2} - 2 + 2\deg t)\sqrt{q} = 38\sqrt{q}$, and $g(X_2(u)) = 0$ is sextic polynomial, we can derive

$$\left| \sum_{\substack{P\in C_2(\mathbb{F}_q) \\ \left(\frac{t(P)}{q}\right)=+1}} (h_2^*\psi^*\chi)(P) \right| \leqslant 26\sqrt{q} + 3.$$

Similarly, $\left| \sum_{\substack{P\in C_3(\mathbb{F}_q) \\ \left(\frac{t(P)}{q}\right)=-1}} (h_3^*\psi^*\chi)(P) \right| \leqslant 26\sqrt{q} + 3.$

Hence $|S_{f_1}(x)| \leqslant 52\sqrt{q} + 6 + \#S_2 + \#S_3 + \#S$. Since $g(X_2(u)) = 0$ and $g(X_3(u)) = 0$ have 2 common roots, then $\#S \leqslant 1 + 6 + 6 - 2 = 11$. Thus $\#S_j \leqslant 2(\#S + 1) \leqslant 24$. Then $|S_{f_1}(x)| \leqslant 52\sqrt{q} + 65$. ∎

$|S_{f_1}(x)| \leqslant 52\sqrt{q} + 65$ implies that f_1 is well-distributed (the definition of well-distributed encoding refers to [30]).

Next, we will prove that the construction $H_1(m) = f_1(h(m))$ is a hash function when h is a classical hash function.

3.4 One-Wayness

One-wayness is defined as follows.

Definition 3. *A hash function h is (t, ϵ) one-way, if any algorithm without precomputation running in time t, when given a random $y \in im(h)$ as input, outputs m such that $h(m) = y$ with probability at most ϵ. If ϵ is negligible for any polynomial time t in the security parameter, then the hash function h is one-way.*

Proposition 1. *If h is a (t, ϵ) one-way function, then H_1 is (t', ϵ') one-way where $H_1(m) = f_1(h(m))$, $\epsilon' = 16\epsilon$ and t' is a polynomial expression of t.*

Proof. Each point has at most $L = 4$ different preimages through f_1. L is a polynomial in the security parameter and h is one way. The main idea of this proof, similar to [13], is to reduce the argument that H_1 is not one-way to absurdity when h is one-way. Taking $L = 4$ in the proof of Lemma 5 in [13], $\epsilon' = L^2\epsilon = 16\epsilon$ and t' is a polynomial expression of t. ■

If ϵ is negligible, then $\epsilon' = 16\epsilon$ also can be negligible. Then if h is one-way, H_1 is one-way. Next, we prove that the construction $H_1(m) = f_1(h(m))$ is collision-resistant where h is collision-resistant.

3.5 Collision-Resistance

The definition of collision-resistance is:

Definition 4. *A family $\mathcal{H}$ of hash functions is (t, ϵ) collision-resistant if any algorithm running in time t when given a random $h \in \mathcal{H}$, outputs m and m' s.t. $h(m) = h(m')$ with probability at most ϵ. If ϵ is negligible for any polynomial time t in the security parameter, then $\mathcal{H}$ is collision-resistant.*

Proposition 2. *If $\mathcal{H} : \{0,1\}^* \to \{0,1\}^k$ is a (t, ϵ) collision-resistant, then H_1' is a (t', ϵ') collision-resistant hash function where $H_1'(m) = f_1(c \cdot h(m) + d)$, $h \in \mathcal{H}$ for $c, d \in \mathbb{F}_q$ selected randomly, $\epsilon' = \epsilon + \frac{2^{2k+2}}{q}$, and t' is a polynomial expression of t.*

Proof. Each point has at most $L = 4$ different preimages through f_1. The main idea of this proof, similar to [13], is to reduce the argument that H_1 is not collision-resistant to absurdity when $\mathcal{H}$ is collision-resistant. Using Theorem 3 [13] for $L = 4$, $\epsilon' = \epsilon + L\frac{2^{2k}}{q} = \epsilon + \frac{2^{2k+2}}{q}$ and t' is a polynomial expression of t. ■

When ϵ is chosen as 2^{-k}, then ϵ' is approximately 5×2^{-k} when the size of q is at least $3k$ bits. If ϵ is negligible, then $\epsilon' = 5\epsilon$ also is negligible. Thus, if h is collision-resistant, H_1' is collision-resistant. When $\epsilon = 2^{-k}$, the size of q is at least $3k$ bits, setting $c = 1$ and $d = 0$, $H_1(m) = f_1(h(m))$ is collision-resistant.

Thus we have proven that the construction $H_1(m) = f_1(h(m))$ is a hash function when h is a classical hash function.

4 Cube Root Encoding

When $q = p^n \equiv 2 \pmod 3$, the function $x \mapsto x^3$ is a bijection with inverse function

$$x \mapsto x^{\frac{1}{3}} = x^{\frac{2p^n-1}{3}} = x^{\frac{2q-1}{3}}.$$

Based on this bijection, our cube root encoding $f_2(u) = (x, y)$ is constructed from $\mathbb{F}_q$ to Jacobi quartic curve, where

$$x = \frac{2\alpha}{u\alpha + \beta}, \qquad y = \frac{2\alpha^2(\alpha - 2a)}{(u\alpha + \beta)^2} - 1, \tag{4}$$

$$\alpha = \frac{4a + u^2}{3} + \sqrt[3]{\beta^2 - \left(\frac{4a + u^2}{3}\right)^3}, \qquad \beta = \frac{4(a^2 - d) - 3\left(\frac{4a+u^2}{3}\right)^2}{2u}. \tag{5}$$

Notice that in Eq. (4), $y = \frac{x^2}{2}(\alpha - 2a) - 1$. Substitute it into the Jacobi quartic equation $y^2 = dx^4 + 2ax^2 + 1$, then $x^4(\alpha - 2a)^2 - 4\alpha x^2 = 4dx^4$. When $x = 0$, the Eq. (4) is right. When $x \neq 0$, it is easy to check that $x^2(\alpha - 2a)^2 - 4\alpha = 4dx^2$. Thus, x and y satisfy the Jacobi quartic equation. $\frac{4a+u^2}{3}$ shows up repeatedly in the representation of α, β to construct the encoding using a cube root.

4.1 Properties of Cube Root Encoding

Lemma 2. *Let $P(x, y)$ be a point on the curve $E_{a,d}$. The solutions u_s of $f_2(u) = P$ are the solutions of the polynomial equation*

$$Q(u; x, y) = x^3u^4 - 4x(ax^2 + 3y + 3)u^2 + 24(ax^2 + y + 1)u + 4x^3(a^2 + 3d) = 0.$$

Proof. Let α and β be defined in Eq. (5).

$$\begin{cases} y^2 = dx^4 + 2ax^2 + 1 \\ x^3u^4 - 4x(ax^2 + 3y + 3)u^2 + 24(ax^2 + y + 1)u + 4x^3(a^2 + 3d) = 0 \end{cases}$$

$$\Leftrightarrow \begin{cases} y^2 = dx^4 + 2ax^2 + 1 \\ x^3\left[4(a^2 - d) - 3\left(\frac{4a+u^2}{3}\right)^2\right] = 4u(2 - xu)(ax^2 + y + 1) \end{cases}$$

$$\Leftrightarrow \begin{cases} y^2 = dx^4 + 2ax^2 + 1 \\ x^3\beta = 2(2 - xu)(ax^2 + y + 1) \end{cases} \quad \Leftrightarrow \begin{cases} y^2 = dx^4 + 2ax^2 + 1 \\ \frac{x\beta}{2-xu} = 2a + \frac{2(y+1)}{x^2} \end{cases}$$

$$\Leftrightarrow \begin{cases} x = \frac{2\alpha}{u\alpha+\beta} \\ x^3\beta = 2(2 - xu)(ax^2 + y + 1) \end{cases} \quad \Leftrightarrow \begin{cases} x = \frac{2\alpha}{u\alpha+\beta} \\ y = \frac{2\alpha^2(\alpha-2a)}{(u\alpha+\beta)^2} - 1 \end{cases}$$

$H_2(m) = f_2(h(m))$ is a hash function if h is a classical hash function whose proof is the same as H_1.

Also, we construct a projective curve $R_{a,d} = \{(u,x,y)|(x,y) \in E_{a,d}, Q(u;x,y) = 0\}$ and a natural morphism from $R_{a,d}$ onto $E_{a,d}$.

4.2 The Genus of $R_{a,d}$

To find the genus of the curve $R_{a,d}$, we have to calculate the number of ramification points. Note that the discriminant of $Q(u;x,y)$ is

$$\begin{aligned} D(x,y) = \; & disc(Q(u;x,y)) = -2^{12}\cdot 3^2 x^6((-72\,x^8d^2 - 36\,ax^6d + 108\,x^4d \\ & +540\,a^2x^4 + 648\,ax^2 + 216 - 96\,x^{10}d^2a - 192\,a^2x^8d + 124\,a^3x^6 \\ & -32\,a^3x^{10}d + 8\,a^4x^8)y + 216 + 864\,ax^2 - 174\,a^2x^8d + 288\,ax^6d \\ & -144\,a^3x^{10}d - 240\,x^{10}d^2a - 64\,x^{12}d^2a^2 + 1080\,a^2x^4 + 51\,a^4x^8 \\ & +216\,x^4d - 45\,x^8d^2 - 48\,x^{12}d^3 + 448\,a^3x^6 - 16\,a^4x^{12}d). \end{aligned} \quad (6)$$

Then the curve ramifies if and only if Eq. (6) vanishes. If $x \neq 0$, then we represent y in Eq. (6) by x and substitute it into $y^2 - dx^4 - 2ax^2 - 1 = 0$, and get an equation of x with degree 12. Each x corresponds a point (x,y) on curve $E_{a,d}$. Hence there are 12 distinct branch points on $E_{a,d}$. If $Q(u;x,y)$ has a triple root, then

$$\begin{cases} E_{a,d} & = 0 \\ Q(u;x,y) & = 0 \\ \dfrac{d}{du}Q(u;x,y) & = 0 \\ \dfrac{d^2}{du^2}Q(u;x,y) & = 0. \end{cases}$$

This branch point has a solution if and only if $4\,a^6 - 45\,da^4 + 270\,d^2a^2 + 27\,d^3 = 0$.

Thus, all 12 branch points have ramification index 2. By Riemann-Hurwitz formula:

$$2g_{R_{a,d}} - 2 = 4(2\cdot 1 - 2) + 12(2-1),$$

hence $g_{R_{a,d}} = 7$.

4.3 Calculating Character Sums on the Curve $R_{a,d}$

Theorem 3. *Let f_2 be a map from $\mathbb{F}_q$ to $E_{a,d}$ defined as Eq. 4. For any non-trivial character χ of $E_{a,d}(\mathbb{F}_q)$, the character sum $S_{f_2}(\chi)$ satisfies:*

$$S_{f_2}(\chi) \leqslant 12\sqrt{q} + 3. \quad (7)$$

Proof. Let $K = F(x,y)$ be the function field of curve $E_{a,d}$, and$L = K[u]/(Q(u;x,y))$. For field inclusions $\mathbb{F}_q(u) \subset L$ and $K \subset L$, we construct birational maps $\varsigma : R_{a,d} \to \mathbb{P}^1(\mathbb{F}_q)$ and $\tau : R_{a,d} \to E_{a,d}$. Then ς is a bijection and $f_2(u) = \tau \circ \varsigma^{-1}(u)$.

Since curve $R_{a,d}$ is of genus 7, by Theorem 1, we have

$$|S_{f_2}(\chi) + \sum_{P\in C(\mathbb{F}_q), s(P)=\infty} \chi\circ h(P)| = |\sum_{P\in C(\mathbb{F}_q)} \chi\circ h(P)| \leqslant (2\cdot 7-2)\sqrt{q} = 12\sqrt{q}.$$

When $x \neq 0$, it can be checked that $Q(u; x, y)$ has 4 finite roots. When $x = 0$, substitute $v = \frac{1}{u}$ into $Q(u; x, y)$, then we have $48v^3 = 0$. Since $q > 3$ and q is odd, then $v^3 = 0$, which implies u has exactly 3 poles on $R_{a,d}$.

Thus $S_f(\chi) \leqslant 12\sqrt{q} + 3$. ■

Since $S_f(\chi) \leqslant 12\sqrt{q} + 3$, then f_2 is well-distributed.

Let F be the algebraic closure of $\mathbb{F}_q$, $K = F(x, y)$ be the function field of curve $E_{a,d}$, L be the function field of $R_{a,d}$. Next, we study Galois group of field extension L/K.

4.4 Galois Group of Field Extension L/K

To estimate the size of the image of f_2, the structure of $Gal(L/K)$ should be investigated.When L/K is a quartic extension, in [32], $Gal(L/K) = S_4$ if and only if

1. $Q(u; x, y)$ is irreducible over K;
2. let $C(u; x, y)$ be the resolvent cubic of $Q(u; x, y)$, then $C(u; x, y)$is irreducible over K;
3. the discriminant of $C(u; x, y)$ is not a square in K.

Hence, we only need to prove Lemmas 3, 4 and 5.

Lemma 3. *The polynomial $Q(u; x, y)$ is irreducible over K.*

Proof. Let σ be the nontrivial Galois automorphism in $Gal(F(x, y)/F(x))$, which maps y to $-y$. We only need to show that $\tilde{Q}(u; x) = Q(u; x, y) \cdot Q(u; x, y)^\sigma$ is irreducible over $F(x)$.

$$\begin{aligned}\tilde{Q}(u;x) = {} & x^6u^8 - 8\,x^4(x^2a+3)u^6 + 48\,x^3(x^2a+1)u^5 + 24\,x^3(x^3a^2 - 5\,x^3d + 4\,xa \\ & -12\,a)u^4 - 192\,x^2(x^3a^2 - 3\,x^3d + 4\,xa - 6\,a)u^3 - 32\,x(3\,ax^5d + a^3x^5 \\ & +27\,x^3d - 15\,x^3a^2 - 36\,xa + 36\,a)u^2 + 192\,x^3(x^2a+1)(a^2+3\,d)u \\ & +16\,x^6(a^2+3\,d)^2. \end{aligned} \tag{8}$$

Suppose that $\tilde{Q}(u; x)$ is reducible over $F(x)$. Note that $\tilde{Q}(u; x)$ can be represented as

$$\begin{aligned}&(24\,u^4a^2 + 96\,a^2d - 8\,u^6a - 120\,u^4d + u^8 - 96\,u^2ad - 32\,u^2a^3 + 16\,a^4 + 144\,d^2)x^3 \\ &+ 48\,u(u^4a - 4\,u^2a^2 + 12\,u^2d + 12\,ad + 4\,a^3)x^2 + 24\,u^2(-u^4 - 8\,au^2 + 20\,a^2 \\ &- 36\,d)x + 48\,u(u^4 + 8\,au^2 + 4\,a^2 + 12\,d). \end{aligned} \tag{9}$$

If $\tilde{Q}(u;x)$ has an irreducible factor $G \in F[u]$, then G is a common factor of all coefficients of Eq. (9). If $a^2+3d=0$, then u is a factor of $\tilde{Q}(u;x)$ and is a factor of $Q(u;x,y)$. Since $\gcd(-u^4-8au^2+20a^2-36d, u^4+8au^2+4a^2+12d)=1$, if u is not a common factor, then all coefficients of Eq. (9) do not have any common factor. Thus $\tilde{Q}(u;x)$ has not a factor in $F[u]$.

Otherwise, $\tilde{Q}(u;x)$ has a factor $G \in F[u,x]$ with $\deg(G,x)=1$. We suppose that $\tilde{Q}(u;x)=(Ax+B)\cdot(Zx^2+Dx+E)=AZx^3+(AD+BZ)x^2+(AE+BD)x+BE, A,B,Z,D,E \in F[u]$. Let $v(H)$ be the lowest degree of $H \in F[u]$. We have

$$\begin{cases} v(A)+v(Z)=0 \\ v(B)+v(E)=1 \\ \min(v(A)+v(D),v(B)+v(Z))=1 \text{ or } v(A)+v(D)=v(B)+v(Z)<1 \\ \min(v(A)+v(E),v(B)+v(D))=2 \text{ or } v(A)+v(E)=v(B)+v(D)<2, \end{cases}$$

which lead to a contradiction for no value satisfies these conditions. ■

Lemma 4. *The resolvent cubic $C(u;x,y)$ is irreducible over K.*

Proof. The resolvent cubic of $Q(u;x,y)$ is

$$\begin{aligned} C(u;x,y) = {} & u^3x^6+4\,x^4\left(x^2a+3\,y+3\right)u^2-16\,x^6\left(a^2+3\,d\right)u-1152\,x^2a \\ & -576\,y^2-1152\,y-576-768\,a^2x^4-192\,ax^6d-64\,a^3x^6 \\ & -192\,ya^2x^4-1152\,x^2ay-576\,ydx^4-576\,x^4d \end{aligned} \tag{10}$$

Similar to Lemma 3, we need to show $\tilde{C}(u;x)=C(u;x,y)\cdot C(u;x,y)^{\sigma}$ is irreducible over $F(x)$. We have

$$\begin{aligned} \tilde{C}(u;x) = {} & x^6\cdot((16\,a^2+8\,ua-144\,d+u^2)(48\,d+16\,a^2-u^2)^2x^6+24\,(48\,d+16\,a^2 \\ & -u^2\cdot(64\,a^3+64\,a^2u+8\,au^2-1344\,ad-u^3+96\,du)x^4+(-4608\,u^3a \\ & -331776\,d^2+73728\,ua^3+221184\,uda+23040\,u^2a^2+13824\,u^2d \\ & -1990656\,da^2-36864\,a^4)x^2-2304\,(-u+8\,a)(48\,d+16\,a^2-u^2)). \end{aligned} \tag{11}$$

Let $z=x^2$, we firstly show that

$$\begin{gathered} (16\,a^2+8\,ua-144\,d+u^2)(48\,d+16\,a^2-u^2)^2z^3+24\,(48\,d+16\,a^2-u^2)\cdot \\ (64\,a^3+64\,a^2u+8\,au^2-1344\,ad-u^3+96\,du)z^2+(-4608\,u^3a-331776\,d^2+ \\ 73728\,ua^3+221184\,uda+23040\,u^2a^2+13824\,u^2d-1990656\,da^2-36864\,a^4)z- \\ 2304\,(-u+8\,a)(48\,d+16\,a^2-u^2) \end{gathered}$$

is irreducible over $F(z)$.

If it is not irreducible, then it can be written as

$$(Az+B)(Zz^2+Dz+E)=AZz^3+(AD+BZ)z^2+(AE+BD)z+BE.$$

Similar to previous proof, we have

$$\begin{cases} \deg(A) + \deg(Z) = 6 \\ \deg(B) + \deg(E) = 3 \\ \max(\deg(A) + \deg(D), \deg(B) + \deg(Z)) = 5, \text{ or} \\ \quad \deg(A) + \deg(D) = \deg(B) + \deg(Z) > 5 \\ \\ \max(\deg(A) + \deg(E), \deg(B) + \deg(D)) = 3, \text{ or} \\ \quad \deg(A) + \deg(E) = \deg(B) + \deg(D) > 3. \end{cases}$$

Then $\deg(A) = \deg(B) + 1$ and this equation set does not have a solution. Thus the resolvent cubic $C(u; x, y)$ is irreducible over K. ■

Lemma 5. *The discriminant of $C(u; x, y)$ is not a square in K.*

Proof. Let $D(x, y)$ be the discriminant of $C(u; x, y)$.

$$\begin{aligned} D(x,y) = &-36864\,x^{12}(135 + 432\,y + 4536\,x^2 + 370244\,x^6 + 1052079\,x^8 \\ &-13720\,x^{12} + 11340\,yx^2 + 852012\,x^{10} + 58842\,x^4 + 6804\,x^2y^2 - 2268\,x^2y^3 \\ &-2268\,x^2y^4 + 383572\,x^6y + 425516\,x^8y + 26082\,x^4y^2 - 43120\,y^2x^8 \\ &-10192\,x^6y^2 - 23520\,x^6y^3 + 54880\,yx^{10} + 102564\,yx^4 - 20160\,y^3x^4 \\ &-2520\,y^4x^4 + 378\,y^2 - 297\,y^4 - 108\,y^5 - 108\,y^3). \end{aligned}$$

We only need to show that

$$\tilde{D}(x) = \frac{D(x,y) \cdot D(x,y)^{\sigma}}{(-36864x^{12})^2}$$

is not a square over $F(x)$. Let $z = x^2$, $\tilde{D}(x)$ can be represented as

$$\begin{aligned} &z^6 \cdot (a^2 - d)^2 \cdot a^2 \cdot (256\,A^2a^6\,(3\,A + 1)^2\,z^6 + 3072\,A^2a^5\,(3\,A + 1)\,z^5 \\ &\quad - 96\,a^4A\left(-72\,A + 9\,A^2 + 7\right)z^4 - 32\,a^3\left(459\,A^2 + 4 + 273\,A\right)z^3 \\ &\quad - 27\,a^2\left(1174\,A + 53 + 309\,A^2\right)z^2 - 864\,a\,(5 + 27\,A)\,z - 1728 - 5184\,A), \end{aligned}$$

where $A = \dfrac{d}{a^2}$.

Then $\tilde{D}$ is a square only if it can be represented as $(z^3(a^2 - d) \cdot a)^2 \cdot (Uz^3 + Vz^2 + Rz + S)^2$ which is impossible. Hence $D(x, y)$ is not a square. ■

Summarize Lemmas 3, 4 and 5, we directly deduce:

Theorem 4. *$Gal(L/K) = S_4$.*

4.5 Calculating the Size of the Images of f_2

Applying Chebotarev density theorem to Jacobi Quartic curves, we give the size of images of f_2.

Theorem 5 (Chebotarev [18]). *Let K be an extension of $\mathbb{F}_q(x)$ of degree $n < \infty$ and L a Galois extension of K of degree $m < \infty$. Assume $\mathbb{F}_q$ is algebraically closed in L, and fix some subset φ of $Gal(L/K)$ stable under conjugation. Let $s = \#\varphi$ and $N(\varphi)$ the number of places v of K of degree 1, unramified in L, such that the Artin symbol $\left(\frac{L/K}{v}\right)$ (defined up to conjugation) is in φ. Then*

$$|N(\varphi) - \frac{s}{m}q| \leqslant \frac{2s}{m}((m+g_L)\cdot q^{1/2} + m(2g_K+1)\cdot q^{1/4} + g_L + nm)$$

where g_K and g_L are genera of the function fields K and L.

we calculating the size of the images of f_2 in Theorem 6.

Theorem 6.

$$|\#Im(f_2) - \frac{5}{8}\#E_{a,d}(\mathbb{F}_q)| < \frac{5}{4}(31q^{1/2} + 72q^{1/4} + 65).$$

Proof. K is the function field of $E_{a,d}$ and is the quadratic extension of $\mathbb{F}_q(x)$. Hence $[K : \mathbb{F}_q(x)] = 2$, and the genus of K is 1 for the genus of elliptic curve is 1. $Gal(L/K) = S_4$, hence $m = \#S_4 = 24$. φ is the subset of $Gal(L/K)$ consisting at least 1 fixed point, which are conjugates of $(1)(2)(3)(4)$, $(12)(3)(4)$ and $(123)(4)$, then $s = 1 + 6 + 8 = 15$. Since the place v of K of degree 1 correspond to the projective unramified points on $E_{a,d}(\mathbb{F}_q)$, hence $|\#Im(f_2) - N(\varphi)| \leq 12$, where 12 represents the number of ramified points. Then we have

$$\begin{aligned}|\#Im(f_2) - \frac{5}{8}q| &\leqslant |\#Im(f) - N(\varphi)| + |N(\varphi) - \frac{5}{8}q| \\ &\leqslant 12 + \frac{5}{4}(31q^{1/2} + 72q^{1/4} + 55) \\ &< \frac{5}{4}(31q^{1/2} + 72q^{1/4} + 65).\end{aligned}$$

■

5 Indifferentiable from Random Oracle

5.1 First Construction

As a consequence of the proof of random oracle, if $f : \mathbb{S} \to \mathbb{G}$ is any weak encoding [29] to a cyclic group $\mathbb{G}$ with generator G, the hash function $H_R : \{0,1\}^* \to \mathbb{G}$ is defined by

$$H_R(m) = f(h_1(m)) + h_2(m)G,$$

where $h_1 : \{0,1\}^* \to \mathbb{F}_p$ and $h_2 : \{0,1\}^* \to \mathbb{Z}_N$ are two hash functions. $H_R(m)$ is indifferentiable from a random oracle in the random oracle model for h_1 and h_2.

If we want to prove $H_{R_i}(m) = f_i(h_1(m)) + h_2(m)G,\ i = 1, 2$, are both indifferentiable from a random oracle in the random oracle model for h_1 and h_2, we only need to prove f_1, f_2 are both weak encodings. From the definition of weak encoding [29], f_1 is an α_1-weak encoding from $\mathbb{F}_q$ to $E_{a,d}(\mathbb{F}_q)$, with $\alpha_1 = 8N/q$, where N is the order of $E_{a,d}(\mathbb{F}_q)$. Similarly, f_2 is an α_2-weak encoding from $\mathbb{F}_q$ to $E_{a,d}(\mathbb{F}_q)$, with $\alpha_2 = 4N/q$. As α_1, α_2 are polynomial functions of the security parameter, f_1, f_2 are both weak encodings. Thus,

$$H_{R_i}(m) = f_i(h_1(m)) + h_2(m)G,\ i = 1, 2 \tag{12}$$

are both indifferentiable from a random oracle in the random oracle model for h_1 and h_2. For more details about hash function indifferentiability from a random oracle, we refer the interested reader to [29,30].

5.2 Second Construction

The second construction is :

$$H_{R_i'}(m) = f_i(h_1(m)) + f_i(h_2(m)),\ i = 1, 2.$$

Using [30]'s Corollary 2: f_i is well-distributed encoding, then $H_{R_i'}(m)$ are well-behaved i.e. $H_{R_i'}(m)$ is indifferentiable from a random oracle.

In Sects. 3.3 and 4.3, we have proved that f_1, f_2 are both a well distributed encodings. Then $H_{R_i'}(m), i = 1, 2$ are both indifferentiable from a random oracle, where h_1, h_2 are regarded as independent random oracles with values in $\mathbb{F}_q$.

6 Time Complexity

For the convenience of making comparisons, we first introduce a birational map which is based on the map in [25] and show that the curve $E_{a,d}$ is birationally equivalent over $\mathbb{F}_q$ to the Weierstrass curve

$$E_W : v^2 = u^3 - \frac{4}{3}(a^2 + 3d)u + \frac{16a}{27}(a^2 - 9d), \tag{13}$$

via maps

$$\begin{aligned}
&\phi : E_{a,d} \to E_W :\\
&(x, y) \mapsto (u, v) = \left(\frac{2dx^2 + 2a(1+y)}{y-1} - \frac{4a}{3}, \frac{4a(dx^2+2a) - 4d(1-y)}{(1-y)^2}x\right),\\
&\psi : E_W \to E_{a,d} :\\
&(u, v) \mapsto (x, y) = \left(\frac{2v}{(u-2a/3)^2 - 4d}, \frac{(u+4a/3)^2 - 4(a^2-d)}{(u-2a/3)^2 - 4d}\right).
\end{aligned} \tag{14}$$

6.1 Theoretical Analysis of Hash

The hash functions $H_i(m) = f_i(h(m))$, $i = 1, 2$ require the calculation of f_i and a classical hash function h. When comparing H_i, we only need to compare f_i.

Let M denote field multiplication, S field squaring, I field inversion, E a modular exponent in a finite field, E_C the cube root, E_S the square root, D a determination of the square residue, and K the security parameter. Suppose that a is a small integer in the Jacobi quartic curve equation. We make the assumptions that $S = M$, $I = 10M$, and $E_C = E_S = E$ for E_C and E_S both need calculating a modular exponent.

The cost of f_1 can be computed as following:

1. Compute u^2 for an S, and that is enough to compute $X_3(u)$.
2. When computing $X_2(u)$, we find the inverse of u^2 and multiply by $-X_3(u)$, for additional $M + I$.
3. Note that since $g(s) = s(s^2 - 4as + 4a^2 - 4d)$, computing $g(s)$ takes $M + S$. Computing $g(X_2(u))$ and $g(X_3(u))$ costs $2M + 2S$.
4. One of the two is a quadratic residue, but we only need to check one, which adds a D, then take square root E_S of whichever is the square. Then we now have new variables s, t.
5. We compute the inverse of $(s - 2a)^2 - 4d$, which requires I, for s^2 and $a \cdot s$ have computed in $g(s)$.
6. Finally, x is $2t$ times that inverse, while y is $s^2 - 4(a^2 - d)$ times the inverse, which adds $2M$ for s^2 have computed in $g(s)$.

Thus, f_1 costs $2I + E_S + 5M + 3S + D = E + 28M + D$. f_2 can be computed as

$$
\begin{aligned}
u\beta &= \frac{4(a^2 - d) - 3\left(\frac{4a+u^2}{3}\right)^2}{2} && 2S \\
u\alpha &= u\frac{4a + u^2}{3} + \sqrt[3]{u(u\beta)^2 - \left(u\frac{4a + u^2}{3}\right)^3} && E_C + 3M + 2S \\
x &= \frac{2u\alpha}{u(u\alpha) + u\beta} = \frac{u\alpha(2u)}{(u(u\alpha) + u\beta)u} && I + 4M \\
y &= \frac{2(u\alpha)^2 u(u\alpha - 2au)}{(u(u\alpha) + u\beta)^2 u^2} - 1 && 2M + S.
\end{aligned}
$$

f_2 costs $I + E_C + 9M + 5S = E + 24M$.

The parameter of Weierstrass equation E_W (Eq. (13)) can be precomputed. Calculating birational equivalence from Weierstrass curve to Jacobi quartic curves in Eq. (14) need $I + 2M + S$. Birational equivalence from Weierstrass curve to Jacobi quartic curves composed with the Ulas' function [16], denoted by f_{Ulas}, costs $(2I + 2M + S + E_S + D) + (I + 2M + S) = 3I + E_S + 4M + 2S + D = E + 36M + D$. Fouque and Tibouchi [18] gave the brief version of Ulas' function, denoted by f_{FT}, costs $(I + 3M + 4S + E_S + D) + (I + 2M + S) = 2I + E_S + 5M + 5S + D = E + 30M + D$. Birational equivalence from Weierstrass curve to Jacobi quartic curves composed with the Icart's function [13], denoted by f_{Icart}, costs $(I + 3M + 3S + E_C) + (I + 2M + S) = 2I + E_C + 5M + 4S = E + 29M$. Alasha's work [21] on deterministic encoding into Jacobi quartic curves over $\mathbb{F}_q$ when $q \equiv 2 \pmod 3$, denoted by f_{Alasha}, is shown as

$$\begin{aligned} m &= u^2 - 2b && S \\ v &= \frac{3m^2 - a}{u} && I + M + S \\ s &= \sqrt[3]{m^3 - u^2 + 2ab} - m && E_C + M \\ x &= \frac{2(b - s)}{us + v} && \\ y &= \frac{s^2 - 2bs + a}{a - s^2} && I + 6M + S. \end{aligned}$$

Total cost is $E_C + 2I + 8M + 3S$.

The cost of different deterministic encodings is summarized in Table 1. f_{Ulas}, f_{FT}, and f_1 work over $\mathbb{F}_q$ when $q \equiv 3 \pmod 4$. f_{Icart}, f_{Alasha}, and f_2 work over $\mathbb{F}_q$ when $q \equiv 2 \pmod 3$.

From Table 1, f_1 saves 8 field multiplications compared with f_{Ulas} and 2 multiplications compared with f_{FT}. f_2 costs 5 field multiplications less than f_{Icart} and 7 multiplications than f_{Alasha}. Thus, the hash functions constructed directly into Jacobi quartic curves are more efficient than birational equivalence from Weierstrass curve to Jacobi quartic curves composed with existing hash functions into short Weierstrass curves.

6.2 Theoretical Analysis of Random Oracle

The constructions $H_{R_i}(m) = f_i(h_1(m)) + h_2(m)G, i = 1, 2$ require one hash function $f_i(h_1(m))$ and one scalar multiplication $h_2(m)G$. Because $H_{R_i}(m), i = 1, 2$ both require calculating $h_1(m)$ and $h_2(m)G$, we only need compare f_i, which is the same as analyzing deterministic encodings.

The constructions $H_{R'_i}(m) = f_i(h_1(m)) + f_i(h_2(m)), i = 1, 2$ require two hash functions $f_i(h_j(m)), j = 1, 2$. Because $H_{R'_i}(m), i = 1, 2$ both require calculating $h_1(m)$ and $h_2(m)G$, we only need compare f_i, which is the same as analyzing deterministic encodings.

Table 1. Time cost of different deterministic encodings

Deterministic encoding	Cost	Converted cost
f_{Ulas}	$3I + E_S + 4M + 2S + D$	$E + D + 36M$
f_{FT}	$2I + E_S + 5M + 5S + D$	$E + D + 30M$
f_1	$2I + E_S + 5M + 3S + D$	$E + D + 28M$
f_{Icart}	$2I + E_C + 5M + 4S$	$E + 29M$
f_{Alasha}	$2I + E_C + 8M + 3S$	$E + 31M$
f_2	$I + E_C + 9M + 5S$	$E + 24M$

Table 2. NIST primes

Prime	Value	Residue (mod 3)	Residue (mod 4)
$P192$	$2^{192} - 2^{64} - 1$	2	3
$P384$	$2^{384} - 2^{128} - 2^{96} + 2^{32} - 1$	2	3

6.3 Practical Implementations

The running times for the various deterministic encodings into Jacobi quartic curves is discussed in the following. The implementation has been done on $\mathbb{F}_{P192}$ and $\mathbb{F}_{P384}$ (Table 2).

Miracl lib [31] was used to implement big number arithmetic. The experiments were tested on an Intel Core 2, 2.66G Hz processor. f_{Ulas}, f_{FT}, f_1, f_{Icart}, f_{Alasha}, and f_2 all ran 1,000,000 times for randomly chosen u on given prime fields $\mathbb{F}_{P192}$, $\mathbb{F}_{P384}$.

Table 3. Time cost (ms) of different methods on NIST

Prime	$P192$	$P384$
f_{Ulas}	0.449	1.212
f_{FT}	0.437	1.190
f_1	0.429	1.180
f_{Icart}	0.360	1.013
f_{Alasha}	0.364	1.018
f_2	0.342	0.992

From the average running times of Table 3, f_1 is faster than F_{Ulas} and f_{FT}. On $\mathbb{F}_{P192}$, it saves 23.8 % running time compared with f_{Ulas}, 21.7 % with f_{FT}, 20.2 % with f_1, 5 % with f_{Icart}, and 5.8 % with f_{Alasha}. On $\mathbb{F}_{P384}$, f_2 is also fastest among these deterministic encodings, but saves less than on $\mathbb{F}_{P192}$.

7 Conclusion

We proposed two deterministic encodings directly from finite field to Jacobi quartic curves, namely, SWU encoding f_0 (its brief version is f_1), and cube root encoding f_2. f_1 is most efficient in all existed methods working over $\mathbb{F}_q$ when $q \equiv 3 \pmod 4$, and f_2 is most efficient when $q \equiv 2 \pmod 3$. In the case $q \equiv 11 \pmod{12}$, both f_1 and f_2 can be applied over $\mathbb{F}_q$ and f_2 is much faster than f_1.

Additionally, Legendre encoding, based on computing Legendre symbols, is proposed in appendix when the j-invariant of Jacobi quartic curves is 1728. The Legendre encoding, from a finite field $\mathbb{F}_q$ to Jacobi quartic curves $y^2 = dx^4 + 1$, costs $E_S + 3M + S + D = E + 4M + D$ when $q \equiv 3 \pmod 4$.

References

1. Boneh, D., Franklin, M.: Identity-based encryption from the Weil pairing. In: Kilian, J. (ed.) CRYPTO 2001. LNCS, vol. 2139, pp. 213–229. Springer, Heidelberg (2001)
2. Baek, J., Zheng, Y.: Identity-based threshold decryption. In: Bao, F., Deng, R., Zhou, J. (eds.) PKC 2004. LNCS, vol. 2947, pp. 262–276. Springer, Heidelberg (2004)
3. Horwitz, J., Lynn, B.: Toward hierarchical identity-based encryption. In: Knudsen, L.R. (ed.) EUROCRYPT 2002. LNCS, vol. 2332, pp. 466–481. Springer, Heidelberg (2002)
4. Boneh, D., Gentry, C., Lynn, B., Shacham, H.: Aggregate and verifiably encrypted signatures from bilinear maps. In: Biham, E. (ed.) EUROCRYPT 2003. LNCS, vol. 2656, pp. 416–432. Springer, Heidelberg (2003)
5. Zhang, F., Kim, K.: Id-based blind signature and ring signature from pairings. In: Zheng, Y. (ed.) ASIACRYPT 2002. LNCS, vol. 2501, pp. 533–547. Springer, Heidelberg (2002)
6. Boyen, X.: Multipurpose identity-based signcryption. In: Boneh, D. (ed.) CRYPTO 2003. LNCS, vol. 2729, pp. 383–399. Springer, Heidelberg (2003)
7. Libert, B., Quisquater, J.-J.: Efficient signcryption with key privacy from gap Diffie-Hellman groups. In: Bao, F., Deng, R., Zhou, J. (eds.) PKC 2004. LNCS, vol. 2947, pp. 187–200. Springer, Heidelberg (2004)
8. Lindell, Y.: Highly-efficient universally-composable commitments based on the DDH assumption. In: Paterson, K.G. (ed.) EUROCRYPT 2011. LNCS, vol. 6632, pp. 446–466. Springer, Heidelberg (2011)
9. Jablon, D.P.: Strong password-only authenticated key exchange. SIGCOMM Comput. Commun. Rev. **26**(5), 5–26 (1996)
10. Boyko, V., MacKenzie, P.D., Patel, S.: Provably secure password-authenticated key exchange using Diffie-Hellman. In: Preneel, B. (ed.) EUROCRYPT 2000. LNCS, vol. 1807, pp. 156–171. Springer, Heidelberg (2000)
11. Boyd, C., Montague, P., Nguyen, K.: Elliptic curve based password authenticated key exchange protocols. In: Varadharajan, V., Mu, Y. (eds.) ACISP 2001. LNCS, vol. 2119, pp. 487–501. Springer, Heidelberg (2001)
12. Shallue, A., van de Woestijne, C.E.: Construction of rational points on elliptic curves over finite fields. In: Hess, F., Pauli, S., Pohst, M. (eds.) ANTS 2006. LNCS, vol. 4076, pp. 510–524. Springer, Heidelberg (2006)
13. Icart, T.: How to hash into elliptic curves. In: Halevi, S. (ed.) CRYPTO 2009. LNCS, vol. 5677, pp. 303–316. Springer, Heidelberg (2009)
14. Skalba, M.: Points on elliptic curves over finite fields. Acta Arith. **117**, 293–301 (2005)
15. Fouque, P.-A., Joux, A., Tibouchi, M.: Injective encodings to elliptic curves. In: Boyd, C., Simpson, L. (eds.) ACISP. LNCS, vol. 7959, pp. 203–218. Springer, Heidelberg (2013)
16. Ulas, M.: Rational points on certain hyperelliptic curves over finite fields. Bull. Polish Acad. Sci. Math. **55**, 97–104 (2007)
17. Fouque, P.-A., Tibouchi, M.: Deterministic encoding and hashing to odd hyperelliptic curves. In: Joye, M., Miyaji, A., Otsuka, A. (eds.) Pairing 2010. LNCS, vol. 6487, pp. 265–277. Springer, Heidelberg (2010)
18. Fouque, P.-A., Tibouchi, M.: Estimating the size of the image of deterministic hash functions to elliptic curves. In: Abdalla, M., Barreto, P.S.L.M. (eds.) LATINCRYPT 2010. LNCS, vol. 6212, pp. 81–91. Springer, Heidelberg (2010)

19. Farashahi, R.R.: Hashing into Hessian curves. In: Nitaj, A., Pointcheval, D. (eds.) AFRICACRYPT 2011. LNCS, vol. 6737, pp. 278–289. Springer, Heidelberg (2011)
20. Yu, W., Wang, K., Li, B., Tian, S.: About hash into montgomery form elliptic curves. In: Deng, R.H., Feng, T. (eds.) ISPEC 2013, LNCS, vol. 7863, pp. 147–159. Springer, Heidelberg (2013)
21. Alasha, T.: Constant-time encoding points on elliptic curve of diffierent forms over finite fields (2012). http://iml.univ-mrs.fr/editions/preprint2012/files/tammam_alasha-IML_paper_2012.pdf
22. Yu, W., Wang, K., Li, B., Tian, S.: Construct hash function from plaintext to C_{34} curves. Chin. J. Comput. **35**(9), 1868–1873 (2012)
23. Billet, O., Joye, M.: The Jacobi model of an elliptic curve and side-channel analysis. In: Fossorier, M., Hoholdt, T., Poli, A. (eds.) AAECC 2003. LNCS, vol. 2643, pp. 34–42. Springer, Heidelberg (2003)
24. Hisil, H., Wong, K.K.H., Carter, G., Dawson, E.: Jacobi quartic curves revisited. In: Boyd, C., Nieto, J. (eds.) ACISP 2009. LNCS, vol. 5594, pp. 452–468. Springer, Heidelberg (2009)
25. Hisil, H.: Elliptic Curves, Group Law, and Efficient Computation. Ph.D. thesis, Queensland University of Technology (2010)
26. Wang, H., Wang, K., Zhang, L., Li, B.: Pairing computation on elliptic curves of jacobi quartic form. Chin. J. Electron. **20**(4), 655–661 (2011)
27. Duquesne, S., Fouotsa, E.: Tate pairing computation on jacobi's elliptic curves. In: Abdalla, M., Lange, T. (eds.) Pairing 2012. LNCS, vol. 7708, pp. 254–269. Springer, Heidelberg (2013)
28. Standards for Efficient Cryptography, Elliptic Curve Cryptography Ver. 0.5 (1999). http://www.secg.org/drafts.htm
29. Brier, E., Coron, J.-S., Icart, T., Madore, D., Randriam, H., Tibouchi, M.: Efficient indifferentiable hashing into ordinary elliptic curves. In: Rabin, T. (ed.) CRYPTO 2010. LNCS, vol. 6223, pp. 237–254. Springer, Heidelberg (2010)
30. Farashahi, R.R., Fouque, P.-A., Shparlinski, I.E., Tibouchi, M., Voloch, J.F.: Indifferentiable deterministic hashing to elliptic and hyperelliptic curves. Math. Comp. **82**, 491–512 (2013)
31. miracl: Multiprecision Integer and Rational Arithmetic Cryptographic Library. http://www.shamus.ie
32. Roman, S.: Field Theory. Graduate Texts in Mathematics 158, 2nd edn. Springer, NewYork (2011)

Cryptanalysis II

Two Generic Methods of Analyzing Stream Ciphers

Lin Jiao[1,2(✉)], Bin Zhang[1,3], and Mingsheng Wang[4]

[1] TCA, Institute of Software, Chinese Academy of Sciences, Beijing 100190, China
jiaolin@tca.iscas.ac.cn
[2] University of Chinese Academy of Sciences, Beijing 100049, China
[3] SKLCS, Institute of Software, Chinese Academy of Sciences, Beijing 100190, China
[4] State Key Laboratory of Information Security,
Institute of Information Engineering, Chinese Academy of Sciences,
Beijing 100093, China

Abstract. Since the security analysis against stream ciphers becomes more difficult nowadays, it is urgent and significant to propose new generic methods. In this work, we introduce guess-and-determine techniques to two traditional analysis methods and make the new approaches methodological for generalization. We show the power of the new methods by analyzing two stream ciphers: Grain-v1 and ACORN. Grain-v1 is one of the finalists selected in the eSTREAM project. We present a time-memory-data tradeoff attack against Grain-v1 by importing the idea of conditional sampling resistance based on the k-linear-normality and a specific guessing path, with the parameters of 2^{61} time online employing a memory of 2^{71} assuming available keystream of 2^{79} and 2^{81} preprocessing time, which are much better than the best tradeoffs in the single key and IV pair setting so far. We transform the parameters into cipher ticks, and all the complexities are lower than $2^{87.4}$ cipher ticks, which is the actual complexity of the brute force attack. We also evaluate the security of another lightweight authenticated cipher ACORN, since there is few security analysis of the recently submitted cipher to CAESAR competition. The analysis against this cipher emphasizes on finding the linear approximations of the output function and the efficiently guessed combination information of the upstate function, and exploiting the integer linear programming problem as a tool to search the optimal complexity. Our attack calls for 2^{157} tests, which estimate the security margin of ACORN.

Keywords: Guess-and-determine · Time-memory-data tradeoff · Linear approximation · Stream cipher · ACORN · Grain-v1

1 Introduction

Additive stream ciphers are an important class of data encryption primitives, whose cores are the pseudo-random keystream generators. They are currently

J. Lopez and C.J. Mitchell (Eds.): ISC 2015, LNCS 9290, pp. 379–396, 2015.
DOI: 10.1007/978-3-319-23318-5_21

used in various aspects of our life, like RC4 on the Internet [25], E0 in Bluetooth [9] and A5/1 in GSM communication [10]. Nowadays, the trend of designing stream ciphers becomes exploiting nonlinear components, which increases the difficulty for analyzing the security of the ciphers using previous traditional analysis methods. Thus to propose new generic approaches to analyze those stream ciphers becomes more urgent and necessary. After massive research, we find that some guess-and-determine techniques seem relatively efficient against the nonlinear components [5,16]. Hence we try to associate some traditional analysis methods with guess-and-determine techniques to further analyze such stream ciphers and make the new methods standard and methodological for general application. We show the power of the new methods by analyzing two additive stream ciphers as follows.

Firstly, we present a time-memory-data tradeoff attack against Grain-v1 by introducing the idea of conditional sampling resistance. Grain-v1 [18] is one of the three remaining candidates in the final hardware portfolio of the eSTREAM project [1]. It uses 80-bit key and 64-bit IV, which consists of two combined registers each has 80-bit state, one NFSR and one LFSR, filtered together by a non-linear function. Grain-v1 has a compact structure with carefully chosen tap positions, feedback functions and output function. The feedback function of NFSR and the filter function make it impossible to take a correlation or a distinguishing attack in time faster than exhaustive search [4,21]. In [12], a slide property in the initialization phase was discovered, and later mounted with several related-key chosen IV attacks [22]. Recently, a near collision attack against Grain-v1 was proposed in [28]. Our work described here is a state recovery attack against Grain-v1 in the single key and IV pair setting using time-memory-data tradeoffs. Firstly, we extend the concept of k-normality into k-linear-normality of Boolean functions. Then we combine the k-linear-normality of the filter function with sampling resistance under the constraints of some state bits, which makes the sampling resistance much longer, and reduces the searching space that supports wider tradeoff parameters. We call this kind of sampling resistance as conditional sampling resistance. For Grain-v1, we find a conditional sampling resistance based on a specific guessing path that by fixing 51 bits of state constraint conditions and guessing 81 bits more of the internal state, the remaining 28 bits of the state can be recovered directly using the first 28 keystream output bits generated from the state, which is 10 bits longer than the sampling resistance given in [8]. According to the conditional sampling resistance, we conduct a time-memory-data tradeoff attack against Grain-v1 with the parameters of $T = 2^{61}$ table look-up operations employing a memory of $M = 2^{71}$ dimensions assuming available keystream length of $D = 2^{79}$ and the preprocessing time of $P = 2^{81}$, which are much better than the best parameters that $T = 2^{71}, D = 2^{53.5}, M = 2^{71}$ and $P = 2^{106.5}$ in the single key and IV pair setting as far as we know, and the preprocessing time can be controlled much lower. We also compare our attack with the brute force attack by transforming the parameters into cipher ticks, and our attack calls for $2^{67.3}$ cipher ticks online after the pre-computation of $2^{87.3}$ ticks, given $2^{78.3}$ bits memory and 2^{79}

keystream bits, while the complexity of brute force attack is actually $2^{87.4}$ cipher ticks which is higher than 2^{80}. Moreover, each brute force attack can only be mounted for each fixed IV, while our precomputation can be applied to the scenario with arbitrary IVs. Furthermore, it is a generic approach to analyze stream ciphers, by bringing the conditional sampling resistance into time-memory-data tradeoff attacks, based on the k-linear-normality of the output function and an efficient guessing path. We also put forward the tradeoff curve of our new time-memory-data tradeoff attacks.

We also present a security evaluation of another stream cipher ACORN. ACORN is a lightweight authenticated cipher recently proposed by H. Wu [27] and submitted to CAESAR (Competition for Authenticated Encryption: Security, Applicability, and Robustness) [2], which is a new competition following a long tradition of focus in secret-key cryptography and expecting to have a tremendous increase in confidence in the security of authentication ciphers. The structure of ACORN is very clear and it has only 128-bit version at present. ACORN-128 contains a 293-bit internal state, and is designed to protect up to 2^{64} bits of associated data and 2^{64} bits of plain-text by using a 128-bit secret key and a 128-bit IV. The authentication tag can be l_{tag} bits, where $64 \leq l_{tag} \leq 128$, and the use of a 128-bit tag is recommended by the designer. The designer requests that in ACORN, each key, IV pair is used to protect only one message and if verification fails, the new tag and the decrypted ciphertext should not be given as output, which are in order to use the cipher securely. Since there is no security analysis against the encryption process of the cipher by the designer and only one analysis proposed in [23], we present a state recovery attack against ACORN to assess the security margin of the cipher. Our method focuses on finding the linear approximations of the output function and the efficiently guessed combination information of the upstate function. Then we exploit the integer linear programming problem to optimize the tradeoff between using better linear approximations and adding fewer feedback steps needed to guess. Concretely, we obtain 265 linear approximation equations and guess 28-bit information, each is a value of a linear function in the internal state variables. Thus we find a system of linear equations in the initial state variables of the cipher, and then we solve the system and deduce the initial state to be verified. The attack calls for 2^{157} tests, which is better than that for 2^{164} tests in [23]. This result can be a security bound of ACORN examined by the method of guess-and-determine attacks so far. Moreover, it provides some insights on such compact stream ciphers and can be viewed as a generic way to evaluate the security of such compact stream ciphers.

This paper is organized as follows. Section 2 contains a time-memory-data tradeoff attack against Grain-v1 using conditional sampling resistance based a guessing path. Section 3 presents a security evaluation of ACORN using linear approximation and guessing strategy. Section 4 concludes the paper.

2 Time-Memory-Data Tradeoff Attack Against Grain-v1 with Conditional Sampling Resistance Based on Guess-and-Determine Strategy

In this section, we present a state recovery attack against Grain-v1 in the single key and IV pair setting. For the self-completeness of the discussion that follows, we first introduce some basic knowledge of time-memory-data tradeoff attacks in Sect. 2.1, and give a brief description concerning Grain-v1 in Sect. 2.2. To analyze Grain-v1, we point out certain weakness of the filter function, and give the conditional sampling resistance of Grain-v1 that employs a specific guessing path in Sect. 2.3. According to the conditional sampling resistance, we conduct the time-memory-data trade-off attack against Grain-v1 in Sect. 2.4, and present the complexity analysis and the comparison with previously reported time-memory-data trade-off attacks against Grain-v1 in Sect. 2.5.

2.1 Introduction of Time-Memory-Data Tradeoff Attack

Time-memory (TM) tradeoffs were first introduced by Hellman [19] in 1980 as a generic way of attacking block ciphers, but can be generalized to the problem of inverting one-way functions. In the case of function inversion, Babbage and Golic [3,15] and later Biryukov, Shamir and Wagner [6] showed that the basic TM tradeoffs can be improved significantly by using several data points, known as time-memory-data (TMD) tradeoffs. For the state recovery of stream ciphers, the one-way function to be inverted is commonly taken to be the map sending an n-bit internal state of the cipher to the first n bits of keystream generated from that state. Several stream ciphers have been broken by TMD tradeoffs, most famously the GSM encryption scheme A5/1 [7].

Using a TMD tradeoff to invert a function can be split into two phases; offline and online phase. In the offline step, the attacker builds large tables relating to the function in question. In the active phase, the attacker obtains a number of actual data points that he wants to invert, and tries to find a preimage of at least one value using the precomputed tables. A TMD tradeoff is thus characterized by five parameters: the size of the search space N, the time consumed by the precomputation P, the amount of memory M used to store the precomputed tables, the time complexity of the online phase T, and the amount of data required D. In particular, the Hellman tradeoff is related by $P = N, TM^2 = N^2, 1 \leq T \leq N$, the Babbage-Golic tradeoff is specified by the relations $P = M, N = TM$ and $T = D$, while the Biryukov-Shamir-Wagner tradeoff is given by $P = N/D, N^2 = TM^2D^2, D^2 \leq T$.

2.2 Description of Grain-v1

Next we describe the parts of Grain-v1, and refer to [18] for the full specification. Grain-v1 is a bit-oriented stream cipher taking an 80-bit key and a 64-bit IV. The internal secret state of Grain-v1 has 160 bits, and it consists of an 80-bit

LFSR, denoted as $(s_0, \cdots, s_{79})$ and an 80-bit NFSR, denoted as $(b_0, \cdots, b_{79})$. The feedback polynomial of the LFSR is a primitive polynomial of degree 80, and the update functions of the LFSR and NFSR are, respectively, described as follows:

$$s_{t+80} = s_{t+62} \oplus s_{t+51} \oplus s_{t+38} \oplus s_{t+23} \oplus s_{t+13} \oplus s_t$$

$$\begin{aligned} b_{t+80} = \; & s_t \oplus b_{t+62} \oplus b_{t+60} \oplus b_{t+52} \oplus b_{t+45} \oplus b_{t+37} \oplus b_{t+33} \oplus b_{t+28} \oplus b_{t+21} \\ & \oplus b_{t+14} \oplus b_{t+9} \oplus b_t \oplus b_{t+63}b_{t+60} \oplus b_{t+37}b_{t+33} \oplus b_{t+15}b_{t+9} \\ & \oplus b_{t+60}b_{t+52}b_{t+45} \oplus b_{t+33}b_{t+28}b_{t+21} \oplus b_{t+63}b_{t+45}b_{t+28}b_{t+9} \\ & \oplus b_{t+60}b_{t+52}b_{t+37}b_{t+33} \oplus b_{t+63}b_{t+60}b_{t+21}b_{t+15} \\ & \oplus b_{t+63}b_{t+60}b_{t+52}b_{t+45}b_{t+37} \oplus b_{t+33}b_{t+28}b_{t+21}b_{t+15}b_{t+9} \\ & \oplus b_{t+52}b_{t+45}b_{t+37}b_{t+33}b_{t+28}b_{t+21}. \end{aligned}$$

During the keystream generation, Grain-v1 outputs a single bit at each clock cycle, which is computed by taking 5 variables from the two registers as input to the filter function $h(x)$, and masking seven more state bits from the NFSR. The output function is defined as

$$z_t = h(s_{t+3}, s_{t+25}, s_{t+46}, s_{t+64}, b_{t+63}) \oplus \sum_{j \in \mathrm{A}} b_{t+j},$$

where $\mathrm{A} = \{1, 2, 4, 10, 31, 43, 56\}$ and the filter function $h(x)$ is given by

$$\begin{aligned} h(x) = \; & x_1 \oplus x_4 \oplus x_0x_3 \oplus x_2x_3 \oplus x_3x_4 \oplus x_0x_1x_2 \oplus x_0x_2x_3 \\ & \oplus x_0x_2x_4 \oplus x_1x_2x_4 \oplus x_2x_3x_4, \end{aligned}$$

where the variables x_0, x_1, x_2, x_3, and x_4 of $h(x)$ correspond to the tap positions $s_{t+3}, s_{t+25}, s_{t+46}, s_{t+64}$ and b_{t+63}, respectively.

Grain-v1 is initialized with a 64-bit IV injected directly into the LFSR (the remaining bits of the LFSR are assigned value one), and an 80-bit key that is loaded into the NFSR. Then the cipher is clocked 160 times without producing any keystream, but feeding the output bits back into both the LFSR and the NFSR. The structures of the Grain-v1 in the keystream generation and key initialization work mode are respectively depicted as shown in Fig. 1. Finally, we note that the state update function of Grain-v1 is invertible both during keystream generation and key initialization. This implies that if we recover the state of the cipher at some time t, we can clock it backwards to recover the key used. Our state recovery attack only focuses on the keystream generation mode after initialization.

2.3 Preliminary Analysis

Firstly, we present some definitions. A Boolean function is said to be k-normal if it is constant on a k-dimensional flat, and the k is referred as the normality

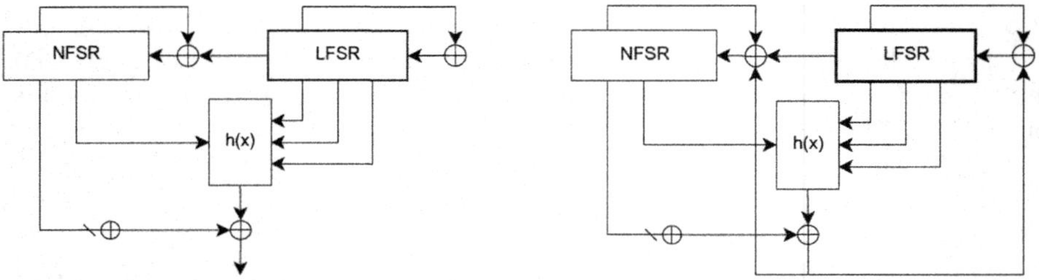

Fig. 1. Grain-v1: Keystream generation mode and Key initialization mode

order of the Boolean function. The notion of normality was introduced in [14] and later on was generalized to k-normality in [11]. In this paper, we further generalize the definition as a Boolean function is said to be k-linear-normal if its restriction is a linear function on a k-dimensional flat. For the filter function $h(x)$ of Grain-v1, we deduce all the linear operation modes by setting certain constraints on the state, shown in Table 1.

Table 1. Linear operation modes of the filter function $h(x)$

Constraint conditions	Linear operation modes
$x_2 = 0, x_3 = 0$	$x_1 \oplus x_4$
$x_2 = 0, x_3 = 1$	$x_0 \oplus x_1$
$x_0 = 0, x_1 = 0, x_2 = 1$	$x_3 \oplus x_4$
$x_0 = 0, x_1 = 1, x_2 = 1$	$x_3 \oplus 1$
$x_0 = 1, x_1 = 0, x_2 = 1$	x_3
$x_0 = 1, x_1 = 1, x_2 = 1$	$x_3 \oplus x_4$
$x_0 = 0, x_1 = 0, x_3 = 0$	x_4
$x_0 = 1, x_1 = 0, x_3 = 1$	x_4
$x_0 = 0, x_1 = 1, x_3 = 1$	$x_2 \oplus 1$
$x_0 = 1, x_1 = 0, x_3 = 1$	1
$x_0 = 1, x_1 = 1, x_3 = 0$	$x_2 \oplus x_4 \oplus 1$
$x_0 = 0, x_1 = 0, x_4 = 1$	$x_3 \oplus 1$
$x_0 = 0, x_1 = 1, x_4 = 1$	$x_2 \oplus x_3$
Continued from previous sheet	
$x_0 = 1, x_1 = 0, x_4 = 0$	x_3
$x_0 = 1, x_1 = 1, x_4 = 0$	$x_2 \oplus x_3 \oplus 1$
$x_1 = 0, x_2 = 0, x_3 = 0$	x_4
$x_1 = 0, x_2 = 0, x_3 = 1$	x_0
$x_1 = 1, x_2 = 0, x_3 = 0$	$x_4 \oplus 1$
$x_1 = 1, x_2 = 0, x_3 = 1$	$x_0 \oplus 1$
$x_1 = 0, x_2 = 1, x_4 = 0$	x_3
$x_1 = 0, x_2 = 1, x_4 = 1$	$x_0 \oplus x_3 \oplus 1$
$x_1 = 1, x_2 = 1, x_4 = 0$	$x_0 \oplus x_3 \oplus 1$
$x_1 = 1, x_2 = 1, x_4 = 1$	$x_3 \oplus 1$
$x_2 = 0, x_3 = 0, x_4 = 0$	x_1
$x_2 = 0, x_3 = 0, x_4 = 1$	$x_1 \oplus 1$
$x_2 = 0, x_3 = 1, x_4 = 0$	$x_0 \oplus x_1$
$x_2 = 0, x_3 = 1, x_4 = 1$	$x_0 \oplus x_1$

Secondly, we bring in the definition of sampling resistance. Consider a stream cipher with n-bit state. If given a value of $n - l$ special state bits of the cipher and the first l bits of the keystream sequence generated from that internal state, the remaining l bits of the internal state can be recovered directly, then the sampling resistance is defined as $R = 2^{-l}$, where l is the maximum value for which this enumeration of special states is possible. Thus, we can associate with each special state a short name of $n - l$ bits, and a short output of $n - l$ bits.

This technique does not modify the Biryukov-Shamir-Wagner tradeoff curve, but offers wider choices of tradeoff parameters.

In many proposed constructions, the sampling resistance depends on the location of the taps and the properties of the filter function. It is usually possible for small values of l, but only leads to improved tradeoff attacks if l is moderately large. The sampling resistance of Grain-v1 was analyzed in [8]. It said that given the value of 133 particular state bits of Grain-v1 and the first 18 keystream bits produced from that state, another 18 internal state bits may be deduced efficiently, and it emphasized that the sampling resistance of Grain-v1 is at most 2^{-18}, which is obviously not enough. Then comes the problem that how to make l longer. Our idea is associating the technique of k-linear-normality with sampling resistance by fixing some state bits. Under the constraints of state bits, the filter function of the remaining bits is linear. Thus we can substitute more state bits with keystream bits, i.e., extend the sampling resistance. Also, we can reduce the space of guessed bits both impacted by the constraints and sampling resistance. We call it conditional sampling resistance, which supports tradeoff parameters in larger range.

Next, we present the specific conditional sampling resistance for Grain-v1, which is based on the following guess-and-determine strategy. At first, we choose two of those linear operation modes according to the k-linear-normality of the filter function $h(x)$:

- Let $x_2 = 0, x_3 = 1$, i.e., $s_{t+46} = 0, s_{t+64} = 1$, then $h(x) = x_0 \oplus x_1 = s_{t+3} \oplus s_{t+25}$,
- Let $x_0 = 1, x_1 = 0, x_2 = 1$, i.e., $s_{t+3} = 1, s_{t+25} = 0, s_{t+46} = 1$, then $h(x) = x_3 = s_{t+64}$.

Combining the linear modes with the output function and the update functions of both LFSR and NFSR, we show the guessing path in Table 2, which is similar with the analysis in [26]. Let us briefly illustrate the procedures in the strategy. For Step 0 to 15, i.e., at $t = 0, \ldots, 15$, we derive

$$b_{t+10} = z_t \oplus b_{t+1} \oplus b_{t+2} \oplus b_{t+4} \oplus b_{t+31} \oplus b_{t+43} \oplus b_{t+56} \oplus s_{t+3} \oplus s_{t+25},$$

under the constraint conditions of $s_{t+46} = 0, s_{t+64} = 1$ by guessing the state bits listing in the table at each step. For Step 16, we derive

$$\underline{b_{29}} = z_{19} \oplus b_{20} \oplus b_{21} \oplus b_{23} \oplus b_{50} \oplus b_{62} \oplus \underline{b_{75}} \oplus \underline{s_{83}},$$

under the constraint conditions of $s_{22} = 1, s_{44} = 0, s_{65} = 1$ at $t = 19$, where underline identifies the state bits unknown to make the explanation more clear. Then we substitute s_{83} with the update function of LFSR

$$s_{83} = s_{65} \oplus s_{54} \oplus s_{41} \oplus s_{26} \oplus s_{16} \oplus s_3,$$

hence we only need to guess b_{75} to compute b_{29}. It is similar for Step 17 and 18. For Step 19, since we have

$$b_{35} = z_{25} \oplus \underline{b_{26}} \oplus \underline{b_{27}} \oplus b_{29} \oplus b_{56} \oplus b_{68} \oplus \underline{b_{81}} \oplus \underline{s_{89}},$$

Table 2. Guess-and-determine strategy

Step	Constraint conditions	Concerned keystream bits	Guessed NFSR and LFSR bits	Recovered bit
0	$s_{46}=0, s_{64}=1$	z_0	$b_1, b_2, b_4, b_{31}, b_{43}, b_{56}, s_3$	b_{10}
1	$s_{47}=0, s_{65}=1$	z_1	$b_3, b_5, b_{32}, b_{44}, b_{57}, s_4$	b_{11}
2	$s_{48}=0, s_{66}=1$	z_2	$b_6, b_{33}, b_{45}, b_{58}, s_5$	b_{12}
3	$s_{49}=0, s_{67}=1$	z_3	$b_7, b_{34}, b_{46}, b_{59}, s_6$	b_{13}
4	$s_{50}=0, s_{68}=1$	z_4	$b_8, b_{35}, b_{47}, b_{60}, s_7$	b_{14}
5	$s_{51}=0, s_{69}=1$	z_5	$b_9, b_{36}, b_{48}, b_{61}, s_8$	b_{15}
6	$s_{52}=0, s_{70}=1$	z_6	$b_{37}, b_{49}, b_{62}, s_9, s_{31}$	b_{16}
7	$s_{53}=0, s_{71}=1$	z_7	$b_{38}, b_{50}, b_{63}, s_{10}, s_{32}$	b_{17}
8	$s_{54}=0, s_{72}=1$	z_8	$b_{39}, b_{51}, b_{64}, s_{11}, s_{33}$	b_{18}
9	$s_{55}=0, s_{73}=1$	z_9	$b_{40}, b_{52}, b_{65}, s_{12}, s_{34}$	b_{19}
10	$s_{56}=0, s_{74}=1$	z_{10}	$b_{41}, b_{53}, b_{66}, s_{13}, s_{35}$	b_{20}
11	$s_{57}=0, s_{75}=1$	z_{11}	$b_{42}, b_{54}, b_{67}, s_{14}, s_{36}$	b_{21}
12	$s_{58}=0, s_{76}=1$	z_{12}	$b_{55}, b_{68}, s_{15}, s_{37}$	b_{22}
13	$s_{59}=0, s_{77}=1$	z_{13}	b_{69}, s_{16}, s_{38}	b_{23}
14	$s_{60}=0, s_{78}=1$	z_{14}	b_{70}, s_{17}, s_{39}	b_{24}
15	$s_{61}=0, s_{79}=1$	z_{15}	b_{71}, s_{18}, s_{40}	b_{25}
16	$s_{22}=1, s_{44}=0, \overline{s_{65}}=1$	z_{19}	b_{75}	b_{29}
17	$s_{23}=1, s_{45}=0, \overline{s_{66}}=1$	z_{20}	b_{76}	b_{30}
18	$s_{24}=1, \overline{s_{46}}=0, \overline{s_{67}}=1$	z_{21}	–	b_{77}
19	$s_{19}=s_{20}=s_{28}=1$	$z_{16}=0$	b_{72}, b_{73}	s_0
	$s_{41}=s_{42}=\overline{s_{50}}=0$	$z_{17}=0$		
	$s_{62}=s_{63}=\overline{s_{71}}=1$	$z_{25}=0$		
20	$\overline{s_{20}}=s_{21}=s_{29}=1$	$z_{17}=0$	b_{74}	s_1
	$\overline{s_{42}}=s_{43}=\overline{s_{51}}=0$	$z_{18}=0$		
	$\overline{s_{63}}=\overline{s_{64}}=\overline{s_{72}}=1$	$z_{26}=0$		
21	$s_{30}=1, \overline{s_{52}}=0, \overline{s_{73}}=1$	z_{27}	–	b_{28}
22	$\overline{s_{21}}=1, \overline{s_{43}}=0, \overline{s_{64}}=1$	z_{18}	–	s_2
23	$\overline{s_{29}}=1, \overline{s_{51}}=0, \overline{s_{72}}=1$	z_{26}	–	b_{27}
24	$\overline{s_{28}}=1, \overline{s_{50}}=0, \overline{s_{71}}=1$	z_{25}	–	b_{26}
25	$s_{25}=1, \overline{s_{47}}=0, \overline{s_{68}}=1$	z_{22}	–	b_{78}
26	$s_{26}=1, \overline{s_{48}}=0, \overline{s_{69}}=1$	z_{23}	–	b_{79}
26	$s_{27}=1, \overline{s_{49}}=0, \overline{s_{70}}=1$	z_{24}	–	b_0

– Overline denotes that this condition has already been assigned under the constraint conditions of $s_{21} = 1, s_{43} = 0, s_{64} = 1$ at $t = 25$, and

$$b_{26} = z_{16} \oplus b_{17} \oplus b_{18} \oplus b_{20} \oplus b_{47} \oplus b_{59} \oplus \underline{b_{72}} \oplus \underline{s_{80}},$$

under the constraint conditions of $s_{19} = 1, s_{41} = 0, s_{62} = 1$ at $t = 16$, and

$$b_{27} = z_{17} \oplus b_{18} \oplus b_{19} \oplus b_{21} \oplus b_{48} \oplus b_{60} \oplus \underline{b_{73}} \oplus \underline{s_{81}},$$

under the constraint conditions of $s_{20} = 1, s_{42} = 0, s_{63} = 1$ at $t = 17$, and also the following update state functions of LFSR and NFSR,

$$\begin{aligned}
s_{80} &= s_{62} \oplus s_{51} \oplus s_{38} \oplus s_{23} \oplus s_{13} \oplus \underline{s_0},\\
s_{81} &= s_{63} \oplus s_{52} \oplus s_{39} \oplus s_{24} \oplus s_{14} \oplus \underline{s_1},\\
s_{89} &= s_{71} \oplus s_{60} \oplus s_{47} \oplus s_{32} \oplus s_{22} \oplus s_9,\\
b_{81} &= \underline{s_1} \oplus b_{63} \oplus b_{61} \oplus b_{53} \oplus b_{46} \oplus b_{38} \oplus b_{34} \oplus b_{29} \oplus b_{22} \oplus b_{15} \oplus b_{10} \oplus b_1 \oplus b_{64}b_{61}\\
&\oplus b_{38}b_{34} \oplus b_{16}b_{10} \oplus b_{61}b_{53}b_{46} \oplus b_{34}b_{29}b_{22} \oplus b_{64}b_{46}b_{29}b_{10} \oplus b_{61}b_{53}b_{38}b_{34}\\
&\oplus b_{64}b_{61}b_{22}b_{16} \oplus b_{64}b_{61}b_{53}b_{46}b_{38} \oplus b_{34}b_{29}b_{22}b_{16}b_{10} \oplus b_{53}b_{46}b_{38}b_{34}b_{29}b_{22},
\end{aligned}$$

we derive

$$\begin{aligned}
\underline{s_0} &= b_{35} \oplus z_{25} \oplus z_{16} \oplus b_{17} \oplus b_{18} \oplus b_{20} \oplus b_{47} \oplus b_{59} \oplus \underline{b_{72}} \oplus s_{62} \oplus s_{51} \oplus s_{38} \oplus s_{23}\\
&\oplus s_{13} \oplus z_{17} \oplus b_{18} \oplus b_{19} \oplus b_{21} \oplus b_{48} \oplus b_{60} \oplus \underline{b_{73}} \oplus s_{63} \oplus s_{52} \oplus s_{39} \oplus s_{24} \oplus s_{14}\\
&\oplus b_{29} \oplus b_{56} \oplus b_{68} \oplus b_{64} \oplus b_{61} \oplus b_{53} \oplus b_{46} \oplus b_{38} \oplus b_{34} \oplus b_{29} \oplus b_{22} \oplus b_{16} \oplus b_{10}\\
&\oplus b_1 \oplus b_{64}b_{61} \oplus b_{38}b_{34} \oplus b_{16}b_{10} \oplus b_{61}b_{53}b_{46} \oplus b_{34}b_{29}b_{22} \oplus b_{64}b_{46}b_{29}b_{10}\\
&\oplus b_{61}b_{53}b_{38}b_{34} \oplus b_{64}b_{61}b_{22}b_{16} \oplus b_{64}b_{61}b_{53}b_{46}b_{38} \oplus b_{34}b_{29}b_{22}b_{16}b_{10}\\
&\oplus b_{53}b_{46}b_{38}b_{34}b_{29}b_{22} \oplus s_{89}.
\end{aligned}$$

That is, we only need to guess b_{72}, b_{73} to compute s_0. It is similar for Step 20. For Step 21, we derive

$$\underline{b_{28}} = z_{27} \oplus b_{29} \oplus b_{31} \oplus b_{37} \oplus b_{58} \oplus b_{70} \oplus \underline{b_{83}} \oplus \underline{s_{91}},$$

under the constraint conditions of $s_{30} = 1, s_{52} = 0, s_{73} = 1$, and then substitute s_{91} and b_{83} with their update functions

$$\begin{aligned}
s_{91} &= s_{73} \oplus s_{62} \oplus s_{49} \oplus s_{34} \oplus s_{24} \oplus s_{11},\\
b_{83} &= s_3 \oplus b_{65} \oplus b_{63} \oplus b_{55} \oplus b_{48} \oplus b_{40} \oplus b_{36} \oplus b_{31} \oplus b_{24} \oplus b_{17} \oplus b_{12} \oplus b_3 \oplus b_{66}b_{63}\\
&\oplus b_{40}b_{36} \oplus b_{18}b_{12} \oplus b_{63}b_{55}b_{48} \oplus b_{36}b_{31}b_{24} \oplus b_{66}b_{48}b_{31}b_{12} \oplus b_{63}b_{55}b_{40}b_{36}\\
&\oplus b_{66}b_{63}b_{24}b_{18} \oplus b_{66}b_{63}b_{55}b_{48}b_{40} \oplus b_{36}b_{31}b_{24}b_{18}b_{12} \oplus b_{55}b_{48}b_{40}b_{36}b_{31}b_{24}.
\end{aligned}$$

We need to guess no more bits to compute b_{28}. For Step 22, we have

$$\underline{s_{82}} = z_{18} \oplus b_{19} \oplus b_{20} \oplus b_{22} \oplus b_{28} \oplus b_{49} \oplus b_{61} \oplus b_{74},$$

under the constraint conditions of $s_{21} = 1, s_{43} = 0, s_{64} = 1$, and substitute s_{82} with the update function

$$s_{82} = s_{64} \oplus s_{53} \oplus s_{40} \oplus s_{25} \oplus s_{15} \oplus \underline{s_2},$$

hence we need to guess no more bits to compute s_2. Step 23 and 24 are similar with Step 21. Step 25 and 26 are similar with Step 16. For Step 27, we have

$$\underline{b_{80}} = z_{24} \oplus b_{25} \oplus b_{26} \oplus b_{28} \oplus b_{55} \oplus b_{67} \oplus b_{34} \oplus \underline{s_{88}},$$

under the constraint conditions of $s_{27} = 1, s_{49} = 0, s_{70} = 1$, and we substitute s_{88} and b_{80} as follows,

$$\begin{aligned} s_{88} &= s_{70} \oplus s_{59} \oplus s_{46} \oplus s_{31} \oplus s_{21} \oplus s_8, \\ b_{80} &= s_0 \oplus b_{62} \oplus b_{60} \oplus b_{52} \oplus b_{45} \oplus b_{37} \oplus b_{33} \oplus b_{28} \oplus b_{21} \oplus b_{14} \oplus b_9 \oplus \underline{b_0} \oplus b_{63}b_{60} \\ &\oplus b_{37}b_{33} \oplus b_{15}b_9 \oplus b_{60}b_{52}b_{45} \oplus b_{33}b_{28}b_{21} \oplus b_{63}b_{45}b_{28}b_9 \oplus b_{60}b_{52}b_{37}b_{33} \\ &\oplus b_{63}b_{60}b_{21}b_{15} \oplus b_{63}b_{60}b_{52}b_{45}b_{37} \oplus b_{33}b_{28}b_{21}b_{15}b_9 \oplus b_{52}b_{45}b_{37}b_{33}b_{28}b_{21}. \end{aligned}$$

We need to guess no more new bits to compute b_0.

In summary, given the guess-and-determine strategy, we derive that by fixing 51 bits of state constraint conditions and guessing 81 bits more of the internal state, the remaining 28 bits of the state can be recovered directly using the first 28 keystream output bits generated from the state, which is 10 bits longer than the sampling resistance given in [8].

2.4 Time-Memory-Data Tradeoff Attack of Grain-v1

Now, we complete the TMD trade-off attack against Grain-v1 according to the conditional sampling resistance given in Sect. 2.3. The attack has two phases: During the preprocessing phase, we explore the general structure of the cryptosystem, and summarize the findings in a large table. During the realtime phase, we are given actual data produced from a particular unknown state, and our goal is to use the precomputed table to find the state as quickly as possible. Procedures in details are as follows.

- Preprocessing Phase:
  1. Choose a fixed string $s \in \{0,1\}^{28}$ as a segment of keystream according to Table 2.
  2. Form a $m \times t$ matrix that tries to cover the whole search space which is composed of all the possible guessed 81 bits of NFSR and LFSR states as follows.
     (a) Randomly choose m startpoints of the chains, each point formed by a vector of 81 bits which is to be an injection into the guessed positions of NFSR and LFSR shown in Table 2.
     (b) Under the constraint conditions of 51 bits shown in Table 2, the remaining 28 bits of the state can be recovered using the segment of keystream s according to the guessing path. Thus, the overall system is obtained. Perform a backward computation of the system and generate the former 81 keystream bits from this moment on. Make it the next point in the chain, i.e., update the injection into NFSR and LFSR with this point.

(c) Iterate Step (b) t times on each startpoint respectively.
(d) Store the pairs of startpoints and endpoints $(SP_j, EP_j), j = 1, \ldots, m$ in a table.

– Realtime Phase:
1. Observe the keystream and find 2^{51} number of 28-bit strings matching with string s. For one such string, let its former 81 bits in the keystream be y.
2. For each y, check if there is EP_j, $j = 1, \ldots, m$ matching with y first. If not, iterate Step (b) w times on y until it matches with one of EP_j, $j = 1, \ldots, m$, where $w = 1, \ldots, t$. When there is a match, jump to the corresponding startpoint, and repeatedly apply Step (b) to the startpoint until the 81-bit keystream vector reaches y again. Then the previous point visited is the 81 guessed state bits of NFSR and LFSR, and the whole initial state of the cipher is recovered by jointing it with the 51 constraint state bits and 28 derived state bits.

There are several points to illustrate. Actually, the one-way function employed here is on the special states from $\{0,1\}^{81}$ to $\{0,1\}^{81}$ under the constraint conditions rather than inverting the full state of the cipher. To reduce the cross points of the chains, we can build more matrices in Step 2 by randomly seeding an 81-bit maximum LFSR and generating a sequence of distinct 81-bit vectors $X_1, X_2, \ldots$, each exclusive-ORs with the original points in each matrix to make the new points.

2.5 Complexity Analysis and Comparison

In this part, we consider the complexity of the attack. Firstly, we focus on the tradeoff parameters. The whole search space is composed of all the possible guessed NFSR and LFSR bits, whose cardinality is $2^{81} = m \cdot t$, which equals the preprocessing time P. The memory complexity is $M = m$ for the storage of the startpoints and endpoints table. Since the matrix is built under the constraint conditions of 51 bits, we expect to encounter a state among the matrix given 2^{51} selected data. Moreover, we need to sample $D = 2^{51} \cdot 2^{28}$ consecutive keystream bits to collect the required 2^{51} strings, since the string s of length 28 bits occurs on average once in 2^{28} keystream bits. For each selected data, we need to calculate Step (b) at most t times, and the time complexity T equals $2^{51} \cdot t$. Here, we choose $m = 2^{71}$ and $t = 2^{10}$, then we get a group of tradeoff parameters as follows, $T = 2^{61}, M = 2^{71}, D = 2^{79}, P = 2^{81}$.

Next, we compare our attack with previously reported TMD trade-off state recovery attacks against Grain-v1 in the single key and IV pair setting. It was first analyzed in the design document [17]. The designers follow the rule stated by Babbage [3] strictly that "if a secret key length of k bits is required, a state size of at least $2k$ bits is desirable". Thus, Grain-v1 is immune to the Babbage-Golic TMD attack. For Biryukov-Shamir-Wagner TMD attack, it needs to obey the tradeoff curve that $P = N/D, N^2 = TM^2D^2, D^2 \leq T$. In this way, the extreme parameters we can obtain is $T = 2^{80}, D = 2^{40}, M = 2^{80}, P = 2^{120}$. Another attack [8] employed the sampling resistance of $R = 2^{-18}$ and finally

Table 3. Comparison of trade-off parameters against Grain-v1 in the single key and IV pair setting

Resource	Time online (T)	Keystream (D)	Memory (M)	Preprocessing time (P)
BSW TMD	2^{80}	2^{40}	2^{80}	2^{120}
[8]	2^{71}	$2^{53.5}$	2^{71}	$2^{106.5}$
new	2^{61}	2^{79}	2^{71}	2^{81}

choose the parameters of $T = 2^{71}, D = 2^{53.5}, M = 2^{71}, P = 2^{106.5}$, which obey the tradeoff curve that $P = N/D, (RN)^2 = TM^2(RD)^2, D^2 \leq T$. Those are all the attacks published against Grain-v1 by time-memory tradeoffs in the case of single key and IV pair. We show the comparison of the parameters in Table 3 and our figures appear as significantly better than the previously reported ones, since the preprocessing time can be controlled much lower.

There are still several TMD attacks against Grain-v1 in the case of multi key and IV pairs, or different initial values. Such as in [17], the designers viewed the initialization process of the stream cipher as a one-way function, i.e., the function taking the key and IV as input and outputting the first $|K| + |IV|$ bits of the keystream. In this case, the search space is $2^{|K|+|IV|}$ and new data is generated by repeated initializations of the cipher. They gave the attack complexities in the tradeoff setting of [20], i.e., $N^2 = TM^2D^2$ and $P = N/D$. Note that D is the number of initializations, rather than keystream here. They presented two parameter choices: $T = 2^{80}, D = 2^{40}, M = 2^{64}, P = 2^{104}$ and $T = 2^{72}, D = 2^{36}, M = 2^{72}, P = 2^{108}$. For the attacks using different initial values, the attackers usually split the samples $D = d \cdot d'$ into d keystreams of length d', each generated by the same key and different IV. Usually, $d \cdot d'$ exceeds 2^{80}, and loading different IVs calls for more complexity for the cipher initialization. For example, it calls for the parameters that $T = 2^{46}, D = 2^{34} \cdot 2^{38}, M = 2^{70}, P = 2^{88}$ in [24], and $T = 2^{69.5}, D = 2^{45.25} \cdot 2^{45.25}, M = 2^{69.5}, P = 2^{69.5}$ in [13].

Next, we transform the tradeoff parameters into cipher ticks. For the preprocessing time, Step (b) needs to run backwards 81 cipher ticks. Thus to cover the space, we need the precomputation of $P = 2^{81} \cdot 81 = 2^{87.3}$ ticks. The memory is $M = m \cdot 81 \cdot 2$ bits for storing the pairs of 81-bit length points. The time online taken is $T = 2^{51} \cdot t \cdot 81$ ticks for Step (b). Since we choose $m = 2^{71}$ and $t = 2^{10}$, the memory is $M = 2^{78.3}$ bits and the time online is $T = 2^{67.3}$ cipher ticks. As a baseline, we analyzed the time complexity of the brute force attack against Grain-v1. Actually, the complexity of brute force attack is higher than 2^{80} ticks. Given a known fixed IV IV and an 80-bit keystream segment w generated by the (K, IV) pair, the goal is to recover K using the exhaustive search strategy. For each enumerated $k_i, 1 \leq i \leq 2^{80} - 1$, the attacker first needs to proceed the initialization phase which needs 160 ticks. During the keystream generation phase, once a keystream bit is generated, the attacker compares it to the corresponding bit in w. If they are equal, the attacker continues to generate the next keystream bit and does the comparison. If not, the attacker searches another key and repeats the previous steps. If each keystream bit is treated as a random independent variable, then for each k_i, the probability that the attacker needs to generate $l (1 \leq l \leq 80)$ keystream bits is 1 for $l = 1$ and $2^{-(l-1)}$ for $l > 1$, which means that the previous $l - 1$ bits are equal to the counter bits in w. Let N_w be the expected number of bits needed to generate for each enumerated key, which is $N_w = \sum_{l=1}^{80} l \cdot 2^{-(l-1)} = 4$. Then, the total time complexity is $(2^{80} - 1) \cdot (160 + 4) = 2^{87.4}$ cipher ticks. Thus the complexities of the new attack are all below that of the brute force attack. Furthermore, the brute force attack can only be mounted for each fixed IV, while our attack can be applied to the scenario with arbitrary IVs.

Moreover, it is a generic approach to analyze stream ciphers, by bringing the conditional sampling resistance in TMD tradeoff attacks, which is based on the k-linear-normality of the output function and an efficient guessing path. To illustrate the universality, we summarize the tradeoff curve of the new TMD attack. Assume a conditional sampling resistance is already obtained, that given r state bits of constraint conditions and g guessed state bits, the remaining l state bits can be recovered by the first l keystream bits generated from the state. Then we have $N = 2^{g+r+l}$, since the search space of the full states can be divided into three parts for constraint, guessing and sampling resistance respectively. The preprocessing time becomes $P = 2^g$, because we have to evaluate all the short states under the fixed constraint conditions. An $m \times t$ matrix is used to cover the enumerated short states, i.e., $P = mt$. The memory required to store the table of start- and end- points is $M = m$. Since we expect to encounter a state among the matrix built under the r constraint conditions, and the rate to find a fixed section of keystream is 2^{-l}, the data required is 2^{r+l}. We need to try 2^r samples online, each iterates up to t times. This product is $t2^r$. Thus we derive the tradeoff curve as follows.

$$\begin{aligned} D &= 2^{r+l} = N/2^g = N/P, \\ TMD &= t2^r m 2^{r+l} = 2^r 2^{g+r+l} = 2^r N. \end{aligned}$$

We believe the method can apply to more stream ciphers and should be taken into considerations when designing the security community of the stream ciphers.

3 Security Evaluation of ACORN Using Linear Approximation and Guessing Strategy

In this section, we present a state recovery attack against the lightweight authenticated cipher ACORN, given some length of keystream and plaintext. We start by giving a short description of the ACORN in Sect. 3.1. Then, we examine the guess-and-determine type of attacks against the cipher combined with the technique of linear approximation in Sect. 3.2. Since the designer did not give any security analysis against the encryption process of the cipher, we think it is necessary to present an evaluation to assess the security margin.

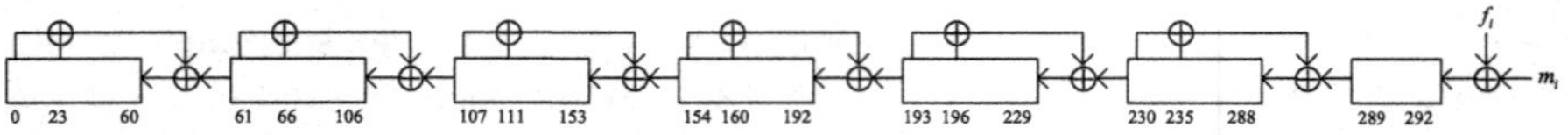

Fig. 2. ACORN-128

3.1 Description of ACORN

A concise description of ACORN is specified here as much detail as needed for the analysis, referred to [27]. ACORN-128 uses a 128-bit key and a 128-bit IV. The associated data length and the plaintext length are required less than 2^{64} bits. The authentication tag length is less than or equal to 128 bits, and the designer strongly recommends the use of a 128-bit tag. The state size is 293 bits, and there are six linear feedback shift registers (LFSRs) being concatenated in ACORN-128, shown in Fig. 2. Two Boolean functions are used in ACORN:

$$maj(x, y, z) = (x\&y) \oplus (x\&z) \oplus (y\&z); \quad ch(x, y, z) = (x\&y) \oplus ((\sim x)\&z),$$

where $\oplus$, & and $\sim$ denote bit-wise exclusive OR, bit-wise AND and bit-wise NOT respectively. Let the j-th bit of the state at the beginning of the i-th step be $S_{i,j}$, the keystream bit generated at the i-th step be ks_i, and the data bit be m_i. One step of ACORN is done as follows:

$$
\begin{aligned}
&S_{i,289} = S_{i,289} \oplus S_{i,235} \oplus S_{i,230};\\
&S_{i,230} = S_{i,230} \oplus S_{i,196} \oplus S_{i,193};\\
&S_{i,193} = S_{i,193} \oplus S_{i,160} \oplus S_{i,154};\\
&S_{i,154} = S_{i,154} \oplus S_{i,111} \oplus S_{i,107};\\
&S_{i,107} = S_{i,107} \oplus S_{i,66} \oplus S_{i,61};\\
&S_{i,61} = S_{i,61} \oplus S_{i,23} \oplus S_{i,0};\\
&ks_i = S_{i,12} \oplus S_{i,154} \oplus maj(S_{i,235}, S_{i,61}, S_{i,193});\\
&f_i = S_{i,0} \oplus (\sim S_{i,107}) \oplus maj(S_{i,244}, S_{i,23}, S_{i,160}) \oplus ch(S_{i,230}, S_{i,111}, S_{i,66})\\
&\qquad \oplus (ca_i \& S_{i,196}) \oplus (cb_i \& ks_i);\\
&\text{for } j := 0 \text{ to } 291 \text{ do } S_{i+1,j} = S_{i,j+1};\\
&S_{i+1,292} = f_i \oplus m_i;
\end{aligned}
$$

where ca_i and cb_i are control bits at the i-th step. Moreover, $ca_i = 1$ and $cb_i = 0$ at each step of the encryption without the separating process, which phase is focused on in our analysis.

The initialization of ACORN-128 consists of loading the key and IV into the state, and running the cipher for 1536 steps. After the initialization, the associated data is used to update the state. Note that even when there is no associated data, it is still needed to run the cipher for 512 steps. After processing the associated data, at each step of the encryption, one plaintext bit p_i is used to update the state, and p_i is encrypted to the ciphertext bit c_i as $c_i = p_i \oplus ks_i$.

Still 256 steps are called for separating the encryption and finalization. At last, the algorithm generates the authentication tag. Note that since the cipher has invertible state-update function, the internal state recovery of the cipher also leads to a key recovery attack.

3.2 Security Evaluation of ACORN

Firstly, we present some observations of the cipher on linear approximations of the filter function and guessed combination information of the upstate function. One observation is that the maj function can be linearly approximated with a big probability. Given the truth table of maj function as follows.

maj	0	0	0	0	1	1	1	1
x	0	0	0	1	0	1	1	1
y	0	0	1	0	1	0	1	1
z	0	1	0	0	1	1	0	1

At each step, the probability that any one of these three variables x, y, z equals the value of the maj function is $\frac{3}{4}$, the probability that any two of these three variables equal the value of the maj function is $\frac{1}{2}$, and the probability that all these three variables equal the value of the maj function is $\frac{1}{4}$. It is true that the probability of only one variable approximation at two steps is bigger than the probability of two variables approximations at only one step, $(\frac{3}{4})^2 > \frac{1}{2}$, and also the probability of only one variable approximation at three steps is bigger than the probability of three variables approximations at only one step, $(\frac{3}{4})^3 > \frac{1}{4}$. Since $ks_i = S_{i,12} \oplus S_{i,154} \oplus maj(S_{i,235}, S_{i,61}, S_{i,193})$, it is easy to get several linear equations before that the nonlinear feedback bits shift into the register and become one tap going into the keystream generating function used for linear approximation. Then there can be 139 steps for linear equations when using the approximation to $S_{i,61}$ of maj, considering the tap of $S_{i,154}$ not loading the feedback bits. Similarly, there can be 100 and 58 steps when using the approximation to $S_{i,193}$ and $S_{i,235}$ of maj, respectively.

To receive more linear equations, we consider the feedback bits. Here comes another observation,

$$\begin{aligned} f_i &= S_{i,0} \oplus (\sim S_{i,107}) \oplus maj(S_{i,244}, S_{i,23}, S_{i,160}) \oplus ch(S_{i,230}, S_{i,111}, S_{i,66}) \oplus S_{i,196} \\ &= S_{i,0} \oplus S_{i,23} \oplus S_{i,66} \oplus (\sim S_{i,107}) \oplus S_{i,196} \oplus (S_{i,23} \oplus S_{i,160}) \& (S_{i,23} \oplus S_{i,244}) \\ &\quad \oplus (S_{i,66} \oplus S_{i,111}) \& S_{i,230}. \end{aligned}$$

Thus we only need to guess two bits combination information of $S_{i,23} \oplus S_{i,160}$ and $S_{i,66} \oplus S_{i,111}$, then the feedback bit becomes liner. Hence, the feedback bit can be used in the approximation accordingly. Moreover, we get another two linear equations. It is easy to transform the state variables at each step into the initial ones linearly.

Thus, there becomes a problem for balancing the number of approximations at each step and the length of steps for approximations regarding to the guesses of feedback bits. Let the number of steps using just one variable approximation, two variables approximations and three variables approximations be a, b and c, respectively. Let the number of feedback steps be f. Then we transform the balancing problem into an integer linear programming problem (ILP) as follows.

$$\text{maximum} \quad Pr := (\frac{3}{4})^a \cdot (\frac{1}{2})^b \cdot (\frac{1}{4})^c \cdot (\frac{1}{2})^{2f}$$

$$\begin{cases} a + 2b + 3c + 2f \geq 293 \\ t := a + b + c = 139 + f \\ b + c \leq 100 + f \\ c \leq 58 + f \\ a, b, c, f \geq 0, \text{ are integers.} \end{cases}$$

We use Maple to solve the optimization problem, and the result is

$$a = 41, b = 112, c = 0, f = 14.$$

Here the goal Pr is about 2^{-157}, and $t = 153$. We explain the transformation briefly. Our objective is to get enough linear equations in the initial state variables as to recover the state with biggest success probability, which is the goal of the ILP problem. The first condition of the ILP problem shows that we have already found an adequate linear system. We have simulated the process, and testified the systems randomly. The results show that those linear systems mostly unisolvent. The following condition denotes the length of steps we used for approximation including 139 steps without feedback and f steps using the feedback bits. The next two conditions confine the steps using two and three variables approximations. At last, all the parameters are the numbers of steps, so they are integers and not negative.

A brief algorithm of the attack is presented as follows.

---

1. Guess 28 bits information to make the feedback bits at 14 steps linear and get 28 linear equations in the initial state variables.
2. For every guess, collect 265 linear approximating equations at optional steps according to the results of the ILP problem.
3. Given 153 keystream bits of ACORN, recover the state of ACORN, and verify the solution.
4. Repeat the loops from Step 1 to 3 until the right initial state is found.

If the solution is not the real state, both the error of linear approximation and the guessed information can lead to the inaccuracy. We expect a right solution among 2^{157} tests according to the success probability. There is also a guess-and-determine attack against ACORN in [23], while its complexity is about 2^{164}

tests, which is larger than ours. This result can be a security bound of ACORN examined by the method of guess-and-determine attacks so far. Moreover, we can view this method a generic way to evaluate the security of stream ciphers, which works as firstly finding the linear approximations of the output function and the efficient guessed combination information of the upstate function, then transforming the bounding problem into an integer linear programming problem for searching the optimal solution.

4 Conclusion

In this paper, we have presented two new generic methods for analyzing stream ciphers. One is a time-memory-data tradeoff attack using the conditional sampling resistance, and its application to Grain-v1 shows that the result is better than the previous ones and lower than the security bound. Another is a security evaluation method using linear approximations, efficiently guessed information and the tool of integer linear programming problem. The result of its application to ACORN gives a security bound of ACORN.

Acknowledgements. This work was supported by the National Grand Fundamental Research 973 Program of China (Grant No. 2013CB338002) and the programs of the National Natural Science Foundation of China (Grant No. 61379142, 60833008, 60603018, 61173134, 91118006, 61272476).

References

1. The ecrypt stream cipher project. eSTREAM Portfolio Revision (2008). http://www.ecrypt.eu.org/stream
2. Caesar (2013). http://competitions.cr.yp.to/index.html
3. Babbage, S.: Improved "exhaustive search" attacks on stream ciphers. In: European Convention on Security and Detection, pp. 161–166. IET (1995)
4. Berbain, C., Gilbert, H., Maximov, A.: Cryptanalysis of grain. In: Robshaw, M. (ed.) FSE 2006. LNCS, vol. 4047, pp. 15–29. Springer, Heidelberg (2006)
5. Biryukov, A., Kizhvatov, I., Zhang, B.: Cryptanalysis of the atmel cipher in SecureMemory, CryptoMemory and CryptoRF. In: Lopez, J., Tsudik, G. (eds.) ACNS 2011. LNCS, vol. 6715, pp. 91–109. Springer, Heidelberg (2011)
6. Biryukov, A., Shamir, A.: Cryptanalytic time/memory/data tradeoffs for stream ciphers. In: Okamoto, T. (ed.) ASIACRYPT 2000. LNCS, vol. 1976, pp. 1–13. Springer, Heidelberg (2000)
7. Biryukov, A., Shamir, A., Wagner, D.: Real time cryptanalysis of A5/1 on a PC. In: Goos, G., Hartmanis, J., van Leeuwen, J., Schneier, B. (eds.) FSE 2000. LNCS, vol. 1978, pp. 1–18. Springer, Heidelberg (2001)
8. Bjørstad, T.: Cryptanalysis of grain using time/memory/data tradeoffs (2008). http://www.ecrypt.eu.org/stream/grainp3.htm
9. Bluetooth, S.: Specification of the bluetooth system, version 1.1 (2001). http://www.bluetooth.com
10. Briceno, M., Goldberg, I., Wagner, D.: A pedagogical implementation of a5/1 (1999). http://jya.com/a51-pi.htm

11. Charpin, P.: Normal boolean functions. J. Complex. **20**(2–3), 245–265 (2004). Festschrift for Harald Niederreiter, Special Issue on Coding and Cryptography
12. De Cannière, C., Küçük, Ö., Preneel, B.: Analysis of grain's initialization algorithm. In: Vaudenay, S. (ed.) AFRICACRYPT 2008. LNCS, vol. 5023, pp. 276–289. Springer, Heidelberg (2008)
13. Ding, L., Jin, C., Guan, J., Qi, C.: New treatment of the BSW sampling and its applications to stream ciphers. In: Pointcheval, D., Vergnaud, D. (eds.) AFRICACRYPT. LNCS, vol. 8469, pp. 136–146. Springer, Heidelberg (2014)
14. Dobbertin, H.: Construction of bent functions and balanced boolean functions with high nonlinearity. In: Preneel, B. (ed.) FES 2000. LNCS, vol. 1008, pp. 61–74. Springer, Heidelberg (1995)
15. Golić, J.D.: Cryptanalysis of alleged A5 stream cipher. In: Fumy, W. (ed.) EUROCRYPT 1997. LNCS, vol. 1233, pp. 239–255. Springer, Heidelberg (1997)
16. Hawkes, P., Rose, G.G.: Guess-and-determine attacks on SNOW. In: Nyberg, K., Heys, H.M. (eds.) SAC 2002. LNCS, vol. 2595, pp. 37–46. Springer, Heidelberg (2003)
17. Hell, M., Johansson, T., Maximov, A., Meier, W.: The grain family of stream ciphers. In: Robshaw, M., Billet, O. (eds.) New Stream Cipher Designs. LNCS, vol. 4986, pp. 179–190. Springer, Heidelberg (2008)
18. Hell, M., Johansson, T., Meier, W.: Grain: a stream cipher for constrained environments. Int. J. Wirel. Mob. Comput. **2**(1), 86–93 (2007)
19. Hellman, M.: A cryptanalytic time-memory trade-off. IEEE Trans. Inf. Theory **26**(4), 401–406 (1980)
20. Hong, J., Sarkar, P.: New applications of time memory data tradeoffs. In: Roy, B. (ed.) ASIACRYPT 2005. LNCS, vol. 3788, pp. 353–372. Springer, Heidelberg (2005)
21. Khazaei, S., Hassanzadeh, M., Kiaei, M.: Distinguishing attack on grain (2005)
22. Lee, Y., Jeong, K., Sung, J., Hong, S.H.: Related-key chosen IV attacks on Grain-v1 and Grain-128. In: Mu, Y., Susilo, W., Seberry, J. (eds.) ACISP 2008. LNCS, vol. 5107, pp. 321–335. Springer, Heidelberg (2008)
23. Liu, M., Lin, D.: Cryptanalysis of lightweight authenticated cipher acorn. Cryptocompetitions mailing list. (2014). https://groups.google.com/forum/#!topic/crypto-competitions/2mrDnyb9hfM
24. Mihaljevic, M., Gangopadhyay, S., Paul, G., Imai, H.: Internal state recovery of grain-v1 employing normality order of the filter function. IET Inf. Secur. **6**(2), 55–64 (2012)
25. Rivest, R.: The rc4 encryption algorithm, rsa data security inc. (1992)
26. Wei, Y., Pasalic, E., Zhang, F., Wu, W.: Key recovery attacks on grain family using bsw sampling and certain weaknesses of the filtering function. Cryptology ePrint Archive, Report 2014/971 (2014). http://eprint.iacr.org/
27. Wu, H.: Acorn: a lightweight authenticated cipher (v1). Submission to CAESAR (2014). http://competitions.cr.yp.to/round1/acornv1.pdf
28. Zhang, B., Li, Z., Feng, D., Lin, D.: Near collision attack on the grain v1 stream cipher. In: Moriai, S. (ed.) FSE 2013. LNCS, vol. 8424, pp. 518–538. Springer, Heidelberg (2014)

Key Recovery Attacks Against NTRU-Based Somewhat Homomorphic Encryption Schemes

Massimo Chenal(✉) and Qiang Tang

APSIA Group, SnT, University of Luxembourg,
6 rue Richard Coudenhove-Kalergi, 1359 Luxembourg, Luxembourg
{massimo.chenal,qiang.tang}@uni.lu

Abstract. A key recovery attack allows an attacker to recover the private key of an underlying encryption scheme when given a number of decryption oracle accesses. Previous research has shown that most existing Somewhat Homomorphic Encryption (SHE) schemes suffer from this attack. In this paper, we propose efficient key recovery attacks against two NTRU-based SHE schemes due to Lopez-Alt et al. (2012) and Bos et al. (2013), which have not gained much attention in the literature. Parallel to our work, Dahab, Galbraith and Morais (2015) have also proposed similar attacks but only for specific parameter settings. In comparison, our attacks apply to all parameter settings and are more efficient.

Keywords: Somewhat homomorphic encryption · Key recovery attack · IND-CCA1 Security

1 Introduction

In the literature, all Somewhat Homomorphic Encryption (SHE) schemes have been developed with the aim of being IND-CPA secure. In [Gen09], Gentry emphasized it as a future work to investigate SHE schemes with IND-CCA1 security (i.e. secure against a non-adaptive chosen-ciphertext attack). Up to now, the only scheme proven IND-CCA1 secure is that by Loftus et al. [LMSV12]. Most works in this direction focus on devising attacks against existing schemes.

It has been shown that most existing SHE schemes suffer from key recovery attacks, which allow an attacker to recover the private key of an underlying encryption scheme when given a number of decryption oracle accesses. It is clear that a key recovery attack is stronger than a typical attack against IND-CCA1 security. Loftus et al. [LMSV12] showed key recovery attacks against SHE schemes from [Gen09, GH11]. Zhang et al. [ZPS12] presented an attack against the SHE scheme in [vDGHV10]. Chenal and Tang [CT14] presented key recovery attacks for all the schemes in [BV11b, BV11a, GSW13, Bra12, BGV12].

Previous analysis has not paid much attention to the NTRU-based SHE schemes. Two representative schemes in this line are those by Lopez-Alt, Tromer

This paper is an extended abstract of the IACR report [CT15].

J. Lopez and C.J. Mitchell (Eds.): ISC 2015, LNCS 9290, pp. 397–418, 2015.
DOI: 10.1007/978-3-319-23318-5_22

and Vaikuntanathan [LATV12] and Jos et al. [BLLN13]. Note that, instead of relying on the original NTRU scheme by Hoffstein, Pipher and Silverman [HPS98], these schemes are based on a variant by Stehle and Steinfeld [SS10]. Parallel to our work in this paper, we noticed that Dahab, Galbraith and Morais [DGM15] constructed key recovery attacks for these schemes from [BLLN13, LATV12].

1.1 Our Contribution

The key recovery attacks by Dahab, Galbraith and Morais [DGM15] work for some tailored parameters for the LTV12 and BLLN13 SHE schemes. For example, they require $6(t^2 + t) < q$ and $B^2 < \frac{q}{36t^2}$ while these conditions are not assumed in [LATV12,BLLN13]. In this paper, we present attacks that work for all parameter settings. Moreover, our attacks are more efficient than theirs, see the following table. Note that n is defined as an integer of power of 2, B is a bound on the coefficient size of error distribution and is much smaller than q, $t \geq 2$ is an integer that partially determines the message space size. More detailed definitions for these parameters can be found in the following sections.

	Our attacks	Attacks from [DGM15]
[LATV12]	$\lfloor \log_2 B \rfloor + n$	$n \cdot \lceil \log_2 B \rceil + n$
[BLLN13] (t is odd)	$\lceil \log_2 (B/t) \rceil$	$n \cdot \lceil \log_2 B \rceil$
[BLLN13] (t is even but not 2)	$\lceil \log_2 (B/t) \rceil + n$	$n \cdot \lceil \log_2 B \rceil$
[BLLN13] ($t = 2$)	$\lceil \log_2 (B/t) \rceil + n$	$n \cdot \lceil \log_2 B \rceil + n$

Our work, together with the results from [LMSV12,ZPS12,CT14], show that most existing SHE schemes, except that from [LMSV12], suffer from key recovery attacks so that they are not IND-CCA1 secure.

1.2 Structure of the Paper

In Sect. 2, we recall some background on SHE schemes. In Sect. 3, we present our attack against the LTV12 SHE scheme. In Sect. 4, we present our attack against the BLLN13 SHE scheme. In Sect. 5, we conclude the paper.

2 Preliminary

Let $\mathbb{N}$ be the set of natural numbers, $\mathbb{Z}$ the ring of integers, $\mathbb{Q}$ the field of rational numbers, and $\mathbb{F}_q$ a finite field with q elements, where q is a power of a prime p. In particular, we will consider often $\mathbb{F}_p = \mathbb{Z}/p\mathbb{Z} = \mathbb{Z}_p$. If $r \in \mathbb{Z}_q$, we indicate as r^{-1} its inverse in $\mathbb{Z}_q$, i.e. that value such that $r^{-1} \cdot r = 1 \mod q$. For a ring R and a (two-sided) ideal I of R, we consider the quotient ring R/I. For a given rational number $x \in \mathbb{Q}$, we let $\lfloor x \rceil$, $\lfloor x \rfloor$ and $\lceil x \rceil$ be respectively the rounding function, the floor function and the ceiling function. For a given integer $n \in \mathbb{N}$, $\lfloor n + 1/2 \rceil = n + 1$. Of course, our attacks work also, with trivial modifications, in the case we define $\lfloor n + 1/2 \rceil = n$. To indicate that an element a is chosen uniformly at random from a set A we use notation $a \xleftarrow{\$} A$. For a set A, we let its cardinality be $|A|$. We denote the map that reduces an integer x modulo q and uniquely represents the result by an element in the interval $(-q/2, q/2]$ by $[\cdot]_q$. Therefore, we will consider the ring $\mathbb{Z}_q$ as $\mathbb{Z}_q := \{-\lfloor \frac{q}{2} \rfloor, -\lfloor \frac{q}{2} \rfloor + 1, \ldots, \lfloor \frac{q}{2} \rfloor\}$. We extend this map to polynomials in $\mathbb{Z}[X]$ and thus also to elements of R by applying it to their coefficients separately; given a polynomial $a(x) \in R$, we define the map $[\cdot]_q : R \to R$, $a(x) = \sum_{i=0}^{n-1} a_i x^i \mapsto \sum_{i=0}^{n-1} [a_i]_q x^i$ Unless otherwise specified, λ will always denote the security parameter. In the asymmetric schemes we are going to discuss, the secret key is denoted as sk, and the public key is pk.

The following definitions are adapted from [Gen09]. We only assume bit-by-bit public-key encryption, i.e. we only consider encryption schemes that are homomorphic with respect to boolean circuits consisting of gates for addition and multiplication mod 2. Extensions to bigger plaintext spaces and symmetric-key setting are straightforward, so that we skip it.

Definition 1 (Homomorphic Encryption). *A public key homomorphic encryption (HE) scheme is a set* $\mathcal{E} = (\mathsf{KeyGen}_\mathcal{E}, \mathsf{Encrypt}_\mathcal{E}, \mathsf{Decrypt}_\mathcal{E}, \mathsf{Evaluate}_\mathcal{E})$ *of four algorithms all of which must run in polynomial time. When the context is clear, we will often omit the index* $\mathcal{E}$.

$\mathsf{KeyGen}(\lambda) = (\mathsf{sk}, \mathsf{pk})$
- input: λ
- output: sk; pk

$\mathsf{Encrypt}(\mathsf{pk}, m) = c$
- input: pk and plaintext $m \in \mathbb{F}_2$
- output: ciphertext c

$\mathsf{Decrypt}(\mathsf{sk}, c) = m'$
- input: sk and ciphertext c
- output: $m' \in \mathbb{F}_2$

$\mathsf{Evaluate}(\mathsf{pk}, C, (c_1, \ldots, c_r)) = c_e$
- input: pk, circuit C, ciphertexts $c_1, \ldots, c_r$, with $c_i = \mathsf{Encrypt}(\mathsf{pk}, m_i)$
- output: ciphertext c_e

Informally, a homomorphic encryption scheme that can perform only a limited number of $\mathsf{Evaluate}$ operations is called a Somewhat Homomorphic Encryption (SHE) scheme.

A public-key encryption scheme is IND-CCA1 secure if a polynomial time attacker can only win the following game with a negligible advantage $\mathrm{Adv}_{\mathcal{A},\mathcal{E},\lambda}^{\text{IND-CCA1}} = |\Pr(b = b') - 1/2|$.

1. $(\mathsf{pk}, \mathsf{sk}) \leftarrow \mathsf{KeyGen}(1^\lambda)$
2. $(m_0, m_1) \leftarrow \mathcal{A}_1^{(\mathsf{Decrypt})}(\mathsf{pk})$
   [Stage 1]
3. $b \leftarrow \{0, 1\}$
4. $c^* \leftarrow \mathsf{Encrypt}(m_b, \mathsf{pk})$
5. $b' \leftarrow \mathcal{A}_2(c^*)$
   [Stage 2]

According to the definition, in order to show that a scheme is not IND-CCA1 secure, we only need to show that an adversary can guess the bit b with a non-negligible advantage given access to the decryption oracle in Stage 1. In comparison, in a *key recovery attack*, an adversary can output the private key given access to the decryption oracle in Stage 1. Clearly, a key recovery attack is stronger and can result in more serious vulnerabilities in practice.

3 Attack Against the LTV12 SHE Scheme

We start by recalling the LTV12 SHE Scheme [LATV12]. Let λ be the security parameter, consider an integer $n = n(\lambda)$ and a prime number $q = q(\lambda) \neq 2$. Consider also a degree-n polynomial $\phi(x) = \phi_\lambda(x)$: following [LATV12], we will use $\phi(x) = x^n + 1$. Finally, let $\chi = \chi(\lambda)$ a $B(\lambda)$-bounded error distribution over the ring $R := \mathbb{Z}[x]/(\phi(x))$. The parameters $n, q, \phi(x)$ and χ are public and we assume that given λ, there are polynomial-time algorithms that output n, q and $\phi(x)$, and sample from the error distribution χ. The message space is $\mathcal{M} = \{0, 1\}$, and all operations on ciphertexts are carried out in the ring $R_q := \mathbb{Z}_q[x]/(\phi(x))$.

$\mathsf{KeyGen}(\lambda)$:
- sample $f', g \leftarrow \chi$
- set $f := 2f' + 1$ so that $f \equiv 1 \bmod 2$
- if f is not invertible in R_q, resample f'
- $\mathsf{pk} := h = 2gf^{-1} \in R_q$
- $\mathsf{sk} := f \in R$

$\mathsf{Encrypt}(\mathsf{pk}, m)$:
- sample $s, e \leftarrow \chi$
- output ciphertext $c := hs + 2e + m \in R_q$

$\mathsf{Decrypt}(\mathsf{sk}, c)$:
- let $\mu = f \cdot c \in R_q$
- output $\mu' := \mu \bmod 2$

Since we don't need the evaluation step, we omit it in the description. In the original paper [LATV12], the somewhat homomorphic encryption scheme is multi-key, i.e. one can use several secret keys $\mathsf{sk}_1 = f_1, \ldots, \mathsf{sk}_M = f_M$ in order to decrypt. By analyzing the original decryption step, one can see that, in order to decrypt the plaintext message, we need to multiply secret keys $\mathsf{sk}_1 = f_1, \ldots, \mathsf{sk}_M = f_M$ together, and then multiply the result with the ciphertext and reduce. For this reason, it is enough to retrieve, as the secret key, the polynomial $f_1 \cdots f_M =: s = s(x) = s_0 + s_1 x + s_2 x^2 + \cdots + s_{n-1} x^{n-1} \in R_q$, with $s_i \in (-q/2, q/2]$ for all $i = 0, 1, \ldots, n-1$. For this reason, it is enough to present the scheme as we saw it, with only one secret key.

Remark 1. In [LATV12], the authors do not explicitly state how the decryption behaves if $\mu \bmod 2$ is not a constant. We consider three scenarios: (1) output directly $\mu \bmod 2$; (2) output the constant of $\mu \bmod R_2$; (3) output an error. In the following, we describe a key recovery attack for scenario (1) and it can be easily extended to scenario (2). It is likely that we can adapt our attack to scenario (3), but we have not succeeded so far.

3.1 Attack Preview

Generally, suppose the secret key is in the form of the polynomial $f = s(x) = s_0 + s_1 x + s_2 x^2 + \cdots + s_{n-1} x^{n-1} \in R_q$. Now, since we assume q odd, and s_i is an integer, we have $-q/2 < s_i < q/2$, and in particular $-\left\lfloor \frac{q}{2} \right\rfloor \leq s_i \leq \left\lfloor \frac{q}{2} \right\rfloor, \quad \forall 0 \leq i \leq n-1$. Each coefficient s_i can have $\left\lfloor \frac{q}{2} \right\rfloor - (-\left\lfloor \frac{q}{2} \right\rfloor) + 1 = q$ possible different values. We remark that there exists a bit representation of the s_i's such that $\#\text{bits}(s_i) = \lfloor \log_2(q-1) \rfloor + 1 =: N$, and $\#\text{bits}(s) = n \cdot$

$\#\text{bits}(s_i) = n \cdot (\lfloor \log_2(q-1) \rfloor + 1)$. The decryption oracle reveals a polynomial $\mu'(x) = \mu(x) \bmod 2 = \mu'_0 + \mu'_1 x + \cdots + \mu'_{n-1} x^{n-1}$, with $\mu'_i \in \{0,1\}$ for $i = 0,1,\ldots,n-1$. Hence, decryption oracle reveals n bits at a time. Therefore, the minimum number of oracle queries needed to recover s is N. As we will see, our attack needs N oracle queries, plus at most $n-1$ oracle queries necessary to determine the signs of the coefficients of the secret key. We remark that the scheme as described in [LATV12] has message space $\mathcal{M} = \{0,1\}$. When the oracle decryption receives an honestly-generated ciphertext, it returns either $0 = \sum_{i=0}^{n-1} 0 \cdot x^i \in R_q$ or $1 = 1 + \sum_{i=1}^{n-1} 0 \cdot x^i \in R_q$. However, in principle the oracle decryption can return any polynomial in $\{0,1\}/(x^n+1)$ and we will use this fact as basis to build our attack.

Here is the workflow of our key recovery attack. First of all, we are going to determine the parity of each coefficient $s_i \in (-q/2, q/2]$. Then, we are going to find s_i by gradually reducing (halving) the interval in which it lies. At some point, s_i will be reduced to belong to some interval with at most two consecutive integers; the absolute value of s_i will be deduced by its (known) parity. At this point, we will know the secret key coefficient s_i in absolute value; in the last step, we are going to query the oracle decryption at most n times in order to recover the sign of the coefficients s_i, for $i = 1,2,\ldots,n-1$, relative to the (unknown) sign of s_0. So in the end, we will end up with two possible candidate secret keys $s_1(x)$ and $s_2(x) = -s_1(x)$. We have then $s(x) = s_1(x)$ or $s(x) = s_2(x)$, and recovering which one of the two is trivial with an extra oracle query.

In our description, we consider the coefficients s_i in the interval $(-q/2, q/2]$ and can recover the private key with at most $\lfloor \log_2 q \rfloor + n$ decryption oracle queries. However, we could consider the stricter interval $[-B, B]$, with B the bound on coefficients given by the distribution χ from which the coefficients are picked from. In this case, we can see that the total number of queries needed to be submitted to the decryption oracle are actually at most $\lfloor \log_2 B \rfloor + n$.

3.2 Detailed Attack

Preliminary Step. Submit to the decryption oracle the "ciphertext" $c(x) = 1 \in R_q$. The oracle will compute and return the polynomial $D(c(x) = 1) = s(x) \bmod 2 = \sum_{i=0}^{n-1}(s_i \bmod 2)x^i$, which tells us the parity of each s_i, $i = 0,1,\ldots,n-1$.

Step 1. Choose and submit to the decryption oracle the "ciphertext" $c(x) = 2 \in R_q$. It will compute and return the polynomial $D(c(x) = 2) = (2s(x) \in R_q) \bmod 2 = \sum_{i=0}^{n-1}[(2s_i \bmod q) \bmod 2]\, x^i$. For all $i \in [0, n-1]$ we have

$$\frac{-q+1}{2} \le s_i \le \frac{q-1}{2}, \text{ and so } -q+1 \le 2s_i \le q-1 \tag{A}$$

For each i, we have two cases to distinguish:

Case A_1: $(2s_i \bmod q) \bmod 2 = 0$. Then, condition (A) implies that $\frac{-q+1}{2} \le 2s_i \le \frac{q-1}{2}$, i.e. $\frac{-q+1}{4} \le s_i \le \frac{q-1}{4}$

$$-q+1 \le 4s_i \le q-1 \tag{A1}$$

Case B_1: $(2s_i \bmod q) \bmod 2 = 1$. Then, condition (A) implies that $\frac{q-1}{2} + 1 \leq 2|s_i| \leq q-1$, i.e. $\frac{q+1}{4} \leq |s_i| \leq \frac{q-1}{2}$

$$q + 1 \leq 4|s_i| \leq 2q - 2 \tag{B1}$$

Step 2. Choose and submit to the decryption oracle the "ciphertext" $c(x) = 4 \in R_q$. It will compute and return the polynomial $D(c(x) = 4) = [s(x) \cdot 4]_q \bmod 2 = \sum_{i=0}^{n-1} [[4s_i]_q \bmod 2]\, x^i$. For each i, we have four cases to distinguish:

Case A_2: In Step 1 case A_1 held, and $[4s_i]_q \bmod 2 = 0$. Then, condition (A1) implies that $\frac{-q+1}{2} \leq 4s_i \leq \frac{q-1}{2}$, i.e. $\frac{-q+1}{8} \leq s_i \leq \frac{q-1}{8}$

$$-q + 1 \leq 8s_i \leq q - 1 \tag{A2}$$

Case B_2: In Step 1 case A_1 held, and $[4s_i]_q \bmod 2 = 1$. Then, condition (A1) implies that $\frac{q-1}{2} + 1 \leq 4|s_i| \leq q - 1$, i.e. $\frac{q+1}{8} \leq |s_i| \leq \frac{q-1}{4}$

$$q + 1 \leq 8|s_i| \leq 2q - 2 \tag{B2}$$

Case C_2: In Step 1 case B_1 held, and $[4s_i]_q \bmod 2 = 0$. Then, condition (B1) implies that $q + 1 + \frac{q-1}{2} \leq 4|s_i| \leq 2q - 2$, i.e. $\frac{3q+1}{8} \leq |s_i| \leq \frac{q-1}{2}$

$$3q + 1 \leq 8|s_i| \leq 4q - 4 \tag{C2}$$

Case D_2: In Step 1 case B_1 held, and $[4s_i]_q \bmod 2 = 1$. Then, condition (B1) implies that $q + 1 \leq 4|s_i| \leq \frac{3q-1}{2}$, i.e. $\frac{q+1}{4} \leq |s_i| \leq \frac{3q-1}{8}$

$$2q + 2 \leq 8|s_i| \leq 3q - 1 \tag{D2}$$

Step 3. Choose and submit to the decryption oracle the "ciphertext" $c(x) = 8 \in R_q$. It will compute and return the polynomial $D(c(x) = 8) = [s(x) \cdot 8]_q \bmod 2 = \sum_{i=0}^{n-1} [[8s_i]_q \bmod 2]\, x^i$. For each i, we have four cases to distinguish:

Case A_3: In Step 2 case A_2 held, and $[8s_i]_q \bmod 2 = 0$. Then, condition (A2) implies that $\frac{-q+1}{2} \leq 8s_i \leq \frac{q-1}{2}$, i.e. $\frac{-q+1}{16} \leq s_i \leq \frac{q-1}{16}$

$$-q + 1 \leq 16s_i \leq q - 1 \tag{A3}$$

Case B_3: In Step 2 case A_2 held, and $[8s_i]_q \bmod 2 = 1$. Then, condition (A2) implies that $\frac{q-1}{2} + 1 \leq 8|s_i| \leq q - 1$, i.e. $\frac{q+1}{16} \leq |s_i| \leq \frac{q-1}{8}$

$$q + 1 \leq 16|s_i| \leq 2q - 2 \tag{B3}$$

Case C_3: In Step 2 case B_2 held, and $[8s_i]_q \bmod 2 = 0$. Then, condition (B2) implies that $\frac{3q+1}{2} \leq 8|s_i| \leq 2q - 2$, i.e. $\frac{3q+1}{16} \leq |s_i| \leq \frac{q-1}{4}$

$$3q + 1 \leq 16|s_i| \leq 4q - 4 \tag{C3}$$

Case D_3: In Step 2 case B_2 held, and $[8s_i]_q \bmod 2 = 1$. Then, condition (B2) implies that $q+1 \leq 8|s_i| \leq \frac{3q-1}{2}$, i.e. $\frac{q+1}{8} \leq |s_i| \leq \frac{3q-1}{16}$

$$2q + 2 \leq 16|s_i| \leq 3q - 1 \tag{D3}$$

Case E_3: In Step 2 case C_2 held, and $[8s_i]_q \bmod 2 = 0$. Then, condition (C2) implies that $\frac{7q+1}{2} \leq 8|s_i| \leq 4q - 4$, i.e. $\frac{7q+1}{16} \leq |s_i| \leq \frac{q-1}{2}$

$$7q + 1 \leq 16|s_i| \leq 8q - 8 \tag{E3}$$

Case F_3: In Step 2 case C_2 held, and $[8s_i]_q \bmod 2 = 1$. Then, condition (C2) implies that $3q+1 \leq 8|s_i| \leq \frac{7q-1}{2}$, i.e. $\frac{3q+1}{8} \leq |s_i| \leq \frac{7q-1}{16}$

$$6q + 2 \leq 16|s_i| \leq 7q - 1 \tag{F3}$$

Case G_3: In Step 2 case D_2 held, and $[8s_i]_q \bmod 2 = 0$. Then, condition (D2) implies that $2q+2 \leq 8|s_i| \leq \frac{5q-1}{2}$, i.e. $\frac{q+1}{4} \leq |s_i| \leq \frac{5q-1}{16}$

$$4q + 4 \leq 16|s_i| \leq 5q - 1 \tag{G3}$$

Case H_3: In Step 2 case D_2 held, and $[8s_i]_q \bmod 2 = 1$. Then, condition (D2) implies that $\frac{5q+1}{2} \leq 8|s_i| \leq 3q - 1$, i.e. $\frac{5q+1}{16} \leq |s_i| \leq \frac{3q-1}{8}$

$$5q + 1 \leq 16|s_i| \leq 6q - 2 \tag{H3}$$

Final Step. We continue in this fashion and finally we obtain integers $s_i' := |s_i| \in [0, \frac{q-1}{2}]$, for $i = 0, 1, \ldots, n-1$. This is obtained in the last step, where all coefficients $|s_i|$, in absolute value, can assume at most only two (consecutive) values; the known parity will then determine $|s_i|$. It is easy to see that in order to achieve this we need $\lfloor \log_2 q \rfloor$ steps.

The strategy now is to find out whether $s_i \cdot s_j < 0$ or $s_i \cdot s_j > 0$ holds, for every i, j with $s_i, s_j \neq 0$. Let s_m be the first non-zero coefficient. This way, we will obtain two possible candidates of the secret key, one with $s_m > 0$ and the other with $s_m < 0$. A trivial query to the oracle decryption will allow us to determine which is the correct secret key.

We have to choose an appropriate "ciphertext" $c(x) = c_0 + c_1 x + \cdots + c_{n-1}x^{n-1}$ to submit to the decryption oracle. Choose $c_0 = 1, c_1 = 1$ and $c_j = 0$ for $j \neq 0, 1$. Oracle decryption will compute and return the polynomial

$$D(c(x)) = s(x) \cdot c(x) = [s_0 - s_{n-1}]_q \bmod 2 + \sum_{i=1}^{n-1} ([s_i + s_{i-1}]_q \bmod 2) x^i$$

Fix $i = 1, 2, \ldots, n-1$ such that $s_i, s_{i-1} \neq 0$. Let $b_i := [s_i + s_{i-1}]_q \bmod 2$ be the coefficient of x^i, and let $b_i' := [s_i' + s_{i-1}']_q \bmod 2$. There are two cases to consider:

1. $s_i' + s_{i-1}' \geq \frac{q+1}{2}$. Then
   - if $b_i = b_i'$, then s_i and s_{i-1} have the same sign;
   - if $b_i \neq b_i'$, then s_i and s_{i-1} have different signs.

2. $0 \le s'_i + s'_{i-1} \le \frac{q-1}{2}$. Then we need to make an extra query to understand whether s_i and s_{i-1} have the same sign or not.

Now, for each one of the i of the previous case (i.e. such that $0 \le s'_i + s'_{i-1} \le \frac{q-1}{2}$, $i = 1, 2, \dots, n-1$, and $s_i, s_{i-1} \neq 0$) we choose and submit to the decryption oracle the polynomial $c(x) = \alpha_i|s_{i-1}| + \alpha_i|s_i|x$, i.e. we choose $c_0 = \alpha_i|s_{i-1}|$, $c_1 = \alpha_i|s_i|$, $c_2 = c_3 = \dots = c_{n-1} = 0$, where α_i is chosen such that

$$\alpha_i|s_{i-1} \cdot s_i| \in \left(\frac{q-1}{4}, \frac{q-1}{2}\right] \tag{1}$$

(it is always possible to find such an α_i). The oracle decryption will return the polynomial $D(c(x)) = s(x) \cdot c(x)$, i.e.

$$[\alpha_i|s_{i-1}|s_0 - \alpha_i|s_i|s_{n-1}]_q \bmod 2 + \sum_{j=1}^{n-1} \left([\alpha_i|s_{i-1}|s_j + \alpha_i|s_i|s_{j-1}]_q \bmod 2\right) x^j$$

Let's focus on the coefficient of x^i, i.e. $\beta_i := [\alpha_i|s_{i-1}|s_i + \alpha_i|s_i|s_{i-1}]_q \bmod 2$. Now, there are two cases:

1. if s_i, s_{i-1} have different signs, then $\beta_i = 0$;
2. if s_i, s_{i-1} have the same sign, then $\beta_i = 1$ (trivial to verify: 1 holds, and therefore $[2\alpha_i \cdot |s_i \cdot s_{i-1}|]_q)$ is odd.

By repeating this idea for every $i = 1, 2, \dots, n-1$ such that $0 \le s'_i + s'_{i-1} \le \frac{q-1}{2}$ we will know which one of the following relations $s_i \cdot s_{i-1} < 0 \quad \vee \quad s_i \cdot s_{i-1} > 0$ holds, for every consecutive non-zero coefficients s_i, s_{i-1}.

Now, we have one more thing to consider: we have to be careful in case one of the coefficient s_i is zero. In this case in fact, no information can be given about the sign of s_{i-1} if we compare it to s_i. To solve this problem, we have to choose and submit to the decryption oracle a polynomial $c(x) = a + bx^j$ for appropriates a, b, j. Let $0 \le m_1 \le n-1$ be an integer such that s_{m_1} is the first non-zero coefficient of the secret key $s(x)$. If there exists $i_1 > m_1$ such that $s_{i_1} = 0$, then let m_2 be the first non-zero coefficient such that $i_1 < m_2 \le n-1$. Then we want to compare the relative signs of s_{m_1} and s_{m_2} by choosing the polynomial $c(x)$ with $c_0 = \alpha|s_{m_1}|$, $c_{m_2-m_1} = \alpha|s_{m_2}|$, $c_j = 0$ for $j \neq 0, m_2 - m_1$. So we have $c(x) = \alpha|s_{m_1}| + \alpha|s_{m_2}|x^{m_2-m_1}$, with α such that $\alpha|s_{m_1}s_{m_2}| \in \left(\frac{q-1}{4}, \frac{q-1}{2}\right]$. The oracle decryption will return the polynomial $D(c(x)) = s(x) \cdot c(x) = \beta_0 + \beta_1 x + \dots + \beta_{n-1}x^{n-1}$. Consider the m_2-th coefficient $\beta_{m_2} = [\alpha|s_{m_1}|s_{m_2} + \alpha|s_{m_2}|s_{m_1}]_q \bmod 2$. As before, we can conclude that if s_{m_1}, s_{m_2} have different signs, then $\beta_{m_2} = 0$, and if s_{m_1}, s_{m_2} have the same sign, then $\beta_{m_2} = 1$.

Now, similar to what just discussed, if there exists $i_2 > m_2$ such that $s_{i_2} = 0$, then let m_3 be the first non-zero coefficient such that $m_3 > i_2$. We will in a similar fashion compare the relative signs of s_{m_1} and s_{m_3}. We keep proceeding this way, and in the end we will know, for every $0 \le i, j \le n-1$ such that $s_i \neq 0, s_j \neq 0$, whether $s_i \cdot s_j > 0$ or $s_i \cdot s_j < 0$ occurs. This allows us to determine two possible candidates for the secret key $s(x)$ (assume s_m is the first non-zero coefficient; then one candidate has $s_m < 0$, the other has $s_m > 0$). A trivial oracle decryption query will reveal which one of the two is the correct secret key. The total number of decryption queries is then at most $\lfloor \log_2 q \rfloor + n$.

4 Attack Against the BLLN13 SHE Scheme

We start by recalling the BLLN13 SHE Scheme [BLLN13]. For a given positive integer $d \in \mathbb{N}_{>0}$, define the quotient ring $R := \mathbb{Z}[x]/(\Phi_d(x))$, i.e. the ring of polynomials with integer coefficients modulo the d-th cyclotomic polynomial $\Phi_d(x) \in \mathbb{Z}[x]$. The degree of Φ_d is $n = \varphi(d)$, where φ is Euler's totient function. As considered by the authors of [BLLN13], for correctness of the scheme, let d be a power of 2; in this case, we have $\Phi_d(x) = x^n + 1$ with n also a power of 2. Therefore $R = \mathbb{Z}[x]/(x^n + 1)$. The other parameters of the [BLLN13] SHE scheme are a prime integer $q \in \mathbb{N}$ and an integer $t \in \mathbb{N}$ such that $1 < t < q$. Let also $\chi_{\text{key}}, \chi_{\text{err}}$ be two distributions on R. The parameters $d, q, t, \chi_{\text{key}}$ and χ_{err} are public and we assume that given λ, there are polynomial-time algorithms that output d, q, t and $\phi(x)$, and sample from the error distributions χ. The message space is $\mathcal{M} = R/tR = \mathbb{Z}_t[x]/(x^n + 1)$, and all operations on ciphertexts are carried out in the ring $R_q := \mathbb{Z}_q[x]/(\phi(x))$.

KeyGen(λ) :

- sample $f', g \leftarrow \chi_{\text{key}}$
- let $f = [tf' + 1]_q$
- if f is not invertible in R_q, resample f'
- set pk $:= h = [tgf^{-1}]_q \in R_q$
- set sk $:= f \in R_q$

Encrypt(pk, m):

- for message $m + tR$, let $[m]_t$ be its representative
- sample $s, e \leftarrow \chi_{\text{err}}$
- output ciphertext $c = [\lfloor q/t \rfloor [m]_t + e + hs]_q \in R_q$

Decrypt(sk, c):

- output $m = \left[\left\lfloor \frac{t}{q} \cdot [fc]_q \right\rceil\right]_t \in R_t$

Since we don't need the evaluation step, we omit it in the description.

4.1 Attack Preview

We are going to recover the secret key $f(x) = f_0 + f_1 x + f_2 x^2 + \cdots + f_{n-1}x^{n-1} \in \frac{\mathbb{Z}_q[x]}{(x^n+1)}$, where f_i is an integer in $(-q/2, q/2]$ for all $i = 0, 1, \ldots, n-1$. In order to recover $f(x)$, we are going to submit specifically-chosen'ciphertexts' of the form $c(x) = c_0 + c_1 x + c_2 x^2 + \cdots + c_{n-1}x^{n-1} \in \frac{\mathbb{Z}_q[x]}{(x^n+1)}$, with integers $c_i \in (-q/2, q/2]$. Choose $c(x) = 1 = 1 + 0x + 0x^2 + \cdots + 0x^{n-1}$. We have

$$\begin{aligned} D(c = 1) &= \left[\left\lfloor \frac{t}{q} \cdot [f \cdot 1]_q \right\rceil\right]_t = \left[\left\lfloor \frac{t}{q} \cdot \left([f_0]_q + [f_1]_q x + \cdots + [f_{n-1}]_q x^{n-1}\right) \right\rceil\right]_t \\ &\stackrel{*}{=} \left[\left\lfloor \frac{t}{q} \cdot \left(f_0 + f_1 x + \cdots + f_{n-1}x^{n-1}\right) \right\rceil\right]_t \\ &= \left[\left\lfloor \frac{t}{q} f_0 \right\rceil + \left\lfloor \frac{t}{q} f_1 \right\rceil x + \cdots + \left\lfloor \frac{t}{q} f_{n-1} \right\rceil x^{n-1}\right]_t \end{aligned}$$

Equality $\stackrel{*}{=}$ holds since the integer coefficients f_i are already reduced modulo q. Now, for every $0 \leq i \leq n-1$ we have $-q/2 < f_i \leq q/2$. We have that $q > 2$ since in [BLLN13] it is claimed that $1 < t < q$, with t, q integers. In particular, q is a prime integer greater than 2, and therefore $q/2 \notin \mathbb{N}$. So we have $-q/2 < f_i < q/2$. In particular we have that $-\frac{t}{2} < \frac{t}{q} \cdot f_i < \frac{t}{2}$. For every $0 \leq i \leq n-1$, let $u_i^{(1)} := \left\lfloor \frac{t}{q} f_i \right\rceil$. We have $\lceil -\frac{t}{2} \rceil \leq u_i^{(1)} \leq \lfloor \frac{t}{2} \rfloor$. Each $u_i^{(1)}$ can have

$$\left\lfloor \frac{t}{2} \right\rfloor - \left\lceil -\frac{t}{2} \right\rceil + 1 = 2\left\lfloor \frac{t}{2} \right\rfloor + 1 = \begin{cases} t & \text{if } t \text{ is odd} \\ t+1 & \text{if } t \text{ is even} \end{cases}$$

possible different values, i.e. $u_i^{(1)}$ can have t different possible values if t is odd, and can have $t+1$ different possible values if t is even. Now, for every $0 \leq i \leq n-1$, we have that $[u_i^{(1)}]_t \in (-t/2, t/2]$ and therefore

- $[u_i^{(1)}]_t \in= [-\frac{t}{2} + \frac{1}{2}, -\frac{t}{2} + \frac{3}{2}, -\frac{t}{2} + \frac{5}{2}, \cdots, \frac{t}{2} - \frac{1}{2}] =: T_1$ if t is odd;
- $[u_i^{(1)}]_t \in \left[-\frac{t}{2} + 1, -\frac{t}{2} + 2, \ldots, \frac{t}{2}\right] =: T_2$ if t is even.

We have that $\#(T_1) = \#(T_2) = t$. Let $v_i^{(1)} := [u_i^{(1)}]_t$ for $0 \leq i \leq n-1$. It is clear that if $u_i^{(1)} = -t/2$, i.e. if $u_i^{(1)} = \lceil -t/2 \rceil$ and t is even, then $v_i^{(1)} = t/2$. We have

$$\begin{aligned} D(c(x) = 1) &= \left[u_0^{(1)} + u_1^{(1)} x + u_2^{(1)} x^2 + \cdots + u_{n-1}^{(1)} x^{n-1}\right]_t \\ &= [u_0^{(1)}]_t + [u_1^{(1)}]_t x + \cdots + [u_{n-1}^{(1)}]_t x^{n-1} \\ &= v_0^{(1)} + v_1^{(1)} x + v_2^{(1)} x^2 + \cdots + v_{n-1}^{(1)} x^{n-1} \end{aligned}$$

where $\forall i = 0, 1, \ldots, n-1$,

$$v_i^{(1)} = \begin{cases} \frac{t}{2} & \text{if } u_i^{(1)} = -\frac{t}{2} (\text{i.e. if } u_i^{(1)} = \lceil -\frac{t}{2} \rceil \text{ and } t \text{ is even}) \\ u_i^{(1)} & \text{otherwise} \end{cases}$$

In particular, if t is odd, then $D(c = 1) = u_0^{(1)} + u_1^{(1)} x + u_2^{(1)} x^2 + \cdots + u_{n-1}^{(1)} x^{n-1}$.
We have, $\forall 0 \leq i \leq n-1$,

$$\text{if } t \text{ is odd}, -\frac{t}{2} + \frac{1}{2} \leq v_i^{(1)} \leq \frac{t}{2} - \frac{1}{2}; \text{if } t \text{ is even}, -\frac{t}{2} + 1 \leq v_i^{(1)} \leq \frac{t}{2}$$

In both cases, $v_i^{(1)}$ *can only have* t *different values. As we saw before, in case of* t *odd we need to perform* $\lceil \log_2(q/t) \rceil + 1$ *oracle decryption queries; in case of* t *even, we need to perform extra oracle decryption queries (at most* $n-1$*) in order to understand which sign are given the coefficients of the secret key. Therefore, the total number of queries to the decryption oracle is at most* $\lceil \log_2(q/t) \rceil + n$. *If we use the actual bound* B *given on the coefficients* s_i *by the distribution* χ, *we have that the total number of queries to the decryption oracle is at most* $\lceil \log_2(B/t) \rceil + n$.

4.2 Detailed Attack in Three Cases

Case 1: t is odd

Step 1: Select $c(x) = 1$. Select "ciphertext" $c(x) = 1$ and submit it to the decryption oracle. Since t is odd and $v_i^{(1)} = u_i^{(1)}$, $\forall 0 \le i \le n-1$, we obtain the polynomial $D(c=1) = u_0^{(1)} + u_1^{(1)}x + u_2^{(1)}x^2 + \cdots + u_{n-1}^{(1)}x^{n-1}$, where $\lceil -\frac{t}{2} \rceil \le u_i^{(1)} \le \lfloor \frac{t}{2} \rfloor$. Every $u_i^{(1)}$ can have only t different values and can be written as $u_i^{(1)} = \lceil -\frac{t}{2} \rceil + k_{i,1}$, with $k_{i,1} \in \{0, 1, \ldots, t-1\}$. Now, it is easy to see that

$$u_i^{(1)} = \left\lceil -\frac{t}{2} \right\rceil + k_{i,1} \Leftrightarrow -\frac{q}{2} + \frac{q}{t}k_{i,1} < f_i < -\frac{q}{2} + \frac{q}{t}(k_{i,1}+1)$$

The polynomial obtained from the decryption oracle can therefore be written as

$$D(c(x)=1) = u_0^{(1)} + u_1^{(1)}x + u_2^{(1)}x^2 + \cdots + u_{n-1}^{(1)}x^{n-1} = \sum_{i=0}^{n-1} \left(\left\lceil -\frac{t}{2} \right\rceil + k_{i,1} \right) x^i$$

Each f_i belongs to the interval $(-q/2, q/2)$. But after this our first query we learn values $k_{i,1} \in [0, 1, \ldots, t-1]$, $0 \le i \le n-1$, such that

$$-\frac{q}{2} + \frac{q}{t}k_{i,1} < f_i < -\frac{q}{2} + \frac{q}{t}(k_{i,1}+1) \qquad \text{(F(0,1))}$$

We have $-\frac{q}{2} + \frac{q}{t}(k_{i+1}+1) - \left(-\frac{q}{2} + \frac{q}{t}k_{i+1}\right) = \frac{q}{t}$. Therefore, we know each integer coefficient f_i with an error up to $\frac{q}{t}$. The idea now is to keep submitting 'ciphertext' to the decryption oracle and obtain values $k_{i,j}$, with $0 \le i \le n-1$ and increasing integers $j = 1, 2, 3, \ldots$, in such a way that we keep reducing the interval in which f_i lies until we know f_i with an error smaller than 1, which determines each f_i completely.

Step 2: Select $c(x) = 2$. Select now "ciphertext" $c(x) = 2 = 2 + 0x + 0x^2 + \cdots + 0x^{n-1}$. Decryption oracle computes and return the polynomial

$$\begin{aligned} D(c=2) &= \left[\left\lfloor \frac{t}{q} \cdot [f \cdot 2]_q \right\rceil \right]_t = \left[\left\lfloor \frac{t}{q} \cdot \left([2f_0]_q + [2f_1]_q x + \cdots + [2f_{n-1}]_q x^{n-1} \right) \right\rceil \right]_t \\ &= \left[\left\lfloor \frac{t}{q} f_0^{(2)} \right\rceil + \left\lfloor \frac{t}{q} f_1^{(2)} \right\rceil x + \cdots + \left\lfloor \frac{t}{q} f_{n-1}^{(2)} \right\rceil x^{n-1} \right]_t \end{aligned}$$

where we have put $f_i^{(2)} := [2f_i]_q$, for every $0 \le i \le n-1$; of course we have $-\frac{q}{2} < f_i^{(2)} < \frac{q}{2}$. Now,

- if $-q/4 < f_i < q/4$, then $-\frac{q}{2} < 2f_i < \frac{q}{2}$ and therefore $f_i^{(2)} = [2f_i]_q = 2f_i$
- if $-q/2 < f_i < -q/4$, then $-q < 2f_i < -\frac{q}{2}$ and therefore $f_i^{(2)} = [2f_i]_q = 2f_i + q$
- if $q/4 < f_i < q/2$, then $\frac{q}{2} < 2f_i < q$ and therefore $f_i^{(2)} = [2f_i]_q = 2f_i - q$

So we have

$$f_i^{(2)} = [2f_i]_q = \begin{cases} 2f_i & \text{if } -\frac{q}{4} < f_i < \frac{q}{4} \\ 2f_i + q & \text{if } -\frac{q}{2} < f_i < -\frac{q}{4}, \text{ and in this case } 0 < f_i^{(2)} < \frac{q}{2} \\ 2f_i - q & \text{if } \frac{q}{4} < f_i < \frac{q}{2}, \text{ and in this case } -\frac{q}{2} < f_i^{(2)} < 0 \end{cases} \tag{2}$$

Let $u_i^{(2)} := \left\lfloor \frac{t}{q} \cdot f_i^{(2)} \right\rceil$. Then $D(c=2) = \left[u_0^{(2)} + u_1^{(2)} x + u_2^{(2)} x^2 + \cdots + u_{n-1}^{(2)} x^{n-1} \right]_t$. As before, $u_i^{(2)}$ can have only t different possible values, and can be written as $u_i^{(2)} = \lceil -\frac{t}{2} \rceil + k_{i,2}$, with $k_{i,2} \in \{0, 1, \ldots, t-1\}$, and also $u_i^{(2)} = \lceil -\frac{t}{2} \rceil + k_{i,2} \Leftrightarrow -\frac{q}{2} + \frac{q}{t} k_{i,2} < f_i < -\frac{q}{2} + \frac{q}{t}(k_{i,2} + 1)$. As before, since $-q/2 < f_i^{(2)} < q/2$ and t is odd, we have $\lceil -\frac{t}{2} \rceil \leq u_i^{(2)} \leq \lfloor \frac{t}{2} \rfloor$, and therefore we can simply write $D(c=2) = u_0^{(2)} + u_1^{(2)} x + u_2^{(2)} x^2 + \cdots + u_{n-1}^{(2)} x^{n-1} = \sum_{i=0}^{n-1} \left(\lceil -\frac{t}{2} \rceil + k_{i,2} \right) x^i$. So now, for each $0 \leq i \leq n-1$, we know $k_{i,1}, k_{i,2}$ such that

$$\begin{cases} -\frac{q}{2} + \frac{q}{t} k_{i,1} < f_i < -\frac{q}{2} + \frac{q}{t}(k_{i,1} + 1) \\ -\frac{q}{2} + \frac{q}{t} k_{i,2} < [2f_i]_q < -\frac{q}{2} + \frac{q}{t}(k_{i,2} + 1) \end{cases}$$

There are 3 cases to distinguish, where $3 = 2^2 - 1$.

$(1/3)_{[c=2]}$. If $-\frac{q}{2} + \frac{q}{t}(k_{i,1}+1) \leq -\frac{q}{4} \wedge -\frac{q}{2} + \frac{q}{t} k_{i,1} \geq -\frac{q}{2}$, which says that $0 \leq k_{i,1} \leq \lfloor \frac{t}{4} - 1 \rfloor$, then we are sure that $f_i \in (-\frac{q}{2}, -\frac{q}{4})$. Therefore, by condition (2), we expect $f_i^{(2)} = [2f_i]_q = 2f_i + q$. Therefore, $-\frac{3q}{4} + \frac{q}{2t} k_{i,2} < f_i < -\frac{3q}{4} + \frac{q}{2t}(k_{i,2} + 1)$

$(2/3)_{[c=2]}$. If $-\frac{q}{2} + \frac{q}{t}(k_{i,1} + 1) \leq \frac{q}{4} \wedge -\frac{q}{2} + \frac{q}{t} k_{i,1} \geq -\frac{q}{4}$, which says that $\lceil \frac{t}{4} \rceil \leq k_{i,1} \leq \lfloor \frac{3t}{4} - 1 \rfloor$, then we are sure that $f_i \in (-\frac{q}{4}, \frac{q}{4})$. Therefore, by condition (2), we expect $f_i^{(2)} = [2f_i]_q = 2f_i$. Therefore, $-\frac{q}{4} + \frac{q}{2t} k_{i,2} < f_i < -\frac{q}{4} + \frac{q}{2t}(k_{i,2} + 1)$

$(3/3)_{[c=2]}$. If $-\frac{q}{2} + \frac{q}{t}(k_{i,1} + 1) \leq \frac{q}{2} \wedge -\frac{q}{2} + \frac{q}{t} k_{i,1} \geq \frac{q}{4}$, which says that $\lceil \frac{3t}{4} \rceil \leq k_{i,1} \leq t-1$, then we are sure that $f_i \in (\frac{q}{4}, \frac{q}{2})$. Therefore, by condition (2), we expect $f_i^{(2)} = [2f_i]_q = 2f_i - q$. Therefore, $\frac{q}{4} + \frac{q}{2t} k_{i,2} < f_i < \frac{q}{4} + \frac{q}{2t}(k_{i,2} + 1)$

Now, we remark that there are values of $k_{i,1}$ for which is not clear to which of the previous cases we are falling in. For instance, if $k_{i,1}$ is such that $-\frac{q}{4} \in \left(-\frac{q}{2} + \frac{q}{t} k_{i,1}, -\frac{q}{2} + \frac{q}{t}(k_{i,1} + 1)\right)$, then we are not sure whether we are in Case $(1/3)_{[c=2]}$ or in Case $(2/3)_{[c=2]}$. This uncertainty happens if $\nexists k_{i,1} \in [0, 1, \ldots, t-1]$ such that $-\frac{q}{2} + \frac{q}{t} k_{i,1} = -\frac{q}{4}$, i.e. such that $k_{i,1} = t/4$. So, if $\nexists k_{i,1} \in [0, 1, \ldots, t-1]$ such that $k_{i,1} = t/4$, i.e. if $4 \nmid t$, then $-\frac{q}{4} \in \left(-\frac{q}{2} + \frac{q}{t} \lfloor \frac{t}{4} \rfloor, -\frac{q}{2} + \frac{q}{t}\left(\lfloor \frac{t}{4} \rfloor + 1\right)\right)$. So, if $k_{i,1} = \lfloor \frac{t}{4} \rfloor$, with $\frac{t}{4} \notin \mathbb{N}$, we have that $f_i \in \left(-\frac{q}{2} + \frac{q}{t} \lfloor \frac{t}{4} \rfloor, -\frac{q}{2} + \frac{q}{t}\left(\lfloor \frac{t}{4} \rfloor + 1\right)\right) =: I$. It is easy to see that

$$-\frac{q}{2} + \frac{q}{t}\left(\left\lfloor \frac{t}{4} \right\rfloor + 1\right) \leq 0, \forall 1 < t < q \tag{3}$$

There are two cases:

1/2: $f_i \in I_1 := I \cap (-q/2, -q/4)$. Then condition (2) implies that $f_i^{(2)} = [2f_i]_q \in (0, q/2)$

2/2: $f_i \in I_2 := I \cap (-q/4, 0)$. Then $f_i^{(2)} = [2f_i]_q \in (-q/2, 0)$

So, to sum up we have that if $k_{i,1} = \lfloor \frac{t}{4} \rfloor$, with $\frac{t}{4} \notin \mathbb{N}$, then

- if $f_i^{(2)} \in (0, q/2)$ then $f_i \in (-q/2, -q/4)$ and apply Case $(1/3)_{[c=2]}$
- if $f_i^{(2)} \in (-q/2, 0)$ then $f_i \in (-q/4, 0)$ and apply Case $(2/3)_{[c=2]}$

Now if $k_{i,1}$ is such that $\frac{q}{4} \in \left(-\frac{q}{2} + \frac{q}{t}k_{i,1}, -\frac{q}{2} + \frac{q}{t}(k_{i,1}+1)\right)$, then similarly to what we have just discussed we are not sure if we are in Case $(2/3)_{[c=2]}$ or in Case $(3/3)_{[c=2]}$ This uncertainty happens when $\nexists k_{i,1} \in [0, 1, \ldots, t-1]$ such that $-\frac{q}{2} + \frac{q}{t}k_{i,1} = \frac{q}{4}$, i.e. such that $k_{i,1} = 3t/4$. So, if $\nexists k_{i,1} \in [0, 1, \ldots, t-1]$ such that $k_{i,1} = 3t/4$, then $\frac{q}{4} \in \left(-\frac{q}{2} + \frac{q}{t}\lfloor \frac{3t}{4} \rfloor, -\frac{q}{2} + \frac{q}{t}\left(\lfloor \frac{3t}{4} \rfloor + 1\right)\right)$. So, if $k_{i,1} = \lfloor \frac{3t}{4} \rfloor$, with $\frac{3t}{4} \notin \mathbb{N}$, we have that $f_i \in \left(-\frac{q}{2} + \frac{q}{t}\lfloor \frac{3t}{4} \rfloor, -\frac{q}{2} + \frac{q}{t}\left(\lfloor \frac{3t}{4} \rfloor + 1\right)\right) =: I$. It is easy to see that

$$-\frac{q}{2} + \frac{q}{t}\left\lfloor \frac{3t}{4} \right\rfloor \geq 0, \forall t, q \tag{4}$$

There are two cases:

1/2: $f_i \in I_1 := I \cap (0, q/4)$. Then $f_i^{(2)} = [2f_i]_q \in (0, q/2)$

2/2: $f_i \in I_2 := I \cap (q/4, q/2)$. Then condition (2) implies that $f_i^{(2)} = [2f_i]_q \in (-q/2, 0)$

So, to sum up we have that if $k_{i,1} = \lfloor \frac{3t}{4} \rfloor$, with $\frac{3t}{4} \notin \mathbb{N}$, then

- if $f_i^{(2)} \in (0, q/2)$ then $f_i \in (-q/4, q/4)$ and apply Case $(2/3)_{[c=2]}$
- if $f_i^{(2)} \in (-q/2, 0)$ then $f_i \in (q/4, q/2)$ and apply Case $(3/3)_{[c=2]}$

We can write now all the 3 cases in a more complete way:

$(1/3)_{[c=2]}$. Suppose that

$$0 \leq k_{i,1} \leq \left\lfloor \frac{t}{4} - 1 \right\rfloor \vee \left(k_{i,1} = \left\lfloor \frac{t}{4} \right\rfloor, \text{with } \frac{t}{4} \notin \mathbb{N} \wedge f_i^{(2)} \in (0, q/2)\right) \tag{K(1,1)}$$

$$\text{Then } f_i \in \left(-\frac{q}{2}, -\frac{q}{4}\right), \quad -\frac{3q}{4} + \frac{q}{2t}k_{i,2} < f_i < -\frac{3q}{4} + \frac{q}{2t}(k_{i,2}+1) \tag{F(1,1)}$$

$(2/3)_{[c=2]}$. Suppose that

$$\begin{aligned} \left\lceil \frac{t}{4} \right\rceil \leq k_{i,1} \leq \left\lfloor \frac{3t}{4} - 1 \right\rfloor \vee \left(k_{i,1} = \left\lfloor \frac{t}{4} \right\rfloor \wedge f_i^{(2)} \in (-q/2, 0)\right) \vee \\ \vee \left(k_{i,1} = \left\lfloor \frac{3t}{4} \right\rfloor \wedge f_i^{(2)} \in (0, q/2)\right) \end{aligned} \tag{K(1,2)}$$

$$\text{Then } f_i \in \left(-\frac{q}{4}, \frac{q}{4}\right), \qquad -\frac{q}{4} + \frac{q}{2t}k_{i,2} < f_i < -\frac{q}{4} + \frac{q}{2t}(k_{i,2}+1) \tag{F(1,2)}$$

$(3/3)_{[c=2]}$. Suppose that

$$\left\lceil \frac{3t}{4} \right\rceil \leq k_{i,1} \leq t-1 \vee \left(k_{i,1} = \left\lfloor \frac{3t}{4} \right\rfloor \wedge f_i^{(2)} \in (-q/2, 0) \right) \qquad \text{(K(1,3))}$$

$$\text{Then } f_i \in \left(\frac{q}{4}, \frac{q}{2}\right), \qquad \frac{q}{4} + \frac{q}{2t} k_{i,2} < f_i < \frac{q}{4} + \frac{q}{2t}(k_{i,2}+1) \qquad \text{(F(1,3))}$$

In all cases, we end up by knowing f_i with an error up to $q/(2t)$.

Generalization and Complexity. At step 3, we select the "ciphertext" $c(x) = 4$; we omit the details of step 3 (for more details, see [CT15]). In general, at each step we keep submitting "ciphertexts" $c(x) := 2^h$, for increasing values $h = 0, 1, 2, \ldots$, i.e. at step $h+1$ we submit ciphertext $c(x) = 2^h$. Suppose we are at step $h+1$. Then we submit to the decryption oracle the 'ciphertext' $c(x) = 2^h$, and the decryption oracle will return us a polynomial

$$D(c = 2^h) = u_0^{(h+1)} + u_1^{(h+1)} x + \cdots + u_{n-1}^{(h+1)} x^{n-1} = \sum_{i=0}^{n-1} u_i^{(h+1)} x^i$$
$$= \sum_{i=0}^{n-1} \left(\left\lceil -\frac{t}{2} \right\rceil + k_{i,h+1} \right) \in R_t$$

from which we learn values $k_{i,h+1}$ for $1 \leq i \leq n-1$. So, at this point, we know $k_{i,j}$, for $0 \leq i \leq n-1$ and $1 \leq j \leq h+1$. These values allow us to distinguish between $m_h := 2^{h+1} - 1$ cases: for each $0 \leq i \leq n-1$, we know that integer f_i belongs to one of the cases:

$(a/2^{h+1} - 1)_{[c=2^h]}$. Suppose that

$$[\text{Condition } (\mathrm{C}(h,a,1)) \text{ holds}] \wedge [\text{Condition } (\mathrm{C}(h,a,2)) \text{ holds}] \qquad \text{(K(h,a))}$$

Then

$$f_i \in (x_{a,h}, y_{a,h}), \qquad \Delta_{h,a} + \frac{q}{2^h t} k_{i,h+1} < f_i < \Delta_{h,a} + \frac{q}{2^h t}(k_{i,h+1}+1) \quad \text{(F(h,a))}$$

where $a \in \{1, 2, \ldots, 2^{h+1} - 1\}$. Since

$$\Delta_{h,a} + \frac{q}{2^h t}(k_{i,h+1}+1) - \left(\Delta_{h,a} + \frac{q}{2^h t} k_{i,h+1} \right) = \frac{q}{2^h t},$$

this allows us to recover, for each $0 \leq i \leq n-1$, the integer f_i with an error up to $\frac{q}{2^h t}$. Therefore, we keep submitting'ciphertexts' $c(x) = 2^h$ for increasing values $h = 0, 1, 2, \ldots$ until h is such that $\frac{q}{2^h t} < 1$, i.e. $h \geq \lceil \log_2(q/t) \rceil$. So, we have to repeat our attack, submitting ciphertexts $c(x) = 1 = 2^0, 2^1, 2^2, 2^3, \ldots, 2^H$, where $H := \lceil \log_2(q/t) \rceil$. Se we repeat our attack $H+1$ times. Now, the secret key is $f(x) = f_0 + f_1 x + \cdots + f_{n-1} x^{n-1}$, where $f_i \in (-q/2, q/2]$, $\forall 0 \leq i \leq n-1$.

So f_i can have q different values. The decryption oracle reveals a polynomial $m(x) = m_0 + m_1 x + \cdots + m_{n-1} x^{n-1}$, where $m_i \in (-t/2, t/2], \forall 0 \leq i \leq n-1$. So m_i can have t different values. Each f_i can be described with at most $\lfloor \log_2(q-1) \rfloor + 1$ bits. So $f(x)$ can be described with $n \cdot (\lfloor \log_2(q-1) \rfloor + 1)$. Oracle decryption reveals $n \cdot (\lfloor \log_2(t-1) \rfloor + 1)$ bits. So the minimum number of oracle queries to determine $f(x)$ is given by $\frac{n \cdot (\lfloor \log_2(q-1) \rfloor + 1)}{n \cdot (\lfloor \log_2(t-1) \rfloor + 1)}$. In order to finish our attack for t odd, we need to give complete description of $\Delta_{h,a}$, Condition C$(h, a, 1)$ and Condition C$(h, a, 2)$, for each $0 \leq h \leq \lceil \log_2(q/t) \rceil = H$ and for each $1 \leq a \leq 2^{h+1} - 1$. Fix $0 \leq h \leq \lceil \log_2(q/t) \rceil$. For a given $1 \leq a \leq 2^{h+1} - 1$ put

$$\delta_{h,a} := \begin{cases} 2^{h-1} & \text{if } a = 2^h \\ \lfloor \frac{a}{2} \rfloor & \text{if } 1 \leq a < 2^h \\ \lceil \frac{a}{2} \rceil & \text{if } 2^h < a \leq 2^{h+1} - 1 \end{cases}, \quad \Delta_{h,a} := -\left(\frac{1}{2} + \frac{1}{2^{h+1}} - \frac{\delta_{h,a}}{2^h} \right) \cdot q$$

Also, put

$$\eta(h, a) := \begin{cases} \lceil \frac{a}{2} \rceil & \text{if } 1 \leq a \leq 2^h \\ \lfloor \frac{a}{2} \rfloor & \text{if } 2^h < a \leq 2^{h+1} - 1 \end{cases}$$

Then

$$\text{Condition } (\mathrm{C}(h, a, 1)) = \text{Condition } (\mathrm{K}(h-1, \eta(h, a)))$$

Remark that, if $h = 0$ or $h = 1$, then Condition (C(h,a,1)) $= \emptyset$ i.e., we don't put any condition at all, vacuous condition.

For Condition C$(h, a, 2)$, remark that if $h = 0$ then Condition $(\mathrm{C}(0, a, 2)) = \emptyset$ i.e., we don't put any condition at all, vacuous condition. One can see that, at step $h + 1$, condition C$(h, a, 2)$ is only one among the following 5:

1. $V_{3,h} := U_{2,1} = U_{1,1} \wedge (r \text{ is even}) = U_{3,1} \wedge (r \text{ is odd})$:

$$0 \leq k_{i,h} \leq \left\lfloor \frac{t}{4} - 1 \right\rfloor \vee \left(k_{i,h} = \left\lfloor \frac{t}{4} \right\rfloor \wedge f_i^{(h+1)} \in \left(0, \frac{q}{2}\right) \right) \tag{$V_{3,h}$}$$

2. $V_{5,h} := U_{2,2}$:

$$\begin{aligned} \left\lceil \frac{t}{4} \right\rceil \leq k_{i,h} \leq \left\lfloor \frac{3t}{4} - 1 \right\rfloor \vee \left(k_{i,h} = \left\lfloor \frac{t}{4} \right\rfloor \wedge f_i^{(h+1)} \in \left(-\frac{q}{2}, 0\right) \right) \vee \\ \vee \left(k_{i,h} = \left\lfloor \frac{3t}{4} \right\rfloor \wedge f_i^{(h+1)} \in \left(0, \frac{q}{2}\right) \right) \end{aligned} \tag{$V_{5,h}$}$$

3. $V_{2,h} := U_{2,3} = U_{1,2} \wedge (r \text{ is odd}) = U_{3,2} \wedge (r \text{ is even})$:

$$\left\lceil \frac{3t}{4} \right\rceil \leq k_{i,h} \leq t - 1 \vee \left(k_{i,h} = \left\lfloor \frac{3t}{4} \right\rfloor \wedge f_i^{(h+1)} \in \left(-\frac{q}{2}, 0\right) \right) \tag{$V_{2,h}$}$$

4. $V_{1,h} := U_{1,1} \wedge (r \text{ is odd}) = U_{3,1} \wedge (r \text{ is even})$:

$$\left\lceil \frac{t}{2} \right\rceil \leq k_{i,h} \leq \left\lfloor \frac{3t}{4} - 1 \right\rfloor \vee \left(k_{i,h} = \left\lfloor \frac{3t}{4} \right\rfloor \wedge f_i^{(h+1)} \in \left(0, \frac{q}{2}\right) \right) \qquad (V_{1,h})$$

5. $V_{0,h} := U_{1,2} \wedge (r \text{ is even}) = U_{3,2} \wedge (r \text{ is odd})$:

$$\left\lceil \frac{t}{4} \right\rceil \leq k_{i,h} \leq \left\lfloor \frac{t}{2} - 1 \right\rfloor \vee \left(k_{i,h} = \left\lfloor \frac{t}{4} \right\rfloor \wedge f_i^{(h+1)} \in \left(-\frac{q}{2}, 0\right) \right) \qquad (V_{0,h})$$

So, suppose we are in case $(a/2^{h+1} - 1)_{[c=2^h]}$. Then we see that we have Therefore, we have

$$C(h,a,2) = \begin{cases} V_{1,h} & \text{if } 1 \leq a \leq 2^h - 2 \wedge a \equiv 1 \bmod 4 \textbf{ or} \\ & \quad \textbf{or } 2^h + 2 \leq a \leq 2^{h+1} - 1 \wedge a \equiv 0 \bmod 4 \\ V_{2,h} & \text{if } 1 \leq a \leq 2^h - 2 \wedge a \equiv 2 \bmod 4 \textbf{ or} \\ & \quad \textbf{or } 2^h + 2 \leq a \leq 2^{h+1} - 1 \wedge a \equiv 1 \bmod 4 \\ & \quad \textbf{or } a = 2^h + 1 \\ V_{3,h} & \text{if } 1 \leq a \leq 2^h - 2 \wedge a \equiv 3 \bmod 4 \textbf{ or} \\ & \quad \textbf{or } 2^h + 2 \leq a \leq 2^{h+1} - 1 \wedge a \equiv 2 \bmod 4 \\ & \quad \textbf{or } a = 2^h - 1 \\ V_{0,h} & \text{if } 1 \leq a \leq 2^h - 2 \wedge a \equiv 0 \bmod 4 \textbf{ or} \\ & \quad \textbf{or } 2^h + 2 \leq a \leq 2^{h+1} - 1 \wedge a \equiv 3 \bmod 4 \\ V_{5,h} & \text{if } a = 2^h \end{cases}$$

Case 2: t is even but not 2

Step 1: Select $c(x) = 1$. Select "ciphertext" $c(x) = 1$ and submit it to the decryption oracle. We obtain the polynomial $D(c(x) = 1) = v_0^{(1)} + v_1^{(1)}x + v_2^{(1)}x^2 + \cdots + v_{n-1}^{(1)}x^{n-1}$. Suppose there exists $v_i^{(1)} = t/2$. This means that either $u_i^{(1)} = \frac{t}{2}$ or $u_i^{(1)} = -\frac{t}{2}$. We want to find out which one among the two above cases holds.

1. If we are in case $u_i^{(1)} = \frac{t}{2}$, then we have $\left\lfloor \frac{t}{q} f_i \right\rceil = \frac{t}{2} \Leftrightarrow \frac{q}{2} - \frac{q}{2t} < f_i < \frac{q}{2}$
2. If we are in case $u_i^{(1)} = -\frac{t}{2}$, then we have $\left\lfloor \frac{t}{q} f_i \right\rceil = -\frac{t}{2} \Leftrightarrow -\frac{q}{2} < f_i < -\frac{q}{2} + \frac{q}{2t}$

To find out which one is the case, we have to wait for the next step.

Now, let's focus on all the other $v_i^{(1)} \neq \frac{t}{2}$. We have in this case, $v_i^{(1)} = u_i^{(1)}$. Now, similarly as before, we have $-\frac{t}{2} + 1 \leq u_i^{(1)} \leq \frac{t}{2}$, and every $u_i^{(1)}$ can have only t different values; it can be written as $u_i^{(1)} = -\frac{t}{2} + 1 + k_{i,1}$, with $k_{i,1} \in \{0, 1, \ldots, t-1\}$. Now, it is easy to see that

$$u_i^{(1)} = -\frac{t}{2} + 1 + k_{i,1} \Leftrightarrow -\frac{q}{2} + \frac{q}{t}(k_{i,1} + \frac{1}{2}) < f_i < -\frac{q}{2} + \frac{q}{t}(k_{i,1} + \frac{3}{2})$$

The polynomial obtained from the decryption oracle can therefore be written as $D(c(x)=1)=\sum_{i=0}^{n-1}\left(-\frac{t}{2}+1+k_{i,1}\right)x^i$. Each f_i belongs to the interval $(-q/2,q/2)$. But after this our first query we learn values $k_{i,1}\in[0,1,\ldots,t-1]$, $0\le i\le n-1$, such that $-\frac{q}{2}+\frac{q}{t}(k_{i,1}+\frac{1}{2})<f_i<-\frac{q}{2}+\frac{q}{t}(k_{i,1}+\frac{3}{2})$. We have that $-\frac{q}{2}+\frac{q}{t}(k_{i+1}+3/2)-\left(-\frac{q}{2}+\frac{q}{t}(k_{i+1}+1/2)\right)=\frac{q}{t}$. Therefore, we know each integer coefficient f_i with an error up to $\frac{q}{t}$.

The idea now is to keep submitting 'ciphertext' to the decryption oracle and obtain values $k_{i,j}$, with $0\le i\le n-1$ and increasing integers $j=1,2,3,\ldots$, in such a way that we keep reducing the interval in which f_i lies until we know f_i with an error smaller than 1, which determines each f_i completely.

Step 2: Select $c(x)=2$. Select now "ciphertext" $c(x)=2=2+0x+\cdots+0x^{n-1}$. Decryption oracle computes and return the polynomial

$$D(c(x)=2)=\left[\left\lfloor\frac{t}{q}\cdot[f\cdot 2]_q\right\rceil\right]_t=\left[\left\lfloor\frac{t}{q}\cdot\left([2f_0]_q+[2f_1]_qx+\cdots+[2f_{n-1}]_qx^{n-1}\right)\right\rceil\right]_t$$

Now, let's focus on $\left[\left\lfloor\frac{t}{q}[2f_i]_q\right\rceil\right]_t x^i$ for each i such that, in the previous step, $v_i^{(1)}=\frac{t}{2}$.

1. We have $\frac{q}{2}-\frac{q}{2t}<f_i<\frac{q}{2}\Leftrightarrow q-\frac{q}{t}<2f_i<q\Leftrightarrow -\frac{q}{t}<[2f_i]_q<0$

$$\Leftrightarrow -1<\frac{t}{q}[2f_i]_q<0\Leftrightarrow -1\le\left[\left\lfloor\frac{t}{q}[2f_i]_q\right\rceil\right]_t\le 0$$

$$\Leftrightarrow\left[\left\lfloor\frac{t}{q}[2f_i]_q\right\rceil\right]_t=\begin{cases}0\text{ or }-1 & \text{if } t>2\\ 0\text{ or }1 & \text{if } t=2\end{cases}$$

2. We have analogously $-\frac{q}{2}<f_i<-\frac{q}{2}+\frac{q}{2t}\Leftrightarrow\left[\left\lfloor\frac{t}{q}[2f_i]_q\right\rceil\right]_t=0\text{ or }1.$

From now on we assume $t>2$; we will consider later the case in which $t=2$. Let $v_i^{(2)}=\left[\left\lfloor\frac{t}{q}[2f_i]_q\right\rceil\right]_t$. We have that

1. if $v_i^{(2)}=-1$, then $u_i^{(1)}=\frac{t}{2}$ and $\frac{q}{2}-\frac{q}{2t}<f_i<\frac{q}{2}$
2. if $v_i^{(2)}=1$, then $u_i^{(1)}=-\frac{t}{2}$ and $-\frac{q}{2}<f_i<-\frac{q}{2}+\frac{q}{2t}$
3. if $v_i^{(2)}=0$, then we can't conclude right now the exact interval in which f_i belongs; this will be considered in the next step.

Remark 2. Suppose we are in the above case 3, i.e. $v^{(2)}=\left\lfloor\frac{t}{q}[2f_i]_q\right\rceil=0$. Then

1. We have $\frac{q}{2}-\frac{q}{2t}<f_i<\frac{q}{2}\quad\wedge\quad\left\lfloor\frac{t}{q}[2f_i]_q\right\rceil=0\Leftrightarrow\frac{q}{2}-\frac{q}{4t}<f_i<\frac{q}{2}$
2. We have $-\frac{q}{2}<f_i<-\frac{q}{2}+\frac{q}{2t}\quad\wedge\quad\left\lfloor\frac{t}{q}[2f_i]_q\right\rceil=0\Leftrightarrow-\frac{q}{2}<f_i<-\frac{q}{2}+\frac{q}{4t}$

We will use this remark in the next step to investigate further the interval in which f_i lies. Now, let's focus on all of the other coefficients. Using the same arguments as in Sect. 4.2, the decryption oracle computes and return the polynomial

$$D(c(x)=2) = \left[\left\lfloor \frac{t}{q} \cdot [f \cdot 2]_q \right\rceil\right]_t = \left[\left\lfloor \frac{t}{q} f_0^{(2)} \right\rceil + \left\lfloor \frac{t}{q} f_1^{(2)} \right\rceil x + \cdots + \left\lfloor \frac{t}{q} f_{n-1}^{(2)} \right\rceil x^{n-1}\right]_t$$
$$= \left[u_0^{(2)} + u_1^{(2)} x + \cdots + u_{n-1}^{(2)} x^{n-1}\right]_t := v_0^{(2)} + v_1^{(2)} x + \cdots + v_{n-1}^{(2)} x^{n-1}$$

As before, suppose there exists $v_i^{(2)} = t/2$. This means that either $u_i^{(2)} = \frac{t}{2}$, or $u_i^{(2)} = -\frac{t}{2}$. We can easily understand which case we are by considering the known value $v_i^{(1)} \neq \frac{t}{2}$. All the other $v_i^{(2)}$ correspond to values $u_i^{(2)} \neq \frac{-t}{2}$. These $u_i^{(2)}$ can then have only t different possible values, and can be written as $u_i^{(2)} = -\frac{t}{2} + 1 + k_{i,2}$, with $k_{i,2} \in \{0, 1, \ldots, t-1\}$, and also

$$u_i^{(2)} = -\frac{t}{2} + 1 + k_{i,2} \Leftrightarrow -\frac{q}{2} + \frac{q}{t}(k_{i,2} + \frac{1}{2}) < f_i < -\frac{q}{2} + \frac{q}{t}(k_{i,2} + \frac{3}{2})$$

So now, for each $0 \leq i \leq n-1$ such that $v_i^{(1)} \neq \frac{t}{2} \vee (v_i^{(1)} = \frac{t}{2} \wedge v_i^{(2)} = 0)$, we know $k_{i,1}, k_{i,2}$ such that

$$\begin{cases} -\frac{q}{2} + \frac{q}{t}(k_{i,1} + \frac{1}{2}) < f_i < -\frac{q}{2} + \frac{q}{t}(k_{i,1} + \frac{3}{2}) \\ -\frac{q}{2} + \frac{q}{t}(k_{i,2} + \frac{1}{2}) < f_i < -\frac{q}{2} + \frac{q}{t}(k_{i,2} + \frac{3}{2}) \end{cases}$$

There are 3 cases to distinguish. These cases can be computed in an analogous way to what seen for the case t odd. We omit the details.

Generalization. We continue in this way, following the blueprint for t odd and taking care of all the coefficients for which $v_i^{(1)} = \frac{t}{2}$ and all subsequents $v_i^{(j)} = 0$ (when we finally find a $j \geq 2$ such that $v_i^{(j)} = 1$ or -1, then we can deduce the original value of $u_i^{(1)} = \frac{t}{2}$ or $-\frac{t}{2}$). If at the last step m we still get $v_i^{(m)} = 0$, then all the values $u_i^{(1)}$ remain undetermined, which also say that all the corresponding coefficients f_i can have only two possible values. At this point, the strategy is to submit to the decryption oracle 'ciphertexts' in order to determine whether $f_i \cdot f_j < 0$ or $f_i \cdot f_j > 0$ holds among all the non-zero coefficients f_i, f_j, in a way similar to what we have already discussed for the attack on the [LATV12] SHE scheme. We omit the details; we will give a description of how to do this in the case $t = 2$; the general case $t > 2$ is then easy to obtain.

Case 3: t = 2

Step 1: Select $c(x) = 1$**.** Choose and submit to the decryption oracle the polynomial $c(x) = 1$. It will compute and return the polynomial

$$D(c(x)=1) = \left[\left\lfloor \frac{2}{q} \cdot [f \cdot 1]_q \right\rceil\right]_2 = \left[\left\lfloor \frac{2}{q} f_0 \right\rceil + \left\lfloor \frac{2}{q} f_1 \right\rceil x + \cdots + \left\lfloor \frac{2}{q} f_{n-1} \right\rceil x^{n-1}\right]_2$$

For every $0 \leq i \leq n-1$, $u_i^{(1)} := \left\lfloor \frac{2}{q} f_i \right\rceil$ is such that $-1 \leq u_i^{(1)} \leq 1$, and so $v_i^{(1)} := [u_i^{(1)}]_2 = 0$ or 1. We have two cases to distinguish:

(1) $v_i^{(1)} = 0 \Leftrightarrow u_i^{(1)} = 0 \Leftrightarrow \left\lfloor \frac{2}{q} f_i \right\rceil = 0 \Leftrightarrow -\frac{1}{2} < \frac{2}{q} f_i < \frac{1}{2} \Leftrightarrow -\frac{q}{4} < f_i < \frac{q}{4}$

(2) $v_i^{(1)} = 1 \Leftrightarrow u_i^{(1)} = -1$ or $u_i^{(1)} = +1 \Leftrightarrow \left\lfloor \frac{2}{q} f_i \right\rceil = -1$ or $\left\lfloor \frac{2}{q} f_i \right\rceil = +1$

$$\Leftrightarrow -\frac{3}{2} < \frac{2}{q} f_i < -\frac{1}{2} \text{ or } \frac{1}{2} < \frac{2}{q} f_i < \frac{3}{2} \Leftrightarrow -\frac{q}{2} < f_i < -\frac{q}{4} \text{ or } \frac{q}{4} < f_i < \frac{q}{2}$$

Step 2: Select $c(x) = 2$. Choose and submit to the decryption oracle the polynomial $c(x) = 2$. It will return the polynomial $D(c(x) = 2) = \sum_{i=0}^{n-1} \left[\left\lfloor \frac{2}{q} [2f_i]_q \right\rceil \right]_2 x^i$ $=: \sum_{i=0}^{n-1} \left[u_i^{(2)} \right]_2 x^i =: \sum_{i=0}^{n-1} v_i^{(2)} x^i$. We have two cases to distinguish:

(1) $v_i^{(2)} = 0$. We have

$$v_i^{(2)} = 0 \Leftrightarrow u_i^{(2)} = 0 \Leftrightarrow \left\lfloor \frac{2}{q} [2f_i]_q \right\rceil = 0 \Leftrightarrow -\frac{1}{2} < \frac{2}{q} [2f_i]_q < \frac{1}{2} \Leftrightarrow -\frac{q}{4} < [2f_i]_q < \frac{q}{4}$$
$$\Leftrightarrow -\frac{q}{4} < 2f_i < \frac{q}{4} \text{ or } -\frac{5q}{4} < 2f_i < -\frac{3q}{4} \text{ or } \frac{3q}{4} < 2f_i < \frac{5q}{4}$$
$$\Leftrightarrow -\frac{q}{8} < f_i < \frac{q}{8} \text{ or } -\frac{q}{2} < f_i < -\frac{3q}{8} \text{ or } \frac{3q}{8} < f_i < \frac{q}{2}$$

We have three cases to distinguish, according to which known interval f_i lies at the end of step 1:

(1.1) If $-\frac{q}{4} < f_i < \frac{q}{4}$, then $-\frac{q}{8} < f_i < \frac{q}{8}$
(1.2) If $-\frac{q}{2} < f_i < -\frac{q}{4}$, then $-\frac{q}{2} < f_i < -\frac{3q}{8}$
(1.3) If $\frac{q}{4} < f_i < \frac{q}{2}$, then $\frac{3q}{8} < f_i < \frac{q}{2}$

(2) $v_i^{(2)} = 1$. We have

$$v_i^{(2)} = 1 \Leftrightarrow u_i^{(2)} = -1 \text{ or } u_i^{(2)} = +1 \Leftrightarrow \left\lfloor \frac{2}{q} [2f_i]_q \right\rceil = -1 \text{ or } \left\lfloor \frac{2}{q} [2f_i]_q \right\rceil = +1$$
$$\Leftrightarrow -\frac{3}{2} < \frac{2}{q} [2f_i]_q < -\frac{1}{2} \text{ or } \frac{1}{2} < \frac{2}{q} [2f_i]_q < \frac{3}{2}$$
$$\Leftrightarrow -\frac{3q}{4} < [2f_i]_q < -\frac{q}{4} \text{ or } \frac{q}{4} < [2f_i]_q < \frac{3q}{4}$$
$$\Leftrightarrow -\frac{3q}{4} < 2f_i < -\frac{q}{4} \text{ or } \frac{q}{4} < 2f_i < \frac{3q}{4}$$
$$\Leftrightarrow -\frac{3q}{8} < f_i < -\frac{q}{8} \text{ or } \frac{q}{8} < f_i < \frac{3q}{8}$$

Now, again we have three cases to distinguish, according to which known interval f_i lies at the end of step 1:

(2.1) If $-\frac{q}{4} < f_i < \frac{q}{4}$, then $-\frac{q}{4} < f_i < -\frac{q}{8}$ or $\frac{q}{8} < f_i < \frac{q}{4}$
(2.2) If $-\frac{q}{2} < f_i < -\frac{q}{4}$, then $-\frac{3q}{8} < f_i < -\frac{q}{4}$
(2.3) If $\frac{q}{4} < f_i < \frac{q}{2}$, then $\frac{q}{4} < f_i < \frac{3q}{8}$

Generalization and the Last Step. We continue in this way, and in the end we will know each coefficient f_i up to the sign. Therefore, we will know a polynomial $f'(x) = f'_0 + f'_1 x + \cdots + f'_{n-1}x^{n-1}$, with $f'_i = |f_i|$ for every i. We proceed similarly to what we have seen for the attack on the [LATV12] scheme, i.e. we query the decryption oracle in order to find out the relations $f_i \cdot f_j < 0$ or $f_i \cdot f_j > 0$ among the coefficients f_i of the secret key $f(x)$. Suppose that the two consecutive coefficients f_i, f_{i-1} are both non-zero. We know their absolute values f'_i, f'_{i-1}. Choose and submit to the decryption oracle the polynomial $c(x) = \alpha|f_{i-1}| + \alpha|f_i|x$, with $\alpha \in (-q/2, q/2]$ such that $[2\alpha|f_{i-1} \cdot f_i|]_q \in \left[\frac{q}{4}, \frac{q}{2}\right]$ (it is always possible to find such an α). Now, the decryption oracle will compute and return the polynomial

$$D(c(x)) = \left[\left\lfloor \frac{2}{q}[\alpha|f_{i-1}|f_0 - \alpha|f_i|f_{n-1}]_q \right\rceil\right]_2 + \sum_{j=1}^{n-1} \left[\left\lfloor \frac{2}{q}[\alpha|f_{i-1}|f_j + \alpha|f_i|f_{j-1}]_q \right\rceil\right]_2 x^j$$

Let's focus on the i-th coefficient $\left[\left\lfloor \frac{2}{q}[\alpha|f_{i-1}|f_i + \alpha|f_i|f_{i-1}]_q \right\rceil\right]_2$. We have two cases:

(1) If f_i, f_{i-1} have different signs, then $\alpha|f_{i-1}|f_i + \alpha|f_i|f_{i-1} = 0$, and therefore the i-th coefficient is zero $\left[\left\lfloor \frac{2}{q}[\alpha|f_{i-1}|f_i + \alpha|f_i|f_{i-1}]_q \right\rceil\right]_2 = 0$

(2) If f_i, f_{i-1} have the same positive sign, we then have $[\alpha|f_{i-1}|f_i + \alpha|f_i|f_{i-1}]_q = [2\alpha|f_i f_{i-1}|]_q \in \left[\frac{q}{4}, \frac{q}{2}\right]$. In case f_i, f_{i-1} are both negative, we have that $[\alpha|f_{i-1}|f_i + \alpha|f_i|f_{i-1}]_q = [-2\alpha|f_i f_{i-1}|]_q \in \left[-\frac{q}{2}, -\frac{q}{4}\right]$).
In both cases, it is easy to see that $\left[\left\lfloor \frac{2}{q}[\alpha|f_{i-1}|f_i + \alpha|f_i|f_{i-1}]_q \right\rceil\right]_2 = 1$.

So we can distinguish whether two consecutive non-zero coefficients f_i, f_{i-1} have the same sign or not. As we saw for the attack on the [LATV12] scheme, this leads us to two candidates for the secret key; to determine which one is the correct one, it is enough to submit an extra query to the decryption oracle.

Remark 3. As we saw for the attack on the [LATV12] scheme, we have to be careful in case one of the coefficient f_i is zero. In this case in fact, no information can be given about the sign of f_{i-1} if we compare it to f_i. To solve this issue, we choose and submit to the decryption oracle a polynomial in the form $c(x) = a + bx^j$, for appropriates a, b, j. We omit the details, which are straightforward from what we have just discussed and from the attack on the [LATV12] scheme.

5 Conclusion

In this paper, we have described efficient key recovery attacks against the SHE schemes from [LATV12,BLLN13]. At this moment, it is still not clear whether we can adapt our attack to the scenario (3) of the LTV12 scheme, as noted in Remark 1 in the beginning of Sect. 3. This is an interesting future work. Up to today, the only known IND-CCA1 SHE scheme is that of Loftus et al. [LMSV12]. It is a wide open problem to design more efficient IND-CCA1 secure SHE schemes, possibly based on standard assumptions such as LWE.

Acknowledgements. Massimo Chenal is supported by an AFR Ph.D. grant from the National Research Fund, Luxembourg. Qiang Tang is partially supported by a CORE (junior track) grant from the National Research Fund, Luxembourg. We thank the IACR ePrint editors for pointing out references for three papers on key recovery attack against NTRUEncrypt and Steven Galbraith for valuable discussions.

References

[BGV12] Brakerski, Z., Gentry, C., Vaikuntanathan, V.: (Leveled) fully homomorphic encryption without bootstrapping. In: ITCS 2012, pp. 309–325. ACM (2012)

[BLLN13] Bos, J.W., Lauter, K., Loftus, J., Naehrig, M.: Improved security for a ring-based fully homomorphic encryption scheme. In: Stam, M. (ed.) IMACC 2013. LNCS, vol. 8308, pp. 45–64. Springer, Heidelberg (2013)

[Bra12] Brakerski, Z.: Fully homomorphic encryption without modulus switching from classical GapSVP. In: Safavi-Naini, R., Canetti, R. (eds.) CRYPTO 2012. LNCS, vol. 7417, pp. 868–886. Springer, Heidelberg (2012)

[BV11a] Brakerski, Z., Vaikuntanathan, V.: Fully homomorphic encryption from Ring-LWE and security for key dependent messages. In: Rogaway, P. (ed.) CRYPTO 2011. LNCS, vol. 6841, pp. 505–524. Springer, Heidelberg (2011)

[BV11b] Brakerski, Z., Vaikuntanathan, V.: Efficient fully homomorphic encryption from (standard) LWE. In: FOCS 2011 (2011)

[CT14] Chenal, M., Tang, Q.: On key recovery attacks against existing somewhat homomorphic encryption schemes. In: Aranha, D.F., Menezes, A. (eds.) LATINCRYPT 2014. LNCS, vol. 8895, pp. 239–258. Springer, Heidelberg (2015)

[CT15] Chenal, M., Tang, Q.: Key recovery attack against an ntru-type somewhat homomorphic encryption scheme (2015). http://eprint.iacr.org/2015/83

[DGM15] Dahab, R., Galbraith, S., Morais, E.: Adaptive key recovery attacks on NTRU-based somewhat homomorphic encryption schemes. In: Lehmann, A., Wolf, S. (eds.) Information Theoretic Security. LNCS, vol. 9063, pp. 283–296. Springer, Heidelberg (2015)

[Gen09] Gentry, C.: Fully homomorphic encryption using ideal lattices. In: STOC 2009, pp. 169–178. ACM (2009)

[GH11] Gentry, C., Halevi, S.: Implementing gentry's fully-homomorphic encryption scheme. In: Paterson, K.G. (ed.) EUROCRYPT 2011. LNCS, vol. 6632, pp. 129–148. Springer, Heidelberg (2011)

[GSW13] Gentry, C., Sahai, A., Waters, B.: Homomorphic encryption from learning with errors: conceptually-simpler, asymptotically-faster, attribute-based. In: Canetti, R., Garay, J.A. (eds.) CRYPTO 2013, Part I. LNCS, vol. 8042, pp. 75–92. Springer, Heidelberg (2013)

[HPS98] Hoffstein, J., Pipher, J., Silverman, J.H.: NTRU: a ring-based public key cryptosystem. In: Buhler, J.P. (ed.) ANTS 1998. LNCS, vol. 1423, pp. 267–288. Springer, Heidelberg (1998)

[LATV12] López-Alt, A., Tromer, E., Vaikuntanathan, V.: On-the-fly multiparty computation on the cloud via multikey fully homomorphic encryption. In: STOC 2012, pp. 1219–1234. ACM (2012)

[LMSV12] Loftus, J., May, A., Smart, N.P., Vercauteren, F.: On CCA-secure somewhat homomorphic encryption. In: Miri, A., Vaudenay, S. (eds.) SAC 2011. LNCS, vol. 7118, pp. 55–72. Springer, Heidelberg (2012)

[SS10] Stehlé, D., Steinfeld, R.: Faster fully homomorphic encryption. In: Abe, M. (ed.) ASIACRYPT 2010. LNCS, vol. 6477, pp. 377–394. Springer, Heidelberg (2010)

[vDGHV10] van Dijk, M., Gentry, C., Halevi, S., Vaikuntanathan, V.: Fully homomorphic encryption over the integers. In: Gilbert, H. (ed.) EUROCRYPT 2010. LNCS, vol. 6110, pp. 24–43. Springer, Heidelberg (2010)

[ZPS12] Zhang, Z., Plantard, T., Susilo, W.: On the CCA-1 security of somewhat homomorphic encryption over the integers. In: Ryan, M.D., Smyth, B., Wang, G. (eds.) ISPEC 2012. LNCS, vol. 7232, pp. 353–368. Springer, Heidelberg (2012)

PUFs and Implementation Security

Bit Error Probability Evaluation of RO PUFs

Qinglong Zhang[1,2,3], Zongbin Liu[1,2], Cunqing Ma[1,2(✉)], and Jiwu Jing[1,2]

[1] Data Assurance and Communication Security Research Center, Beijing, China
{qlzhang,zbliu,cqma,jing}@is.ac.cn
[2] State Key Laboratory of Information Security,
Institute of Information Engineering, CAS, Beijing, China
[3] University of Chinese Academy of Sciences, Beijing, China

Abstract. Physically unclonable functions (PUFs) are crucial to the implementations of secure key protection and authentication protocol. Ring oscillator PUF (RO PUF) is popular for its nice properties of neat structure. The power of an accurate model to describe the characteristics of a physical system is beyond doubt. However, there are few publications to quantitatively analyze the source of RO PUF's bit error probability and give out a calculation model for the bit error probability. In this paper, based on the characteristics of RO and the pairwise comparison of different oscillations, we quantitatively describe the factors to affect RO PUF's bit error rate, including the process variation, sampling interval and temperature. Experiments are conducted to demonstrate the validation of our calculating model. Our work allows the studying of RO PUF's bit error probability in full detail, and strengthen the evaluation scheme of RO PUF. What's more, it is an important tool for designers to construct more efficient RO PUF-based systems.

Keywords: Physically unclonable functions · Ring oscillators · Bit error probability · Evaluation model

1 Introduction

In recent years, researchers have given the high degree of attention on PUF technology. Utilizing the intrinsic process variation inside the electronic circuits, PUF is steadily solving problems of secure key storage, secure boot and so on [1]. Since PUFs don't operate on their own but are usually embedded into systems, it is invaluable to design PUF architectures which can meet the high reliability and security constraints imposed by such systems. Generally, the fundamental physical security of PUF-based system is derived from the PUF implementation and post-processing is involved to make the PUF implementation suitable for some applications, e.g., key storage. The essential characteristic of PUF implementation is the probabilistic behavior of the PUF itself since the private keys

Cunqing Ma is the contact author of this paper. The work is supported by a grant from the National Natural Science Foundation of China (No.61402470).

J. Lopez and C.J. Mitchell (Eds.): ISC 2015, LNCS 9290, pp. 421–436, 2015.
DOI: 10.1007/978-3-319-23318-5_23

cannot tolerate any bit error. However, PUF's original response almost may contain bit errors. Therefore, fuzzy extractors [2] is necessary for almost all PUF implementations, but more complex the fuzzy extractor is, more resources it consumes. As a result, constructing a PUF system becomes an intricate work because it should make a trade-off between the reliability, resource consumption and efficiency.

There are many types of PUFs, like SRAM PUF [3,4], Glitch PUF [5], Arbiter PUF [6], RO PUF [7–10] and so on. SRAM PUF takes advantage of SRAM cell's uninitialized value after power on, but traditionally in the commercial FPGA products, SRAM cells are forcibly reset to a known value. Therefore, SRAM PUF is not usually available on FPGA platform. Glitch PUF exploits glitches that behave non-linearly from delay variation between gates and the characteristic of pulse propagation of each gate. For its complex architecture, Glitch PUF is not easy to be implemented. Arbiter PUF utilizes the variation of two symmetrical signal propagation path, but on FPGA platform, it is difficult to place two symmetrical paths and it is also not easy to implement Arbiter PUF on FPGAs. The construction of RO PUF is simple and it is easy to be implemented on both ASIC and FPGA platforms, so the researches on the characteristics evaluation of RO PUF is significant as the guide to design RO PUF-based systems.

Up to now, since bit error rate is one of the most essential characteristics, there are many researches [4,11,12] on the probabilistic behavior. Some researches are focused on the design of the optimized error correction module. In CHES 2008, Bösch et al. [11] propose an efficient helper data key extractor on FPGAs and in CHES 2009, Maes et al. [12] propose a soft decision helper data algorithm for error correction. If the error correction module is powerful enough, all the response's errors can be corrected. However, in the practical design of PUF-based systems, we should make balancing typically opposing goals between reliability and design's complexity. What's more, there are also researches on the PUF architectures which are error-free. In J.Cryptol.2011, Maiti et al. [8] describe a configurable RO PUF which can generate responses nearly without errors over varying environmental conditions. The configurable RO PUF consumes extra resource and the selection of the most stable RO pairs also consumes extra resource. If we can master the bit error probability in advance, it can be the guide to construct the error correction module or error-free construction with high efficiency and appropriate resource consumption.

In order to analyze the source of RO PUF's bit error probability, we propose an error bit analysis model which is based on the RO's classical oscillation model. Considering three main factors, process variation, sampling interval and environmental conditions, to affect the RO's bit error probability, we quantitatively analyze their effects on the bit error probability. From the results of our model calculation and experiments, it achieves high consistency to demonstrate the validity of our model to describe the probabilistic behavior of RO PUF. In summary, in this paper, we make the following contributions.

- According to the basic RO's characteristics, we describe our bit error calculation model which can quantitatively calculate the bit error probability with basic oscillation parameters.
- We conduct experiments to show the effects of sampling interval and environmental conditions on the bit error probability. The results demonstrate the validity of our analysis model.
- Our work contributes to the evaluation scheme of RO PUF and can help designers efficiently construct RO PUF with an acceptant bit error rate.

Structure. In Sect. 2, we briefly describe the concept of RO PUF and its basic notations. Section 3 presents our proposed bit error analysis model based on RO's classical oscillation mode and show the simulation design for our new model. In Sect. 4, we conduct experiments to demonstrate the validity of our model. Section 5 gives a simple evaluation of the sampling interval and temperature on the bit error probability. In Sect. 6, we conclude this paper.

2 Preliminaries

A ring oscillator consists of odd number of inverters. The frequency of this ring depends on the propagation delay of all the inverters. During the period of manufacture, slight difference of an inverter's propagation delay appears among different rings. The slight difference is called process variation and it can not be avoidable.

2.1 Ring Oscillator PUF

Ring oscillator architecture is a typical method to construct PUFs. The ring oscillator PUF is first proposed by Suh and Devadas [7], and an RO PUF is composed of n identical ROs, RO_1 to RO_n, with frequencies, f_1 to f_n, respectively. Generally, RO PUF also contains two counters and two n-to-1 multiplexers that control which ROs are currently applied to both counters. Due to process variation, the frequencies of these two selected ROs, f_i and f_j, tend to be different from each other. One bit response r_{ij} can be extracted from two ring oscillators by using a simple comparison of their frequencies as follows.

$$r_{ij} = \begin{cases} 1 & if \;\; f_i > f_j, \\ 0 & otherwise. \end{cases} \tag{1}$$

Since all the process variation and other noises have influence on the frequencies, the resulting comparison bit will be random and device-specific. The above comparison operation is a basic form of compensated measurement, which is proposed by Gassend et al. [1]. The compensated measurements based on the ratio of two frequencies is particularly effective because the environmental changes and systematic noises can affect both frequencies of two ring oscillators simultaneously.

2.2 The Evaluation Scheme of RO PUF

In brief, the evaluation scheme of RO PUF has three parts: reliability, uniqueness and security [8].

- Reliability shows the ability to generate stable responses as the environmental variables changing.
- Uniqueness evaluates how uniquely to distinguish different PUF instances based on the generated responses.
- Security indicates the ability to prevent an adversary from stealing the PUF's secrets.

In this paper, we focus on the reliability because the bit error probability is the same with reliability. Reliability is measured by intra-distance. For a particular challenge, the intra-distance between two evaluations on one PUF instantiation is the hamming distance between the responses resulting from applying this challenge twice to the PUF. Due to the environmental changes and systematic noises, the reproducing probability of a PUF response can not be 100 %. To evaluate the bit error probability, we get one n-bit response $m+1$ times at some environmental condition and select one response as the reference response R_0 and the other responses as R_j $(1 \leq j \leq m)$. The average intra-distance μ_{intra} can be calculated as follows.

$$\mu_{intra} = \frac{1}{m}\sum_{j=1}^{m}\frac{HD(R_0, R_j)}{n} \times 100\,\% \tag{2}$$

3 Analysis of the Model on Bit Error Probability

In order to simplify the scenario for our analysis, we assume that except Gaussian noise there are no other disturbance signals to affect RO's oscillations. Therefore, there are two random variables in our model, one is process variation and the other is Gaussian noise.

3.1 Notations

On the observation of previous researches on ROs, there are some important notations in oscillator-based architecture, like TRNGs [13–15]. The period of one ring oscillation between two rising edges is X_k, and X_k is affected by two parts, intrinsic manufacturing factor and Gaussian noise. In CHES 2014, Ma et al. [14] give one assumption that X_k is i.i.d. The mean and variance of X_k is denoted as μ and σ^2. Amaki et al. [15] point out that for an oscillator jitter characteristic of gates is important factor and define variance constant r as the variance of X_k divided by the mean of X_k. Due to this definition, the variance constant of an oscillator composed of n gates with r variance constants is conveniently equal to r and we describe the variance constant r as follows.

$$r = var(X_k)/mean(X_k) = \sigma^2/\mu \tag{3}$$

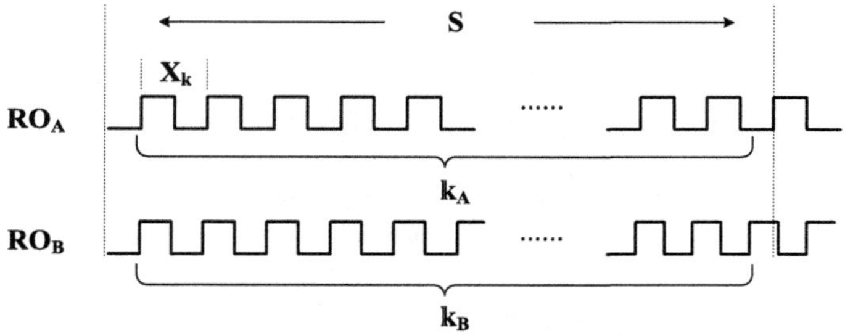

Fig. 1. The oscillators in sampling interval S

From the results of [16] in CHES 2008 on the probability calculation of sampling bits, there are some basic preliminary of our work.

As shown in Fig. 1, in an equal sampling interval S, RO_A has k_a periods and RO_B has k_b periods. The relationship between the number of period and the sampling interval is described as follows.

$$T_{k_a} = X_{A,1} + X_{A,2} + \cdots + X_{A,k_a} < S \tag{4}$$

$$T_{k_a+1} = X_{A,1} + X_{A,2} + \cdots + X_{A,k_a+1} > S \tag{5}$$

$$T_{k_b} = X_{B,1} + X_{B,2} + \cdots + X_{B,k_b} < S \tag{6}$$

$$T_{k_b+1} = X_{B,1} + X_{B,2} + \cdots + X_{B,k_b+1} > S \tag{7}$$

From formula (4)~(7), let $N_i = max\{ k \mid T_k < S \}$ and N_i is the number of periods in sampling interval S. The probability $Prob(N_i = k)$ is calculated as shown in formula (8).

$$Prob(N_i = k) = Prob(T_k \leq S) - Prob(T_{k+1} \leq S) \tag{8}$$

The distribution of T_k is derived from the central-limit theorem (CLT), so it can be deduced as follows.

$$Prob(\frac{T_k - k\mu}{\sigma\sqrt{k}} \leq x) \to \Phi(x), k \to \infty. \tag{9}$$

where $\Phi(x)$ denotes the cumulative distribution function of the standard normal distribution $N(0,1)$. Based on formulas (8) and (9), calculate $Prob(N_i = k)$ as follows.

$$\begin{aligned} &Prob(N_i = k) \\ &= Prob(T_k \leq S) - Prob(T_{k+1} \leq S) \\ &\approx \Phi((v-k) \cdot \frac{\mu}{\sigma\sqrt{k}}) - \Phi((v-k-1) \cdot \frac{\mu}{\sigma\sqrt{k+1}}) \end{aligned} \tag{10}$$

In formula (10), $v = S/\mu$ is denoted as the frequency ratio.

3.2 Analysis on RO PUF

A one-bit response of RO PUF is derived from the comparison of pairwise ROs. In a sampling interval, record two oscillation counting values and get the one-bit response by comparing these two counting values. However, the existence of noise may result in the difference of the one bit response in multiple response extractions. For example, one bit response has preference to be zero, but sometimes the measured response changes to be zero. This unexpected bit error is not friendly to extract private information because if there is no error correction module, bit errors from extraction lead to the uncertainty of private information. Based on the assumptions and notations proposed in previous section, we analyze the bit error probability of RO PUF from the ROs' jitter characteristics.

As shown in Fig. 1, the mean of X_{RO_A} is μ_A and that of X_{RO_B} is μ_B. Here we regard the difference of μ_A and μ_B as the process variation between RO_A and RO_B. If there are no noise influences, the one-bit response from RO_A and RO_B is only determined by the comparison of μ_A and μ_B, and the one-bit response keeps stable forever. However, in practice the noise influences exist. Moreover, if the difference value between μ_A and μ_B is not enough large, the cumulative influence of noise may be larger than that of process variation. The basic model for our analysis is described as follows.

$$d_{Loop} = d_{AVG} + d_{RAND_P} + d_{NOISE} \tag{11}$$

where d_{AVG} is the nominal delay that is the same for all identical ROs. d_{RAND_P} is the delay variation due to the random process variation. d_{NOISE} is the delay variation because of the noise influence.

Based on formula (11), the bit error probability $Prob_{error}$ is that the sign of the difference value between μ_a and μ_B is the same to the sign of the difference value between k_a and k_b.

$$Prob_{error} = Prob(k_a > k_b \mid \mu_a > \mu_b) \tag{12}$$

From formula (10), we can get both the probability of $Prob(N_A = k_a)$ and $Prob(N_B = k_b)$. Then we can calculate the probability $Prob(k_a > k_b)$ as follows.

$$Prob(k_a > k_b) = \sum_{i=0}^{+\infty}\{Prob(k_a = i) \cdot [\sum_{j=0}^{i-1} Prob(k_b = j)]\} \tag{13}$$

Let $Prob(k_b = 0) = 0$, and the calculation of $\sum_{j=0}^{i-1} Prob(k_b = j)$ can be deduced as follows.

$$\sum_{j=0}^{i-1} Prob(k_b = j)$$

$$\approx \{\sum_{j=0}^{i-1} \Phi((v_b - k_b) \cdot \frac{\mu_b}{\sigma_b\sqrt{k_b}}) - \Phi((v_b - k_b - 1) \cdot \frac{\mu_b}{\sigma_b\sqrt{k_b + 1}})\}$$

$$= \Phi((v_b - (i-1)) \cdot \frac{\mu_b}{\sigma_b \sqrt{i-1}}) - \Phi((v_b - i) \cdot \frac{\mu_b}{\sigma_b \sqrt{i}}) + \Phi((v_b - (i-2)) \cdot \frac{\mu_b}{\sigma_b \sqrt{i-1}})$$

$$-\Phi((v_b - (i-1)) \cdot \frac{\mu_b}{\sigma_b \sqrt{i-1}}) + \Phi((v_b - (i-3)) \cdot \frac{\mu_b}{\sigma_b \sqrt{i-3}}) - \Phi((v_b - (i-2)) \cdot \frac{\mu_b}{\sigma_b \sqrt{i-2}})$$

$$+ \cdots\cdots + \Phi((v_b - (1)) \cdot \frac{\mu_b}{\sigma_b \sqrt{1}}) - \Phi((v_b - (2)) \cdot \frac{\mu_b}{\sigma_b \sqrt{2}}) + Prob(k_b = 0)$$

$$= \Phi((v_b - 1) \cdot \frac{\mu_b}{\sigma_b \sqrt{1}}) - \Phi((v_b - i) \cdot \frac{\mu_b}{\sigma_b \sqrt{i}}) \tag{14}$$

Based on formulas (13) and (14), $Prob(k_a > k_b)$ is deduced that

$$Prob(k_a > k_b)$$

$$= \sum_{i=0}^{+\infty} \{[\Phi((v_a - (i-1)) \cdot \frac{\mu_a}{\sigma_a \sqrt{i-1}}) - \Phi((v_a - i) \cdot \frac{\mu_a}{\sigma_a \sqrt{i}})] \cdot [\sum_{j=0}^{i-1} Prob(k_b = j)]\}$$

$$= \sum_{i=0}^{+\infty} \{[\Phi((v_a - (i-1)) \cdot \frac{\mu_a}{\sigma_a \sqrt{i-1}}) - \Phi((v_a - i) \cdot \frac{\mu_a}{\sigma_a \sqrt{i}})] \cdot$$
$$[\Phi((v_b - 1) \cdot \frac{\mu_b}{\sigma_b \sqrt{1}}) - \Phi((v_b - i) \cdot \frac{\mu_b}{\sigma_b \sqrt{i}})]\} \tag{15}$$

According to formula (15), the probability $Prob(k_a > k_b)$ is involved with the process variation parameter μ_i, the sampling interval S and the variance constant r. As described in [15], the variance constant of an oscillator composed of n gates with r variance constants is conveniently equal to r, namely, in RO PUF, the variance constants of different n-stage ROs are the same under the same environmental conditions. Therefore, transform formula (15) to another form as (16).

$$Prob(k_a > k_b)$$

$$= \sum_{i=0}^{+\infty} \{[\Phi((S/\mu_a - (i-1)) \cdot \frac{\sqrt{\mu_a}}{\sqrt{r}\sqrt{i-1}}) - \Phi((S/\mu_a - i) \cdot \frac{\sqrt{\mu_a}}{\sqrt{r}\sqrt{i}})] \cdot$$
$$[\Phi((S/\mu_b - 1) \cdot \frac{\sqrt{\mu_b}}{\sqrt{r}\sqrt{1}}) - \Phi((S/\mu_b - i) \cdot \frac{\sqrt{\mu_b}}{\sqrt{r}\sqrt{i}})]\} \tag{16}$$

According to formula (16), we can utilize the contingent probabilities $Prob(k_a > k_b \mid \mu_a > \mu_b)$ or $Prob(k_a < k_b \mid \mu_a < \mu_b)$ to estimate the bit error probability.

First, we assume that the process variation distribution is a normal distribution with the probability density function

$$f_{PV}(x) = \frac{1}{\sqrt{2\pi}\sigma_{PV}} exp(-\frac{(x-\mu_{PV})^2}{2\sigma_{PV}^2}),$$

and for m ROs, the mean of the RO_i period is μ_i, $1 \leq i \leq m$.

$$(\mu_1, \mu_2, \cdots\cdots, \mu_{m-1}, \mu_m) \sim N(\mu_{PV}, \sigma_{PV}^2) \tag{17}$$

Then from formulas (16) and (17), we can achieve that

$$Prob(k_a > k_b \mid \mu_a > \mu_b)$$

$$= \int_0^{+\infty} \int_0^{x} \sum_{i=0}^{+\infty} \{[\Phi((S/x-(i-1)) \cdot \frac{\sqrt{x}}{\sqrt{r}\sqrt{i-1}}) - \Phi((S/x-i) \cdot \frac{\sqrt{x}}{\sqrt{r}\sqrt{i}})] \cdot$$

$$[\Phi((S/y-1) \cdot \frac{\sqrt{y}}{\sqrt{r}\sqrt{1}}) - \Phi((S/y-i) \cdot \frac{\sqrt{y}}{\sqrt{r}\sqrt{i}})]\} \cdot f_{PV}(x) \cdot f_{PV}(y) \cdot dy \cdot dx$$

$$= \sum_{i=0}^{+\infty} \int_0^{+\infty} \int_0^{x} \{[\Phi((S/x-(i-1)) \cdot \frac{\sqrt{x}}{\sqrt{r}\sqrt{i-1}}) - \Phi((S/x-i) \cdot \frac{\sqrt{x}}{\sqrt{r}\sqrt{i}})] \cdot$$

$$[\Phi((S/y-1) \cdot \frac{\sqrt{y}}{\sqrt{r}\sqrt{1}}) - \Phi((S/y-i) \cdot \frac{\sqrt{y}}{\sqrt{r}\sqrt{i}})]\} \cdot f_{PV}(y) \cdot f_{PV}(x) \cdot dy \cdot dx \tag{18}$$

Therefore if the sampling interval S, RO PUF's variance constant r and the probability density function $f_{PV}(x)$ is known, we can calculate the expectation of the bit error probability.

3.3 Simulation Design for the Bit Error Probability Estimation

In order to verify the model for bit error probability estimation, we utilize Matlab to simulate this test scenario and there are 1024 ROs in this simulation. First, the inherent characteristic of RO is resulted from the process variation and instantiate these ROs by 1024 sampling values $(\mu_{RO_1}, \mu_{RO_2}, \cdots, \mu_{RO_{1024}})$ from the process variation's normal distribution $N(\mu_{PV}, \sigma_{PV}^2)$. Second, the variance constant is mainly affected environmental variables (such as temperature and supply voltage) and manufacturing technology, and in our simulation, we assume that the variance constant r is the same for these 1024 ROs. Therefore, we can have the period distributions for these 1024 ROs as follows.

$$for\ RO_1, \qquad T_{RO_1} \sim N(\mu_{RO_1}, \sigma_{RO_1}^2) = N(\mu_{RO_1}, r \cdot \mu_{RO_1}) \tag{19}$$

$$for\ RO_2, \qquad T_{RO_2} \sim N(\mu_{RO_2}, \sigma^2_{RO_2}) = N(\mu_{RO_2}, r \cdot \mu_{RO_2}) \tag{20}$$

$$\cdot \quad \cdot \quad \cdot \quad \cdot \quad \cdot \quad \cdot \tag{21}$$

$$for\ RO_{1024}, \quad T_{RO_{1024}} \sim N(\mu_{RO_{1024}}, \sigma^2_{RO_{1024}}) = N(\mu_{RO_{1024}}, r \cdot \mu_{RO_{1024}}) \tag{22}$$

Based on these 1024 period distributions, we can simulate to complete one record of the oscillation's numbers in sampling interval S.

$$for\ RO_1, \quad max\{\ k_1 \mid t_{RO_1,1} + t_{RO_1,2} + \cdots + t_{RO_1,k_1-1} + t_{RO_1,k_1} < S\} \tag{23}$$

$$for\ RO_2, \quad max\{\ k_2 \mid t_{RO_2,1} + t_{RO_2,2} + \cdots + t_{RO_2,k_2-1} + t_{RO_2,k_2} < S\} \tag{24}$$

$$\cdot \quad \cdot \quad \cdot \quad \cdot \quad \cdot \quad \cdot \tag{25}$$

$$for\ RO_{1024}, \quad max\{\ k_{1024} \mid t_{RO_{1024},1} + t_{RO_{1024},2} + \cdots + t_{RO_{1024},k_{1024}} < S\} \tag{26}$$

Then when we get these 1024 oscillations' values, we can extract a 512-bit response by comparing the oscillations' values pairwise. Finally, repeat the step shown in formula (26) 1000 times and we can get 1000 512-bit responses. From the 1000 512-bit responses, we can extract the bit error probability based on the calculation of intra-distance. We briefly summarize the steps as follows.

1. According to the process variation's normal distribution, get 1024 ROs' inherent characteristics.
2. Based on the variance constant r, get the 1024 period distributions of different ROs.
3. Select the sampling interval S and record the oscillations' values.
4. Extract the response and calculate the bit error probability.

Among the parameters for simulation, the process variation distribution and the variation constant are fixed if the environmental variables and the manufacturing technology for simulation are unchanged. However, the sampling interval S is not fixed and on the observation of formula (18), the sampling interval S is one factor to affect the bit error probability. Therefore, in our simulation design, it is necessary to consider the influence of sampling interval on the bit error probability.

Keep variance constant r and the process variation distribution's parameters unchanged, and only change the value of sampling interval S, the simulation results can show the influence of the sampling interval on the bit error probability. We select the different sampling intervals as

$$S_0, \quad S_1 = 2 \cdot S_0, \quad S_2 = 2^2 \cdot S_0, \quad \cdots \quad , \quad S_n = 2^n \cdot S_0,$$

and achieves different sampling intervals' bit error probabilities respectively.

The basic parameters for simulation, like the variance constant and the process variation's normal distribution, can be extracted from practical experiments. In next section, we conduct practical experiments to extract these parameters and verify the consistency of the simulation and experiments on the estimation of bit error probability.

4 Experiment Design for Model Verification

In this section, we conduct practical experiments on fifteen Xilinx Virtex-5 XC5VLX110T-1ff1136 FPGA boards and there are 1024 ROs deployed in every board. The evaluation system is shown in Fig. 2. A 50 MHz crystal oscillator onboard is used for the setting of the sampling interval S. For example, if the sampling interval is S_0, the reference counting value N_0 based on this 50 MHz crystal oscillator is setting as that

$$N_0 = S_0/20 \cdot 10^9$$

The architecture of the ROs in our experiments is shown in Fig. 3 and they are 15-stage ROs composed of 16 LUTs. One of these LUTs is configured as a 2-input AND gate and one input is used as the enable signal for ROs. The other 15 LUTs are configured as inverters. These LUTs are deployed in 4 slices and we utilize the Hard Macro technique to deploy these 1024 ROs. In order to record every RO's oscillations, we utilize the UART to transmit all the counting values to PC for analysis. Our configured ROs' frequencies are about 132 MHz on Virtex-5 FPGA.

4.1 Parameter Extraction

When we conduct our experiments, there are two modes to complete the record of counting values. One mode is that use our configured RO to drive reference counter. When our configured RO has N oscillations, record the counter N_{CO} which is driven by the 50 MHz crystal oscillator. The feature of this mode is that we can fix the number of configured RO's oscillations and get the time consumption's distribution. Although in this mode we can extract all the parameters for our simulation, the disadvantage of this mode is that every record of counting value only records the time consumption for one RO and it is not very efficient to extract the process variation's distribution function. The other mode is that use the 50 MHz crystal oscillator to drive reference counter. When the crystal oscillator has N_{CO} oscillations, record the counters that driven by all our configured ROs. The feature of this mode is that we can record all the ROs' counting values in a sampling interval. However, this mode can only roughly extract the process variation's distribution function, and it can not extract the variance constant r. Therefore, we extract the variance constant r in mode one, and extract the parameters for process variation's distribution in mode two.

First, we conduct experiments in mode two to extract the parameters for the process variation's distribution function. In order to record the counting values, we select $N_{CO} = 2^{11}$ and the sampling interval S_0 is 40.96 μs. In every FPGA board, we record counting values 1000 times in the same sampling interval S_0. The different counting values of 1024 ROs are shown as Fig. 4 and these counting values is one of these 1000 records. From Fig. 3, the frequencies of these 1024 ROs are about 130.86 MHz to 134.27 MHz. The histogram of the counting values seems a normal distribution. From these records, we extract the process

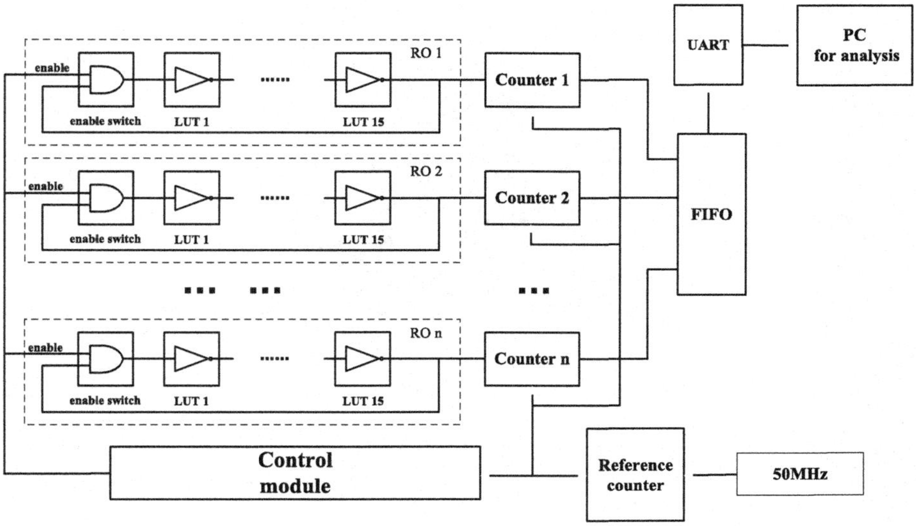

Fig. 2. The basic architecture of this evaluation system

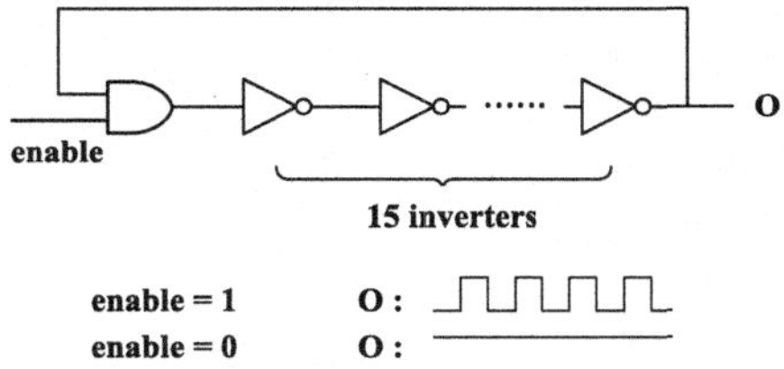

Fig. 3. The architecture of the ROs in our experiments

variation distribution function. Notice the assumption that for one RO_i, its oscillation period is a normal distribution $N(\mu_i, \sigma_i^2)$ and every oscillation period is i.i.d.. Therefore, in sampling interval S_0, based on the average number of RO_i's oscillations, $\overline{k_i}$, we can extract the parameter $\mu_i = \frac{S_0}{\overline{k_i}}$. Because in a sampling interval, there are $\overline{k_i}$ oscillations, it can be roughly regarded that $\overline{k_i}$ variables are added up and the sum's distribution is $N(\overline{k_i}\mu_i, \overline{k_i}\sigma_i^2)$. The same calculation is used for all the 1024 ROs, and we get 1024 process variation parameters. The description of process variation distribution function's extraction is shown in Algorithm 1. The distributions of these 1024 process variation parameters μ_i, $1 \le i \le 1024$, are shown in Fig. 5. Therefore, we can get the mean value of the process variation distribution function is $7.5627 * 10^{-9}$ s and the variance of this distribution is $9.8156 * 10^{-22}$.

Then, in mode one we extract the variance constant r. We select the reference counting value for our configured RO RO_i, $N = 2^{11}$. Once RO_i has N oscillations, record the counter of the crystal oscillator's oscillations. Repeat the

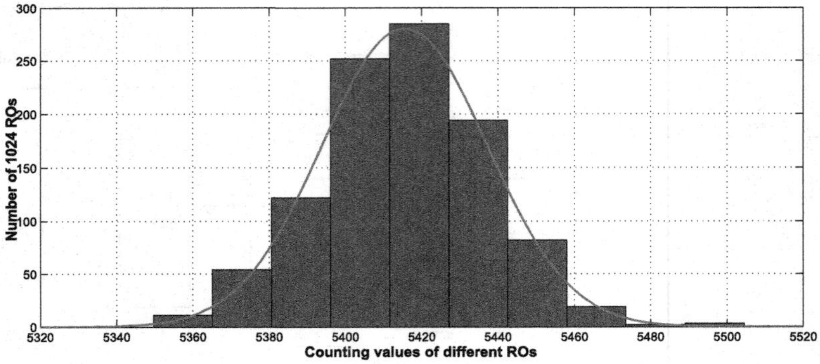

Fig. 4. The histogram of the counting values

Algorithm 1. Process variation distribution function's extraction

---

Input: · $Counter_{i,j}$ is the jth counting value of RO_i.
· $1 \le i \le N_{ROs}$ and $1 \le j \le T_{Record}$.
· N_{ROs} denotes ROs' quantity and T_{Record} denotes records' quantity.
Output: Process variation distribution function's parameters.
σ_{PV}^2 denotes the standard deviation.
μ_{PV} denotes the mean value.

for i = 1 to N_{ROs} **do**
  for j = 1 to T_{Record} **do**
    $Temp = Temp + Counter_{i,j}$;
  end for
  $\overline{k_i} = Temp/T_{Record}$;
end for
for i = 1 to N_{ROs} **do**
  $\mu_i = S_0/\overline{k_i}$
end for
$N(\mu_{PV}, \sigma_{PV}^2) \longleftarrow (\mu_1, \mu_2, \cdots, \mu_{N_{ROs}})$

---

operation 1000 times and we can get the RO_i's time consumption distribution function, and calculate the variance constant r. And then repeat the calculation of the variance constant multiple times to extract an average variance constant for the simulation.

Based on the variance constant r and the process variation's parameters, theoretically we can utilize formula (18) to calculate the bit error probability with a pre-set sampling interval S.

4.2 Simulation

During the process of simulation, there are complicated calculation which is mainly from the formula (18) and when we conduct simulations with Matlab, the calculation of formula (18) takes a large amount of time. In order to improve the

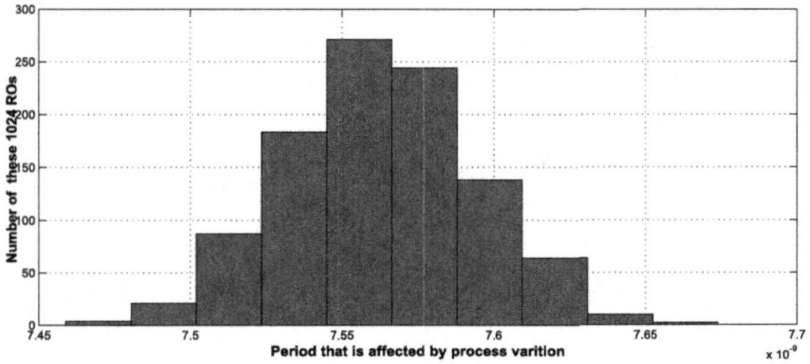

Fig. 5. The histogram of 1024 ROs' period affected by process variation

simulation's efficiency, we can appropriately simplify the calculation of formula (18). First, in the integrating range there are some values which are so small that we can ignore these values. As shown in Fig. 5, the RO's period is about $7.45 * 10^{-9}$ s to $7.7 * 10^{-9}$ s, but the integrating ranges of formula (18) are

$$(0, +\infty) \qquad and \qquad (0, x)$$

So we can use the smaller integrating ranges

$$(7.3 * 10^{-9},\ 7.7 * 10^{-9}) \qquad and \qquad (7.3 * 10^{-9},\ x)$$

to replace previous integrating ranges. Second, in a pre-set sampling interval, because the number of oscillations just fluctuates a bound which is affected by the sampling interval, we can change the i's from $Bound_{down}$ to $Bound_{up}$. B_{down} and B_{up} can be approximately computed from the sampling interval S. Third, the continuous integration may lead to time-consuming calculations and we can take advantage of the trapezoidal numerical integration to calculate formula (18). As long as the interval for the trapezoidal numerical integration is small enough, it works well to take place of the continuous integration. The new efficient calculation of bit error probability is shown as follows.

$$Prob(k_a > k_b \mid \mu_a > \mu_b)$$

$$= \sum_{i=B_{down}}^{B_{up}} \sum_{x=7.3*10^{-9}}^{7.7*10^{-9}} \sum_{y=7.3*10^{-9}}^{x} \{[\Phi((S/x-(i-1)) \cdot \frac{\sqrt{x}}{\sqrt{r}\sqrt{i-1}}) - \Phi((S/x-i) \cdot \frac{\sqrt{x}}{\sqrt{r}\sqrt{i}})] \cdot$$

$$[\Phi((S/y-1) \cdot \frac{\sqrt{y}}{\sqrt{r}\sqrt{1}}) - \Phi((S/y-i) \cdot \frac{\sqrt{y}}{\sqrt{r}\sqrt{i}})]\} \cdot f_{PV}(y) \cdot f_{PV}(x) \cdot \Delta y \cdot \Delta x \quad (27)$$

4.3 Results From Simulations and Practical Experiments

We assess the validity of the bit error probability calculating model by the results from simulations and practical experiments. In our practical experiments of RO

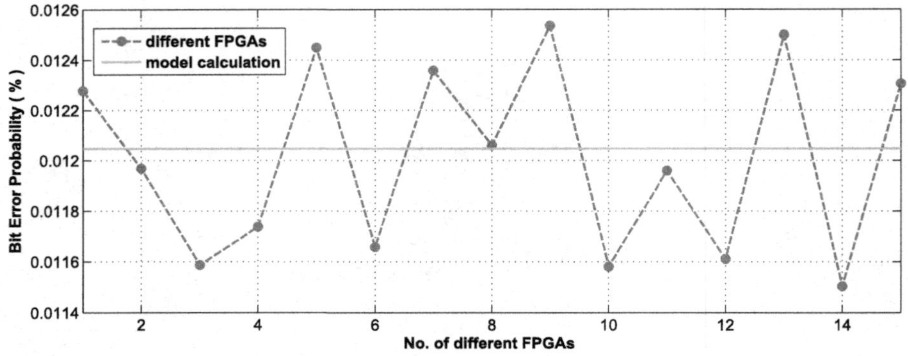

Fig. 6. The results from simulations and experiments

PUFs on Virtex-5 FPGA, the sampling time interval is 2^{11} oscillations of 50 MHz crystal oscillator, and the environmental variables, such as temperature and supply voltage, are both in normal conditions. All the ROs have 15 inverters and an extra enable switch. Because on every FPGA board there are 1024 ROs and ROs are compared pairwise to generate a 512-bit response, we reproduce the 512-bit response 1000 times and calculate the bit error probability. Figure 6 shows the bit error probabilities which are the results of data statistic from every FPGA board and the bit error probability calculated by model simulation. The dot denotes the bit error probabilities from practical experiments and the line denotes the bit error probability calculated from formula 27. The parameters for the model calculation is that the sampling time interval is 40.96 μs which is the same for the experiments on FPGAs. The process variation distribution function has the mean value, $7.5627 * 10^{-9}$ s, and the variance, $9.8156 * 10^{-22}$ s. The variance constant for simulation is $5*10^{-13}$. As shown in Fig. 6, the error bit probabilities of experiments on different FPGAs are near the bit error probability calculated from the RO PUF model.

Therefore, according to Fig. 6, it indicates that with these environmental variables and sampling interval, the model simulation and the practical experiments achieve high consistency. However, from the formula (18) or (27), besides the process variation, the magnitude of sampling interval and the variance constant also affect the bit error probability. The variance constant is involved with temperature, supply voltage and the number of RO's stage.

5 Further Discussion

As is mentioned above, the sampling interval and variance constant affect the bit error probability. In this section, first we conduct experiments with different sampling intervals. Then perform a rough temperature test in which we can get the relationship between variance constant and the bit error probability since the variance constant is mainly affected by environmental temperature.

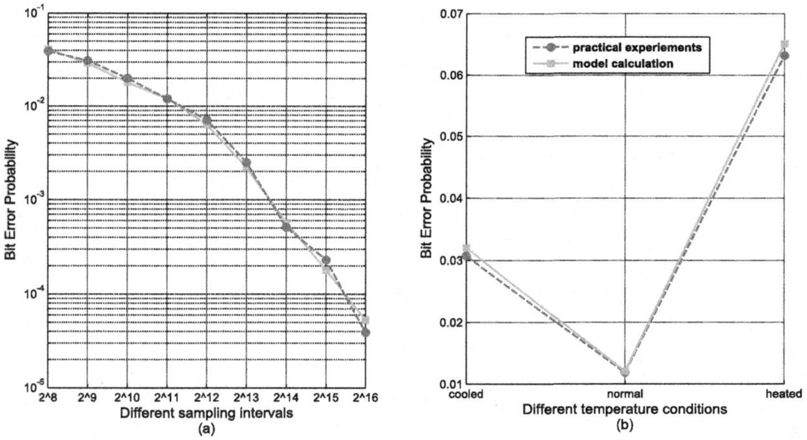

Fig. 7. The results of further discussions (Color figure online)

The sampling interval is set from 2^8 to 2^{16} and Fig. 7 shows the results from both model calculation and practical experiments. With the sampling interval becoming larger, the bit error probability decreases. When the sampling interval is larger than 2^{15}, the bit error probability is less than 10^{-4}.

Using a thermoelectric device, we heat the FPGA board to make its temperature to be about 70°C and cool it to about 5°C. We measure the variance constant r under different temperature conditions. In Fig. 7(b), the blue squares denote the bit error rate from experiments and the red dots denote the bit error probability from model calculation. In a word, our model calculation is suitable for the bit error probability evaluation of RO PUFs.

6 Conclusion

In this paper, we propose a bit error analysis model which utilizes the RO's oscillating characteristics. Therefore, this model can be more accurate and quantitatively describe the source of bit error probability. In this bit error analysis model, we consider three factors, process variation, sampling interval and temperature. The effect of temperature is transformed to the influence of the oscillation's jitter and the variance constant is used to denote the jitter. Experiments are conducted to demonstrate the validity of this new model. Therefore, the evaluation scheme is further improved based on our work and our bit error analysis model can serve as a guide to construct RO PUF with a prescient bit error probability and design the RO PUF-based system efficiently and economically.

References

1. Gassend, B., Clarke, D., Van Dijk, M., Devadas, S.: Silicon physical random functions. In: Proceedings of the 9th ACM conference on Computer and communications security, pp. 148–160 (2002)
2. Dodis, Y., Reyzin, L., Smith, A.: Fuzzy extractors: how to generate strong keys from biometrics and other noisy data. In: Cachin, C., Camenisch, J.L. (eds.) EUROCRYPT 2004. LNCS, vol. 3027, pp. 523–540. Springer, Heidelberg (2004)
3. Guajardo, J., Kumar, S.S., Schrijen, G.-J., Tuyls, P.: FPGA intrinsic PUFs and their use for IP protection. In: Paillier, P., Verbauwhede, I. (eds.) CHES 2007. LNCS, vol. 4727, pp. 63–80. Springer, Heidelberg (2007)
4. Maes, R., Tuyls, P., Verbauwhede, I.: A soft decision helper data algorithm for sram pufs. In: IEEE International Symposium on Information Theory, ISIT 2009, pp. 2101–2105 (2009)
5. Suzuki, D., Shimizu, K.: The glitch PUF: a new delay-PUF architecture exploiting glitch shapes. In: Mangard, S., Standaert, F.-X. (eds.) CHES 2010. LNCS, vol. 6225, pp. 366–382. Springer, Heidelberg (2010)
6. Tajik, S., Dietz, E., Frohmann, S., Seifert, J.-P., Nedospasov, D., Helfmeier, C., Boit, C., Dittrich, H.: Physical characterization of arbiter PUFs. In: Batina, L., Robshaw, M. (eds.) CHES 2014. LNCS, vol. 8731, pp. 493–509. Springer, Heidelberg (2014)
7. Suh, G.E., Devadas, S.: Physical unclonable functions for device authentication and secret key generation. In: Proceedings of the 44th annual Design Automation Conference, pp. 9–14 (2007)
8. Maiti, A., Schaumont, P.: Improved ring oscillator puf: an fpga-friendly secure primitive. J. Cryptology **2**, 375–397 (2011)
9. Rahman, T., Forte, D., Fahrny, J., Tehranipoor, M.: Aro-puf: an aging-resistant ring oscillator puf design. In: Proceedings of the Conference on Design, Automation & Test in Europe (2014)
10. Merli, D., Heyszl, J., Heinz, B., Schuster, D., Stumpf, F., Sigl, G.: Localized electromagnetic analysis of ro pufs. In: 2013 IEEE International Symposium on Hardware-Oriented Security and Trust (HOST) (2013)
11. Bösch, C., Guajardo, J., Sadeghi, A.-R., Shokrollahi, J., Tuyls, P.: Efficient helper data key extractor on FPGAs. In: Oswald, E., Rohatgi, P. (eds.) CHES 2008. LNCS, vol. 5154, pp. 181–197. Springer, Heidelberg (2008)
12. Maes, R., Tuyls, P., Verbauwhede, I.: Low-overhead implementation of a soft decision helper data algorithm for SRAM PUFs. In: Clavier, C., Gaj, K. (eds.) CHES 2009. LNCS, vol. 5747, pp. 332–347. Springer, Heidelberg (2009)
13. Fischer, V., Lubicz, D.: Embedded evaluation of randomness in oscillator based elementary TRNG. In: Batina, L., Robshaw, M. (eds.) CHES 2014. LNCS, vol. 8731, pp. 527–543. Springer, Heidelberg (2014)
14. Ma, Y., Lin, J., Chen, T., Xu, C., Liu, Z., Jing, J.: Entropy evaluation for oscillator-based true random number generators. In: Batina, L., Robshaw, M. (eds.) CHES 2014. LNCS, vol. 8731, pp. 544–561. Springer, Heidelberg (2014)
15. Amaki, T., Hashimoto, M., Mitsuyama, Y., Onoye, T.: A worst-case-aware design methodology for noise-tolerant oscillator-based true random number generator with stochastic behavior modeling. IEEE Trans. Inf. Forensics Secur. **8**, 1331–1342 (2013)
16. Killmann, W., Schindler, W.: A design for a physical RNG with robust entropy estimators. In: Oswald, E., Rohatgi, P. (eds.) CHES 2008. LNCS, vol. 5154, pp. 146–163. Springer, Heidelberg (2008)

Extracting Robust Keys from NAND Flash Physical Unclonable Functions

Shijie Jia[1,2,3], Luning Xia[1,2](✉), Zhan Wang[1,2], Jingqiang Lin[1,2], Guozhu Zhang[1,2,3], and Yafei Ji[1,2,3]

[1] Data Assurance and Communication Security Research Center, Chinese Academy of Sciences, Beijing, China
{jiashijie,halk,zwang,linjq,zhangguozhu,jiyafei12}@is.ac.cn
[2] State Key Laboratory of Information Security, Institute of Information Engineering, Chinese Academy of Sciences, Beijing, China
[3] University of Chinese Academy of Sciences, Beijing, China

Abstract. Physical unclonable functions (PUFs) are innovative primitives to extract secret keys from the unique submicron structure of integrated circuits. PUFs avoid storing the secret key in the nonvolatile memory directly, providing interesting advantages such as physical unclonability and tamper resistance. In general, Error-Correcting Codes (ECC) are used to ensure the reliability of the response bits. However, the ECC techniques have significant power, delay overheads and are subject to information leakage. In this paper, we introduce a PUF-based key generator for NAND Flash memory chips, while requiring no extra custom hardware circuits. First, we present three methods to extract raw PUF output numbers from NAND Flash memory chips, namely partial erasure, partial programming and program disturbance, which are all based on the NAND Flash Physical Unclonable Function (NFPUF). Second, we use a bit-map or a position-map to select the cells with the most reliable relationship of the size between raw NFPUF output numbers. Only the selected cells are used for key generation. Finally, we describe the practical implementations with multiple off-the-shelf NAND Flash memory chips, and evaluate the reliability and security of the proposed key generator. Experimental results show that our NFPUF based key generator can generate a cryptographically secure 128-bit key with a failure rate $< 10^{-6}$ in 93.83 ms.

Keywords: Physical Unclonable Functions (PUFs) · NAND flash · Process variation · Secret keys · Error correction

1 Introduction

As electronic devices have become interconnected and ubiquitous, people are increasingly depending on electronic devices to perform sensitive tasks and to

This work was partially supported by the National 973 Program of China under award No. 2013CB338001 and the Strategic Priority Research Program of Chinese Academy of Sciences under Grant XDA06010702.

J. Lopez and C.J. Mitchell (Eds.): ISC 2015, LNCS 9290, pp. 437–454, 2015.
DOI: 10.1007/978-3-319-23318-5_24

handle sensitive information. As a result of the merits of NAND Flash memory, such as small size, low power consumption, light weight, high access speed, shock/temperature resistance and mute characteristics [12], now virtually all portable electronic devices such as smartphones, SD cards, USB memory sticks and tablets use NAND Flash memory as nonvolatile storage.

Now many electronic devices of embedded systems have become to contain more confidential information, and many applications need to identify and authenticate users. Therefore, the secret keys used by the devices and the applications should be protected to ensure the security of the communication system. However, in the real world implementations of cryptosystems, the cryptographic keys are recently revealed from nonvolatile memories by sophisticated tampering methods [3,9,21,25]. Based on the above situation, we leverage the special virtue of NAND Flash to avoid storing the secret key in the nonvolatile memory directly.

In order to prevent both the invasive and noninvasive physical attacks, Physical Unclonable Functions (PUFs) have been attracting wider attention and studied intensively in recent years. Due to the advantages of physical unclonability and tamper proof, PUFs are used to avoid storing actual bits of the secret keys in the storage memory. Generally, PUFs are engaged in two typical classes of applications, namely authentication and secret key generation. In the authentication applications, the responses of the PUFs can be designed to tolerate a certain amount of errors. While in the secret key generation applications, the responses of the PUFs need to be consistent [16]. The conventional method to ensure the robustness and the reliability of the responses is to utilize fuzzy extractors [6,14]. Traditionally, fuzzy extractors employ an Error-Correcting Code (ECC) and a cryptographic hash function. There have been several state of the art papers cite the use of ECC with PUFs to generate cryptographic keys [2,10,15,29]. However, ECC is not viable for resource constrained electronic devices. First, the error rates for PUFs across environmental variations can be as high as 25 % [5], making a straightforward use of ECC infeasible [15,24], namely the codeword sizes required will be too large in practice. Second, ECC is generally performed by specialised hardware chips, which not only requires tremendous area and power overheads, but also scales up as the number of bits of correction increases [1]. Third, ECC requires additional helper or syndrome data to be publicly stored for the regeneration of the key. The helper data of ECC reveals information about the outputs of the PUFs [23]. Thus, error reduction techniques can be applied to reduce the cost of ECC and to ensure the reliability and security of the responses.

In this work, we focus on the NFPUF and the error reduction techniques to generate cryptographic robust keys. First, we present three methods (partial erasure, partial programming and program disturbance) to extract raw NFPUF output numbers from NAND Flash memory chips. Second, we introduce two methods (the bit-map and position-map) to select the cells with the most reliable relationship of the size between raw NFPUF output numbers. In other words, the size relationship of the raw NFPUF output numbers from the selected cell pairs is almost constant during the whole lifetime of the NAND Flash chips. At last, we evaluate the reliability and security of the proposed key generator. The proposed key generator can get reliable and robust keys for the electronic devices with limited hardware resources, meanwhile it reduces the implementation costs and hardware overheads of ECC significantly.

Our Contributions. In this paper we introduce a robust key generator based on NFPUF for NAND Flash memory chips. The main contributions of this paper are as follows:

- We present the first implementation of secret key generator from unmodified commercial NAND Flash memory chips. Most importantly, the proposed key generator can be applied to any NAND Flash memory chips, extending the functionality of NAND Flash memory chips, while requiring no extra hardware circuits overheads.
- We describe three specific methods to extract raw NFPUF output numbers from NAND Flash memory chips. Particularly, the partial erasure method is proposed for the first time.
- We present two methods to select the NAND Flash memory cells with the most reliable relationship of the size between raw NFPUF output numbers. It reduces the system overheads of ECC significantly, and it is feasible for the electronic devices even with constrained hardware resources.
- We evaluate the reliability and security of our proposed key generator with multiple NAND Flash memory chips from different manufacturers by plenty of experiments.

Organization of the Paper. The organization of the rest of this paper is as follows. Section 2 introduces the related work. The background is introduced in Sect. 3. We will present the specific secret key generator in Sect. 4. Implementation details and evaluations are shown in Sect. 5. Finally, conclusions are given in Sect. 6.

2 Related Works

Pappu et al. introduced the Physical One-Way Functions (POWFs) in [18]. They used a transparent optical medium with a three-dimensional micro-structure as a POWF. The concept of silicon PUFs was introduced in [7,8]. Silicon PUFs have substantial challenge-response pairs (CRPs) owing to the manufacturing process variations, so it is impossible for the attacker to clone all the potential CRPs [13]. Our key generator also takes advantage of the manufacturing process variations.

Škorić et al. presented a key extraction method from the bit-string extraction of noisy optical PUFs in [20]. Different PUFs circuit designs based on ring oscillators were introduced in [13,15,23,29]. The first construction of a PUF intrinsic based on the power-up state of SRAM memory on current FPGAs was presented in [10] to solve the IP protection problem. An efficient helper data key extractor technique was introduced to generate secret keys on FPGAs in [2], which leverages several complicated concatenated codes (repetition code and ECC) to ensure the reliability of the keys. Our key generator does not require a power cycle or the special circuit designs that the prior PUFs need, while it can be done by any electronic devices with commercial off-the-shelf NAND Flash memory chips as nonvolatile storage.

Xu et al. introduced the sources of variations in Flash memory for PUFs in [28]. It points out that the uniqueness and robustness of the NFPUF are indeed universally applicable, rather than just a phenomenon presented in the limited selection. In general, NFPUF distributions are translated to threshold voltage distributions via tunneling current during programming and erasing operations to analyze its physical origins. Our key generator also leverages the threshold voltage distributions of the NAND Flash memory cell transistors to extract the raw NFPUF output numbers.

Prabhu et al. evaluated seven techniques to extract unique signatures from Flash devices based on the observable effects of process variations as device fingerprints [17]. They exploited formal correlation metric (the Pearson correlation) to distinguish whether the extracted signatures were from the same page or different pages, then they could uniquely identify individual Flash devices. Yang et al. took advantage of the uncertainty of Random Telegraph Noise (RTN) from Flash memory to provide two security functions: true random number generation and digital fingerprinting [27]. As a result of the high uncertainty of the random numbers and device fingerprints, neither the techniques they proposed could be used to extract unique and reproduceable secret keys with a tiny bit error rate. Our key generator leverages the specific physical characteristics of the NAND Flash memory cells to extract numbers, then we select the cells with the most reliable relationship of the size between the extracted numbers during the whole life of the chip for key generation. Our key generator ensures the reproducibility and the reliability of the key, meanwhile, it avoids the costly overheads of ECC.

3 Background

The secret keys generator that we will describe in Sect. 4 bases on the composition of NAND Flash memory cells, and how NAND Flash memory chip organize the cells into memory arrays. The specific NAND Flash memory cell composition and array organization lead to noises exist in NAND Flash memory cells universally. This section summarizes the primary characteristics of NAND Flash memory chips that we rely on for this study.

3.1 Uncertain States of NAND Flash Memory Cells

As Fig. 1(a) shows, a NAND Flash memory cell is based on a floating gate metal oxide semiconductor (MOS) transistor. There are two gates in a floating gate transistor. The top one is called control gate, which is capacitive coupled. The bottom one is the floating gate, which is surrounded by dielectrics. The special

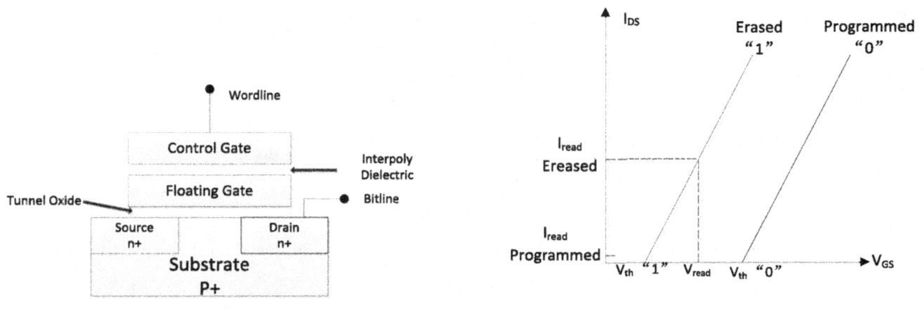

(a) (b)

Fig. 1. (a) NAND flash memory cell structure. (b) Threshold voltage schematic diagram.

property of dielectrics makes the NAND Flash memory nonvolatile. In general, a triple layer of oxide-nitride-oxide isolates the two gates. In addition, the thin oxide between the floating gate and transistor channel is known as tunnel oxide. The source and drain electrodes are heavy doped, and they are electron-rich (n-type). While the substrate is less doped, and it is electron-deficient (p-type).

Both programming and erasing operations of common-ground NAND Flash memory cells are by Fowler-Nordheim (FN) tunneling, which is a quantum-mechanical tunneling mechanism induced by the electric field [19]. The presence or absence of trapped charge on the floating gate is expressed as logical state "0" or logical state "1" respectively.

The trapped charge affects the threshold voltage (V_{th}) of the transistor [28]. When an electron charge is stored in the floating gate, the threshold voltage of this transistor increases, and the increase amplitude is proportional to the stored charge. As illustrated in Fig. 1(b), the charge stored in the floating gate discourages the presence of current in the transistor channel, then the cell is sensed and translated into logical state "0", thus the NAND Flash memory cell will be in the programmed state. On the contrary, when the floating gate has no electron charge, then it forms a conductive path between the drain and the source electrodes, creating a current (I_{DS}) in the transistor channel, and hence the cell will be sensed and translated into logical state "1", then the NAND Flash memory cell will be in the erased state. In conclusion, by applying an appropriate voltage to the control gate and measuring the current flow through the transistor channel of the target cell, a NAND Flash memory chip can effectively measure the threshold voltage of the cells, and determine the logical states of the cells.

However, on account of variations in manufacturing processes, the threshold voltages of ($V_{th``1"}$) and ($V_{th``0"}$) vary from cell to cell. When the threshold voltage is not shifted sufficiently from the programmed state to the erased state, and vice versa, then the cell will be in an uncertain state. In these cases, the cell can be expressed as either logic state "0" or logic state "1". In this paper, we propose the partial erasure and the partial programming methods, both the methods exploit the uncertain states of the NAND Flash memory cells to extract raw NFPUF output numbers.

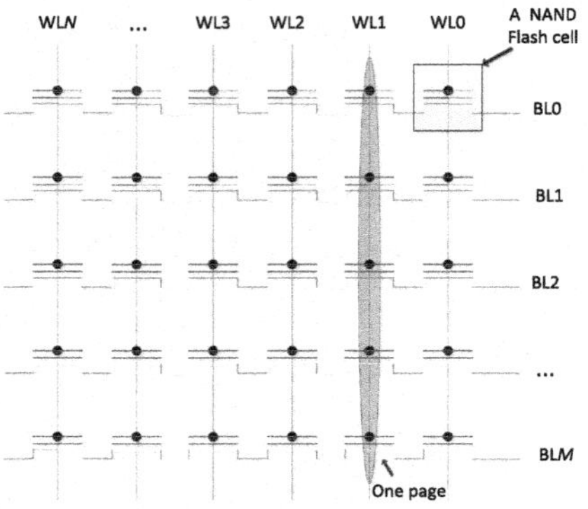

Fig. 2. NAND flash memory array organization.

3.2 Disturbance Related to NAND Flash Memory Array Organization

As Fig. 2 illustrates, the NAND Flash memory cells are arranged in a coherent and structured manner, normally in arrays, to achieve high density. Due to the array organisation, each cell can be accessed by its specific row and column address. In general, the NAND Flash memory cells are grouped into pages (typically 512 bytes-4 KB) and blocks. A block contains dozens of (typically 32–128) adjacent pages [22]. Thousands of independent blocks make up a NAND Flash memory chip. The common drain connection that the rows of cells share is called a bitline (BL), while the common poly-2 gate connection that the columns of cells share is called a wordline (WL) [9]. A single wordline connects the gates on all the transistors in a page or more than one page, and the latter case is particularly general for multi-level cells (MLC) NAND Flash memory chips. Programming and reading operations are performed on the unit of a page, whereas erasing operation must be performed on an entire block. What is more, the pages in a block must be programmed sequentially. The programmed sequence is designed to minimize the programming disturbance between the neighboring pages, which aims to avoid undesired voltage shifts in the pages despite not being selected.

However, although the array organization of the NAND Flash memory is specially designed, there still exists electrical influence between adjacent NAND Flash memory cells. During the programming and reading operations, a high voltage is applied to the wordlines of the selected pages, meanwhile producing an intermediate gate voltage to the neighbouring wordline. After multiple repeating operations, the intermediate gate voltage makes the according adjacent NAND Flash memory cells flip, which is a process of quantitative change to qualitative change. In particular, as the result of the capacitive coupling between the selected wordline and the physically adjacent wordline, the effect of programming operations is much stronger [30]. In this paper, we propose the program disturbance method, which leverages the effects between the adjacent pages to extract raw NFPUF output numbers.

4 Robust Key Generation

As Sect. 3.1 discussed the physical sources of variations in NAND Flash memories for NFPUF, not only the initial and after-erase voltages, but also the initial and after-program voltages for a same NAND Flash memory chip may vary from cell to cell due to the manufacturing process variations. Section 3.2 discussed the disturbance related to the specific NAND Flash memory array organization. The repeating programming operations to a same page, resulting in its neighbouring page unexpected bit variations.

Both the above phenomena are due to the maximum density of NAND Flash memory cells. Since process variations are beyond the manufacturers' control, small variations in tunnel oxide thicknesses and control gate coupling ratio may make a big difference in the threshold voltage of the floating gate transistor [28]. So even an adversary who has the detail information of the NFPUF principle still cannot clone the NFPUF.

4.1 Extracting Raw NFPUF Output Numbers

In this paper, we propose three methods to extract raw NFPUF output numbers, namely partial erasure, partial programming and program disturbance. We will discuss the details of the three methods in order.

Partial Erasure. This method exploits the feature of NAND Flash memory chip that the initial and after-erase voltages vary from cell to cell due to the manufacturing process variations. Algorithm 1 provides the pseudo-code. First, we erase the selected block ($BlockNum$), then we program all the cells of the selected page ($PageNum$) belonging to the selected block to logic state "0". Second, we perform fixed number ($PENum$) of partial erasure operations to the selected page. The time of each partial erasure operation (T_e) is also fixed. After each partial erasure operation, some cells in the selected page will have been erased enough to flip their states from logic state "0" to logic state "1". Therefore, we record the number of partial erasure operations that the selected cells need to flip. Third, after the fixed number of partial erasure operations, some cells may have not flipped, then the value of $PENum$ plus 1 is assigned to these cells. At last, we extract the raw NFPUF output numbers by repeating partial erasure operations from the specific block and page.

Partial Programming. Similar with the partial erasure method, the partial programming method leverages the feature of NAND Flash memory chip that the initial and after-program voltages vary from cell to cell due to the manufacturing process variations. First, we erase the selected block. Second, we perform fixed number ($PPNum$) of partial programming operations to the selected page, the time of each partial programming operation (T_p) is also fixed. After each partial programming operation, some cells may have been programmed enough to flip their states from logic state "1" to logic state "0". Therefore, we record the number of partial programming operations that the selected cells need to flip. Third, after the fixed number of partial programming operations, some cells

Algorithm 1. Partial erasure: Extract the raw NFPUF output numbers by repeating partial erasure operations to the specific block and page.

Require:
The number of block to erase ($BlockNum$);
The number of page to read ($PageNum$);
The number of cells to record ($CellsNum$);
The time of each partial erasure operation (T_e);
The number of partial erasure operations ($PENum$);

Ensure:
The number of partial erasure operations of each NAND Flash memory cell need to reach the erased state ($RawPuf[CellsNum]$).

```
Erase(BlockNum);
Program(PageNum, 0);
for i = 1; i <= PENum; i++ do
    PartiallyErase (T_e, BlockNum);
    Read(PageNum);
    for All the selected cells to record do
        if The first observation of the cell flips from 0 to 1 then
            RawPuf[The Position of the cell in the selected cells ]= i;
        end if
    end for
end for
for The cells have not flipped after PENum partial erasure operations do
    RawPuf[The Position of the cell in the selected cells]= PENum + 1;
end for
```

may have not flipped, then the value of $PPNum$ plus 1 is assigned to these cells. At last, we extract the raw NFPUF output numbers by repeating partial programming operations from the specific block and page.

Program Disturbance. Unlike the above two methods, this method is based on the disturbance between the adjacent pages due to the specific NAND Flash memory array organization. The repeating programming operations to a same page, resulting in its neighbouring page unexpected bit variations. First, we erase the selected block. Second, we perform fixed number ($PDNum$) of programming operations to the selected page. After each programming operation, some cells in its physically adjacent page will have been programmed enough to flip their states from logic state "1" to logic state "0". Therefore, we record the number of programming operations that the selected cells in its physically adjacent page need to flip. Third, after the fixed number of programming operations, some cells may have not flipped, then the value of $PDNum$ plus 1 is assigned to these cells. At last, we extract the raw NFPUF output numbers by repeating programming operations from the specific block and pages.

4.2 Extracting Robust Keys from the Raw NFPUF Output Numbers

If the PUFs are measured repeatedly, the cell-wise extracted numbers apparently will have non-negligible fluctuations as a result of noises. Therefore, the raw PUFs output numbers are not fit as secret key directly [4]. In general, fuzzy extractors are used to ensure the reliability of the PUFs response outputs. Fuzzy extractors employ an ECC and a cryptographic hash function. As a result of the tremendous raw bits and helper data overheads are needed in real system implementations of ECC, it is expensive to implement in electronic devices with limited hardware resources [5,24].

In this work, our objective is to extract robust keys from raw NFPUF output numbers with a tiny bit error rate, meanwhile reducing the costly overheads in the implementations of ECC. Therefore, it will be feasible for NAND Flash devices even with constrained hardware resources to generate robust keys.

Due to the layout and spatial variations of NAND Flash memory chips, a consistent systematic variation exists among the average page NFPUF output numbers and the average block NFPUF output numbers [28]. Typically NAND Flash memory can withstand 100,000 program and erase (P/E) cycles for single-level cell (SLC) type and 10,000 for MLC type [26]. Repetitive P/E cycles can alter the raw extracted numbers of the cells due to cyclic endurance aging effects [27]. As Fig. 3(a) illustrates, the raw NFPUF output numbers from the cells of the same page present an irregular distribution and have great difference. However, as shown in Fig. 3(b), although the raw NFPUF output numbers slightly decreased over P/E cycles, the relationship between the size of raw NFPUF output numbers extracted from different cells is relatively stable during the whole lifetime of the NAND Flash memory chips.

Our key generator is to find the NAND Flash memory cells with the most reliable and stable relationship between the size of raw NFPUF output numbers. We translate the size relationship of the raw NFPUF output numbers into binary numbers as a robust secret key, meanwhile we record the according cell position information as helper data for key regeneration.

We introduce two methods to extract secret keys by selecting the NAND Flash memory cells with the most reliable relationship of the size between raw NFPUF output numbers, namely the bit-map method and position-map method. Only the selected cells are used for generation of the key.

Figure 4 describes an example of our two key extraction methods. We extract ten raw NFPUF output numbers ($RawPuf$) from ten cells, and the position of the cells starts from 0×0065 to 0×006e (due to the page size of the commercial off-the-shelf NAND Flash memory chip is from 512 bytes to 4 KB generally, so we use 16 bits to represent an address of a cell). The specific quantities of the raw NFPUF output numbers and the according bit number of the extracted key are related to the bit error rate of the key, which will be discussed in detail in the experimental section.

Bit-Map Method. First, we compare the adjacent raw NFPUF output numbers in pairs and record the absolute values of the corresponding D-values

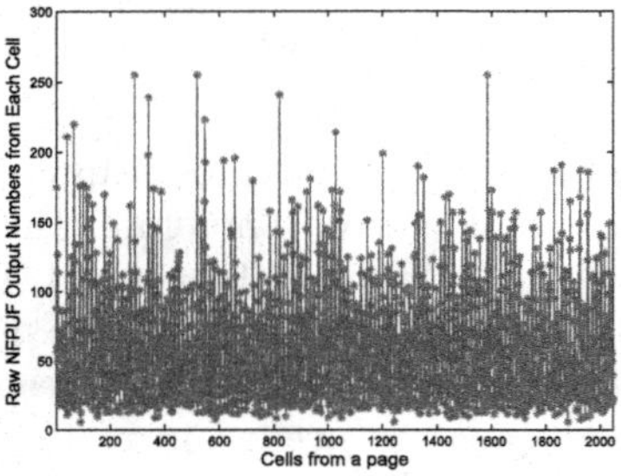

(a) Raw NFPUF output numbers from a page.

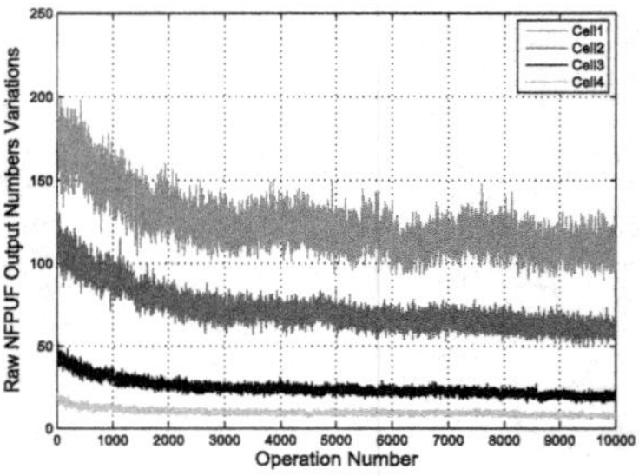

(b) Fluctuations of raw NFPUF output numbers from four cells of a MLC type NAND Flash memory chip.

Fig. 3. The distributions of raw NFPUF output numbers.

(*ADvalue*). Second, we sort the the recorded *ADvalues* from small to large, and only a part of the cell pairs with the top largest *ADvalues* are selected to generate a key (here three cell pairs are selected). Third, we assign "1" to the selected cell pairs, and assign "0" to the rest as helper data (*BitMap*). Fourth, if the former raw NFPUF output number is bigger than the latter one in the selected cell pairs, we allocate "1" to the key (*Key*1), if not, then we allocate "0" to the key. At last, we obtain the secret key (*Key*1) and store the *BitMap* for regeneration of the key.

Position-Map Method. First, we sort the extracted raw NFPUF output numbers from small to large (*SortedRawPuf*). Second, we select a part of the top smallest and the top largest cells to make up the selected cell pairs (here both the cells with the top three smallest and the top three largest raw NFPUF output numbers are selected). Then we sort the raw NFPUF output numbers of each cell pairs according to the cell positions (*PairRawPuf*). Third, we record the cell position of the selected NAND Flash memory cells as helper data (*PositionMap*). Fourth, we compare the raw NFPUF output numbers of the selected cells. If the former is bigger than the latter one, we allocate "1" to the key (*Key*2), if not, we allocate "0" to the key. At last, we obtain the secret key (*Key*2) and store the *PositionMap* for regeneration of the key.

Note that both the *BitMap* and *PositionMap* just represent the location of the selected NAND Flash memory cells, and they have nothing to do with the relationship of the size between the raw NFPUF output numbers. Therefore, the helper data carries no information about the polarity of the bits in the key. Hence the helper data does not leak any information about the key, unless there is a location-based correlation found in the numbers generated from the NFPUF. As the proposed key generator is based on the manufacturing process variations, which is a random process, and hence the polarity of the bits of the key is also random. Therefore, the helper data of this study is significantly more resilient to information leakage as compared to the helper data in conventional ECC.

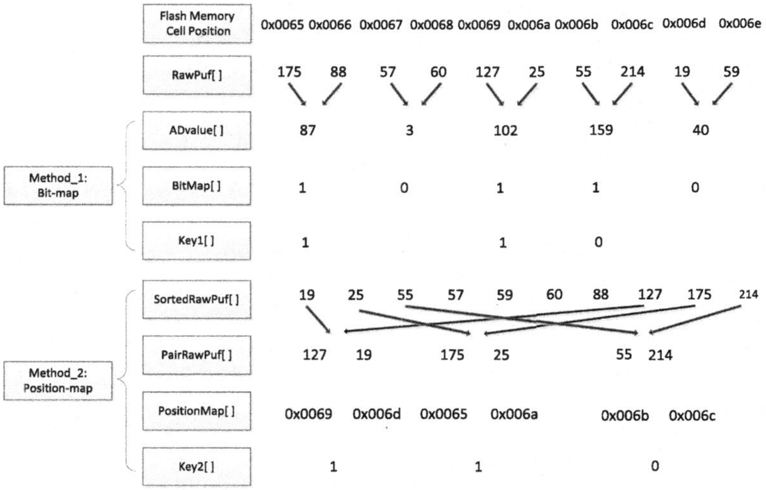

Fig. 4. Schematic diagram of the two key extraction methods.

5 Implementation and Evaluation

In this section, we present the experimental facilities used in this study, and evaluate the primary characteristics of reliability and security of the key generator based on NFPUF.

5.1 Tested Device

To extract raw NFPUF output numbers from NAND Flash memory chips, we use a custom PCB test board that contains the STM32F103VCT6 controller, which has a common ARM Cortex-M3 32-bit RISC core. With the controller, we can send program, read, and erase operations to the tested NAND Flash memory chips at will. This common device shows that our robust key generator can be applied to commercial off-the-shelf NAND Flash memory devices with no extra integrated circuits.

We evaluate NFPUF with a set of NAND Flash memory chips from different manufacturers. Table 1 shows the chips we use in this study.

5.2 Experimental Results and Evaluation

In order to evaluate the performance of the proposed key generator based on NFPUF, we need to analyse the primary characteristics of the security and reliability of the generated keys, such as speed (for performance), reproducibility (for reliability), uniqueness (for security), and randomness (for high-entropy).

Speed. Both the generation and regeneration processes of our proposed key generator need multiple program or erase operations, so the throughput of the

Table 1. Tested NAND flash memory chips.

Chip	Manufacturer	Part number	Capacity	Quantity	Technology
A	Samsung	K9K8G08U0M	8 Gbit	10	90 nm SLC
B	Samsung	K9F2G08U0B	2 Gbit	10	SLC
C	Micron	MT29F4G08ABA DAWP:D	4 Gbit	5	34 nm SLC
D	Micron	MT29F16G08CB ACAWP:C	16 Gbit	5	MLC
E	Intel	JS29F64G08AAME1	64 Gbit	5	MLC
F	Hynix	HY27UF084G2B	4 Gbit	5	SLC
G	Numonyx	NAND04GW3B2DN6	4 Gbit	5	57 nm SLC

proposed key generator varies significantly depending on the program and erase characteristics of the selected NAND Flash memory chips. Table 2 shows the parameters of the proposed three raw NFPUF output numbers extraction methods of the selected NAND Flash memory chips.

First, we find out the typical block erase time (t_{BERS}) and the typical page program time (t_{PROG}) from the datasheet of each NAND Flash memory chip. Second, we determine the time of each partial erasure operation (T_e) and the time of each partial programming operation (T_p) by trial and error until we get obviously diacritical outputs from each cell. The vast experimental results show that the T_e should be about the 1/12 of t_{BERS}, and the T_p should be about the 1/20 of t_{PROG}. To determine the number of partial erasure operations ($PENum$), we can partially erase the specific block and page with the selected T_e repeatedly until 99 % cells are erased in a page. In the same way, we can determine the number of partial programming operations ($PPNum$) by repeatedly partial programming operations with the determined T_p until 99 % cells are programmed in a page. At last, we can determine the number of repeated programming operations ($PDNum$) by normal programming operations to a specific page until 99 % cells are programmed in its adjacent page.

With the determined parameters in Table 2, we obtain the average throughput of the raw NFPUF output numbers with the NAND Flash memory chips from five manufactures, and it is shown in Table 3. The average throughput ranges from 7.35 Kbits/second to 22.38 Kbits/second. On account of the operation time, the average throughput of the partial programming method shows the highest speed, the next one is the partial erasure method, while the program disturbance method shows the slowest speed comparatively.

To get a 128-bit key with the bit error rate $< 10^{-6}$, we need 18.28 Kbits raw NFPUF output numbers for the bit-map method, so we can get a 128-bit key in 816.8 ms to 2.48 s. While we only need 2.1 Kbits raw NFPUF output numbers for the position-map method, so we can get a 128-bit key in 93.83 ms to 285.7 ms.

In our experiments, the average throughput is largely limited by the timing of the asynchronous interface, which is controlled by an ARM microcontroller with CPU frequency of 72 MHz and the 8-bit bus of the NAND Flash memory chips.

Table 2. Parameters setting of the raw NFPUF output numbers extraction methods.

Chip	$t_{BERS}(\mu s)$	$T_e(\mu s)$	$PENum$	$t_{PROG}(\mu s)$	T_p (μs)	$PPNum$	$PDNum$
A	1500	180	250	200	10	245	3912
B	1500	180	250	200	10	245	4002
C	700	85	200	200	11	200	3017
D	700	85	200	200	11	200	3123
E	3000	358	230	1200	58	220	3438
F	1500	179	240	200	10	250	3621
G	1500	181	235	200	11	240	3419

Table 3. The average throughput of the raw NFPUF output numbers(Kbits/second).

Method	Sumsung	Micron SLC	Micron MLC	Intel MLC	Hynix	Numonyx
Partial erasure	14.67	13.92	13.87	12.21	14.52	13.68
Partial programming	22.38	17.47	16.32	14.45	18.36	17.18
Program disturbance	9.72	8.78	7.89	7.35	9.74	8.29

The throughput performance can be much higher if the data can be transferred more quickly through the controller interface.

Reproducibility. In order to indicate the reproducibility of the PUF outputs, we evaluate the intra-chip variation, namely the number of bits changes when regenerated from a single PUF with or without environmental changes [23]. Ideally, the intra-chip variation should be 0 %.

To reduce the cost of ECC, we propose the bit-map and position-map methods to select the NAND Flash memory cells with the most reliable relationship of the size between raw NFPUF output numbers. For our reference implementation, we aim to obtain a 128-bit key with intra-chip variation $< 10^{-6}$, which means that our proposed key generator is applicable and reliable during the whole lifetime of the NAND Flash memory chips.

As Fig. 5 illustrates, we evaluate the average intra-chip variation with temperature and aging variations. We extract Y bits key from X raw NFPUF output numbers. The x-axis represents the ratio of Y in the X (as we use the relationship of the size between two raw NFPUF output numbers to extract a bit for the key, so the maximum of Y is $X/2$). The y-axis shows the according average intra-chip variation of the tested NAND Flash memory chips.

The variations of temperature influence thermal noise amplitude, while RTN amplitude stays almost the same [27]. Since we extract raw NFPUF output numbers based on RTN primarily, as Fig. 5(a) and (c) show, there is little difference across different temperatures.

NAND Flash memory chips wear-out over time due to program/erase (P/E) operations are performed. The average page NFPUF decreases slightly as P/E cycles increases [28]. However, we test the chips in Table 1 under different temperature conditions with our test board to verify the relationship of the size between raw NFPUF output numbers, all the chips show the same result as Fig. 3(b), namely the relationship of the size between raw NFPUF output numbers is rather stable. Therefore, we can see in the Fig. 5(b) and (d), the aging influence between different P/E cycles is also unconspicuous.

In the bit-map method, the intra-chip variation decreases from 14.42 % to less than 10^{-6} as the ratio of Y/X decreases from 0.5 to 0.055. In this case, when the ratio of Y/X is 0.055, now 2340 NAND Flash memory cells are needed to generate a 128-bit robust key, and the length of helper data is 2340/2=1.14 Kbits.

In the position-map method, the intra-chip variation decreases from 2.3×10^{-5} to less than 10^{-6} as the ratio of Y/X decreases from 0.5 to 0.474. In this case, when the ratio of Y/X is 0.474, now 270 NAND Flash memory cells are needed to generate a 128-bit robust key, and the length of helper data is $128\times16\times2$=4Kbits.

In conclusion, to generate a 128-bit key with the bit error rate $< 10^{-6}$, we can select 2340 or 270 NAND Flash memory cells for the bit-map method and the position-map method, respectively. Then we just choose the top 128 cell pairs with the maximal difference of the raw NFPUF output numbers to generate the key. Comparatively, the bit-map method needs more NAND Flash memory cells and less helper data, while the position-map method requires much less cells and more helper data. Therefore, we can select the appropriate method according to the specific implementation requirement.

ECC is too complex and expensive to implement for efficient PUF-based key generation [5,15,24]. To generate a 128-bit key with a targeted key error rate $< 10^{-6}$, ECC implementations typically require 3 K-10 K PUF raw response bits (with bit error rate of 15 %) to generate the key, and the helper data generated for this case will be typically 3 K-15 K bits [1]. What is more, the error correcting capability of a specific ECC technology is fixed, if the number of error bits are beyond its fixed ability, the ECC would be useless. Therefore, our key generator is much more flexible, and it can achieve a 128-bit key with error rate $< 10^{-6}$ by using much less overheads compared with ECC.

Uniqueness. Uniqueness is a measure of how uncorrelated the PUFs response numbers are across different chips [1]. We evaluate the inter-chip variation, namely the number of bits which are different between two keys extracted from different PUF numbers. If the PUF produces uniformly distributed and independent random bits, the Hamming distance (HD) of a k-bit response from ideally unique chips should follow a binomial distribution with parameters $N = k$ and $p = 0.5$, and the mean of the HD distribution should be equal to $k/2$, namely the inter-chip variation should be 50 % on average [1].

The inter-chip variations of the three proposed extraction methods to generate 128-bit keys are shown in Fig. 6. The x-axis represents the different bit

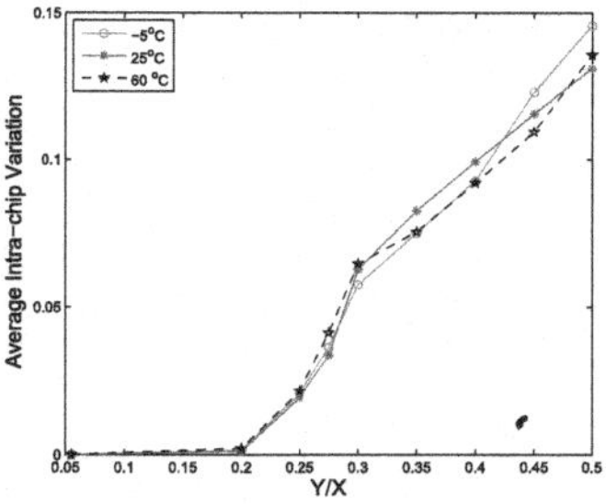

(a) Bit-map: Temperature variations influence.

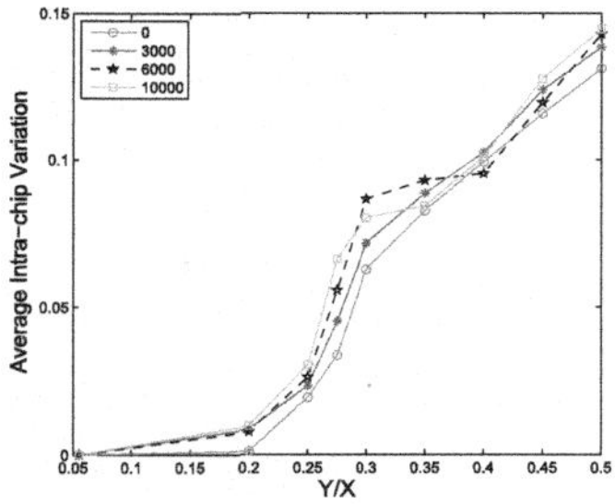

(b) Bit-map: Aging variations influence.

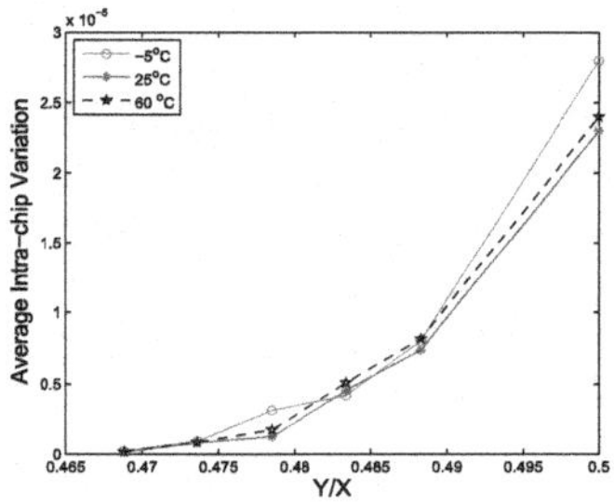

(c) Position-map: Temperature variations influence.

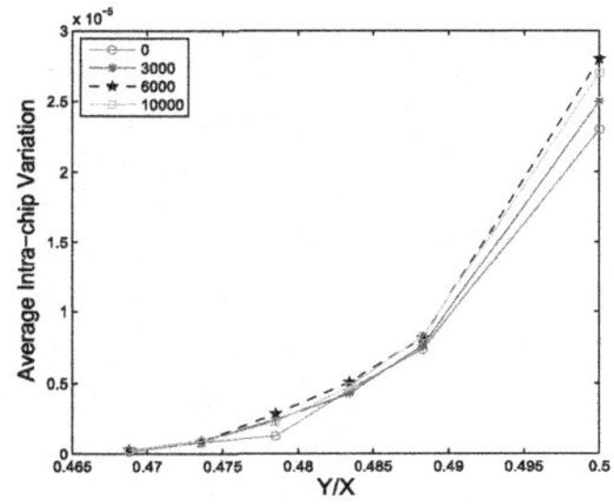

(d) Position-map: Aging variations influence.

Fig. 5. The intra-chip variations with environmental changes.

number of the 128-bit key, and the y-axis represents the according probability. Here, the bars (blue) show the experimental results from 10000 pair-wise comparisons, and the lines (read) show a binomial distribution. As shown in the Fig. 6, the average different bits out of 128 bits are 63.91, 63.94, and 59.98, respectively. The average inter-chip variations of the three methods are 49.93 %, 49.95 % and 46.86 % respectively. The results are all pretty close to the ideal average of 50 %.

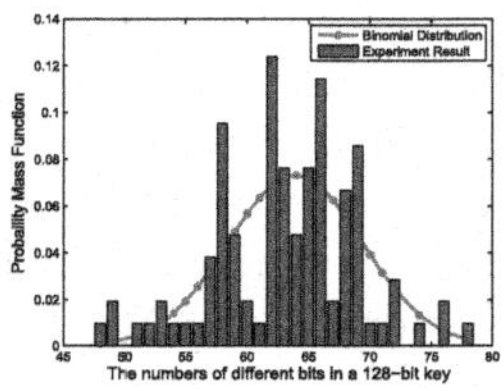

(a) Partial Programming.

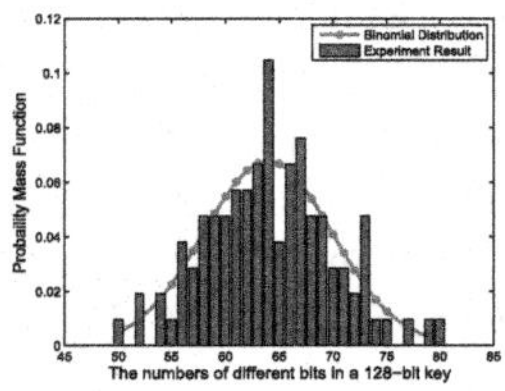

(b) Partial Erasure.

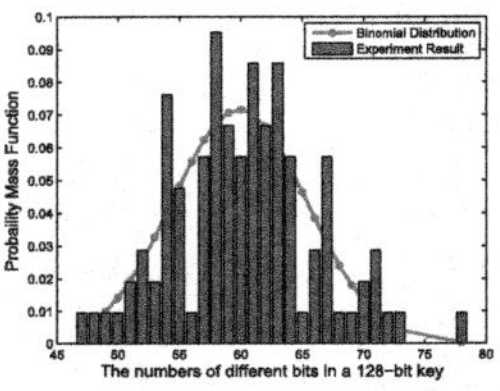

(c) Program Disturbance.

Fig. 6. The inter-chip variations of the three proposed extraction methods.

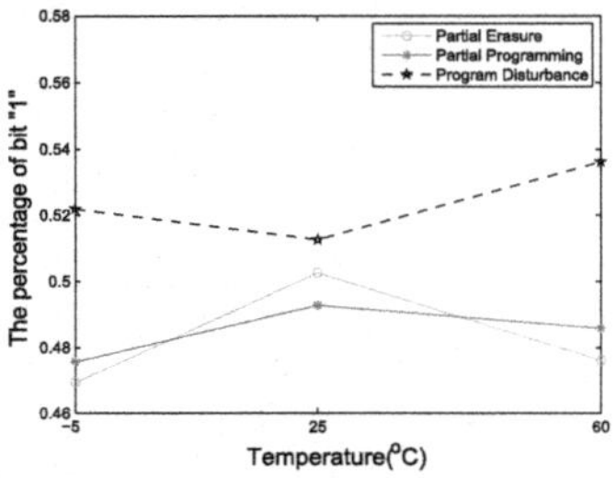

Fig. 7. The percentage of bit "1" with temperature changes.

Randomness. To ensure that the generation of the keys does not favor bits with a certain polarity, we compute the percentage of bit "1" in 10000 groups of 128-bit keys under three temperatures. As Fig. 7 shows, we find that the percentage of bit "1" ranges from 46.94 % to 53.62 %, and it is quite close to ideal 50 % with temperature variations. What's more, we can just leverage Von Neumann skew-correction algorithm to generate uniformly random bits, and use a hash function [11] to ensure the high entropy requirement.

6 Conclusion

In this work, we showed that common NAND Flash memory chips could be used to generate robust keys based on NFPUF. First, we proposed three methods to extract raw NFPUF output numbers from NAND Flash memory chips. Second, we utilized the bit-map or position-map method to select the NAND Flash memory cells with the most reliable relationship of the size between raw NFPUF output numbers. Only the selected cells are used for key generation. At last, we evaluated the primary characteristics of the generated key in various temperature and aging conditions. To our knowledge, this is the first time that a key generator based NFPUF implementation has been evaluated. Our key generator could generate a 128-bit key with a bit error rate $< 10^{-6}$ in 93.83 ms. The bit error rate ensures our key generator is reliable during the whole lifetime of the NAND Flash memory chips. Such low bit error rate is conventionally only achievable using powerful, but costly, error correction codes (ECC). Our key generator eschews the costly ECC overheads to generate robust and error-free keys. This study extends the functionality of NAND Flash memory chips, while requires no hardware change. Due to the widespread use of NAND Flash memory chips, the proposed robust key generator is potential to be widely applied to any electronic encryption devices, as long as the device leverages NAND Flash memory chip as nonvolatile storage.

References

1. Bhargava, M., Mai, K.: An efficient reliable PUF-based cryptographic key generator in 65 nm CMOS. In: Proceedings of the Conference on Design, p. 70. European Design and Automation Association, Automation and Test in Europe (2014)
2. Bösch, C., Guajardo, J., Sadeghi, A.-R., Shokrollahi, J., Tuyls, P.: Efficient helper data key extractor on FPGAs. In: Oswald, E., Rohatgi, P. (eds.) CHES 2008. LNCS, vol. 5154, pp. 181–197. Springer, Heidelberg (2008)
3. Breeuwsma, M., De Jongh, M., Klaver, C., Van Der Knijff, R., Roeloffs, M.: Forensic data recovery from flash memory. Small Scale Digital Device Forensics J. **1**(1), 1–17 (2007)
4. Delvaux, J., Verbauwhede, I.: Attacking PUF-based pattern matching key generators via Helper data manipulation. In: Benaloh, J. (ed.) CT-RSA 2014. LNCS, vol. 8366, pp. 106–131. Springer, Heidelberg (2014)
5. Devadas, S., Yu, M.: Secure and robust error correction for physical unclonable functions. IEEE Des. Test Comput. **27**(1), 48–65 (2010)
6. Dodis, Y., Ostrovsky, R., Reyzin, L., Smith, A.: Fuzzy extractors: how to generate strong keys from biometrics and other noisy data. SIAM J. Comput. **38**(1), 97–139 (2008)
7. Gassend, B., Clarke, D., Van Dijk, M., Devadas, S.: Silicon physical random functions. In: Proceedings of the 9th ACM Conference on Computer and Communications Security, pp. 148–160. ACM (2002)
8. Gassend, B.L.: Physical random functions. Ph.D. thesis, Massachusetts Institute of Technology (2003)
9. Handschuh, H., Trichina, E.: Securing flash technology. In: Fault Diagnosis and Tolerance in Cryptography, FDTC 2007, pp. 3–20. IEEE (2007)
10. Guajardo, J., Kumar, S.S., Schrijen, G.-J., Tuyls, P.: FPGA intrinsic PUFs and their use for IP protection. In: Paillier, P., Verbauwhede, I. (eds.) CHES 2007. LNCS, vol. 4727, pp. 63–80. Springer, Heidelberg (2007)
11. Krawczyk, H.: LFSR-based hashing and authentication. In: Desmedt, Y.G. (ed.) CRYPTO 1994. LNCS, vol. 839, pp. 129–139. Springer, Heidelberg (1994)
12. Lee, J., Heo, J., Cho, Y., Hong, J., Shin, S.Y.: Secure deletion for nand flash file system. In: Proceedings of the 2008 ACM Symposium on Applied Computing, pp. 1710–1714. ACM (2008)
13. Lim, D., Lee, J.W., Gassend, B., Suh, G.E., Van Dijk, M., Devadas, S.: Extracting secret keys from integrated circuits. In: IEEE Transactions on Very Large Scale Integration (VLSI) Systems, vol. 13, no. 10, pp. 1200–1205 (2005)
14. Linnartz, J.P., Tuyls, P.: New shielding functions to enhance privacy and prevent misuse of biometric templates. In: Kittler, J., Nixon, M.S. (eds.) AVBPA 2003. LNCS, vol. 2688, pp. 393–402. Springer, Heidelberg (2003)
15. Maes, R., Van Herrewege, A., Verbauwhede, I.: PUFKY: a fully functional PUF-based cryptographic key generator. In: Prouff, E., Schaumont, P. (eds.) CHES 2012. LNCS, vol. 7428, pp. 302–319. Springer, Heidelberg (2012)
16. Paral, Z., Devadas, S.: Reliable and efficient PUF-based key generation using pattern matching. In: IEEE International Symposium on Hardware-Oriented Security and Trust (HOST), pp. 128–133. IEEE (2011)
17. Prabhu, P., Akel, A., Grupp, L.M., Yu, W.-K.S., Suh, G.E., Kan, E., Swanson, S.: Extracting device fingerprints from flash memory by exploiting physical variations. In: McCune, J.M., Balacheff, B., Perrig, A., Sadeghi, A.-R., Sasse, A., Beres, Y. (eds.) Trust 2011. LNCS, vol. 6740, pp. 188–201. Springer, Heidelberg (2011)

18. Ravikanth, P.S.: Physical one-way functions. Ph.D. thesis, Massachusetts Institute of Technology (2001)
19. Selmi, L., Fiegna, C.: Physical aspects of cell operation and reliability. In: Flash Memories, pp. 153–239. Springer, USA (1999)
20. Škorić, B., Tuyls, P., Ophey, W.: Robust key extraction from physical uncloneable functions. In: Ioannidis, J., Keromytis, A.D., Yung, M. (eds.) ACNS 2005. LNCS, vol. 3531, pp. 407–422. Springer, Heidelberg (2005)
21. Skorobogatov, S.: Flash memory 'Bumping' attacks. In: Mangard, S., Standaert, F.-X. (eds.) CHES 2010. LNCS, vol. 6225, pp. 158–172. Springer, Heidelberg (2010)
22. Subha, S.: An algorithm for secure deletion in flash memories. In: 2nd IEEE International Conference on Computer Science and Information Technology, ICCSIT 2009, pp. 260–262. IEEE (2009)
23. Suh, G.E., Devadas, S.: Physical unclonable functions for device authentication and secret key generation. In: Proceedings of the 44th Annual Design Automation Conference, pp. 9–14. ACM (2007)
24. Suh, G.E., O'Donnell, C.W., Devadas, S.: Aegis: a single-chip secure processor. Inf. Secur. Tech. Rep. **10**(2), 63–73 (2005)
25. Wang, A., Li, Z., Yang, X., Yu, Y.: New attacks and security model of the secure flash disk. Math. Comput. Model. **57**(11), 2605–2612 (2013)
26. Wang, C., Wong, W.F.: Extending the lifetime of nand flash memory by salvaging bad blocks. In: Proceedings of the Conference on Design, Automation and Test in Europe, pp. 260–263. EDA Consortium (2012)
27. Wang, Y., Yu, W.k., Wu, S., Malysa, G., Suh, G.E., Kan, E.C.: Flash memory for ubiquitous hardware security functions: true random number generation and device fingerprints. In: IEEE Symposium on Security and Privacy (SP), pp. 33–47. IEEE (2012)
28. Xu, S.Q., Yu, W.k., Suh, G.E., Kan, E.C.: Understanding sources of variations in flash memory for physical unclonable functions. In: IEEE 6th International Memory Workshop (IMW), pp. 1–4. IEEE (2014)
29. Yu, M.-D.M., M'Raihi, D., Sowell, R., Devadas, S.: Lightweight and secure PUF key storage using limits of machine learning. In: Preneel, B., Takagi, T. (eds.) CHES 2011. LNCS, vol. 6917, pp. 358–373. Springer, Heidelberg (2011)
30. Zambelli, C., Chimenton, A., Olivo, P.: Reliability issues of nand flash memories. In: Inside NAND Flash Memories, pp. 89–113. Springer, Netherlands (2010)

On Security of a White-Box Implementation of SHARK

Yang Shi and Hongfei Fan(✉)

School of Software Engineering, Tongji University, Shanghai, China
{shiyang, fanhongfei}@tongji.edu.cn

Abstract. In a white-box attack context, an attacker has full visibility of the implementation of a cipher and full control over its execution environment. As a countermeasure against the threat of a key exposure in this context, a white-box implementation of the block cipher SHARK, i.e., the white-box SHARK, was proposed in a piece of prior work in 2013. However, based on our observation and investigation, it has been derived that the white-box SHARK is insufficiently secure, where the hidden key and external encodings can be extracted with a work factor of approximately 1.5 * (2 ^ 47).

Keywords: White-box attack contexts · Symmetric encryption · Key exposure · SHARK · Cryptanalysis

1 Introduction

Symmetric encryption is one of the most frequently used techniques for encrypting information. It is a class of algorithms for cryptography that use the same cryptographic keys for both the encryption of plaintexts and the decryption of ciphertexts.

Cryptographic models based on standard symmetric encryption algorithms assume that endpoints such as hosts are secure. However, if those endpoints reside in potentially hostile environments, crackers are able to extract the keys whenever they are used by actively monitoring standard cryptographic functions or memory dumps. Hence, the standard design and implementation of symmetric encryption algorithms are not suitable for "white-box" environments and contexts, where their executions on devices can be observed. In fact, the contexts of adversaries to attack cryptosystems can be categorized in three types [1, 2] as follows.

First is the ***black-box attack context***. It is a traditional attack context where an adversary only has access to the functionality of a cryptosystem. Second is the ***gray-box attack context***, which refers to a model where a leakage function is present. In this attack context, the adversary can deploy side-channel cryptanalysis techniques such as fault analysis, electromagnetic analysis, and power analysis. Because of the large variety of leakage functions, several gray-box models can be defined. However, in many applications, a gray-box attack model is not realistic. The ***white-box attack context*** (WBAC) is the third one where the adversary has total visibility of the software implementation of the cryptosystem and full control over its execution platform. One could refer to the white-box attack context as the worst protection case. The white-box attack context is used to analyze algorithms that are running in a non-trustable

J. Lopez and C.J. Mitchell (Eds.): ISC 2015, LNCS 9290, pp. 455–471, 2015.
DOI: 10.1007/978-3-319-23318-5_25

environment, that is, an environment in which applications are subject to attacks from the execution platform. There exist distinct cryptanalysis techniques applicable to each of the three attack contexts respectively. For black-box attack contexts, the most commonly used cryptanalysis techniques against block ciphers include differential and linear attacks. Typical side-channel cryptanalysis (gray-box attacks) includes time analysis, fault injection, power analysis, and electromagnetic radiation attack. In contrast, powerful cryptanalysis techniques such as memory inspection, CPU call interception, debugging, reverse-engineering, code tampering and entropy attack are available in WBAC.

Secure computing in a WBAC is more challenging than in other two attack contexts because WBAC is commonly subject to three assumptions [3, 4]: (1) fully-privileged attack tools run on the same host as the cryptographic software, which allows the attack software complete access to the implementation of algorithms; (2) dynamic executions (with instantiated cryptographic keys) can be observed; and (3) internal details of cryptographic algorithms are both completely visible and alterable.

Nowadays, with the rapid development of networking, distributed computing, mobile computing, and ubiquitous computing, more and more white-box attack contexts can be discovered in various IT systems and computing devices. Some typical white-box attack contexts include: (1) a server or an endpoint for which a hacker has got the "root" or "admin" privilege; (2) a malicious host where mobile agents are running [5, 6]; (3) an outdoor node of a wireless sensor network captured by an attacker [7, 8]; (4) Digital Rights Management (DRM) components in TV set-top boxes or IPTV equipments [9, 10]; (5) On Board Units (OBUs) and Road Side Units (RSUs) in a VANet (e.g., the device suffering from the so-called "Industrial Insiders" attack in [11], the "on-board tampering" attack in [12], or even the "Malware attack" in [13]); and (6) mobile devices (e.g., smart phones and tablets) captured by an attacker [14].

To protect implementations of symmetric encryption algorithms in WBACs, specialized obfuscating techniques are urgently demanded. A white-box encryption algorithm (WBEA) is a particular form or implementation of a block cipher with strong security in terms of preventing attackers from extracting the embedded cryptographic key, even if the attackers are positioned in a WBAC. Since it was introduced in 2002 by Chow et al. [4], many WBEAs were proposed, such as in [3–5, 15–20]. All of these WBEAs have been found insecure except the white-box SHARK in [5].

In this paper, we demonstrate that white-box SHARK is not sufficiently secure. By a set of theoretical analysis on the design of the white-box SHARK components, it is found that the cryptographic key can be extracted from a white-box implementation with a work factor of 1.5×2^{47} at most.

The remainder of this paper is organized as follows. In Sect. 2, we briefly review recent advances in white-box cryptography and cryptanalysis. The block cipher SHARK and the white-box SHARK proposed in the piece of prior work [5] are reviewed and sketched out in Sect. 3. In Sect. 4 we provide the theoretical analysis of the white-box SHARK implementation. In Sect. 5, we propose an approach that can extract the key and even external encodings from a white-box implementation of SHARK. Finally, concluding remarks are summarized and presented in Sect. 6.

2 Recent Advances in White-Box Cryptography and Cryptanalysis

WBEAs can be classified into two categories: light-weight WEBAs and heavy-weight WBEAs. Light-weight WBEAs are used to provide time-limited protection only in resource-constrained scenarios such as wireless sensor networks. Heavy-weight WBEAs commonly require large space and lead to relatively low efficiency, but they have been designed for providing long-term protection. Existing research on heavy-weight WBEAs has been mainly concerned with white-box implementations of classical symmetric encryption algorithms, such as DES and AES.

Chow et al. [4] proposed a white-box implementation of DES by interleaving affine transformations and using de-linearization techniques. Chow et al. [3] implemented white-box AES by representing it with a set of key-dependent look-up tables. They suggested the use of these two white-box encryption algorithms in DRM applications to protect digital information content and the associated usage rights from unauthorized access, use, and dissemination. These two works form the foundation of almost all white-box encryption papers. Many attacks are proposed against [3, 4], and these two algorithms are insecure now. The next two paragraphs introduce attacks on [3, 4].

Jacob et al. [21] proposed a fault injection based attack, where an attacker injects errors into the environment during program execution, to defeat some obfuscation methods. They presented a cryptanalysis of the naked variant of the Chow et al.'s white-box DES, that is, a variant without external encodings. Similar to Chow et al.'s white-box DES, Link et al. [18] implemented white-box DES and white-box triple-DES with alterations that improved the security of the key. Their system is secure against the previously published attacks on Chow et al.'s white-box DES implementation and their own adaptation of a statistical bucketing attack. In 2007, Wyseur et al. [22] and Goubin et al. [23], independently of each other, cryptanalyzed all existing obfuscation methods of DES. These attacks were based on a truncated differential cryptanalysis. Goubin, et al. presented an attack that analyzed the first rounds of the white-box DES implementations, while Wyseur et al. presented an attack that works on the internal information. Hence, none of proposed white-box DES implementations are secure.

Billet et al. [24] presented an efficient practical attack against the obfuscated AES implementation proposed by Chow et al., with negligible memory and worst time complexity of 2^{30}. In 2009, Michiels et al. [25] generalized the attack so that it can be deployed on a generic class of white-box implementations. The most time-consuming part of Billet et al.'s attack [24] is finding the used byte permutation up to an affine mapping, which takes a time-complexity of 2^{24} in the worst situation. In 2012, Tolhuizen [26] provided a variation on this part of the attack, reducing the time complexity to at most 2^{14}. With the help of this improvement, the overall worst time complexity of breaking Chow et al.'s white-box AES in [3] is reduced from 2^{30} to 2^{20}.

Bringer et al. [20] proposed a solution to address the issue of white box representation of block ciphers by applying perturbation on algebraic structures of a cipher. The authors claimed that the techniques can be applied to a variant of the AES block cipher for which the S-Boxes are all different.

Unfortunately, De Mulder et al. [27] showed that the perturbation procedures and random equation calculations can be separated from the implementation, and the linear input and output encodings can be removed. Accordingly, attackers can transform the white-box implementation into a simple but functionally equivalent implementation and then retrieve a set of keys that are equivalent to the original key. The cryptanalysis of the white-box AES in [20] has a worst-case work factor of 2^{17}.

Xiao et al. [15] proposed a white-box AES implementation after a detailed analysis of attack techniques in [24]. The size of this implementation is considerably large to achieve a higher security level. In Xiao et al.'s scheme, the obfuscation works on at least two cells of an AES state; moreover, the attacker cannot divide it into smaller units (e.g. one cell of an AES state) and remove it using the attack techniques proposed in [24]. The time complexity of the Xiao-Lai white-box AES implementation is 2^{24}, which is slower than the Chow et al.'s implementation in [3] (2^{20}), and the size is 20502 KB.

De Mulder et al. [28] presented a practical cryptanalysis of Xiao et al.'s white-box AES. They applied the linear equivalence algorithm presented by Biryukov et al. [29] as a building block in their key-extraction algorithm. The cryptanalysis efficiently extracts the AES key with a work factor of about 2^{32}.

Another white-box implementation for AES is proposed by Karroumi [19]. This implementation makes InvSubBytes and InvMixColumns operations variable by using additional sets of coefficients taken from dual representations of AES. Karroumi claimed that the expected security level is raised from 2^{30} to 2^{91}. However, an algebraic analysis [30] was proposed in 2013 and Karroumi's implementation can be easily broken.

The two key factors of a white-box encryption algorithm are size and security level. Unfortunately, in many cases, the two key factors form a tradeoff and cannot be achieved simultaneously. The goal of light-weight WBEAs proposed in prior work [16, 17] is to make the size of implementations as small as possible in order to meet the requirements on computing with mobile agents and sensor nodes. In general, these white-box implementations only intend to support time-limited black box security [31].

As summarized in Table 1, white-box implementations designed for providing long term protections (i.e., heavy-weight WBEAs) are listed in the left column with the corresponding attacks listed in the right column. The latest published heavy-weight WBEA (i.e., the white-box SHARK [5]) is also included.

3 SHARK and its White-Box Implementation

SHARK [32] is a six round substitution-permutation-network that alternates a key mixing stage with linear and non-linear transformation layers. Each round of the SHARK encryption algorithm can be split into three distinct layers: a non-linear confusion layer of S-Boxes, a linear diffusion layer and a key addition (XOR) layer.

Let $S : GF(2^8) \rightarrow GF(2^8), x \mapsto S[x]$ be the mapping of S-Boxes. Then the substitution layer can be defined as $\gamma : GF(2^8)^8 \rightarrow GF(2^8)^8$, $\gamma(a) = b \Leftrightarrow b_i = S[a_i]$, $0 \leq i \leq 7$.

Table 1. Heavy-weight WBEAs and cryptanalysis

WBEA	Cryptanalysis
White-box DES [4] (Chow et al., 2002)	Jacob, Boneh and Felten in [21], 2002
	Wyseur, Michiels, Gorissen and Preneel in [22], 2007
	Goubin, Masereel, and Quisquater in [23], 2007
White-box DES [18] (Link and Neumann, 2005)	Wyseur, Michiels, Gorissen and Preneel in [22], 2007
	Goubin, Masereel, and Quisquater in [23], 2007
White-box AES [3] (Chow et al., 2002)	Billet, Gilbert and Ech-Chatbi in [24], 2004
	Tolhuizen in [26], 2012 (an improvement of [24])
	Lepoint, Rivain, De Mulder, Roelse and Bart Preneel in [30], 2013 (an improvement of [24])
A generic construction base on [3]	Michiels, Gorissen and Hollmann in [25], 2008
Perturbated White-Box AES [20] (Bringer et al., 2006)	De Mulder, Wyseur, and Preneel in [27], 2010
White-box AES [15] (Xiao and Lai, 2009)	De Mulder, Roelse and Preneel in [28], 2013
White-box AES with dual ciphers [19] (Karroumi, 2011)	Lepoint, Rivain, De Mulder, Roelse and Preneel in [30], 2013
White-box SHARK [5] (Shi et al., 2013)	This paper

Let $\lambda : GF(2^8)^8 \to GF(2^8)^8$ denote the linear transformation corresponding to the linear diffusion layer. There exists a matrix H such that $\lambda(a) = b \Leftrightarrow b = a \cdot H$.

Furthermore, let K^r be the round key of the rth round and let $\sigma[K^r] : GF(2^8)^8 \to GF(2^8)^8$ be the key exclusive or mapping.

Then, the encryption algorithm of SHARK with cryptographic key K can be defined by:

$$SHARK[K] = \lambda^{-1} \circ \left(\mathop{\circ}_{r=1}^{6} \sigma[K^r] \circ \lambda \circ \gamma \right) \circ \sigma[K^0] \tag{1}$$

An overall illustration of the SHARK workflow is shown in Fig. 1.

In constructing the white-box SHARK according to [5], the following two steps are applied. Firstly, for each round r, the rth round function of the block cipher SHARK is transformed into a random "dual version" by applying a transformation Δ_r. Δ_r is an isomorphic transformation that takes the description of (a round of) the cipher under the standard irreducible polynomial to another description with a different irreducible polynomial. Secondly, several operations of each dual round function are merged into lookup tables which are blended by randomly generated mixing bijections. It should be noted that each Δ_r is kept secret from attackers and corresponds to 8160 possible choices. Hence, the authors of [5] claimed that Δ_r makes an attack almost 8160 times harder than a white-box encryption algorithm being protected by the second step only.

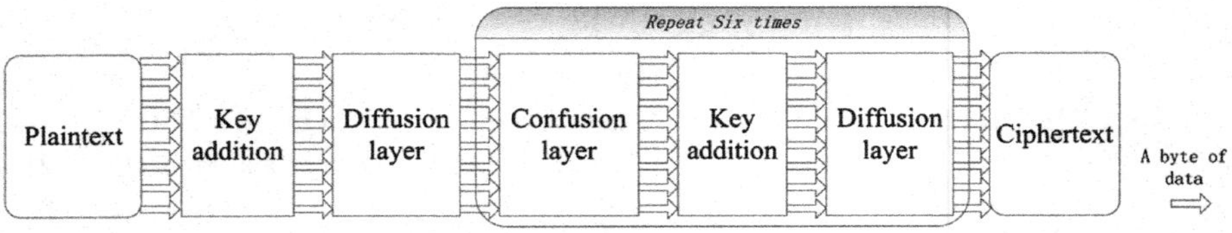

Fig. 1. The workflow of SHARK

Let M_{64} be the set that consists of all 64 × 64 nonsingular binary matrices and M_{16} be the set that consists of all 16 × 16 nonsingular binary matrices. In addition, the symbol $\in_{\$}$ denotes a randomly selected element of a set. Components of the white-box SHARK are defined as follows.

Let $Q^{(r,i)} \in_{\$} M_{16},\ r = 1, \ldots, 6, i = 0, \ldots, 3,$ and $N^{(r)} \in_{\$} M_{64},\ r = 0, \ldots, 6$. For each $i \in \{0, 1, 2, 3\}$, let $L_i^{(0)} = \Delta_0 \circ \left(\cdot Q^{(0,i)}\right)$ and ss be pre-round mixing bijections.

Let H be the 64 × 64 binary matrix that corresponds to the diffusion operation of SHARK. The post-round diffusion-mixing bijections are given by (2) and (3) as follows.

$$P^{(r)} = \Delta_r(H) \cdot N^{(r)} \stackrel{def}{=} \begin{bmatrix} P_0^{(r)} \\ P_1^{(r)} \\ P_2^{(r)} \\ P_3^{(r)} \end{bmatrix}, r = 0, \ldots, 5 \tag{2}$$

$$P^{(6)} = \Delta_6\left(H^{-1}\right) \cdot N^{(6)} \stackrel{def}{=} \begin{bmatrix} P_0^{(6)} \\ P_1^{(6)} \\ P_2^{(6)} \\ P_3^{(6)} \end{bmatrix} \tag{3}$$

Let $k_{2i}^{(r)} \Big\| k_{2i+1}^{(r)}$ be the $(2i)^{th}$ and $(2i+1)^{th}$ bytes of the rth round-key, and K be the cipher key. Lookup tables of the white-box SHARK are generated by using (4) and (5) as follows:

$$\rho_W[r, i, K](x) = \left((S\|S)_{\Delta r}\left(\left(L_i^{(r)}(x)\right) \oplus \left(\Delta_r\left(k_{2i}^{(r)} \Big\| k_{2i+1}^{(r)}\right)\right)\right)\right) P_i^{(r)}, r = 0, \ldots, 5;\ i = 0, \ldots, 3 \tag{4}$$

$$\rho_W[6, i, K](x) = \left(\left(L_i^{(6)}(x)\right) \oplus \left(\Delta_6\left(k_{2i}^{(6)} \Big\| k_{2i+1}^{(6)}\right)\right)\right) P_i^{(6)}, i = 0, \ldots, 3, \tag{5}$$

where $S\|S$ denotes two SHARK S-Boxes (Substitution-Boxes) operating in parallel and $(S\|S)_{\Delta r}$ is given by (6):

$$(S\|S)_{\Delta r}\colon GF(2^8)^2 \to GF(2^8)^2, x \mapsto \Delta_r\left((S\|S)\left(\Delta_r^{-1}(x)\right)\right). \tag{6}$$

Let $M^{(r)}$ be a 64×64 nonsingular binary matrix, defined for $r = 1, \ldots, 6$ as in (7).

$$M^{(r)} = \left(N^{(r-1)}\right)^{-1} \cdot \begin{bmatrix} \left(Q^{(r,0)}\right)^{-1} & & & \\ & \left(Q^{(r,1)}\right)^{-1} & & \\ & & \left(Q^{(r,2)}\right)^{-1} & \\ & & & \left(Q^{(r,3)}\right)^{-1} \end{bmatrix} \tag{7}$$

Based on the components described above, the structure of the round function of white-box SHARK is shown in Fig. 2; the workflow of white-box SHARK is shown in Fig. 3; and the encryption algorithm is then presented following the two figures. Note that the structure of the last round function is slightly different from others, and readers may refer to [5] for details.

Algorithm $SHARK_W[K]$ (on input x):

```
1:  r ← 0
2:      (x0, x1, x2, x3) ← x
3:      i ← 0
4:          yi ← TBox_{r,i}[xi]     //Lookup in a T-Box
5:          i ← i+1
6:          if (i < 4)  goto(4);  else  goto(7)
7:      x ← y0 ⊕ y1 ⊕ y2 ⊕ y3
8:      if (r < 7)  goto(9);  else  goto(11)
9:      x ← x·M^(r)
10:     r ← r+1;  goto(2)
11: output x
```

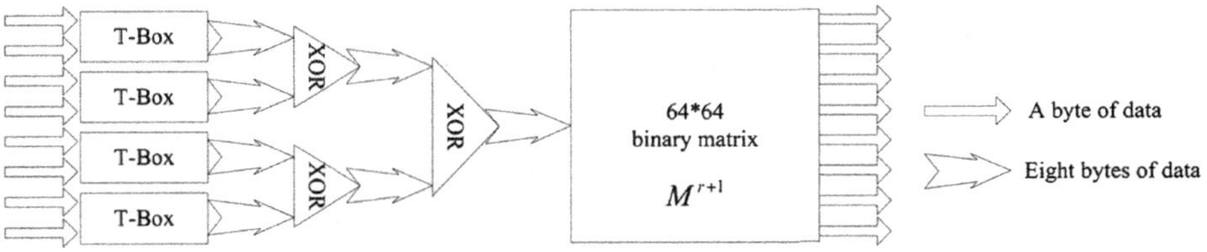

Fig. 2. The structure of the round function of the white-box SHARK

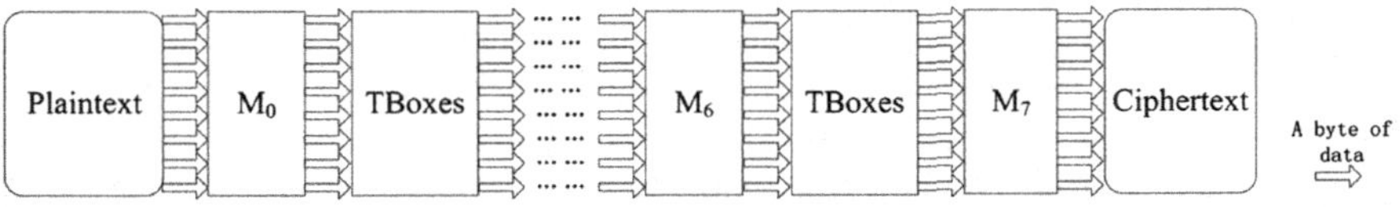

Fig. 3. The workflow of the white-box SHARK

Note that each $TBox_{r,i}$ in the algorithm is a lookup table in correspondence to the functionality of $\rho_W[r, i, K]$ that is given by (4) or (5).

4 Theoretical Analysis of T-Boxes

The security of white-box SHARK partially relies on the secret isomorphic transformation Δ_r that is used to transform each round into a dual version. One major step in breaking the white-box SHARK is to defeat the extra protection provided by Δ_r. By means of Theorem 1 below, we found that the transformation Δ_r can be "merged into" the secret transformation embedded inside the T-Boxes. Consequently, the secret transformation Δ_r cannot effectively protect the implementation in WBACs.

Suppose that

$$H = \begin{bmatrix} H_0 \\ H_1 \\ H_2 \\ H_3 \end{bmatrix} \tag{8}$$

and

$$H^{-1} = \begin{bmatrix} H'_0 \\ H'_1 \\ H'_2 \\ H'_3 \end{bmatrix}, \tag{9}$$

where H_0, H_1, H_2, H_3 and H'_0, H'_1, H'_2, H'_3 are 16×64 block binary matrices, and we present the following theorem with the proof.

Theorem 1.

$$\forall r \in \{0,1,2,3,4,5\}, \forall i \in \{0,1,2,3\}, \exists A^{(r,i)} \in \mathrm{M}_{16} \wedge \exists B^{(r)} \in \mathrm{M}_{64}, st.$$

$$\rho_W[r,i,K](x) = \left((S\|S)\left(x \cdot A^{(r,i)} \oplus \left(k_{2i}^{(r)} \middle\| k_{2i+1}^{(r)}\right)\right)\right) \cdot H_i \cdot B^{(r)} \tag{10}$$

and

$$\forall i \in \{0,1,2,3\}, \exists A^{(6,i)} \in \mathrm{M}_{16} \wedge \exists B^{(6)} \in \mathrm{M}_{64}, st.$$

$$\rho_W[6,i,K](x) = \left((S\|S)\left(x \cdot A^{(6,i)} \oplus \left(k_{2i}^{(6)} \middle\| k_{2i+1}^{(6)}\right)\right)\right) \cdot H'_i \cdot B^{(6)} \tag{11}$$

Proof. The isomorphic mapping Δ transforms the description of the cipher under the standard irreducible polynomial into another description under a distinct irreducible polynomial. Transparently, the mapping is linear. Hence, there exists a 16×16 non-singular binary matrix D_r corresponding to the isomorphic transformation Δ_r, $r = 0, 1, \ldots, 6$.

For $r = 1, \ldots, 5$, we have

$$
\begin{aligned}
\rho_W[r,i,K](x) &= \left((S\|S)_{\Delta r}\left(\left(L_i^{(r)}(x)\right) \oplus \left(\Delta_r\left(k_{2i}^{(r)}\middle\|k_{2i+1}^{(r)}\right)\right)\right)\right)P_i^{(r)} \\
&= \left(\Delta_r\left((S\|S)\left(\Delta_r^{-1}\left(\left(L_i^{(r)}(x)\right) \oplus \left(\Delta_r\left(k_{2i}^{(r)}\middle\|k_{2i+1}^{(r)}\right)\right)\right)\right)\right)\right)P_i^{(r)} \\
&= \left(\Delta_r\left((S\|S)\left(\Delta_r^{-1}\left(\left(\Delta_r \circ \Delta_{r-1}^{-1} \circ \left(\cdot Q^{(r,i)}\right)(x)\right) \oplus \left(\Delta_r\left(k_{2i}^{(r)}\middle\|k_{2i+1}^{(r)}\right)\right)\right)\right)\right)\right)P_i^{(r)} \\
&= \left((S\|S)\left(\Delta_r^{-1}\left(\left(x \cdot Q^{(r,i)} \cdot D_{r-1}^{-1} \cdot D_r\right) \oplus \left(k_{2i}^{(r)}\middle\|k_{2i+1}^{(r)}\right) \cdot D_r\right)\right)\right) \cdot D_r \cdot P_i^{(r)} \\
&= \left((S\|S)\left(\left(\left(\left(x \cdot Q^{(r,i)} \cdot D_{r-1}^{-1}\right) \oplus \left(k_{2i}^{(r)}\middle\|k_{2i+1}^{(r)}\right)\right) \cdot D_r\right) \cdot D_r^{-1}\right)\right) \cdot D_r \cdot P_i^{(r)} \\
&= \left((S\|S)\left(x \cdot Q^{(r,i)} \cdot D_{r-1}^{-1} \oplus \left(k_{2i}^{(r)}\middle\|k_{2i+1}^{(r)}\right)\right)\right) \cdot D_r \cdot P_i^{(r)}
\end{aligned}
$$

and

$$
\begin{aligned}
&P^{(r)} \\
&= (\Delta_r(H)) \cdot N^{(r)} \\
&= \begin{bmatrix} H_{0,0} & H_{0,1} & H_{0,2} & H_{0,3} \\ H_{1,0} & H_{1,1} & H_{1,2} & H_{1,3} \\ H_{2,0} & H_{2,1} & H_{2,2} & H_{2,3} \\ H_{3,0} & H_{3,1} & H_{3,2} & H_{3,3} \end{bmatrix} \cdot \begin{bmatrix} D_r & & & \\ & D_r & & \\ & & D_r & \\ & & & D_r \end{bmatrix} \cdot N^{(r)}, \\
&= \begin{bmatrix} H_0 \\ H_1 \\ H_2 \\ H_3 \end{bmatrix} \cdot \begin{bmatrix} D_r & & & \\ & D_r & & \\ & & D_r & \\ & & & D_r \end{bmatrix} \cdot N^{(r)}
\end{aligned}
$$

where each $H_{i,j}$ is a 16×16 block matrix.

Let

$$
B^{(r)} = \begin{bmatrix} D_r & & & \\ & D_r & & \\ & & D_r & \\ & & & D_r \end{bmatrix} \cdot N^{(r)}
$$

and

$$
\begin{aligned}
&A^{(r)} \\
&= \begin{bmatrix} A^{(r,0)} & & & \\ & A^{(r,1)} & & \\ & & A^{(r,2)} & \\ & & & A^{(r,3)} \end{bmatrix}
\end{aligned}
$$

$$\stackrel{def}{=}\begin{bmatrix} Q^{(r,0)}\cdot D_{r-1}^{-1} & & & \\ & Q^{(r,1)}\cdot D_{r-1}^{-1} & & \\ & & Q^{(r,2)}\cdot D_{r-1}^{-1} & \\ & & & Q^{(r,3)}\cdot D_{r-1}^{-1} \end{bmatrix}$$

Thus,

$$\begin{aligned}
&\rho_W[r,i,K](x)\\
&=\left((S\|S)\left(x\cdot Q^{(r,i)}\cdot D_{r-1}^{-1}\oplus\left(k_{2i}^{(r)}\middle\|k_{2i+1}^{(r)}\right)\right)\right)\cdot D_r\cdot P_i^{(r)}\\
&=\left((S\|S)\left(x\cdot A^{(r,i)}\oplus\left(k_{2i}^{(r)}\middle\|k_{2i+1}^{(r)}\right)\right)\right)\cdot H_i\cdot B^{(r)}
\end{aligned}$$

For $r=0$, we have

$$\begin{aligned}
&\rho_W[0,i,K](x)\\
&=\left((S\|S)_{\Delta 0}\left(\left(L_i^{(0)}(x)\right)\oplus\left(\Delta_0\left(k_{2i}^{(0)}\middle\|k_{2i+1}^{(0)}\right)\right)\right)\right)P_i^{(0)}\\
&=\left(\Delta_0\left((S\|S)\left(\Delta_0^{-1}\left(\left(L_i^{(0)}(x)\right)\oplus\left(\Delta_0\left(k_{2i}^{(0)}\middle\|k_{2i+1}^{(0)}\right)\right)\right)\right)\right)\right)P_i^{(0)}\\
&=\left(\Delta_0\left((S\|S)\left(\Delta_0^{-1}\left(\left(\Delta_0\circ\left(\cdot Q^{(0,i)}\right)(x)\right)\oplus\left(\Delta_0\left(k_{2i}^{(0)}\middle\|k_{2i+1}^{(0)}\right)\right)\right)\right)\right)\right)P_i^{(0)}\\
&=\left((S\|S)\left(x\cdot Q^{(0,i)}\oplus\left(k_{2i}^{(0)}\middle\|k_{2i+1}^{(0)}\right)\right)\right)\cdot D_0\cdot P_i^{(0)}
\end{aligned}$$

Hence, the matrices given by (12) and (13) would satisfy (10) when $r=0$.

$$B^{(0)}=\begin{bmatrix} D_0 & & & \\ & D_0 & & \\ & & D_0 & \\ & & & D_0 \end{bmatrix}\cdot N^{(0)} \tag{12}$$

$$A^{(0,i)}=\begin{bmatrix} Q^{(0,0)} & & & \\ & Q^{(0,1)} & & \\ & & Q^{(0,2)} & \\ & & & Q^{(0,3)} \end{bmatrix} \tag{13}$$

As to the final round, similarly, we have

$$\begin{aligned}
&\rho_W[6,i,K](x)\\
&=\left(\left((S\|S)_{\Delta 6}\left(\left(L_i^{(6)}(x)\right)\oplus\left(\Delta_6\left(k_{2i}^{(6)}\middle\|k_{2i+1}^{(6)}\right)\right)\right)\right)P_i^{(6)}\right.\\
&=\left((S\|S)\left(x\cdot Q^{(6,i)}\cdot D_5^{-1}\oplus\left(k_{2i}^{(6)}\middle\|k_{2i+1}^{(6)}\right)\right)\right)\cdot D_6\cdot P_i^{(6)}
\end{aligned}$$

and

$$\begin{aligned}
&P^{(6)}\\
&=\left(\Delta_6\left(H^{-1}\right)\right)\cdot N^{(6)}\\
&=\begin{bmatrix} H_{0,0} & H_{0,1} & H_{0,2} & H_{0,3}\\ H_{1,0} & H_{1,1} & H_{1,2} & H_{1,3}\\ H_{2,0} & H_{2,1} & H_{2,2} & H_{2,3}\\ H_{3,0} & H_{3,1} & H_{3,2} & H_{3,3}\end{bmatrix}^{-1}\cdot\begin{bmatrix} D_6 & & & \\ & D_6 & & \\ & & D_6 & \\ & & & D_6\end{bmatrix}\cdot N^{(6)}\\
&=\begin{bmatrix} H'_0\\ H'_1\\ H'_2\\ H'_3\end{bmatrix}\cdot\begin{bmatrix} D_6 & & & \\ & D_6 & & \\ & & D_6 & \\ & & & D_6\end{bmatrix}\cdot N^{(6)}
\end{aligned}$$

where each $H_{i,j}$ is a 16×16 block matrix.
Let

$$B^{(6)}=\begin{bmatrix} D_6 & & & \\ & D_6 & & \\ & & D_6 & \\ & & & D_6\end{bmatrix}\cdot N^{(6)}$$

and

$$\begin{aligned}
&A^{(6)}\\
&=\begin{bmatrix} A^{(6,0)} & & & \\ & A^{(6,1)} & & \\ & & A^{(6,2)} & \\ & & & A^{(6,3)}\end{bmatrix}\\
&\stackrel{def}{=}\begin{bmatrix} Q^{(6,0)}\cdot D_5^{-1} & & & \\ & Q^{(6,1)}\cdot D_5^{-1} & & \\ & & Q^{(6,2)}\cdot D_5^{-1} & \\ & & & Q^{(6,3)}\cdot D_5^{-1}\end{bmatrix}
\end{aligned}$$

Thus, $\rho_W[6,i,K](x)=\left((S\|S)\left(x\cdot A^{(6,i)}\oplus\left(k_{2i}^{(6)}\middle\|k_{2i+1}^{(6)}\right)\right)\right)\cdot H'_i\cdot B^{(6)}$.

This ends the proof.

Based on Theorem 1, an attacker can regard the white-box SHARK as a series of protected lookup tables which directly correspond to the (round functions of) standard SHARK, and the protection are only provided by means of secret linear mappings $A^{(r,i)}$ and $B^{(r)}$ in fact.

5 Extracting the Embedded Key

To demonstrate that the vulnerability identified by the theoretic analysis in the previous section can be utilized to break the white-box SHARK, we propose an algorithm for extracting the key and even external encodings from a white-box implementation of SHARK. As the proposed algorithm incorporates an extended version of the linear equivalence (LE) algorithm as a building block, we briefly review the LE algorithm and its extension first.

For a pair of given S-Boxes S' and S'', if there exists a pair of linear mappings A and B such that $A \circ S' \circ B = S''$, we say that S' and S'' are linear equivalent. Clearly, a brutal-force implementation would start by guessing the mapping B and then computing the mapping A using the equation $A = S'' \circ (S' \circ B)^{-1}$. If A turns out to be linear as well, then we have found a solution; if not, we try again with a different guess. The worst-case work factor is $n^3 \times 2^{n^2}$ for n-bit to n-bit S-Boxes and thus the attack would be impossible.

Fortunately, the LE algorithm proposed in [29] is a powerful and efficient tool for cryptanalysis with a much smaller work factor at approximately $n^3 \times 2^n$. The general idea of the LE algorithm is to guess the linear mapping A by the fewest input points possible, and then use the linearity of A and B to continue with these guesses as far as possible. The inputs of the algorithm are S' and S'', whereas the output is either a linear equivalence (A, B) in case that S' and S'' are linearly equivalent, or a message that such a linear equivalence does not exist. The LE process starts by selecting two distinct non-zero input values $x1$ and $x2$, and guesses the values of $A(x1)$ and $A(x2)$. Then, the two equations $B(yi) = S''(xi)$ and $yi = S'(A(xi))$ are utilized to retrieve the information about B. Furthermore, new information about A is obtained according to the equation $A(x3) = S'^{-1}(y1 \oplus y2)$ where $x3 = S''^{-1}(B(y1) \oplus B(y2))$. This process is applied iteratively, and the linearity of the partially determined mappings A and B can be verified by a Gaussian elimination. The entire process is briefly illustrated in Fig. 4; for in-depth descriptions of the LE algorithm, please refer to [29].

The original linear equivalence algorithm LE terminates upon finding one single linear equivalence which adequately proves that both given S-Boxes, i.e. S' and S'', are linearly equivalent. However, as suggested in [28], by executing LE over all possible guesses (i.e., both initial guesses and possible additional guesses made during the execution of LE), other linear equivalences (A, B) can also be found. Therefore, we can adopt the techniques in [28] to perform further cryptanalysis. Such an extended variant of LE can be denoted as ELE (Extended LE). Consequently, the work factor of ELE is at least $n^3 \times 2^{2n}$, i.e., a Gaussian elimination (n^3) for each possible pair of initial guesses (2^{2n}).

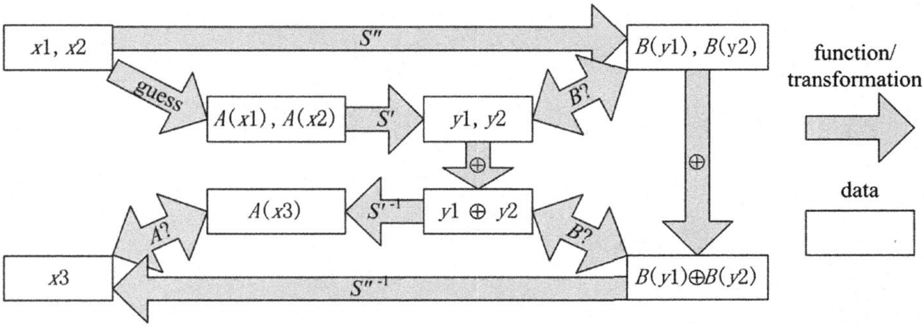

Fig. 4. The process of LE

One remaining task for breaking the white-box SHARK is to propose an efficient approach for constructing a pair of known S-Boxes that are linearly equivalent. The theoretic foundation of such an approach is provided by Theorem 2 below. Obviously, $\exists s_0 \in GF(2)^8, S(s_0) = 0$. Let $S'' = (S||S) \circ (\oplus_{s_0||s_0})$. Theorem 2 demonstrates that for each pair $\{r, i\}$, an S-Box which is linearly equivalent to S'' can be derived from $\rho_W[r, i, K]$.

Theorem 2. Suppose that $\rho_W[r, i, K]\left(z_i^{(r)}\right) = 0$ where $z_i^{(r)} \in GF(2)^{16}$.

Let $S'_{r,i}(x) = \rho_W[r, i, K]\left(x \oplus z_i^{(r)}\right)$. $\forall i \in \{0, \cdots, 3\}$, we have

$$S'_{r,i} = \left(\cdot A^{(r,i)}\right) \circ S'' \circ \left(\cdot H_i \cdot B^{(r)}\right), r = 0, \cdots, 5 \tag{14}$$

and

$$S'_{6,i} = \left(\cdot A^{(6,i)}\right) \circ S'' \circ \left(\cdot H'_i \cdot B^{(6)}\right) \tag{15}$$

Proof. We only prove (14) whereas the proof of (15) is omitted, because it can be acquired by similar deductions.

$$0 = \rho_W[r, i, K]\left(z_i^{(r)}\right) = \left((S||S)\left(z_i^{(r)} \cdot A^{(r,i)} \oplus \left(k_{2i}^{(r)} \middle\| k_{2i+1}^{(r)}\right)\right)\right)$$
$$s_0||s_0 = z_i^{(r)} \cdot A^{(r,i)} \oplus \left(k_{2i}^{(r)} \middle\| k_{2i+1}^{(r)}\right)$$

Hence,

$$
\begin{aligned}
&S'_{r,i}(x)\\
&= \left((S\|S)\left(\left(x \oplus z_i^{(r)}\right) \cdot A^{(r,i)} \oplus \left(k_{2i}^{(r)} \middle\| k_{2i+1}^{(r)}\right)\right)\right) \cdot H_i \cdot B^{(r)}\\
&= \left((S\|S)\left(\left(x \cdot A^{(r,i)} \oplus z_i^{(r)} \cdot A^{(r,i)}\right) \oplus \left(k_{2i}^{(r)} \middle\| k_{2i+1}^{(r)}\right)\right)\right) \cdot H_i \cdot B^{(r)}\\
&= \left((S\|S)\left(\left(x \cdot A^{(r,i)} \oplus (s_0\|s_0) \oplus \left(k_{2i}^{(r)} \middle\| k_{2i+1}^{(r)}\right)\right) \oplus \left(k_{2i}^{(r)} \middle\| k_{2i+1}^{(r)}\right)\right)\right) \cdot H_i \cdot B^{(r)}\\
&= \left((S\|S)\left(x \cdot A^{(r,i)} \oplus (s_0\|s_0)\right)\right) \cdot H_i \cdot B^{(r)}
\end{aligned}
$$

This ends the proof.

According to Theorem 2, when $S'_{r,i}$ and S'' are analyzed by applying the ELE algorithm, $A^{(r,i)}$ and $B^{(r)}$ are obtained with work factor of 2^{44}, i.e., $n^3 \times 2^{2n}$ for $n = 16$. It is feasible to break a round by running the ELE algorithm four times.

From the specification of SHARK, we know that the 128-bit cipher key is used as input of SHARK with a public-known encryption key (PEK) and a public-known initial value (PIV) in 64-bit cipher feedback (CFB) mode. Then the output bits are used as the actual round keys for the encryption of the message. Hence, the cipher key can be recovered from the first two round-keys. Moreover, the external encodings can be recovered by analyzing the first and the last rounds. Thus, the overall work factor is dominated by running the ELE algorithm 12 times, i.e., $12 \times 2^{44} = 1.5 \times 2^{47}$. Although the work factor is rather large, the time efficiency could be optimized by executing the ELE algorithm programs in parallel (multi-threading) on a workstation with a group of multi-core processors. For example, on a medium-level workstation with 2 hexa-core processors or 4 quad-core processors, the work factor would decrease to 2^{44}. To achieve higher efficiency, the attacking software could be implemented based on CUDA or OpenCL by utilizing GPU computational techniques and capabilities. In such case, the Gaussian eliminations could be significantly accelerated [33, 34].

In addition to the ELE algorithm, three more algorithms are used in the attack process. $SHARK.Dec_{CFB}$ is the SHARK decryption algorithm which works in the CFB mode, where the input arguments are the ciphertext, the key, and the initial value in sequence. $SHARK.Enc_{ECB}$ is the SHARK encryption algorithm which works in the ECB (Electronic Codebook) mode, where the input arguments are the ciphertext and the key in sequence. An algorithm named SearchKey takes a pair of matrices, namely A and B, and the round number r as the input arguments, which performs a brutal force searching of the round key in $GF(2)^{16}$ and returns the key as output.

The attack first analyzes the T-Boxes of Round 0, 1 and 6 by using the ELE algorithm. The output of the analysis is the complete set of possible linear transformations used in the 12 T-Boxes, i.e. $P_{r,i}$ for $r = 0, 1, 6$ and $i = 0, 1, 2, 3$. Next, the "real" transformations used in the first two rounds are extracted by testing whether they match the corresponding inter-round matrices. The algorithm SearchKey is then used to find the values of round keys of the first two rounds. With the round keys found, the cipher key can be acquired by inversing the key-scheduling process of SHARK. Finally, the "real" transformations used in the last round are extracted by comparing the results from (a) the encryption using the white-box encryption algorithm and (b) the encryption using the standard SHARK algorithm plus external encodings.

The integrated algorithm that implements the above attack process is presented as follows.

Algorithm Extract
BEGIN
 For $r = 0,1,6$
  For $i = 0,1,2,3$
   $P_{r,i} \leftarrow \text{ELE}(S'_{r,i}, S")$
  EndFor
  Generate:
$$P_r = \left\{ \langle A_r, B_r \rangle \middle| \begin{array}{l} A_r = diag\{A^{(r,0)}, A^{(r,1)}, A^{(r,2)}, A^{(r,3)}\} \\ \wedge \forall i \in 0,1,2,3, \langle A^{(r,i)}, B_r \rangle \in P_{r,i} \end{array} \right\}$$
 EndFor

  For each $\langle A_0, B_0 \rangle \in P_0$
   For each $\langle A_1, B_1 \rangle \in P_1$
    IF $(H^{-1} \cdot B_0 = A_1^{-1})$
     $\langle A^{(0)}, B^{(0)} \rangle \leftarrow \langle A_0, B_0 \rangle$
     $\langle A^{(1)}, B^{(1)} \rangle \leftarrow \langle A_1, B_1 \rangle$
     break;
   EndFor
  EndFor
  $k^{(0)} \leftarrow \text{SearchKey}(A^{(0)}, B^{(0)}, 0)$
  $k^{(1)} \leftarrow \text{SearchKey}(A^{(1)}, B^{(1)}, 1)$
  $key \leftarrow SHARK.Dec_{CFB}(k^{(0)} \| k^{(1)}, PEK, PIV)$
  $C_W \leftarrow SHARK_W(0^{64})$ // 64 bits '0'
  $C \leftarrow SHARK.Enc_{ECB}(0^{64}, key)$
  For each $\langle A_6, B_6 \rangle \in P_6$
   $C' \leftarrow C \cdot H \cdot B_6$
   IF $(C_W = C')$
    $B^{(6)} \leftarrow H \cdot B_6$ *break*;
  EndFor
  Return $(key, A^{(0)}, B^{(6)})$
 END

6 Conclusions

Theoretical analysis in this paper has proved that, in essence, each embedded round key fragment hidden in a lookup table of the white-box SHARK is only protected by a pair of secret linear mappings. As a result, the unknown isomorphic transformations are not capable of providing effective protections on the tables that contain embedded keys in WBACs. It has been demonstrated that the white-box SHARK can be broken by the proposed attack, where the hidden key and external encodings can be extracted with a work factor of approximately 1.5×2^{47}. Although the computational efficiency of the attack procedure could be potentially improved, it is sufficient to indicate the insecurity of the white-box SHARK.

Acknowledgments. This research has been supported by the National Natural Science Foundation of China (No. 61202382), the Fundamental Research Funds for the Central Universities, and the Scientific Research Foundation for the Returned Overseas Chinese Scholars. The authors would like to extend their appreciation to the PC members and anonymous reviewers for their valuable comments and suggestions.

References

1. Wyseur, B.: White-box cryptography. Katholieke Universiteit, Doctoral Dissertation, B-3001 Heverlee (Belgium) (2009)
2. Michiels, W.: Opportunities in white-box cryptography. IEEE Secur. Priv. **8**, 64–67 (2010)
3. Chow, S., Eisen, P., Johnson, H., Van Orschot, P.C.: White-box cryptography and an AES implementation. In: Nyberg, K., Heys, H.M. (eds.) SAC 2002. LNCS, vol. 2595, pp. 250–270. Springer, Heidelberg (2003)
4. Chow, S., Eisen, P., Johnson, H., van Oorschot, P.C.: A white-box DES implementation for DRM applications. In: Feigenbaum, J. (ed.) DRM 2002. LNCS, vol. 2696, pp. 1–15. Springer, Heidelberg (2003)
5. Shi, Y., Liu, Q., Zhao, Q.: A secure implementation of a symmetric encryption algorithm in white-box attack contexts. J. Appl. Math. **2013** Article ID 431794, 9 p. (2013). doi:10.1155/2013/431794
6. Shi, Y., Xiong, G.Y.: An undetachable threshold digital signature scheme based on conic curves. Appl. Math. Inform. Sci. **7**, 823–828 (2013)
7. Babamir, F.S., Norouzi, A.: Achieving key privacy and invisibility for unattended wireless sensor networks in healthcare. Comput. J. **57**, 624–635 (2014)
8. Tague, P., Li, M.Y., Poovendran, R.: Mitigation of control channel jamming under node capture attacks. IEEE Trans. Mob. Comput. **8**, 1221–1234 (2009)
9. Hwang, S.O.: Content and service protection for IPTV. IEEE Trans. Broadcast. **55**, 686 (2009)
10. Nishimoto, Y., Imaizumi, H., Mita, N.: Integrated digital rights management for mobile IPTV using broadcasting and communications. IEEE Trans. Broadcast. **55**, 419–424 (2009)
11. Razzaque, M.A., Ahmad Salehi, S., Cheraghi, S.M.: Security and privacy in vehicular ad-hoc networks: survey and the road ahead. In: Khan, S., Khan Pathan, A.-S. (eds.) Wireless Networks and Security. SCT, vol. 2, pp. 107–132. Springer, Heidelberg (2013)
12. Yang, W.: Security in vehicular Ad Hoc networks (VANETs). In: Chen, L., Ji, J., Zhang, Z. (eds.) Wireless Network Security, pp. 95–128. Springer, Heidelberg (2013)
13. Mejri, M.N., Ben-Othman, J., Hamdi, M.: Survey on VANET security challenges and possible cryptographic solutions. Veh. Commun. **1**, 53–66 (2014)
14. He, S., Lin, L., Letong, F., Yuan Xiang, G.: Introducing code assets of a new white-box security modeling language. In: 2014 IEEE 38th International Computer Software and Applications Conference Workshops (COMPSACW), pp. 116–121 (2014)
15. Xiao, Y., Lai, X.: A secure implementation of white-box AES. In: 2nd International Conference on Computer Science and its Applications, CSA 2009, pp. 1–6. IEEE (2009)
16. Shi, Y., Lin, J., Zhang, C.: A white-box encryption algorithm for computing with mobile agents. J. Internet Technol. **12**, 981–993 (2011)

17. Shi, Y., He, Z.: A lightweight white-box symmetric encryption algorithm against node capture for WSNs. In: 2014 IEEE Wireless Communications and Networking Conference (WCNC), pp. 3058–3063. IEEE (2014)
18. Link, H.E., Neumann, W.D.: Clarifying obfuscation: improving the security of white-box DES. In: ITCC 2005: International Conference on Information Technology: Coding and Computing, vol. 1, pp. 679–684 (2005)
19. Karroumi, M.: Protecting white-box AES with dual ciphers. In: Rhee, K.-H., Nyang, D. (eds.) ICISC 2010. LNCS, vol. 6829, pp. 278–291. Springer, Heidelberg (2011)
20. Bringer, J., Chabanne, H., Dottax, E.: White Box Cryptography: Another Attempt (2006)
21. Jacob, M., Boneh, D., Felten, E.W.: Attacking an obfuscated cipher by injecting faults. In: Feigenbaum, J. (ed.) DRM 2002. LNCS, vol. 2696, pp. 16–31. Springer, Heidelberg (2003)
22. Wyseur, B., Michiels, W., Gorissen, P., Preneel, B.: Cryptanalysis of white-box DES implementations with arbitrary external encodings. In: Adams, C., Miri, A., Wiener, M. (eds.) SAC 2007. LNCS, vol. 4876, pp. 264–277. Springer, Heidelberg (2007)
23. Goubin, L., Masereel, J.-M., Quisquater, M.: Cryptanalysis of white box DES implementations. In: Adams, C., Miri, A., Wiener, M. (eds.) SAC 2007. LNCS, vol. 4876, pp. 278–295. Springer, Heidelberg (2007)
24. Billet, O., Gilbert, H., Ech-Chatbi, C.: Cryptanalysis of a white box aes implementation. In: Handschuh, H., Hasan, M.A. (eds.) SAC 2004. LNCS, vol. 3357, pp. 227–240. Springer, Heidelberg (2004)
25. Michiels, W., Gorissen, P., Hollmann, H.D.: Cryptanalysis of a generic class of white-box implementations. In: Avanzi, R.M., Keliher, L., Sica, F. (eds.) SAC 2008. LNCS, vol. 5381, pp. 414–428. Springer, Heidelberg (2009)
26. Tolhuizen, L.: Improved cryptanalysis of an AES implementation. In: Proceedings of the 33rd WIC Symposium on Information Theory in the Benelux, Boekelo, The Netherlands, 24–25 May 2012. WIC (Werkgemeenschap voor Inform.-en Communicatietheorie) (2012)
27. De Mulder, Y., Wyseur, B., Preneel, B.: Cryptanalysis of a perturbated white-box AES implementation. In: Gong, G., Gupta, K.C. (eds.) INDOCRYPT 2010. LNCS, vol. 6498, pp. 292–310. Springer, Heidelberg (2010)
28. De Mulder, Y., Roelse, P., Preneel, B.: Cryptanalysis of the Xiao – Lai white-box AES implementation. In: Kn udsen, L., Wu, H. (eds.) SAC 2012. LNCS, vol. 7707, pp. 34–49. Springer, Heidelberg (2013)
29. Biryukov, A., Cannière, C.D., Braeken, A., Preneel, B.: A toolbox for cryptanalysis: linear and affine equivalence algorithms. In: Biham, E. (ed.) EUROCRYPT 2003. LNCS, vol. 2656, pp. 33–50. Springer, Heidelberg (2003)
30. Lepoint, T., Rivain, M., De Mulder, Y., Roelse, P., Preneel, B.: Two attacks on a white-box AES implementation. In: Lange, T., Lauter, K., Lisoněk, P. (eds.) SAC 2013. LNCS, vol. 8282, pp. 265–286. Springer, Heidelberg (2014)
31. Hohl, F.: Time limited blackbox security: protecting mobile agents from malicious hosts. In: Vigna, G. (ed.) Mobile Agents and Security. LNCS, vol. 1419, pp. 92–113. Springer, Heidelberg (1998)
32. Rijmen, V., Daemen, J., Preneel, B., Bosselaers, A.: The cipher SHARK. In: Gollmann, D. (ed.) FSE 1996. LNCS, vol. 1039, pp. 99–111. Springer, Heidelberg (1996)
33. Doerksen, M., Solomon, S., Thulasiraman, P.: Designing APU oriented scientific computing applications in OpenCL. In: 2011 IEEE 13th International Conference on High Performance Computing and Communications (HPCC), pp. 587–592 (2011)
34. Manocha, D.: General-purpose computations using graphics processors. Computer **38**, 85–88 (2005)

GPU-Disasm: A GPU-Based X86 Disassembler

Evangelos Ladakis[1(✉)], Giorgos Vasiliadis[1], Michalis Polychronakis[2], Sotiris Ioannidis[1], and Georgios Portokalidis[3]

[1] FORTH-ICS, Heraklion, Greece
{ladakis,gvasil,sotiris}@ics.forth.gr
[2] Stony Brook University, New York, USA
mikepo@cs.stonybrook.edu
[3] Stevens Institute of Technology, Hoboken, USA
gportoka@stevens.edu

Abstract. Static binary code analysis and reverse engineering are crucial operations for malware analysis, binary-level software protections, debugging, and patching, among many other tasks. Faster binary code analysis tools are necessary for tasks such as analyzing the multitude of new malware samples gathered every day. Binary code disassembly is a core functionality of such tools which has not received enough attention from a performance perspective. In this paper we introduce GPU-Disasm, a GPU-based disassembly framework for x86 code that takes advantage of graphics processors to achieve efficient large-scale analysis of binary executables. We describe in detail various optimizations and design decisions for achieving both inter-parallelism, to disassemble multiple binaries in parallel, as well as intra-parallelism, to decode multiple instructions of the same binary in parallel. The results of our experimental evaluation in terms of performance and power consumption demonstrate that GPU-Disasm is twice as fast than a CPU disassembler for linear disassembly and 4.4 times faster for exhaustive disassembly, with power consumption comparable to CPU-only implementations.

1 Introduction

Code disassemblers are typically used to translate byte code to assembly language, as a first step in understanding the functionality of binaries when source code is not available. Besides software debugging and reverse engineering, disassemblers are widely used by security experts to analyze and understand the behaviour of malicious programs [8,12], or to find software bugs and vulnerabilities in closed-source applications. Moreover, code disassembly forms the basis of various add-on software protection techniques, such as control-flow integrity [24] and code randomization [16].

Most previous efforts in the area have primarily focused on improving the accuracy of code disassembly [9,13,24]. Besides increasing the accuracy of code disassembly, little work has been performed on improving the speed of the actual disassembly process. As the number of binary programs that need to be analyzed is growing rapidly, improving the performance of code disassembly is vital for

J. Lopez and C.J. Mitchell (Eds.): ISC 2015, LNCS 9290, pp. 472–489, 2015.
DOI: 10.1007/978-3-319-23318-5_26

coping with the ever increasing demand. For instance, mobile application repositories contain thousands of applications that have to be analyzed for malicious activity [17]. To make matters worse, most of these applications are updated quite frequently, resulting in large financial and time costs for binary analysis workloads. At the same time, antivirus and security intelligence vendors need to analyze a multitude of malware samples gathered every day from publicly available malware scanning services and deployed malware scanners.

In this work, we focus on improving the performance of code disassembly and propose to offload the disassembly process on graphics processing units (GPUs). We have designed and implemented *GPU-Disasm*, a GPU-based disassembly engine for x86 code that takes advantage of the hundreds of cores and the high-speed memory interfaces that modern GPU architectures offer, to achieve efficient large-scale analysis of binary executables. GPU-Disasm achieves both inter-parallelism, by disassembling many different binaries in parallel, as well as intra-parallelism, by decoding multiple instructions of the same binary in parallel. We discuss in detail the challenges we faced for achieving high code disassembly throughput.

GPU-Disasm can be the basis for building sophisticated analysis tools that rely on instruction decoding and code disassembly. We chose to focus on the x86 instruction set architecture for several reasons. First, x86 and x86-64 are the most commonly used CISC architectures. Second, building a disassembler for a CISC architecture poses more challenges compared to RISC, due to much larger set of instructions and the complexity of the instruction decoding process. Third, it is easier to apply the proposed GPU-based design decisions to a RISC code disassembler than the other way around.

We have experimentally evaluated GPU-Disasm in terms of performance and power consumption with a large set of Linux executables. The results of our evaluation demonstrate that GPU-Disasm is twice as fast compared to a CPU disassembler for linear disassembly, and 4.4 times faster for exhaustive disassembly, with power consumption comparable to CPU-only implementations.

In summary, the main contributions of this paper are:

1. We present the first (to our knowledge) GPU-based code disassembly framework, aiming to improve the performance of the instruction decoding process.
2. We present techniques that exploit the GPU memory hierarchy for optimizing the read and write throughput of the decoding process. Such memory optimizations can be applied in tools with similar memory I/O operations.
3. We evaluate and compare our GPU-based disassembly library with a CPU-based approach in terms of performance, cost, and power consumption.

2 Background

2.1 General Purpose Computing on GPUs (GPGPU)

While GPUs are traditionally used for computer graphics, they can also be used for general-purpose computation. Due to the massive parallelism they offer, they can achieve significant performance boosts to certain types of computation. GPUs typically contain hundreds (or even thousands) of streaming cores, organized in multiple *stream multiprocessors* (SM). GPU Threads are divided in groups of 32, called *warps*, with each core hosting one warp. Each warp executes the same block of code, meaning that the threads within a warp do not execute independently, but all of them run the same instruction concurrently. Consequently, code containing control flow statements that lead to different threads following divergent execution paths, cannot fully utilize the available cores. When some threads within a warp diverge, because a branch follows a different path than the rest of them (*branch divergence*), they are stalled. Consequently, the tasks that can truly benefit from the massively parallel execution of GPUs are the ones that do not exhibit branch divergence. Among many domains, GPUs have been used in scientific computing [2], cracking passwords [1], machine learning [5], and network traffic processing [20–22].

GPUs have a distinct memory model. Each multiprocessor has a set of 64 K registers, which are the fastest GPU memory component. Registers are assigned to threads and are privately scoped. The scheduler is responsible for ensuring that register values are saved and restored during context switches of threads. Each multiprocessor has its own Level 1 (L1) cache and shared memory, which are shared by all the threads running on it, and are part of the same physical memory component. This allows for choosing at run time (before spawning the GPU threads) how to distribute memory between cache and shared memory. The L1 cache is organized in data cache lines of 128 bytes. Shared memory is as fast as L1 cache but is programmable, which means that it can be statically allocated and used in GPGPU programs.

GPUs also include *global memory*, which is equivalent to the host's RAM. It is the slowest memory interface, but has the largest capacity. Global memory is available to all SMs and data from the host to the device and vice versa can be transfered only through this part of memory. Interestingly, global memory also hosts *local memory*, which is used by threads to spill data when they run out of registers or shared memory. Finally, global memory also includes constant memory, a region where programs can keep read-only data, allowing for fast access when threads use the same location repeatedly.

A Level 2 (L2) cache is shared between all SMs and has a larger capacity than L1. Every read/write from and to the global memory passes through the L2 cache. A GPU multiprocessor can fetch 128 byte lines. The driver keeps this alignment in global memory and in cache lines to achieve increased throughput for read and write operations. The maximum transfer throughput to global memory is 180 GB/s.

There are two frameworks commonly used to program GPUs for general purpose computations, both using C API extensions. The first is CUDA [14], a programming framework developed by NVIDIA (which we use in this work), and the second is OpenCl [19], which is a generic framework for programming co-processors with general purpose computational capabilities.

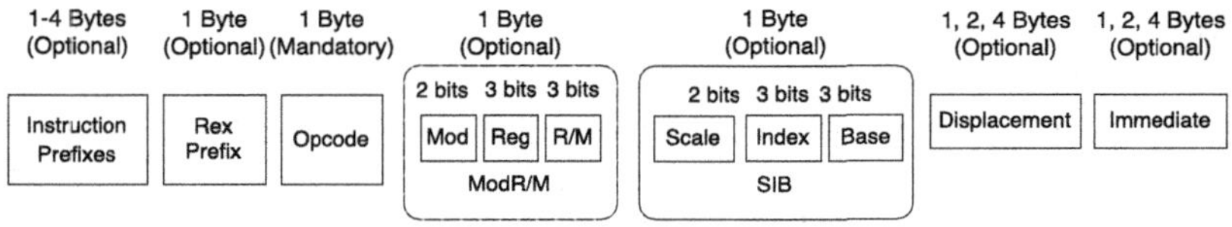

Fig. 1. x86 Instruction format.

2.2 x86 Architecture

The x86 and x86-64 architectures are probably the most widely used CISC (Complex Instruction-Set Computing) architectures [7]. Their instruction sets are rich and complex, and most importantly they support instructions of varying length. Instruction lengths range from just *one* byte (i.e., instructions comprising just an opcode) to 15 bytes. Generally, instructions consist of optional prefix bytes, which extend the functionality of the instruction, the opcode, which defines the instruction, the ModR/M and SIB bytes, which describe the operands, followed by an immediate value, that is also optional. The overall format of an x86 instruction is depicted in Fig. 1.

Due to the extensive instruction set and the variable size if its instructions, it is very easy for disassemblers to be confused, decoding arbitrary bytes as instructions [3], e.g., because data may be interleaved with instructions, or because the beginning of a block of instructions is not correctly identified.

2.3 Code Disassemblers

There are two widely used code disassembly techniques, *linear* and *recursive* disassembly [6]. In linear disassembly, a segment of bytes is disassembled by decoding instructions from the beginning of the of the segment until the end is reached. Linear disassembly typically does not apply any heuristics to distinguish between code and data, and consequently, it is easy to get "confused" and produce erroneous results. For example, compilers emit data and patching bytes for function alignment, which a linear disassembler decodes as instructions, along with the actual code. Thus, when disassembling the whole text segment of a binary, the output of linear disassembly is likely to contain erroneous parts that correspond to embedded data and alignment byte Binaries may also contain unreachable functions that are included during compilation, e.g., due to the static linkage of libraries, which will also be included in the output of linear disassembly.

Recursive disassemblers use a different approach that eliminates the erroneous assembly produced by linear disassembly, but with its own disadvantages. The decoding process starts from an address out of a set of entry points (exported functions, entry points) and linearly disassembles the byte code. Whenever the disassembler encounters control flow instructions, it adds all targets to the set of entry points. The disassembly process stops when it finds indirect (computed) branches which cannot be followed statically. The process continues recursively by decoding from a new target out of the set of entry points. They main drawback of recursive disassembly is that it cannot reach code segments that are accessible only through indirect control flow transfer instructions.

3 Architecture

In this section, we describe the overall architecture of our system. Our aim is to design a GPU-based disassembly engine that is able to process a large number of binaries in parallel. The key factors for achieving good performance are: (i) exploit the massively parallel computation features of the GPU, (ii) optimize PCIe transfers and pipeline all components for keeping the hardware utilized at all times, and (iii) design optimization heuristics for exploiting further capabilities of the hardware.

The basic operations of our approach include: (i) *Pre-processing*: loading of the binaries from disk to properly aligned buffers of the host's memory space, (ii) *Host-to-device*: transfer of the input data buffers to the memory space of the GPU, (iii) *Disassembly*: the actual parallel code disassembly of the inputs on the GPU, and storage of the decoded instructions into pre-allocated output data buffers, (iv) *Device-to-host*: transfer of the output buffers to the host's memory space, and finally (v) *Post-processing*: delivery of the disassembled output and initialization of the pointers to the next chunk of bytes of each binary, if any, that will be fed to the GPU for disassembly. Once processing of all binaries has completed, input buffers are loaded with the next binaries to be analyzed.

3.1 Transferring Input Binaries to the GPU

The operation to consider is how input binary files will be transferred from the host to the memory space of the GPU. The simplest approach would be to transfer each binary file directly to the GPU for processing. However, due to the overhead associated with data transfer operations to and from the GPU, grouping many small transfers into a larger one achieves much better performance than performing each transfer separately. Thus, we have chosen to copy the binary files to the GPU in batches. In addition, the input file buffer is allocated as a special type of memory, called page-locked or "pinned down" memory, in order to prevent it from being swapped out to secondary storage. The copy from page-locked memory to the GPU is performed using DMA, without occupying the CPU. This allows for higher data transfer throughput compared to the use of pageable memory, e.g., using traditional memory allocation functions such as `malloc()`.

3.2 Disassembling x86 Code on the GPU

Instruction Decoding and Linear Disassembly. Linear disassembly blindly decodes a given sequence of bytes from the beginning to the end without applying any further heuristics or logic. Initially, the GPU decoder dispatches the instruction prefixes (if present), which always come before the opcode of x86 instructions. Afterwards, the decoder dispatches the next byte of the instruction which is the actual opcode we are interested in. The decoder shifts the opcode bytes to bring them in a form that it can easily use them as an index for a look-up table. After decoding the opcode, we determine if the instruction has operands or not, by decoding the ModR/M byte. The operands can be registers or immediate values. If the operands are registers, they can be either implicit, as part of the instruction, or explicit, defined by the following bytes. If the instruction uses indexed addressing, then the next decoded byte corresponds to the SIB (Scale Index Base) which determines the addressing mode of the array. Lastly, the disassembler decodes the displacement and immediate bytes.

The disassembly process can fail while decoding an instruction. Depending on the failure reason, the disassembler handles it in a different way. When more bytes than available are expected based on the last decoded opcode, the instruction decoding process stops and an appropriate error is reported. When invalid instructions are encountered, the disassembler marks them and continues the decoding process from the following byte.

Each GPU thread is assigned to disassemble a single chunk of an input binary at a time. Consequently, the total GPU kernel execution time is equal to the time of the slowest (last finished) thread. Note that the overall performance would drop in case some threads remained under-utilized, i.e., they were assigned smaller workloads. To avoid this, we assign fixed-sized input buffers (chunks) to all threads, which minimizes the possibility of having idle threads. However, as all input binaries do not have the same size, some imbalance unavoidably happens as the processing of smaller input files completes. Our current prototype does not handle such imbalances, but their effect can be minimized by selecting input file batches based on file sizes, so that each batch includes files of similar sizes.

Having fixed size chunks leads to more complex data splitting, when a binary may not fit inside the buffer all at once. Therefore, we have to divide the binary in several chunks and perform the disassembly process on batches. Due to the nature of the `x86` instruction set (Sect. 2.2) we have to carefully choose the starting point of the next chunk of bytes for decoding, otherwise any split instructions will generate incorrect disassembly.

Exhaustive Disassembly. We have also implemented an exhaustive disassembly mode, which applies linear disassembly by starting from each and every byte of the input, i.e., by decoding all possible (valid) instructions contained in the input. Further analysis of the output can be then performed to identify function boarders, basic blocks, and even obfuscated code constructs. For instance, Bao et al. [4] use exhaustive disassembly to generate all possible outputs, and then apply machine learning techniques to find instruction sequences that correspond to function entry and exit points. Other approaches [8,13] disassemble the same regions of a binary from different indexes and apply heuristics to identify basic blocks and reconstruct the control flow graph.

For exhaustive disassembly, we transfer the input buffer to the GPU memory space and spawn as many threads as the bytes of the binary. Each thread starts the decoding of the same input from a different index. Although each thread decodes only one instruction, this approach is effective in quickly extracting all possible instructions contained in the input.

3.3 Transferring the Results to the Host

After an instruction is decoded, the corresponding data is stored in the GPU memory. As storing extensive data for all decoded instructions from all threads can easily deplete the memory capacity, we chose to save only basic information about each decoded instruction, which though is enough for further analysis. Specifically, we store the relative address of the instruction within the input file, its opcode, the group to which it belongs (e.g., indirect control flow transfer, arithmetic operation, and so on), and all explicit operands such as registers and immediate values. The above extracted information can fully describe each decoded instruction, and can be easily used for further static analysis, compared to more verbose storing of raw fields, such as ModR/M bits. Information such as implicit operands and the size of the instruction mnemonic can be easily extracted from the stored metadata. For example, the size of the instruction can be calculated from the distance between the relative addresses of the current and the next instruction.

The decoded instructions are stored in a pre-allocated array with enough space for all instructions of the input. As shown in Fig. 4 (discussed in more detail in Sect. 5.1), only less than 20 % of the encountered instructions on average are a single-byte long, so the number of decoded instructions in typically much smaller than the size of the input in bytes. Consequently, we safely set the number of slots in the array as half the size of the input buffer in bytes.

The GPU disassembly engine saves the decoded instructions on GPU memory and transfers them back to the host for further analysis. After the device to host transfer has completed, the system evaluates the extracted information as part of a post-process phase. This includes checks for errors due to any misconfiguration of the GPU threads, and for each thread, whether there are pending bytes for disassembly for the current input binary being processed. Then, the pointer for the next chunk to be processed is set according to the last successfully decoded instruction, so that the disassembly process is not corrupted. If a thread has finished disassembling an input binary, the pointer is set to NULL so that a new binary will be assigned to it, after the processing of the whole batch is completed.

3.4 Pipeline

After optimizing the basic operations, we have to design the overall architecture in such a way that will keep every hardware component utilized. The GPGPU API supports running computations using streams. Thus, we can parallelize data transfers with the disassembly process and eliminate idle time for the PCIe bus and the GPU multiprocessors. We use double buffers for both input and output, so that when the GPU processes a buffer, the system can transfer the output data and fill the next input buffers with new binaries for disassembly. With the proper usage of streams, we can keep the CPU, the PCIe bus, and the GPU utilized concurrently at all times.

The GPU can handle the synchronization of GPU operations internally. However, before the host proceeds with output analysis, it needs to synchronize the GPU operations. The host is unable to know if the device has finished processing until the driver receives a signal from the GPU that denotes completion. Ideally, we would like to keep the GPU utilized without blocking for synchronization. The architecture can be designed so that synchronization is kept to a minimum, just for one of the operations. By placing all input values (binaries, sizes, memory addresses) and all output data into a single buffer, as described above, requires invoking the synchronization process only after the copy of the output from device to host, eliminating in this way any intermediate serialization points.

4 Optimizations

4.1 Access to Global Memory

Due to the linear nature of the disassembly process, we enforce both reads and writes to the input and output buffers to be performed only once for each decoded instruction. As mentioned, the instruction sizes of the `x86` ISA vary significantly, ranging between 1 and 15 bytes. According to the alignment property that GPUs follow for the memory accesses, different sequences of instructions with different sizes may result in misaligned accesses, consequently resulting in degraded memory access throughput.

We describe the improvement of the reading process in Sect. 4.3. Regarding the improving the write throughput of the disassembly output to global memory, GPU best practices [15] propose that data structures on the GPU should be placed as structs of arrays. In most cases, this results in improved data throughput from global memory. However, in our case we observed lower performance due to the drop of the writing throughput back to global memory. We tackled this issue and achieved a better throughput by having a struct with the decoded information per instruction, instead of separate arrays for each field.

4.2 Constant Memory

A crucial part of the disassembler are the look-up tables with the decoding information that are hardcoded in the instruction decoder. These tables are used as dispatchers for the decoding process. They hold information about each instruction, such as the opcode, whether there are operands and how many to expect, the type of the instruction, the group of the architecture extension of an instruction, and so on. The look-up tables are constants and shared through all threads. Therefore, we can use the constant memory of the GPU in order to have fast access to these tables. The constant memory though is limited in size, and the look-up tables can easily exceed the available memory. To strike a balance between performance and accuracy, we measured the most used tables and placed them to the GPU constant memory, and kept the more rarely used tables in the (slower) global device memory. Furthermore, global variables such as function pointers that are being assigned by the initialization process, are placed to the shared memory of each multiprocessor, which can be initialized at run time.

4.3 Access to L2 Cache

Read and write data accesses pass through the L2 cache, which is a shared memory interface for all multiprocessors as the global memory. The L2 cache memory is n-associative [23], which means that data lines are placed depending on the least significant bits of the accessed address. When assigning large input buffers to each thread, memory divergence increases, and consequently, line collisions inside the L2 cache occur more frequently as well. On the other hand, having small input buffers will result in under-utilization of the GPU threads, and an overall drop in performance.

Taking in consideration this trade-off, we sought a solution that combines the benefits of both approaches. Each read access to the global memory from a multiprocessor fetches a 128-byte line of data. Consequently, we chose to divide large buffers into smaller ones (as shown in Fig. 2) with a size aligned to the access line of the GPU, and place them within the larger buffer in such a way that threads access the buffer as a group. We evaluated buffer sizes of 16, 32 and 64 bytes, and the results of our experiments showed that beyond 32 bytes, the L2 hit ratio from the L1 cache dropped due to line collisions (Table 1). For every 32 bytes of the input buffer, we place in the first 16 bytes the previous 16 decoded bytes, and in the following 16 bytes the new bytes that have to be decoded. The repeated bytes are needed for correcting the decoding alignment, in case of out-of-bounds errors of a previous disassembly. In that case, we continue the decoding process from the byte where the previous disassembly stopped at, until the end of the 32 bytes. Furthermore, this optimization forces the disassembler to make fixed read accesses to global memory, which achieves better throughput.

4.4 Data in GPU Registers

We take advantage of the GPU registers to store statically allocated data that is frequently used by the decoder. Typically, instruction operands are dynamically allocated for each instruction, due to the fact that the number of operands am `x86` instruction uses is not known in advance. We changed the list of operands to a static array, which eventually the compiler keeps in registers. As mentioned earlier, operands may be either explicit or implicit. Due to memory capacity limitations, we decided to keep in registers only the explicit operands (three or less). Implicit operands depend on the instruction opcode, and therefore can be easily inferred.

Keeping operands into registers instead of shared memory is preferable because the latter would affect the L1 cache of each multiprocessor, which corresponds to the same hardware, and therefore would drop the read access

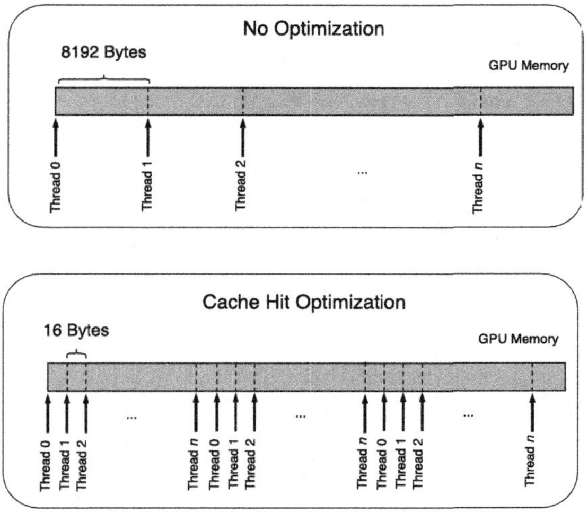

Fig. 2. Reading inputs from GPU global memory with L2 cache optimization.

throughput of the input binaries. Also, the shared memory would have to be divided according to the number of threads for each multiprocessor, imposing an upper-bound on the number of threads that could be spawn due to the size of temporary list of operands for each thread.

Another use of registers is related to improving the read throughput of the input buffers. Traditionally, read requests pass from global memory through the L2 cache, and finally the data are fetched to the L1 cache of the corresponding multiprocessor. In order to avoid reading from the L1 cache, or even worse to overwrite the cache line where decoded bytes are stored, we save the 32 byte lines into a *uint4_t* statically declared array, which is translated at compile time in register storage. Although excessive use of registers can result in register spilling to local memory, any incurred latencies can be hidden by spawning more threads. Our experiments show that stall instructions due to local data accesses are rare.

5 Evaluation

In order to evaluate our GPU-based disassembler, we create a corpus of 32,768 binaries from the `/usr/bin/` directory of a vanilla Ubuntu 12.04 installation, allowing duplicates to reach the desired set size. The sizes of the binaries vary between 30 KB and 40 KB. Our testbed consists of a PC equipped with an Intel i7-3770 CPU at 3.40 GHz and 8 GB of RAM, and an NVIDIA GeForce GTX 770 GPU with 1536 cores and 4 GB of memory.

5.1 Performance Analysis

The performance evaluation examines both the system as a whole, as well as its sub-parts (e.g., the decoding engine and data transfers). We also test existing

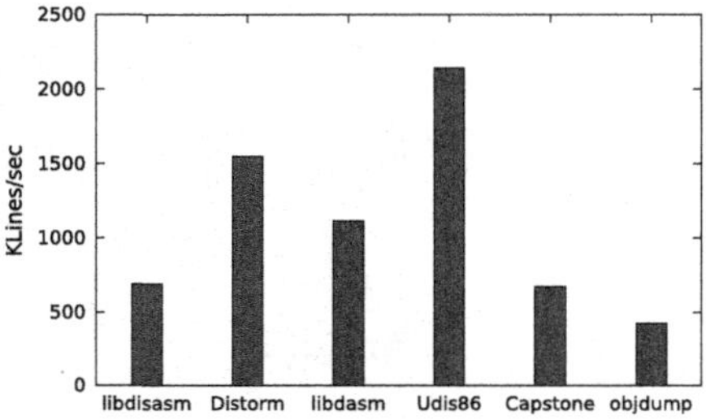

Fig. 3. Performance evaluation of various open-source linear disassemblers.

CPU-only disassemblers for comparison. We report the throughput of the disassembly process as the number of assembly lines (or decoded instructions) produced per second. As the size of instructions in `x86` ISA varies, it would be misleading to measure the number of bytes processed per time.

Performance Analysis of Open-Source Disassemblers. As a first step, we evaluate several popular open-source linear disassemblers to estimate the throughput of conventional CPU-based disassemblers. In order to eliminate any I/O overhead, we redirect the output of the tools to `/dev/null`. Figure 3 depicts the average disassembly rate for various disassemblers in thousands of assembly lines (KLines) per second, when utilizing a single CPU thread. The faster disassembler is *Udis86*, which achieves a throughput of 2142.2 KLines/sec and the slower is the *objdump* utility, which processes 423.664 KLines/sec. The differences in throughput are mostly due to the data produced for disassembled instruction; the more information generated by a disassembler, the lower its throughput. For instance, some tools record only the opcodes and the corresponding operands for each instruction, while others include information such as its instruction group, relative virtual addresses, etc.

Data Transfer Costs. In this experiment, we measure the data transfer rate between CPU and GPU over PCIe for different block sizes of data. Figure 5 shows the results in GB/sec including standard error bars for transferring data from host to GPU memory and vice versa. The maximum theoretical transport bandwidth for PCIe 3.0 is 16 GB/s, however, in this experiment the maximum achieved rate is 12 GB/s, when transferring blocks of 16 MB.

GPU Instruction-Decoding Performance. In this section, we evaluate the decoding performance of the GPU, excluding any data transfers, and pre- and post-processing occurring on the CPU (e.g., opening files and preparing data exchanges). In this experiment, we use three different inputs: (i) linear disassembly of synthetic binaries, (ii) linear disassembly of binaries corpus, and (iii) exhaustive disassembly of a subset of the corpus.

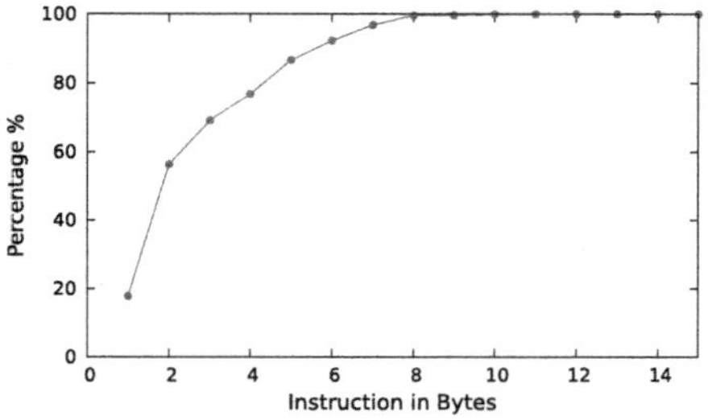

Fig. 4. CDF of the x86 Instructions sizes found on GNU Binutils.

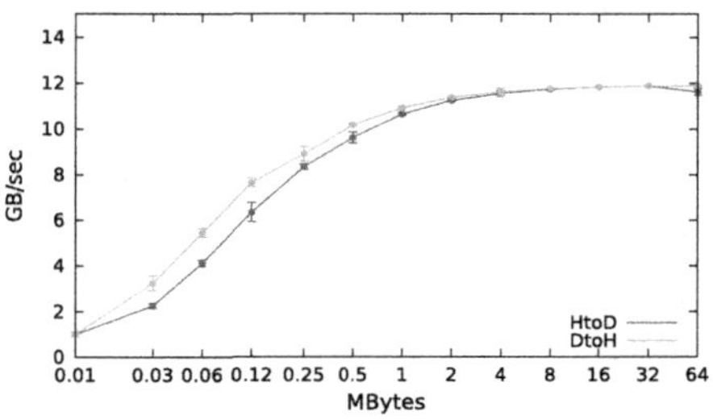

Fig. 5. PCIe 3.0 transfer throughput.

Synthetic Binaries. In this experiment, we aim to evaluate our various optimizations and the effect of instruction-size. First, we generate buffers including 2-byte instructions, which is the most common instruction length (about 38.54 % in our dataset, see Fig. 4), and measure how the buffer size used in decoding affects the

Table 1. Average hit rate at L2 cache for all read requests from L1 cache, when decoding 2-byte instructions in the GPU.

Cache hit rate in L2	
Buffer optimized size	Average hit rate %
16 Bytes	58.70
32 Bytes	53.65
64 Bytes	45.26

Table 2. Impact of *Access To Global* optimizations, when decoding 2-byte instructions in the GPU.

Optimization	MLines/Sec.	Performance gained %
No optimization	52.05	-
Improve cache hits	65.51	+25.85 %
Array of structs	43.85	-15.75 %

Table 3. Effect of instruction sizes in decoding.

Instruction size	MLines/Sec. CPU	Performance dropped CPU %	MLines/Sec. GPU	Performance dropped GPU %
1	35.90	-	100.91	-
2	14.12	60.6	66.67	33.93
4	12.63	64.81	59.53	41.00
8	9.96	72.25	46.32	54.09

Table 4. Exhaustive disassembly of 101 binaries from the corpus. GPU speedup results compared to the CPU. (Includes only instruction decoding.)

Exhaustive disassemble results	
Description	Speedup
Average	4.411
Standard deviation	0.928
Maximum	7.122
Minimum	2.729

Table 5. End-to-end disassembly of binaries coprus. Overall Performance in MLines/sec. (Data transfer buffer is 8192 Bytes)

Threads	Performance MLines/sec
512	3.096
1024	4.857
2048	9.335
4096	17.548
8192	28.053
16384	28.085
CPU performance: 13.933	

L2 cache hit rate, when using 4096 threads. Table 1 shows that the optimal buffer size is 16 bytes. Table 2 shows the performance gained in accessing global memory by each of the optimizations described in Sect. 4.

As mentioned in Sect. 2.2, the size of an `x86` instruction can be between 1–15 bytes. Figure 4 shows the cumulative distribution function (CDF) of instruction sizes in the binaries used in the evaluation. In order to understand how binaries containing a mix of instructions with different sizes will affect performance, we decode files containing instructions with different sizes, with each file containing only a single length of instructions. We also compare decoding throughput by running our decoder both on the CPU and GPU, and try different numbers of threads on the GPU. We again use 4096 threads in the GPU, as we found that is optimal in the synthetic binaries scenario. Table 3 lists the results of this experiment.

Linear Disassembly of Binaries. In this experiment, we evaluate the GPU performance on disassembling the binaries in our corpus. This is likely to be the common use case of our prototype on large scale binary analysis. Each thread is assigned a different binary for disassembling. In Fig. 6, we plot the speedup gained when offloading the disassembling process to the GPU. We evaluate several configurations, i.e., bytes per thread and number of threads, in order to find the best configuration. We can see that the GPU reaches maximum performance on different number of threads (8192) than with the synthetic binaries (4096). We also observe that the performance on different binaries drops to 28.4 MLines/sec compared to decoding all 8-byte instructions in the GPU (46.32 MLines/sec). This performance loss happens due to the different memory stalls that occur to each thread at a given moment. Threads decode different sizes of instructions when disassembling binaries, as a consequence they do misaligned accesses to the global memory and the cache misses increase. Still, by increasing the threads per multiprocessor we can hide some of these stalls and therefore the disassembler scales up to 8,192 threads. From the other hand, just spawning threads is not enough for hiding all the stalls. Spawning more threads arises more races to the caches and more cache misses for the concurrent cache lines. Lastly, by decoding different instructions, we slightly increase the branch divergence that also

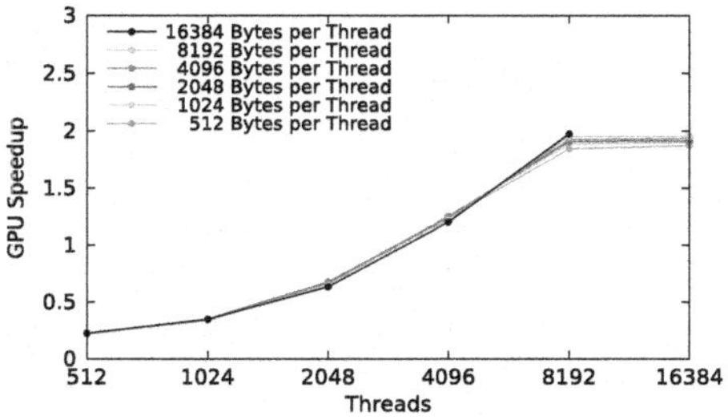

Fig. 6. GPU-Disassembler speed up compared to the CPU on different set ups. Comparing only the disassemble process without the transfers.

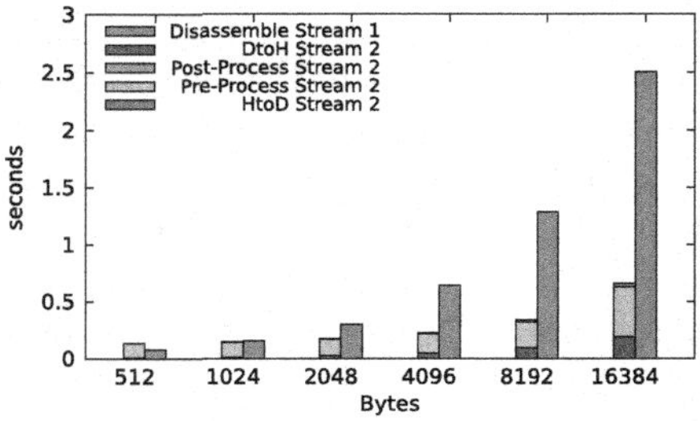

Fig. 7. The disassembly components of the GPU pipelined using streams. Focused on 8192 Threads.

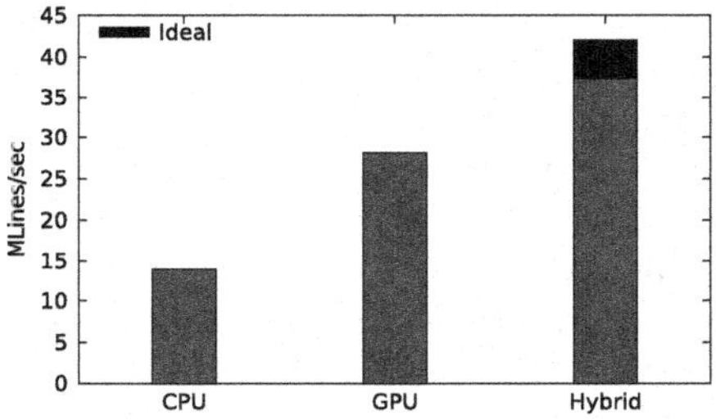

Fig. 8. The overall performance on CPU, GPU and on both processors. Focused on 8192 Threads for the GPU.

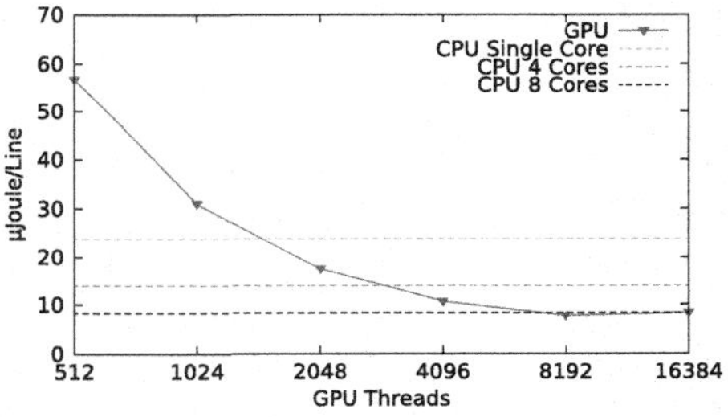

Fig. 9. Power Consumption per decoded line.

creates stalls. As we can see in Fig. 6 the GPU was ≈2 times faster on the disassembly process than a relevant high-end CPU. Performance stops scaling after 8192 threads which we can safely state that this is the optimum configuration for the disassembly process.

Exhaustive Disassembly of Binaries. In this experiment we disassemble each binary starting from each byte in order to find all possible instructions included in the binary. The evaluated prototype is the one described in Sect. 3.2. We evaluate the prototype using several number of threads in order to find the optimal for this case. The best performance is reached when we spawn 131,072 threads. Therefore, the exhaustive prototype, shall perform better, if we disassemble binaries of size bigger than the threads we spawn. In case the binary is smaller than the optimal amount of threads we spawn as many threads as the size of the binary. As we saw the disassemble performance differs among different sizes of instructions. In order to be accurate, we exhaustive disassemble a set of 101 binaries and evaluate the achieved performance. In Table 4 we can see the results of the experiment described, on disassembling binaries exhaustively. The average speedup we gained is 4.411 with a standard deviation of 0.928.

Overall Performance. In this section, we evaluate our prototype in an end-to-end scenario. As mentioned in Sect. 3, we use streams in order to pipeline the operations and hide communication costs. We measure the time spent for each component in isolation. For all subsequent experiments with use 8,192 threads, as this configuration achieves the best performance, as we have shown in Sect. 5.1. In Fig. 7 we can see the raw times of the corresponding components stacked in the order they execute in a given stream, pipelined with the current disassembly process of the previous stream. When the number of threads is lower than 1024 we can see that the bottleneck operation is pre-processing. However, after 1024 bytes per thread we can see that the disassembly component becomes the bottleneck of the whole process. Therefore, pipelining does not reduce performance. In Table 5 we demonstrate the raw performance in MLines/sec of the GPU in several threads with the size of the input buffer at 8192 bytes per thread.

Hybrid Model: We also evaluate the performance of utilizing all CPU cores and the GPU to massively disassemble binaries. Despite the fact that the GPU is an independent processing system, it still requires interaction with the CPU for transferring data, spawning the GPU kernel for execution, etc. Therefore, when we over-utilize the CPU with workload, we increase the probability of having threads stalled due to context switching. At the evaluation process, by overloading the CPU we experience an increase in the pre- and post-processing overhead and so, we wasted time by having idle the GPU and decrease the overall performance. In order to evaluate properly the hybrid model we assigned one CPU thread to the GPU processes (pre, post, GPU invocation and interaction) and the rest for disassembly on the CPU. In Fig. 8 we can see the performance on different devices and the hybrid model as described. The hybrid model achieved the performance of 37.336 MLines/sec which is 2.67 times faster than having only the CPU utilized and 1.32 times faster than the GPU implementation. The divergence of the hybrid model from the ideal performance is due to the assigned thread to the GPU controlling processes.

5.2 Power Consumption and Cost

Power Consumption per line. In this experiment we measure the power consumption of our prototype at a given moment, with the components pipelined, when disassembling binaries. For the comparison we define the metric Joules consumed per decoded line. We evaluate the watts consuming per second and the performance of the tool as defined in previous sections (Lines/sec). By dividing these values we come up with Joules consumed per decoded line (Joules/line). In Fig. 9 we demonstrate the power consumption efficiency for the GPU and CPU in different threads. For the measurement of the power consumption we used sensors that can measure the power consumption of the CPU, the PCI bus, the RAM and peripherals. For each set up, we sum up the power consumed at a given moment and then we calculate the power consumed per decode line. Both of the devices perform similar in terms of power consumption per decoded line. GPU consumes 8.34 μJ at the best configuration for decoding an instruction.

Lines per Dollar. For our hardware setup, we have selected relatively high-end devices; for the CPU we used an Intel(R) i7-3770 which costs around \$305, and the NVIDIA GTX 770 graphics card with similar cost at \$396.[1,2] These are the prices at the time this work was published. The total system cost is around \$1120 with the current values of the components. Our prototype performs with an overall cost of 23.36 KLines/\$.

6 Related Work

The improvement of the disassembly process for the `x86` and `x86-64` architecture is still an open issue. There are various publications that address disassembly correctness and effectively differentiate code from alignment patching bytes inside the text section of the binaries. Most of these publications, are based on a similar approach. They use the targets of control-flow instructions in order to recognize the regions of basic blocks and functions borders. They make several disassembly passes on these code regions until the given conditions of correctness are satisfied. Finally, they construct the final call graph and discard the unreachable regions [9,13,18,24]. However, there is also a dynamic approach that leverages machine learning techniques [4,11]. This approach uses decision trees, that are constructed by feeding binaries, compiled from various compilers and optimization flags as training sets. They perform exhaustive disassembly on the binary to produce all the possible assembly output. Lastly, they use the constructed tree to match and recognize the entry and exit points of functions.

GPUs continuously become more powerful and with extended computational capabilities that can support more applications. In the scientific community, there are several security analysis tools that exploit the parallelism offered by GPUs for fast processing such as network packet processing [10,20,21].

7 Limitations

The implementation of our prototype comes with limitations. The size of the decoded instructions for all the threads can be enormous and as a result, we can easily run out of memory. Also, memory constrains occur on the fast memory interfaces such as constant memory, shared and register usage per thread. Furthermore, GPU limitations with regards to dynamic memory allocation, forces us to use static allocation and requires rewriting of the dynamic parts of the disassembler.

We are unable to further exploit GPU parallelization due to memory stalls that occur at decoding time. GPU threads, make arbitrary accesses to memory at the decoding process which under-utilize the access throughput. Although, we can hide memory stalls by spawning more threads, there is a limit on how the cuda-process scales. The GPU hides stalled threads by context-switching to

[1] Cpu benchmarks: Intel core i7-3770 @ 3.40ghz. http://www.cpubenchmark.net/.

[2] Videocard benchmarks: Geforce gtx 770. http://www.videocardbenchmark.net/.

threads that are ready to execute. However, complex programs, that have high needs in resources and frequently access memory, can generate more stalls when excessively utilizing threads. Thus, it is not trivial to determine the optimal number of threads for a GPU-Disassembler; it really depends on the implementation and the disassembly algorithm (linear, exhaustive, etc.).

8 Conclusion

GPUs are powerful co-processors, which we can use to accelerate computationally intensive tasks like binary disassembly through parallelization. In this work we have built a GPU based x86-disassembler that exploits the hardware features offered from GPUs to accelerate disassembly. We evaluate our GPU-based x86-disassembler in terms of performance and cost. Our prototype performs two times faster in linear disassembly and 4.4x faster in exhaustive disassembling of the same binary compared to a CPU implementation. In terms of performance over power consumption; GPU performs similar with a full utilized CPU at 8.34 μJ/Line.

Acknowledgments. We want to express our thanks to the anonymous reviewers for their valuable comments. This work was supported by the General Secretariat for Research and Technology in Greece with the Research Excellence grant GANDALF, and by the projects NECOMA, SHARCS, funded by the European Commission under Grant Agreements No. 608533 and No. 644571. This work was also partially supported by the US Air Force through contract AFRL-FA8650-10-C-7024. Any opinions, findings, conclusions or recommendations expressed herein are those of the authors, and do not necessarily reflect those of the US Government or the Air Force.

References

1. New 25 GPU Monster Devours Passwords In Seconds. http://securityledger.com/new-25-gpu-monster-devours-passwords-in-seconds/
2. Preis, T., Virnau, P., Paul, W., Schneider, J.J.: GPU accelerated Monte Carlo simulation of the 2D and 3D ising model. J. Computat. Phy. **228**(12), 4468–4477 (2009)
3. Balakrishnan, G., Reps, T.: Wysinwyx: what you see is not what you execute. ACM Trans. Program. Lang. Syst. (TOPLAS) **32**(6), 23 (2010)
4. Bao, T., Burket, J., Woo, M., Turner, R., Brumley, D.: Byteweight: learning to recognize functions in binary code. In: Proceedings of USENIX Security 2014 (2014)
5. Catanzaro, B., Sundaram, N., Keutzer, K.: Fast support vector machine training and classification on graphics processors. In: Proceedings of the 25th International Conference on Machine Learning, ICML 2008, pp. 104–111 (2008)
6. Eagle, C.: The IDA Pro Book: The Unofficial Guide to the World's Most Popular Disassembler. No Starch Press, San Francisco (2008)
7. Intel Intel. and ia-32 architectures software developer's manual, volume 3b: System programming guide. Part, 1:2007, 64

8. Kapoor, A.: An approach towards disassembly of malicious binary executables. PhD thesis, University of Louisiana at Lafayette (2004)
9. Kinder, J.: Static analysis of x86 executables (2010)
10. Koromilas, L., Vasiliadis, G., Manousakis, I., Ioannidis, S.: Efficient software packet processing on heterogeneous and asymmetric hardware architectures. In: Proceedings of the 10th ACM/IEEE Symposium on Architecture for Networking and Communications Systems, ANCS (2014)
11. Krishnamoorthy, N., Debray, S., Fligg, K.: Static detection of disassembly errors. In: 16th Working Conference on Reverse Engineering 2009, WCRE 2009, pp. 259–268. IEEE (2009)
12. Kruegel, C., Kirda, E., Mutz, D., Robertson, W., Vigna, G.: Polymorphic worm detection using structural information of executables. In: Valdes, A., Zamboni, D. (eds.) RAID 2005. LNCS, vol. 3858, pp. 207–226. Springer, Heidelberg (2006)
13. Kruegel, C., Robertson, W., Valeur, F., Vigna, G.: Static disassembly of obfuscated binaries. In: USENIX Security Symposium vol. 13, p. 18 (2004)
14. NVIDIA. CUDA C Programming Guide, Version 5.0
15. CUDA NVidia. C best practices guide. NVIDIA, Santa Clara, CA (2012)
16. Pappas, V., Polychronakis, M., Keromytis, A.D.: Smashing the gadgets: hindering return-oriented programming using in-place code randomization. In: Proceedings of the 33rd IEEE Symposium on Security and Privacy (S&P), May 2012
17. Petsas, T., Papadogiannakis, A., Polychronakis, M., Markatos, E.P., Karagiannis, T.: Rise of the planet of the apps: A systematic study of the mobile app ecosystem. In: Proceedings of the 2013 Conference on Internet Measurement Conference, pp. 277–290. ACM (2013)
18. Schwartz, E.J., Lee, J., Woo, M., Brumley, D.: Native x86 decompilation using semantics-preserving structural analysis and iterative control-flow structuring. In: Proceedings of the USENIX Security Symposium, p. 16 (2013)
19. Stone, J.E., Gohara, D., Shi, G.: Opencl: a parallel programming standard for heterogeneous computing systems. Comput. Sci. Eng. **12**(1–3), 66–73 (2010)
20. Vasiliadis, G., Antonatos, S., Polychronakis, M., Markatos, E.P., Ioannidis, S.: Gnort: high performance network intrusion detection using graphics processors. In: Lippmann, R., Kirda, E., Trachtenberg, A. (eds.) RAID 2008. LNCS, vol. 5230, pp. 116–134. Springer, Heidelberg (2008)
21. Vasiliadis, G., Koromilas, L., Polychronakis, M., Ioannidis, S.: GASPP: a GPU-accelerated stateful packet processing framework. In: Proceedings of the USENIX Annual Technical Conference (ATC), June 2014
22. Vasiliadis, G., Polychronakis, M., Ioannidis, S.: MIDeA: a multi-parallel intrusion detection architecture. In: Proceedings of the 18th ACM Conference on Computer and Communications Security (CCS), October 2011
23. Wong, H., Papadopoulou, M.-M., Sadooghi-Alvandi, M., Moshovos, A.: Demystifying GPU microarchitecture through microbenchmarking. In: 2010 IEEE International Symposium on Performance Analysis of Systems and Software (ISPASS), pp. 235–246. IEEE (2010)
24. Zhang, M., Sekar, R.: Control flow integrity for cots binaries. In: USENIX Security, pp. 337–352 (2013)

Key Generation, Biometrics and Image Security

Reasoning about Privacy Properties of Biometric Systems Architectures in the Presence of Information Leakage

Julien Bringer[1], Hervé Chabanne[1,2], Daniel Le Métayer[3], and Roch Lescuyer[1](✉)

[1] Morpho, Issy-Les-Moulineaux, France
roch.lescuyer@morpho.com
[2] Télécom ParisTech, Paris, France
[3] Inria, Université de Lyon, Lyon, France

Abstract. Motivated by the need for precise definitions of privacy requirements, foundations for formal reasoning, and tools for justifying privacy-preserving design choices, a recent work introduces a formal model for the description of system architectures and the formal verification of their privacy properties. A subsequent work uses this framework to reason about privacy properties of biometric system architectures. In these studies, the description of an architecture specifies each component, their computations and the communications between them. This static approach makes it possible to reason about design choices at the very architectural level, leaving aside the implementation details. Although it is important to express privacy properties at this level, this approach fails to catch some leakage which may result from the system runtime. In particular, in the case of biometric systems, known attacks allow to recover some biometric information following a black-box approach, without breaking any part of the system. In this paper, we extend the existing formal model in order to deal with such side-channel attacks and we apply the extended model to analyse biometric information leakage in several variants of a biometric system architecture.

Keywords: Formal methods · Biometric systems · Privacy by design

1 Introduction

The privacy-by-design approach requests that the privacy properties of a given system or service should be considered from the early design steps. The draft of the Data Protection Regulation adopted by the European parliament in March 2014 [11] introduces privacy-by-design and privacy-by-default as legal obligations. The concrete consequences of this requirement on system engineering, however, are far from obvious. In particular, precise definitions of privacy properties are lacking, as well as a clear view of the tensions between functional, security and privacy requirements.

J. Lopez and C.J. Mitchell (Eds.): ISC 2015, LNCS 9290, pp. 493–510, 2015.
DOI: 10.1007/978-3-319-23318-5_27

The framework introduced in [1] is a first step towards a rigorous treatment of this issue. It relies on a formal approach to describe system architectures and to express their privacy and integrity requirements. Location of data and computations are first identified, then the use of formal methods allows designers to make the trust assumptions between stakeholders explicit, and to supply rigorous justifications of architectural choices.

A subsequent work [6] uses the framework of [1] to describe different biometrics system architectures and to reason about their privacy properties. Several constructions have been proposed to protect biometric data, based for instance on secure sketches [9], fuzzy vaults [17, 35], existing cryptographic tools [23], or secure hardware solutions [29]. Each proposal has been designed separately and comparisons between different solutions are not straightforward. The merit of [6] is to supply a unified framework for the precise comparison of existing solutions.

The authors of [1] advocate the definition of privacy requirements and design choices at the architecture level. Formal architectural descriptions include the specification of the exchanges between components and the location of the computations. Details of protocols are hidden from the architecture description, and the cryptographic building blocks are assumed to be correct. However, in general side-channel information may leak from the system runtime. It is especially true for biometric systems, where the comparison between biometric data is inherently fuzzy. The result of a matching between a fresh biometric data and an enrolled data inherently gives some information, even if the latter is protected and the building cryptographic blocks assumed to be correct (cf. [7, 30, 32]). The formal model of [1] fails to catch this leakage in its architecture analysis.

Our contribution. In this work, building upon the formalism of [1], we propose a formal model in which such runtime leakage can be expressed and analyzed. In particular, this model makes it possible to describe in the formal language the leakage of information through several executions of the same protocol. We also apply the extended model to analyse biometric information leakage in several variants of a biometric system architecture.

Related work. Privacy properties for biometric authentication have been addressed in [34]. [19, 21, 22] also formalize privacy properties for a particular representation of biometric templates, close to error correcting codes. Privacy by design is advocated by several authors [15, 20, 24, 26, 27, 33], who also underline the complexity of the engineering needed to ensure privacy properties. [13] defines a language for expressing privacy properties of computations that are ensured by the compilation process. Studies about the use of formal methods to ensure privacy properties of protocols include dedicated privacy languages, such as [3, 4], applied π-calculus [8], and methods for the design of algorithms meeting privacy metrics [25]. Privacy metrics, such as k-anonymity, l-diversity, or ϵ-differential privacy [10], supply a quantitative estimation of the level of privacy provided by an algorithm. Most previous work address privacy properties at the protocol level. In contrast, following the approach advocated in [1, 6], we study privacy properties at the architecture level. Because they abstract away unnecessary details and focus on critical issues, architectural descriptions enable a more systematic exploration of the design space and make it easier to prove high level privacy properties of a system.

ORGANIZATION OF THE PAPER. Section 2 introduces a formalism to describe architectures, and a privacy logic used to express privacy properties of architectures. Section 3 presents an architecture exibiting black-box leakage of biometric data and shows how the model of Sect. 2 makes it possible to integrate this leakage in the formal analysis.

2 A Formalism for Architectures and Privacy Properties

This section introduces a formalism to describe system architectures, based on the work of [1]. From the logical perspective, privacy properties are expressed with a dedicated epistemic logic [12], in order to avoid the so-called *logical omniscience problem* [16].

2.1 Architecture Syntax

We assume that a *functionality* achieved by an architecture is described by a set $\Omega = \{X = T\}$ of equations over the term language $\mathcal{T}$ defined as follows.

$$
\begin{aligned}
T &::= \tilde{X} \mid \ c \ \mid F(\tilde{X}_1, \ldots, \tilde{X}_m, c_1, \ldots, c_q) \mid \odot F(X) \\
\tilde{X} &::= X \mid X[k]
\end{aligned}
$$

$\tilde{X}$ is a (possibly indexed) variable, X a variable ($X \in Var$), k an variable index ($k \in Index$), c a constant ($c \in Const$) and F a function ($F \in Fun$). If X is an array, $Range(X)$ denotes its size. Each variable X is associated with at most one index k. In this case k is exclusive to X. $\odot F(X)$ denotes the iterative application of the binary function F to each element of the array X. $\tilde{X} \in T$ denotes that the variable $\tilde{X}$ appears in the term T. $V(\tilde{X})$ denotes either the array to which the variable belongs, if $\tilde{X} = X[k]$ is the entry of an array, or the variable itself in the other case.

Let $\mathcal{L}_A$ be the following language used to described architectures.

$$
\begin{aligned}
\mathcal{A} &::= \{R\} \\
R &::= Has_i^{(n)}(X) \mid Has_i(c) \mid Receive_{i,j}^{(n)}(\{St\}, \{X\} \cup \{c\}) \mid Check_i^{(n)}(\{Eq\}) \\
&\quad\ \mid Trust_{i,j} \quad \mid Reset \quad \mid Compute_G^{(n)}(X = T) \qquad\quad \mid Verify_i^{(n)}(\{St\})
\end{aligned}
$$

$$
\begin{aligned}
St &::= Pro \mid Att & Att &::= Attest_i(\{Eq\}) \\
Pro &::= Proof(\{P\}) & Eq &::= Pred(T_1, \ldots, T_m) \\
P &::= Att \mid Eq & &
\end{aligned}
$$

$\{Z\}$ is a set of elements of category Z. *Pred* is a set of predicates, depending on the architectures to be considered in practice. An architecture defines not only a set of primitives, but may also give a bound on the number of times a primitive can be used. The superscript notation $^{(n)}$ denotes that a primitive can be carried out at most $n \in (\mathbb{N} \setminus \{0\}) \cup \{\infty\}$ times by the component(s) – where ($\forall n' \in \mathbb{N}$: $n' < \infty$). To be consistent, we assume that n never equals 0. $\mathsf{mul}(\alpha)$ denotes the multiplicity (n) of the primitive α, if any. Each architecture $\mathcal{A}$ is built from a set of components $\mathcal{C} = \{C_1, \ldots, C_{|\mathcal{C}|}\}$. For the sake of conciseness we often use indices such as i or j to denote the components themselves (C_i or C_j). $G \subseteq \mathcal{C}$ is a subset of components. We note $\tilde{X} \in Eq := Pred(T_1, \ldots, T_m)$ if ($\exists l \in [1, n]$: $\tilde{X} \in T_l$); $\tilde{X} \in E := \{Eq_1, \ldots, Eq_n\}$ if ($\exists l \in [1, n]$: $\tilde{X} \in Eq_l$).

Consistency assumptions are made about the architectures to avoid meaningless definitions. For instance we require that components carry out computations only on the values that they have access to (either through *Has*, *Compute*, or *Receive*). We also require that all multiplicities (n) specified by the primitives are identical in a consistent architecture. As a result, a consistent architecture $\mathcal{A}$ is parametrized by an integer $n \geq 1$ (we note $\mathcal{A}(n)$ when we want to make this integer explicit). Let *Arch* be the set of consistent architectures.

2.2 Traces of Events

A trace is a sequence of events occurring in the system. Let $\mathcal{T}^\epsilon$ be the following extension of $\mathcal{T}$, used to manage computations using different values of the same variables from different sessions and to instantiate indices by specific integer values.

$$
\begin{aligned}
T^\epsilon &::= \tilde{X} \mid c \mid F(\tilde{X}_1^{(n_1)}, \ldots, \tilde{X}_m^{(n_m)}, c_1, \ldots, c_q) \mid \odot F(X) \\
\tilde{X} &::= X \mid X[K] \\
K &::= k \mid Ck
\end{aligned}
$$

When a function is applied to some variables, each variable is given with a multiplicity $^{(n)}$, for some $n \geq 1$. $F(\tilde{X}_1, \ldots, \tilde{X}_m, c_1, \ldots, c_q) \in \mathcal{T}^\epsilon$ is a short-cut for $F(\tilde{X}_1^{(1)}, \ldots, \tilde{X}_m^{(1)}, c_1, \ldots, c_q)$. Ck is a index constant (an integer).

Let $\mathcal{L}_\mathcal{T}$ be the following language.

$$
\begin{aligned}
\theta &::= \mathsf{Seq}(\epsilon) \\
\epsilon &::= Has_i(X : V) \mid Has_i(c) \mid Receive_{i,j}(\{St\}, \{X : V\} \cup \{c\}) \\
&\quad \mid Session \mid Reset \mid Compute_G(X = T^\epsilon) \\
&\quad \mid Check_i(\{Eq\}) \mid Verify_i(\{St\})
\end{aligned}
$$

In particular, an event can instantiate variables X with specific values V. Constants always map to the same value. Let Val be the set of values the variables and constants can take. The set $Val_\perp$ is defined as $Val \cup \{\perp\}$ where $\perp \notin Val$ is a specific symbol used to denote that a variable or a constant has not been assigned yet.

$\epsilon \cdot \theta$ denotes a trace where the first event is ϵ and the remaining of the trace is θ. $\epsilon \in \theta$ is a short-cut for ($\theta = \epsilon_1 \cdots \epsilon_t$ and $\exists a \in [1, t]$: $\epsilon_a = \epsilon$). For two events $\epsilon_a, \epsilon_b \in \theta$, we note $\epsilon_b \geq^\theta \epsilon_a$ if $b \geq a$. $\theta_{<a} = \epsilon_1 \cdots \epsilon_{a-1}$ is the restriction of θ to the events that precede ϵ_a, and $\theta_{|<a}$ is the restriction of θ to the current session up to ϵ_a. We note $\theta_{|l}$ the restriction of θ to the current session. For an event

$\epsilon_a \in \theta$, $\mathsf{s}(a)$ denotes the current session, *i.e.* the session to which ϵ_a belongs. $\theta_{|\mathsf{s}(a)}$ is the restriction of θ to the session that contains ϵ_a. Likewise $\mathsf{R}(a)$ denotes the current sub-trace between two *Reset* events.

The following correspondence relation expresses the link between traces and architectures. The events (session markers excepted) are instantiations of the architectural primitives (trust relations excepted). The *correspondence relation*, defined over $Event \times \mathcal{A}$ and noted $\mathsf{C}(\epsilon, \alpha)$, holds if, except for the *Session* event, the event ϵ can be obtained from the architectural primitive α, by assigning values to variables measured and received, multiplicities to variables in right-hand side of computations, and integers to indices.

Consistency assumptions are made about the traces, as for architectures. In particular, we require that, between two *Reset* events, all $Has_i(c)$ events follow the first *Reset* event.

A trace θ of events is said *compatible* with a consistent architecture $\mathcal{A}(n)$ if all events in θ (except the computations) can be obtained by instantiation of some architectural primitive from $\mathcal{A}$, and if the number of events between two *Reset* events corresponding to a given primitive respects the bound n specified by the architecture: $((\forall \epsilon \in \theta\colon \text{if } \epsilon \neq Compute_G(X = T^{\epsilon}), \text{then } \exists \alpha \in \mathcal{A}\colon \mathsf{C}(\epsilon, \alpha))$, and $(\forall \alpha \in \mathcal{A}\colon \text{if } \mathsf{mul}(\alpha) = n, \text{then } (\forall \epsilon \in \theta\colon |\{\epsilon' \in \theta_{|\mathsf{R}(\epsilon)} \mid \mathsf{C}(\epsilon', \alpha)\}| \leq n)))$.

Due to the exclusion of computation from the definition, a compatible trace can contain additional computations from an adversary trying to obtain information about the variables. It may in particular use information leaking through several executions of a protocol. A trace can also contain erroneous computations from the adversary, leading to the reception of erroneous values by components. Let $T(\mathcal{A})$ be the set of traces which are compatible with an architecture $\mathcal{A} \in Arch$.

2.3 Architecture Semantics

To define the semantics of an architecture, each component is associated with a state. Each event in a trace of events affects the state of each component involved by the event. Then, the semantics of an architecture is defined as the set of states reachable by compatible traces.

The state of a component is either the *Error* state or a pair consisting of: (i) a variable state assigning values to variables, and (ii) a property state defining what is known by a component.

$$\begin{aligned} State_{\perp} &= (State_V \times State_P) \cup \{Error\} \\ State_V &= Var \cup Const \rightarrow \mathsf{List}(Val_{\perp}) \\ State_P &= \{Eq\} \cup \{Trust_{i,j}\} \end{aligned}$$

The data structure List over a set S denotes the finite ordered lists of elements of S. $\mathsf{size}(L)$ denotes the size of the list L. $()$ is the empty list, its size is 0. For a non-empty list $L = (e_1, \ldots, e_n) \in S^n$ where $\mathsf{size}(L) = n \geq 1$, $L[m]$ denotes the element e_m for $1 \leq m \leq n$, $\mathsf{last}(L)$ denotes $L[n]$, and $\mathsf{append}(L, e)$ denotes the list $(e_1, \ldots, e_n, e) \in S^{n+1}$.

$\sigma := (\sigma_1, \ldots, \sigma_{|\mathcal{C}|})$ denotes the global state (*i.e.* the list of states of all components) defined over $(State_\perp)^{|\mathcal{C}|}$. σ_i^v and σ_i^{pk} denotes, respectively, the variable and the knowledge state of the component $C_i \in \mathcal{C}$. The variable state assigns values to variables, but also to constants (constants however are either undefined or takes a single specific value). $\sigma_i^v(X)[m]$ (resp. $\sigma_i^v(c)[m]$) denotes the m-th entry of the variable state of $X \in Var$ (resp. $c \in Const$), and contains a value. The initial state of an architecture $\mathcal{A}$ is noted $Init^{\mathcal{A}} = \langle Init_1^{\mathcal{A}}, \ldots, Init_{|\mathcal{C}|}^{\mathcal{A}} \rangle$ where: $\forall C_i \in \mathcal{C}$: $Init_i^{\mathcal{A}} = (Empty, \{Trust_{i,j} \mid \exists C_j \in \mathcal{C}\colon Trust_{i,j} \in \mathcal{A}\})$. *Empty* associates to each variable and constant a list of a single undefined value ($\perp$). We assume that, in the initial state, the system lies in the first session. Alternatively, we could set empty lists in the initial state and assume that every consistent trace begins with a *Session* event.

Let $S_T : Trace \times (State_\perp)^{|\mathcal{C}|} \to (State_\perp)^{|\mathcal{C}|}$ and $S_E : Event \times (State_\perp)^{|\mathcal{C}|} \to (State_\perp)^{|\mathcal{C}|}$ be the following two functions. S_T is defined recursively by iteration of S_E: for all state $\sigma \in (State_\perp)^{|\mathcal{C}|}$, event $\epsilon \in Event$ and consistent trace $\theta \in Trace$, $S_T(\langle\rangle, \sigma) = \sigma$ and $S_T(\epsilon \cdot \theta, \sigma) = S_T(\theta, S_E(\epsilon, \sigma))$. The modification of a state is noted $\sigma[\sigma_i/(v, pk)]$ the variable and knowledge states of C_i are replaced by v and pk respectively. $\sigma[\sigma_i/Error]$ denotes that the *Error* state is reached for component C_i. We assume that a component reaching an *Error* state gets no longer involved in any later action (until a reset of the system). The function S_E is defined event per event.

The effect of $Has_i(X : V)$ and $Receive_{i,j}(S, \{(X : V)\})$ on the variable state of component C_i is the replacement of the last value of the variable X by the value V: $\mathsf{last}(\sigma_i^v(X)) := V$. This effect is denoted by $\sigma_i^v[X/V]$:

$$S_E(Has_i(X : V), \sigma) = S_E(Receive_{i,j}(S, \{X : V\}), \sigma) = \sigma[\sigma_i/(\sigma_i^v[X/V], \sigma_i^{pk})].$$

In the case of constants, the value V is determined by the interpretation of c (as in the function symbols in the computation).

The effect of $Compute_G(X = T^\epsilon)$ is to assign to X, for each component $C_i \in G$, the value V produces by the evaluation (denoted ε) of T^ϵ. The new knowledge is the equation $X = T$. A computation may involve values of variables from different sessions. As a result, some consistency conditions must be met, otherwise an error state is reached:

$$S_E(Compute_G(X = T^\epsilon), \sigma) = \begin{cases} \sigma[\forall C_i \in G : \sigma_i/(\sigma_i^v[X/V], \sigma_i^{pk} \cup \{X = T\})] \\ \text{ if the condition on the computation holds,} \\ \sigma[\sigma_i/Error] \text{ otherwise,} \end{cases}$$

where $V := \varepsilon\left(T^\epsilon, \cup_{C_i \in G}\sigma_i^v\right)$. For each $\tilde{X}^{(n)} \in T^\epsilon$, the evaluation of T^ϵ is done with respect to the n last values of $\tilde{X}$ that are fully defined. An error state is reached if n such values are not available. The condition on the computation is then: $\forall C_i \in G, \tilde{X}^{(n)} \in T^\epsilon$: $\mathsf{size}\left(\left\{m \mid \sigma_i^v\left(V(\tilde{X})\right)[m] \text{ is fully defined}\right\}\right) \geq n$.

Semantics of the verification events are defined according to the (implicit) semantics of the underlying verification procedures. In each case, the knowledge state of the component is updated if the verification passes, otherwise the component reaches an $Error$ state. The variable state is not affected.

$$S_E(Check_i(E), \sigma) = \begin{cases} \sigma[\sigma_i/(\sigma_i^v, \sigma_i^{pk} \cup E)] \text{ if the check holds,} \\ \sigma[\sigma_i/Error] \text{ otherwise,} \end{cases}$$

$$S_E(Verify_i(Proof(E)), \sigma) = \begin{cases} \sigma[\sigma_i/(\sigma_i^v, \sigma_i^{pk} \cup new_{Proof}^{pk})] \text{ if the proof is valid,} \\ \sigma[\sigma_i/Error] \text{ otherwise,} \end{cases}$$

$$S_E(Verify_i(Attest_j(E)), \sigma) = \begin{cases} \sigma[\sigma_i/(\sigma_i^v, \sigma_i^{pk} \cup new_{Attest}^{pk})] \\ \text{ if the attestation is valid,} \\ \sigma[\sigma_i/Error] \text{ otherwise.} \end{cases}$$

The new knowledge new_{Proof}^{pk} and new_{Attest}^{pk} are defined as: $new_{Proof}^{pk} := \{Eq \mid Eq \in E \vee (\exists j' : Attest_{j'}(E') \in E \wedge Eq \in E' \wedge Trust_{i,j'} \in \sigma_i^{pk})\}$ and $new_{Attest}^{pk} := \{Eq \mid Eq \in E \wedge Trust_{i,j} \in \sigma_i^{pk}\}$.

In the session case, the knowledge state is reinitialized and a new entry is added in the variable states:

$$S_E(Session, \sigma) = \sigma[\forall i \in \mathcal{C} : \sigma_i/(upd^v, \{Trust_{i,j} \mid \exists C_j \in \mathcal{C} : Trust_{i,j} \in \mathcal{A}\})],$$

where the new variable state upd^v is such that $\sigma_i^v(X) := \mathsf{append}(\sigma_i^v(X), \perp)$ for all variables $X \in Var$, and $\sigma_i^v(c) := \mathsf{append}(\sigma_i^v(c), \mathsf{last}(\sigma_i^v(c)))$ for all constants $c \in Const$. The session event is not local to a component, all component states are updated. As a result, we associate to each global state σ a unique number, noted $\mathsf{s}(\sigma)$, which indicates the number of sessions. In the initial state, $\mathsf{s}(\sigma) := 1$, and at each $Session$ event, $\mathsf{s}(\sigma)$ is incremented.

In the reset case, all values are dropped and the initial state is restored: $S_E(Reset, \sigma) = Init^{\mathcal{A}}$.

The semantics $S(\mathcal{A})$ of an architecture $\mathcal{A}$ is defined as the set of states reachable by compatible traces: $S(\mathcal{A}) = \{\sigma \in (State_\perp)^{|\mathcal{C}|} \mid \exists \theta \in T(\mathcal{A}) : S_T(\theta, Init^{\mathcal{A}}) = \sigma\}$. Let $S_i(\mathcal{A}) = \{\sigma \in S(\mathcal{A}) \mid \sigma_i \neq Error\}$ denote the restriction $S(\mathcal{A})$ to well-defined states with respect to component C_i.

2.4 Privacy Properties of Architectures

Following the approach of [1], we define a dedicated epistemic logic to express privacy properties of architectures. Let $\mathcal{L}_P$ be the following language.

$$\begin{aligned} \phi &::= Has_i(X^{(n)}) \mid Has_i(c) \mid Has_i^{none}(X) \mid Has_i^{none}(c) \mid K_i(Eq) \mid \phi_1 \wedge \phi_2 \\ Eq &::= Pred(T_1, \ldots, T_m) \end{aligned}$$

The formula Has_i represents $n \geq 1$ accesses by C_i to some variable X. The *knowledge operator* K_i represents the knowledge a component C_i can derive, following the "deductive algorithmic knowledge" approach [12,31]. More precisely, the knowledge of a component C_i is defined as the set of properties that this component can actually derive using its own information and the deductive system $\triangleright_i$. The deductive relation $\triangleright_i$ of a component C_i is defined as a set of relations: $\{Eq_1, \ldots, Eq_n\} \triangleright_i Eq_0$. The notation Eq in $\mathcal{L}_P$ is an overloaded notation of the Eq definition in the language architecture $\mathcal{A}$.

In addition, each component $C_i \in C$ is associated with a dependence relation Dep_i, catching dependencies between variables. In particular, several values of the same variables from different sessions can give information about other variables. For a variable Y (or a constant c), a set $\mathcal{X}$ of variables (and eventually constants), each variable being given with a multiplicity, $Dep_i(Y, \mathcal{X})$ means that a value of Y can be obtained by the component C_i if it gets access to the n values of X, for each $X^{(n)} \in \mathcal{X}$, and to the value of the constants, if any.

The *semantics* $S(\phi)$ *of a property* $\phi \in \mathcal{L}_P$ is defined as the set of architectures where ϕ is satisfied. The fact that ϕ is satisfied by a (consistent) architecture $\mathcal{A} \in Arch$ is defined as follows.

$\mathcal{A}$ satisfies $Has_i(X^{(n)})$ if there is a reachable state in which X is fully defined (at least) $n \geq 1$ times:
$\exists \sigma \in S(\mathcal{A})$: $\exists (1 \leq m_1 < m_2 < \cdots < m_n)$: $(\forall l \in [1, n]$: $[(\sigma_i^v(X)[m_l] \neq \bot)$ and (if $\sigma_i^v(X)[m_l] = \langle v_1, \ldots, v_k \rangle$, then $\forall l' \in [1, k]$: $(v_{l'} \neq \bot))])$.

$\mathcal{A}$ satisfies $Has_i(c)$ if there is a reachable state in which c is fully defined:
$\exists \sigma \in S(\mathcal{A}), \exists m \geq 1$: $((\sigma_i^v(c)[m] \neq \bot)$ and (if $\sigma_i^v(c)[m] = \langle v_1, \ldots, v_k \rangle$, then $\forall l' \in [1, k]$: $(v_{l'} \neq \bot)))$.

$\mathcal{A}$ satisfies $Has_i^{none}(X)$ (resp. $Has_i^{none}(c)$) if no compatible trace leads to a state in which C_i assigns a value to X (resp. c):
(resp. $\forall \sigma \in S(\mathcal{A})$: $\forall m \geq 1$: (if $\mathsf{size}(\sigma_i^v(c)) \leq m$, $(\sigma_i^v(c)[m] = \bot)))$.

$\mathcal{A}$ satisfies $K_i(Eq)$ if for all reachable states, there exists a state in the same session in which C_i can derive Eq:
$\forall \sigma' \in S_i(\mathcal{A})$: $\exists \sigma \in S_i(\mathcal{A})$: $\sigma \geq_i \sigma'$ and $\mathsf{s}(\sigma) = \mathsf{s}(\sigma')$ and $\sigma_i^{pk} \triangleright_i Eq$.

$\mathcal{A}$ satisfies $\phi_1 \wedge \phi_2$ if $\mathcal{A}$ satisfies ϕ_1 and $\mathcal{A}$ satisfies ϕ_2.

A set $\mathcal{D}$ of *deductive rules* for $\mathcal{L}_P$ is given Fig. 1. $\mathcal{A} \vdash \phi$ denotes that ϕ can be derived from $\mathcal{A}$ thanks to $\mathcal{D}$ (*i.e.* there exists a derivation tree such that all steps belong to $\mathcal{D}$ and such that the leaf is $\mathcal{A} \vdash \phi$). $\mathcal{D}$ is sound and complete with respect to the above semantics of properties.

Theorem 1 (Soundness). *For all $\mathcal{A} \in Arch$, if $\mathcal{A} \vdash \phi$ then $\mathcal{A} \in S(\phi)$.*

Theorem 2 (Completeness). *For all $\mathcal{A} \in Arch$, if $\mathcal{A} \in S(\phi)$ then $\mathcal{A} \vdash \phi$.*

Due to the length of the proofs and the lack of place, we only give sketch for these proofs in Appendix A. Soundness is proved by induction on the derivation tree. For each theorem $\mathcal{A} \vdash \phi$, one can find traces satisfying the claimed property, or show that all traces satisfy the claimed property (depending on the kind of property). Completeness is shown by induction on the property ϕ. For each property belonging to the semantics, one can exhibit a tree that derives it from the architecture.

3 Application to Biometric Systems Architectures

In this section, we analyse a particular biometric system architecture, introduced in [5], with the formal model of the precedent section. This architecture extends the Match-On-Card (MOC) technology, used for authentication purposes, to the identification paradigm. We first describe it within the framework of [6]. We then note that this framework is insufficient to consider some dynamic leakage of information and show how to use our extended formalism to integrate this leakage into the architecture analysis.

3.1 Extension of the MOC Technology to Biometric Identification

In a biometric system, users are registered during an *enrolment* phase and the biometric templates, aka the references, are recorded in a database. During the *identification*, a user presents a fresh biometric trait to a terminal, equipped

$$\textbf{H1}\ \frac{Has_i^{(n)}(X) \in \mathcal{A}}{\mathcal{A} \vdash Has_i(X^{(n)})} \qquad \textbf{H2}\ \frac{Receive_{i,j}^{(n)}(S,E) \in \mathcal{A} \qquad X \in E}{\mathcal{A} \vdash Has_i(X^{(n)})}$$

$$\textbf{H1'}\ \frac{Has_i(c) \in \mathcal{A}}{\mathcal{A} \vdash Has_i(c)} \qquad \textbf{H2'}\ \frac{Receive_{i,j}^{(n)}(S,E) \in \mathcal{A} \qquad c \in E}{\mathcal{A} \vdash Has_i(c)}$$

$$\textbf{H3}\ \frac{Compute_G^{(n)}(X = T) \in \mathcal{A} \qquad C_i \in G}{\mathcal{A} \vdash Has_i(X^{(n)})} \qquad \textbf{H4}\ \frac{\mathcal{A} \vdash Has_i(X^{(n)}) \qquad 1 \leq m \leq n}{\mathcal{A} \vdash Has_i(X^{(m)})}$$

$$\textbf{H5}\ \frac{Dep_i(Y, \mathcal{X}) \qquad \forall X^{(n)} \in \mathcal{X}\colon \mathcal{A} \vdash Has_i(X^{(n)}) \qquad \forall c \in \mathcal{X}\colon \mathcal{A} \vdash Has_i(c)}{\mathcal{A} \vdash Has_i(Y^{(1)})}$$

$$\textbf{H5'}\ \frac{Dep_i(c, \mathcal{X}) \qquad \forall X^{(n)} \in \mathcal{X}\colon \mathcal{A} \vdash Has_i(X^{(n)}) \qquad \forall c' \in \mathcal{X}\colon \mathcal{A} \vdash Has_i(c')}{\mathcal{A} \vdash Has_i(c)}$$

$$\textbf{HN}\ \frac{\mathcal{A} \nvdash Has_i(X^{(1)})}{\mathcal{A} \vdash Has_i^{none}(X)} \qquad \textbf{HN'}\ \frac{\mathcal{A} \nvdash Has_i(c)}{\mathcal{A} \vdash Has_i^{none}(c)}$$

$$\textbf{K1}\ \frac{Compute_G^{(n)}(X = T) \in \mathcal{A} \qquad C_i \in G}{\mathcal{A} \vdash K_i(X = T)} \qquad \textbf{K}\wedge\ \frac{\mathcal{A} \vdash K_i(Eq_1) \qquad \mathcal{A} \vdash K_i(Eq_2)}{\mathcal{A} \vdash K_i(Eq_1 \wedge Eq_2)}$$

$$\textbf{K2}\ \frac{Check_i^{(n)}(E) \in \mathcal{A} \qquad Eq \in E}{\mathcal{A} \vdash K_i(Eq)} \qquad \textbf{K}\triangleright\ \frac{E \triangleright_i Eq_0 \qquad \forall Eq \in E\colon \mathcal{A} \vdash K_i(Eq)}{\mathcal{A} \vdash K_i(Eq_0)}$$

$$\textbf{K3}\ \frac{Verify_i^{(n)}(Proof(E)) \in \mathcal{A} \qquad Eq \in E}{\mathcal{A} \vdash K_i(Eq)} \qquad \textbf{I}\wedge\ \frac{\mathcal{A} \vdash \phi_1 \qquad \mathcal{A} \vdash \phi_2}{\mathcal{A} \vdash \phi_1 \wedge \phi_2}$$

$$\textbf{K4}\ \frac{Verify_i^{(n)}(Proof(E)) \in \mathcal{A} \qquad Attest_k(E') \in E \qquad Eq \in E' \qquad Trust_{i,k} \in \mathcal{A}}{\mathcal{A} \vdash K_i(Eq)}$$

$$\textbf{K5}\ \frac{Verify_i^{(n)}(Attest_j(E)) \in \mathcal{A} \qquad Trust_{i,j} \in \mathcal{A} \qquad Eq \in E}{\mathcal{A} \vdash K_i(Eq)}$$

Fig. 1. Set $\mathcal{D}$ of deductive rules

with a sensor. A fresh biometric template is extracted from the trait and is compared with the templates stored in the database. The identification procedure accepts if there is a match between the fresh template and the database. In the *authentication* paradigm, the fresh biometric template is compared to a single reference template, instead of a whole database. In the *Match-On-Card* (MOC) technology [14,28,29] (also called *comparison-on-card*), the reference biometric template is stored on a smart-card. During the verification, the fresh template is sent to the card, the comparison with the reference template is done inside the card. In practice, since an entire database cannot be stored on a smart-card, the MOC technology is used for authentication purpose.

In order to extend the MOC technology to the identification paradigm, the main idea of [5] is to store a *quantized* version of the database inside a secure module (playing the role of the card in the MOC case). From each biometric reference template, a *quantization* is computed, using typically a secure sketch scheme [9,18]. The reference database is encrypted and stored outside the secure module, whereas the quantizations of the templates are stored inside.

The verification step is processed as follows. Suppose one wants to identify himself in the system. A terminal captures the fresh biometrics, extracts a template, computes its quantization and sends them to the secure module. Then, the module proceeds to a comparison between the fresh quantization and all enrolled quantizations. The c nearest quantizations, for some parameter c of the system, are the c potential candidates for the identification. Then, the module queries the c corresponding (encrypted) templates to the database, decrypts them, and compares them with the fresh template. The module finally sends its response to the terminal: 1 if one of the enrolled templates is close enough to the fresh template, 0 otherwise.

The Fig. 2 gives a graphical representation of the resulting architecture. We focus on the verification, and assume that users are already enrolled. A component per actor is introduced, denoted by upper case sans serif letters S, T, *etc.* The component U represents the user, and T the terminal. The issuer I enrols

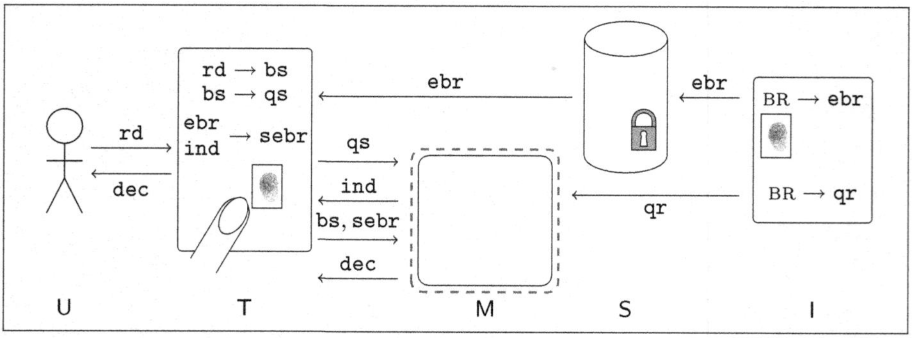

Fig. 2. Architecture of the extension of the Match-On-Card technology to biometric identification. The dotted red line indicates the location of the comparison.

users and certifies their templates. Moreover it computes the quantizations and sends them into the module. The server S stores the reference templates inside a database. Finally a module M stores the quantizations, performs the matching and eventually takes the decision.

Type letters bs, qs, *etc.* denote variables and capital letters THR, BR, *etc.* denote constants. BR is a biometric reference templates database, built during the enrolment phase. ebr is the encrypted version of BR. rd is a raw biometric data provided by the user to the terminal during the verification phase. bs is a fresh biometric template derived from the raw data rd. THR is a threshold used as a closeness criterion for the biometric templates. The output dec of the verification is the result of the matching between the fresh template bs and the enrolled templates BR, considering the threshold THR. N denotes the size of the database (*i.e.* the number of enrolled users), Q the size of the quantizations, and C the number of indices asked by the card. The ranges are $Range(\text{BR}, \texttt{ebr}, \texttt{qr}) = \text{N}$, $Range(\texttt{rd}, \text{THR}, \texttt{bs}, \texttt{qs}, \texttt{dec}) = 1$, and $Range(\texttt{ind}, \texttt{sebr}, \texttt{sbr}) = \text{C}$.

The set *Fun* of functions contains the extraction procedure *Extract*, the encryption and decryption procedures *Enc* and *Dec*, the (non-invertible) quantization *Quant* of the biometric templates, the comparison of the quantizations *QComp*, which takes as inputs two sets of quantizations and the parameter C, the selection of the encrypted templates *EGet*, and finally the matching μ, which takes as arguments two biometric templates and the threshold THR.

The biometric reference templates are enrolled by the issuer ($Has_\mathsf{I}(\text{BR})$). A verification process is initiated by the terminal T receiving as input a raw biometric data rd from the user U. T extracts the fresh biometric template bs from rd using the function $Extract \in Fun$. The architecture then contains, as other biometric systems, $Receive_{\mathsf{T},\mathsf{U}}(\{\}, \{\texttt{rd}\})$ and $Compute_\mathsf{T}(\texttt{bs} = Extract(\texttt{rd}))$ and the Dep_T relation is such that $(\texttt{bs}, \{\texttt{rd}\}) \in Dep_\mathsf{T}$. The user receives the final decision dec from the terminal: $Receive_{\mathsf{U},\mathsf{T}}(\{\}, \{\texttt{dec}\})$. To sum up, the architecture is described as follows in the framework of [6]:

$$\begin{aligned}
\mathcal{A}^{\mathsf{mi}} := \{&Has_\mathsf{I}(\text{BR}), Has_\mathsf{U}(\texttt{rd}), Has_\mathsf{M}(\text{C}), Has_\mathsf{M}(\text{THR}),\\
&Compute_\mathsf{I}(\texttt{ebr} = Enc(\text{BR})), Compute_\mathsf{I}(\texttt{qr} = Quant(\text{BR})),\\
&Compute_\mathsf{T}(\texttt{bs} = Extract(\texttt{rd})), Compute_\mathsf{T}(\texttt{sebr} = EGet(\texttt{ebr}, \texttt{ind})),\\
&Compute_\mathsf{T}(\texttt{qs} = Quant(\texttt{bs})), Compute_\mathsf{M}(\texttt{ind} = QComp(\texttt{qs}, \texttt{qr}, \text{C})),\\
&Compute_\mathsf{M}(\texttt{sbr} = Dec(\texttt{sebr})), Compute_\mathsf{M}(\texttt{dec} = \mu(\texttt{sbr}, \texttt{bs}, \text{THR})),\\
&Receive_{\mathsf{S},\mathsf{I}}(\{Attest_\mathsf{I}(\texttt{ebr} = Enc(\text{BR}))\}, \{\texttt{ebr}\}), Receive_{\mathsf{T},\mathsf{U}}(\{\}, \{\texttt{rd}\}),\\
&Receive_{\mathsf{T},\mathsf{S}}(\{Attest_\mathsf{I}(\texttt{ebr} = Enc(\text{BR}))\}, \{\texttt{ebr}\}), Receive_{\mathsf{M},\mathsf{T}}(\{\}, \{\texttt{qs}\}),\\
&Receive_{\mathsf{M},\mathsf{I}}(\{Attest_\mathsf{I}(\texttt{qr} = Quant(\text{BR}))\}, \{\texttt{qr}\}), Receive_{\mathsf{T},\mathsf{M}}(\{\}, \{\texttt{ind}\}),\\
&Receive_{\mathsf{M},\mathsf{T}}(\{\}, \{\texttt{sebr}, \texttt{bs}\}), Receive_{\mathsf{T},\mathsf{M}}(\{\}, \{\texttt{dec}\}),\\
&Trust_{\mathsf{T},\mathsf{I}}, Trust_{\mathsf{M},\mathsf{I}}, Trust_{\mathsf{T},\mathsf{M}}, Verify_\mathsf{T}(Attest_\mathsf{I}(\texttt{ebr} = Enc(\text{BR}))),\\
&Verify_\mathsf{T}(\{Attest_\mathsf{M}(\texttt{dec} = \mu(\texttt{sbr}, \texttt{bs}, \text{THR}))\}),\\
&Verify_\mathsf{M}(Attest_\mathsf{I}(\texttt{qr} = Quant(\text{BR}))), Verify_\mathsf{T}(\{Attest_\mathsf{M}(\texttt{sbr} = Dec(\texttt{ebr}))\})\}
\end{aligned}$$

The issuer encrypts the templates and computes the quantizations, which is expressed by the dependencies: $Dep_{\mathsf{I}}^{\mathsf{mi}} := \{(\mathtt{ebr}, \{\mathrm{BR}\}), (\mathtt{qr}, \{\mathrm{BR}\})\}$. The terminal and module computations are reflected in the dependencies as well: $Dep_{\mathsf{T}}^{\mathsf{mi}} := \{(\mathtt{bs}, \{\mathtt{rd}\}), (\mathtt{qs}, \{\mathtt{bs}\})\}, (\mathtt{sebr}, \{\mathtt{bs}, \mathtt{ind}\})\}$. The dependency relation of the module reflects its ability to decrypt the templates: $Dep_{\mathsf{M}}^{\mathsf{mi}} := \{(\mathtt{ind}, \{\mathtt{qs}, \mathtt{qr}, \mathrm{C}\}), (\mathtt{sbr}, \{\mathtt{sebr}\}), (\mathtt{dec}, \{\mathtt{sbr}, \mathtt{bs}, \mathrm{THR}\}), (\mathrm{BR}, \{\mathtt{ebr}\})\}$. The absence of such a relation in other dependencies prevents the corresponding components to get access to the plain references, even if they get access to the ciphertexts.

3.2 Learning of the Protected Quantizations

The formalism of [1] is insufficient to consider the leakage of the sensitive biometric data stored inside the module. Let us now discuss this point. In $\mathcal{A}^{\mathsf{mi}}$, we would like that the terminal gets no access to the quantizations: $\mathcal{A}^{\mathsf{mi}} \in Has_{\mathsf{T}}^{none}(\mathtt{qr})$. It is indeed possible to derive $\mathcal{A}^{\mathsf{mi}} \vdash Has_{\mathsf{T}}^{none}(\mathtt{qr})$, thanks to the (**HN**) rule. According to the notations of [1], where $Has_i^{all}(X)$ stands for $Has_i(X^{(1)})$ in this paper, we have:

$$\mathbf{HN}\ \frac{\begin{array}{cc}\nexists X : Dep_{\mathsf{T}}(\mathtt{qr}, X) \in \mathcal{A}^{\mathsf{mi}} & \nexists j, S : Receive_{\mathsf{T},j}(S, \{\mathtt{qr}\}) \in \mathcal{A}^{\mathsf{mi}} \\ Has_{\mathsf{T}}(\mathtt{qr}) \notin \mathcal{A}^{\mathsf{mi}} & \nexists T : Compute_{\mathsf{T}}(\mathtt{qr} = T) \in \mathcal{A}^{\mathsf{mi}}\end{array}}{\dfrac{\mathcal{A} \nvdash Has_{\mathsf{T}}^{all}(\mathtt{qr})}{\mathcal{A} \vdash Has_{\mathsf{T}}^{none}(\mathtt{qr})}}$$

This corresponds to the intuition saying that quantizations are protected since they are stored in a secure hardware element.

However, an attack (described in [7]) shows that, in practice, quantizations can be learned if a sufficient number of queries to the module is allowed. The attack roughly proceeds as follows (we drop the masks for sake of clarity). The attacker maintains a $\mathrm{N} \times \mathrm{Q}$ table (say T) of counters for each bit to be guessed. All entries are initialized to 0. Then it picks Q-bits random vector Q (in a non-adaptive fashion) and sends it to the module. The attacker observes the set of indices $\mathtt{ind} \subseteq [1, \mathrm{N}]$ corresponding to the encrypted templates asked by the module. It updates its table T as follows, according to its query Q and the response $\mathtt{ind}$: for each $i \in [1, \mathrm{N}]$ and $j \in [1, \mathrm{Q}]$, it decrements the entry $T[i][j]$ if $Q[j] = 0$, and increments it if $Q[j] = 1$. At the end of the attack, the N quantizations are guessed from the signs of the counters.

We experimentally verified what is claimed in [7]: our experiments show that one can retrieve all the stored quantizations, with overwhelming probability, if a linear number of attempts is allowed to the module (linear in the number of bits to be guessed). For instance, if 10 quantizations of 128 bits are stored within the module, around 1000–2000 queries are sufficient to correctly guess all the bits.

3.3 Architecture Description in our Extended Framework

The number of query made to the module is the crucial point in the attack above (and generally in other black-box attacks against biometric systems [7]).

The extended model of Sect. 2 enables to introduce a bound on the number of actions allowed to be performed. We now use this model to integrate such a bound in the formal architecture description. Let $\mathcal{A}^{\mathsf{mi\text{-}e}}(n)$ be the following architecture, for some $n \geq 1$:

$$\begin{aligned}
\mathcal{A}^{\mathsf{mi\text{-}e}}(n) := \{&Has_{\mathsf{I}}(\mathrm{BR}), Has_{\mathsf{U}}^{(n)}(\mathtt{rd}), Has_{\mathsf{M}}(\mathrm{C}), Has_{\mathsf{M}}(\mathrm{THR}),\\
&Compute_{\mathsf{I}}^{(n)}(\mathtt{ebr} = Enc(\mathrm{BR})), Compute_{\mathsf{I}}^{(n)}(\mathtt{qr} = Quant(\mathrm{BR})),\\
&Compute_{\mathsf{T}}^{(n)}(\mathtt{bs} = Extract(\mathtt{rd})), Compute_{\mathsf{T}}^{(n)}(\mathtt{sebr} = EGet(\mathtt{ebr}, \mathtt{ind})),\\
&Compute_{\mathsf{T}}^{(n)}(\mathtt{qs} = Quant(\mathtt{bs})), Compute_{\mathsf{M}}^{(n)}(\mathtt{ind} = QComp(\mathtt{qs}, \mathtt{qr}, \mathrm{C})),\\
&Compute_{\mathsf{M}}^{(n)}(\mathtt{sbr} = Dec(\mathtt{sebr})), Compute_{\mathsf{M}}^{(n)}(\mathtt{dec} = \mu(\mathtt{sbr}, \mathtt{bs}, \mathrm{THR})),\\
&Receive_{\mathsf{S,I}}^{(n)}(\{Attest_{\mathsf{I}}(\mathtt{ebr} = Enc(\mathrm{BR}))\}, \{\mathtt{ebr}\}), Receive_{\mathsf{T,U}}^{(n)}(\{\}, \{\mathtt{rd}\}),\\
&Receive_{\mathsf{T,S}}^{(n)}(\{Attest_{\mathsf{I}}(\mathtt{ebr} = Enc(\mathrm{BR}))\}, \{\mathtt{ebr}\}), Receive_{\mathsf{M,T}}^{(n)}(\{\}, \{\mathtt{qs}\}),\\
&Receive_{\mathsf{M,I}}^{(n)}(\{Attest_{\mathsf{I}}(\mathtt{qr} = Quant(\mathrm{BR}))\}, \{\mathtt{qr}\}), Receive_{\mathsf{T,M}}^{(n)}(\{\}, \{\mathtt{ind}\}),\\
&Receive_{\mathsf{M,T}}^{(n)}(\{\}, \{\mathtt{sebr}, \mathtt{bs}\}), Receive_{\mathsf{T,M}}^{(n)}(\{\}, \{\mathtt{dec}\}),\\
&Trust_{\mathsf{T,I}}, Trust_{\mathsf{M,I}}, Trust_{\mathsf{T,M}}, Verify_{\mathsf{T}}^{(n)}(Attest_{\mathsf{I}}(\mathtt{ebr} = Enc(\mathrm{BR}))),\\
&Verify_{\mathsf{T}}^{(n)}(\{Attest_{\mathsf{M}}(\mathtt{dec} = \mu(\mathtt{sbr}, \mathtt{bs}, \mathrm{THR}))\}),\\
&Verify_{\mathsf{M}}^{(n)}(Attest_{\mathsf{I}}(\mathtt{qr} = Quant(\mathrm{BR}))),\\
&Verify_{\mathsf{T}}^{(n)}(\{Attest_{\mathsf{M}}(\mathtt{sbr} = Dec(\mathtt{ebr}))\})\}
\end{aligned}$$

In addition to the dependence of $\mathcal{A}^{\mathsf{mi}}$, the dependence relations indicates that the leakage is conditioned by a specific link mapping between the outsourced ciphertexts and the stored quantizations: $Dep_{\mathsf{T}}^{\mathsf{mi\text{-}e}}(\mathtt{qr}, \{\mathtt{ind}^{(\mathrm{N}\cdot\mathrm{Q})}, \mathtt{qs}^{(\mathrm{N}\cdot\mathrm{Q})}\})$. Furthermore, the module may learn the entire database ebr in a number of queries depending on the size of the database and the number of indices asked by the module: $Dep_{\mathsf{M}}^{\mathsf{mi\text{-}e}}(\mathtt{ebr}, \{\mathtt{sebr}^{(\lceil \mathrm{N}/\mathrm{C} \rceil)}\})$.

3.4 Variants of the Architecture

Now, based on some counter-measures of the attacks indicated in [7], we express several variants of the architecture $\mathcal{A}^{\mathsf{mi\text{-}e}}$. For each variants, the deductive rules $\mathcal{D}$ for the property language $\mathcal{L}_P$ are used to show that, for some conditions on the parameters, the quantizations qr are protected.

Variant 1. As a first counter-measure, the module could ask the entire database at each invocation. It is rather inefficient, and, in some sense, runs against to initial motivation of its design. However, this can be described within the language $\mathcal{L}_A$, and, in practice, can be manageable for small databases. This architecture, denoted $\mathcal{A}^{\mathsf{mi\text{-}e1}}$, is given by $\mathcal{A}^{\mathsf{mi\text{-}e}}(n)$ for some $n \geq 1$, except that $Dep_{\mathsf{T}}^{\mathsf{mi\text{-}e1}} := Dep_{\mathsf{T}}^{\mathsf{mi}}$. It is now possible to prove that the quantizations are protected, even in presence of several executions of the protocols. Since the relations Dep_{T} no longer contains a dependence leading to $\mathtt{qr}$, an application of (**HN**) becomes possible and gives the expected property.

$$\cfrac{\begin{array}{cc} \nexists X : Dep_{\mathsf{T}}(\mathtt{qr}, X) \in \mathcal{A}^{\mathsf{mi\text{-}e1}} & \nexists j : Receive^{(n)}_{\mathsf{T},j}(S, \{\mathtt{qr}\}) \in \mathcal{A}^{\mathsf{mi\text{-}e1}} \\ Has^{(n)}_{\mathsf{T}}(\mathtt{qr}) \notin \mathcal{A}^{\mathsf{mi\text{-}e1}} & \nexists T : Compute^{(n)}_{\mathsf{T}}(\mathtt{qr} = T) \in \mathcal{A}^{\mathsf{mi\text{-}e1}} \end{array}}{\mathbf{HN}\ \cfrac{\forall n : \mathcal{A} \nvdash Has_{\mathsf{T}}(\mathtt{qr}^{(n)})}{\mathcal{A} \vdash Has^{none}_{\mathsf{T}}(\mathtt{qr})}}$$

VARIANT 2. In the precedent variant, the effect of the counter-measure is the withdrawal of the dependence relation. We now consider architectures where such a dependency is still given, but where counter-measures are used to prevent a critical bound on the number of queries to be reached.

A first measure is to block the number of attempts the terminal can make. The module can detect it and refuse to respond. This architecture, denoted $\mathcal{A}^{\mathsf{mi\text{-}e2}}$, is given by $\mathcal{A}^{\mathsf{mi\text{-}e}}(\mathrm{B})$, for some $\mathrm{B} \ll \mathrm{N}\cdot\mathrm{Q}$. As a result, the $Has^{none}_{i}(\mathtt{qr})$ property can be derived. In particular one must show that $\mathcal{A}^{\mathsf{mi\text{-}e2}} \nvdash Has_{\mathsf{T}}(\mathtt{ind}^{(\mathrm{N}\cdot\mathrm{Q})})$, in order to prevent the dependence rule **H5** to be applied.

$$\cfrac{\begin{array}{cc} & Has^{(\mathrm{B})}_{\mathsf{T}}(\mathtt{ind}) \in \mathcal{A}^{\mathsf{mi\text{-}e2}} \qquad \mathrm{B} < \mathrm{N}\cdot\mathrm{Q} \\ \nexists S : Receive^{(\mathrm{B})}_{\mathsf{T},\mathsf{M}}(S, \{\mathtt{ind}\}) \in \mathcal{A}^{\mathsf{mi\text{-}e2}} & \nexists T : Compute^{(\mathrm{B})}_{\mathsf{T}}(\mathtt{ind} = T) \in \mathcal{A}^{\mathsf{mi\text{-}e2}} \end{array}}{\mathcal{A}^{\mathsf{mi\text{-}e2}} \nvdash Has_{\mathsf{T}}(\mathtt{ind}^{(\mathrm{N}\cdot\mathrm{Q})})}$$

An application of **HN** enables to conclude.

$$\cfrac{\begin{array}{cc} & Has^{(\mathrm{B})}_{\mathsf{T}}(\mathtt{qr}) \notin \mathcal{A}^{\mathsf{mi\text{-}e2}} \\ Dep^{\mathsf{mi\text{-}e2}}_{\mathsf{T}}(\mathtt{qr}, \{\mathtt{ind}^{(\mathrm{N}\cdot\mathrm{Q})}\}) \in \mathcal{A}^{\mathsf{mi\text{-}e2}} & \nexists j : Receive^{(\mathrm{B})}_{\mathsf{T},j}(S, \{\mathtt{qr}\}) \in \mathcal{A}^{\mathsf{mi\text{-}e2}} \\ \mathcal{A}^{\mathsf{mi\text{-}e2}} \nvdash Has_{\mathsf{T}}(\mathtt{ind}^{(\mathrm{N}\cdot\mathrm{Q})}) & \nexists T : Compute^{(\mathrm{B})}_{\mathsf{T}}(\mathtt{qr} = T) \in \mathcal{A}^{\mathsf{mi\text{-}e2}} \end{array}}{\mathbf{HN}\ \cfrac{\mathcal{A}^{\mathsf{mi\text{-}e2}} \nvdash Has_{\mathsf{T}}(\mathtt{qr}^{(1)})}{\mathcal{A}^{\mathsf{mi\text{-}e2}} \vdash Has^{none}_{\mathsf{T}}(\mathtt{qr})}}$$

VARIANT 3. In the precedent variant, the terminal cannot accumulate enough information since he cannot query the module enough times to derive a useful knowledge. We now describe a variant where the terminal has no bound on the number of times it asks the module, but where the systems is regularly reinitialised, so that the accumulated information becomes useless.

The leakage of the system runtime is dependent on some association between the quantizations $\mathtt{qr}$ and the encrypted database $\mathtt{ebr}$; namely the association π that maps the quantization $\mathtt{qr}[i] = Quant(\mathrm{BR}[\pi(i)])$ to the encrypted template from which it has been computed $\mathtt{ebr}[\pi(i)] = Enc(\mathrm{BR}[\pi(i)])$. Once this mapping is changed, the information is cancelled. For instance the database can be randomly permuted after B queries to the secure module.

Formally, this is caught by adding a *Reset* primitive to the architecture. Let $\mathcal{A}^{\mathsf{mi\text{-}e3}}$ be the architecture defined as $\mathcal{A}^{\mathsf{mi\text{-}e3}} := \mathcal{A}^{\mathsf{mi\text{-}e2}} \cup \{Reset\}$. The semantics of the *Reset* events ensures that no more than B values of $\mathtt{ind}$ will be gathered by the terminal for a fixed mapping. The proof that $\mathcal{A}^{\mathsf{mi\text{-}e3}} \vdash Has^{none}_{\mathsf{T}}(\mathtt{qr})$ is as the proof that $\mathcal{A}^{\mathsf{mi\text{-}e2}} \vdash Has^{none}_{\mathsf{T}}(\mathtt{qr})$.

4 Conclusion

Previous work on the use of formal methods to reason about privacy properties at the architecture level fail to consider the information leakage through system sessions. When applied to biometric systems, this leakage must be considered to analyze the confidentiality of biometric data. Indeed, several attacks against biometric systems do not defeat the cryptographic building blocks, assumed to remain secure, but exploit information leakage related to the distances between biometrics templates. The goal of this paper was to address this limitation.

A computer aided privacy engineering tool called CAPRIV [2] has been developed to help non-expert designers to build architectures following a trust based strategy. CAPRIV implements an iterative design procedure in which the key decision to be taken by the designer is the type of trust which can be accepted by the components. The design of the tool makes it possible to hide the formal aspects of the model to the designer who does not want to be exposed to mathematical notations. This is mainly achieved through the use of a graphical user interface (GUI) and natural language statements. However CAPRIV implements the formal framework introduced in [1] and needs to be extended to include the extensions proposed in this paper.

Approaches based on π-calculus or privacy metrics are complementary to the work described in this paper. We leave to a future work to bridge the gap between analysis done at the architectural level and the formal methods used at the protocol level.

Acknowledgements. This work has been partially funded by the French ANR-12-INSE-0013 project BIOPRIV and the European FP7-ICT-2013-1.5 project PRIPARE.

A Sketch of Proof for Completeness and Correctness

A trace is said to be a *covering trace* if it contains an event corresponding to each primitive specified in an architecture $\mathcal{A}$ (except trust relations) and if for each primitive it contains as much events as the multiplicity $^{(n)}$ of the primitive. As a first step to prove soundness, it is shown that for all consistent architecture $\mathcal{A}$, there exists a consistent trace $\theta \in T(\mathcal{A})$ that covers $\mathcal{A}$.

Then the soundness is shown by induction on the depth of the tree $\mathcal{A} \vdash \phi$.

Let us assume that $\mathcal{A} \vdash Has_i(X^{(n)})$, and that the derivation tree is of depth 1. By definition of $\mathcal{D}$, such a proof is obtained by application of (**H1**), (**H2**) or (**H3**). In each case, it is shown (thanks to the existence of covering traces) that an appropriate trace can be found in the semantics of $\mathcal{A}$, hence $\mathcal{A} \in S(Has_i(X^{(n)}))$. The case of $\mathcal{A} \vdash Has_i(c)$ is very similar.

Let us assume that $\mathcal{A} \vdash K_i(Eq)$, and that the derivation tree is of depth 1. By definition of $\mathcal{D}$, such a proof is obtained by application of (**K1**), (**K2**), (**K3**), (**K4**) or (**K5**). In each case, starting from a state $\sigma' \in S_i(\mathcal{A})$ such that $\mathsf{s}(\sigma') \geq n$, it is first shown that there exists a covering trace $\theta \geq \theta'$ that extends θ' and that contains n corresponding events $Compute_G(X = T^{\epsilon}) \in \theta$ in n distinct sessions (for the **K1** case, and other events for the other rules). Then by the properties of the deductive algorithmic knowledge, it is shown that the semantics of the property $\mathcal{A} \in S(K_i(X = T))$ holds.

Let us assume that $\mathcal{A} \vdash Has_i(X^{(n)})$, and that the derivation tree is of depth strictly greater than 1. By definition of $\mathcal{D}$, such a proof is obtained by application of (**H4**) or (**H5**).
In the first case, by the induction hypothesis and the semantics of properties, there exists a reachable state $\sigma \in S(\mathcal{A})$ and n indices $i_1, \ldots, i_n$ such that $\sigma_i^v(X)[i_l]$ is fully defined for all $l \in [1, n]$. This gives, *a fortiori*, $\mathcal{A} \in S(Has_i(X^{(m)}))$ for all m such that $1 \leq m \leq n$.
In the second case, we have that $(Y, \{X_1^{(n_1)}, \ldots, X_m^{(n_m)}, c_1, \ldots, c_q\}) \in Dep_i$, that $\forall l \in [1, m] : \mathcal{A} \vdash Has_i(X_l^{(n_l)})$ and $\forall l \in [1, q] : \mathcal{A} \vdash Has_i(c_l)$. of a covering trace that contains an event $Compute_G$ $(Y = T)$ (where $i \in G$), allowing to conclude that $\mathcal{A} \in S(Has_i(Y^{(1)}))$.
Again, the corresponding cases for constant are very similar.

A derivation for Has^{none} is obtained by application of (**HN**). The proof assume, towards a contradiction, that $A \notin S(Has_i^{none}(X))$. It is shown, by the architecture semantics, that there exists a compatible trace that enable to derive $\mathcal{A} \vdash Has_i^{(1)}(X)$. However, since (**HN**) was applied, we have $\mathcal{A} \nvdash Has_i^{(1)}(X)$, hence a contradiction.

The last case (the conjunction $\wedge$) is fairly straightforward.

The completeness is proved by induction over the definition of ϕ.

Let us assume that $\mathcal{A} \in S(Has_i(X^{(n)}))$. By the architecture semantics and the semantics of traces, it is shown that the corresponding traces either contain events where X is computed, received or measured, or that some dependence relation on X exists. In the first case, we have $\mathcal{A} \vdash Has_i(X^{(n)})$ by applying (respectively) (**H1**), (**H2**), or (**H3**) (after an eventual application of (**H4**)). In the last case, the proof shows how to exhibit a derivation tree to obtain $\mathcal{A} \vdash Has_i(X^{(n)})$ (the (**H5**) rule is used).

Let us assume that $\mathcal{A} \in S(Has_i^{none}(X))$. By the semantics of properties, this means that in all reachable states, X does not receive any value. The proof shows that $\mathcal{A} \nvdash S(Has_i(X^{(1)}))$, otherwise $\mathcal{A} \in S(Has_i^{none}(X))$ would be contradicted. So as a conclusion, $\mathcal{A} \vdash Has_i^{none}(X)$ by applying (**HN**).

The constant cases $\mathcal{A} \in S(Has_i(c)$ and $\mathcal{A} \in S(Has_i^{none}(c))$ case are similar to the variable cases.

Let us assume that $\mathcal{A} \in S(K_i(Eq))$. By the semantics of properties this means that for all reachable states, there exists a later state in the same session where the knowledge state enables to derive Eq. By the semantics of architecture, we can exhibit a compatible trace that reaches a state where Eq can be derived. By the semantics of compatible traces, the proof shows, by reasoning on the events on the traces, that $\mathcal{A} \vdash K_i(Eq)$ by applying either (**K1**), (**K2**), (**K3**), (**K4**) or (**K5**).

Finally the conjunctive case is straightforward.

References

1. Antignac, T., Le Métayer, D.: Privacy architectures: reasoning about data minimisation and integrity. In: Mauw, S., Jensen, C.D. (eds.) STM 2014. LNCS, vol. 8743, pp. 17–32. Springer, Heidelberg (2014)
2. Antignac, T., Le Métayer, D.: Trust driven strategies for privacy by design. In: Damsgaard Jensen, C., Marsh, S., Dimitrakos, T., Murayama, Y. (eds.) IFIPTM 2015. IFIP AICT, vol. 454, pp. 60–75. Springer, Heidelberg (2015)
3. Barth, A., Datta, A., Mitchell, J.C., Nissenbaum, H.: Privacy and contextual integrity: framework and applications. In: IEEE Symposium on Security and Privacy, S&P 2006, pp. 184–198. IEEE Computer Society (2006)
4. Becker, M.Y., Alexander, M., Laurent, B.: S4P: A generic language for specifying privacy preferences and policies. Technical report, Microsoft Research/IMDEA Software/EMIC (2010)
5. Bringer, J., Chabanne, H., Kevenaar, T.A.M., Kindarji, B.: Extending match-on-card to local biometric identification. In: Fierrez, J., Ortega-Garcia, J., Esposito, A., Drygajlo, A., Faundez-Zanuy, M. (eds.) BioID MultiComm2009. LNCS, vol. 5707, pp. 178–186. Springer, Heidelberg (2009)
6. Bringer, J., Chabanne, H., Le Métayer, D., Lescuyer, R.: Privacy by design in practice: reasoning about privacy properties of biometric system architectures. In: Bjørner, N., de Boer, F. (eds.) FM 2015. LNCS, vol. 9109, pp. 90–107. Springer, Heidelberg (2015)
7. Bringer, J., Chabanne, H., Simoens, K.: Blackbox security of biometrics (invited paper). In: Conference on Intelligent Information Hiding and Multimedia Signal Processing, IIH-MSP 2010, pp. 337–340. IEEE Computer Society (2010)
8. Delaune, S., Kremer, S., Ryan, M.: Verifying privacy-type properties of electronic voting protocols: a taster. In: Chaum, D., Jakobsson, M., Rivest, R.L., Ryan, P.Y.A., Benaloh, J., Kutylowski, M., Adida, B. (eds.) Towards Trustworthy Elections. LNCS, vol. 6000, pp. 289–309. Springer, Heidelberg (2010)
9. Dodis, Y., Reyzin, L., Smith, A.: Fuzzy extractors: how to generate strong keys from biometrics and other noisy data. In: Cachin, C., Camenisch, J.L. (eds.) EUROCRYPT 2004. LNCS, vol. 3027, pp. 523–540. Springer, Heidelberg (2004)
10. Dwork, C.: Differential privacy. In: Bugliesi, M., Preneel, B., Sassone, V., Wegener, I. (eds.) ICALP 2006. LNCS, vol. 4052, pp. 1–12. Springer, Heidelberg (2006)
11. European Parliament. European parliament legislative resolution of 12 March 2014 on the proposal for a regulation of the European parliament and of the council on the protection of individuals with regard to the processing of personal data and on the free movement of such data. General data protection regulation, ordinary legislative procedure: first reading (2014)
12. Fagin, R., Halpern, J., Moses, Y., Vardi, M.: Reasoning about Knowledge. MIT Press, Cambridge (2004)
13. Fournet, C., Kohlweiss, M., Danezis, G., Luo, Z.: ZQL: a compiler for privacy-preserving data processing. In: USENIX 2013 Security Symposium, pp. 163–178. USENIX Association (2013)
14. Govan, M., Buggy, T.: A computationally efficient fingerprint matching algorithm for implementation on smartcards. In: Biometrics: Theory, Applications, and Systems, BTAS 2007, pp. 1–6. IEEE Computer Society (2007)
15. Gürses, S., Troncoso, C., Díaz, C.: Engineering privacy by design. In: Privacy and Data Protection Conference, Presented at the Computers (2011)
16. Halpern, J.Y., Pucella, R.: Dealing with logical omniscience. In: Conference on Theoretical Aspects of Rationality and Knowledge, TARK 2007, pp. 169–176 (2007)

17. Juels, A., Sudan, M.: A fuzzy vault scheme. Des. Codes Crypt. **38**(2), 237–257 (2006)
18. Juels, A., Wattenberg, M.: A fuzzy commitment scheme. In: ACM Conference on Computer and Communications Security, CCS 1999, pp. 28–36. ACM Press (1999)
19. Kanak, A., Sogukpinar, I.: BioPSTM: a formal model for privacy, security, and trust in template-protecting biometric authentication. Secur. Commun. Netw. **7**(1), 123–138 (2014)
20. Kerschbaum, F.: Privacy-preserving computation. In: Preneel, B., Ikonomou, D. (eds.) APF 2012. LNCS, vol. 8319, pp. 41–54. Springer, Heidelberg (2014)
21. Lai, L., Ho, S.-W., Vincent Poor, H.: Privacy-security trade-offs in biometric security systems - part I: single use case. IEEE Trans. Inf. Forensics Secur. **6**(1), 122–139 (2011)
22. Li, L., Ho, S.-W., Vincent Poor, H.: Privacy-security trade-offs in biometric security systems - part II: multiple use case. IEEE Trans. Inf. Forensics Secur. **6**(1), 140–151 (2011)
23. Li, H., Pang, L.: A novel biometric-based authentication scheme with privacy protection. In: Conference on Information Assurance and Security, IAS 2009, pp. 295–298. IEEE Computer Society (2009)
24. Maffei, M., Pecina, K., Reinert, M.: Security and privacy by declarative design. In: IEEE Symposium on Computer Security Foundations, CSF 2013, pp. 81–96. IEEE Computer Society (2013)
25. McSherry, F.: Privacy integrated queries: an extensible platform for privacy-preserving data analysis. In: ACM Conference on Management of Data, SIGMOD 2009, pp. 19–30. ACM Press (2009)
26. Le Métayer, D.: Privacy by design: a formal framework for the analysis of architectural choices. In: ACM Conference on Data and Application Security and Privacy, CODASPY 2013, pp. 95–104. ACM Press (2013)
27. Mulligan, D.K., King, J.: Bridging the gap between privacy and design. Univ. Pennsylvania J. Const. Law **14**, 989–1034 (2012)
28. National Institute of Standards and Technology (NIST). MINEXII - an assessment of match-on-card technology (2011). http://www.nist.gov/itl/iad/ig/minexii.cfm
29. International Standard Organization. International standard ISO/IEC 24787:2010, information technology - identification cards - on-card biometric comparison (2010)
30. Pagnin, E., Dimitrakakis, C., Abidin, A., Mitrokotsa, A.: On the leakage of information in biometric authentication. In: Meier, W., Mukhopadhyay, D. (eds.) INDOCRYPT 2014. Lecture Notes in Computer Science, vol. 8885, pp. 265–280. Springer, LNCS (2014)
31. Pucella, R.: Deductive algorithmic knowledge. J. Log. Comput. **16**(2), 287–309 (2006)
32. Simoens, K., Bringer, J., Chabanne, H., Seys, S.: A framework for analyzing template security and privacy in biometric authentication systems. IEEE Trans. Inf. Forensics Secur. **7**(2), 833–841 (2012)
33. Spiekermann, S., Faith Cranor, L.: Engineering privacy. IEEE Trans. Softw. Eng. **35**(1), 67–82 (2009)
34. Tang, Q., Bringer, J., Chabanne, H., Pointcheval, D.: A formal study of the privacy concerns in biometric-based remote authentication schemes. In: Chen, L., Mu, Y., Susilo, W. (eds.) ISPEC 2008. LNCS, vol. 4991, pp. 56–70. Springer, Heidelberg (2008)
35. Uludag, U., Pankanti, S., Jain, A.K.: Fuzzy vault for fingerprints. In: Kanade, T., Jain, A., Ratha, N.K. (eds.) AVBPA 2005. LNCS, vol. 3546, pp. 310–319. Springer, Heidelberg (2005)

Improvement of Multi-bit Information Embedding Algorithm for Palette-Based Images

Anu Aryal(✉), Kazuma Motegi, Shoko Imaizumi, and Naokazu Aoki

Graduate School of Advanced Integration Science, Chiba University, 1–33 Yayoicho, Inage-ku, Chiba-shi, Chiba 263-8522, Japan
{anu,afha3739,imaizumi}@chiba-u.jp

Abstract. We propose a new approach that is an improvement on the conventional of a multi-bit information embedding algorithm for palette-based images. The proposed method embeds secret information by changing the pixel values for each of the $2^{k-2}+1$ pixels. Hence, our scheme can embed more information compared to the conventional method. Furthermore, the developed algorithm does not drastically change the pixel values with large color differences between before and after the changes. Therefore, it reduces the degradation of the image quality. Our experimental results show that the proposed scheme is superior to the conventional method.

Keywords: Data embedding · Steganography · Palette-based images · Capacity enhancement · Quality improvement

1 Introduction

With the development of the Internet, a vast amount of information has been exchanged through different communication channels. Hence, security for such information has been required. Accordingly, many researches on steganography techniques and its security have been studied. Digital steganography [1,2] is a security technique to protect digital information. The main purpose of it is to hide the existence of embedded confidential information. The confidential information is imperceptibly embedded into digital media (hereinafter referred to as cover data) and is transmitted through open channels without creating noticeable artifacts. Therefore, the risk of information leakage is reduced because unauthorized recipients are totally unaware of the existence of the message that is embedded into the cover data.

Compared to full-color images, palette-based images possess a limited number of colors, which leads to a reduction in the amount of data needed. For this reason, these images are widely used in many multimedia applications, websites, and so forth. Here, the color of each pixel is managed by a number (referred to as the index). Each index corresponds to a color in the color palette.

For palette-based images, there are two methods to embed data by controlling entries in the steganographic schemes. One of the methods changes the colors of

J. Lopez and C.J. Mitchell (Eds.): ISC 2015, LNCS 9290, pp. 511–523, 2015.
DOI: 10.1007/978-3-319-23318-5_28

the entries slightly to embed the message [3–12] whereas the other one retains the colors of the entries and reorders the entries in the palette [13]. If the pixel values are changed, the quality of the image is also reduced compared to that of the original image. In addition, the degree of image quality degradation increases as the amount of embedded data increases. However, the suppression of degradation of image quality is demanded while embedding information in steganography. Fridrich [8] proposed a method for hiding a message bit in the parity bit of each neighboring color. Similarly, techniques to improve color difference calculation [9,10] and methods of suppression of image quality degradation [4,7] have been proposed. However, if the amount of embedded data per pixel is increased, the image quality degradation also increases in these methods. Therefore, Ozawa et al. [11] extended the embedded unit from a pixel to a pixel matrix and proposed a technique to suppress the image quality degradation.

In this paper, we propose a new information embedding method using a limited color image as cover data. The conventional method [11] reduces the number of pixels used in the unit of a pixel matrix compared to their previous work [12] but might lead to large color differences between before and after the changes. The proposed method, which is an improvement on the conventional method [11], can extend the embedded information capacity and suppress the degradation of image quality.

2 Conventional Method

In Ozawa et al.'s method [11], each k-bit message w ($w = \{w_n | w_n \in \{0, 1, \ldots, 2^k - 1\}, n = 0, 1, 2, \ldots\}$ is embedded into the pixels $t_l(n)$ of a 2×2 pixel matrix m_n ($t_{l(n)} \in m_n$, $l = 0, 1, 2, 3$) as shown in Fig. 1. In this case, each $t_{l(n)}$ has an index value of $d_{l(n)}$.

2.1 Sorting of Color Palette

First, we temporarily sort the entries C_i (i is a positive number) of the color palette in accordance with the following steps. As a result, the new indices can be assigned to the entries as C'_j (j is a positive number).

Step 1. Calculate a_i for each C_i in the original palette using the following equation

$$a_i = \left(256^2 r_i + 256^1 g_i + 256^0 b_i\right), \quad i = 0, 1, 2, \cdots, \tag{1}$$

and determine

$$C'_0 = \arg\min a_i. \tag{2}$$

Step 2. The j-th color is determined by calculating arg min $\delta_{j-1,j}$, where C'_{j-1} is the color that was assigned the index in the last process. Here, δ denotes the color difference using the Euclidean distance. The color difference $\delta_{p,q}$ between the colors C_p and C_q is expressed by,

$$\delta_{p,q} = \sqrt{(r_p - r_q)^2 + (g_p - g_q)^2 + (b_p - b_q)^2}. \tag{3}$$

Step 3. Step 2 is repeated until the indices are assigned to all the entries.

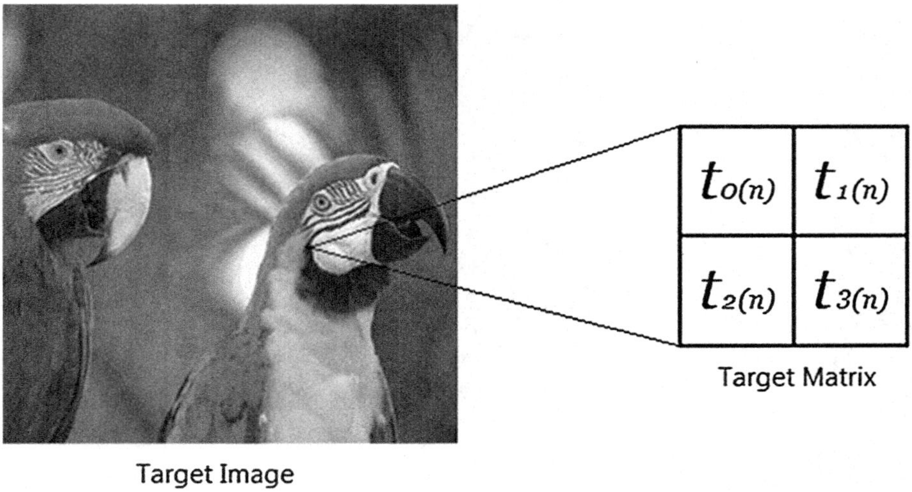

Fig. 1. Embedded unit of Ozawa et al.'s method [11].

2.2 Embedding of Message

In Ozawa et al.'s method [11], the message w_n is embedded into each 2×2 pixel of matrix m_n. In this section, the length of a message is considered to be 3-bit, i.e., $k = 3$.

Step 1. As shown in Fig. 1, a 2×2 matrix m_n is selected from the target image.
Step 2. Calculate the parity S_n as

$$S_n = \sum_{l=0}^{3} d_{l(n)} \mod 2^k, \tag{4}$$

and take the difference x between S_n and w_n as

$$x = S_n - w_n. \tag{5}$$

Here, the number of pixels to be changed a_n is determined as

$$a_n = \begin{cases} \min(x, -x + 2^k) & (S_n > w_n) \\ \min(-x, x + 2^k) & (S_n < w_n) \\ 0 & (S_n = w_n). \end{cases} \tag{6}$$

When $a = 0$, **Steps 3–5** are not performed; thus, no pixel is changed.

Step 3. Choose a_n of the pixels $t_{l(n)}$ in ascending order corresponding to the Euclidean distance $D_{l(n)}$ between the entry C'_j and the neighboring entry C'_{j+1} or C'_{j-1}. The distance $D_{l(n)}$ between the two entries is given as

$$D_{l(n)} = \begin{cases} \delta_{j,j-1}, & \text{if } a_n = x(S_n > w_n) \text{ or } a_n = x + 2^k(S_n < w_n) \\ \delta_{j,j+1}, & \text{if } a_n = -x(S_n < w_n) \text{ or } a_n = -x + 2^k(S_n > w_n). \end{cases} \tag{7}$$

Fig. 2. Embedded unit of proposed method ($k = 3$)

Step 4. Replace a_n of the indices $d_{l(n)}$ for the pixels $t_{l(n)}$, which are chosen by **Step 3**, with $j - 1$ or $j + 1$.
Step 5. Steps 1–4 are repeated n times.

2.3 Extraction of Message

By calculating w_n, that is, S_n of matrix m_n, and concatenating w_n in the order of n, the entire message w is obtained. The embedding order of the message is determined by the stego-key, which is a seed value of the pseudo-random sequence. The sender of the message gives the stego-key and message length k to the receiver to extract the message.

2.4 Problems of the Conventional Method

In Ozawa et al.'s method [11], four pixels are used to embed a 3-bit message. The maximum embedded amount is smaller than that of the methods that embed one bit message into one pixel [7–10]. In addition, there is also a tendency for the color differences among the neighboring colors in the reordered color palette to become large because the difference is based on the Euclidean distance. This is one of the causes of the image degradation.

3 Proposed Method

We propose an efficient method to embed k-bit messages $w(w = \{w_n | w_n \in \{0, 1, \ldots, 2^k - 1\},\ n = 0, 1, 2, \ldots\})$ into a limited color image by controlling

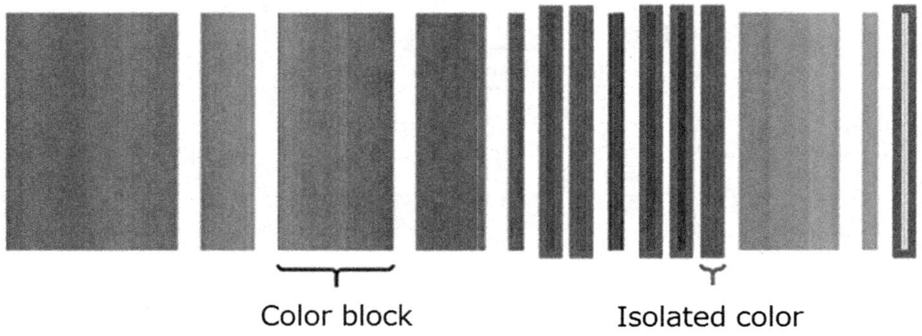

Fig. 3. Color blocks and isolated colors (Color figure online)

the index values. An embedded matrix m_n consists of the pixels of $p_{(n)}$ and $t_{l(n)}(t_{l(n)} \in m_n,\ l = 0, 1, \ldots, 2^{k-2} - 1)$.

3.1 Sorting of Color Palette

The color palette of a limited color bitmap image is represented using RGB. Since we sort the colors using CIEDE2000 [14] (hereinafter referred to as ΔE_{00}), it is necessary to convert the RGB color space to the $L^*a^*b^*$ color space.

The entries C_i in the color palette (i is a positive integer less than 256) are temporarily sorted to change the index. Then, the entries C'_j (j is a positive integer less than 256) are generated with a new index j. The sorting procedure is described as follows.

Step 1. $C'_0(j = 0)$, which is the darkest color in C_i, is obtained by calculating $\arg\min L^*_i$.
Step 2. Set $j = j + 1$.
Step 3. Calculate ΔE_{00} between C'_{j-1} and the initial entries C_i that have not been assigned an index. The entry C_i with the minimum ΔE_{00} is assigned j and becomes C'_j.
Step 4. Steps 2 and **3** are repeated until the index is assigned to all the entries.

Figure 3 represents the latter part of the sorted palette. Each block has been generated by delimiting the colors when $\Delta E_{00} > 5.0$. The colors with thick frames are isolated colors, that is, they do not have any other colors in their color blocks. For the pixels represented by the isolated colors, the color differences after changing the indices would become incredibly large. Therefore, we cannot embed the message into the pixels with the isolated colors.

3.2 Embedding of Message

In the proposed method, the k-bit messages are embedded into each $2^{k-2} + 1$ pixel matrix. In this section, the message length is considered to be 3 bits ($k = 3$).

Table 1. Example of P_n, S_n and embedded information w_n

P_n	S_n	Embedded information w_n
0	3	7
0	2	6
0	1	5
0	0	4
1	0	3
1	1	2
1	2	1
1	3	0

Step 1. As shown in Fig. 2, a 1×3 pixel matrix is selected from the target image. Note that it must be confirmed that a k-bit message has not yet been embedded into the selected matrix.

Step 2. The parity S_n is calculated as

$$S_n = d_{0(n)} + d_{l(n)} \mod 4, \tag{8}$$

where $d_{l(n)}$ indicates the index of the pixel $t_{l(n)}$.

Step 3. The value of S_n can be controlled by changing the indices of $p_{(n)}$, $t_{0(n)}$, and/or $t_{1(n)}$ by either +1 or −1. Table 1 shows an example of the relation among P_n, S_n, and w_n. Here, we define $P_n = 0$ when the index of $p_{(n)}$ is an even number and $P_n = 1$ when the index of $p_{(n)}$ is an odd number. The indices of $p_{(n)}$, $t_{0(n)}$, and/or $t_{1(n)}$ may be changed to be the corresponding values of the embedded message w_n in accordance with Table 1.

(a) Change of P_n: If the value of P_n before the embedding process is different from that after the process, the index of $p_{(n)}$ is changed by +1 or −1. In this process, we should choose either +1 or −1 such that it makes ΔE_{00} be a smaller value.

(b) Change of S_n: If the difference between the values of S_n before and after the embedding process is 2, the indices of $d_{0(n)}$ and $d_{1(n)}$ of $t_{0(n)}$ and $t_{1(n)}$ are changed by +1 or −1. In this process, we should choose either +1 or −1 such that it makes the average of ΔE_{00} be a smaller value.

If the difference between the values of S_n before and after the embedding process is 1, either $d_{0(n)}$ or $d_{1(n)}$ is changed by +1 or −1. In this process, we should choose either $d_{0(n)}$ or $d_{1(n)}$ and either +1 or −1 such that it makes ΔE_{00} be the smallest value.

If the differences between the values of S_n before and after the embedding process are zero, neither $d_{0(n)}$ nor $d_{1(n)}$ is changed.

The proposed method performs the embedding process only when all the ΔE_{00} values for the pixels of the matrix become 5.0 or less. If any of the ΔE_{00} values for the pixels exceeds 5.0, the embedding process is not performed on the matrix.

Step 4. Steps 1–3 are repeated until all the messages are embedded.

a) Balloon

b) Parrot

Fig. 4. Original images

3.3 Extraction of Message

We can extract each embedded message w_n in accordance with Table 1 after calculating P_n and S_n. It is possible to restore message w by sorting w_n. To restore the message, the embedding position and the message length k are required.

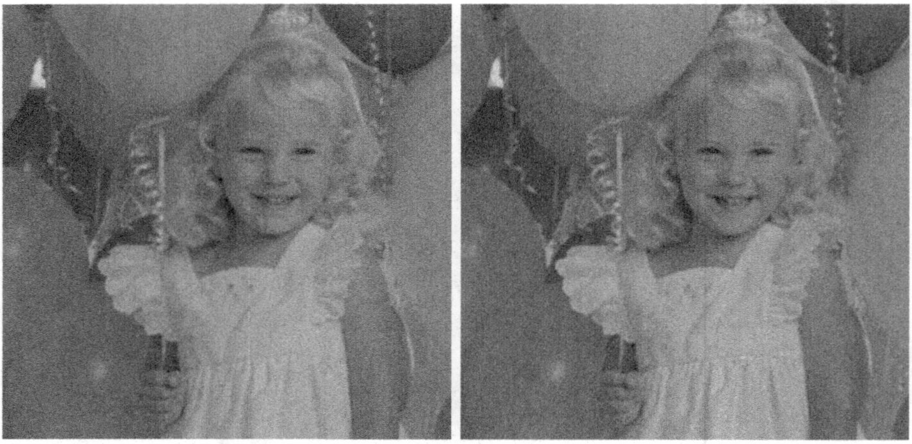

a) Proposed method b) Conventional method [11]

Fig. 5. Embedding of 10,800-bit message (Balloon)

a) Proposed method b) Conventional method [11]

Fig. 6. Embedding of 10,800-bit message (Parrot)

4 Experimental Results

In this section, the proposed method is compared to the conventional method [11] in terms of image quality and the maximum amount of embedding data. We performed our experiments on twelve different 256 × 256 palette-based images (8-bit color bitmap images).

a) Proposed method b) Conventional method [11]

Fig. 7. Embedding of 21,600-bit message (Balloon)

a) Proposed method b) Conventional method [11]

Fig. 8. Embedding of 21,600-bit message (Parrot)

4.1 Evaluation of Image Quality

The original images of Balloon and Parrot are shown in Fig. 4. The simulation results for embedding a 10,800-bit message where $k = 3$ for the proposed method and the conventional method [11] are depicted in Figs. 5 and 6. Similarly, the simulation results of embedding a 21,600-bit message where $k = 3$ for the proposed method and the conventional method [11] are demonstrated in Figs. 7 and 8. Furthermore, Tables 2 and 3 indicate the PSNR and the SSIM [15] values of the simulation results of the proposed method and the conventional method [11],

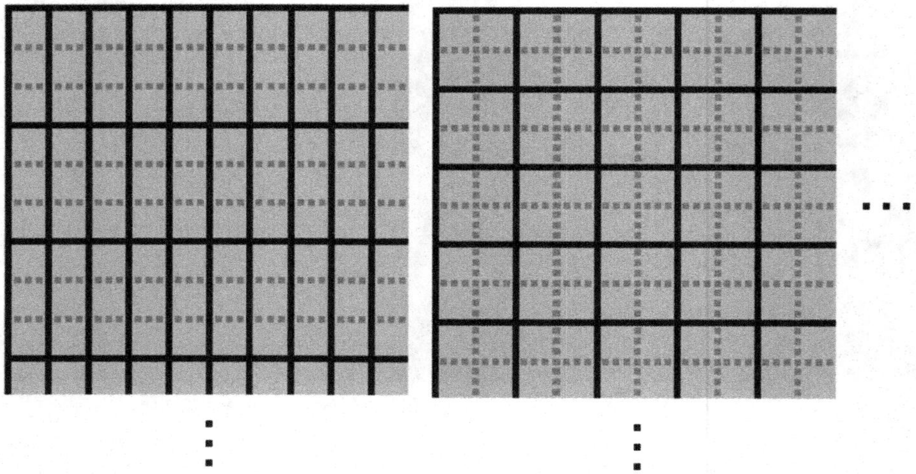

a) Proposed method b) Conventional method [11]

Fig. 9. Matrix arrangement for maximum amount of embedded bits

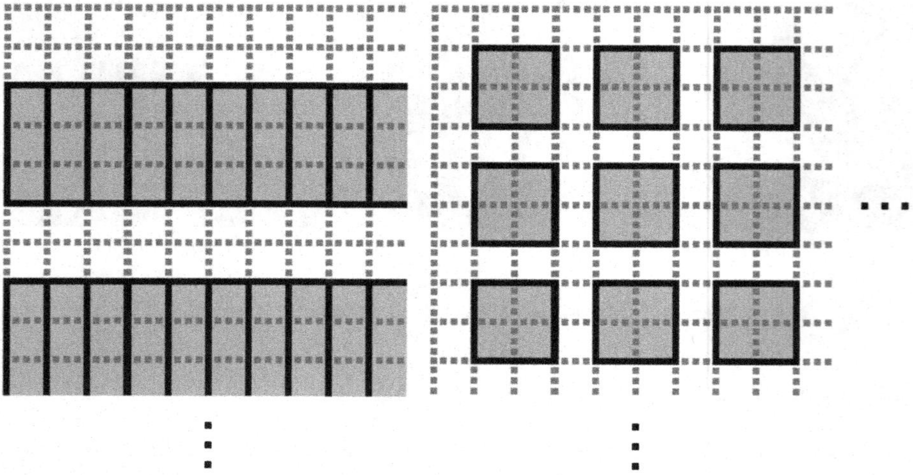

a) Proposed method b) Conventional method [11]

Fig. 10. Matrix arrangement for minimum amount of embedded bits

Table 2. Evaluation using PSNR

	10,800 Bits		21,600 Bits	
	Proposed	Conventional [11]	Proposed	Conventional [11]
Aerial	41.96	39.61	38.87	36.81
Airplane	43.13	40.54	40.15	37.76
Balloon	43.49	40.73	40.62	37.68
Couple	41.87	39.18	38.77	35.98
Earth	45.59	43.38	42.54	40.54
Girl	40.34	36.23	37.30	32.88
Lena	43.03	40.65	40.00	37.67
Mandrill	39.06	36.38	35.73	33.25
Milkdrop	42.93	40.30	39.83	37.24
Parrots	39.33	35.77	36.30	32.48
Pepper	40.34	36.58	37.19	33.67
Sailboat	41.32	38.57	38.41	35.68

Table 3. Evaluation using SSIM

	10,800 Bits		21,600 Bits	
	Proposed	Conventional [11]	Proposed	Conventional [11]
Aerial	0.902	0.862	0.801	0.729
Airplane	0.846	0.825	0.709	0.670
Balloon	0.867	0.822	0.751	0.674
Couple	0.896	0.849	0.799	0.710
Earth	0.882	0.828	0.765	0.665
Girl	0.840	0.722	0.706	0.585
Lena	0.871	0.827	0.749	0.667
Mandrill	0.878	0.830	0.753	0.667
Milkdrop	0.871	0.835	0.753	0.683
Parrots	0.854	0.825	0.723	0.681
Pepper	0.871	0.831	0.754	0.699
Sailboat	0.863	0.827	0.755	0.685

respectively. The values of PSNR and SSIM demonstrate that the image quality of the proposed method is superior to that of the conventional method [11].

4.2 Evaluation of Maximum Possible Amount of Embedding Bits

Next, we discuss the maximum amount of embedded bits in our proposed method. Figure 9 represents the matrix arrangement for embedding the maximum

Table 4. Maximum and minimum values of embedded bits

	Proposed	Conventional [11]
Maximum amount of embedded bits	65,280	49,152
Minimum amount of embedded bits	39,168	21,675

possible amount of message bits for the proposed method and the conventional method [11], respectively. The squares surrounded by dashed lines represent each pixel whereas the matrix is represented as a set of multiple pixels with thick lines. It should be noted that the coordinates of each matrix to be embedded in the messages are separately determined by pseudo-random numbers. We also show the arrangements when dealing with the minimum amount of embedded bits for the proposed method and the conventional method in Fig. 10. Note that when the number of vertical pixels is not divisible by k in the proposed method and the number of vertical and/or horizontal pixels is not divisible by 2 in the conventional method [11], the excess pixels are considered to be outside the embedding object.

Table 4 denotes the maximum and minimum amounts of embedded bits for an image of 256×256 pixels in the proposed method and the conventional method [11], respectively. It can be seen that in the proposed method, the maximum possible amount of embedded bits is increased by 1.3 times whereas the minimum number of embedded bits is increased by 1.8 times compared to that of the conventional method [11].

5 Conclusion

We have proposed an improvement on the conventional multi-bit information embedding algorithm for steganography of palette-based images. This method enhances data embedding with larger capacity and improves the image quality as well. The values of PSNR and SSIM in our experiments have been improved in the proposed method. This indicates that image quality degradation is suppressed in our method. Furthermore, by reducing the size of the embedded matrix, the maximum amount of embedded bits is 1.3 times more than that of the conventional method, and the minimum amount is 1.8 times more.

Acknowledgments. This work was supported by JSPS KAKENHI Grant Number 26820138.

References

1. Kahn, D.: The history of steganography. In: Anderson, R. (ed.) IH 1996. LNCS, vol. 1174, pp. 1–5. Springer, Heidelberg (1996)
2. Chandramouli, R., Kharrazi, M., Memon, N.D.: Image steganography and steganalysis: concepts and practice. In: Kalker, T., Cox, I., Ro, Y.M. (eds.) IWDW 2003. LNCS, vol. 2939, pp. 35–49. Springer, Heidelberg (2004)

3. Wang, X., Yao, Z., Li, C.-T.: A palette-based image steganographic method using colour quantisation. In: Proceedings of IEEE ICIP, pp. II-1090–II-1093 (2005)
4. Niimi, M., Noda, H., Kawaguchi, E.: High capacity and secure digital steganography to palette-based images. In: Proceedings of IEEE ICIP, pp. II-917–II-920 (2002)
5. Zhao, H., Wang, H., Khan, M.K.: Steganalysis for palette-based images using generalized difference image and color correlogram. Sig. Process. **91**(11), 2595–2605 (2011)
6. Zhang, X., Wang, S., Zhou, Z.: Multibit assignment steganography in palette images. IEEE Signal Proc. Lett. **15**, 553–556 (2008)
7. Tzeng, C.-H., Yang, Z.-F., Tsai, W.-H.: Adaptive data hiding in palette images by color ordering and mapping with security protection. IEEE Trans. Commun. **52**(5), 791–800 (2004)
8. Fridrich, J.: A new steganographic method for palette-based image. In: Proceedings of IS&T PICS, pp. 285–289 (1999)
9. Huy, P.T., Thanh, N.H., Thang, T.M., Dat, N.T.: On fastest optimal parity assignments in palette images. In: Pan, J.-S., Chen, S.-M., Nguyen, N.T. (eds.) ACIIDS 2012, Part II. LNCS, vol. 7197, pp. 234–244. Springer, Heidelberg (2012)
10. Inoue, K., Hotta, S., Takeichi, Y., Urahama, K.: A steganographic method for palette-based images. Trans. Inst. Electron. Inf. Commun. Eng. A **82**(11), 1750–1751 (1999). (in Japanese)
11. Imaizumi, S., Ozawa, K.: Palette-based image steganography for high-capacity embedding. Bull. Soc. Photogr. Image Jpn. **25**(1), 7–11 (2015)
12. Imaizumi, S., Ozawa, K.: Multibit embedding algorithm for steganography of palette-based images. In: Klette, R., Rivera, M., Satoh, S. (eds.) PSIVT 2013. LNCS, vol. 8333, pp. 99–110. Springer, Heidelberg (2014)
13. Kwan, M.: Gifshuffle (2003). http://www.darkside.com.au/gifshuffle/
14. Colorimetry - Part 6: CIEDE2000 Colour-difference formula. ISO/CIE 11664-6 (2014)
15. Wang, Z., Bovik, A.C., Sheikh, H.R., Simoncelli, E.P.: Image quality assessment: from error visibility to structural similarity. IEEE Trans. Image Process. **13**(4), 600–612 (2004)

Efficient Ephemeral Elliptic Curve Cryptographic Keys

Andrea Miele(✉) and Arjen K. Lenstra

EPFL, Lausanne, Switzerland
andrea.miele@epfl.ch

Abstract. We show how any pair of authenticated users can on-the-fly agree on an elliptic curve group that is unique to their communication session, unpredictable to outside observers, and secure against known attacks. Our proposal is suitable for deployment on constrained devices such as smartphones, allowing them to efficiently generate ephemeral parameters that are unique to any single cryptographic application such as symmetric key agreement. For such applications it thus offers an alternative to long term usage of standardized or otherwise pre-generated elliptic curve parameters, obtaining security against cryptographic attacks aimed at other users, and eliminating the need to trust elliptic curves generated by third parties.

Keywords: Elliptic curve cryptography · Complex multiplication method

1 Introduction

Deployment of elliptic curve cryptography (ECC) [32,40] is becoming more common. A variety of ECC parameters has been proposed or standardized [1,5,9,13,16,17,39,58], with or without all kinds of properties that are felt to be desirable or undesirable, and as reviewed in Sect. 2. All these proposals and standards contain a fixed number of possible ECC parameter choices. This implies that many different users will have to share their choice, where either choice implies trust in the party responsible for its construction. Notwithstanding a variety of design methods intended to avoid trust issues (cf. [6]) and despite the fact that parameter sharing is generally accepted for discrete logarithm cryptosystems, recent allegations [28,52] raise questions. As extensively discussed at the recent Workshop on Elliptic Curve Cryptography Standards [45], currently the main challenge in curve selection lies in re-establishing users' trust in ECC which vacillated after the above allegations were announced, and keep being followed by a continuing string of disconcerting information security related mishaps. Relying on choices made by others, either related to elliptic curve parameter selection or to any other personal choice related to information security, parameter sharing and long term usage of any type of cryptographic key material, may have to be reconsidered. In this paper we consider what can realistically

J. Lopez and C.J. Mitchell (Eds.): ISC 2015, LNCS 9290, pp. 524–547, 2015.
DOI: 10.1007/978-3-319-23318-5_29

be done if this reconsideration is taken to the extreme and trust in other parties' contributions is reduced to a minimum: never rely on choices made by others, avoid parameter sharing as much as possible, and refresh key material as often as feasible.

Specifically, we consider an approach that is diametrically different from current common practice, namely selection of *personalized*, *short-lived* ECC parameters. By personalized we mean that no party but the party or parties owning or directly involved in the usage of parameters should be responsible for their generation:

- for a certified public key, **only** the owner of the corresponding private key should be responsible for the selection of **all** underlying parameters;
- in the Diffie-Hellman protocol, as there is no a priori reason for the parties to trust each others' public key material other than for mutual authentication, **both** parties, and no other party, should be equally responsible for the construction of the group to be used in the key agreement phase.

Personalization excludes parameter choice interference by third parties with unknown and possibly contrary incentives. It also avoids the threats inherent in parameter sharing.

By short-lived, or *ephemeral*, we mean that parameters are refreshed (and possibly recertified) as often as feasible and permitted by their application; for the Diffie-Hellman protocol it means that a group is generated and used for just a single protocol execution and discarded after completion of the key agreement phase. Ephemeral parameters minimize the attack-window before the parameters are discarded. Attacks after use cannot be avoided for any type of public key system. But the least we can do is to avoid using parameters that may have been exposed to cryptanalysis for an unknown and possibly extended period of time before their usage.

In this paper we discuss existing methods for personalized, short-lived ECC parameter generation. Even with current technology, each end-user can in principle refresh and recertify his or her ECC parameters on a daily basis (cf. Sect. 2): "in principle" because user-friendly interfaces to the required software are not easily available to regular users. But it allows arbitrary, personalized choices – within the restrictions of ECC of course – in such a way that no other party can control or predict any of the newly selected parameters (including a curve parameterization and a finite field that together define an elliptic curve group, cf. below). Personalization isolates each user from attacks against other users, and using keys for a period of time that is as short as possible reduces the potential attack pay-off. Once personalized, short-lived ECC (public, private) key pairs are adopted at the end-user level, certifying parties may also rethink their sometimes decades-long key validities.

To satisfy the run time requirements of the Diffie-Hellman protocol, it should take at most a fraction of a second (jointly on two consumer-devices) to construct a personalized elliptic curve group suitable for the key agreement phase, that will be used for just that key agreement phase, and that will be discarded right after its usage – never to be used or even met again. In full generality this is not yet possible, as far as we know, and a subject of current research. However, for the moment the method from [34] can be used if one is willing to settle for partially personalized parameters: the finite field and thus the elliptic curve group cardinality are still fully personalized and unpredictable to any third party, but not more than eight choices are available for the Weierstrass equation used for the curve parameterization. Although the resulting parameters are not in compliance with the security criteria adopted by [9] and implied by [39], we point out that there is no indication whatsoever that either of these eight choices offers inadequate security: citing [9] "there is no evidence of serious problems". The choice is between being vulnerable to as yet unknown attacks – as virtually all cryptographic systems are – or being vulnerable to attacks aimed at others by sharing parameters, on top of trusting choices made by others. Given where the uncertainties lie these days, we opt for the former choice.

An issue that we are not paying attention to in this paper is the performance of the elliptic curve cryptosystem itself, once the parameters have been generated, or a comparison between the curves as generated here and the standardized ones. This is not because the issue is not of interest, but mostly because for either type of curve perfectly adequate runtimes can easily be achieved using generally available software. Also, our main point of concern in this paper is not performance optimization but minimization of trust in other parties.

After introductory sections on elliptic curves and their selection for ECC (Sect. 2) and complex multiplication (Sect. 3.1) we provide an explanation (in Sect. 3.2) how the "class number one" Weierstrass equations proposed in [34] were derived and how that same method generalizes to slightly larger class numbers. As a result we expand, also in Sect. 3.2, the table from [34] with eleven more Weierstrass equations, thereby more than doubling the number of equations available. In Sect. 3.3 we show how our methods can be further generalized, and why practical application of these ideas may not be worthwhile. In Sect. 4 we describe a new method for partially personalized ECC parameter generation that is substantially faster than the one from [34] and that also allows generation of *Montgomery friendly* primes and, at non-trivial overhead, of *twist-secure* curves. We demonstrate the effectiveness of our approach with an implementation on an Android Samsung Galaxy S4 smartphone. It generates a unique 128-bit secure elliptic curve group in about 50 milliseconds on average and thus allows efficient generation and ephemeral usage of such groups during Diffie-Hellman key agreement. Security issues (including the one mentioned above) are discussed in Sect. 5. In the concluding Sect. 6 we briefly discuss extension of our method to genus 2.

Our source code will be made available. As selecting ECC parameters on the fly adds more complexity to Diffie-Hellman key agreement, users should use the open-source code we will provide, after extensive testing and potential improvement, to minimize the probability of failure. Techniques to make the selection process more robust like additional sanity checks may be applied.

2 Preliminaries

Elliptic Curves. We fix notation and recall some well known facts. For a finite field K of characteristic larger than 3, a pair $(a, b) \in K^2$ with $4a^3 + 27b^2 \neq 0$ defines an *elliptic curve* $E_{a,b} = E$ over K, to be thought of as the coefficients of the Weierstrass equation

$$y^2 = x^3 + ax + b.$$

The set of pairs $(x, y) \in K^2$ that satisfy this equation along with a *point at infinity* O is the *set of points* $E(K)$ of E over K. This set has the structure of an abelian group (with O acting as the identity element) and is thus also referred to as the *group of points* of E over K or simply the *elliptic curve group*. Traditionally, this group is written additively. See [53, Chap. III] for a more general and formal introduction to this material, including effective ways to perform the group operation in a constant number of operations in K.

For $g \in E(K)$ the *discrete logarithm* problem with respect to g is the problem to find, given $h \in \langle g \rangle$, an integer n such that $h = ng$. For properly chosen E, the fastest published methods to solve this problem require on the order of $\sqrt{q}$ operations in the group $E(K)$ (and thus in K), where q is the largest prime dividing the order of g. If $k \in \mathbf{Z}$ is such that $2^{k-1} \leq \sqrt{q} < 2^k$, the discrete logarithm problem in $E(K)$ is said to offer *k-bit security*.

With $K = \mathbf{F}_p$ the finite field of cardinality p for a prime $p > 3$, and a randomly chosen elliptic curve E over $\mathbf{F}_p$, the order $\#E(\mathbf{F}_p)$ behaves as a random integer close to $p+1$ (see [38] for the precise statement) with $|\#E(\mathbf{F}_p) - p - 1| \leq 2\sqrt{p}$. For ECC at k-bit security level it therefore suffices to select a $2k$-bit prime p and an elliptic curve E for which $\#E(\mathbf{F}_p)$ is prime (or *almost prime*, i.e., up to an ℓ-bit factor, at an $\frac{\ell}{2}$-bit security loss, for a small ℓ), and to rely on the alleged hardness of the discrete logarithm with respect to a generator (of a large prime order subgroup) of $E(\mathbf{F}_p)$. How suitable p and E should be constructed is the subject of this paper. For reasons adequately argued elsewhere (cf. [7, Sect. 4.2]), for cryptographic purposes we explicitly exclude from consideration elliptic curves over extension fields.

Depending on the application, *twist-security* may have to be enforced as well: not just $\#E_{a,b}(\mathbf{F}_p) = p + 1 - t$ must be (almost) prime (where $|t| \leq 2\sqrt{p}$), but also $p + 1 + t$ must be (almost) prime. This number $p + 1 + t$ is the cardinality of the group of points of a *(quadratic) twist* $\widetilde{E} = E_{r^2a, r^3b}$ of $E = E_{a,b}$, where r is any non-square in $\mathbf{F}_p$.

Generating Elliptic Curves for ECC. The direct approach is to first select, for k-bit security, a random $2k$-bit prime p and then to randomly select elliptic curves E over $\mathbf{F}_p$ until $\#E(\mathbf{F}_p)$ is (almost) prime. Because of the random behavior of $\#E(\mathbf{F}_p)$, the expected number of elliptic curves to be selected is linear in k and can be halved by considering $\#\widetilde{E}(\mathbf{F}_p)$ as well (and replacing E by $\widetilde{E}$ if a prime $\#\widetilde{E}(\mathbf{F}_p)$ is found first). Because $\#E(\mathbf{F}_p)$ can be computed in time polynomial in k using the Schoof-Elkies-Atkin algorithm (SEA) [50], the overall expected effort is polynomial in k. This method is referred to as *SEA-based*

Table 1. Timings of cryptographic parameter generation on a single 2.7 GHz Intel Core i7-3820QM, averaged over 100 parameter sets, for prime elliptic curve group orders and 80-bit, 112-bit, and 128-bit security. For RSA these security levels correspond, roughly but close enough, to 1024-bit, 2048-bit, and 3072-bit composite moduli, for DSA to 1024-bit, 2048-bit, and 3072-bit prime fields with 160-bit, 224-bit, and 256-bit prime order subgroups of the multiplicative group, respectively. The timings in the last two rows have been obtained using a C implementation of the method we present in this paper.

	80-bit security	112-bit security	128-bit security
RSA	80 milliseconds	0.8 s	2.5 s
DSA	0.2 s	1.8 s	8 s
Random ECC (MAGMA)	12 s	47 s	120 s
Same, but twist-secure	6 min	37 min	83 min
Low discriminant curves over random prime fields	2 milliseconds	5 milliseconds	6 milliseconds
Same, but twist secure	10 milliseconds	27 milliseconds	45 milliseconds

ECC parameter selection. Generating twist-secure curves in this way is slower by a factor linear in k.

Table 1 lists actual ECC parameter generation times, for $k \in \{80, 112, 128\}$. Using primes p with special properties (such as being *Montgomery friendly*, i.e., $p \equiv \pm 1 \bmod 2^{32}$ or 2^{64}) has little or no influence on the timings. For comparison, key generation times are included for traditional non-ECC asymmetric cryptosystems at approximately the same security levels. The ECC parameter generation timings – in particular the twist-secure ones – may explain why the direct approach to ECC parameter generation is not considered to be a method that is suitable for the general public. Although this may have to be reconsidered and end-users could in principle – given appropriate software – (re)generate their ECC parameters and key material on a daily basis, the current state-of-the-art of the direct approach does not allow fast enough on-the-fly ECC parameter generation in the course of the Diffie-Hellman protocol. Table 1 also lists timings obtained using the method presented in this paper.

Pre-selected Elliptic Curves. We briefly discuss some of the elliptic curves that have been proposed or standardized for ECC. As mentioned above, we do not consider any of the proposals that involve extension fields (most commonly of characteristic two).

With two notable exceptions that focus on $\approx$ 125-bit security, most proposals offer a range of security levels. Although 90-bit security [11] is still adequate, it is unclear why parameters that offer less than 112-bit security (the minimal security level recommended by NIST [43]) should currently still be considered, given that the $\approx$ 125-bit security proposals offer excellent performance. With 128-bit security more than sufficient for the foreseeable future, it is not clear either what purpose is served by higher security levels, other than catering to "TOP SECRET" 192-bit security from [44]. In this context it is interesting to note that 256-bit AES, also prescribed by [44] for "TOP SECRET", was introduced only to still have a 128-bit secure symmetric cipher in the post-quantum world (cf. [55]), and that 192-bit security was merely a side-effect that resulted from the calculation $\frac{128+256}{2}$ (cf. [55]). In that world ECC is obsolete anyhow.

In [16] eleven different primes are given, all of a special form that makes modular arithmetic somewhat easier than for generic primes of the same size, and ranging from 112 to 521 bits. They are used to define fifteen elliptic curves of eight security levels from 56-bit to 260-bit, four with $a = 0$ and b small positive ("Koblitz curves"), the other eleven "verifiably at random" but nine of which with $a = p - 3$, and all except two with prime group order (two with cofactor 4 at security levels 56 and 64). Verifiability means that a standard pseudo random number generator when seeded with a value that is provided, results in the parameters a (if $a \neq p - 3$) and b. The arbitrary and non-uniform choice for the seeds, however, does not exclude the possibility that parameters were aimed for that have properties that are unknown to the users. This could easily have been avoided, but maybe this was not a concern at the time when these curves were generated (i.e., before the fall of the year 2000). Neither was twist-security a design criterion back then; indeed some curves have poor twist security (particularly so the 96-bit secure curve), whereas the single 192-bit secure curve is perfectly twist-secure. If one is willing to use pre-selected curves, there does not seem to be a valid argument, at this point in time, to settle for anything less than optimal twist-security: for general applications they are arguably preferable and their only disadvantage is that they are relatively hard to find, but this is done just once and thus no concern. Surprisingly, in the latest (2013) update of the federal information processing standards ("FIPS") for digital signatures (cf. [58]) only two out of five twists of the curves at security level 96 or higher and with $a = p - 3$ (all "recommended for federal government use") satisfy the group-cardinality margins allowed by [58].

The use of special primes was understandable back in 2000, because at that time ECC was relatively slow and any method to boost its performance was welcome, if not crucial, for the survival of ECC. The trend to use special primes persists to the present day, in a seemingly unending competition for the fastest ECC system. However, these days also regular primes without any special form offer more than adequate ECC performance. This is reflected in one of the proposals discussed below.

The proposals [5,7] each contain a single twist-secure curve of (approximately) 125-bit security, possibly based on the sensible argument that there is no need to settle for less if the performance is adequate, and no need to require more (cf. above). All choices are deterministic given the design criteria, easily verifiable, and have indeed been verified. For instance, the finite field in [5] is defined by the largest 255-bit prime, where the choice 255 is arguably optimal given the clever field arithmetic. The curve equation is the "first" one given the computationally advantageous curve parameterization and various requirements on the group orders. Another, but similarly rigidly observed, design criterion (beyond the scope of the present paper) underlies the proposal in [7].

The curves from [5,7] are perfectly adequate from a security-level and design point of view. If the issue of sharing pre-selected curves is disregarded they should suffice to cater to all conceivable cryptographic applications (with the exception of pairing-based cryptography, cf. below). Nevertheless, their design approach triggered two follow-up papers by others. In [1] they are complemented with their counterparts at approximate security levels 112, 192, and 256. In [13] the scope of [7] is broadened by allowing more curve parameterizations and more types of special primes, while handling exceptions more strictly. This leads to eight new twist-secure curves of (approximately) 128-bit security, in addition to eight and ten twist-secure curves at approximate security levels 192 and 256, respectively.

The seven Brainpool curves [39] at seven security levels from 80-bit to 256-bit revert to the verifiably pseudo random approach from [16], while improving it and thereby making it harder to target specific curve properties (but see [6]). The primes p have no special form (except that they are 3 mod 4) and are deterministically determined as a function of a seed that is chosen in a uniform manner based on the binary expansion of $\pi = 3.14159\ldots$. The curves use $a = p - 3$ and a quadratic non-residue $b \in \mathbf{F}_p$ (deterministically determined as a function of a different seed, similarly generated based on $e = 2.71828\ldots$) for which the orders of the groups of the curve and its twist are both prime. As an additional precaution, curves are required to satisfy $\#E_{a,b}(\mathbf{F}_p) < p$. In [37] it is shown how usage of constants such as π and e can be avoided while still allowing verifiable and trustworthy random parameter generation.

The SafeCurves project [9] specifies a set of criteria to analyze elliptic curve parameters aiming to ensure the security of ECC and not just the security (i.e., the difficulty) of the elliptic curve discrete logarithm problem, and analyzes many proposed parameter choices, including many of those presented above, with respect to those criteria. This effort represents a step forward towards better security for ECC. For this paper it is relevant to mention that the SafeCurves security criteria include the requirement that the complex-multiplication field discriminant (cf. below) must be larger than 2^{100} in absolute value. Aside from the lack of argumentation for the bound, this requirement seems to be unnecessarily severe (and considerably larger than the rough 2^{40} requirement implied by [39]), not just because it is not supported by theoretical evidence, but also because the requirement cannot be met by pairing-based cryptography, considered by many as a legitimate and secure application of elliptic curves. On the other hand, [9] does not express concerns about the trust problem inherent in the usage of (shared) parameters pre-selected by third parties.

Attacking Multiple Keys. We conclude this section with a brief summary of results concerning the security of multiple instances of the "same" asymmetric cryptographic system. Early successes cannot be expected, or are sufficiently unlikely (third case).

1. *Multiple RSA moduli of the same size.* It is shown in [19, Sect. 4] that after a costly size-specific precomputation (far exceeding the computation and storage cost of an individual factoring effort), any RSA modulus of the proper size can be factored at cost substantially less than its individual factoring effort. This is not a consequence of key-sharing (as RSA moduli should not be shared), it is a consequence of the number field sieve method for integer factorization [35].
2. *Multiple discrete logarithms all in the same multiplicative group of a prime field.* Finding a single discrete logarithm in the multiplicative group of a finite field is about as hard as finding any number of discrete logarithms in the same multiplicative group. Sharing a group is common (cf. DSA), but once a single discrete logarithm has been solved, subsequent ones in the same group are relatively easy.
3. *Multiple discrete logarithms all in the same elliptic curve group.* Solving a single discrete logarithm problem takes on the order of $\sqrt{q}$ operations, if the group has prime order q, and solving k discrete logarithm problems takes effort $\sqrt{kq}$. Thus, the average effort is reduced for each subsequent key that uses the same group.
4. *Multiple discrete logarithms in as many distinct, independent groups.* Solving k distinct discrete logarithm problems in k groups that have no relation to each other requires in general solving k independent problems. With the proper choice of groups, no savings can be obtained.

The final two cases most concern us in this paper. In the third case, with k users, an overall attack effort $\sqrt{kq}$ leads to an average attack effort per user of "just" $\sqrt{q/k}$. This may look disconcerting, but if q is properly chosen in such a way that effort $\sqrt{q}$ is infeasible to begin with, there is arguably nothing to be concerned about. Compared to the rather common second case (i.e., shared DSA parameters), the situation is actually quite a bit better. Nevertheless, existing users cannot prevent that new users may considerably affect the attack incentives. In the final case such considerations are of no concern. However, given the figures from Table 1, realizing the final case for ECC with randomly chosen parameters is not feasible yet for all applications. The next best approach that we are aware of is further explored below.

3 Special Cases of the Complex Multiplication Method

Our approach is based on and extends [34]. It may be regarded as a special case, or a short-cut, of the well known *complex multiplication* (CM) method of which many variants have been published and which appears under a variety of names in the literature (such as "Atkin-Morain" method). As no explanation is provided in [34], we first sketch one approach to the CM method and describe how it leads to the method from [34]. We then use this description to get a more general method, and indicate how further generalizations can be obtained.

3.1 The CM Method

We refer to [3, Chap. 18], [48], and the references therein for all details of the method sketched here. In the SEA-based ECC parameter selection described in Sect. 2 one selects a prime field $\mathbf{F}_p$ and then keeps selecting elliptic curves over $\mathbf{F}_p$ until the order of the elliptic curve group has a desirable property. Checking the order is relatively cumbersome, making this type of ECC parameter selection a slow process. Roughly speaking, the CM method switches around the order of some of the above steps, making the process much faster at the expense of a much smaller variety of resulting elliptic curves: first primes p are selected until a trivial to compute function of p satisfies a desirable property, and only then an elliptic curve over $\mathbf{F}_p$ is determined that satisfies one's needs.

The standard CM method works as follows. Let $d \neq 1, 3$ be a square-free positive integer and let $H_d(X)$ be the Hilbert class polynomial[1] of the imaginary quadratic field $\mathbf{Q}(\sqrt{-d})$. If $d \equiv 3 \bmod 4$ let $m = 4$ and $s = 1$, else let $m = 1$ and $s = 2$. Find integers u, v such that $u^2 + dv^2$ equals mp for a suitably large prime p and such that $p + 1 \pm su$ satisfies the desired property (such as one of $p + 1 \pm su$ prime, or both prime for perfect twist security). Compute a root j of $H_d(X)$ modulo p, then the pair $\left(\frac{-27j}{4(j-12^3)}, \frac{27j}{4(j-12^3)}\right) \in \mathbf{F}_p^2$ defines an elliptic curve E over $\mathbf{F}_p$ such that $\#E(\mathbf{F}_p) = p + 1 \pm su$ (and $\#\widetilde{E}(\mathbf{F}_p) = p + 1 \mp su$). Finally, use scalar multiplications with a random element of $E(\mathbf{F}_p)$ to resolve the ambiguity. For $d \equiv 3 \bmod 4$ the case $u = 1$ should be excluded because it leads to anomalous curves.

The method requires access to a table of Hilbert class polynomials or their on-the-fly computation. Either way, this implies that only relatively small d-values can be used, thereby limiting the resulting elliptic curves to those for which the "complex-multiplication field discriminant" (namely, d) is small. The degree of $H_d(X)$ is the class number h_{-d} of $\mathbf{Q}(\sqrt{-d})$. Because $h_{-d} = 1$ precisely for $d \in \{1, 2, 3, 7, 11, 19, 43, 67, 163\}$ (assuming square-freeness), for those d-values the root computation and derivation of the elliptic curve become a straightforward one-time precomputation that is independent of the p-values that may be used. This is what is exploited in [34], as further explained, and extended to other d-values for which h_{-d} is small, in the remainder of this section.

3.2 The CM Method for Class Numbers at Most Three

In [34] a further simplification was used to avoid the ambiguity in $p+1\pm u$. Here we follow the description from [56, Theorem 1], restricting ourselves to $d > 1$ with $\gcd(d, 6) = 1$, and leaving $d \in \{3, 8\}$ from [34] as special cases. We assume that $d \equiv 3 \bmod 4$ and aim for primes $p \equiv 3 \bmod 4$ to facilitate square root computation in $\mathbf{F}_p$. It follows that $\left(\frac{-1}{p}\right) = -1$.

[1] Obviously, we could have used Weber polynomials instead. Here we explain and generalize the method from [34] and therefore use Hilbert polynomials because those were the ones used in that paper.

Let $H_d(X)$ be as in Sect. 3.1. If $d \equiv 3 \bmod 8$ let $s = 1$, else let $s = -1$. As above, find integers $u > 1, v$ such that $u^2 + dv^2$ equals $4p$ for a (large) prime $p \equiv 3 \bmod 4$ for which the numbers $p + 1 \pm u$ are (almost) prime, and for which

$$a = 27d\sqrt[3]{j} \quad \text{and} \quad b = 54sd\sqrt{d(12^3 - j)}$$

are well-defined in $\mathbf{F}_p$, where j is a root of $H_d(X)$ modulo p. Then for any non-zero $c \in \mathbf{F}_p$, the pair $(c^4a, c^6b) \in \mathbf{F}_p^2$ defines an elliptic curve E over $\mathbf{F}_p$ such that $\#E(\mathbf{F}_p) = p + 1 - \left(\frac{2u}{d}\right)u$ (and $\#\widetilde{E}(\mathbf{F}_p) = p + 1 + \left(\frac{2u}{d}\right)u$).

As an example, let $d = 7$, so $s = -1$. The Hilbert class polynomial $H_7(X)$ of $\mathbf{Q}(\sqrt{-7})$ equals $X + 15^3$, which leads to $j = -15^3$, $a = -3^4 \cdot 5 \cdot 7$, and $b = -54 \cdot 7\sqrt{7(12^3 + 15^3)} = -2 \cdot 3^6 \cdot 7^2$. With $c = \frac{1}{3}$ we find that the pair $(-35, -98)$ defines an elliptic curve E over any prime field $\mathbf{F}_p$ with $4p = u^2 + 7v^2$ and that $\#E(\mathbf{F}_p) = p + 1 - \left(\frac{2u}{7}\right)u$.

Similarly, $H_{11}(X) = X + 2^{15}$ for $d = 11$. With $s = 1$ this leads to $j = -2^{15}$, $a = -2^5 \cdot 2^3 \cdot 11 = -9504$, and $b = 2 \cdot 3^3 \cdot 11\sqrt{11(12^3 + 2^{15})} = 365904$. For any $p \equiv 3 \bmod 4$ the pair $(-9504, 365904)$ defines an elliptic curve E over $\mathbf{F}_p$ for which $\#E(\mathbf{F}_p) = p + 1 - \left(\frac{2u}{11}\right)u$, where $4p = u^2 + 11v^2$. This is the twist of the curve for $d = 11$ in [34].

The elliptic curves corresponding to the four d-values with $h_{-d} = 1$ and $d > 11$ are derived in a similar way, and are listed in Table 2. The two remaining cases with $h_{-d} = 1$ listed in Table 2 are dealt with as described in [2, Theorem 8.2] for $d = 3$ and [47] for $d = 8$.

For $d = 91$, the class number h_{-91} of $\mathbf{Q}(\sqrt{-91})$ equals two and $H_{91}(X) = X^2 + 2^{17} \cdot 3^3 \cdot 5 \cdot 227 \cdot 2579X - 2^{30} \cdot 3^6 \cdot 17^3$ has root $j = \left(-2^4 \cdot 3(227 + 3^2 \cdot 7\sqrt{13})\right)^3$. It follows that $a = -2^4 \cdot 3^4 \cdot 7 \cdot 13(227 + 3^2 \cdot 7\sqrt{13})$ and $b = 2^4 \cdot 3^6 \cdot 7^2 \cdot 11 \cdot 13(13 \cdot 71 + 2^8\sqrt{13})$ so that with $c = \frac{1}{3}$ we find that the pair $(-330512 - 91728\sqrt{13}, 103479376 + 28700672\sqrt{13})$ defines an elliptic curve E over any prime field $\mathbf{F}_p$ with $p \equiv 3 \bmod 4$ and $\left(\frac{13}{p}\right) = 1$, and that $\#E(\mathbf{F}_p) = p + 1 - \left(\frac{2u}{91}\right)u$ where $4p = u^2 + 91v^2$.

Table 2 lists nine more d-values for which $h_{-d} = 2$, all with $d \equiv 3 \bmod 4$: for those with $\gcd(d, 6) = 1$ the construction of the elliptic curve goes as above for $d = 91$, the other three (all with $\gcd(d, 6) = 3$) are handled as shown in [30]. The other d-values for which $h_{-d} = 2$ also have $\gcd(d, 6) \neq 1$ and were not considered (but see [30]). The example for $h_{-d} = 3$ in the last row of Table 2 was taken from [30].

3.3 The CM Method for Larger Class Numbers

In this section we give three examples to illustrate how larger class numbers may be dealt with, still using the approach from Sect. 3.2. For each applicable d with $h_{-d} < 5$ a straightforward (but possibly cumbersome) one-time precomputation suffices to express one of the roots of $H_d(X)$ in radicals as a function of the coefficients of $H_d(X)$, and to restrict to primes p for which the root exists in $\mathbf{F}_p$. For larger h_{-d} there are in principle two obvious approaches (other possibilities

Table 2. Elliptic curves for fast ECC parameter selection. Each row contains a value d, the class number h_{-d} of the imaginary quadratic field $\mathbf{Q}(\sqrt{-d})$ with discriminant $-d$, the root used (commonly referred to as the j-invariant), the elliptic curve $E = E_{a,b}$, the constraints on the prime p and the values u and v, the value s such that $\#E(\mathbf{F}_p) = p+1-su$, and with γ and $\tilde{\gamma}$ denoting fixed factors of $\#E(\mathbf{F}_p)$ and $\#\widetilde{E}(\mathbf{F}_p)$, respectively.

h_{-d}	d	j-invariant	a, b	$p, u, v \in \mathbf{Z}_{>0}$	s	$\{\gamma\} \cup \{\tilde{\gamma}\}$
1	3	0	$0, 16$	$u^2+3v^2=4p$, $p \equiv 1 \bmod 3$, $u \equiv 1 \bmod 3$, $v \equiv 0 \bmod 3$	-1	$\{1,9\}$
	8	20^3	$-270, -1512$	$u^2+2v^2=p$, $u \equiv 1 \bmod 4$ if $p \equiv 3 \bmod 16$, $u \equiv 3 \bmod 4$ if $p \equiv 11 \bmod 16$	2	$\{2\}$
	7	-15^3	$-35, -98$	$u^2+dv^2=4p$, $u>1$	$\left(\frac{2u}{d}\right)$	$\{8\}$
	11	-32^3	$-9504, 365904$			$\{1,9\}$
	19	-96^3	$-608, 5776$			$\{1\}$
	43	-960^3	$-13760, 621264$			
	67	-5280^3	$-117920, 15585808$			
	163	-640320^3	$-34790720, 78984748304$			
2	91	$-48^3(227+63\sqrt{13})^3$	$-330512-91728\sqrt{13}$, $103479376+28700672\sqrt{13}$			
	115	$-48^3(785+351\sqrt{5})^3$	$-1444400-645840\sqrt{5}$, $944794000+422522880\sqrt{5}$			
	187	$-240^3(3451+837\sqrt{17})^3$	$-51626960-12521520\sqrt{17}$, $+201921077072+48973056000\sqrt{17}$			
	235	$-528^3(8875+3969\sqrt{5})^3$	$-367070000-164157840\sqrt{5}$, $3828113058000+1711984189440\sqrt{5}$			
	403	$-240^3(2809615+779247\sqrt{13})^3$	$-90581987600-25122923280\sqrt{13}$, $1399216(10605743499+2941504000\sqrt{13})$			
	427	$-5280^3(236674+30303\sqrt{61})^3$	$-177865244480-22773310560\sqrt{61}$, $1099951(37121542375+4752926464\sqrt{61})$			
	51	$-48^3(4+\sqrt{17})^2(5+\sqrt{17})^3$	$-245616-59568\sqrt{17}$, $66257296+16069760\sqrt{17}$			$\{1,3\}$
	123	$-480^3(32+5\sqrt{41})^2(8+\sqrt{41})^3$	$-580796160-90705120\sqrt{41}$, $7619012947280+1189889913856\sqrt{41}$			
	267	$-240^3(500+53\sqrt{89})^2(625+53\sqrt{89})^3$	$-12015034710000-1273591132080\sqrt{89}$, $9968(2274273163768531+241072473215000\sqrt{89})$			
	35	$-16^3(15+7\sqrt{5})^3$	$-226800-105840\sqrt{5}$, $60858000+27095040\sqrt{5}$			$\{1,9\}$
3	243	$-160^3(151022371885959 +104713064226304\sqrt[3]{3} -72603983653110\sqrt[3]{9})$	$-1560+720\sqrt[3]{9}$, $32258-11124\sqrt[3]{3}-7704\sqrt[3]{9}$		$\left(\frac{-2\alpha}{p}\right)\left(\frac{2u}{243}\right)$ $\alpha = 2-\sqrt[3]{9}$	

exist, but we do not explore them here). One approach would be to exploit the solvability by radicals of the Hilbert class polynomial [29] for any d, to carry out the corresponding one-time root calculation, and to restrict, as usual, to primes modulo which a root exists. The other approach is to look up $H_d(X)$ for some appropriate d, to search for a prime p such that $H_d(X)$ has a root modulo p, and to determine it. In our application, the precomputation approach leads to relatively lightweight online calculations, which for the last approach quickly become more involved. We give examples for all three cases, with run times obtained on a 2.7 GHz Intel Core i7-3820QM.

For $d = 203$ we have $h_{-203} = 4$ and $H_{203}(X) = X^4 + 2^{18} \cdot 3 \cdot 5^3 \cdot 739 \cdot 378577789X^3 - 2^{30} \cdot 5^6 \cdot 17 \cdot 1499 \cdot 194261303X^2 + 2^{54} \cdot 5^9 \cdot 11^6 \cdot 4021X + 2^{66} \cdot 5^{12} \cdot 11^6$

with root $-2^{14} \cdot 5^3 j'$ where

$$j' = 3357227832852 + 623421557759\sqrt{29} + 3367\sqrt{29(68565775894279681 + 12732344942060216\sqrt{29})}.$$

This precomputation takes an insignificant amount of time for any polynomial of degree at most four. With $c = 2^4 \cdot 3^3 \cdot 203$ it follows that the pair $\left(-5c\sqrt[3]{4j'}, c\sqrt{203(3^3 + 2^8 \cdot 5^3 j')}\right)$ defines an elliptic curve E over any prime field $\mathbf{F}_p$ that contains the various roots, and that $\#E(\mathbf{F}_p) = p + 1 - \left(\frac{2u}{203}\right)u$ where $4p = u^2 + 203v^2$. The online calculation can be done very quickly if the choice of p is restricted to primes for which square and cube roots can be computed using exponentiations modulo p.

As an example of the second approach, for $d = 47$ the polynomial $H_{47}(X)$ has degree five and root $25j'$, with the following expression by radicals for j':

$$\begin{aligned}
&13^3(7453991996007968795256512 - 2406037696832339815\sqrt{5} + A(408914360902374 16B \\
&- 28095336077279242712004810905 5211\sqrt{5}/B))/(2^{3/5}C) - 13(5364746311921861372 \\
&- 856800988085\sqrt{5} - A(29162309591B - 135009745365087109801596264\sqrt{5}))/(2C^2)^{1/5} \\
&+ (3861085845907 - 1237935\sqrt{5})/(2 \cdot 13^3 C^{1/5}) - 18062673 + 13C^{1/5}/2^{2/5},
\end{aligned}$$

where

$$A = \frac{67206667}{827296299281}, \quad B = \sqrt{47(119957963395745 + 21781710063898\sqrt{5})}$$

and

$$C = -2071374628128425156312708988 1529 + 1665551744948633926890917 5\sqrt{5} - \frac{D}{B}$$

for

$$D = 5^2 \cdot 11^2 \cdot 19 \cdot 23 \cdot 29 \cdot 31 \cdot 41 \cdot 47(2069683334124917088 47 - 461495327025091583738 45\sqrt{5}).$$

This one-time precomputation took 0.005 s (using Maple 18). Elliptic curves and group orders follow easily, for properly chosen primes. In principle such root-expressions can be tabulated for any list of d-values one sees fit, but obtaining them, in general and for higher degrees, may be challenging.

As an example of the final approach mentioned above, for $d = 5923$ the polynomial $H_{5923}(X)$ has degree seven and equals

$$\begin{aligned}
&X^7 + 2^{15} \cdot 3^3 \cdot 5^3 \cdot 7 \cdot 31 \cdot 127 \cdot 2429520931 \cdot 136238689771578256215972490257607347497085841560925219572863881662960257476074094637X^6 \\
&-2^{30} \cdot 3^7 \cdot 5^6 \cdot 7 \cdot 62983 \cdot 1112240226499 \cdot 192924280073389856473204919112650 71 \cdot 171556657076224699685934416851052653070777X^5 \\
&+2^{45} \cdot 3^9 \cdot 5^9 \cdot 7 \cdot 53 \cdot 97 \cdot 769 \cdot 259381 \cdot 4437462560116423 \cdot 976042195206305867192519561 83 \cdot 271475671651404722645770221908783 51X^4 \\
&-2^{60} \cdot 3^{12} \cdot 5^{1}2 \cdot 7 \cdot 31 \cdot 99208777 \cdot 340691724206563027829933344798 69 \cdot 211581900590194937311557316394276049622142 4793X^3 \\
&+2^{75} \cdot 3^{16} \cdot 5^{15} \cdot 7 \cdot 11^3 \cdot 10477 \cdot 47581 \cdot 240853 \cdot 104531840353 \cdot 10353927562807 \cdot 355302735176948792722753488988566621288 31X^2 \\
&-2^{90} \cdot 3^{18} \cdot 5^{18} \cdot 7 \cdot 11^6 \cdot 47^3 \cdot 727 \cdot 7603931 \cdot 88452227997949 \cdot 174930707434708830526362836641919931158 9957X \\
&+(2^{35} \cdot 3^7 \cdot 5^7 \cdot 11^3 \cdot 17 \cdot 23 \cdot 41 \cdot 47^2 \cdot 71 \cdot 593 \cdot 659 \cdot 1103 \cdot 1109)^3.
\end{aligned}$$

Given $H_{5923}(X)$ and 128-bit security, we look for 123-bit integers u and v such that $4p = u^2 + 5923v^2$ for a prime p for which $H_{5923}(X)$ has a root j modulo p

and such that $\sqrt[3]{j}$ and $\sqrt{j}$ exist in $\mathbf{F}_p$ and can easily be calculated. For the present case it took 0.11 s (using Mathematica 9) to find

$$u = 97989548965232974261222572203796365 84,$$

$$v = 679415845702168995816816244342227 1774$$

which leads to the 256-bit prime

$$p = 68376297247017003283970261221870401697343820120616991149309517708508634100051$$

and

$$j = 54243655991109505677097612140273606931478183421749872324499965496758684443312.$$

Because $p \equiv 2 \bmod 3$ all elements of $\mathbf{F}_p$ have a cube root (in particular $\sqrt[3]{j} = j^{\frac{2p-1}{3}} \bmod p$), $\left(\frac{j}{p}\right) = 1$ and $p \equiv 3 \bmod 4$. The elliptic curve and group order follow in the customary fashion.

From our results and run times it is clear that none of these approaches (one-time root precomputations, or online root calculation) is compatible with the requirements on the class number (at least 10^6 in [39]) or the discriminant (at least 2^{100} in [9]). In the remainder of this paper we focus on the approach from Sect. 3.2. Our approach thus does not comply with the class number or discriminant requirements from [9,39], security requirements that are, as far as we know, not supported by published evidence.

4 Ephemeral ECC Parameter Generation

We describe how to use Table 2 to online generate ephemeral ECC parameters, improving the speed of the search for a prime p and curve E over $\mathbf{F}_p$ compared to the method from [34, Sect. 3.2], and while allowing an additional security requirement to the ones from [34] (without explicitly mentioning the ones already in place in [34]; refer to Sect. 5 for details). In the first place, on top of the trivial modifications to handle the extended table and determination of a base point as mentioned in [34, Sect. 3.6], we introduce the following additional search criteria:

1. *Efficiency considerations.*
   (a) *Montgomery friendly modulus.* The prime p may be chosen as -1 modulo 2^{64} or modulo 2^{32} to allow somewhat faster modular arithmetic.
   (b) *Conversion friendly curve.* A small positive factor f may be prescribed that must divide $\#E(\mathbf{F}_p)$ (such as for instance $f = 4$ to allow conversion to a Montgomery curve).
2. *Twist security.* Writing $\#E(\mathbf{F}_p) = fcq$ and $\#\widetilde{E}(\mathbf{F}_p) = \tilde{c}\tilde{q}$, with $f \in \mathbf{Z}_{>0}$ as above, cofactors $c, \tilde{c} \in \mathbf{Z}_{>0}$, and primes q and $\tilde{q}$, independent upper bounds ℓ and $\tilde{\ell}$ on the total security loss may be specified such that $fc < 2^{\ell}$ and $\tilde{c} < 2^{\tilde{\ell}}$. The roles of E and $\widetilde{E}$ may be reversed to meet these requirements faster (with f always a factor of the "new" $\#E(\mathbf{F}_p)$, which is automatically the case if $p \equiv 3 \bmod 4$ and $f = 4$).

These new requirements still allow a search as in [34, Sect. 3.2] where, based on external parameters and a random value, an initial pair (u_0, v_0) is chosen and the pairs $(u, v) \in \{(u_0, v_0 + i) : i \in [0, 255]\}$ are inspected on a one-by-one basis for each of the eight rows of [34, Table 1] until a pair is found that corresponds to a satisfactory p and E. If the search is unsuccessful (after trying $256 * 8$ possibilities), the process is repeated with a fresh random value and new initial pair (u_0, v_0). With $m = 1$, $c = 32$, and no restrictions on $\#\widetilde{E}(\mathbf{F}_p)$, it required on average less than ten seconds on a 133 MHz Pentium processor to generate a satisfactory ECC parameter set at the 90-bit security level. Though this performance was apparently acceptable at the time [34] was published, it does not bode well for higher security levels and, in particular, when twist security is required as well. This is confirmed by experiments (cf. runtimes reported in Table 3 below).

Sieving-Based Search. Secondly, we show how the performance of the search can be considerably improved compared to [34]. Because, for a fixed d, the prime p and both group orders are quadratic polynomials in u and v, sieving with a set P of small primes can be used to quickly identify (u, v) pairs that do not correspond to a satisfactory p or E. The remaining pairs, for which the candidates for the prime and for the group order(s) do not have factors in P, can then be subjected to more precise inspection, similar to the search from [34]. We sketch our sieving-based search for ECC parameters as in Table 2 where we assume that $\min(2^{\ell} - 1, 2^{\tilde{\ell}} - 1) = f$ and $\max(2^{\ell} - 1, 2^{\tilde{\ell}} - 1) \in \{f, \infty\}$, i.e., we settle for perfect twist security (except for the factor f) or no twist security at all. More liberal choices require a more cumbersome approach to the sieving; we do not elaborate.

Let (u_0, v_0) be chosen as above, but restricted to certain residue classes modulo small primes to satisfy a variety of divisibility criteria depending on the above choices of f, ℓ, and $\tilde{\ell}$, and with respect to Montgomery friendliness. We found it most convenient to fix u_0 and to sieve over regularly spaced $(v_0 + i)$-values, again restricted to certain residue classes for the same reasons (including divisibility of $\#E(\mathbf{F}_p)$ by f in case $f > 1$), but using a much larger range of i-values than in [34]. Fixing u_0, the first at most sixteen compatible d-values from Table 2 are selected; only ten d-values may remain and depending on the parity of u_0 the value $d = 7$ may or may not occur. Let $d_0, d_1, \ldots, d_{k-1}$ be the selected d-values, with $10 \leq k \leq 16$. With I the set of distinct i-values to be considered, we initialize for all $i \in I$ the sieve-location s_i as $2^k - 1$ (i.e., all "one"-bits in the k bit-positions indexed from 0 to $k - 1$), while leaving the constant difference between consecutive i-values unspecified for the present description. We mostly used difference 16, using difference 4 only for $d = 8$, and using a substantially larger value if the prime p must be Montgomery friendly.

For each d_j and each sieving-prime $\varsigma \in P$ up to six roots $r_{j\varsigma}$ modulo ς of up to three quadratic polynomials are determined (computing square roots using $\frac{\varsigma+1}{4}$-th powering for $\varsigma \equiv 3 \bmod 4$ and using the Tonelli-Shanks algorithm [20, 2.3.8] otherwise); the polynomials follow in a straightforward fashion from Table 2. To sieve for d_j the following is done for all $\varsigma \in P$ and for all roots $r_{j\varsigma}$: all sieve-locations s_i with $i \in (r_{j\varsigma} + \varsigma\mathbf{Z}) \cap I$ are replaced by $s_i \wedge 2^k - 2^j - 1$ (thus setting a possible "one"-bit at bit-position j in s_i to a "zero"-bit, while not changing the bits at the other $k-1$ bit-positions in s_i).

A "one"-bit at bit-position j in s_i that is still "one" after the sieving (for all indices, all sieving primes, and all roots) indicates that discriminant $-d_j$ and pair $(u_0, v_0 + i)$ warrants closer inspection because all relevant related values are free of factors in P. If the search is unsuccessful (after considering $k|I|$ possibilities), the process is repeated with a new sieve. If for all indices j and all $\varsigma \in P$ all last visited sieve locations are kept (at most $6k|P|$ values), recomputation of the roots can be avoided if the same (u_0, v_0) is re-used with the "next" interval of i-values.

Some savings may be obtained, in particular for small ς values, by combining the sieving for identical roots modulo ς for distinct indices j. Or, one could make just a single sieving pass per ς-value but simultaneously for all indices j and all roots $r_{j\varsigma}$ modulo ς, by gathering (using "$\wedge$"), for that ς, all sieving information (for all indices and all roots) for a block of ς consecutive sieve locations, and using that block for the sieving.

Parameter Reconstruction. A successful search results in an index j and value i such that d_j and the prime corresponding to the (u, v)-pair $(u_0, v_0 + i)$ leads to ECC parameters that satisfy the aimed for criteria. Any party that has the information required to construct (u_0, v_0) can use the pair (j, i) to instantaneously reconstruct (using Table 2) those same ECC parameters, without redoing the search. It is straightforward to arrange for an additional value that allows easy (re)construction of a base point.

Implementation Results. We implemented the basic search as used in [34] and the sieving based approach sketched above for generic x86 processors and for ARM/Android devices. To make the code easily portable to other platforms as well we used the GMP 6.0 library [24] for multi-precision integer arithmetic after having verified that modular exponentiation (crucial for an efficient search) offers good performance on ARM processors. Making the code substantially faster would require specific ARM processor dependent optimization. We used the Java native interface [46] and the Android native development kit [26] to allow the part of the application written in Java to call the GMP-based C-routines that underlie the compute intensive core. To avoid making the user interface non-responsive and avoid interruption by the Android run-time environment, a background service (*IntentService* class) [27] is instantiated to run this core independently of the thread that handles the user interface.

Table 3 lists detailed results for the 128-bit security level, using empirically determined (and close to optimal, given the platform) sieving bounds, lengths, etc. The implementations closely followed the description above, but we omit many details that were used to obtain better performance, such as precomputations and extra conditions, and to make sure that a variety of security requirements is met (more on this in Sect. 5). Table 4 shows average timings in milliseconds for different security levels in two cases: prime order non twist-secure generation and perfect twist security. The x86 platform is an Intel Core

Table 3. Performance results in milliseconds for parameter generation at the 128-bit security level, with ℓ, $\tilde{\ell}$, f, P, and I as above, the "MF"-column to indicate Montgomery friendliness, and μ the average and σ the standard deviation.

					x86, over 10 000 runs						ARM, over 3000 runs					
					basic		sieving				basic		sieving			
ℓ	$\tilde{\ell}$	$\{\ell\} \cup \{\tilde{\ell}\}$	f	MF	μ	σ	μ	σ	$\|P\|$	$\|I\|$	μ	σ	μ	σ	$\|P\|$	$\|I\|$
					not twist secure:											
		$\{6, \infty\}$			8.2	4.8	7.8	3.6	100	2^{10}	64	47	50	30	150	2^{12}
6					9.6	6.2	8.6	3.8	200	2^{10}	72	58	59	35	250	2^{12}
		$\{6, \infty\}$		✓	8.3	5.0	7.8	3.7	100	2^{10}	64	44	49	29	200	2^{12}
6				✓	9.7	6.4	8.7	3.8	200	2^{10}	71	55	60	33	250	2^{12}
		$\{6, \infty\}$	4		8.4	5.2	7.9	4.0	100	2^{10}	64	49	54	35	200	2^{12}
6			4		9.7	6.4	8.8	4.7	200	2^{10}	71	57	61	36	250	2^{12}
		$\{6, \infty\}$	4	✓	8.6	5.2	7.9	3.8	100	2^{10}	62	48	50	29	200	2^{12}
6			4	✓	9.7	6.4	8.6	3.7	200	2^{10}	72	58	56	35	250	2^{12}
		$\{1, \infty\}$			8.8	5.4	8.0	4.0	100	2^{10}	65	47	53	32	200	2^{12}
1					10.4	7.1	8.9	4.0	200	2^{10}	77	61	58	36	250	2^{12}
		$\{1, \infty\}$		✓	8.8	5.5	8.0	3.9	100	2^{10}	65	50	50	31	200	2^{12}
1				✓	10.4	7.0	8.8	3.9	200	2^{10}	76	62	57	35	250	2^{12}
					twist secure:											
		$\{6\}$			148	143	46	33	700	2^{14}	1280	1271	357	304	750	2^{15}
1	6				167	162	55	44	800	2^{14}	1432	1392	410	335	750	2^{15}
		$\{1, 6\}$			160	151	49	34	800	2^{14}	1350	1341	392	326	750	2^{15}
		$\{1\}$			180	177	49	40	800	2^{14}	1433	1372	390	325	750	2^{15}
		$\{6\}$		✓	143	139	50	36	700	2^{14}	1301	1270	390	311	750	2^{15}
1	6			✓	165	161	51	38	800	2^{14}	1428	1321	409	315	750	2^{15}
		$\{1, 6\}$		✓	154	148	49	35	800	2^{14}	1327	1300	380	316	750	2^{15}
		$\{1\}$		✓	172	168	48	36	800	2^{14}	1491	1428	378	326	750	2^{15}
		$\{6\}$	4		162	158	49	34	700	2^{14}	1307	1245	390	319	750	2^{15}
		$\{6\}$	4	✓	165	159	50	38	700	2^{14}	1287	1253	385	318	750	2^{15}

i7-3820QM, running at 2.7 GHz under OS X 10.9.2 and with 16 GB RAM. The ARM device is a Samsung Galaxy S4 smartphone with a Snapdragon 600 (ARM v7) running at 1.9 GHz under Android 4.4 with 2 GB RAM. Key reconstruction takes around 1.5 (x86) and 10 (ARM) milliseconds.

5 Security Criteria

In this section we review security requirements that are relevant in the context of ECC. Most are taken from [9], the order and keywords of which we roughly follow

Table 4. Performance results in milliseconds for parameter generation at different security levels: 80-bit, 112-bit, 128-bit, 160-bit, 192-bit and 256-bit. In comparison with Table 3 the Montgomery friendliness option is always disabled and $f = 1$.

		x86					ARM				
		basic	sieving			runs	basic	sieving			runs
k	$\{\ell\} \cup \{\tilde{\ell}\}$	μ	μ	$\|P\|$	$\|I\|$		μ	μ	$\|P\|$	$\|I\|$	
80	$\{1, \infty\}$	3	3	50	2^9	10000	22	19	100	2^{11}	100
	$\{1\}$	31	10	200	2^{12}	1000	197	61	450	2^{12}	100
112	$\{1, \infty\}$	6	6	100	2^9	10000	47	38	200	2^{10}	100
	$\{1\}$	114	30	800	2^{14}	1000	981	214	650	2^{14}	100
128	$\{1, \infty\}$	9	8	100	2^{10}	10000	65	53	250	2^{12}	3000
	$\{1\}$	180	49	800	2^{14}	10000	1433	390	750	2^{15}	3000
160	$\{1, \infty\}$	19	16	300	2^{11}	1000	143	87	200	2^{10}	100
	$\{1\}$	474	95	800	2^{14}	1000	5425	808	750	2^{15}	100
192	$\{1, \infty\}$	36	25	400	2^{12}	1000	265	169	20	2^{10}	100
	$\{1\}$	1144	222	1200	2^{16}	1000	10785	2231	900	2^{17}	20
256	$\{1, \infty\}$	105	70	400	2^{13}	1000	14543	575	450	2^{11}	100
	$\{1\}$	4635	994	1200	2^{16}	1000	50 s	10 s	1200	2^{17}	10

for ease of reference, and some are from [22]. We discuss to what extent these requirements are met by the parameters generated by our method. Generally speaking our approach is to focus on existing threats, as dealing with non-existing ones only limits the parameter choice while not serving a published purpose.

ECDLP Security. For the security of ECC, the discrete logarithm problem in the group of points of the elliptic curve must be hard. In this first category of security requirements one attempts to make sure that elliptic curve groups are chosen in such a way that this requirement is met.

- **Pollard rho attack** becomes ineffective if the group is chosen in such a way that a sufficiently large prime factor divides its order. This is a straightforward "key-length" issue (cf. [36]). Using a 128-bit prime field cardinality with $\ell \leq 5$, as suggested by Table 3, is more than sufficient.
- **Transfers** refer to the possibility to embed the group into a group where the discrete logarithm problem is easy, as would be the case for "anomalous curves" and for curves with a low "embedding degree". For the former, the

elliptic curve group over the finite field $\mathbf{F}_p$ has cardinality p and can be effectively embedded in the additive group $\mathbf{F}_p$, allowing trivial solution of the elliptic curve discrete logarithm problem (cf. [49,51,54]). By construction our method avoids these curves.
For the latter, the group can be embedded in the multiplicative group $\mathbf{F}_{p^k}^\times$ of $\mathbf{F}_{p^k}$ for a low embedding degree k. To avoid those curves, we follow the approach from [34] which ties the smallest permissible value for k to the published difficulty of finding discrete logarithms in $\mathbf{F}_{p^k}^\times$. It would be trivial, and would have negligible effect on our performance results, to adopt the "overkill" approach favored by [9,13,39], but we see no good reason to do so.

- **Complex-multiplication field discriminants** refers to the concern that for small values of the discriminant ($-d$ in our case) there are endomorphism-based speedups for the Pollard rho attack [25,61]. For instance, the first row of Table 2 leads to groups with the same *automorphism group* [53, Chap. III.10] as the *pairing-friendly* groups proposed in [4] and thereby to an additional speedup of the Pollard rho attack by a factor of $\sqrt{3}$. We refer to [14,21] for a discussion of the practical implications and note that such speedups are of no concern for 128-bit prime field cardinalities with $\ell \leq 5$.
Despite the fact that the authors of [9] agree with this observation (cf. their quotation cited in the introduction), and as already mentioned in Sect. 2, [9] chooses a lower bound of 2^{100} for the absolute value of the complex-multiplication field discriminant while [39] settles for roughly 2^{40}. Neither bound can be satisfied by out method, as amply illustrated in Sect. 3.3. Until a valid concern is published, we see no reason to abandon our approach.
- **Rigidity** is the security requirement that the entire parameter generation process must be transparent and exclude the possibility that malicious choices are targeted. Assuming a transparent process to generate the initial pair (u_0, v_0) (for instance by following the approach described in [34]) the process proposed here is fully deterministic, fully explained, and leaves no room for trickery. Note also that a third party is excluded and that the affected parties (the public key owner or the two communicating parties engaging in the Diffie-Hellman protocol) are the only ones involved in the parameter generation process.

ECC Security. Properly chosen groups can still be used in insecure ways. Here we discuss a number of precautions that may be taken to avoid some attacks that are aimed at exploiting the way ECC may be used.

- **Constant-time single-coordinate scalar multiplication** ("Ladders" in [9]) makes it harder to exploit timing differences during the most important operation in ECC, the multiplication of a group element by a scalar that usually needs to be kept secret, as such differences may reveal information about the scalar (where it should be noted that the "single-coordinate" part is just for efficiency and ease of implementation). For all Weierstrass curve parameterizations used here constant-time single-coordinate scalar multiplication can be achieved using the method from [15]. If efficiency is a bigger concern than freedom of choice, one may impose the requirement that the group order is divisible by four ("$f = 4$" in Table 3) as it allows conversion to Montgomery form [42] and thereby a more efficient constant-time single-coordinate scalar multiplication [41].

- **Invalid-point attacks** ("Twists" in [9]) refer to attempts to exploit a user's omission to verify properties of alleged group elements received. They are of no concern if the proper tests are consistently performed (at the cost of some performance loss) or if a closed software environment can be relied upon. Some are also thwarted if the curve's twist satisfies the same ECDLP security requirements as the curve itself, an approach that thus avoids implementation assumptions while replacing recurring verification costs by one-time but more costly parameter generation: for one-time parameter usage one-time verification is less costly (than relatively expensive generation of twist secure parameters), for possibly repeated usage (as in certified keys) twist secure parameters may be preferred. Our parameter selection method includes the twist security option and thus caters to either scenario. Below we elaborate on the various attack possibilities.

  Small-Subgroup Attacks. If the group order is not prime but has a relatively small factor h, an attacker may send a group element of order h (as opposed to large prime order), learn the residue class modulo h of the victim's secret key, and thus obtain a speedup of the Pollard rho attack by a factor of $\sqrt{h}$. It suffices to ascertain that group elements received do not have order dividing h, or to generate the parameters such that the group order is prime (one of our options).

  Invalid-Curve Attacks. An attacker may send elements of different small prime orders belonging to different appropriately selected elliptic curve groups, all distinct from the proper group. Each time the targeted victim fails to check proper group membership of elements received the attacker learns the residue class modulo a new small prime of the victim's secret key, ultimately enabling the attacker to use the Chinese remainder theorem to recover the key [10]. This attack cannot be avoided at the parameter selection level, but is avoided by checking that each element received belongs to the right group (at negligible cost). Also, using parameters just once renders the attack ineffective.

  Twist Attacks Against Single-Coordinate Scalar Multiplication. Usage of single coordinates goes a long way to counter the above invalid-curve attacks, because each element that does not belong to the group of the curve automatically belongs to the group of the twist of the curve. Effective attacks can thus be avoided either by checking membership of the proper group (i.e., not of the group of the twist) or by making sure that the group of the twist of the curve satisfies the same security requirements as the group of the curve itself (at a one-time twist secure parameter generation cost, avoiding the possibly recurring membership test). As mentioned above, it depends on the usage scenario which method is preferred; for each scenario our method offers a compatible option.

- **Exceptions in scalar multiplication** ("Completeness" in [9]). Depending on the curve parameterization, the implementation of the group law may distinguish between adding two distinct points and doubling a point. Using addition where doubling should have been used may be leveraged by an attacker to learn information about the secret key [31]. Either a check must be included (while maintaining constant-time execution, as in [13]) or a "complete" addition formula must be used, i.e., one that works even if the two points are not distinct. This leads to a somewhat slower group law for our Weierstrass curve parameterizations, but if they are used along with $f = 4$ in Table 3 the parameterization can be converted to Edwards or Montgomery form, which are both endowed with fast complete formulae for the group law [5,8].
- **Indistinguishability** of group elements and uniform random strings is important for ECC applications such as censorship-circumvention protocols [9], but we are not aware of its importance for the applications targeted in this paper. We refer to [7,23] for ways to achieve indistinguishability using families of curves in Montgomery, Edwards or Hessian form and to [57] for a solution that applies to the Weierstrass curve parameterization (which, however, doubles the lengths of the strings involved). Either way, our methods can be made to deal with this issue as well.
- **Strong Diffie-Hellman problem** (not mentioned in [9]). In [18] it is shown that for protocols relying on the ECC version of the strong Diffie-Hellman problem the large prime q dividing the group order must be chosen such that $q - 1$ and $q + 1$ both have a large prime factor. Although several arguments are presented in [17, Sect. B.1] why this attack is "unlikely to be feasible", [17] nevertheless continues with "as a precautionary measure, one may want to choose elliptic curve domain parameters that resist Cheon's attack by arranging that $q - 1$ and $q + 1$ have very large prime factors". Taking this precaution, however, would add considerable overhead to the parameter generation process. Our methods can in principle be adapted to take this additional requirement into account, but doing so will cause the parameter generation timings to skyrocket. The attack is not considered in [9], and none of the standardized parameter choices that we inspected take the precaution recommended in [17].

Side-channel attacks are physical attacks on the device executing the parameter generation process or the cryptographic protocols. Most of these attacks require multiple runs of the ECC protocol with the same private key (cf. [22, Table 1]) and are thus of no concern in an ephemeral key agreement application. There are three attacks for which a single protocol execution suffices:

Simple power analysis (SPA) attacks are avoided when using a scalar multiplication algorithm ensuring that the sequence of operations performed is independent of the scalar.

Fault induced invalid curve attacks can be expected to require several trials before a weak parameter choice is hit, and can be prevented by enforcing more sanity checks in the scalar multiplication [22].

Template attacks may recover a small number of bits of the secret key and can be avoided using one of the randomization techniques mentioned in [22].

6 Conclusions and Future Work

We showed how communicating parties can efficiently generate fresh ECC parameters every time they need to agree on a session key, generalizing and improving the method from [34]. Our major modifications consist of the use of sieving to speed up the generation process, a greater variety of security and efficiency options, and the inclusion of eleven more curve equations. Furthermore, we explained how to further generalize our method and showed that doing so may have limited practical value. We demonstrated the practical potential of our method on constrained devices, presented performance figures of an implementation on an ARM/Android platform, and discussed relevant security issues.

Future work could include further efficiency enhancements by targeting specific ARM processors, direct inclusion of Montgomery and Edwards forms, extension to genus 2 hyperelliptic curves and, much more challenging and important, improving elliptic curve point counting methods to allow on-the-fly generation of ephemeral random elliptic curves over prime fields. Unfortunately, we do not know yet how to approach the latter problem, but genus 2 extension of our methods seems to be quite within reach. We conclude with a few remarks on this issue.

Extension to Genus 2 Hyperelliptic Curves. Jacobians of hyperelliptic curves of genus 2 allow cryptographic applications similar to elliptic curves [33] and, as recently shown in [12], offer comparable or even better performance. Genus 2 hyperelliptic curves may thus be a worthwhile alternative to elliptic curves and, in particular given the lack of a reasonable variety of standardized genus 2 curves, generalization of our methods to the genus 2 case may have practical appeal. In [60] it is described how this could work. The imaginary quadratic fields are replaced by *quartic* CM fields and the j-invariant (a root of the Hilbert class polynomial) is replaced by three j-invariants which are usually referred to as Igusa's invariants. In [59] a table is given listing equations with integer coefficients of genus 2 hyperelliptic curves having complex multiplication by class number one quartic CM fields and class number two quartic CM fields. The three algorithms presented at the beginning of [60, Sect. 8] can then be used to easily compute the orders of the Jacobians of these curves over suitably chosen prime fields. The main remaining problem seems to be to resolve the ambiguity between the order of the Jacobian of the hyperelliptic curve and of its quadratic twist other than by using scalar multiplication. We leave the solution of this problem – and implementation of the resulting genus 2 parameter selection method – as future work.

Acknowledgement. Thanks to Adrian Antipa for bringing the strong Diffie-Hellman security requirement and additional precaution from [17, Sect. B.1] to our attention, and to René Schoof for inspiring this paper by providing the original table in [34].

References

1. Aranha, D.F., Barreto, P.S.L.M., Geovandro, C.C.F.P., Ricardini, J.E.: A note on high-security general-purpose elliptic curves. IACR Cryptology ePrint Archive, 2013:647 (2013)
2. Atkin, A.O.L., Morain, F.: Elliptic curves and primality proving. Math. Comput. **61**, 29–68 (1993)
3. Avanzi, R.M., Cohen, H., Doche, C., Frey, G., Lange, T., Nguyen, K., Vercauteren, F.: Handbook of Elliptic and Hyperelliptic Curve Cryptography. Chapman & Hall/CRC, Boca Raton (2006)
4. Barreto, P.S.L.M., Naehrig, M.: Pairing-friendly elliptic curves of prime order. In: Preneel, B., Tavares, S. (eds.) SAC 2005. LNCS, vol. 3897, pp. 319–331. Springer, Heidelberg (2006)
5. Bernstein, D.J.: Curve25519: new diffie-hellman speed records. In: Yung, M., Dodis, Y., Kiayias, A., Malkin, T. (eds.) PKC 2006. LNCS, vol. 3958, pp. 207–228. Springer, Heidelberg (2006)
6. Bernstein, D.J., Chou, T., Chuengsatiansup, C., Hülsing, A., Lange, T., Niederhagen, R., van Vredendaal, C.: How to manipulate curve standards: a white paper for the black hat. Cryptology ePrint Archive, Report 2014/571 (2014). http://eprint.iacr.org/2014/571
7. Bernstein, D.J., Hamburg, M., Krasnova, A., Lange, T.: Elligator: elliptic-curve points indistinguishable from uniform random strings. In: Proceedings of the 2013 ACM SIGSAC Conference on Computer & Communications Security, CCS 2013, pp. 967–980. ACM, New York (2013)
8. Bernstein, D.J., Lange, T.: Faster addition and doubling on elliptic curves. In: Kurosawa, K. (ed.) ASIACRYPT 2007. LNCS, vol. 4833, pp. 29–50. Springer, Heidelberg (2007)
9. Bernstein, D.J., Lange, T.L.: Safecurves: choosing safe curves for elliptic-curve cryptography
10. Biehl, I., Meyer, B., Müller, V.: Differential fault attacks on elliptic curve cryptosystems. In: Bellare, M. (ed.) CRYPTO 2000. LNCS, vol. 1880, p. 131. Springer, Heidelberg (2000)
11. Blaze, M., Diffie, W., Rivest, R.L., Schneier, B., Shimomura, T., Thompson, Wiener, M.: Minimal key lengths for symmetric ciphers to provide adequate commercial security, January 1996. http://www.schneier.com/paper-keylength.pdf
12. Bos, J.W., Costello, C., Hisil, H., Lauter, K.: Fast cryptography in genus 2. In: Johansson, T., Nguyen, P.Q. (eds.) EUROCRYPT 2013. LNCS, vol. 7881, pp. 194–210. Springer, Heidelberg (2013)
13. Bos, J.W., Costello, C., Longa, P., Naehrig, M.: Selecting elliptic curves for cryptography: an efficiency and security analysis. Cryptology ePrint Archive, Report 2014/130 (2014). http://eprint.iacr.org/
14. Bos, J.W., Costello, C., Miele, A.: Elliptic and hyperelliptic curves: a practical security analysis. In: Krawczyk, H. (ed.) PKC 2014. LNCS, vol. 8383, pp. 203–220. Springer, Heidelberg (2014)
15. Brier, E., Joye, M.: Weierstraß elliptic curves and side-channel attacks. In: Naccache, D., Paillier, P. (eds.) PKC 2002. LNCS, vol. 2274, p. 335. Springer, Heidelberg (2002)
16. Certicom Research. Standards for efficient cryptography 2: Recommended elliptic curve domain parameters. Standard SEC2, Certicom (2000)

17. Certicom Research. Standards for efficient cryptography 1: Elliptic curve cryptography (version 2.0). Standard SEC1, Certicom (2009)
18. Cheon, J.H.: Security analysis of the strong diffie-hellman problem. In: Vaudenay, S. (ed.) EUROCRYPT 2006. LNCS, vol. 4004, pp. 1–11. Springer, Heidelberg (2006)
19. Coppersmith, D.: Modifications to the number field sieve. J. Cryptology **6**(3), 169–180 (1993)
20. Crandall, R., Pomerance, C.: Prime Numbers: A Computational Perspective. Lecture Notes in Statistics, 2nd edn. Springer, New York (2005)
21. Duursma, I.M., Gaudry, P., Morain, F.: Speeding up the discrete log computation on curves with automorphisms. In: Lam, K.-Y., Okamoto, E., Xing, C. (eds.) ASIACRYPT 1999. LNCS, vol. 1716, pp. 103–121. Springer, Heidelberg (1999)
22. Fan, J., Verbauwhede, I.: An updated survey on secure ECC implementations: attacks, countermeasures and cost. In: Naccache, D. (ed.) Cryphtography and Security: From Theory to Applications. LNCS, vol. 6805, pp. 265–282. Springer, Heidelberg (2012)
23. Fouque, P.-A., Joux, A., Tibouchi, M.: Injective encodings to elliptic curves. In: Boyd, C., Simpson, L. (eds.) ACISP. LNCS, vol. 7959, pp. 203–218. Springer, Heidelberg (2013)
24. Free Software Foundation, Inc. GMP: The GNU Multiple Precision Arithmetic Library (2014). http://www.gmplib.org/
25. Gallant, R.P., Lambert, R.J., Vanstone, S.A.: Improving the parallelized Pollard lambda search on anomalous binary curves. Math. Comput. **69**(232), 1699–1705 (2000)
26. Google. Android NDK. https://developer.android.com/tools/sdk/ndk/index.html
27. Google. Android SDK guide. http://developer.android.com/guide/index.html
28. Hales, T.C.: The NSA back door to NIST. Not. AMS **61**(2), 190–192 (2013)
29. Hanrot, G., Morain, F.: Solvability by radicals from an algorithmic point of view. In: Proceedings of the 2001 International Symposium on Symbolic and Algebraic Computation, ISSAC 2001, pp. 175–182. ACM, New York (2001)
30. Ishii, N.: Trace of frobenius endomorphism of an elliptic curve with complex multiplication. Bull. Aust. Math. Soc. **70**, 125–142 (2004)
31. Izu, T., Takagi, T.: Exceptional procedure attack on elliptic curve cryptosystems. In: Desmedt, Y.G. (ed.) PKC 2003. LNCS, vol. 2567, pp. 224–239. Springer, Heidelberg (2002)
32. Koblitz, N.: Elliptic curve cryptosystems. Math. Comput. **48**(177), 203–209 (1987)
33. Koblitz, N.: Hyperelliptic cryptosystems. J. Cryptology **1**(3), 139–150 (1989)
34. Lenstra, A.K.: Efficient identity based parameter selection for elliptic curve cryptosystems. In: Pieprzyk, J.P., Safavi-Naini, R., Seberry, J. (eds.) ACISP 1999. LNCS, vol. 1587, pp. 294–302. Springer, Heidelberg (1999)
35. Lenstra, A.K., Lenstra Jr., H.W. (eds.): The Development of the Number Field Sieve. Lecture Notes in Mathematics, vol. 1554. Springer, Heidelberg (1993)
36. Lenstra, A.K., Verheul, E.R.: Selecting cryptographic key sizes. J. Cryptology **14**(4), 255–293 (2001)
37. Lenstra, A.K., Wesolowski, B.: A random zoo: sloth, unicorn, and trx. Cryptology ePrint Archive, Report 2015/366 (2015). http://eprint.iacr.org/2015/366
38. Lenstra Jr., H.W.: Factoring integers with elliptic curves. Ann. Math. **126**(3), 649–673 (1987)
39. Lochter, M., Merkle, J.: Elliptic curve cryptography (ECC) brainpool standard curves and curve generation. RFC 5639 (2010)

40. Miller, V.S.: Use of elliptic curves in cryptography. In: Williams, H.C. (ed.) CRYPTO 1985. LNCS, vol. 218, pp. 417–426. Springer, Heidelberg (1986)
41. Montgomery, P.L.: Speeding the pollard and elliptic curve methods of factorization. Math. Comput. **48**(177), 243–264 (1987)
42. Morain, F.: Edwards curves and cm curves. Technical report (2009)
43. National Institute of Standards and Technology. Special publication 800–57: Recommendation for key management part 1: General (revised). http://csrc.nist.gov/publications/nistpubs/800-57/sp800-57-Part1-revised2_Mar08-2007.pdf
44. National Security Agency. Fact sheet NSA Suite B Cryptography (2009). http://www.nsa.gov/ia/programs/suiteb_cryptography/index.shtml
45. NIST. Workshop on Elliptic Curve Cryptography Standards 2015, June 2015. http://www.nist.gov/itl/csd/ct/ecc-workshop.cfm
46. Oracle. Java native interface
47. Rajwade, A.R.: Certain classical congruences via elliptic curves. J. Lond. Math. Soc. **2**(8), 60–62 (1974)
48. Rubin, K., Silverberg, A.: Choosing the correct elliptic curve in the cm method. Math. Comput. **79**(269), 545–561 (2010)
49. Satoh, T., Araki, K.: Fermat quotients and the polynomial time discrete log algorithm for anomalous elliptic curves. Commentarii Mathematici Univ. Sancti Pauli **47**(1), 81–92 (1998)
50. Schoof, R., Schoof, P.R.E.: Counting points on elliptic curves over finite fields (1995)
51. Semaev, I.A.: Evaluation of discrete logarithms in a group of p-torsion points of an elliptic curve in characteristic p. Math. Comput. **67**, 353–356 (1998)
52. Shumov, D., Ferguson, N.: On the Possibility of a Back Door in the NIST SP800-90 Dual EC PRNG (2007)
53. Silverman, J.H.: The Arithmetic of Elliptic Curves. Gradute Texts in Mathematics, vol. 106. Springer, New York (1986)
54. Smart, N.P.: The discrete logarithm problem on elliptic curves of trace one. J. Cryptology **12**(3), 193–196 (1999)
55. Snow, B.: Private communication, June 2014
56. Stark, H.: Counting points on *cm* elliptic curves. Rocky Mt. J. Math. **26**(3), 1115–1138 (1996)
57. Tibouchi, M.: Elligator squared: Uniform points on elliptic curves of prime order as uniform random strings. IACR Cryptology ePrint Archive 2014:43 (2014)
58. U.S. Department of Commerce/National Institute of Standards and Technology. Digital Signature Standard (DSS). FIPS-186-4 (2013). http://nvlpubs.nist.gov/nistpubs/FIPS/NIST.FIPS.186-4.pdf
59. van Wamelen, P.B.: Examples of genus two cm curves defined over the rationals. Math. Comput. **68**(225), 307–320 (1999)
60. Weng, A.: Constructing hyperelliptic curves of genus 2 suitable for cryptography. Math. Comput. **72**(241), 435–458 (2003)
61. Wiener, M.J., Zuccherato, R.J.: Faster attacks on elliptic curve cryptosystems. In: Tavares, S., Meijer, H. (eds.) SAC 1998. LNCS, vol. 1556, pp. 190–200. Springer, Heidelberg (1999)

Distributed Parameter Generation for Bilinear Diffie Hellman Exponentiation and Applications

Aggelos Kiayias[1], Ozgur Oksuz[2(✉)], and Qiang Tang[2]

[1] National and Kapodistrian University of Athens, Athens, Greece
aggelos@cse.uconn.edu
[2] University of Connecticut, Mansfield, USA
ozgur.oksuz@engr.uconn.edu, qiang@cse.uconn.edu

Abstract. Distributed parameter and key generation plays a fundamental role in cryptographic applications and is motivated by the need to relax the trust assumption on a single authority that is responsible for producing the necessary keys for cryptographic algorithms to operate. There are many well-studied distributed key generation protocols for the discrete logarithm problem. In this paper, building upon previous distributed key generation protocols for discrete logarithms, we provide two new building blocks that one can use them in a sequential fashion to derive distributed parameter generation protocols for a class of problems in the bilinear groups setting, most notably the n-Bilinear Diffie Hellman Exponentiation problem. Based on this we present new applications in distributed multi-party oriented cryptographic schemes including decentralized broadcast encryption, revocation systems and identity based encryption.

1 Introduction

The n-Bilinear Diffie Hellman Exponentiation (n-BDHE) problem introduced in [7] has found many applications in the design of multi-party encryption schemes, notably broadcast encryption (BE) schemes [8]. In these schemes there is an authority or trustee that produces a suitable public key that may be used subsequently by the participants to communicate securely. Specifically, a sender can use the public parameters to transmit to any subset of the group members a private message. The trustee is supposed to issue private key information to the members of the group. Naturally, the trustee is a single point of failure and it is highly desirable to develop a mechanism to distribute its operation.

The n-BDHE assumption allows substantial efficiency gains (e.g., it enabled the first constant size ciphertext BE to be developed [8]) and for this reason it is highly structured. An n-BDHE instance contains a series of powers g^{α^i}, with all values in the set $i \in \{1, \ldots, n, n+2, \ldots, 2n\}$ public. The n-BDHE assumption postulates that the value $g^{\alpha^{n+1}}$ is indistinguishable from a random group element for any probabilistic polynomial time observer (the assumption was shown to be true in the generic bilinear group model, [7]). The specialized structure of the n-BDHE public parameter complicates its distributed generation: specifically,

J. Lopez and C.J. Mitchell (Eds.): ISC 2015, LNCS 9290, pp. 548–567, 2015.
DOI: 10.1007/978-3-319-23318-5_30

any set of entities possessing the shares of α should be able to produce all intermediate powers g^{α^i} for $i \neq n+1$, while at the same time no information about $g^{\alpha^{n+1}}$ should be leaked. No existing discrete-logarithm based distributed key generation protocols can handle this task.

Our Contribution. Motivated by the above we build upon previous distributed key generation protocols for discrete logarithms and we provide two new building blocks that one can use them in a sequential fashion to derive a distributed parameter generation protocol for the n-BDHE problem (as well as other related assumptions).

Using our distributed parameter generation protocol for n-BDHE we show how to apply our protocols to [8] and we develop a decentralized broadcast encryption (DBE) with constant size ciphertext and private key. Furthermore, we comment how one can easily apply our protocols to enable other primitives to be distributed such as identity-based encryption (IBE) schemes [6,32] and revocation systems [27].

Related Work on Distributed Key Generation. A number of previous works focused on distributed generation of cryptographic keys and parameters by a set of parties so that the parties equally share the trust of key generation. This line of work starts with the seminal works of Shamir [31] and Blakley [5], followed by the works of Benaloh and Yung, [4], Desmedt, [15] and a great number of others, e.g., [19]. Desmedt et al. [16] proposed a black box threshold secret sharing scheme that secret is chosen from any Abelian group. To share a single element in the black box group among n players, each player receives n elements from the group. The result of [16] is exploited in [14] to obtain an efficient and secure solution for sharing any function. Later, Cramer et al. [12] improved the result of [16] by decreasing the expansion factor from n to $O(\log n)$. The work of [12] was also improved by Cramer et al. [13] where they provide a computationally efficient scheme. Notably, for the discrete-logarithm problem, Gennaro et al. [20] present a distributed key generation protocol for discrete log based cryptosystems that parties jointly generate a system public/secret key pair. Canny et al. [9] provided a protocol with asymptotically better efficiency. Their protocol reduces communication and computation complexity from $O(n \log n)$ [20] to $O(\log^3 n)$ assuming $t = \Theta(\log n)$ is the threshold parameter and n the total number of parties running the protocol. Kate et al. [25] introduced a private distributed key generation protocol for IBE schemes in the multi-server setting.

These well-known distributed key generation protocols do not solve our main problem to generate distributively an instance for the n-BDHE problem: specifically, using the above techniques, parties that are given shares of α can easily produce $g^{\alpha^{i+1}}$ given g^{α^i}, however it is not straightforward how to generate $g^{\alpha^{n+2}}$ given g^{α^n} without leaking any information about $g^{\alpha^{n+1}}$. A possible strategy is to invoke a protocol that enables the squaring of a shared secret (or more generally the multiplication of two shared secret values). Unfortunately, a straightforward multiplication of the shares would not allow the simulation of the protocol to go through and the polynomial degree doubles (when standard polynomial secret sharing is used). A number of previous works consider a similar problem in other settings. Notably using the techniques of Ben-Or et al. [3], or those of [10,22] it is possible to achieve the stated goal, however the resulting generic protocols are quite inefficient for multiplication in the exponent especially for the n-BDHE parameters that require elliptic curve operations. In a more targeted work, Gennaro et al. [21] propose a way for an efficient share multiplication in the exponent but the security of the construction is based on Decisional Diffie-Hellman (DDH) assumption. Since we are in the bilinear group setting (where DDH does not hold), their protocol is not useful. Abe [1] presents a robust protocol that does not use zero-knowledge proofs and enables multiplication of shared secrets very efficiently. We are inspired by this idea and combine [1] with *Pedersen* [29] and *Feldman* [17] verifiable secret sharing protocols to obtain a squaring protocol that yields the solution.

Related Work on Decentralized Broadcast Encryption (DBE) Schemes. BE was first introduced by Fiat et al. [18] to enable a distributor to privately transmit information to a set of parties. It had been the major open problem of BE since then to achieve a scheme with constant size ciphertext, until Boneh et al. exploited the structure of n-BDHE and constructed the first constant size ciphertext BE [8]. Phan et al. [30] recently proposed a DBE, in which there is no need for a trusted party to generate public and secret keys for parties, and instead parties themselves can jointly create key pairs. They provide two schemes that use complete subset (CS) and subset difference (SD) thus the ciphertext size depends on the number of revoked parties. They achieve ciphertext size $O(r \log n/r)$ for CS and $O(2r-1)$ for the SD scheme variant where r is number of revoked parties and n is the total number of parties; in contrast the decentralized BE that we construct with our protocols has only constant size ciphertexts.

There are some other works that use the asymmetric group key agreement primitive to produce DBE schemes. Wu et al. [33], introduced such a scheme that has $O(n)$ size public/secret key per member and does not allow a sender to do revocation thus it is not suitable, as it is, for a DBE. The work of [36] extends the setting of [33] so that interaction is permitted and members may be excluded at will. However, this protocol requires $O(n^2)$ public and $O(n)$ secret key size per member. They introduce a tradeoff between key and ciphertext size and the resulting BE has $O(n^{2/3})$ public key and ciphertext size and $O(n^{1/3})$ private key size. Wu et al. [35] suggest to use multiple dealers in a BE scheme. The dealers interactively generate a common system public key and share the master secret key. Each party needs to interact with at least t dealers in order to obtain its secret decryption key. Every party needs $O(n)$ elements from each dealer and the total size of private-key is $O(tn)$ (where t is the threshold parameter) for the scheme of [35]. The public information is of length $O(n)$ and the scheme offers constant size ciphertexts. They also provide a tradeoff between key and ciphertext size and the resulting public key, private key and ciphertext size are $O(n^{1/2})$. A more closely related notion is that of ad hoc broadcast encryption (AHBE) which was proposed in [34] with a construction of $O(n^2)$ public key size, $O(n)$ secret key size per party and constant size ciphertext but without a formal security proof. They also provide a way to achieve a tradeoff between ciphertexts and public keys; the resulting AHBE has sub-linear complexity $O(n^{2/3})$ for both public keys and ciphertexts and $O(n^{1/3})$ for private keys.

None of the above schemes have the same efficiency profile in terms of size of ciphertext, secret key and public key as [8]. Recall that [8] is the first construction for (plain) BE that has constant size ciphertext and secret key and linear size (in the total number of parties in the system) public key. With our distributed parameter generation protocol for the n-BDHE problem, we can construct a decentralized BE on top of [8] that inherits all the good properties of [8].

2 Definitions

In this section, we introduce definitions of the underlying blocks. Specifically, we will define several correlated sub-protocols which will later be composed in a sequential fashion to distributively generate n-BDHE parameters. For the ease of presentation, we use the terminology of a suite of distributed protocols introduced in [26]. A suite of n-party protocols $Suite = (PROT_0, PROT_1, ..., PROT_s)$ is a set of protocols that can be executed sequentially and use some joint states. In particular after the initialization protocols ($PROT_0$ and $PROT_1$) are performed, subsequent protocols executions can be sequentially composed in an arbitrary way. Each party takes the current state as input for each protocol, in that way each party can easily keep track the protocols. An execution of n-party protocol $Suite$ denoted by $\Upsilon_{\mathcal{A}}^{Suite}(\kappa, n)$ proceeds as follows: (1) First the adversary $\mathcal{A}$ selects a set of at most t parties to corrupt subject to the constraint $t < n/3$; (2) The initialization protocols $prot_0$ and $prot_1$(execution of the program of the protocols $PROT_0$ and $PROT_1$ with a single trusted party) is executed with adversary participation on behalf of the corrupted parties; note that this protocol requires no private inputs for any of the parties and its public input is security parameter κ and number of parties n; the private outputs of honest parties are maintained in a local state of the execution that is inaccessible to the adversary; (3) The adversary may provide a public input and ask the honest parties to execute together with the corrupted parties under adversary's control any of the protocols in the $Suite$. This execution can be repeated sequentially as many times as the adversary commands.

Definition 1. *[26] A suite of $n-$party protocols Suite $= PROT_0, \ldots, PROT_s$ is called t-distribution safe if for all adversaries $\mathcal{A}$ corrupting no more than t parties and $t < n/3$, it holds that there exists an expected PPT simulator $\mathcal{S}$, $|\Pr[\Upsilon_{\mathcal{A}}^{Suite}(1^\kappa, 1^n) = 1] - \Pr[\mathcal{S}^{proto_0, prot_1, \ldots, prot_s}(1^\kappa) = 1]| = negl(\kappa)$, where $(proto_0, prot_1, ..., prot_s)$ expresses the execution of the functionality of ($PROT_0$, $PROT_1, ..., PROT_s$) by a trusted party.*

The intuition behind this definition is that the adversary's knowledge gain (view) can be simulated by a sequential execution of the same set of protocols with a single trusted party (represented by the set of oracles available to the simulator $\mathcal{S}$). Given that anything the adversary can compute in the corrupted setting with controlling t parties, it can also compute while interacting with a single trusted party (This follows from the standard simulation based security of the protocols, e.g., [9,20]). We can conclude that the protocol *Suite* is t-distribution safe or t-secure. The adversary's view consists of his local coins, inputs and received messages from other honest parties in those protocols. As a note that we will use this definition for all of the sub-protocols that we introduce.

Next, we present definitions of several sub-protocols which will be used in a suite. We will apply them to generate the parameters of the n-BDHE problem, and they can later be used in other applications as well. Each party i in a protocol suite $(PROT_0, PROT_1, ..., PROT_s)$ keeps a local state st_i^k for each $PROT_k$ sub-protocol which consists some of the communication transcript in $PROT_k$. We also use I_i^k, O_i^k to denote for the party i's inputs and outputs of $PROT_k$ respectively. As a note that since each honest party's output is the same as the protocol's output, we can write $O^k = O_i^k$ for $PROT_k$ for each honest party i.

Definition 2. *[Parameter Generation (ParGen)] This is an n-party protocol and $ParGen = PROT_0$. It inputs the security parameter κ and number of parties n, outputs the public parameter set $PP = (\kappa, n, t, p, q, g, h, e(,))$, where p, q are two large prime numbers, q divides $p-1$, g and h are two group elements in G, t is the threshold parameter and $e(,)$ is description of a bilinear group.*

Definition 3. *$\mathcal{DKG}$ is a t-secure n-party protocol that is run in $PROT_1$ as well as any protocol in suite. Suppose it is run in $PROT_k$, where $1 \leq k \leq s$. Every party $i \in \{1, .., n\}$ inputs $I_i^k = (PP)$ (it is the public parameter set that output by ParGen), outputs g^x for a random $x \in Z_q$. The local states of $\mathcal{DKG}$ is $st_i^k = [(x_i, x_i'), \mathcal{V}_k, \{\tau_{ji}^k\}_{j \in \mathcal{V}_k}]$, where $\mathcal{V}_k$ is the valid party set, τ_{ji}^k are the messages received from party $j \in \mathcal{V}_k$ in $PROT_k$, (x_i, x_i') values are computed as a function of $\{\tau_{ji}^k\}_{j \in \mathcal{V}_k}$ by the party i and they are called party i's shared of secret x and random x'. A t-secure $\mathcal{DKG}$ sub-protocol satisfies the following properties as follows [20]: (1) All subsets of $t+1$ shares $\{x_j\}_{j \in \mathcal{V}_k}$ provided by honest parties define the same unique secret key x, even if up to t shares are submitted by faulty parties; (2) All honest parties have the same value of public key g^x; (3) x is uniformly distributed in Z_q.*

Definition 4. *$\mathcal{REC}$ is an n party protocol that has to run after a $\mathcal{DKG}$ sub-protocol. Suppose $\mathcal{REC}$ is for $PROT_l$. We assume the parties agree on one $\mathcal{DKG}$ instance (say $PROT_k$ with $O_k = g^x$) if multiple $\mathcal{DKG}$ sub-protocols were run before $PROT_l$. Each party $i \in \mathcal{V}_{l-1}$ takes $I_i^l = \left(st_i^{l-1}, st_i^k, PP, y\right)$ as inputs, where $y \in G$ is any public value. It outputs $O_i^j = y^x$, if $|\mathcal{V}_l| \geq t+1$. Party i's local state for $\mathcal{REC}$ sub-protocol is $st_i^l = \left(\mathcal{V}_l, \{\tau_{ji}^l\}_{j \in \mathcal{V}_l}\right)$.*

Definition 5. *$\mathcal{RECSQ}$ is an n party protocol that has to run after a $\mathcal{DKG}$ sub-protocol. Suppose $\mathcal{RECSQ}$ is for $PROT_l$. We assume the parties agree on one $\mathcal{DKG}$ instance (say $PROT_k$ with $O_k = g^x$) if multiple $\mathcal{DKG}$ sub-protocols were run before $PROT_l$. Each party $i \in (\mathcal{V}_{l-1} \subseteq \mathcal{V}_k)$ takes $I_i^l = \left(st_i^{l-1}, st_i^k, PP, y\right)$ as inputs, where $y \in G$ is any public value. It outputs $O_i^l = y^{x^2}$, if $|\mathcal{V}_l| \geq 2t+1$. Party i's local state for $\mathcal{RECSQ}$ is $st_i^l = \left(\mathcal{V}_l, \{\tau_{ji}^l\}_{j \in \mathcal{V}_l}\right)$.*

Definition 6. *A t-secure decentralized broadcast encryption is a protocol that n parties jointly generate the public keys and secret keys of the parties and one can send encrypted message to any subset of these parties. In a static corruption model, a t-secure decentralized broadcast encryption scheme is* static *chosen plaintext attack (*static-*CPA) secure if in the following game between a PPT adversary $\mathcal{A}$ and a challenger $\mathcal{C}$ holds: (1) $\mathcal{A}$ submits corrupted party set S with $|S| \leq t$ and $\mathcal{C}$ returns $\mathcal{A}$ all private inputs of the parties in S; (2) $\mathcal{A}$ submits the target set $\mathcal{T} \subseteq \{1,..,n\}$ (static adversary); (3) $\mathcal{C}$ interacts with $\mathcal{A}$ to jointly generate the public keys and secret keys of the system. In addition, $\mathcal{A}$ receives all secret keys of parties that are not in $\mathcal{T}$; (4) $\mathcal{A}$ submits two messages M_0, M_1 and receives a challenge chooses a bit b and encrypts M_b to $\mathcal{T}$; (5) $\mathcal{A}$ receives the challenge ciphertext and outputs a bit b'.*

For any PPT $\mathcal{A}$, $|\Pr[b' = b] - \frac{1}{2}| \leq \epsilon$ where ϵ is a negligible function.

3 Building Blocks

In this section, we will describe three building blocks which are t-secure sub-protocols, named as $\mathcal{DKG}, \mathcal{REC}, \mathcal{RECSQ}$ respectively. $\mathcal{DKG}$ is the well-studied distributed key generation for discrete log problem and $\mathcal{REC}$ is a simple extension to raise the same exponent to a different base, while $\mathcal{RECSQ}$ is a new building block that allows the parties to jointly square at the exponent and leaks nothing more. They can be used together to distributively generate n-BDHE problem parameters [8]; in turn, we can have distributed version of cryptographic primitives constructed using those algebraic structures, and we will present some examples of applications using our t-secure sub-protocols in Sect. 4. Note that those applications are by no means exhaustive, one can freely combine our building blocks sequentially to derive new distributed parameter generation protocols for other problem instances.

3.1 Sub-protocols

$\mathcal{DKG}$ **Sub-protocol.** This is the well-known distributed key generation protocol in which n parties jointly generate g^x, for public parameters PP, and the protocol satisfies (i) x is uniform; (ii) more than t parties together can recover x via polynomial interpolation. The main difficulty in the $\mathcal{DKG}$ is to guarantee that the exponent in the output is uniformly distributed, and how to build a simulator to show the security via simulation. Each party executes one Pedersen verifiable secret sharing (PVSS) [29] and one Feldman verifiable secret sharing (FVSS) [17] of their individual secrets so that guarantees uniform output distribution and $t+1$ receivers could jointly recover all the secrets, thus simulator could recover the secret inputs of the corrupted parties, then simulate a transcription identically distributed as in the real protocol. As a note that we assume that the parties already agree with the discrete log parameters (g, p, q) and $h \in G$ before the $PROT_1 = \mathcal{DKG}$. This parameters can be distributively generated the same way as in [26] or [24]. We can simply use the previous results on this, e.g. [9,20], so we do not present full construction of it in this paper and omit the details.

$\mathcal{REC}$ Sub-protocol. This sub-protocol is carried out after a $\mathcal{DKG}$ as each party will take states from the $\mathcal{DKG}$ as inputs. Since parties keep local states about the protocols' computation history, they can agree on which $\mathcal{DKG}$ instance to use if there were multiple such sub-protocols, and then start $\mathcal{REC}$. Suppose $\mathcal{REC}$ is the $PROT_l$ and it will be run under $\mathcal{DKG} = PROT_k$ for some $k < l$. Each valid party $i \in \mathcal{V}_{l-1}$ inputs its states st_i^{l-1} from the $PROT_{l-1}$, st_i^k from the $PROT_k$ (outputs $O_k = g^x$), public parameter set PP, and y, then jointly generates y^x. st_i^{l-1} is needed because the party i needs to know the valid party set $\mathcal{V}_{l-1} \in st_i^{l-1}$ in order to avoid interacting with malicious party j that $j \in \mathcal{V}_k$ but $j \notin \mathcal{V}_{l-1}$. Observe that in the $\mathcal{DKG}$, at least $t+1$ honest parties together can recover x via polynomial interpolation. Suppose $x_i \in st_i^k$ is the piece of information party i can contribute for the interpolation, and $x = \sum_{i=1}^{t+1} \lambda_i x_i$, where λ_i is the Lagrange interpolation coefficient. We would use this fact to build a simple $\mathcal{REC}$ to put x into the exponent of y. The details are presented in Fig. 1.

Each party $i \in \mathcal{V}_{l-1}$ on inputs $\left(\mathcal{V}_{l-1}, (g^{x_j})_{j\in\mathcal{V}_k}, PP, y, x_i\right)$, where valid set $\mathcal{V}_{l-1} \in st_i^{l-1}$; x_i and $(g^{x_j})_{j\in\mathcal{V}_k}$ values are computed from the transcript $\{\tau_{ji}^k\}_{j\in\mathcal{V}_k} \in (st_i^k)$; and $y \in G$. Then the party i does the followings: (1) computes y^{x_i} and broadcasts it; (2) checks if $j \in (\mathcal{V}_{l-1} \subseteq \mathcal{V}_k)$ and $e(g, y^{x_j}) = e(g^{x_j}, y)$. Then, (3) computes $y^x = \prod_{j\in\mathcal{V}_z l}(y^{x_j})^{\lambda_j}$ and outputs $O_i^l = y^x$ and stores $st_i^l = \left(\mathcal{V}_l, \{\tau_{ji}^l\}_{j\in\mathcal{V}_l}\right)$ to the local state, where $\lvert\mathcal{V}_l\rvert \geq t+1$.

Fig. 1. $\mathcal{REC}$ sub-protocol

$\mathcal{RECSQ}$ Sub-protocol. This sub-protocol is also run after a $\mathcal{DKG}$, with the transcript from the $\mathcal{DKG}$ generating g^x, this sub-protocol allows $2t+1$ parties jointly generate y^{x^2} for another value $y \in G$ without leaking extra information. One may think that we can run $\mathcal{REC}$ twice and the second one use y^x as the base value, however, this trivially leaks the value of y^x. Alternatively, if x is the shared constant term of two degree-t polynomials $f(\cdot)$ and $g(\cdot)$ (they could be the same polynomials), then x^2 will be the constant term of a $2t$-degree polynomial $f \cdot g$. Suppose the shares for the party i for x from f, g are x_i, y_i respectively, then one can recover x^2 from $2t+1$ shares of $x_i \cdot y_i$ from interpolation. If we implement this idea straightforwardly as in [21], the security analysis could not go through. With t adversaries' secret shares $\left(g^{x_i^2}\right)_{i=1,..,t}$ recovered, it is not clear how the simulator can produce consistent share $g^{x_j^2}$ for each honest party j so that any $2t+1$ shares will interpolate to the same target value g^{x^2}. If we use a similar simulation procedure from [21], that the simulator chooses honest parties shares at random (say, the $t+1,..,2t$ parties) (e.g. for party $j \in \mathcal{G}$, it chooses random $g^{r_j^2}$ from G) then it recovers other $2t+1,..,3t$ honest parties share one by one by simply doing a reverse interpolation. However, the adversary can easily check the consistency of these values (the ones that are randomly chosen by the simulator) by computing $e\left(g^{x_j}, g^{x_j}\right) = e\left(g^{r_j^2}, g\right)$ since g^{x_j} value is public. Abe proposed a secure multiparty multiplication protocol in [1] that uses couple of PVSS [29] on the product of their shares with degree reduction technique that allow $t+1$ of parties to compute xy. We leverage the idea of Abe for secure multiplication [1] together with one PVSS and FVSS protocols to compute xy (in our case x^2, where $x = y$) in the exponent. Using one more PVSS and FVSS guarantees that the parties share their secrets correctly. This method is also used in [20]. In this case, the simulator perfectly simulates the transcript with the degree reduction technique. Since the simulator recovers t corrupted parties secret shares during the PVSS, each honest party's value (in the exponent) will be recovered based on the target value. As any $t+1$ of them interpolate to the target value on the exponent, no consistency problem will appear in the simulation. Similar to the $\mathcal{REC}$, since each valid party keeps every sub-protocol's states that he participated in, every valid party can jointly run $\mathcal{RECSQ}$ with other valid parties based on a $\mathcal{DKG}$ instance (if there are multiple instances, they can agree one of them before running $\mathcal{RECSQ}$). If $\mathcal{RECSQ}$ is the $PROT_l$ and is run under $\mathcal{DKG} = PROT_k$. Each party $j \in \mathcal{V}_{l-1}$ inputs states st_j^{l-1} from last $PROT_{l-1}$, st_j^k from $PROT_k$, PP and a value y, then jointly generates y^{x^2} without obtaining y^x. The details are given in Fig. 2.

3.2 Security Analysis

In this section, we will analyze the security of the sub-protocols. In particular, we will show when these sub-protocols are arbitrarily and sequentially executed (composed) in a suite as in Definition 1. Then we sequentially compose them to derive the security for the distributed key generation protocol for the n-BDHE problem.

Since $\mathcal{REC}$ and $\mathcal{RECSQ}$ sub-protocols are both based on $\mathcal{DKG}$ sub-protocol, we assume $\mathcal{DKG}$ is only run once at the beginning, (the case that has multiple independent running of $\mathcal{DKG}$ can be easily handled as the simulation can be done by the same way for each). For any sub-protocol $PROT_i$ if it is $\mathcal{REC}$, the simulation of the joint view with $\mathcal{DKG}$ can be proceeded by running the simulator of $\mathcal{S}^{\mathcal{DKG}}$ (simulator for $\mathcal{DKG}$ sub-protocol) first, and use the transcript of $\mathcal{S}^{\mathcal{DKG}}$ to simulate corresponding $\mathcal{REC}$ sub-protocol. Note that $\mathcal{S}^{\mathcal{DKG}}$ is able to recover all the t corrupted parties' private inputs (x_i, $i = 1, .., t$), and thus to guarantee consistency, the $\mathcal{REC}$ simulator can simply do a reverse interpolation to recover the private inputs of honest parties (x_j, $j = t+1, .., 3t$) one by one by hitting the desired simulator's input which is output of $\mathcal{REC}$ sub-protocol (as the exponent can be derived using $t+1$ secret inputs through interpolation).

If $PROT_i$ is $\mathcal{RECSQ}$, the simulation is a little bit more involved. The crucial part of the simulation is to produce honest parties' share of squared secret in the exponent. The difficulty of this was explained in Sect. 3.1 using the idea [21], due to having bilinear map structure. In order to provide consistent shares, the $\mathcal{S}^{\mathcal{RECSQ}}$ (simulator for $\mathcal{RECSQ}$) controls the honest parties and chooses random messages for the malicious parties in behalf of honest parties by simulating

DM-1 [1]: Each party $i \in \mathcal{V}_{l-1}$ on input $\left(\mathcal{V}_{l-1}, \left(g^{x_j} h^{x'_j}\right)_{j \in \mathcal{V}_k}, PP, y, x_i, x'_i\right)$, where $\mathcal{V}_{l-1} \in st_i^{l-1}$; (x_i, x'_i) and $\left(g^{x_j} h^{x'_j}\right)_{j \in \mathcal{V}_k}$ values are computed from $\{\tau_{ji}^k\} \in st_i^k$; and $y \in G$. Then party $i \in \mathcal{V}_{l-1}$ does the followings;

- randomly picks t-degree polynomials f_{i1}, d_{i1}, d_{i2} from $Z_q[X]$ so that $f_{i1}(0) = x_i, d_{i1}(0) = x'_i$. Let r_i randomly chosen free term of $d_{i2}(.)$. Party i shares x_i twice as
  - $PVSS(x_i, x'_i)(g, h) \xrightarrow{f_{i1}, d_{i1}} (x_{ij}, x'_{ij})(\langle X_i \rangle, X_{i1}, ..., X_{it})$
  - $PVSS(x_i, r_i)\left(g^{x_i} h^{x'_i}, h\right) \xrightarrow{f_{i1}, d_{i2}} (\langle x_{ij} \rangle, r_{ij})(Y_{i0}, Y_{i1}, ..., Y_{it})$.
- Party $i \in \mathcal{V}_{l-1}$ then selects two random polynomials f_{i2}, d_{i3} that satisfy $f_{i2}(0) = x_i^2, d_{i3}(0) = x_i x'_i + r_i \bmod q$ and performs: $PVSS\left(x_i^2, x_i x'_i + r_i\right)(g, h) \xrightarrow{f_{i2}, d_{i3}} \left(c_{ij}, c'_{ij}\right)(\langle Y_{i0} \rangle, Z_{i1}, ..., Z_{it})$.

DM-2 [1]: Party $j \in \mathcal{V}_{l-1}$ verifies messages received from party i as follows:

- checks if $i \in (\mathcal{V}_{l-1} \subseteq \mathcal{V}_k)$. It the check fails, the messages from party i are discarded. Otherwise party j checks if
  $g^{x_{ij}} h^{x'_{ij}} = X_i \prod_{k=1}^t X_{ik}^{j^k}$; $\left(g^{x_i} h^{x'_i}\right)^{x_{ij}} h^{r_{ij}} = \prod_{k=0}^t Y_{ik}^{j^k}$; $g^{c_{ij}} h^{c'_{ij}} = Y_{i0} \prod_{k=1}^t Z_{ik}^{j^k}$.

If a check fails, party j declares so and goes to the disqualification protocol described in figure 3.

DM-3 [1]: Let $\mathcal{I} \subseteq \mathcal{V}_{l-1}$ and $|\mathcal{I}| \geq 2t+1$. Each party $j \in \mathcal{I}$ then computes $c_j := \sum_{i \in \mathcal{I}} \lambda_i c_{ij}$; $c'_j := \sum_{i \in \mathcal{I}} \lambda_i c'_{ij}$; $Z_k := \prod_{i \in \mathcal{I}} Z_{ik}^{\lambda_i}$ for $k = 0, .., t$, where $Z_{i0} = Y_{i0}$.

DM-4 [New]: Each party $i \in (\mathcal{I} \subseteq \mathcal{V}_{j-1})$ broadcasts y^{c_i} and performs one $PVSS$ and one $FVSS$ below:

- $PVSS(c_i, c'_i)(g, h) \xrightarrow{f_{i3}, d_{i4}} (e_{ij}, e'_{ij})(\langle D_{i0} \rangle, D_{i1}, ..., D_{it})$, where $D_{i0} = g^{c_i} h^{c'_i} = \prod_{k=0}^t Z_k^{i^k}$ that can be computed from **DM-3**.
- $FVSS(c_i)(y) \xrightarrow{f_{i3}} (\langle e_{ij} \rangle)(E_{i0}, E_{i1}, ..., E_{it})$.

DM-5 [New]: Each party $j \in \mathcal{I} \subseteq \mathcal{V}_{j-1}$ verifies everything received from party i in **DM-4** as follows:

- $g^{e_{ij}} h^{e'_{ij}} = D_{i0} \prod_{k=1}^t D_{ik}^{j^k}$; $y^{e_{ij}} = \prod_{k=0}^t E_{ik}^{j^k}$
  If the checks do not fail, the party j saves local state $st_j^l = \left(\mathcal{V}_l, \{\tau_{ij}^l\}_{i \in \mathcal{V}_l}\right)$, where $\mathcal{I} = \mathcal{V}_l$. If a check fails, party j declares so and goes to the disqualification protocol described in figure 3.

It there are $t+1$ party set $\mathcal{Q} \subseteq \mathcal{I}$ satisfies **DM-5** above, the party j computes, $y^{x^2} = \prod_{i \in \mathcal{Q}} (y^{c_i})^{\lambda_i}$.

Fig. 2. $\mathcal{RECSQ}$ sub-protocol: Note that in **DM-1** $g^{x_i} h^{x'_i}$ is used as the base of the commitments in the second sharing and angle bracket means that the value can be publicly computed by the parties or it has been sent before.

couple PVSS schemes using degree reduction technique. Using this method, the simulator recovers the malicious parties' shares (t of them). Because of the degree reduction technique, $t+1$ honest parties are sufficient for recovering the squared secret (x^2). To recover honest parties' shares in the exponent, $\mathcal{RECSQ}$ simulates one FVSS and one PVSS. This step provides consistency between shares

DQ-1: Party $i \in \mathcal{V}_{l-1} \subseteq \mathcal{V}_k$ is requested to broadcast all the data that he privately sent to party j, which is $(x_{ij}, x'_{ij}, r_{ij}, c_{ij}, c'_{ij}, e_{ij}, e'_{ij})$. (If party i keeps silent, he is disqualified immediately.)
DQ-2: If *t+1* or more parties decide that those shares are faulty, party i is disqualied.

Fig. 3. Protocol *disqualification*

that [21] does not provide in Sect. 3.1. The simulation process starts with simulating the $\mathcal{S}^{\mathcal{DKG}}$, then the simulator simulates the corresponding $\mathcal{RECSQ}$ sub-protocol using the transcript from $\mathcal{DKG}$ and gets the malicious parties (t of them) shares of x^2, where x is the random secret value. After getting the malicious parties' shares of x^2, it basically does reverse interpolation to recover each honest party's share of x^2 in the exponent by hitting the output of the $\mathcal{RECSQ}$ sub-protocol which is y^{x^2} since $t+1$ shares are enough to recover y^{x^2} through interpolation. However, the task of recovering the malicious party inputs requires at least $2t+1$ honest parties.

The above simulation procedure can be done for any i, as any of $\mathcal{REC}, \mathcal{RECSQ}$ is independent to other instances (except $\mathcal{DKG}$), thus the joint view can be simulated with a concatenation of each simulated view. The joint simulated view is still indistinguishable from real as long as each $PROT_i$ is handled as above and the correlation with the same $\mathcal{DKG}$ is preserved. Now, we will present a theorem that after the $\mathcal{DKG}$ sub-protocol, all the sub-protocols above, (e.g., $\mathcal{REC}, \mathcal{RECSQ}$) can be arbitrarily composed in a sequential fashion.

Theorem 1. *An n-party protocol GSuite* $= (PROT_0, PROT_1, ..., PROT_s)$, *where* $PROT_0 = ParGen, PROT_1 = \mathcal{DKG}, PROT_k \in \{\mathcal{REC}, \mathcal{RECSQ}\}$ *for* $k = 2, .., s$ *is t-secure in the sense of Definition 1, when* $t < \frac{n}{3}$.

Proof. In order to prove the protocol $GSuite$ is t-secure, we need to show that for every adversary $\mathcal{A}$ there exists a simulator $\mathcal{S}$ such that $\{\Upsilon_{\mathcal{A}}^{GSuite=PROT_0,..,PROT_s}\} \approx \{\mathcal{S}^{proto_0,..,prot_s}\}$ computationally indistinguishable. We need to build such $\mathcal{S}$ from $\mathcal{S}^{\mathcal{DKG}}, \mathcal{S}^{\mathcal{REC}}$, and $\mathcal{S}^{\mathcal{RECSQ}}$ that each of them simulates adversary's view for the corresponding real protocol $PROT$ by using input and output of the $PROT$. W.l.o.g., we assume that the adversary $\mathcal{A}$ controls the set $\mathcal{B} = \{1, 2, ..., t\}$ and honest parties controlled by the simulator $\mathcal{S}$ are $\mathcal{G} = \{t+1, t+2, ..., n\}$. We need to show that for each protocol $PROT_i$, where $i = 1, .., s$ adversary's view $\left(view_j^i\right)_{j=1,..,t}$ needs to be simulated by $\mathcal{S}^{\mathcal{DKG}}$ (a simulator for $\mathcal{DKG}$ sub-protocol), $\mathcal{S}^{\mathcal{REC}}$, and $\mathcal{S}^{\mathcal{RECSQ}}$ sequentially. For $PROT_1 = \mathcal{DKG}$, a simulator $\mathcal{S}^{\mathcal{DKG}}$ takes public parameter set PP and desired output g^x which is output of the protocol $\mathcal{DKG}$ and simulates the adversary's view in $PROT_1$. The view of the $\mathcal{A}$ for $PROT_1$ is $view_{\mathcal{A}}^1 = \left(view_j^1\right)_{j=1,..,t}$ consists of $(\alpha_i(j), \alpha'_i(j), C_{ik}, A_{ik}) \in \tau_{ij}^1$, where $j \in \mathcal{B}, i \in \mathcal{G}, k = 0, .., t$. As explained in Sect. 3.1, the simulator would reconstruct the secret inputs $((\alpha_i, \alpha'_i)_{i=1,..,t})$ and corresponding secret outputs $((x_i, x'_i)_{i=1,..,t})$ where of the t corrupted parties from the shares of the PVSS received from them, and manipulate one of the simulated honest values based on the final output. The view of the $\mathcal{A}$ can be

$\mathcal{S}^{\mathcal{REC}}$ takes $\left((x_j)_{j\in\mathcal{J}}, PP, y, y^x\right)$ as inputs and does the followings:

– Computes y^{x_j}, for all $j \in \mathcal{J}$ and y^{x_i} for each $i \in \mathcal{G} \subseteq \mathcal{V}_1$ one by one as $y^{x_i} = \left(\frac{y^x}{\prod_{j\in\mathcal{Q}_i\setminus\{i\}}\left(y^{(x_j)}\right)^{\lambda_j}}\right)^{(\lambda_i)^{-1}}$, where λ_js are appropriate Lagrange coefficients and $Q_i = \mathcal{J} \cup \{i\}$ and broadcast y^{x_i} for all $i \in \mathcal{G} \subseteq \mathcal{V}_1$.

As a result, y^{x_i}, where $i \in \mathcal{G}$, is obtained.

Fig. 4. Simulator $\mathcal{S}^{\mathcal{REC}}$

perfectly simulated. Due to space constraint and since we use $\mathcal{DKG}$ from [20], we do not present the correctness and secrecy of $\mathcal{DKG}$ in this paper.

If $PROT_2 = \mathcal{REC}$, we will construct a simulator $\mathcal{S}^{\mathcal{REC}}$ to simulate the view of the adversary from only inputs and outputs from $PROT_2$ so that any distinguisher could not tell the difference between the real protocol and the simulated protocol. This means the protocol does not leak more information other than the input and output of the protocol. The view of the $\mathcal{A}$ $view^2_{\mathcal{A}} = \left(view^2_j\right)_{j=1,..,t}$ consists of $(y^{x_i})_{i\in\mathcal{G}} \in st^2_j$ that any $t+1$ of them recovers x in the exponent (y^x). Recall that $\mathcal{REC}$ is run after a $\mathcal{DKG}$, and the simulator of $\mathcal{DKG} = PROT_1$ where recovers the secret inputs of the $\mathcal{A}$ $(x_i)_{i=1,..,t}$ during the simulation, and manipulates one simulated honest value according to the output, thus the $\mathcal{S}^{REC}$ can take those transcripts from corresponding simulator $\mathcal{S}^{DKG}$ as inputs and manipulate the corresponding simulated honest value. Assume $\mathcal{J}$ is the set of corrupted parties, and $\mathcal{V}_1$ the valid party set from simulated protocol $PROT_1$, $\mathcal{S}^{REC}$ continues the simulation as in Fig. 4.

If $PROT_2 = \mathcal{RECSQ}$, in this case to show that $\mathcal{A}$ is not able to learn any extra information about the secret x from the y^{x^2} which is output of the protocol $PROT_2$ including y^x, we create a simulator $\mathcal{S}^{\mathcal{RECSQ}}$ and show that the view of the $\mathcal{A}$ interacts with $\mathcal{S}^{RECSQ}$ on inputs $y^{(x^2)}$ is the same as the view of $\mathcal{A}$ in a real sub-protocol that outputs the given y^{x^2}. The view of the adversary from real sub-protocol $PROT_2$ is $view^2_{\mathcal{A}} = \left(view^2_j\right)_{j=1,..,t}$ and consisting of $\left(x_{ij}, x'_{ij}, r_{ij}, c_{ij}, c'_{ij}, X_{ik}, Y_{ik}, Z_{ik}, e_{ij}, e'_{ij}, D_{ik}, E_{ik}\right) \in st^2_j$, where $j \in \mathcal{B}$ $i \in \mathcal{G}, k = 0, .., t$. With the help of the corresponding $\mathcal{S}^{\mathcal{DKG}}$, the $\mathcal{S}^{\mathcal{RECSQ}}$ simulates the transcript taking inputs and outputs of the $\mathcal{DKG} = PROT_1$. During the simulation, the simulator has to simulate the transcript for a couple of FVSS and PVSS protocols. To present the proof in a clearer way, we describe the simulation of these two secret sharing schemes in Fig. 5.

The details of $\mathcal{S}^{\mathcal{RECSQ}}$ in Figs. 6 and 7 are shown. In step **SIM-4** in Fig. 7, $\mathcal{G}$ is the honest party set, Q_i is the set of all corrupted parties plus a honest party $i \in \mathcal{G}$.

Now, we consider for any l, where $3 \leq l \leq s$;

$FVSS^{BB}_{\mathcal{S}}\left([r_{ij}]_{j\in[1,t]}, \langle PP, V_{i0}\rangle\right) \stackrel{f_i(.)}{\rightarrow} (V_{ik})$
– takes publicly available (or known before) $t, g \in PP$, V_{i0} values, chooses $r_{ij} \in Z_q$ randomly for each server $i \in \mathcal{G}$ and $j \in \mathcal{B}$, where $i = t+1, ..., n$ and $j = 1, .., t$.
– computes V_{ik} where $i = t+1, ..., n$ and $k = 1, .., t$.
– sends r_{ij} to $j \in \mathcal{A}$ and broadcasts $V_{ik} = g^{a_{ik}}$ where a_{ik} kth coefficient of f_i.
$PVSS^{BB}_{\mathcal{S}}\left([r_{ij}, r'_{ij}]_{j\in[1,t]}, \langle PP, V_{i0}\rangle\right) \xrightarrow{f_i(.), f'_i(.)} (V_{ik})$
– takes publicly available (or known before) $t, g, h \in PP$, V_{i0} values, chooses $r_{ij}, r'_{ij} \in Z_q$ randomly for each server $i \in \mathcal{G}$ and $j \in \mathcal{B}$, where $i = t+1, ..., n$ and $j = 1, .., t$.
– computes V_{ik} where $i = t+1, ..., n$ and $k = 1, .., t$
– sends r_{ij}, r'_{ij} to $j \in \mathcal{A}$ and broadcasts $V_{ik} = g^{a_{ik}} h^{b_{ik}}$ where a_{ik}, b_{ik} are th k–th coefficient of f_i, f'_i for $k = (1, .., t)$.

Fig. 5. Simulation of BlackBox *FVSS* and *PVSS*: Square brackets mean that the variable(s) are random values and chosen by the simulator $\mathcal{S}$.

The simulator $\mathcal{S}^{\mathcal{RECSQ}}$ takes PP, y, y^{x^2}, bad parties' share of secrets x_j, x'_j $(j = 1, .., t)$ that follows the protocol from $\mathcal{S}^{\mathcal{DKG}}$ where $j \in \mathcal{V}_1$ and does the following:

SIM-1 : $PVSS^{BB}_{\mathcal{S}}\left([x_{ij}, x'_{ij}]_{j\in[1,t]}, \langle PP, X_{i0}\rangle\right) \xrightarrow{f_{i1}(.), d_{i1}(.)} (X_{ik})$.

SIM-2 : $PVSS^{BB}_{\mathcal{S}}\left([\{r_{ij}\}_{j\in[1,t]}, Y_{i0}], \langle \{x_{ij}\}_{j\in[1,t]}, PP, X_{i0}\rangle\right) \xrightarrow{f_{i1}(.), d_{i2}(.)} (Y_{ik})$.

SIM-3 : $PVSS^{BB}_{\mathcal{S}}\left([c_{ij}, c'_{ij}]_{j\in[1,t]}, \langle PP, Y_{i0}\rangle\right) \xrightarrow{f_{i2}(.), d_{i3}(.)} (Z_{ik})$.

As a result, $x_{ij}, x'_{ij}, r_{ij}, c_{ij}, c'_{ij}, X_{ik}, Y_{ik}, Z_{ik}$ where $i \in \mathcal{G}, j \in \mathcal{B}, k = 0, ..., t$ obtained.

Fig. 6. Algorithm of the $\mathcal{S}^{\mathcal{RECSQ}}$-(First Part): Note that the value Y_{i0} is chosen randomly by $\mathcal{S}^{\mathcal{RECSQ}}$ in the **SIM-2** above.

From the first part of the proof, $S^{\mathcal{RECSQ}}$ has enough information to calculate: (1) $c_j := \sum_{i\in\mathcal{I}} \lambda_i c_{ij}$, (2) $c'_j := \sum_{i\in\mathcal{I}} \lambda_i c'_{ij}$, (3) $Z_k := \prod_{i\in\mathcal{I}} Z_{ik}^{\lambda_i}$ for $k = 0, .., t$, where $Z_{i0} = Y_{i0}$, where $j = 1, .., t$, $i \in \mathcal{I} \subseteq \mathcal{V}_k$ and $|\mathcal{I}| \geq 2t+1$.

SIM-4 : Then $S^{\mathcal{RECSQ}}$ calculates $E_{i0} = y^{c_i}$ for each $i \in \mathcal{G}$ one by one as $y^{c_i} = \left(\frac{y^{x^2}}{\prod_{j\in\mathcal{Q}_i\setminus\{i\}}\left(y^{(c_j)}\right)^{\lambda_j}}\right)^{(\lambda_i)^{-1}}$, where λ_js are appropriate Lagrange coefficients.

SIM-5 : $PVSS^{BB}_{\mathcal{S}}\left([e_{ij}, e'_{ij}]_{j\in[1,t]}, \langle t, g, h, D_{i0}\rangle\right) \xrightarrow{f_{i3}(.), d_{i4}(.)} (D_{ik})$.
($D_{i0} = g^{c_i} h^{c'_i} = \prod_{k=0}^{t} Z_k^{i^k}$ can be calculated by the simulator from step 3 above.)

SIM-6 : $FVSS^{BB}_{\mathcal{S}}\left(\langle \{e_{ij}\}_{j\in[1,t]}, t, y, E_{i0}\rangle\right) \stackrel{f_{i3}(.)}{\rightarrow} (E_{ik})$.

As a result, $e_{ij}, e'_{ij}, D_{ik}, E_{ik}$ where $i \in \mathcal{G}, j \in \mathcal{B}, k = 0, ..., t$ obtained.

Fig. 7. Algorithm of $\mathcal{S}^{\mathcal{RECSQ}}$-(Second Part)

if $PROT_l = \mathcal{REC}$, the $\mathcal{S}^{\mathcal{REC}}$ takes the corresponding simulated $\mathcal{DKG}$ transcript, the output and input values from $PROT_l$ to simulate view of the $\mathcal{A}$ in $PROT_l$ as described in Fig. 4.

If $PROT_l = \mathcal{RECSQ}$, the $\mathcal{S}^{\mathcal{RECSQ}}$ inputs the corresponding simulated $\mathcal{DKG}$ transcript, the output and input values from $PROT_l$ to simulate view of the adversary in $PROT_l$ as described in Figs. 6 and 7. As a result, the $\mathcal{S}$ consists of sequence of simulators that each of them simulates adversary's corresponding real protocol views $(view_j^k)$ sequentially, where $j = 1, .., t$ and $k = 1, .., s$. □

4 Distributed Parameter Generation for the n-BDHE Problem and Applications

We will show how to combine these sub-protocols introduced above to distributively generate n-BDHE problem parameters. We then show how to use this to construct a t-secure DBE without a trusted authority that enjoys constant size ciphertext, secret key overhead and present some other possible applications.

4.1 t-Secure Distributed n-BDHE Problem Parameters Generation Protocol

Recall that the parameters of the n-BDHE problem have the form of $g_1, g_2, .., g_n, g_{n+2}, ..., g_{2n}$, where $g_i = g^{\alpha^i}$ and for a random $\alpha \in Z_q$. Having the three sub-protocols, one can compose them and distributively generate the n-BDHE problem parameters. We describe a suite protocol for n-BDHE parameters as $(n{-}BDHE)^{Suite} = PROT_0, PROT_1, ..., PROT_{2n-1}$, where $PROT_0 = ParGen$, $PROT_1 = DKG$,$PROT_2 = REC_1, .., PROT_n = REC_{n-1}, PROT_{n+1} = RECSQ$, $PROT_{n+2} = REC_n, .., PROT_{2n-1} = REC_{2n-3}$ and $\mathcal{REC}_j$ is the j

Generating g_1: Each party i runs $\mathcal{DKG}$ with input PP, outputs g^α and keeps its local state as $st_i^1 = (\mathcal{V}_1, \{\tau_{ji}^1\}_{\mathcal{V}_1})$.
Generating $g_2, \ldots, g_n$: For $1 \leq k \leq n-1$, each party $i \in \mathcal{V}_k$ runs $\mathcal{REC}$ with inputs $\left(\mathcal{V}_k, (g^{x_j})_{j \in \mathcal{V}_1}, PP, g_k, x_i\right)$, outputs (g_k^α) and keeps its local state as $st_i^{k+1} = (\mathcal{V}_{k+1}, \{\tau_{ji}^{k+1}\}_{\mathcal{V}_{k+1}})$.
Generating g_{n+2}: Each party i runs $\mathcal{RECSQ}$ with input $\left(\mathcal{V}_n, \left(g^{x_j} h^{x'_j}\right)_{j \in \mathcal{V}_1}, PP, g_n, x_i, x'_i\right)$, outputs $(g_n)^{\alpha^2}$ and keeps its local state as $st_i^{n+1} = (\mathcal{V}_{n+1}, \{\tau_{ji}^{n+1}\}_{\mathcal{V}_{n+1}})$.
Generating $g_{n+3}, \ldots, g_{2n}$: For $n+2 \leq l \leq 2n-1$, each party $i \in \mathcal{V}_{n+1}$ runs $\mathcal{REC}$ with input $\left(\mathcal{V}_{n+1}, (g^{x_j})_{j \in \mathcal{V}_1}, PP, g_l, x_i\right)$, outputs (g_l^α) and keeps its local state as $st_i^l = (\mathcal{V}_l, \{\tau_{ji}^l\}_{\mathcal{V}_l})$.

Fig. 8. n-BDHE problem parameters generation protocol

th run of the sub-protocol $\mathcal{REC}$. The details for generating these parameters are given in Fig. 8. The security follows easily from Theorem 1, and we summarize them into the following theorem, for the detailed definition and analysis for the security of the n-BDHE problem, we refer to Appendix A.2.

Theorem 2. *The protocol described in Fig. 8 is a t-secure distributed n-BDHE problem protocol when $t < n/3$.*

4.2 t-Secure DBE with Constant Size Ciphertext and Secret Key

Now we proceed to construct a $t-$secure DBE scheme by allowing parties to distributively generate the public and secret keys of the Boneh-Gentry-Waters (BGW) BE scheme [8]. We will inherit the good property of constant size secret keys and ciphertexts. Recall that the public keys for BGW BE is in the form of $\mathcal{PK} = (g, g_1, .., g_n, g_{n+2}, .., g_{2n}, v = g^{\gamma})$, while secret key for party i is in the form of $d_i = g_i^{\gamma}$. As a note that our $t-$secure DBE scheme's *Encryption* and *Decryption* algorithms are the same as in BGW scheme.

Decentralized BGW Public Key Generation. It is easy to see that one can generate the public keys using our $t-$secure distributed n-BDHE parameter generation protocol (Fig. 8) plus an independent $\mathcal{DKG}$ to generate $v = g^{\gamma}$. Basically, sequentially composable n party $Suite = (PROT_0, .., PROT_s)$ protocol can generate all the public key instances, where $(PROT_0 = ParGen, PROT_1 = \mathcal{DKG}, (PROT_k)_{k=2,..,n,n+2,..,2n} = \mathcal{REC}_{k-1}, PROT_{n+1} = \mathcal{RECSQ}, PROT_{2n+1} = \mathcal{DKG}_2)$ and $s = 2n + 1$. The proof also easily follows from Theorem 1.

Decentralized BGW Secret Key Generation. Generating the secret keys are slightly different, as for public keys, every party can publicly broadcast their messages and compute the final output from the received messages. While for each secret key, only a single party should be able to compute the final value without referring to a secure channel. We let the party i select a random mask so that one can compute a blinded secret key but only he can remove the random mask, and further derive the secret key. The details are given in Fig. 9 below.

- party i chooses a random r_i, calculates $g^{r_i}, g_i^{r_i}$ and broadcasts them.
- Each party j first checks if the chosen $\mathcal{PK}$ instance by party i is g_i by checking, $e(g^{r_i}, g_i) = e(g_i^{r_i}, g)$. If this is correct, each party $j \in \mathcal{V}_{2n+1}$ runs $\mathcal{REC}$ on inputs $\left(\mathcal{V}_{2n+1}, \left(g^{k_i}\right)_{i\in\mathcal{V}_{2n+1}}, PP, g_i^{r_i}, k_j\right)$ and party i gets the value $(g_i^{r_i})^{\gamma}$ following the steps in Fig. 1, where valid set $\mathcal{V}_{2k+1} \in st_j^{2n+1}$, k_j and $\left(g^{k_i}\right)_{i\in\mathcal{V}_{2n+1}}$ values are computed from the transcript $\{\tau_{ij}^k\}_{i\in\mathcal{V}_{2n+1}} \in \left(st_j^{2n+1}\right)$.
- party i gets his decryption key by computing $d_i = ((g_i^{r_i})^{\gamma})^{1/r_i} = g_i^{\gamma}$ if it satisfies $e(g^{\gamma}, g_i) = e(g_i^{\gamma}, g)$.

Fig. 9. Decentralized BGW secret key generation protocol

Theorem 3. *Our t-secure DBE is* static-*CPA secure under static corruption.*

Proof. After adversary $\mathcal{A}$ submits the corrupted party set S to control and the target set $\mathcal{T}$ that wishes to be challenged on, the simulator $\mathcal{S}^{BE}$ needs to generate public keys in a way that he can answer secret key queries, specifically, he wants to set $v = g^u \left(\prod_{j \in \mathcal{T}} g_{n+1-j}\right)$, where $u \in Z_q$. This can be done by running the simulator of the $\mathcal{DKG}$ $\mathcal{S}^{\mathcal{DKG}}$ as any value can be set as a target value as long as the simulator controls a majority of honest parties. While other parts of the public key instances are in the form of the n-BDHE problem, thus they can be simulated by running the simulator of the n-BDHE protocol. Since the simulator $\mathcal{S}^{BE}$ knows the value u, he can generate the private keys for every $i \notin \mathcal{T}$ as $d_i = g_i^u \left(\prod_{j \in \mathcal{T}} g_{n+1-j+i}\right)^{-1}$ and gives it to $\mathcal{A}$. The other phases (challenge and guess) will proceed identically as in the proof of [8]. □

4.3 Other Applications

Similar as we construct the DBE, our n-BDHE parameters generation protocol can be applied directly to many other primitives to make a distributive version of the corresponding primitive when they have the public/private keys with the structure of the n-BDHE problem. To name a few, p-signature from [23], forward-secure and searchable broadcast encryption with short ciphertexts and private keys from [2], conditional proxy broadcast re-encryption from [11], and TMDS: Thin-Model Data Sharing Scheme Supporting Keyword Search in Cloud Storage from [28].

It is not hard to see that our building blocks can also be composed sequentially and distributively to generate parameters for other problems. The most immediate one would be the Diffie-Hellman type of problems, e.g., the structure of g^α, g_1^α in the bilinear map setting for which we can run the $\mathcal{REC}$. One of the examples is the revocation system from Lewko et al. [27], the public key is in the form of $(g, g^b, g^{b^2}, h^b, e(g,g)^\alpha)$. The suite protocol for generating public key for this example is $Suite = ParGen, \mathcal{DKG}_1, \mathcal{DKG}_2, \mathcal{REC}_1, \mathcal{REC}_2$. The authorities (servers) can run $\mathcal{DKG}$ twice to generate $g^b, e(g,g)^\alpha$, and run $\mathcal{REC}$ twice to generate $(g^b)^b, h^b$. The secret key for a user in [27] is in the form of $D_0 = g^\alpha g^{b^2 t}, D_1 = (g^{bID} h)^t, D_2 = g^{-t}$ and can be distributively generated as follows: (1) The servers first run the $\mathcal{DKG}_3$ to generate g^t and compute $(g^b)^{ID} h$; (2) The servers use $(g^b)^{ID} h$ as a base to run $\mathcal{REC}_3$ to get D_1; (3) since each server has the share of α from $\mathcal{DKG}_2$ and t from the $\mathcal{DKG}_3$, they simply send the user $g^{\alpha_i}(g^{b^2})^{t_i}$ and the user can interpolate to recover D_0. D_2 is basically the inverse of g^t.

Another example is Waters IBE [32], in which the master public key is in the form of $g, g_1 = g^\alpha, g_2, u'$ and an n length random vector $U = (u_i)$, while the master secret key is g_2^α, where $g, g_1, g_2, u', (u_i) \in G$ and $\alpha \in Z_q$. The suite protocol for generating public key for this example is $Suite = ParGen, \mathcal{DKG}_1, .., \mathcal{DKG}_{n+3}$. Servers run $\mathcal{DKG}$ $n+3$ times to generate $g_1 = g^\alpha, g_2, u', (u_i)$ ($u', (u_i)$ can be generated using coin flipping protocols). As a note that each server has the share of g_2^α (without running the $\mathcal{REC}$) from $\mathcal{DKG}_1$. The secret key for a user with identity ID is in the form of $d_v = (g_2^\alpha \cdot (u' \prod_{i \in V} u_i)^r, g^r)$, where v is an n bit string representing an identity, V is the set of indicies for which the bitstring v is set to 1, it can be distributively generated as follows: (1) The servers run the $\mathcal{DKG}_{n+4}$ to generate g^r, now they have the shares of $(u' \prod_{i \in V} u_i)^r$ without running the $\mathcal{REC}$. Thus (2) each server j can send $g_2^{\alpha_j}(u' \prod_{i \in V} u_i)^{r_j}$, then the user can interpolate to get his secret key by computing $\prod_{j \in \mathcal{Q}}(g_2^{\alpha_j}(u' \prod_{i \in V} u_i)^{r_j})^{\lambda_j} = \prod_{j \in \mathcal{Q}}(g_2^{\alpha_j \lambda_j}(u' \prod_{i \in V} u_i)^{r_j \lambda_j}) = g_2^\alpha \cdot (u' \prod_{i \in V} u_i)^r$, where $|\mathcal{Q}| \geq t+1$.

Note: In these schemes [27,32], the secret keys are given to the users via secure channels. These channels can be omitted by using random mask that we showed in Fig. 9. The analysis of above examples are very similar to that of our building blocks and the n-BDHE protocol, and we omit the details here.

5 Conclusion

We construct a cryptographic tool set consisting of three sub-protocols $\mathcal{DKG}$, $\mathcal{REC}$, and $\mathcal{RECSQ}$ that are used in a sequential fashion to produce distributed parameter generation protocols. Our main application is a protocol that generates public parameters for schemes based on the n-BDHE problem. As a result, we distribute the BGW broadcast encryption scheme [8] that has constant size secret keys and ciphertexts. In addition, we show some other applications that they are based on the bilinear Diffie Hellman (BDH) problem such as identity based encryption (IBE) and revocation systems that can be similarly distributed.

Acknowledgment. The first author was supported by the ERC project CODAMODA and the project FINER of the Greek Secretariat of Research and Technology.

A Appendix

A.1 Preliminaries

Parties (Servers and an Adversary): Let $\mathcal{P}$ is a set of parties $\mathcal{P} = \{1, .., n\}$. Party $i \in \mathcal{P}$ is assumed to be probabilistic polynomial time Turing Machine. Among those parties, there are up to t corrupt parties completely controlled by a static adversary and the adversary is active.

Input and Output: Each party is given private and public input. The input of each party includes the number of parties n. At the end of the computation each party will produce private and public output that should be equal among all honest parties (global public output). The private input of corrupted servers as well as the public input is given to the adversary at the start of the protocol.

Communication Model: We assume that the communication is synchronous and protocol execution proceeds in rounds. In each round the each party using its current state and all history of communication from all rounds produces two types of messages to be delivered to other parties: (1) private messages that are sent to other parties by using private channel network where a message is assured of being delivered in a fixed period. The network is assumed to be secure and complete, that is every pair of parties is connected by an untappable and mutually authenticated channel; (2) broadcast message that will be delivered to all parties at the beginning of the next round. At each round a party produces private messages for all other parties as well as a public broadcast message.

Adversarial Operation: Each round, after adversary sees all broadcast messages and secret messages from honest parties that are received by corrupted parties, he sends public and private messages depending on those received messages (public and private) as well as all information that the corrupted parties have had from previous rounds.

Computational Assumption: We use the large primes p, q that satisfy $q|p-1$. We represent by G the subgroup of elements of order q in Z_p^*. It is assumed that solving the discrete logarithm problem in G is intractable.

Feldman's Verifiable Secret Sharing (FVSS): FVSS [17] allows a malicious adversary which corrupts up to $\frac{(n-1)}{2}$ parties including the dealer. The dealer generates a random t-degree polynomial $f(.)$, where $f(0) = x$ which is the secret value, and sends to each party i a share $s_i = f(i) \bmod q$. The dealer also broadcasts values $V_k = g^{a_k}$, where a_k is the kth coefficient of $f(.)$. This will allow the parties to check if the values s_i really define a secret by checking that $g^{s_i} = \prod_{k=0}^{t} V_k^{i^k} \bmod p$ (**Eq. 1**), where $k = 0, ..., t$. If this equation is not satisfied, party i complains and asks the dealer to reveal his share. If more than t parties complain then the dealer is clearly bad and he is disqualified. Otherwise, he reveals the share s_i matching Equation **Eq. 1** for each complaining i. Equation **Eq. 1** also allows detection of incorrect shares s_i' at reconstruction time. Notice that the value of the secret is only computationally secure, e.g., the value $g^{a_0} = g^x$ is leaked. However, it can be shown that an adversary that learns t or less shares cannot obtain any information on the secret x beyond what can be derived from g^x. We will use the following notation to describe the execution of a FVSS protocol: $FVSS(x)(g) \xrightarrow{f,n,t} (s_i)(V_k)$, $k = 0, ..., t$.

Pedersen's Verifiable Secret Sharing (PVSS): We now recall a VSS protocol that provides information theoretic secrecy for the shared secret. This is in contrast to FVSS protocol which leaks the value of g^x. PVSS [29] uses the parameters p, q, g as defined for FVSS. In addition, it uses an element $h \in Z_p^*$ such that h belongs to the subgroup generated by g and the discrete log of h in base g is unknown (and assumed hard to compute). The dealer first chooses two t-degree random polynomials $f(.), f'(.)$, with random coefficients over Z_q, subject to $f(0) = x$, which is the secret. The dealer sends to each party i the values $x_i = f(i) \bmod q$ and $x_i' = f'(i) \bmod q$. The dealer then commits to each coefficient of the polynomials f and f' publishing the values $V_k = g^{a_k} h^{b_k}$, where a_k (resp. b_k) is the kth coefficient of f (resp. f'). This allows the parties to verify the received shares by checking that $g^{s_i} h^{s_i'} = \prod_{k=0}^{t} V_k^{i^k} \bmod p$ (**Eq. 2**). If the shares that do not satisfy the equation **Eq. 2** broadcast a complaint. If more than t parties complain the dealer is disqualified. Otherwise the dealer broadcasts the values x_i and x'_i matching the equation for each complaining party i. At reconstruction time the parties are required to reveal both x_i and x'_i and Equation **Eq. 2** is used to validate the shares. Indeed in order to have an incorrect share t_i accepted at reconstruction time, it can be shown that party i has to compute the discrete log of h in base g. Notice that the value of the secret is unconditionally protected since the only value revealed is $V_0 = g^s h^r$ (it can be seen that for any value x' there is exactly one value r' such that $V_0 = g^{s'} h^{r'}$ thus V_0 gives no information on s). We will use the following notation to denote an execution of PVSS: $PVSS\,(x, x')\,(g, h) \xrightarrow{f, f', n, t} (x_i, x'_i)\,(V_k)$ (**Eq. 3**), $k = 0, ..., t$.

Bilinear Maps: (1) G and G' are two multiplicative cyclic groups of prime order q; (2) g is a generator of G; (3) $e : G \times G \to G'$. Let G and G' be two groups as above. A bilinear map is a map $e : G \times G \to G'$ with the following properties: (1) for all $u, v \in G$ and $a, b \in Z$, we have $e\left(u^a, v^b\right) = e\,(u, v)^{ab}$; (2) the map is not degenerate, i.e., $e\,(g, g) \neq 1$

A.2 Proof of Theorem 2

Definition 7 (t-Secure Distributed n-BDHE Protocol). *$D^{n-\mathrm{BDHE}}$ is an n-party sequentially composable $2n$ protocols (each protocol generates one instance of n-BDHE parameter). Each party takes public parameter set PP as input, and sequentially outputs $n - \mathrm{BDHE} = (g_1, .., g_n, g_{n+2}, .., g_{2n})$, where $g_i = g^{x^i}$ for some random value x with the presence of at most t corrupted parties. t-Secure Distributed n-BDHE protocol satisfies the following properties from [20]:*

Correctness: *(1) x is uniformly distributed in Z_q; (2) All subsets of $t + 1$ shares provided by honest players define the same unique secret key x; (3) All honest parties have the same public values $g_1, .., g_n, g_{n+2}, .., g_{2n}$; (4) If at least $2t + 1$ parties follow the protocol, shares are accepted with probability 1.*

Secrecy: *No information on x can be learned by the adversary except for what is implied by the values $g_1, .., g_n, g_{n+2}, .., g_{2n}$. More formally, we state this condition in terms of simulatability: for every PPT adversary $\mathcal{A}$ that corrupts up to t parties, there exists a PPT simulator $\mathcal{S}$, such that on input an elements $g_1, .., g_n, g_{n+2}, .., g_{2n}$, produces an output distribution which is polynomially indistinguishable from $\mathcal{A}$'s view (Definition 1) of a run of the n-BDHE protocol that ends with $g_1, .., g_n, g_{n+2}, .., g_{2n}$ as its public key output.*

Proof. Correctness: The correctness properties (1), (2), (3) for g_1 can be shown by following [20], the other instances $g_2, .., g_n$ can be obtained by the presence of at least $t+1$ honest parties that use their share of secret x_i and recover the value x in the exponent sequentially. Basically, they raise sequentially their shares (x_i) to recover x value in the exponent using Lagrange interpolation. The share $g_i^{x_j}$, where $i = 1, 2, .., n$ can be verified publicly using bilinear map. To show the value

g_{n+2} is obtained from g_n by any $t+1$ honest parties, at least $2t+1$ parties are needed that they follow $\mathcal{RECSQ}$ sub-protocol. The process can be followed in a similar way in [1] (Lemma 2). The difference is that we have x^2 in the exponent. To do that parties need to run one more PVSS and one more FVSS to show they share the correct value of their c_is in the exponent using g and g_n as the bases. The other instances $g_{n+2}, .., g_{2n}$ also can be obtained as the same way with at least $t+1$ honest parties.

Secrecy: It follows from Theorem 1 since it is the special protocol of $\Upsilon_{\mathcal{A}}^{GSuite}$. □

References

1. Abe, M.: Robust distributed multiplication without interaction. In: Wiener, M. (ed.) CRYPTO 1999. LNCS, vol. 1666, pp. 130–147. Springer, Heidelberg (1999)
2. Attrapadung, N., Furukawa, J., Imai, H.: Forward-secure and searchable broadcast encryption with short ciphertexts and private keys. In: Lai, X., Chen, K. (eds.) ASIACRYPT 2006. LNCS, vol. 4284, pp. 161–177. Springer, Heidelberg (2006)
3. Ben-Or, M., Goldwasser, S., Wigderson, A.: Completeness theorems for non-cryptographic fault-tolerant distributed computation. In: STOC 1988 (1988)
4. Benaloh, J.C., Yung, M.: Distributing the power of a government to enhance the privacy of voters. In: PODC 1986, pp. 52–62. ACM, New York (1986)
5. Blakley, G.: Safeguarding cryptographic keys. In: AFIPS National Computer Conference, pp. 313–317. AFIPS Press, Monval (1979)
6. Boneh, D., Boyen, X.: Efficient selective-ID secure identity-based encryption without random oracles. In: Cachin, C., Camenisch, J.L. (eds.) EUROCRYPT 2004. LNCS, vol. 3027, pp. 223–238. Springer, Heidelberg (2004)
7. Boneh, D., Boyen, X., Goh, E.-J.: Hierarchical identity based encryption with constant size ciphertext. IACR Cryptology ePrint Archive 2005, 15 (2005)
8. Boneh, D., Gentry, C., Waters, B.: Collusion resistant broadcast encryption with short ciphertexts and private keys. In: Shoup, V. (ed.) CRYPTO 2005. LNCS, vol. 3621, pp. 258–275. Springer, Heidelberg (2005)
9. Canny, J., Sorkin, S.: Practical large-scale distributed key generation. In: Cachin, C., Camenisch, J.L. (eds.) EUROCRYPT 2004. LNCS, vol. 3027, pp. 138–152. Springer, Heidelberg (2004)
10. Chaum, D., Crépeau, C., Damgard, I.: Multiparty unconditionally secure protocols. In: STOC 1988, pp. 11–19. ACM, New York (1988)
11. Chu, C.-K., Weng, J., Chow, S.S.M., Zhou, J., Deng, R.H.: Conditional proxy broadcast re-encryption. In: Boyd, C., González Nieto, J. (eds.) ACISP 2009. LNCS, vol. 5594, pp. 327–342. Springer, Heidelberg (2009)
12. Cramer, R., Fehr, S.: Optimal black-box secret sharing over arbitrary abelian groups. In: Yung, M. (ed.) CRYPTO 2002. LNCS, vol. 2442, pp. 272–287. Springer, Heidelberg (2002)
13. Cramer, R., Fehr, S., Stam, M.: Black-box secret sharing from primitive sets in algebraic number fields. In: Shoup, V. (ed.) CRYPTO 2005. LNCS, vol. 3621, pp. 344–360. Springer, Heidelberg (2005)
14. De Santis, A., Desmedt, Y., Frankel, Y., Yung, M.: How to share a function securely. In: Proceedings of the Twenty-Sixth Annual ACM Symposium on Theory of Computing, STOC 1994, pp. 522–533. ACM, New York (1994)
15. Desmedt, Y.G., Frankel, Y.: Threshold cryptosystems. In: Brassard, G. (ed.) CRYPTO 1989. LNCS, vol. 435, pp. 307–315. Springer, Heidelberg (1990)

16. Desmedt, Y.G., Frankel, Y.: Perfect homomorphic zero-knowledge threshold schemes over any finite abelian group (1994)
17. Feldman, P.: A practical scheme for non-interactive verifiable secret sharing. In: FOCS 1987. IEEE Computer Society (1987)
18. Fiat, A., Naor, M.: Broadcast encryption. In: Stinson, D.R. (ed.) CRYPTO 1993. LNCS, vol. 773, pp. 480–491. Springer, Heidelberg (1994)
19. Frankel, Y., MacKenzie, P.D., Yung, M.: Robust efficient distributed rsa-key generation. In: STOC 1998, pp. 663–672. ACM, New York (1998)
20. Gennaro, R., Jarecki, S., Krawczyk, H., Rabin, T.: Secure distributed key generation for discrete-log based cryptosystems. J. Cryptol. **20**(1), 51–83 (2007)
21. Gennaro, R., Raimondo, M.D.: Secure multiplication of shared secrets in the exponent. Inf. Process. Lett. **96**(2), 71–79 (2005)
22. Goldreich, O., Micali, S., Wigderson, A.: Proofs that yield nothing but their validity or all languages in np have zero-knowledge proof systems. J. ACM **38**(3), 691–729 (1991)
23. Izabachène, M., Libert, B., Vergnaud, D.: Block-wise P-signatures and non-interactive anonymous credentials with efficient attributes. In: Chen, L. (ed.) IMACC 2011. LNCS, vol. 7089, pp. 431–450. Springer, Heidelberg (2011)
24. Jarecki, S.: Efficient Threshold Cryptosystems. Ph.D. thesis, MIT (2001)
25. Kate, A., Goldberg, I.: Distributed private-key generators for identity-based cryptography. In: Garay, J.A., De Prisco, R. (eds.) SCN 2010. LNCS, vol. 6280, pp. 436–453. Springer, Heidelberg (2010)
26. Kiayias, A., Xu, S., Yung, M.: Privacy preserving data mining within anonymous credential systems. In: Ostrovsky, R., De Prisco, R., Visconti, I. (eds.) SCN 2008. LNCS, vol. 5229, pp. 57–76. Springer, Heidelberg (2008)
27. Lewko, A., Sahai, A., Waters, B.: Revocation systems with very small private keys. In: SP 2010, pp. 273–285. IEEE Computer Society, Washington, DC (2010)
28. Liu, Z., Li, J., Chen, X., Yang, J., Jia, C.: TMDS: thin-model data sharing scheme supporting keyword search in cloud storage. In: Susilo, W., Mu, Y. (eds.) ACISP 2014. LNCS, vol. 8544, pp. 115–130. Springer, Heidelberg (2014)
29. Pedersen, T.P.: Non-interactive and information-theoretic secure verifiable secret sharing. In: Feigenbaum, J. (ed.) CRYPTO 1991. LNCS, vol. 576, pp. 129–140. Springer, Heidelberg (1992)
30. Phan, D.H., Pointcheval, D., Strefler, M.: Decentralized dynamic broadcast encryption. In: Visconti, I., De Prisco, R. (eds.) SCN 2012. LNCS, vol. 7485, pp. 166–183. Springer, Heidelberg (2012)
31. Shamir, A.: How to share a secret. Commun. ACM **22**(11), 612–613 (1979)
32. Waters, B.: Efficient identity-based encryption without random oracles. In: Cramer, R. (ed.) EUROCRYPT 2005. LNCS, vol. 3494, pp. 114–127. Springer, Heidelberg (2005)
33. Wu, Q., Mu, Y., Susilo, W., Qin, B., Domingo-Ferrer, J.: Asymmetric group key agreement. In: Joux, A. (ed.) EUROCRYPT 2009. LNCS, vol. 5479, pp. 153–170. Springer, Heidelberg (2009)
34. Wu, Q., Qin, B., Zhang, L., Domingo-Ferrer, J.: Ad hoc broadcast encryption. In: CCS 2010 (2010)
35. Wu, Q., Qin, B., Zhang, L., Domingo-Ferrer, J.: Fully distributed broadcast encryption. In: Boyen, X., Chen, X. (eds.) ProvSec 2011. LNCS, vol. 6980, pp. 102–119. Springer, Heidelberg (2011)
36. Wu, Q., Qin, B., Zhang, L., Domingo-Ferrer, J., Farràs, O.: Bridging broadcast encryption and group key agreement. In: Lee, D.H., Wang, X. (eds.) ASIACRYPT 2011. LNCS, vol. 7073, pp. 143–160. Springer, Heidelberg (2011)

Author Index

Printed in the United States
By Bookmasters